Modern Developments in Structural Interpretation, Validation and Modelling

Geological Society Special Publications

Series Editor A. J. FLEET

GEOLOGICAL SOCIETY SPECIAL PUBLICATION NO. 99

Modern Developments in Structural Interpretation, Validation and Modelling

EDITED BY

P. G. BUCHANAN

Oil Search Ltd
Port Moresby, Papua New Guinea

and

D. A. NIEUWLAND

Shell Research BV
Rijswijk, Netherlands

1996

Published by
The Geological Society
London

THE GEOLOGICAL SOCIETY

The Society was founded in 1807 as the Geological Society of London and is the oldest geological society in the world. It received its Royal Charter in 1825 for the purpose of 'investigating the mineral structure of the Earth'. The Society is Britain's national society for geology with a membership of 7500. It has countrywide coverage and approximately 1000 members reside overseas. The Society is responsible for all aspects of the geological sciences including professional matters. The Society has its own publishing house, which produces the Society's international journals, books and maps, and which acts as the European distributor for publications of the American Association of Petroleum Geologists, SEPM and the Geological Society of America.

Fellowship is open to those holding a recognized honours degree in geology or cognate subject and who have at least two years' relevant postgraduate experience, or who have not less than six years' experience in geology or a cognate subject. A Fellow who has not less than five years' relevant postgraduate experience in the practice of geology may apply for validation and, subject to approval, may be able to use the designatory letters C Geol (Chartered Geologist).

Further information about the Society is available from the Membership Manager, The Geological Society, Burlington House, Piccadilly, London W1V 0JU, UK. The Society is a Registered Charity, No. 210161

Published by the Geological Society from:
The Geological Society Publishing House
Unit 7
Brassmill Enterprise Centre
Brassmill Lane
Bath BA1 3JN
UK
(*Orders*: Tel 01225 445046
Fax 01225 442836)

First published 1996

British Library Cataloguing in Publication Data

A catalogue record for this book is available from the British Library ISBN 1-897799-43-8

Typeset by EJS Chemical Composition,
Midsomer Norton, Bath, Avon

Printed by The Alden Press, Osney Mead, Oxford, UK

Distributors

USA
AAPG Bookstore
PO Box 979
Tulsa
OK 74101-0979
USA
(*Orders*: Tel (918) 584-2555
Fax (918) 548-0469)

Australia
Australian Mineral Foundation
63 Conyngham Street
Glenside
South Australia 5065
Australia
(*Orders*: Tel (08) 379-0444
Fax (08) 379-4634)

India
Affiliated East-West Press PVT Ltd
G-1/16 Ansari Road
New Delhi 110 002
India
(*Orders*.: Tel (11) 327-9113
Fax (11) 326-0538

Japan
Kanda Book Trading Co.
Tanikawa Building
3–2 Kanda Surugadai
Chiyoda-Ku
Tokyo 101
Japan
(*Orders*: Tel (03) 3255-3497
Fax (03) 3255-3495)

Contents

Regional analyses and remote sensing

Introduction

D. A. NIEUWLAND[1] & P. G. BUCHANAN[2]

[1] *Shell Research BV, KSEPL, PO Box 60, Rijswijk, 2280 AB, The Netherlands*

[2] *Oil Search Ltd, NIC Haus, PO Box 1031, Champion Parade, Port Morseby, Papua New Guinea*

The scope of this volume is captured in the title, we have aimed to present a comprehensive overview of the latest developments in structural interpretation, validation and modelling techniques.

The role and importance of structural geology in exploration and production of hydrocarbons is increasing with the tendency towards the development of smaller, more complex structural prospects and satellite fields. Despite advances in seismic acquisition and processing which facilitate the visualization of sub-surface structural form, there is still a need to understand often complex kinematics and to refine trap definition. In the last few years, the availability of high powered computer hardware and sophisticated software has facilitated a quantum leap in the accessibility of structural restoration, modelling and visualisation techniques and made the rapid and accurate manipulation of large data sets possible. Many papers in this volume fully reflect use of this latest technology and demonstrate a broad range of capabilities. In areas where seismic is unobtainable, structural imaging is one of the few remaining tools available to the earth scientist. Companies operating in such areas are reliant on the development of techniques in making accurate predictions about the sub-surface.

The intention of this book is to provide anyone wishing to undertake structural interpretation in complex areas with a comprehensive review of the latest techniques available. Each of the techniques has limitations and makes assumptions, which should always be considered and accounted for. It should also be stated, that the advent of new models and technology to manipulate them, is not to be considered a substitute for creative thinking supported by field based studies. There is still a very large requirement to validate the modelling with real examples and one should not be carried away with the power of the computer. Also the integration of various techniques is strongly recommended, as the strengths of one can compensate the weaknesses of another. Through increased communication and integration between different disciplines, the validation process has moved beyond the realms of pure structural geology. One of the underlying themes of the book is to highlight the fact that natural processes are not mutually exclusive and a comprehensive multi-disciplinarian approach to interpretation is both possible and desirable.

Although the chapters in this book are organized on the basis of the methods and techniques used, most authors have combined two or more approaches to achieve their goals. As geological processes are by nature ‘multidisciplinary’, it is evident that integration of all the appropriate and available techniques is a basic requirement for unravelling complex geological systems. This special publication contains some elegant examples illustrating this point. Although the majority of the papers contain original work, some contributions are primarily review papers (**Buchanan, Coward, McClay and Insley**).

The single most important technique that has enabled enormous advances in structural interpretation is undoubtedly 3D seismic. Although some structures can be satisfactory approximated by a series of 2D sections (e.g. a series of parallel cross-sections through a graben, or an axi-symmetric approximation of a salt diapir), geological structures are 3D phenomena and can only be fully understood by using 3D data sets. However, access to 3D seismic is not always available and in many situations 2D data is all that can be relied on. Techniques applicable to 2D are also relevant to 3D and several papers cover this theme.

Pickering *et al.* go into the use of 2D seismic and the estimation of sub-seismic strain, **Horscroft & Bain** promote integration of 3D seismic with other geophysical tools such as gravity and magnetics. **Walsh *et al.*** also concern themselves with problems associated with sub-seismic strain in their discussion of ductile strain effects in the analysis of seismic interpretation of normal fault systems, reducing ductility to a function of the scale of observation.

Seismic is not the only technique to have helped the advance of structural geology. Palinspastic reconstructions have always been a very useful

From Buchanan, P. G. & Nieuwland, D. A. (eds), 1996, *Modern Developments in Structural Interpretation,Validation and Modelling*, Geological Society Special Publication No. 99, pp. 1–3.

cross-section validation method, but notoriously laborious and time consuming. It is only since the availability of sufficient computer power and user-friendly software, that the technique has been developed into a practical tool, rapidly gaining recognition and general use. This is also born out by the relatively large number of contributions to this volume (eight in total).

The first two papers on palinspastic reconstruction and forward modelling, are review papers. **Buchanan** deals with the application of reconstruction to hydrocarbon exploration and production and **Coward** evaluates the problems encountered with balancing sections in inversion structures. A case history of palinspastic reconstruction in inverted basins is presented by **Hill & Cooper** who advocate the integration of thermochronology with palinspastic reconstruction and substantiate their case with examples. The wider use of palinspastic reconstructions has also led to a better appreciation of the many pitfalls and traps of which the user needs to be aware. Fortunately, computer programs are not regarded as black boxes by today's critical geological community. The basic assumption of plain strain deformation, as a necessary requirement for 2D reconstructions, is well known. An example of a pragmatic approach to this problem, using 2D reconstruction techniques, is treated for areas of salt tectonics (**Rowan**), where out-of-plane movements are notorious. Other workers have paid attention to different fundamental problems, such as the discrepancy between the mechanisms of rock deformation and those chosen for palinspastic reconstructions (**Hauge & Gray**), the use of the lost area and strain method in extensional cross sections (**Groshong**), or errors in extension estimates in rifts (**Morley**). Hand in hand with palinspastic reconstruction goes its counterpart, forward modelling, an example of which is presented from the Sirte Basin in Libya (**Skuce**).

Detailed geometric analyses of faults and the study of fault populations has also benefited from the advances made in seismic, but in these subjects, good old field geology has contributed significantly. **Kerr & White** present a method for calculating fault geometries based on seismic, whereas **Needham *et al.*** discuss the analysis of fault geometries and displacement patterns based on seismic, following an alternative approach to that of Kerr and White. A detailed study on the growth and linkage of faults is presented by **Cartwright *et al.***, based on fieldwork in the Canyonlands, Utah.

Analogue modelling has been around for more than a century, since Cadell presented his 'Experimental researches in mountain building' in 1889. In this volume, **McClay** discusses recent advances in analogue modelling studies. **Verschuren *et al.*** present multilayer sandbox models of thrust tectonics and analyse the effects of the viscous and non-viscous interlayers by means of palinspastic reconstruction.

Mathematical modelling of geological structures is a much more recent development, which has been helped by increasing computing power and the advances made in numerical techniques. The increasing interest in this field is demonstrated by no less than five contributions to this volume. It is encouraging that various scales of structural geology and tectonics are being covered. **Barnichon & Charlier** take a closer look at finite-element modelling of the development of shear-bands, **Hardy & Poblet** study a scale larger by modelling growth strata associated with fault-related fold structures. Studies by **ter Voorde & Cloetingh** and **van Wees *et al.*** and **Beekman *et al.*** examine progressively larger scales by going from basin stratigraphy (**ter Voorde & Cloetingh**) to the role of pre-existing weak zones in basin evolution (**van Wees *et al.***) and finally, in his study of a crustal scale relation between fault reactivation and lithospheric folding, **Beekman *et al.*** introduce one of the first published examples of 3D flexure modelling.

We feel that a book on structural geology would not be complete without examples of 'real' geology and we are fortunate to be able to include two excellent regional geological studies, each using a totally different approach. The paper by **Insley *et al.*** gives a comprehensive introduction of the use of satellite imagery, they also demonstrate the added value of integrating various techniques by combining satellite imagery with seismic, gravity, magnetic and field measurements. In his paper on gravity driven nappes and their relation to paleobathymetry, **Turner** uses examples from Africa and Europe and combines gravity, seismic and regional geological information to unravel the structural geological histories of these areas.

This volume stems from a three-day conference hosted at the Geological Society in London on 21–23 February 1994. Speakers from the United States and Europe gave a total of 41 papers. Some were invited to contribute to this special publication and we solicited further contributions to extend the coverage of the volume. The conference provided the forum for cross-pollination of ideas and stimulated lively debate. In part this was due to the quality of the presentations, but also to a perceptive and discerning audience. The conference would not have been possible without the tireless support of the Geological Society of London. We are also indebted to the following organisations who in one way or the other, gave their support: British Gas, BD Exploration, CogniSeis development, OMV and Shell Research. We would also like to thank the numerous referees who kindly donated significant time to reviewing the manuscripts and returning them in a timely manner.

The following referees are gratefully acknowledged for their reviewing efforts: Maurice Bamford, Jilles v.d.Beukel, Dan Bishop, Peter v.d. Bogert, James Buchanan, Joe Cartwright, Tim Chapman, Jean Chery, Richard Cooper, John Cosgrove, Patience Cowie, Ian Davison, Joop Dessauvagie, Gloria Eisenstadt, Jack Filbrandt, Raymond Franssen, Bertrand Gauthier, Gary Gray, Rick Groshong, Jake Hossak, Colin Howard, Sara Indrelid, David Kessler, Charles Kluth, Bob Lillie, Lidia Lonergan, Joe Mcnutt, Mike Naylor, Frank Nieuwland, Giles Pickering, Pascal Richard, Mark Rowan, William Sassi, Jonathan Turner, Jennifer Urquhart, Mark Verschuren, John Walsh, Jan Diedrik van Wees, Manuel Willemse, Matthew Willis, Reiny Zoetemeyer.

Reference

CADELL, H. M. 1889 Experimental Researches in Mountain Building. *Transactions of the Royal Society of Edinburgh*, **1**, 339–343, 1889.

Validation of seismic data processing and interpretation with integration of gravity and magnetic data

T. R. HORSCROFT & J. E. BAIN

LCT Ltd, 2nd floor, 83–84 George Street, Richmond, Surrey TW9 1HE, UK

Abstract: integration of gravity and magnetic data with seismic interpretations provides a definitive means of section validation. Traditionally this has been achieved as a two stage and separate process – seismic interpretation in the time domain and the potential fields model in the depth domain.

Recent advances in software technology now enables interactive simultaneous real-time modelling of seismic gravity and magnetic data. In addition to conventional interpretations, modern uses of modelling now includes: quality control of seismic depth conversion, control of the potential field model using offset raytracing; derivation of a high quality depth model as a starting point for processes like pre-stack depth migration or turning wave analyses. Integration too can be used to differentiate between geometrically similar but lithologically different seismic structures.

This paper describes the advantages of dynamically linking velocity, density and susceptibility in the modelling process illustrated by three different North Sea geological examples.

(a) Depth–velocity modelling of salt structures in the Central Graben.
(b) Structural interpretations of the Rattray volcanic rocks in the Outer Moray Firth.
(c) Shallow sand channels and velocity models from high resolution aeromagnetic data.

Geophysical data integration can at its simplest mean having one's maps at the same scale, enabling visual inspection and comparison of the data. Quantitative integration however involves a significant component of modelling in both 2D and 3D. Traditionally this has been used as an aid to structural interpretation of seismic data.

Many of the modern seismic processing techniques have a structural model as the starting premise. Tuning of this model is then performed and used as a refinement on the processing parameter selection. Other independently measured geophysical techniques, which respond to the same (or linked to) lithological parameters as the seismic method, can be used to provide important input to the selection of these starting point models.

Modern uses of integrated modelling include:

- Interpretation

  Depth model verification
  Fault analysis: position, type, throw
  Salt/shale: volume; structure
  'Reef' versus 'volcanic rocks'
  Sub-salt/sub-shale, or deep seismic structure.

- Processing

  Test seismic processing velocities – depth conversion parameters
  High resolution data for seismic statics corrections
  Test seismic dip parameters using magnetics
  Reduce number of iterations for pre-stack depth, turning-wave analysis.

Fundamental to the integration process is the link between physical rock properties. Density is the fundamental rock parameter in gravity interpretation and is a component of both seismic reflection strength and p-wave velocity. Gravity interpretation can involve direct inversion to density (and therefore to velocity). This is an important parameter in, for instance, AVO analyses. Conversely seismic inversion to velocity can be used as constraint on variation in density in for instance a 3D geological model. This can further enable seismically controlled gravity stripping (Bain et al. 1993). Susceptibility can be used to infer lithology and hence impose constraints on seismic velocity.

Integrated modelling techniques

Sensitivity model

An initial depth model is constructed from seismic and other data: refraction, resistivity, well and geological data. Rock properties are derived from well logs including velocity logs, resistivity logs, density and susceptibility logs as well as from rock samples, cores and tables. The starter model is digitized and then modified interactively in order

From Buchanan, P. G. & Nieuwland, D. A. (eds), 1996, *Modern Developments in Structural Interpretation,Validation and Modelling*, Geological Society Special Publication No. 99, pp. 5–9.

that the observed gravity and/or magnetic data may be satisfied. This approach to modelling is commonly used to test sensitivity of the model at various parts and in consideration of various 'what if' scenarios.

Iterative model

The depth and time verification of the interpretation relies upon a link between the time seismic interpretation and the potential fields model in depth.

Indirect
The interpretation is verified with seismic modelling. The depth model once adjusted to fit the gravity and magnetic data can via offset raytracing be compared with the input data. Geometric and physical rock properties are adjusted iteratively until the model converges, satisfying both the seismic and potential field data.

Direct
The ability to directly compare and modify seismic interpretations in time (or depth) simultaneously and in real-time with depth models constrained by gravity and magnetic data provides the most efficient and precise modelling available.

Examples of integration

Depth–velocity modelling

Although the conversion of time to depth in seismic interpretation is considered routine for certain geological conditions, seldom is the process error free, especially in areas of complex geology having velocity gradients, diffractions and geometric ambiguities. One efficiently imposed constraint is the use of gravity and magnetics data to substantiate or refine the assumed time to depth function and resultant velocity distribution.

Areas of the North Sea affected by mobilized salt present particular problems for seismic interpretation and depth conversion. The example shown in Fig. 1a, b is from the central North Sea where the exploration objective is the subsalt Permian Rotliegendes section. Uncertainties in the volume of salt present and complex velocity variations cause the simple layer cake depth conversion method to be suspect (Marsden, 1989).

The methods developed by Zijlstra *et al.* (1991) exploit seismic velocity modelling as an aid to depth conversion in this area. Input time picks and stacking velocities are utilized in a velocity modelling program with interval velocities calculated using the Dix formula. Forward modelling is then performed using offset raytracing to yield reflection travel times and associated stacking velocities for comparison with the input data. The starting interval velocity field is then improved through inverse raytracing, ending when a reasonable convergence to within the data uncertainties is reached. The final seismic-derived depth conversion and velocity field are shown in Fig. 1a.

In this case the resultant seismic-derived depth model is input directly into LCT's gravity and magnetic modelling system, 2MOD^{TM}. Synthetic density values are obtained from the velocity field obtained from velocity modelling and these are calibrated with logged density from well data. The theoretical gravity and magnetics fields are then computed for the model and compared with the observed fields, as shown in Fig. 1b. The sensitivity of the potential fields model to changes in geometry is tested. This sensitivity is governed by the density or susceptibility contrast, the depth of the structure and scale of the adjustment. For this model structural variations of the order of 5% of the depth are significant.

Of course, if the time interpretation, depth conversion and lithological parameters were perfect, the computed and observed fields would match. Differences between the fields can be reconciled in real time by altering the depth model, which inherently requires an update to the velocity field if the time interpretation is deemed correct. Once an altered model is derived, the depth model can, with offset raytracing be compared with the input seismic section. This process is continued until a combined earth model is derived, which satisfies all of the available input data constraints and concepts.

This example demonstrates that a higher degree of confidence is gained when a time interpretation is converted to depth, the velocity field tested and updated using inverse raytracing, and the result further verified or altered using the gravity and magnetic constraint.

Fault interpretation/thickness of volcanics

The following example is from the Outer Moray Firth region of the UK North Sea. Seismic investigations of the area are hampered by the presence of a high velocity basaltic layer – the Rattray Formation (Boldy & Brealey 1990). This area has recently been surveyed using a multi-sensor marine vessel to collect 3D seismic data and high resolution gravity data. In addition, a high resolution total intensity aeromagnetic survey was flown over the area, extended to include a large halo around the marine survey.

The primary use of the gravity and magnetic data was to determine:

(a) the structure of the Rattray Formation (mid-Jurassic volcanic rocks);

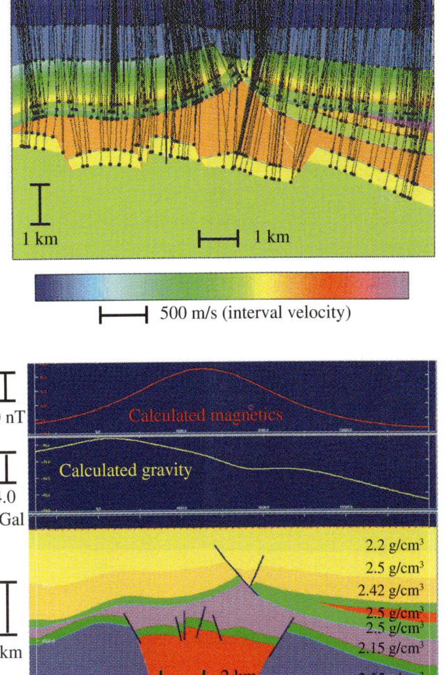

Fig. 1. (**a**) Final depth conversion/velocity field. Results generated using Jason Geosystems' DSite™. (**b**) Dynamically linked seismic/gravity/magnetics modelling. Results generated using Jason Geosystems' DSite™ and LCT's 2MOD™.

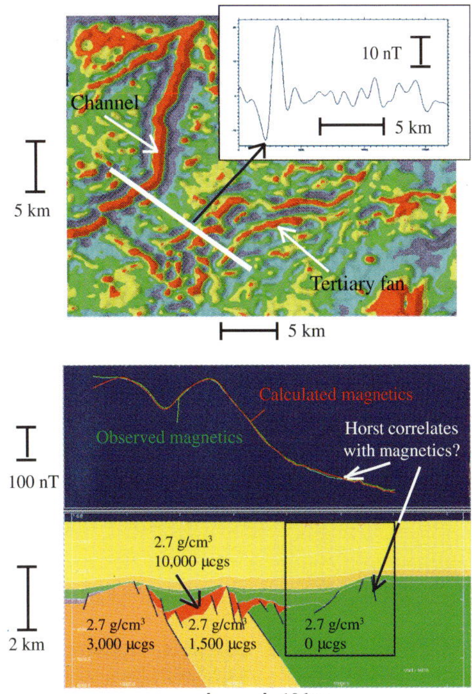

Fig. 2. (**a**) Magnetic effect of near-surface channel. (**b**) Narrow horst versus shallow channel.

(b) the nature and age of the pre-Rattray Formation sediments;
(c) the structure of the crystalline basement.

Figure 3a illustrates the initial seismic interpretation along a N–S oriented line which crosses the Renee Ridge in the south (left hand side). This model (interpreted to top Rattray level) was depth converted and analysed using the gravity and magnetics data. Susceptibilities in the range of 10 000 micro cgsu were assigned to the basaltic Rattray Formation rocks. The total magnetic intensity field is dominated by two high amplitude (100–150 nT) oval, positive magnetic anomalies. The Renee Ridge is characterized by one positive anomaly separated by a strong negative gradient coincident with the location of the Renee Graben. The second magnetic high to the north has a strong negative gradient on the northern side.

A regional magnetic field was derived via application of a 30 km low-pass filter applied to the pole reduced field. This was based upon an analysis of the power spectrum and a 3D model of the magnetic basement. The regional field is characterized by a large triaxial positive anomaly centred to the north of the Renee Ridge. The cylindrical symmetry of this anomaly is typical of what might be expected from a large dome (possible high level magnetic core). Alternatively this 'high' could be explained by the localized thickening of a broad and shallow flat lying body (thick volcanic model). The latter model provides a less conclusive explanation however for the strong negative gradient around the northern margin of the anomaly, evident as a strong gradient in the profile.

The magnetics data were used to identify variations in basement structure and lithology, with a high susceptibility 'basaltic' complex in the centre, rimmed by lower susceptibility 'acidic' components as shown in Fig. 2b. Of particular note was the low angle basement contact on the right side of the model, as interpreted from the magnetics data. Much of the primary faulting in the area appears to be coincident with or related to these large apparent basement contact zones, confirming that the younger structuring may be a reactivation of older basement faulting.

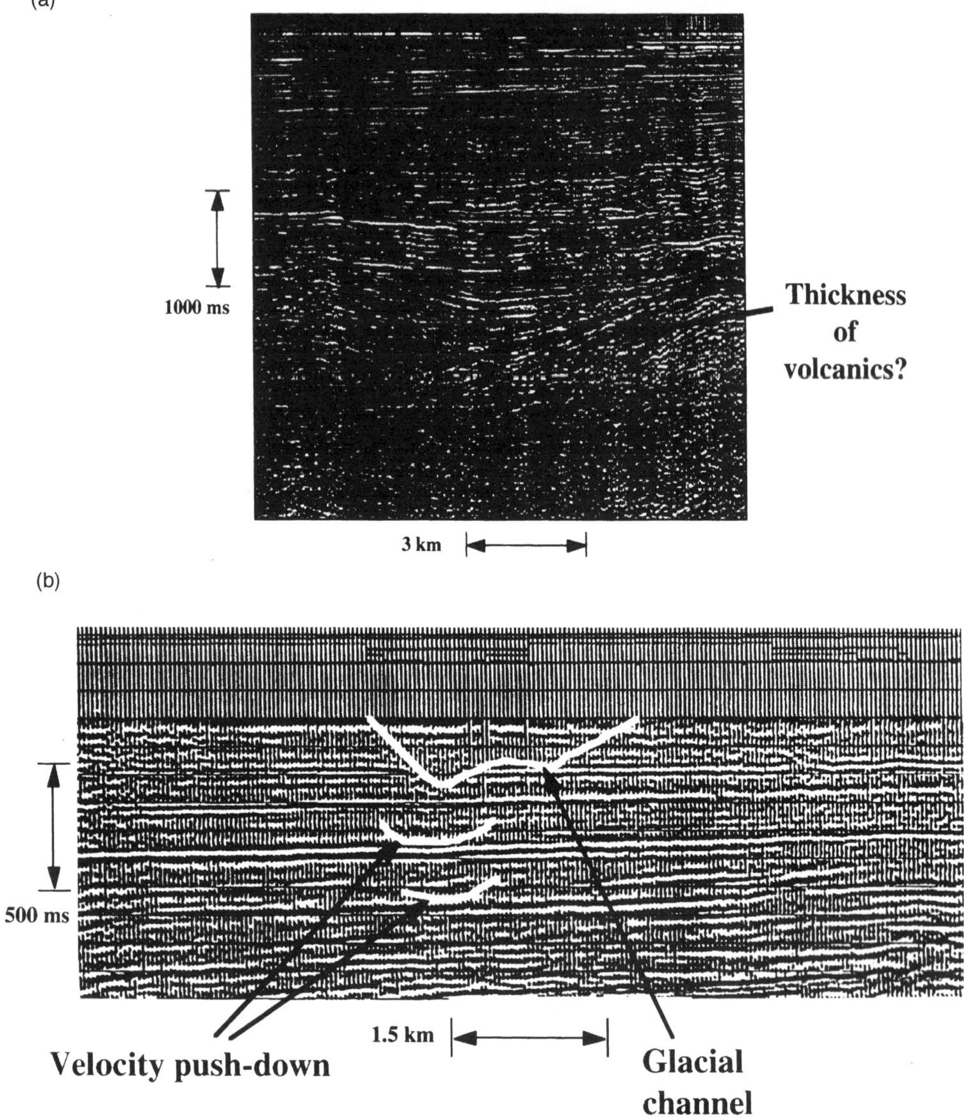

Fig. 3. (**a**) Preliminary seismic interpretation, Central Graben, North Sea. (**b**) Dynamically linked seismic/gravity/magnetics glacial channel model.

The key features of the 2D model include:

(a) the volcanic sequence north of the Renee Ridge is preserved in discreet fault bounded blocks of variable thickness;
(b) the model demonstrates progressive thinning of the Rattray Formation at the northernmost margin of the area;
(c) the model demonstrates that (with a contrast susceptibility) the Rattray Formation is thicker on the downthrown side of the boundary faults and may have evolved during a period of syngenetic growth faulting.

This example demonstrates that improved structural understanding is gained using a multidisciplinary seismic, gravity and magnetics approach.

Shallow sand channel and velocity effects with seismic/magnetic modelling

The magnetization of shallow sediments typically varies between 0–300 micro cgsu, with 70 micro cgsu being a useful global average (Saad 1993).

Modern low level (100 m) high density (200 m line spacing) aeromagnetic surveys are of sufficiently high resolution to enable an interpretation of these low susceptibility sediments at depth.

Several such surveys have been flown in the North Sea which have, as a general characteristic, a very high frequency, high relief character. Whilst interesting, any qualitative interpretation of these high frequency events needs to be made with caution as the following example demonstrates.

Figure 2a illustrates the magnetic field over a part of the north Central Graben region of the North Sea. The signature is easily correlated with shallow channels evident in the seismic data and then used as an additional guide for linking up the seismic events. The magnetic data reveal sediments of both glacial (linear channels) and fluvial (braided river or fan identified as a birds' foot anomaly) origin. The associated magnetic highs have amplitudes in the range of 0.5–10 nT (gammas), with excellent correlation between the anomaly gradients and the margins of the channels. Interesting to note is the particular character of the anomalies observed over these channels. Each anomaly is typically asymmetric with pronounced negative side lobes characteristic of a source geometry with restricted depth extent. Most channels generate positive anomalies indicative of normal magnetization (sedimentary magnetization or strong overprinting) in the present day field. In this case the dominant wavelengths very between 0.5 km and 2 km.

The model in Fig. 2b shows an interesting correlation between a typical high frequency magnetic anomaly and a small basement horst. It is easily demonstrated that the maximum depth the magnetic anomaly can be sourced from is substantially shallower than the horst feature. Detailed analysis of the magnetics data suggests source depths for the magnetic features in the range of 100 to 300 m. Close inspection of the shallow seismic data suggests glacial channels may be present which would tend to be slightly higher in magnetic susceptibility than the surrounding sediments. Figure 3b contains the expanded area of the apparent glacial channel, which is simultaneously modelled using seismic, gravity and magnetics data. The seismic data in this area indicate a possible velocity push-down resulting from the occurrence of somewhat lower velocities in the channel than assumed for the shallow section.

This example shows that high resolution magnetics can be helpful in the corroboration and delineation of sand channels and similar stratigraphic features. Shallow seismic processing can be improved by identifying low density, slightly higher susceptibility channels, which may be corrupting the shallow seismic processing.

Conclusions

The results discussed herein demonstrates that we can, in certain cases, use gravity and magnetics to:

(a) validate the seismic depth conversion process;
(b) resolve seismic ambiguities and improve structural interpretation;
(c) improve stratigraphic interpretation and seismic processing parameters.

References

BAIN, J. E., WEYLAND, J., HORSCROFT, T. T., SAAD, A. H. & BULLING, D. N. 1993. *Complex salt features resolved by integrating seismic, gravity and magnetics.* EAEG/EAPG conference, Stavanger, Norway.

BOLDY, S. A. R. & BREALEY, S. 1990. Timing, nature and sedimentary result of Jurassic tectonism in the Outer Moray Firth. *In*: HARDMAN, R. F. P. & BROOKS, J. (eds) *Tectonic Events Responsible for Britain's Oil and Gas Reserves.* Geological Society, London, Special Publications, **55**, 259–279.

MARSDEN, D. 1989. Layer cake depth conversation. *The Leading Edge*, **8**, 10–14.

SAAD, A. H. 1993. *Interactive Intergrated Interpretation of Gravity, Magnetic and Seismic Data: Tools and Examples.* Presented at the 25th Annual OTC in Houston, Texas.

ZIJLSTRA, D. M., VAN DER MADE, P. M., BUSSEMAKER, F., VAN RIEL, P. 1991. *Effective Depth Conversion – A North Sea Case Study.* Applications to Exploration Problems, Jason Geosystems.

Scaling of fault displacements and implications for the estimation of sub-seismic strain

G. PICKERING, J. M. BULL & D. J. SANDERSON

Geomechanics Research Group, Department of Geology, University of Southampton, Southampton SO17 1BJ, UK

Abstract: Fault displacement populations have been shown to follow a power-law scaling relationship characterized by an exponent D. This relationship can be used to make predictions of the sub-seismic fault population from data derived from seismic surveys. Although fault populations exist in three dimensions the use of section data is recommended. *D*-values derived from sections can be applied directly to several problems, and are also related to the *D*-value for the fault set in higher dimensions. Accurate determination of *D* requires proper consideration of the scale range and sample size limitations of available data. The most common technique of using a cumulative frequency graph often leads to an upwards bias. An iterative correction procedure is proposed. Discrete frequency methods avoid this bias, but as a standard linear interval graph has other associated problems, a log-interval graph method is preferred. Simulations of these methods, applied to random computer generated samples from power-law distributions, have been made to examine the accuracy of D-values derived from typical data. Equations to estimate the confidence intervals for these D-values have been derived from a synthesis of the results. The application of the techniques is shown using fault data measured on seismic sections from the Southern North Sea and the Inner Moray Firth. Where local differences in D are shown to be significant, there is usually a marked change in structural style. Fault data are used to make improved estimates of crustal extension (ß) by extrapolating the derived power-law relationship. A value of ß = 1.20 is calculated for the Inner Moray Firth. Applications predicting the intersection of horizontal wells with 'large' sub-seismic faults and quality control of fault interpretation on seismic sections are also described.

Over the past decade or so there has been considerable research into fractal or power-law distributions and their application to fault populations. However, there remains some confusion and not a little scepticism as to the validity of such models and how they may be applied in practice. A power-law distribution is defined,

$$N = cu^{-D} \tag{1}$$

where N is the cumulative number of values $\geq u$, c is a constant and D is the power-law exponent or D-value. The traditional way of testing this model for fault populations has been to plot log$N$ against log$u$ where u is the displacement or component of displacement. This graph should have a linear relationship with slope of $-D$. There are problems with this method which will be discussed later, but it is often sufficient to show whether the power-law model is appropriate.

The best evidence to support a power-law model comes from the common geological experience of scale-invariance, in which phenomena appear similar at different scales. Faulting is common over many scale ranges, and the power-law is only quantifying this observation. Other distributions, such as log-normal or negative exponential, have a characteristic size and predict very low fault densities above or below the scale of observation. These distributions do not match geological reality. The question that remains is whether there is a single systematic relationship between the numbers of faults and the scale at which they occur. There is considerable evidence to confirm this relationship as power-law, particularly for fault populations in extensional basins. Firstly, there are many published data sets which fit the power-law model, e.g. Kakimi (1980), Childs *et al.* (1990), Heffer & Bevan (1990), Walsh *et al.* (1991, 1994), Jackson & Sanderson (1992) and Pickering *et al.* (1994). Secondly, where predictions of sub-seismic faults using power-laws have been tested against core or field data, the results have shown a good correspondence, e.g. Walsh *et al.* (1991), Yielding *et al.* (1992) and Fig. 1.

There are clearly limits to this scale invariance. At the small scale, other structures may contribute significantly to the strain, such as tensile fractures. This may alter the stress field introducing some scale dependent effect. The particulate nature of the rock may also impose some fixed scale, breaking the scale invariance. As most data on faults is limited to millimetre sized displacements, the exact

From Buchanan, P. G. & Nieuwland, D. A. (eds), 1996, *Modern Developments in Structural Interpretation, Validation and Modelling,* Geological Society Special Publication No. 99, pp. 11–26.

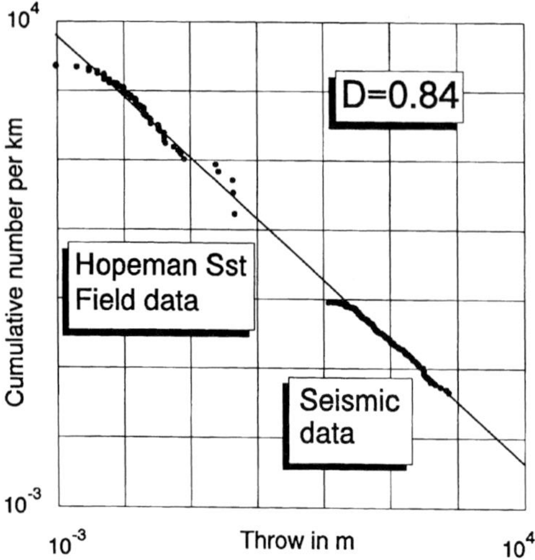

Fig. 1. Fault throw plotted against cumulative number per km on log axes. The two data sets were measured from Triassic rocks located in the Inner Moray Firth (IMF A). The seismic sections used were from the offshore SSL-MF89 survey and were located in UKCS Quad 17 north of Lossiemouth. The field data were measured on several cliff sections 5 km east of Lossiemouth. These exposures are continuous with the offshore basin. The good fit of both data sets to the line shown proves the fault population in this area follows a power-law distribution with a *D*-value of 0.84. Further details of this study are in Pickering *et al.* (1994).

limit of the scale invariance is difficult to estimate, but values of around a millimetre have often been assumed (Walsh & Watterson 1992). At the largest scale, there is a fundamental geometrical distinction between faults that are within the brittle crust (intra-crust) and those that span the crust. A fault which spans the brittle crust only occupies a two-dimensional shell, rather than a three-dimensional space (Marrett & Allmendinger 1991). This geometrical change introduces a change in the scaling relationships of both earthquakes and the dimension and displacement of the faults (Pacheo *et al.* 1992; Westaway 1994). However, all the fault data in this paper are taken from intra-crustal faults, and no geometrically controlled scaling change is expected. References to small scale and large scale faulting relate to size differences within this population.

This paper first presents the methods of analysis of power-law distributions to introduce them to the non-specialist. These are then illustrated using examples based on seismic sections from the North Sea basins. Applications of the technique to hydrocarbon exploration and production are then outlined.

Data and methods of analysis

Fault displacement data

This paper will concentrate on the analysis of fault displacement data from sections, however much of the theory is applicable to any power-law distributed data. There are several reasons for concentrating on section data. Structural interpretation is often made using sections, therefore such fault data are readily available and often of good quality. Section data also avoid some interpretational and censoring problems associated with fault maps (see Yielding et al. 1992) and as sections provide a projection through the fault set, the D-value found is directly related to the D-value describing the full set in three dimensions (Marrett & Allmendinger 1991; Walsh et al. 1994). It is important that the property measured is proportional to the displacement. Dip-slip faults or oblique-slip faults with a large dip-slip component, can be properly measured on vertical sections. Strike-slip faults must be measured using the lateral offset of markers (e.g. dykes) as shown on maps (e.g Jackson & Sanderson 1992). In much of the published work a component of displacement has been used. With seismic sections vertical separation or throw is usually the most accurately determined parameter and will be proportional to displacement as long as the above requirement for the use of vertical sections is satisfied. The examples given in this paper are all based on throw measurements.

Unfortunately most single sections do not resolve large numbers of faults. A large sample size is needed to reasonably estimate the *D*-value as will be shown later, consequently many researchers have combined measurements from multiple traverses, so called multi-line data sets (Yielding *et al.* 1992). If the lines intersect different faults then the multi-line sample will be effectively equivalent to single-line data. This is usually the case for the majority of the sample, as smaller faults tend to have smaller dimensions (Gillespie *et al.* 1992; Cowie & Scholz 1992). One of the disadvantages of 2D seismic grids is that correctly correlating small faults is impossible as they often intersect only one line. Large faults will be intersected by many lines, especially if the lines are close spaced. Therefore, the large scale data often represents the distribution of displacement along faults, rather than the distribution of displacement amongst the fault population.This problem has been recognized (Walsh *et al.* 1991, 1994) and leads to a steep right-hand tail on cumulative graphs. Their approach has been to graph all the data but only fit a line to the data outside of the tail, on the assumption that this will give a representative *D*-value. The validity of this method depends on a number of factors, which will be discussed after considering the problems with single-line or effective single-line data. These must be understood before a full discussion of multi-line data is possible.

Any data derived from a section or map are scale limited. Many authors have identified the problem of small-scale or left-hand truncation, but there is also a problem of large-scale or right-hand truncation in most data sets which is less often considered. Left-hand truncation (LHT) is caused by a lower limit to the resolution, often due to the sampling method. A common example is the limit of resolution on a seismic section due to the band-width of the seismic signal. There is often a loss of data as this limit is approached, as the actual limit will depend upon the noise level, which is likely to vary across the section. This is called the LHT 'fall-off'. This problem has been widely discussed with regard to fracture studies (e.g. Einstein & Baecher 1983) and some corrective methods have been derived. Right-hand truncation (RHT) can be caused in two ways. First the large values may be excluded from the sample. On an outcrop the hanging wall or foot wall cutoff may not be visible or maybe lost due to erosion. A basin bounding fault may be beyond the areal coverage of a seismic survey. Secondly, as they are few in number, the largest faults always have a low probability of being sampled. The effect that these problems have on measuring *D*-values is dependent upon the method used and will now be discussed.

Cumulative frequency graph

This is the most common method employed for the measurement of *D*-values and is directly derived from the definition of the power-law given in equation (1). By taking logs of (1) a plot of log$N$ against log$u$ should plot giving a straight line with

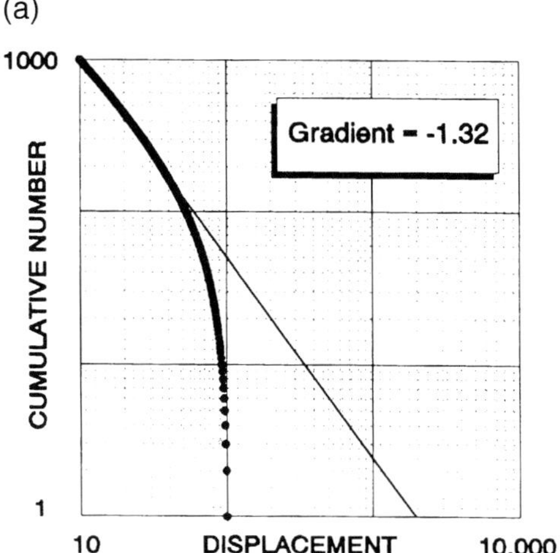

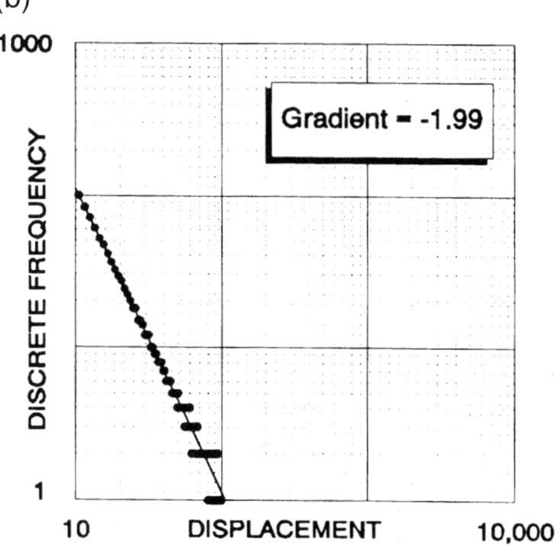

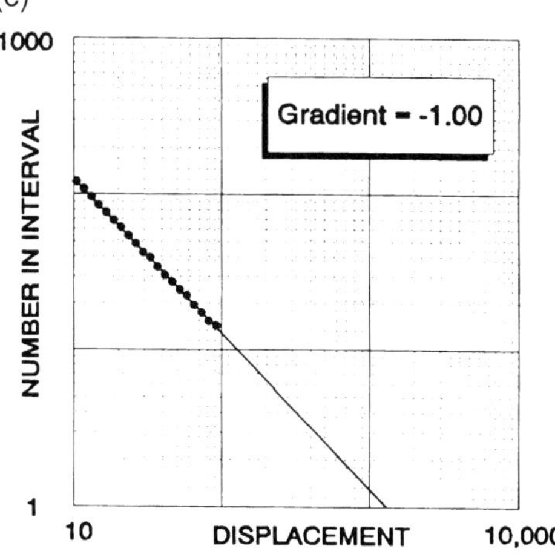

Fig. 2. Three graphs of the same idealized data set of 1000 values from a power-law distribution with a *D*-value of 1.0. The scale range of the data is only one order of magnitude.
(**a**) Log–log cumulative number graph showing finite-range deviation and giving a biased *D*-value estimate of 1.32.
(**b**) Linear interval discrete frequency graph plotted on log–log axes giving a slope of –1.99, which is equivalent to $D = 0.99$. The breakdown of the method can be seen around the larger values in the sample.
(**c**) Log–interval discrete frequency graph, with equal intervals on the log scale of 0.05. The number counted in each interval is plotted against the lower bound of the interval. This graph shows no bias or breakdown and gives a *D*-value equal to that of the distribution.

a slope of $-D$. The LHT fall-off is easily detected on this graph, showing a clear deviation from the expected linear trend at the small scale. Such deviations can be seen on both of the data sets shown in Fig. 1, particularly in the case of the field data. If this part of the data is excluded from the analysis, then the D-value derived will not require any correction. At the large scale there is often another deviation from linearity, which can occur on both single-line and multi-line data and is different from the multi-line effect described above. This is introduced by the RHT, but is also dependent on other factors. This deviation has been called 'censoring', e.g. Jackson & Sanderson (1992), Pickering *et al.* (1994), however this has lead to confusion with other sampling effects, and we feel that a more appropriate term is the finite range effect. An example graph is shown in Fig. 2a.

Finite range effect

Although almost all data-sets have a RHT, not all show a deviation at the large scale. Consider the data-sets shown in Fig. 3. This graph shows three idealized samples with no LHT fall-off, derived from a power-law distribution with $D = 1.0$, and a maximum value of 10 000. Sample A, of size 100, has a maximum value of 100 and a minimum at 1.

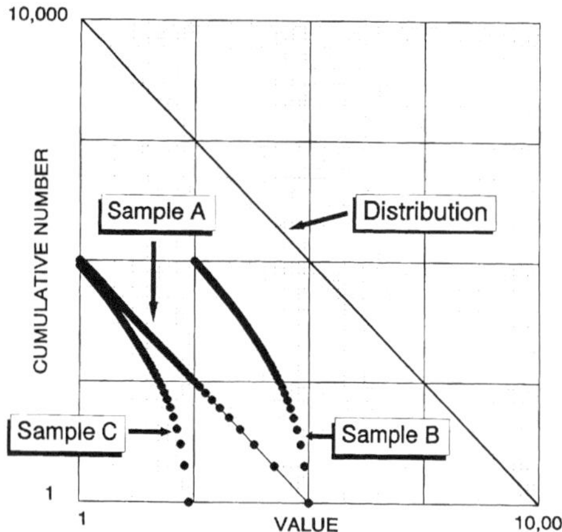

Fig. 3. Log–log graphs of cumulative number against value for three idealized data sets, taken from a power-law distribution. This distribution has a maximum at 10 000 and a D-value of 1.0. Samples A & B contain 100 values, with scale ranges of two and one order of magnitude respectively. Sample C contains 90 values, and is equivalent to Sample A with the largest 10 values removed. The deviation from linearity shown by samples B and C is due to the finite range effect (see text).

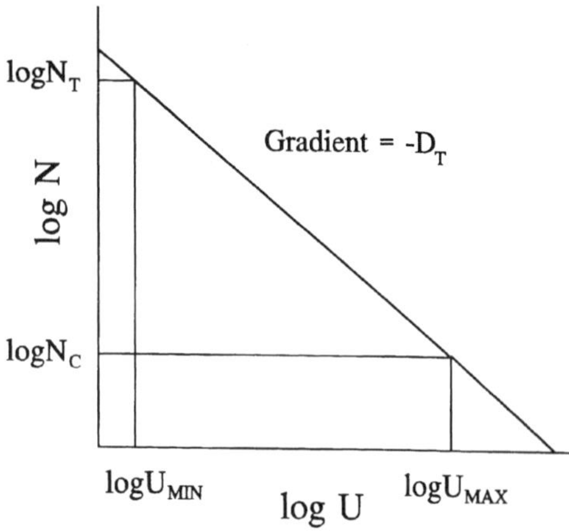

Fig. 4. Log–log cumulative graph of a self-similar sample from a power-law distribution with exponent D_T. U is in arbitrary units and N is cumulative number. If a sub-sample from U_{MIN} to U_{MAX} is taken then N must take values from N_C to N_T, but in practice data will be plotted from 1 to $N_T - N_C$. The geometry of the plot may be used to estimate N_C – see text.

This does plot giving a slope of one. Sample B has the same RHT at 100, but a minimum at 10 and is clearly distorted. More importantly a graph of this sample would lead to an over-estimate of D. The essential difference between the two is the way that the scale range and size have been reduced in comparison with the original distribution. For sample A both have been reduced by a factor of 1/100, while for sample B the size has been reduced by 1/100, but the scale range has been reduced by 1/1000. Therefore, sample A is 'self-similar' to the distribution, whereas sample B is not.

There is a solution for this problem. Consider sample C, this is sample A with the ten largest values removed, but could equally well have been derived from the original distribution. If the largest value was assigned a cumulative number of 11, the distortion would be removed. This 'correction' can be predicted using the following derivation for a general power-law.

Figure 4 shows the line of a self-similar sample from a power law distribution on a log–log plot, which shows no finite range effect. The slope of the line is $-D_T$. The scale range of the sub-sample taken from this is U_{MIN} to U_{MAX}. To make this sub-sample self-similar to the distribution $(N_C - 1)$ extra faults with displacement $>U_{MAX}$ are needed, where N_C is the cumulative number assigned to U_{MAX} in the original self-similar sample.

From Fig. 4

$$\log N_T - \log N_C = -D_T(\log U_{MIN} - \log U_{MAX}) \quad (2)$$

where N_T is the sample size plus the correction ($N_C - 1$). Therefore:

$$\log N_C = \log N_T - D_T(\log U_{MAX} - \log U_{MIN}) \quad (3)$$

For the examples given in Fig. 2, the D-value was one, and the size and scale range had to be equal to preserve self-similarity. In the general case the two will scale differently. This is determined by the D-value, hence the D_T term in equation (3). In the case of a D-value of 0.5, for example, the size of the sample must be reduced by two orders of magnitude for every one order of magnitude reduction in scale range.

With real data only U_{MAX}, U_{MIN} and N_E, the sample size, are known. The slope of the distorted graph, D_E, can be used to estimate D_T, and if we assume that $N_C \ll N_T$ then

$$N_E = N_T - (N_C - 1) \approx N_T \quad (4)$$

or

$$\log N_E \approx \log N_T \quad (5)$$

substituting into equation (3) gives

$$\log N_C \approx \log N_E - D_E(\log U_{MAX} - \log U_{MIN}) \quad (6)$$

N_C can now be estimated and used to correct the bias in the log–log cumulative plots. As the distribution D-value is not known, the correction is approximate, and will become increasingly more inaccurate as N_E and D_E become poor estimates of N_T and D_T respectively. Returning to Fig. 3 and sample C, $D_E \approx 1.2$, $N_E = 90$, giving $N_C = 6$, where as the true value of N_C is 11. An iterative approach can be used, where the D-value from a graph corrected using the first estimate of N_C ($N_C^{\#1}$) is substituted for D_E and $N_E + N_C^{\#1}$ is substituted for N_E. This approach is shown for sample C, previously displayed in Fig. 3, in Fig. 5, where the first uncorrected estimate of D (1.31) is biased. As N_C can only take integer values, once the majority of the bias is removed a small change in D-value will lead to the same value of N_C and therefore no further change in D. In this case after six iterations the method converges on a D-value of 1.05 which is still slightly biased but is much closer to the true value.

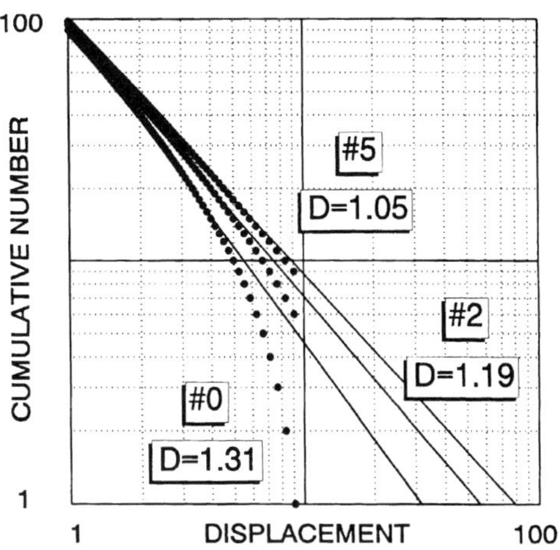

Fig. 5. A series of plots of sample C from Fig. 2 illustrating the finite-range correction procedure. Plot #0 is the uncorrected data which gives a D-value of 1.31, compared to the original distribution D-value of 1.0. After two iterations (#2) D = 1.19. At five iterations (#5) no further correction is predicted giving a final result of 1.05.

Discrete frequency graph

The discrete frequency distribution (n) can be derived for a power-law using equation (1):

$$n = \delta N = -cDu^{-(D+1)}\delta u \quad (7)$$

A plot of n against u on log axes should give a slope of $-(D + 1)$. An example using idealized data is shown in Fig. 2b. As the method makes no assumptions about the nature of the population beyond any RHT, there is no finite range effect. The LHT fall-off is inherent in the data and is more pronounced on discrete frequency graphs. However, if the affected intervals are excluded, the calculated D-value will be representative. The derivation of n assumes that small changes in u (i.e. $u + \partial u$) cause a change in n. This is only reasonable at the small scale, where the data are concentrated. At the large scale the assumption breaks down and the intervals either become empty or contain only one value. Consequently only those intervals that contain a significant proportion of the data can be used to find D.

Log-interval graph

The log-interval graph retains the advantage of a discrete frequency method, but the data is spread more evenly among the intervals. If we take two cumulative numbers, $N_2 > N_1$ where

$$N_2 = cu_2^{-D}$$
$$N_1 = cu_1^{-D} \quad (8)$$

then

$$N_2 - N_1 = c(u_2^{-D} - u_1^{-D})$$
$$- cu_2^{-D}\left[1 - \left(\frac{u_1}{u_2}\right)^{-D}\right] \quad (9)$$

taking logs

$$\log(N_2 - N_1) = \log c - D\log u_2 + \log\left[1 - \left(\frac{u_1}{u_2}\right)^{-D}\right] \quad (10)$$

where $N_2 - N_1$ is the number in the interval $u_1 \rightarrow u_2$. If we take a series of intervals, keeping u_1/u_2 constant, then a plot of (10) will give a slope of $-D$. The effect of keeping u_1/u_2 constant is to make the intervals a constant size on a log scale, producing a discrete frequency plot of the logs of values. This is similar to a discrete frequency plot of earthquake magnitudes, as the magnitude is a log function of the size of the earthquake (e.g. Main & Burton, 1989). An example of this method is shown in Fig. 2c.

Multi-line data

The description of the methods so far has been made on the assumption that the data was single-line or effectively single-line. Data sets that contain repeated measurements from the same fault behave somewhat differently. The right-hand tail seen on cumulative graphs will be represented as a peak at the large scale on the discrete graphs (see Fig. 6). If those intervals within this peak are excluded from the analysis then the remaining data can be used to define D. The approach is less straight forward for the cumulative graph. In many cases the steep right-hand segment is equivalent to correcting the remaining part of the graph for the finite range effect. This is why D-values derived from these graphs in the past have often been reasonable. In some cases the number contained in this segment may be significantly different from the actual correction required, which will lead to

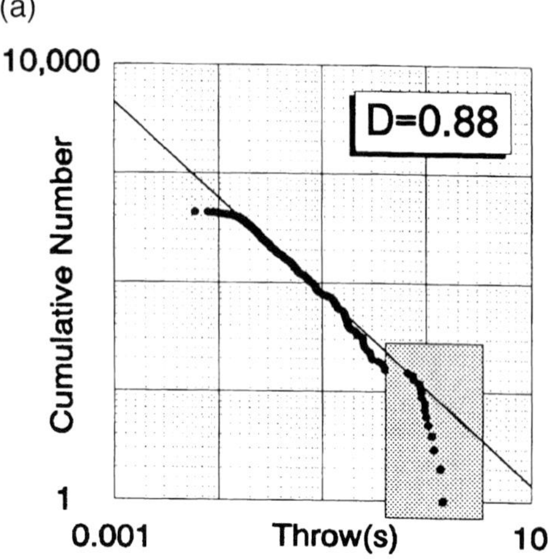

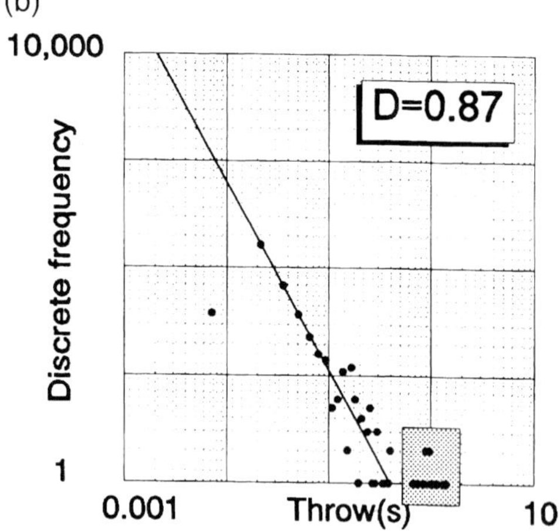

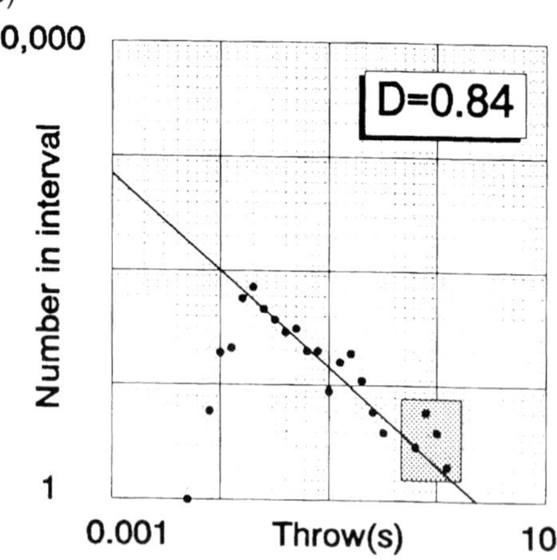

Fig. 6. Graphs of the same multi-line data set using the three analysis methods (cf. Fig. 2). The data are throw measurements from steeply dipping normal faults, made on the dip lines of a seismic survey.
(**a**) The largest fault in the area has been intersected by all of the lines. The similar throw values measured for this fault on each intersection with a seismic line plot as a steep right-hand tail on the cumulative graph as marked by the grey shading. Although the line is only fitted to the rest of the data, the inclusion of this tail affects where the other data fall on the cumulative graph.
(**b**) On the discrete frequency curve this right-hand tail is seen as a small peak at the large scale end of the data. This peak has no effect on the remaining data, which can be fitted to give the D-value.
(**c**) A similar peak is seen on the log–interval graph, and again this peak is excluded from the line fitting.

under- or over-corrected graphs. Clearly if the right-hand segment is removed then the finite range effect is likely to cause distortion. This problem can be solved by either correction or by using a discrete method. Although Walsh *et al.* (1994) did not address the problem of RHT and the consequent effect that this has, they did offer several methods to analyse these edited multi-line data sets. Their methods were all based on averaging the D-values derived from single lines. These smaller data sets were unaffected by the finite range effect as the size and scale range were usually similar. However, due to the small sample size the D-values derived are prone to large random error, and even when averaged this may lead to a consequently large error in D.

Simulations of power-law data and analysis

There are two reasons for simulating the process of sampling power-law distributions and measuring D-values. The first is to test the effectiveness of the methods on random samples. The estimates of D derived from the samples can be checked against the known distribution D-value for bias and random error. The second is to try to estimate confidence limits for these sample D-values. These aims will not be achieved unless the simulation is a reasonable model of the actual data and analysis involved. Producing samples from a power-law is relatively simple. The power-law distribution is defined in terms of its cumulative-frequency distribution, therefore the Inverse Transform Method (Rubinstein, 1981) can be used. This method uses the cumulative number N, picked at random, to calculate a value of u which for power-law distributions can be achieved by using equation (1). A scale-limited sample can be produced by restricting the values of N.

The results presented in this paper are all based on simulations of single-line samples with no LHT fall-off. Although almost all real data sets suffer from a LHT fall-off problem, only the unaffected part of the data is used for analysis, so using unaffected data in the simulation is valid. The same reasoning applies to multi-line data, where if the large scale 'tail' part is removed the data is effectively single-line.

At each sample size and scale range, 1000 simulated data sets were produced in order to characterize the distribution of sample D-values. These samples were analysed using the same methods as those used for any real data set. However, some automation is required in order to process enough samples to make the results statistically significant. The simulation of each method is discussed separately. A more detailed discussion of the program is included in Pickering *et al.* (in press).

Cumulative frequency method

As there was no LHT fall-off, all the data points in each sample were used to fit a line to the graph. For those samples where a finite-range effect would be predicted, the mean D-value showed the expected upwards bias. Also as expected, the larger the difference between the sample size and D_T x the scale range (see equation (3)), the greater the bias. As the D-value of the distribution is known, the effectiveness of the correction can be tested. If the correction is calculated using the known distribution D-value then the bias is successfully removed. For real data this is not possible, and the iterative process described above (Fig. 4) is required. This process was tested by allowing the program to iterate, applying a correction and re-calculating D, until the D-value remained constant. The results of this simulation depended critically on the scale range and numerical size of the samples. When the scale-range is two orders of magnitude or more a good estimate of D is obtained, however when the range is only one order of magnitude and the sample size is small (< 100) the mean D of the samples is a 5–10% under-estimate of the true D-value, with some samples apparently over-corrected. This is an artifact of the simulation, due to the removal of the operator from the decision process. For real data the operator would stop the process once the deviation was removed even if it had not converged. However, the lack of control on the D-value caused by the narrow scale-range can lead to over-correction unless care is taken.

The distribution of sample D-values produced in the simulation, can be used to derive confidence limits for D-values derived from real data (Press *et al.* 1986). Sample sizes from 100 to 1000 were used with scale ranges of one to three orders of magnitude. The analysis used the iterative correction procedure. The distribution D-value was set between 0.5 and 1.5. These bracket typical parameters for fault populations. A synthesis of the results from these simulations gave the basic form for an empirical equation estimating these confidence intervals, i.e.

$$\text{(a) } D \geq 1: \ \sigma = kD\sqrt{\frac{1}{\textit{Sample size}}} \qquad \text{(b) } D < 1: \ \sigma = k\sqrt{\frac{D}{\textit{Sample size}}} \tag{11}$$

The value of k is dependent upon the scale range of the data and the confidence level desired. The k

Table 1. k *values used in equation (11)*

	Cumulative frequency graph	Log-interval discrete frequency graph
Sample scale range (order of magnitude)	*k*	*k*
One	1.9	2.0
Two	1.2	1.2
Three	1.1	1.2

The cumulative graphs were corrected for the finite range effect using the iterative correction procedure. The log interval values are from smoothed graphs using interval sizes of 0.1 (see text).

values for the 68% interval are shown in Table 1, these should be doubled for the 95% interval. For two orders of magnitude the value of k is *c.* 1.2, where as for one order of magnitude k is *c.* 1.9. The widening of the interval for the smaller scale range reflects both the reduced control on *D* and the consequent problem with deriving a reasonable finite-range correction.

Discrete frequency method

This method is difficult to simulate as a high level of interpretation of the graphs is required. The analysis always breaks down at the large scale end, but the value at which this occurs depends on each individual data set. The program simplifies the process by only fitting a line to those intervals which contain at least two values. In general, the method consistently underestimated the true value of *D*, with a random error 5–10 times that seen in the simulation of the cumulative method. This large error is to some extent predictable. As most of the data are concentrated in a small proportion of the intervals at the small scale, these are the only ones that can be used to calculate *D*. Consequently the value found is highly dependent on the exact position of the interval boundaries compared to the random variations in the distribution of data within the sample. This problem can be reduced by smoothing the graph, using a three point running average: (a + 2b + c)/4. However, the log-interval discrete method will usually give better results, and this method is not recommended.

Log-interval method

If the data does not have a LHT fall-off all the points on a log-interval graph can be used to fit a line. There are several options when using this method, for example choice of interval, smoothing algorithm and line fitting routine. A weighted least squares gives the best results, as those intervals containing more data will give points less affected by noise, and should be given more weight. The weighting is proportional to the square root of the number in each interval. Smoothing the graphs appears to make little difference to the spread of the sample *D*-value distribution, but does reduce the tendency of the graph to under-estimate *D*. A simple three point running average, using the rule (a + 2b + c)/4 is suitable. The edge intervals should be excluded as these may partially cover ranges beyond the truncations of the data. The choice of interval is a trade-off between accuracy and precision. An interval of 0.1 is most common in earthquake studies. Reducing the size, and thereby increasing the number, of the intervals reduces the spread, but also increases the bias, particularly at the smaller sample sizes. Given that the improvement in precision for intervals smaller than 0.1 is marginal compared to the increase in bias, an interval size of 0.1 or more is recommended. The general rule is that data should be divided between 10–20 intervals, and therefore as scale range increases so should the interval size. The confidence intervals follow the same relationship as for the cumulative graph (equation (11)), and typical values of *k* are shown in Table 1.

Choosing an analysis method

Of the three methods tested, the discrete frequency method can be rejected as far too inaccurate. Choosing between the other two is more sample specific, with sample size and scale range as the determining factors. For small samples (<100) over narrow scale ranges (less than two orders of magnitude) the log-interval method is best, with an interval of 0.1. If the sample is small but covers a wider scale range, then the cumulative frequency graph is a better choice. For larger sample sizes, the differences in *k* become less significant and the chance of bias is reduced (assuming the cumulative graph is corrected) and so choice of method is less important. The log-interval method does have the advantage of not requiring correction, but care should be taken to choose the best interval size given the data.

Data examples

Determination of D

The data shown in Fig. 7a were derived from the dip lines of a 2D seismic survey over part of a block in the Southern North Sea basin (Area A). The measurements are all vertical separation/throw measured in TWT at the Top Rotliegendes horizon. Conversion to depth difference with a constant velocity would have no effect on the *D*-value

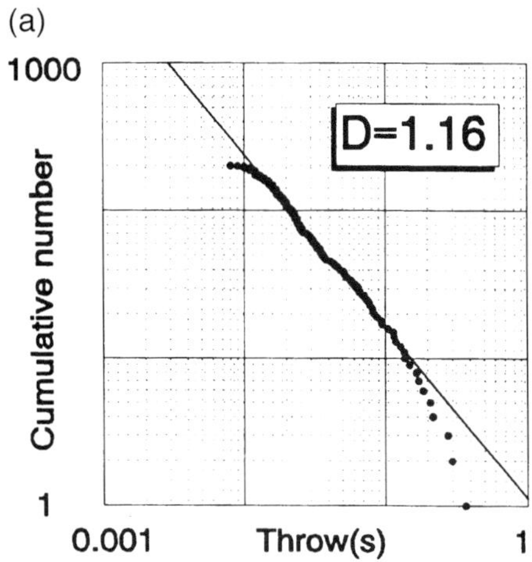

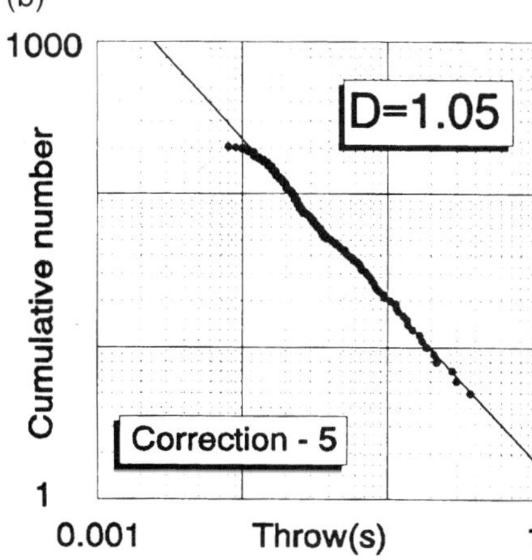

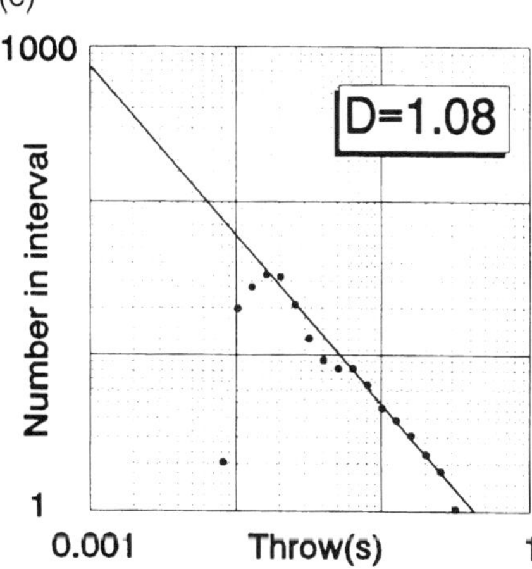

Fig. 7. Graphs of the same data set of throw values measured on a set of seismic sections from part of Quad 53 in the Southern North Sea (SNS A).
(**a**) Cumulative graph showing the expected finite-range effect deviation, due to the difference between the size and scale range of the data.
(**b**) Cumulative graph of the data after the iterative correction procedure has been applied.
(**c**) Log–interval graph of the data.

calculated as it would just cause a bulk shift of the data when plotted on log axes. A very large lateral velocity change may cause distortion on a graph of TWT measures, however even a factor of two is not a large variation (Walsh *et al.* 1994; Pickering *et al.* 1994). The line spacing is 1 km, but the lines are offset along strike and there is no multi-line right-hand tail. As the data set contains 200 values, but only covers ~1.5 orders of magnitude there is a finite range effect deviation. This was corrected using the iterative process, giving a *D*-value of 1.05 ± 0.1 (Fig. 7b). The data were also graphed using a smoothed log-interval graph with an interval size of 0.1 (Fig. 7c). This gave a *D*-value of 1.08 ± 0.1. As expected these unbiased estimates are in good agreement. The pronounced downward curve at the small scale on this graph is a combination of LHT fall-off and the problem identified earlier of intervals covering scale ranges beyond the LHT.

Significance of D-value variation

An increase in *D* across a survey would indicate a higher degree of small scale fracturing and might be used to aid well location within a field. The confidence interval equations derived earlier can be used to test the significance of changes in *D*. Figure 8 shows plots of four multi-line data sets with the right-hand tail segment removed. Each plot has been iteratively corrected for the finite range effect. Plots (a) and (b) are from measurements of throw made on the Top Triassic horizon from adjoining areas in the Inner Moray Firth and show a variation in *D*. The associated 68% confidence intervals are plotted in Fig. 9a, which show that this difference is not significant. Figure 8b and c are plots of fault throws from the Top Rotliegendes horizon in the Southern North Sea. The faults are believed to be mostly of late Triassic or Jurassic age. These measurements are also from adjoining blocks but display a greater variation. The confidence intervals (Fig 9b) indicate that this variation is significant. Line drawings of sections taken from each of the areas are shown in Fig. 10. In area B the faulting at Top Rotliegendes is generally decoupled from higher levels by salt in

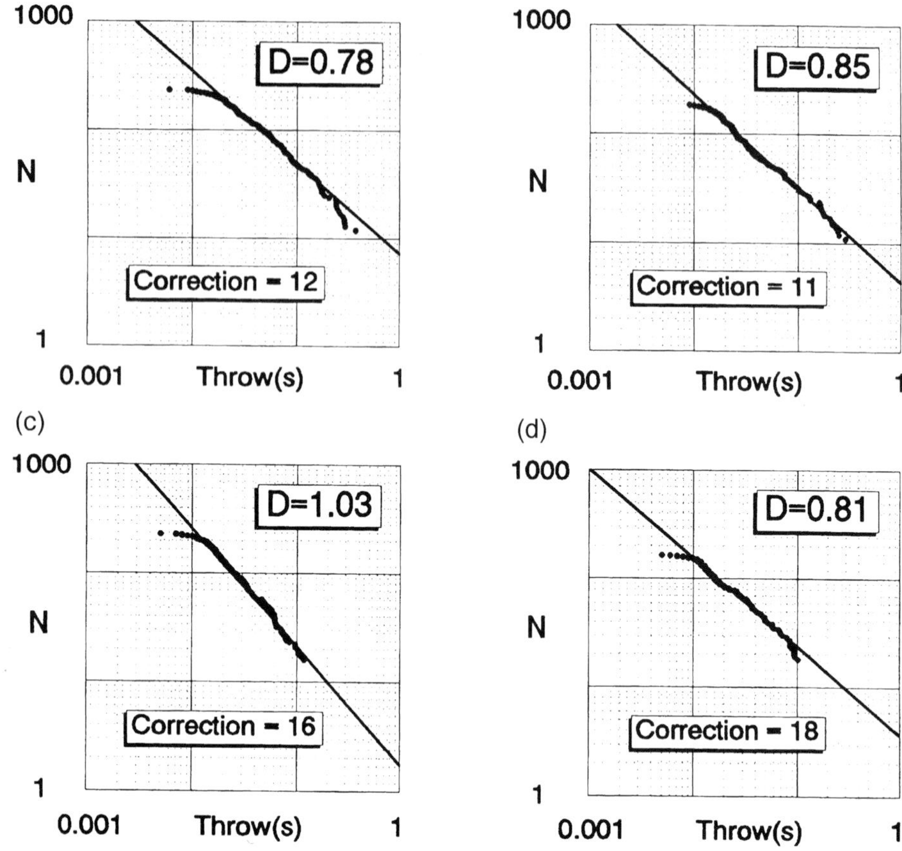

Fig. 8. Four corrected cumulative graphs of fault throw measured from seismic sections located in the Inner Moray Firth (UKCS Quad 12) and the Southern North Sea (UKCS Quad 53). (**a**) IMF B – Top Triassic. (**b**) IMF C – Top Triassic. (**c**) SNS B – Top Rotliegendes. (**d**) SNS C – Top Rotliegendes.

the Zechstein sequence. Area C is nearer the basin margin and the Triassic has not been decoupled from the Permian. This has resulted in a lower *D*-value for the fault population, due to the geological or mechanical differences between the two areas. In the Moray Firth case there was no discernable change in the structural geology between the two areas and the fault population distribution is likely to be the same in both, as suggested by the confidence intervals.

Applications

Tectonic extension estimates

It is well established that the total extension measured by summing the heaves of seismically visible faults, underestimates the true value of tectonic extension, e.g. Wood & Barton (1983). Adding in the total extension on faults below seismic resolution can provide more realistic estimates of extension. There are several methods of estimating the sub-seismic contribution. Cowie & Scholz (1990) proposed integrating the discrete frequency distribution of the fault population, giving a total heave measurement:

$$\int_{U_{min}}^{U_{max}} DCu^{-D}\,du = \left[\frac{DC}{(1-D)}U^{1-D}\right]_{U_{min}}^{U_{max}} \qquad (12)$$

The choice of U_{min} is somewhat arbitrary, however 1 mm is believed to be reasonable as below this the bulk material properties relevant at the large scale are unlikely to be retained by most rocks (Walsh & Watterson 1992). This is also the limit of most small scale data sets (e.g. Fig. 1) and consequently below this the validity of the power-law model cannot be tested. Equation (12) tends to underestimate the additional extension as the integration incorrectly assumes *N* to be a

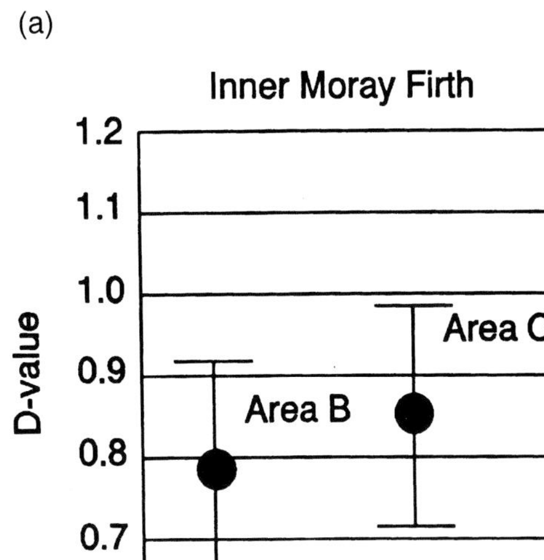

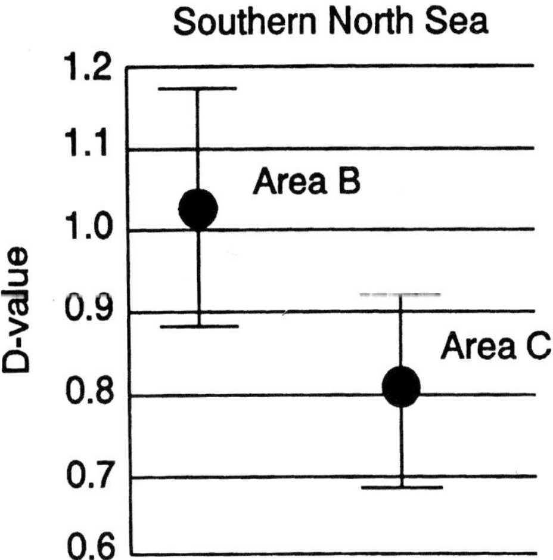

Fig. 9. 68% Confidence intervals for the D-values derived from the data sets shown in Fig. 8. These were calculated using equation (11) (see text), with k-values set in (**a**) at 1.6, and in (**b**) at 1.9 reflecting the scale ranges of the data sets.

continuous function. Marrett & Allmendinger (1991) suggested an improvement on this equation:

$$u_e = u_N \frac{D}{1-D}(N+1)\left(\frac{N}{N+1}\right)^{1/D} \quad (13)$$

where u_N is the smallest measured fault, and u_e is the total unmeasured heave. By subtracting from this the total heave from $U_{MIN} \rightarrow 0$ the summation to zero heave is avoided. Both equations can only be used when $D \neq 1$. The results are quite similar for D-values close to or greater than 1.0, but diverge as D decreases.

The main control on the significance of the sub-seismic contribution is the D-value. Figure (11) shows the extension estimates that would be made from an ideally power-law data set according to the D-value of the population and the minimum resolution. For each D-value the true extension is 10 km forming a basin 50 km wide giving a β-factor of 1.25. When $D = 0.5$ the β-factor measured from a section with a resolution limit of 20 m would be 1.24 and the sub-seismic contribution can be neglected. However if the D-value is higher, say 0.7, then the estimate is only 1.2, and for $D = 0.9$ less than half of the extension would be seen on the seismic section. Therefore for fault populations with high D-values the sub-seismic contribution is critical to an accurate extension estimate.

An example of using equation (13) to add the sub-seismic contribution is shown in Fig. 12. This was made using heave measurements from one of the seismic lines in area A, and the D-value derived from the combined field and seismic graph (Fig 1). The D-value for heave and throw are the same as the distribution of dips does not change with scale. In this case, as the faults cut several horizons, the heave measurements were reasonably accurate and could be used to measure the seismically visible extension. This measured extension gave a β-factor of 1.11. Adding the sub-seismic contribution down to 1 mm increases this to 1.20, which agrees better with estimates made using a seismic refraction experiment (Smith & Bott, 1975). Also displayed on this graph are the error estimates for this β-factor, if only the seismic data had been used to define D. In this case the 68% confidence interval for D would be $0.78 < D < 0.90$, giving an interval for β of $1.17 < \beta < 1.29$. The rapid increase in the sub-seismic estimate with D-value shows the importance of ensuring un-biased estimates of D by using the proper correction for the finite range effect. Given the somewhat arbitrary limit of 1 mm, the estimates gained by extending the minimum down to 10 μm or zero are also shown. For D-values of 0.8 or less this leads to little change to the β estimate. However for higher D-values there is a considerable increase especially for D-values ≥ 0.9. These large estimates are clearly incorrect illustrating the importance of taking a reasonable lower limit when making this calculation for high D-values.

A similar calculation was performed by Roberts *et al.* (1993) for the Viking Graben using the

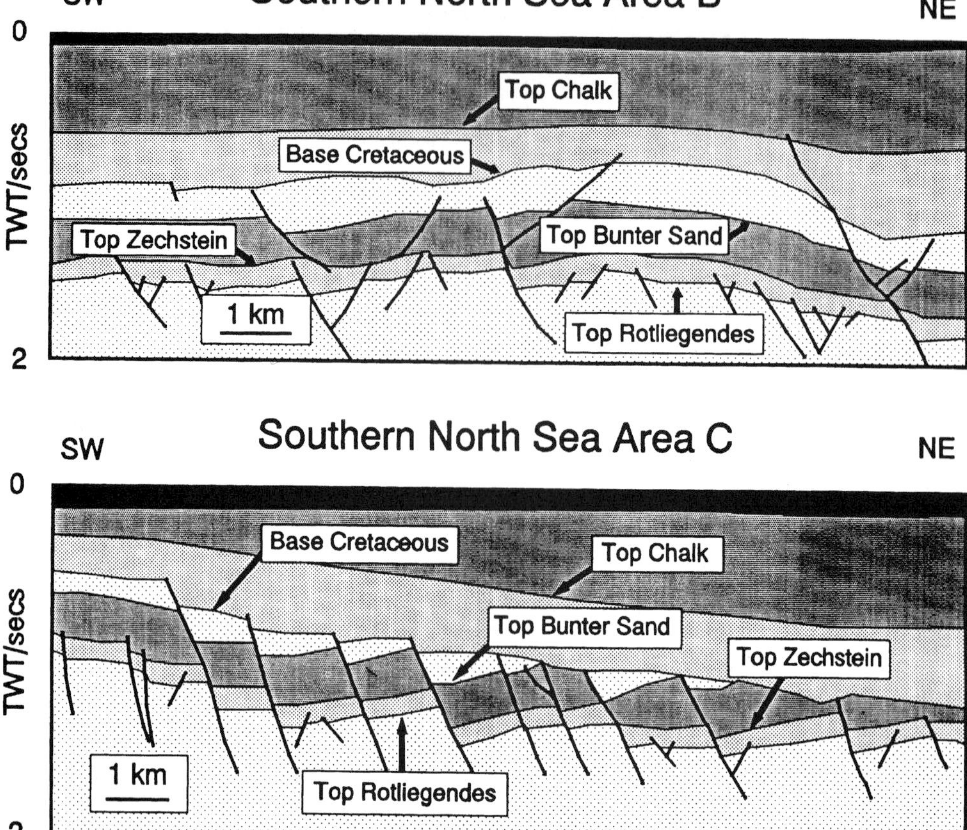

Fig. 10. Line drawings of two seismic sections from Southern North Sea areas B and C. The significant difference between the *D*-values for the two fault populations (Fig. 9) is reflected in a distinct change in structural style. In area B the Post Zechstein sequences are decoupled by salt in the Zechstein sequence. In area C the majority of the faults penetrate up to the Triassic, due to the thinning of the Zechstein, particularly the salt layers.

'missing percentage' method, e.g. Walsh *et al.* (1991), Marrett & Allmendinger (1992). This method also assumes that the fault population is power-law. The *D*-value is derived from the fault displacement data and the distribution is then based on this value and the maximum fault seen on the section. By using either equation (12) or (13) the contribution of the sub-seismic faulting is calculated as a percentage of the seismically visible extension. This is then added to the measured heaves to give an improved β estimate. Roberts *et al.* (1993) found a good agreement between the β-factors from using this method and those derived from basin modelling using the flexural-cantilever model (Marsden *et al.* 1990)

Predicting "large" sub-seismic faults on sections

'Large' sub-seismic faults, i.e. those in the range 5–20 m are likely to have a significant effect on reservoir behaviour. They may affect the sealing properties and inter-connectivity of the reservoir or offset thin reservoir formations above or below the course of horizontal wells. As these are only just below seismic resolution, extrapolation of the power-law distribution is likely to give good predictions of their density. The specific location of the faults is of course unknown, however there is evidence that they may be expected to cluster around those that can be seen on the seismic (Gillespie *et al.* 1994). Heffer & Bevan (1990)

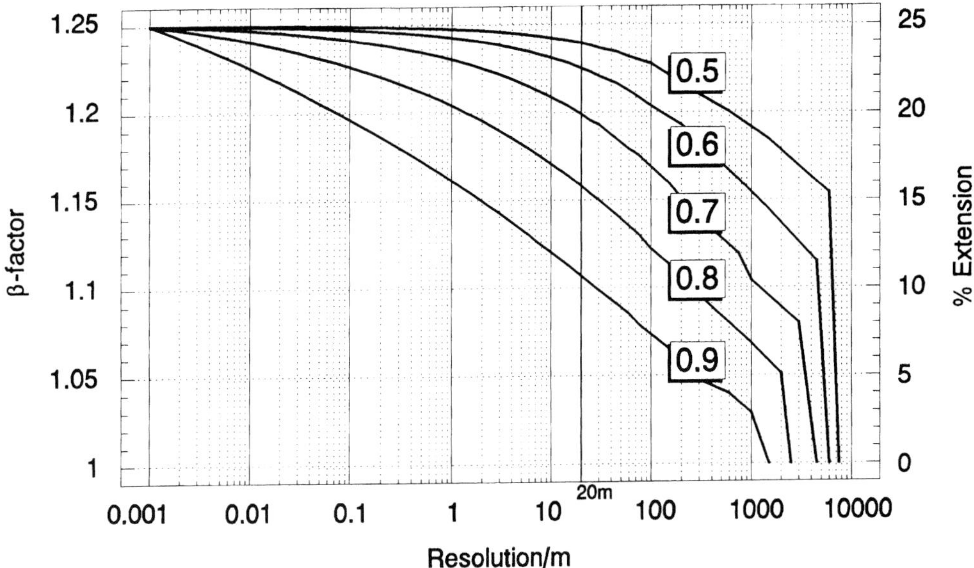

Fig. 11. Graphs of the measured extension of a basin by summing fault heaves against the lower-limit of resolution of these measurements, for different ideally power-law populations of faults. The distributions give the same total extension when summed down to 1mm of 25% or $\beta = 1.25$. The total heave down to each resolution was calculated using equation (13) (see text). The decrease in the maximum fault with increasing D-value reflects the increase in the smaller scale contribution. For $D = 0.5$ almost all of the extension is taken up by faults greater than 20 m, where as for $D = 0.9$ there contribution is less than half of the total.

derived a method of predicting the number density for a horizontal well by extrapolating the scaling relationship for fault trace lengths. Then by calculating the probability of intersecting these faults and assuming a displacement-length relationship a number density can be estimated. The horizontal well is a 1D transect and is equivalent to the traverses used in this paper to derive D-values for the fault population. Therefore, this D-value can be used directly to predict the number density without having to assume any particular displacement–length relationship, over which there is still considerable debate, e.g. Cowie & Scholz (1992), Gillespie *et al.* (1992). If the cumulative number is divided by section length then the number per km can be measured directly from the cumulative frequency graph. Using the data shown in Fig. 1 for example, $N(u \geq 5\text{ m}) = 3.4\text{ km}^{-1}$, and $N(u \geq 20\text{ m}) = 1.1\text{ km}^{-1}$, therefore $N(5\text{ m} \leq u \leq 20\text{ m}) = 2.3\text{ km}^{-1}$. If the seismic data alone are used then the predictions are $1.8\text{ km}^{-1} \leq N(5\text{ m} \leq u \leq 20\text{ m}) \leq 3.3\text{ km}^{-1}$. Therefore, a 1 km well has a good chance of intersecting at least one, but probably two large sub-seismic faults in this area.

Quality control of fault interpretation on sections

Fault interpretation on seismic sections is clearly critical to proper mapping of the sub-surface. There are two ways in which these can be quality controlled by using the power-law relationship. First, the distribution derived from a set of seismic sections can be analysed and the D-value compared to neighbouring surveys or other interpretations. If differences are significant and there is no good geological reason, the interpretation should be checked. Secondly the plots of the data would give the effective size limit above which all faults were picked. This may significantly differ from the expected limit or the minimum fault size identified. Once this limit has been found the power-law relationship could be extrapolated to give a better indication of the true fault density in the typical 10–50 m cut-off region for seismic data.

Conclusions

This paper has confirmed previous studies that there is good evidence to support a power-law

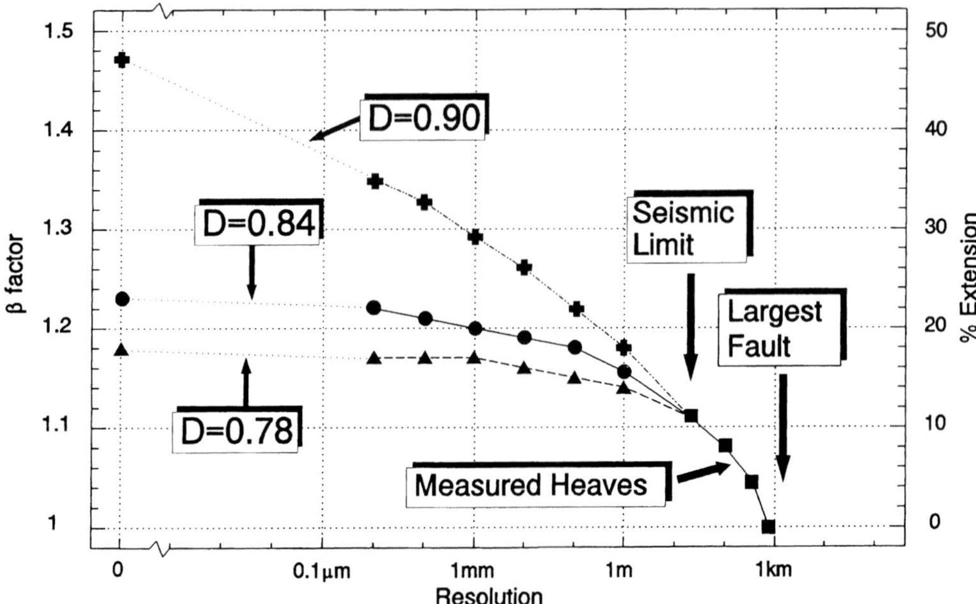

Fig. 12. Graph showing the use of the *D*-value, derived from seismically visible faults, to estimate the sub-seismic contribution to extension. The seismic data used is from a line over Inner Moray Firth area A. The faults measured on the seismic lines gave an extension. The seismic data used is from a line over Inner Moray Firth area A. The faults measured on the seismic lines gave an extension estimate of $\beta = 1.11$. The sub-seismic contribution is calculated from 20 m, down to resolutions varying from 1 m to 0. The calculation is shown for the *D*-value found from Fig. 1 (0.84) and the upper and lower confidence limits for this value (0.78–0.90), had it been derived solely from the seismic data. Setting the limit of the population at 1 mm gives β-factors varying from 1.17 to 1.29, with the expected *D*-value of 0.84 giving $\beta = 1.20$.

distribution for fault displacement populations, particularly in extensional basins. Fault throw data sets derived from sections can be used to derive the power-law exponent (*D*) of this fault population, whose values usually range from 0.5–1.5. This is the preferred source of data as long as the measurements made are proportional to the true displacement of the fault. Measurements made on multiple sections can be combined to increase the sample size of the data. However, if one fault is significantly multi-sampled these measurements should be removed.

Three methods of analysis are available: (i) cumulative frequency graph, (ii) discrete frequency graph with linear intervals, and (iii) log-interval discrete frequency graph.

D-values derived using (i) will be biased due to the finite range effect. This is likely to be serious if the size of the data set is significantly greater than the scale range over which it has been sampled. A correction for this bias can be made, and an iterative procedure has been described. This bias is only found on the cumulative graph, and the two discrete methods are unaffected.

Simulations of the three methods were made using random samples from power-law distributions. As the distribution *D*-value is known, this value can be compared to those derived from the samples. The sample *D*-values derived from method (i) showed the expected bias due to the finite range effect. The correction procedure successfully removed this bias. The results from the simulation of method (ii) suggest that the method can give highly inaccurate results, and is not recommended. Method (iii) gave unbiased results, with a spread of sample *D*-values similar to method (i). This spread can be quantified by estimating confidence intervals. Empirical equations for these intervals were derived from a synthesis of the results of the simulation. As expected, the greater the size of the data set and the wider the scale range the greater the accuracy of the *D*-value derived from the sample. These intervals can be used to assess the significance of any change in *D*-value over an area. Examples from the Inner Moray Firth and the Southern North Sea suggest that significant differences are usually related to a distinct change in structural style.

Although the model can be used as an additional quality control on fault interpretation on seismic sections, the main application of the power-law model is to estimate sub-seismic fault density. For example, the model can be used to predict the number density of 'large' faults just below the seismic resolution, i.e. 5–20 m. These predictions are particularly important in hydrocarbon production as faults in this range may significantly affect the performance of a reservoir. If the power-law model is extrapolated down to the likely limit of the fault population, the total tectonic extension due to fault heave can be calculated. Estimates made in this way will significantly improve upon those made using only the seismically visible faults if the *D*-value is greater than *c.* 0.8. Results from the Viking Graben and the Inner Moray Firth show that the β-values calculated are comparable with those found from other independent methods.

The authors are grateful to Seismograph Service Ltd & SHELF for permission to use the SSL-MF89 seismic survey. A review by John Walsh helped to improve the final version. This research is funded by a studentship from Mobil North Sea Ltd.

References

CHILDS, C., WALSH, J. J. & WATTERSON, J. 1990. A method for estimation of the density of fault displacements below the limits of seismic resolution in reservoir formations. *In*: BULLER, A. T. (ed.) *North Sea Oil and Gas Reservoirs II.* Graham & Trotman, London, 309–318.

COWIE, P. A. & SCHOLZ, C. H. 1992. Displacement-length scaling relationship for faults: data synthesis and discussion. *Journal of Structural Geology*, **14**, 1149–1156.

EINSTEIN, H. H. & BAECHER, G. B. 1983. Probabilistic and Statistical Methods in Engineering Geology, Specific Methods and Examples Part I: Exploration. *Rock Mechanics and Rock Engineering*, **16**, 39–72.

GILLESPIE, P. A., WALSH, J. J. & WATTERSON, J. 1992. Limitations of dimension and displacement data from single faults and the consequences for data analysis and interpretation. *Journal of Structural Geology*, **14**, 1157–1172.

——., HOWARD, C. B., WALSH, J. J. & WATTERSON, J. 1994 Measurement and characterization of spatial distributions of fractures. *Tectonophysics*, **226**, 113–141.

HEFFER, K. & BEVAN, T. 1990. Scaling relationships in natural fractures – data, theory and applications. *Proceedings of the European Petroleum Conference*, 2, 367-376 (SPE paper No. 20981).

JACKSON, P. & SANDERSON, D. J. 1992. Scaling of fault displacements from the Badajoz-Cordoba shear zone, SW Spain. *Tectonophysics*, **210**, 179–190.

KAKIMI, T. 1980. Magnitude-frequency relation for displacement of minor faults and its significance in crustal deformation. *Bulletin of the Geological Survey of Japan*, **31**, 467–487.

MAIN, I. G. & BURTON, P. W. 1989 Seismotectonics and the earthquake frequency magnitude distribution in the Aegean area. *Geophysical Journal*, **98**, 575–586.

MARRETT, R. & ALLMENDINGER, R. W. 1991. Estimates of strain due to brittle faulting: sampling of fault populations. *Journal of Structural Geology*, **13**, 735–738.

—— & —— 1992. Amount of extension on 'small faults': An example from the Viking Graben. *Geology*, **20**, 47–50.

MARSDEN, G., YIELDING, G., ROBERTS, A. & KUSZNIR, N. J. 1990. Application of a flexural cantilever simple-shear/pure-shear model of continental lithosphere extension to the formation of the northern North Sea Basin. *In*: BLUNDELL, D. J. & GIBBS, A. D. (eds) *Tectonic Evolution of the North Sea Rifts.* Oxford University Press, Oxford, 241–261.

PACHECO, J. F., SCHOLZ, C. H. & SYKES, L. R. 1992. Changes in frequency-size relationship from small to large earthquakes. *Nature*, **355**, 71–73.

PICKERING, G., BULL, J. M. & SANDERSON, D. J. in press. Sampling power-law distributions. *Tectonophysics.*

——, ——, —— & Harrison, P. V. 1994. Fractal Fault Displacements: A Case Study from the Moray Firth, Scotland. *In*: Kruhl, J. H. (ed.) *Fractals and Dynamic Systems in Geosciences.* Springer-Verlag, Frankfurt, 105–119.

PRESS, W. H., FLANNERY, B. P., TEUKOLSKY, S. A. & VETTERLING, W. T. 1986. *Numerical Recipes: The Art of Scientific Computing.* Cambridge University Press, Cambridge.

ROBERTS, A. M., YIELDING, G., KUSZNIR, N. J., WALKER, I. & DORN-LOPEZ, D. 1993 Mesozoic extension in the North Sea: constraints from flexural backstripping, forward modelling and fault populations. *In*: PARKER, J. R. (ed.) *Petroleum Geology of Northwest Europe: Proceedings of the 4th Conference.* The Geological Society, London, 1123–1136

RUBINSTEIN, R. Y. 1981. *Simulation and the Monte Carlo method.* John Wiley & Sons, New York.

SCHOLZ, C. H. & COWIE, P. A. 1990. Determination of total strain from faulting using slip measurements. *Nature*, **346**, 837–839.

SMITH, P. J. & BOTT, M. H. P. 1975. Structure of the Crust Beneath the Caledonian Foreland and Caledonian Belt of the North Scottish Shelf Region. *Geophysical Journal of the Royal Astronomy Society*, **40**, 187–205.

WALSH, J. J. & WATTERSON, J. 1992. Populations of faults and fault displacements and their effects on estimates of fault-related regional extension. *Journal of Structural Geology*, **14**, 701–712.

——., —— & Yielding, G. 1991. The importance of small scale faulting in regional extension. *Nature*, **351**, 391–393.

——, —— & —— 1994. Determination and interpretation of fault size populations: procedures and problems. *North Sea Oil and Gas Reservoirs – III.* Norwegian Institute of Technology, 141–155

WESTAWAY, R. 1994. Quantitative analysis of populations of small faults. *Journal of Structural Geology*, **16**, 1259–1273.

WOOD, R. & BARTON, P. 1983. Crustal Thinning and subsidence in the North Sea. *Nature*, **302**, 134–136.

YIELDING, G., WALSH, J. & WATTERSON, J. 1992. The prediction of small-scale faulting in reservoirs. *First Break*, **10**, 449–460.

Ductile strain effects in the analysis of seismic interpretations of normal fault systems

J. J. WALSH, J. WATTERSON, C. CHILDS & A. NICOL

Fault Analysis Group, Department of Earth Sciences, University of Liverpool, Brownlow Street, Liverpool L69 3BX, UK

Abstract: The limited vertical resolution of reflection seismic data results in sub-resolution structures accommodating appreciable ductile strains which should be taken into account in structural restoration and balancing and in analysis of fault displacements. Non-ductile strains are expressed as discontinuities, such as fault offsets. A fault map of an intensively mined coal-seam is used to demonstrate the role of structures which would not be detected by a typical offshore seismic survey. Four examples are given of interpretations of seismic datasets in which significant ductile strain can be demonstrated; (i) an array of normal faults in which significant extension is accommodated by sub-seismic faults, (ii) a relay zone in which ductile shear strain accommodates significant displacement and change of bed-lengths, (iii) an intersection zone between conjugate normal faults which is effectively a zone of ductile pure shear strain with horizontal extension, and (iv) a hanging-wall fold which accommodates appreciable ductile displacement and ductile bed extension. In all these cases the ductile strains and displacements can be estimated from the seismic interpretation. Unless ductile strains are incorporated in restorations bed lengths will not balance. Balancing nevertheless remains the prime method for detection of grossly invalid sections.

Interpretations of seismic data now provide the basis for a high proportion of geometric analyses of normal fault systems, but the systematic resolution limitations of seismic data have to be considered when applying analytical methods originally developed for other types of data, although resolution effects occur with all types of data. Bed-length and/or area balance are convenient geometrical assumptions used in the construction, validation and restoration of cross-sections. Although the role of section balancing varies with the type of exercise to which it is applied, its usefulness as a tool is always dependent on the basic premise that 'valid sections' must balance. Valid results for section balancing or restoration and fault displacement analysis based on seismic interpretations may require quantification and explicit incorporation of ductile strains associated with faults and arising from limited resolution, in addition to the discontinuous displacements on seismically imaged faults. We define ductile strain as a change in shape produced by structures which are too small to be imaged individually by a particular technique and/or too small to be represented individually on a given map or cross-section. Non-ductile strains are expressed by discontinuities, such as fault offsets. The ductile strain threshold varies with the scale of the map or cross-section and is therefore a scale dependent, or resolution dependent, property. The deformation processes which give rise to ductile strains include sub-resolution faulting and plastic deformation processes, including compaction, that do not produce geometrical discontinuities other than on the microscopic scale. Homogeneous ductile shear strains can be simply related to ductile displacements but displacements may be more difficult to estimate when the ductile shear is heterogeneous and gives rise to folds. The ductile strains that are necessary to accommodate displacement variations or changes of fault surface geometry on faults and that are an integral part of the faulting process are not considered here (Verrall 1982; White *et al.* 1986; Gibson *et al.* 1989; Anders & Schlische 1994). Five examples of structures in which ductile strain is either significant or in which it would be if the structures were to be imaged seismically are reviewed; two of the examples are more fully described elsewhere (Nicol *et al.* 1995; Watterson *et al.* in press). These examples are used to briefly consider how ductile strains may effect the veracity of section restoration and validation.

Regional ductile strain

All maps and cross-sections derived from seismic interpretations have resolution, or truncation, values below which faults are not explicitly represented (Walsh & Watterson 1991). Truncation value

From Buchanan, P. G. & Nieuwland, D. A. (eds), 1996, *Modern Developments in Structural Interpretation, Validation and Modelling*, Geological Society Special Publication No. 99, pp. 27–40.

is determined by the quality of the seismic data and its interpretation; good quality North Sea seismic data, for example, rarely permit interpretation of all fault throws down to *c.* 20 m on pre-Cretaceous horizons.

Sub-resolution fault populations, and their fault-related ductile strains, can be predicted in many cases because the fault size distributions within many fault systems are systematic (Childs *et al.* 1990; Marrett & Allmendinger 1991; Jackson & Sanderson 1992; Walsh & Watterson 1992), allowing the population in one size range to be inferred from data for a different size range (Yielding *et al.* 1992). Extensional tectonic fault systems show the highest degree of ordering with fault size distributions conforming to a power-law such that,

$$\log N = a - d\log(S)$$

or

$$N = \log(a) \,.\, S^{d},$$

where S is fault size, measured as either displacement or dimension, N is the cumulative number of faults with size greater than or equal to a given value, $-d$ is the power-law exponent and a is a constant depending on sample size. d is often referred to as the fractal dimension of the size distribution. Higher values of fractal dimension indicate higher proportions of small faults relative to larger faults (Walsh *et al.* 1991). Even in systems that do not have such a simple scaling law the fault size distributions usually indicate substantial numbers of faults below seismic resolution limits. Only datasets measured from outcrops of finely laminated sequences include the smallest faults in a system.

Coalfield dataset

The ductile strain contribution of sub-resolution faults is illustrated by reference to a fault map of a 12 km^2 area in the South Yorkshire coalfield (Fig. 1). The map was constructed from 1:2500 seam abandonment plans for the Lidgett seam in North Gawber and Woolley Collieries. The faults are of Carboniferous age and form a near orthogonal system of NE and NW striking normal fault sets. These data are a sub-set of a larger dataset constructed for investigation of the scaling properties of faults (Watterson *et al.* in press).

Because the population systematics of different fault sets should, where possible, be measured separately (Childs *et al.* 1990; Gauthier & Lake 1993), we consider only the main fault set in the area, i.e. the NE striking normal faults. The fault population is measured using a line sampling method in which the throws are recorded for all faults intersecting an array of parallel sample lines (Walsh *et al.* 1994). The throw population curve for 7 sample lines oriented normal to fault strike shows three main segments (Fig. 2a): a steep right-hand segment with length corresponding to the number of sample lines, an approximately straight line central segment which merges with a left-hand segment of low slope at a throw value of *c.* 30 cm. This value is therefore taken as the truncation value so the range of throws fully represented is 30 cm to 30 m. The central straight segment of the population curve is representative of the size distribution of the fault set and its slope, *c.* –0.55, corresponds to a fractal dimension of 0.55. This value lies at the lower end of the range of fractal dimensions, 0.5–1.0, established from other studies (Yielding *et al.* 1992), indicating a lower proportion of small to large faults than in many tectonic fault systems.

The proportion of extensional strain accommodated by small faults is nevertheless still significant (Fig. 2b). For example, fault throws between 30 cm and 3 m, which represent half the valid range of throw of the dataset, accommodate *c.* 14% of the cumulative throw on the NE fault set. Therefore, with a truncation value of 3 m the ductile extension accommodated by sub-resolution faults with throws down to 30 cm would be *c.* 14% of the extension accommodated by fault throws between 30 cm and 30 m. The proportion of extension represented by sub-seismic faults is dependent on the minimum and maximum fault sizes which are taken into account (Fig. 2c). The smallest fault size in the systematic population for this coalfield dataset will be in the range 1 mm–10 cm. The largest fault size will increase with size of the area considered up to the maximum sized fault in the fault system. This point is illustrated by throw population data from an 8 km long sample line which extends beyond but includes the study area and on which the largest throw intersected is 110 m (Fig. 2c) as opposed to the 30 m maximum in the study area. The fractal dimension of this population curve is also relatively low (*c.* 0.4). If the throw resolution limit along this sample line were 20 m, i.e. a value typical of high quality offshore seismic data, sub-resolution faults could accommodate a ductile extensional strain representing as much as 15% of the total fault-related extension.

The significance of resolution effects on strain estimates is easily appreciated by comparison of fault maps produced by removing fault traces with throws below a specified truncation value. Figure 1 shows truncated derivatives of the primary fault map for 5 m, 10 m and 20 m throw truncations. Degradation of the data resolution reduces the apparent compartmentalisation. At 20 m truncation, only one fault remains and the proportion of

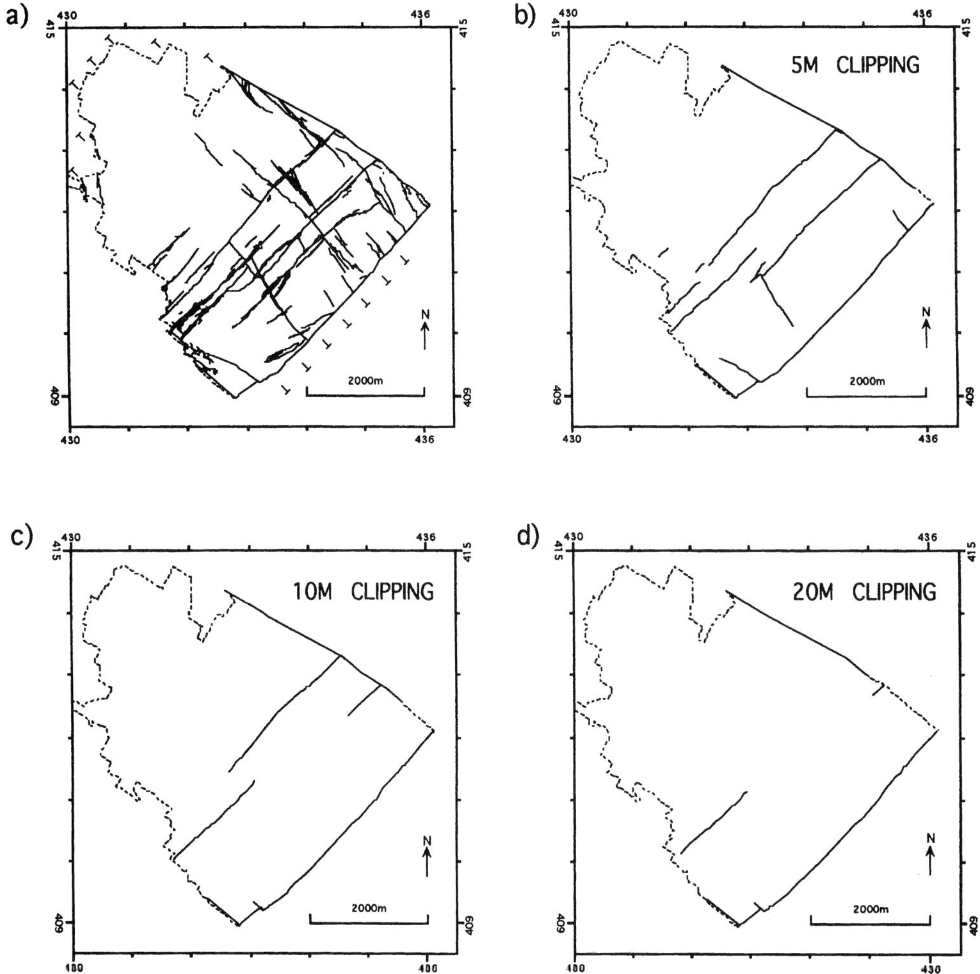

Fig. 1. Fault maps for the Lidgett seam in Woolley and North Gawber Collieries, South Yorkshire. All mapped faults are shown in (**a**), and maps showing only fault traces (or parts of fault traces) with throws greater than a given throw truncation value in (**b-d**): (**b**) 5 m, (**c**) 10 m, (**d**) 20 m. Faults are shown as solid lines and the boundary to the mapped area is defined either by the limits of the seam workings (broken line) or by faults with throws greater than 30 m (solid lines). The locations of NW-oriented sample lines ($n = 7$) for the multi-line throw population curve in Fig. 2a are shown in (a).

extension accommodated by sub-resolution faults (20 m to 30 cm fault throw) is *c.* 70%. Although such exercises reproduce, in a general way, fault maps appropriate to seismic data of different vertical resolutions, the effects of lateral seismic resolution are not incorporated. Limited lateral seismic resolution will effectively aggregate throws of closely spaced faults and therefore effectively image fault throws which would be sub-resolution if more widely spaced. The effect on Fig. 1b of incorporating lateral resolution would be slight and there would no effect on Figs 1c and d.

The conclusions to be drawn from the coal-mine data are that the vertical resolution limit of the data is a principal factor in determining ductile strain due to sub-resolution faults. With highly ordered size population systematics this ductile strain can be assessed and a reasonable estimate made even when the fault size distribution is non-fractal, as is sometimes the case for both gravity driven and tectonic fault systems (Gillespie *et al.* 1993). These estimates of ductile extension should be explicitly incorporated in restorations.

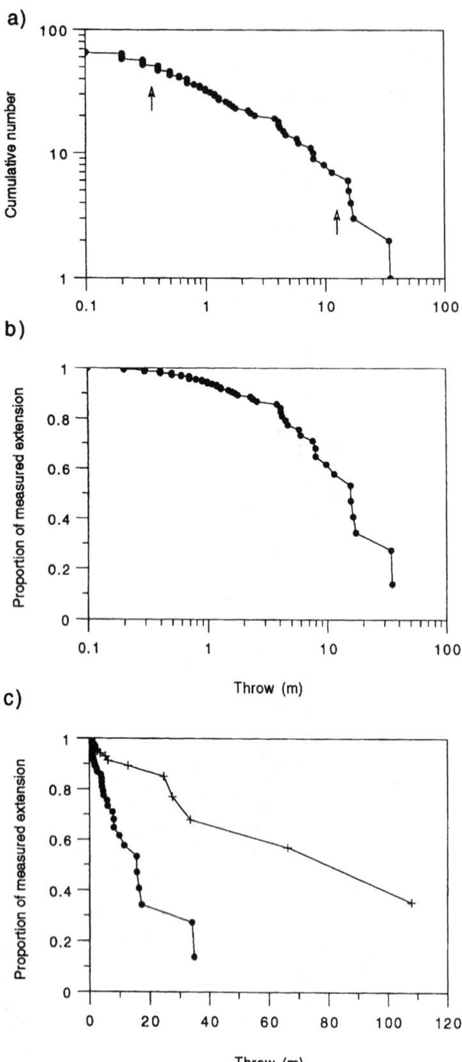

Fig. 2. Fault population curves from South Yorkshire coalfield. (**a**) Logarithmic plot of fault throw v. cumulative number for multi-line ($n = 7$) throw population of NE striking faults on the 12 km^2 fault map shown in Fig. 1. The limits of the central segment of the population curve are arrowed. (**b**) Throw v. proportion of measured extension for the fault data in (a). (**c**) Throw v. proportion of measured extension, for NE striking faults intersecting an 8 km sample line extending SE from the mapped area in Fig. 1. The curve (crosses) shows the contributions of faults of decreasing throw to the total extension accommodated by faults with throws between 30 cm and 110 m. The curve from (b) is shown for comparison.

Seismic dataset

Several studies have shown that fault populations of seismically imaged faults from offshore oil and gas fields are either fractal or otherwise highly systematic (Childs *et al.* 1990; Yielding *et al.* 1992; Gauthier & Lake 1993). These systematics provide the basis for determining the relative numbers of seismic and sub-seismic faults and for estimating the ductile strain contribution of the sub-seismic faults (Walsh & Watterson 1992). We use fault data from a 2D seismic dataset for a 900 km^2 area on the south-east edge of the Cartier Trough within the Timor Sea to illustrate this point (Figs 3 & 4). Seismic coverage consists of a rectangular array of seismic lines (32 NW–SE oriented dip-lines, with 0.5 km to 1 km spacing, and 11 strike-lines). The faults are assigned to 2 main phases of extension: Upper Jurassic–Base Cretaceous and Plio-Pleistocene (Woods 1992; Nicol *et al.* 1995). The most recent phase of extension is represented by a conjugate system of ENE–WSW striking normal faults (Fig. 4), some of which are reactivated Late Jurassic structures. Seismic data quality is very good within a strongly reflective Palaeocene–Recent sequence of shelf carbonates (Fig. 3), but deteriorates rapidly below the Top Cretaceous.

The contribution of faults of different sizes to fault-related Plio-Pleistocene extension is shown on a plot of fault throw versus cumulative β value (Fig. 5) using fault data from 3 pre-faulting horizons on a single sample line across the study area (Fig. 3). Seismically resolved faults show an approximately linear increase in β with decrease in the smallest included fault throw from 200 m to *c.* 20 m (Fig. 4). The decrease in curve slope at throws of *c.* 10–20 m marks the effective limit of resolution of the seismic data. Extrapolation of the straight-line segments (which span throws 20–200 m) of the curves to 1m throw shows that faults in the range 1–20 m, many of which are sub-seismic, may accommodate a ductile extension of 2–3%, representing *c.* 40% of the total extension (Fig. 5). The validity of linear extrapolation beyond the data remains untested as no quantitative well data are available. Inspection of the population curve (Fig. 5a) suggests however that even if a linear extrapolation is inappropriate, small faults (i.e. with displacements of 1–20 m) are likely to accommodate a significant component of the extension. Although this ductile extensional strain is a significant proportion of the whole, it will be represented by a thinning of the faulted sequence which is insignificant in comparison to likely compaction-induced thickness changes. The relative insignificance of fault-related thinning of the sequence will apply also to rifting events with larger beta values, as in the Late Jurassic

(a)

NW SE

(b)

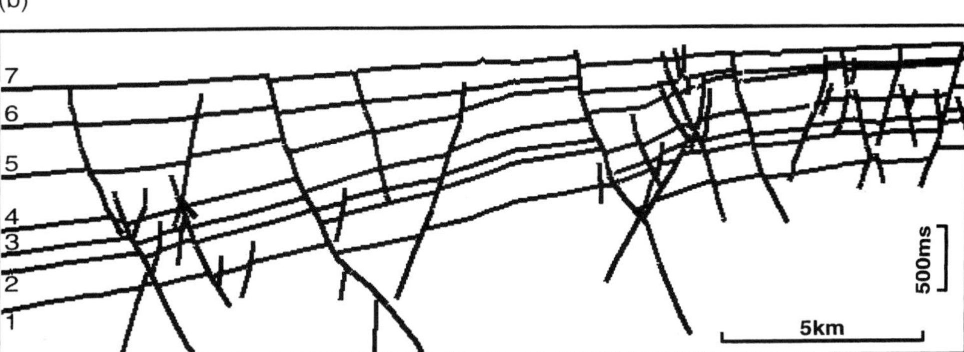

Fig. 3. Seismic (**a**) and interpretation (**b**) of 24 km section from the south-east edge of the Cartier Trough, Timor Sea. Interpreted horizons are: (1) Top Cretaceous, (2) Top Palaeocene, (3) Middle Eocene, (4) Base Miocene, (5) Top Miocene, (6) *c.* Middle Pliocene and (7) *c.* Lower Pleistocene. Vertical exaggeration is *c.* 2 (1 ms = *c.* 1.4 m). A carbonate mound is responsible for the seismic image distortion towards the centre of (a) (see Houland *et al.* 1994).

extension of the North Sea (*c.* $\beta < 1.2$)(Roberts *et al.* 1993).

The resolution sensitivity of these fault data is illustrated graphically using fault maps with different throw truncations (Fig. 4). If this Timor Sea dataset had a throw resolution of 40 m, typical of poorer quality 2D and 3D North Sea seismic datasets, the numbers of imaged faults would decrease and those imaged would have shorter traces. For the seismic line shown in Fig. 3, a 40 m throw truncation would reduce the estimated β value from 1.035 to 1.027 unless account were taken of the ductile extensional strain. Seismically resolved faults in this dataset are not strongly clustered on scales relevant to lateral resolution effects (< 300 m), but some aggregation of small and large throws would be expected.

Restored bed lengths in datasets of this type will be inaccurate unless allowance is made for ductile extension. It might be argued that as the ductile extensions of all horizons is the same the true values are not required. If ductile strains are homogeneous then section restoration and validation may be valid. If however, the spatial distributions of sub-seismic faults, and therefore ductile strain, is heterogeneous, as is usually the case, then section restoration and validation will be subject to error. In addition, where some combination of pre-, syn- and post-faulting horizons is included in a restoration the ductile extension will be neither the same on every horizon nor be a constant proportion of the total extension on each horizon (Fig. 5b). The significance of the errors arising from these effects depends entirely on the use which is made of the results.

Localized ductile strain

Ductile strains accommodated by sub-seismic faults and/or plastic processes occur where there is

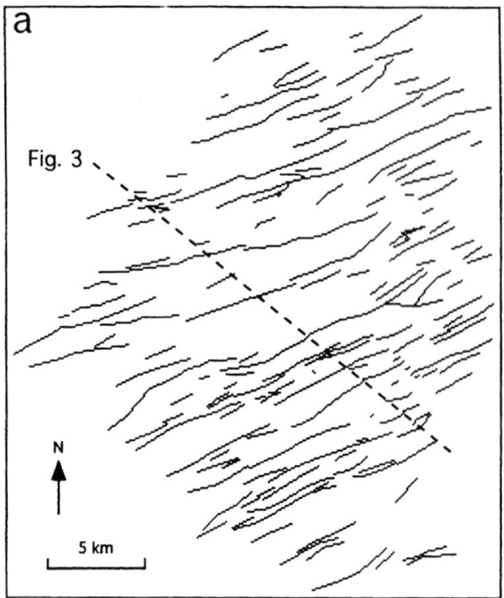

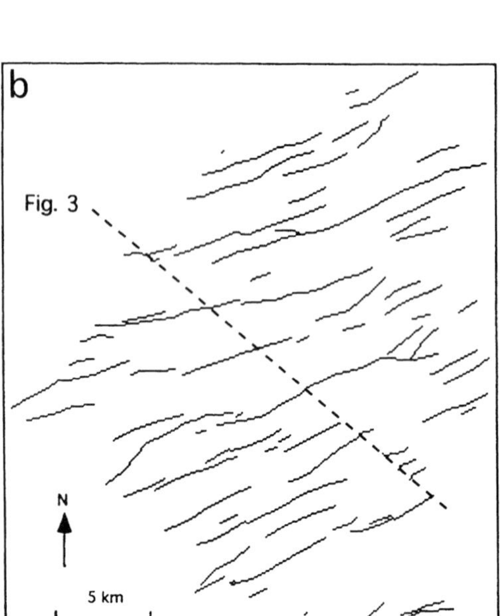

Fig. 4. Fault maps for a pre-faulting Base Miocene horizon (horizon 4, Fig. 3) in an area from the southeast edge of the Cartier Trough, Timor Sea. (**a**) All interpreted fault traces (i.e. all faults correlated on two or more seismic lines). (**b**) Fault traces with throws >40 m. Northwest oriented seismic lines are spaced at 0.5–1 km, and the location of the seismic line in Fig. 3 is shown (broken line).

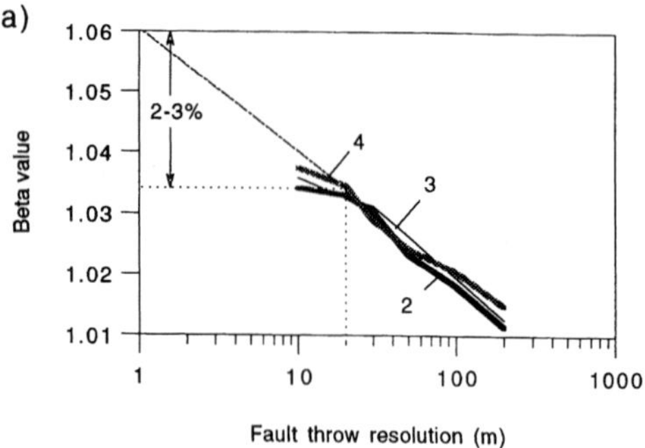

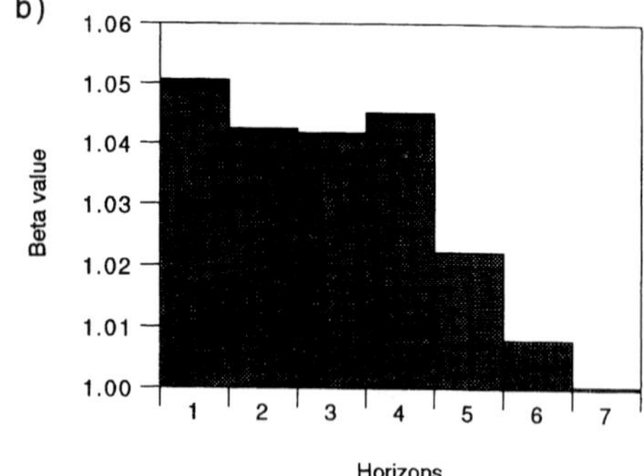

Fig. 5. (**a**) Fault throw resolution v. cumulative β value for three pre-faulting horizons (2, 3 and 4) on section shown in Fig. 3b. The seismic vertical resolution of *c.* 20 m is reflected in the shallowing of the left-hand segments of the curves. The cumulative β value at 1 m throw is indicated by the extrapolated straight segment of the data curve (broken line). The β value at 20 m throw is shown by the horizontal dotted line. (**b**) Histogram showing the extension accommodated by seismically imaged faults (stipple) and ductile strain at conjugate intersections (black) for seven horizons on the cross-section in Fig. 3b. Horizons 5–7 are syn-faulting and are less extended than the pre-faulting horizons.

interaction between adjacent coeval faults or where there is fault-related folding. Where the interacting faults have the same dip direction the commonest form of interaction is that associated with relay zones. Where the faults have opposed dips interaction is concentrated at conjugate intersections. Examples of ductile strains associated with these two circumstances are described below.

Relay zones

Relay zone geometries require high shear strains in the plane of the overlapping faults which are manifested as bed rotations and relay ramps (Fig. 6a). There may also be an element of shear strain in

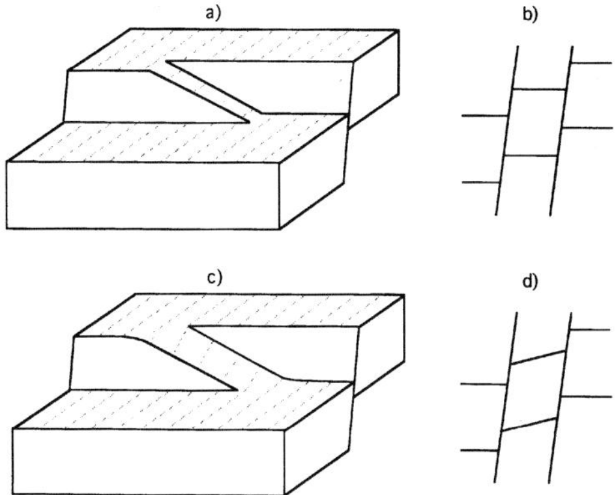

Fig. 6. Schematic block diagrams and cross-sections showing displacement and horizon geometries of relay zones without (**a** & **b**) and with (**c** & **d**) a component of shear displacement in the fault-normal plane within the relay.

the plane normal to the overlapping faults resulting in horizon dips towards the mutual hangingwall of the overlapping faults (Fig. 6c & d). Shear strain in the fault-normal plane is not a geometrically essential feature of relay zones but appears usually to be present (Peacock & Sanderson 1991, 1994; Huggins *et al.* in press). Fault-parallel and fault-normal shear strains together effect a non-plane strain deformation within a relay zone. However, it is shear strain in the fault-normal plane (Fig. 6c & d) which is of most significance in the context of section balancing and restoration, and in the analysis of fault displacements.

Figure 7a shows structural contours on a 1.5 km wide relay zone which provides lateral closure to a structural hydrocarbon trap lying to the west. The relay zone is bounded by two faults (A and B in Fig. 7a) with throws of up to *c.* 200 ms (≈ 300 m). Within the relay zone the rate of lateral change of throw on fault A is *c.* 5 times greater than elsewhere on this fault; no information is available on fault B beyond the ramp zone. Reorientation and tightening of the horizon depth contours within the relay give a contour azimuth 60° anti-clockwise of the regional ENE–WSW structural trend. On the cross-section in Fig. 7b the apparent dip of horizons within the relay zone is in the same direction as the fault dip. This direction of apparent dip is exceptional within this survey area where horizon apparent dips in the fault-normal plane are elsewhere opposed to the dip direction of the large faults.

The horizon geometry within the ramp zone is shown on the horizon separation diagram in Fig. 8a. The hangingwall cutoff of fault A and the footwall cutoff of B are both straight and approximately horizontal. Within the ramp zone the footwall cutoff of A and the hangingwall cutoff of B are parallel to one another and plunge *c.* 10° towards the east. Figure 8b shows the throw profiles for the two faults within the ramp zone. The profiles are projected horizontally onto a vertical plane parallel to the strike of the overlapping faults. As imaged, there is clearly very little actual overlap between these two faults and their aggregate throw profile has a central low (Fig. 8b).

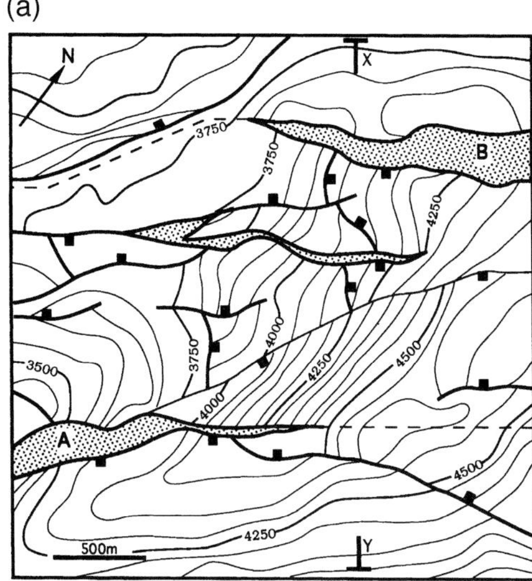

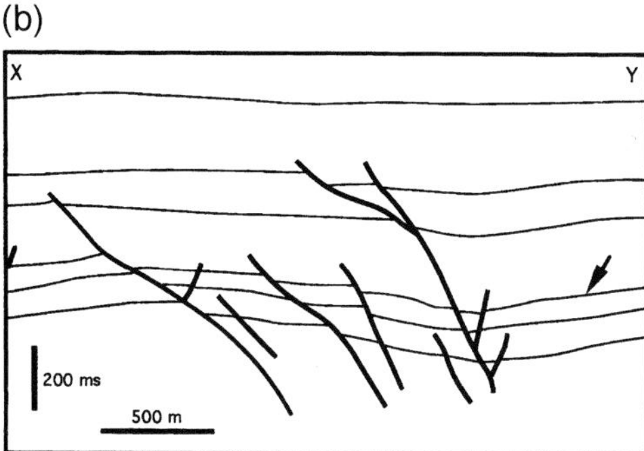

Fig. 7. (**a**) Structure contour map of a large relay zone from the North Sea: depth contours are in feet. Heave polygons of major faults are shown (stippled). The relay zone is bounded by faults A and B. Broken lines are the traces of cross-sections used to estimate differences in bed elevation across the relay zone (see Fig. 8b).
(**b**) Cross-section along X–Y (location shown in (a)). Mapped horizon in (a) is identified (arrow). The sequence above this horizon is syn- and post-faulting. Data are from seismic lines with 100m spacings oriented parallel to the vertical edge of (a).

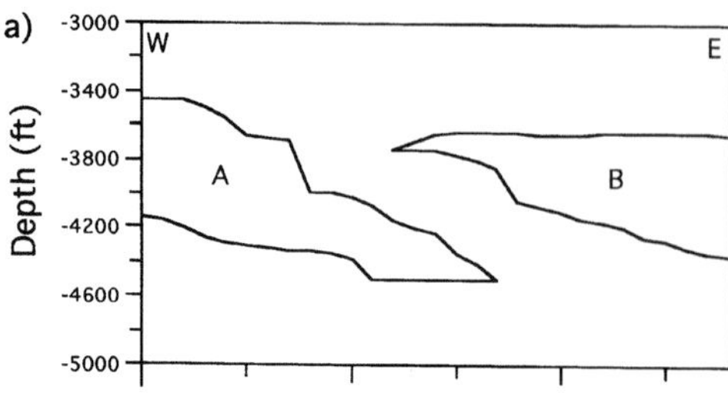

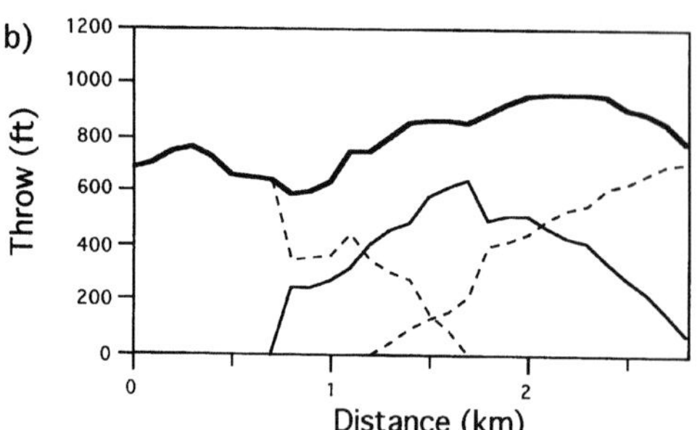

Fig. 8. (**a**) Horizon separation diagram for relay bounding faults A and B (see Fig. 7a), showing the traces of the mapped horizon cutoffs. Distance (km) is measured from the left-hand margin of the map in Fig. 7a and depths are in feet. (b) Throw profiles for (i) faults A and B (broken lines), (ii) elevation changes of horizon due to combined effect of throws on minor faults and bed rotations within the relay zone (light solid line; values represent a minimum estimate and are calculated from the depth difference between the two fault parallel sections shown in Fig. 7a), and (iii) the aggregate of curves (i) and (ii) (heavy solid line). Throws are in feet.

The horizon apparent dip towards the mutual hangingwall (Fig. 7b) is partly accommodated by small faults and partly by bed rotation. The apparent dip and its variation along strike can be estimated from differences in horizon elevations on lines bounding the relay zone and parallel to its margins. The map traces of the two sections are shown in Fig. 7a and the differences in horizon elevations on the two sections are shown in Fig. 8b. The depth difference between the two sections provides a minimum estimate of the combined effects of bed rotation and of throws on the minor faults, as in the profile in Fig. 8b. The imaged minor faults within the ramp are responsible for *c.* 40% of the difference in elevation across the relay zone. The bed rotation and minor faulting together largely compensate for the low aggregate throws on faults A and B, as shown by the more regular aggregate profile (Fig. 8b). Throw is therefore conserved across the ramp region but is partly accommodated by the ductile shear strain giving rise to the horizon rotation and increased bed length across the relay zone. Only after the significance and spatial distribution of this ductile strain is established can a 'valid' restoration of a given cross-section across this relay be performed using a method which explicitly incorporates the ductile strain (e.g. synthetic simple shear method; Rowan & Kligfield 1989). Displacement analysis of this structure must also take account of the displacements accommodated by the ductile shear strain.

Conjugate faults

Intersecting conjugate faults are common in areas where the fault density and the numbers of adjacent opposed-dipping faults are relatively high. Cross-sections of intersecting faults show conjugate, or hourglass, structures with opposed-dipping faults or fault arrays (Horsfield 1980; Woods 1992; Nicol *et al.* 1995). The notion of contemporaneous

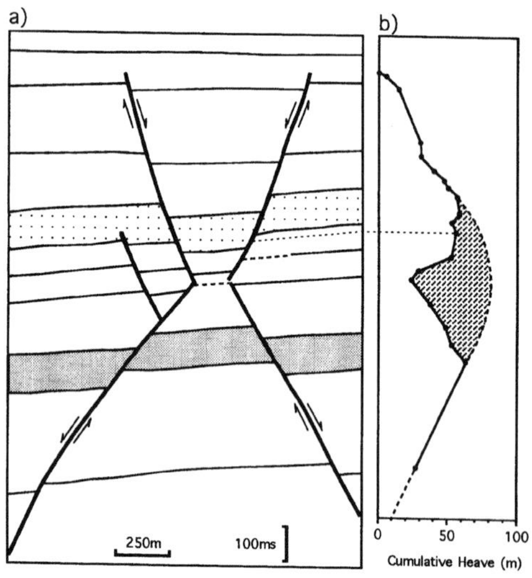

Fig. 9. (**a**) Line drawing of interpretation of a seismic section from the Timor Sea dataset, showing conjugate fault structure (see Nicol *et al.* 1995). The bounding faults cannot be traced across the intersection zone, because their displacements are believed to be accommodated by a more distributed network of sub-seismic faults. The horizons within the intersection zone are therefore continuous at the resolution of the seismic data (see text for details). The sequence above the light stippled unit is syn-faulting. Vertical and horizontal scales are approximately equal. (**b**) Aggregate heave profile for seismically imaged faults within the conjugate structure in (a). Heaves are aggregated along each horizon. Heavy broken line shows the extension (heave) profile expected for two faults with central displacement maxima, while the shaded area shows the inferred extension due to ductile deformation.

movements, over protracted geological time (millions of years), on mutually cross-cutting faults is difficult to envisage from a single cross-section of a conjugate structure (Fig. 9). However, where a conjugate structure is seen to be restricted to the lateral tip regions of the constituent faults with most of their surfaces not intersecting one another, contemporaneous movement is plausible (Fig. 10). Nicol *et al.* (1995) have shown that conjugate structures formed at overlapping tip regions of opposed-dipping faults of similar size, change polarity along the overlap region with an approximately symmetrical conjugate structure only towards the centre of the overlap (Fig. 10).

Significant interaction between faults and a structurally complex intersection zone are predictable consequences of synchronous movement on intersecting faults. Figure 9 shows a cross-section of a small conjugate structure within the highly reflective post-Base Cretaceous sequence of the Timor Sea (see Fig. 3). The bounding faults, which cannot be traced through the conjugate intersection zone, have displacements which decrease both upwards and downwards from the approximate centre of each fault trace (Fig. 9a). The aggregate heave profile (Fig. 9b) shows increasing heave upwards and downwards from the intersection zone but with a sharp decrease centred on the intersection; the expected heave, or extension, profile is shown by the broken line in Fig. 9b. Nicol *et al.* (1995) have shown that the fault displacement variations are consistent with an increasing component of pure shear ductile strain towards the intersection zone. This ductile strain is expressed as a thinning and extension of the stratigraphic units between the conjugate faults by an amount approximately equivalent to the shortfall in aggregate extension on the observed heave profile (Fig. 9b). Although conjugate structures are common within the survey area the ductile strains associated with their intersections account for only a small proportion of the regional extension (Fig. 5b).

The decrease in imaged fault displacements towards the intersection zones of the conjugate structures is interpreted as being complemented by an increase in ductile extension arising from the changes in the fault slip geometries within the intersection zone (Nicol *et al.* 1995). Whereas the fault surfaces beyond the intersection zones

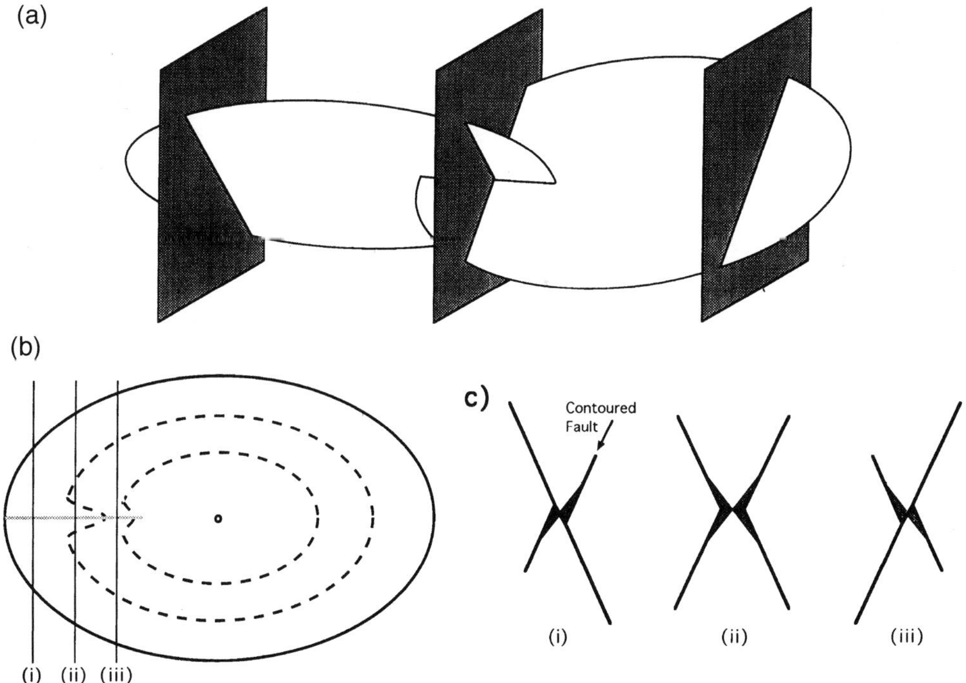

Fig. 10. (**a**) Schematic diagram showing a conjugate structure at the lateral overlap between ideal elliptical fault surfaces. (**b**) Throw contour pattern on right hand fault in (a), with regularity of throw contours (broken lines) modified by apparent decreases in throw towards the intersection line (stippled) of the two fault surfaces. (**c**) Cross-sections along lines (i), (ii) & (iii) in (b) showing symmetry. (a) and (b) modified from Nicol *et al.* (1995).

accumulated many increments of slip on essentially the same fault surfaces, which is the usual method of fault growth, the progressive offset of each fault by the other requires generation of numerous new slip surfaces to maintain continuity of each fault surface and related displacements. Thus within the intersection zones the cumulative fault displacements are believed to be high but accommodated by an array of sub-seismic structures which either splay from the bounding faults or form a more distributed network of minor faults. Deformation associated with this complex system of opposed-dipping sub-seismic faults is seen as a ductile strain. In these circumstances, bed length balancing will not be possible but, if no volume loss occurs, pre- and post-faulting areas will balance. Figure 11 shows a relatively simple outcrop example of the type of structure expected at the intersection of the seismically imaged conjugate faults.

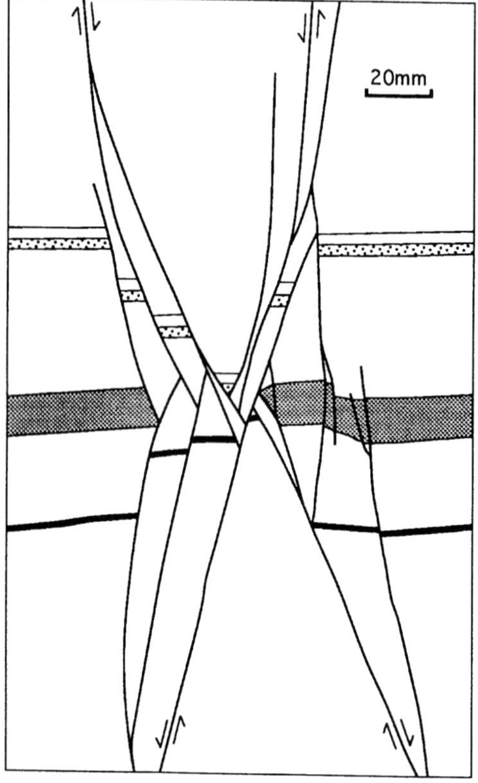

Fig. 11. Outcrop example of conjugate normal faults exposed in an outcrop of Pleistocene fluvio-glacial sands, Germany, showing the type of structure expected at the intersection of the seismically imaged conjugate faults. Drawing traced from a photograph provided by Dietmar Meier.

The concept of area balance associated with conjugate structures provides the basis for a simple geometric model (Fig. 12) which predicts heave variations on bounding faults similar to those described above and which could be used in restorations. The model assumes that the intersection point of conjugate faults bounding a graben lies at the top of a notional layer which is uniformly extended by ductile strain during extension. Area balance is maintained within the conjugate graben and initially horizontal horizons remain so during extension (Fig. 12a & b): the mutual hangingwall area extends by pure shear and no deformation occurs outwith the graben. Pre-faulting sequences are overlain by syn-faulting units which are modelled by assuming constant rates of both footwall (i.e. regional) sedimentation and fault slip: the effects of compaction are neglected. Calculated heave profiles are shown in Fig. 12c, for a model with parameters (extension, sequence thicknesses and fault dips) similar to those of the Timor Sea conjugates (Fig. 9; Nicol *et al.* 1995). Within the pre-faulting sequence, fault displacements decrease downwards from the surface to the intersection point (Fig. 12c), compensated for by an increase in ductile strain at the base of the mutual hangingwall wedge. With increasing extension, the extension accommodated by ductile strain increases. Heave profiles are similar to those of seismically imaged conjugates, mainly because the downward increase in ductile strain towards the intersection point is consistent with the notion that discrete bounding faults are replaced towards the intersection by an array of sub-seismic structures.

Bed lengths across conjugate structures do not balance, because the extension across each structure is accommodated by seismically imaged faults in some parts and by sub-seismic faults in other parts. However, if there is no volume change, area balance criteria can be used both for validating cross-sections and for estimating extensions (e.g. Rowan & Kligfield 1989).

Ductile folding

Distinction between different types of fault-related folds is critical to structural evaluation of fault systems. The fold types associated with normal faults include (see also Schlische in press) the following.

(i) Fault propagation folds which develop at fault tip-lines and are believed to be largely responsible for the presence of 'normal drag' adjacent to normal faults. 'Forced folds' within cover sequences above reactivated basement normal faults (Withjack *et al.* 1989) are an accentuated form of propagation fold. Normal drag folds

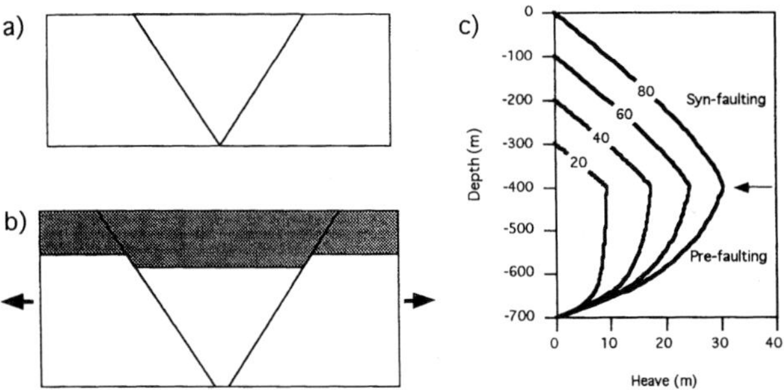

Fig. 12. Simple geometric model for conjugate faults. (**a**) Faults prior to extension. The intersection is at the top of a layer which is assumed to extend ductilely during extension. (**b**) Geometry after extension and sedimentation. The pre-faulting unit within the graben area balances: there is no deformation of the footwall blocks. The syn-faulting unit (stippled) infills the fault generated topography. (**c**) Curves of heave (m) vs. depth (m) for a model bounding fault in a model with parameters consistent with those of simple conjugates in the Timor Sea: fault dip = 75°, pre-faulting sequence thickness = 300 m, syn-faulting hangingwall sequence thickness = 400 m. Heave profiles for extensions of 20 m, 40 m, 60 m and 80 m are shown, based on a ratio of 1:10 between rate of footwall sedimentation and rate of extension.

accommodate a component of continuous displacement which is complementary and additional to the discontinuous component of displacement represented by the fault offset (Walsh & Watterson 1990). Faults which are surrounded by a cluster of numerous smaller sub-resolution synthetic faults may appear to be bordered by normal drag folds on seismic sections.

(ii) Reverse drag folds within the volume surrounding an individual fault accommodate reducing displacement with distance from a fault surface (Hamblin 1965; Barnett *et al.* 1987; Williams & Vann 1987). The wavelengths of 'reverse drag' folds are often an order of magnitude larger than 'normal drag' folds.

(iii) Compaction synclines in which the compaction of hangingwall sediments forms a syncline adjacent to the fault (White *et al.* 1986; Gibson *et al.* 1989).

Fault offsets and normal drag folds accommodate the discontinuous and the continuous, or ductile, components of fault-related displacement respectively (Walsh & Watterson 1990). The continuous component of displacement is no less significant than the discontinuous component for displacement analysis and for section balancing or restoration purposes. Indeed what is seen on one scale as folding may be faulting when seen on another scale. In this section we describe an example of changes in the proportions of discontinuous and continuous components of displacement which would not be apparent from examination of a single cross-section.

Figure 13 shows two cross-sections across a Middle–Upper Jurassic fault interpreted from a Northern North Sea 3D seismic dataset. Faulting initiated during the Brent with stratigraphic growth across the fault. Apart from some fault reactivation at the northern end of the mapped area, fault movement had largely ceased by the Base Cretaceous. Along the length of the fault, displacement is locally partitioned onto 3 sub-parallel, laterally discontinuous, synthetic hanging-wall faults (Fig. 14a). At the Base Brent these minor faults, which may branch from the main fault below the seismic data window, occur between the main fault and the axial trace of a hanging-wall syncline of variable amplitude (Fig. 13b & c). The main fault shows very irregular lateral variations in throw (Fig. 14b). Aggregation of throws on the main fault and the minor faults shows throw variations which are only slightly more regular (Fig. 14b): throws are aggregated along the regional extension direction, but aggregation along lines perpendicular to the local fault strike provides similar results. If the continuous throw component represented by the fold is included in the throw aggregation, the throw variations are regular (Fig. 14b): throws accommodated by folding are calculated by extrapolating the horizon trends beyond the fold hinge, up to the fault. Figure 13b shows that the proportion of displacement accommodated by folding can be as high as 50%.

This exercise suggests that the monocline is not primarily a result either of compaction or of fault surface irregularity at depth. The fold cannot be a fault-propagation structure, because the oldest syn-rift sequence does not increase in thickness towards the axis of the syncline. Instead, the systematic partitioning of displacements between faults and fold (i.e. discontinuous and continuous displacements), indicates that they are coeval and

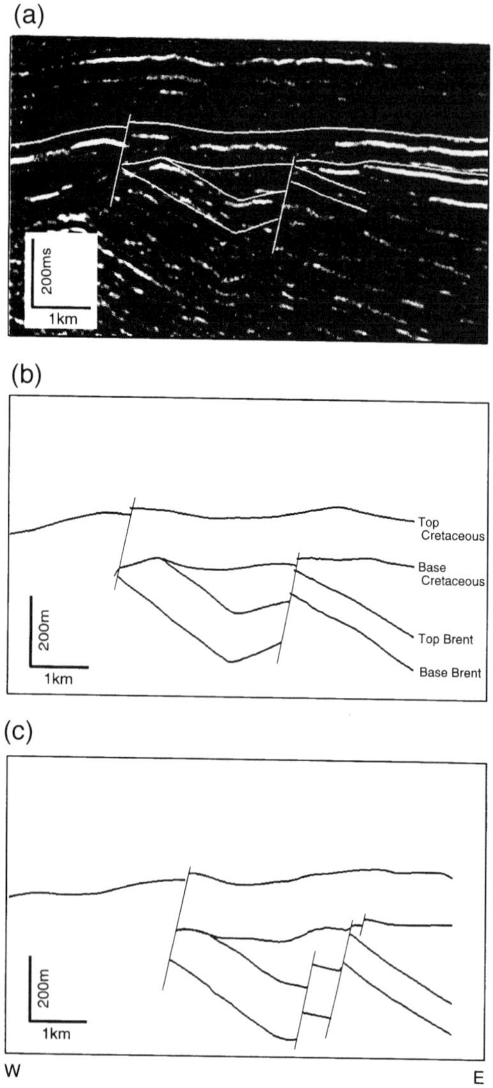

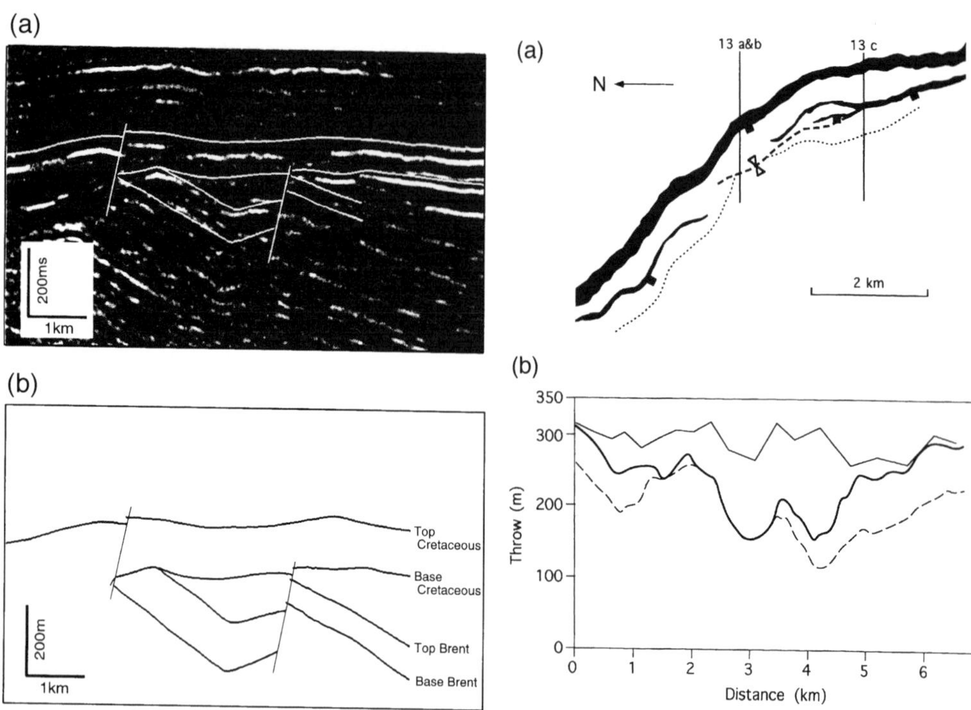

(c)

200m

1km

W E

Fig. 13. (**a**) Seismic section across a fault from the Northern North Sea, showing four interpreted horizons (Base Brent, Top Brent, Base Cretaceous and Top Cretaceous). Brent sequence thickens in the hangingwall as a result of syn-faulting sedimentation. (**b**) Depth-converted interpretation of section in (a). Vertical exaggeration is *c.* 6. (**c**) Depth-converted interpretation of section *c.* 2.3 km along strike from (b). All sections are viewed from the south and Fig. 14a shows their locations.

represent a kinematically coherent system. Whether or not the folding is accommodated by sub-seismic faults cannot be determined from the seismic data. Conclusions based on bed length restoration and or extension estimates based only on the discontinuous components of throw would depend on which cross-section(s) was analysed. Only by analysing the structure in 3D can the origin and displacement contribution of the fault-related fold be assessed.

Fig. 14. (**a**) Fault map of the Base Brent horizon showing locations of sections in Fig. 13 which extend beyond the map area. The axial trace (heavy broken line) of a hangingwall syncline and the trace of the outer limit of the fault-related fold (light broken line) are also shown. (**b**) Throw profiles for the Base Brent horizon along the length of the fault array shown in (a): profile for main fault – broken line; aggregate profile for main fault and hangingwall minor faults – heavy solid line, aggregate throw profile for all faults and fold – light solid line. Errors on throw measurements are up to *c.* 20 m.

Discussion

The restoration of depth-converted interpretations of seismic sections represents one of the principal methods used in the structural analysis of extensional basins. The technique provides a basis for both construction of geologically valid cross-sections and estimation of extensions. The two main approaches to section balancing, i.e. line length and area balancing, were originally applied to thrust belt tectonics (Bally *et al.* 1966; Dahlstrom 1969) and later modified for application to extensional tectonics (Verrall 1982; Gibbs 1983).

Explicit incorporation of ductile strain in balancing of contractional sections has sometimes been possible where outcrop data are available (Hossack 1979; Cooper & Trayner 1986). In extensional fault systems, the primary data generally comprise 2D or 3D seismic surveys, supplemented by well control, so the estimation of ductile strains requires different methods. In areas characterized by gravity-driven listric faulting (e.g. Gulf of Mexico and Niger Delta), various restoration techniques (incorporating fault shape and a variety of geometric constructions; see Williams & Vann 1987) can be used to provide estimates of the ductile strain accommodating hangingwall rollover geometries, but the choice of technique to apply can be difficult and often appears to be arbitrary. None of these techniques can be applied where faults are known not to be listric or where they accommodate regional extension or where 2nd order faults are not necessarily rollover accommodation structures. In this article we have described several examples of structures which come into this category, but for which ductile strain can be estimated by displacement and fault size population analysis techniques. The examples show that significant ductile strains occur on both regional and local scales and can be heterogeneous, varying both laterally and vertically. Line length balancing cannot be simply applied to sections within which there is heterogeneous stretching or shear. Where the ductile strain is homogeneous across a section, the section will line length balance without taking account of the ductile strain but extension estimates will be minimum values; the contribution of sub-resolution faulting should be estimated. Area balancing methods are more robust, particularly where a well constrained pre-deformation stratigraphic template with several mapped horizons is available. However, where the pre-deformation stratigraphy is poorly constrained, as is usually the case for syn-rift sequences, the results of area balancing are equivocal. Our examples suggest that validation or modification of seismic interpretations on the basis of conventional section restoration methods should be avoided. Even where several restoration techniques are utilized, selection of the most appropriate technique is subjective given the variety and complexity of possible fault-related ductile strains. In such circumstances, displacement analysis represents a supplementary, or alternative, method for testing the validity of 3D fault-related geometries.

Conclusions

(1) Ductile strains accommodated by sub-resolution faults can represent a significant proportion (*c*. 40–50%) of the total strain.

(2) The significance of fault-related ductile strain increases in the vicinity of interacting faults (e.g. intersecting conjugate faults and within relay zones).

(3) Displacements accommodated by ductile strain (i.e. continuous, as opposed to discontinuous displacements) can vary along the length of a fault system.

(4) The amount of fault-related ductile strain can be assessed from fault displacement and fault size population analysis. The coalfield map and other data suggest that prediction of sub-seismic fault numbers and throws is possible.

(5) While bed length and area balancing are useful methods for identification of grossly invalid sections, they should be supplemented by displacement and ductile strain analysis when used for more refined testing and reconstruction of sections.

The authors thank Ampolex Ltd, BP Exploration Operating Company and Norsk Hydro for providing the seismic data on which this work is based. They are grateful to Chris Bonson, Peter Bretan, Marie Eeles, Dan Ellis and Isabel Jones for preparation of diagrams and other members of the Fault Analysis Group for much useful discussion. They also wish to thank Dietmar Meier for providing the photograph from which Fig. 11 was traced. Reviews by Jack Filbrandt and Mark Rowan improved the paper. This research was part funded by Amoco (UK) Production Company (contract RAD 59 (90)), OSO/NERC Hydrocarbon Reservoirs LINK Programme (project 827/7053), NERC Petroleum Earth Sciences Programme (grant D1/G1/189/03) and E.C. JOULE II Hydrocarbons Programme (contracts JOU2-CT92-0182 and JOU2-CT92-0099).

References

ANDERS, M. H. & SCHLISCHE, R. W. 1994. Overlapping faults, intrabasin highs and the growth of normal faults. *Journal of Geology*, **102**, 165–180.

BALLY, A. W., GORDY, P L. & STEWART, G. A. 1966. Structure, seismic data, and orogenic evolution of southern Canadian Rocky Mountains. *Bulletin of Canadian Petroleum Geology,* **14,** 337–381.

BARNETT, J. A. M., MORTIMER, J., RIPPON, J., WALSH, J. J. & WATTERSON, J. 1987. Displacement geometry in the volume containing a single normal fault. *American Association of Petroleum Geologists Bulletin*, **71,** 925–937.

CHILDS, C., WALSH, J. J. & WATTERSON, J. 1990. A method for estimation of the density of fault displacements below the limits of seismic resolution in reservoir formations. *In*: BULLER, A. T., BERG, E., HJELMELAND, O., KLEPPE, J., TORSÆTER, O. and AASEN, J. O. (eds) *North Sea Oil and Gas*

Reservoirs II. Graham & Trotman, London, 309–318.

COOPER, M. A. & TRAYNER, P. M. 1986. Thrust-surface geometry: implications for thrust belt evolution and section balancing techniques. *Journal of Structural Geology*, **8**, 305–312.

DAHLSTROM, C. D. A. 1969. Balanced cross sections. *Canadian Journal of Earth Sciences*, **6**, 743–757.

GAUTHIER, B. D. M. & LAKE, S. D. 1993. Probabilistic modelling of faults below the limit of seismic resolution in the Pelican Field, North Sea, offshore U.K. *American Association of Petroleum Geologists Bulletin*, **77**, 761–771.

GIBBS, A. D. 1983. Balanced cross-section construction from seismic sections in areas of extensional tectonics. *Journal of Structural Geology*, **5**, 153–160.

GIBSON, J. R., WALSH, J. J. & WATTERSON, J. 1989. Modelling of bed contours and cross-sections adjacent to planar normal faults. *Journal of Structural Geology*, **11**, 317–328.

GILLESPIE, P. A., HOWARD, C. B., WALSH, J. J. & WATTERSON, J. 1993. Measurement and characterisation of spatial distributions of fractures. *Tectonophysics*, **226**, 113–141.

HAMBLIN, W. K. 1965. Origin of 'reverse drag' on the downthrown side of normal faults. *Bulletin of the Geological Society of America*, **76**, 1145–1164.

HORSFIELD, W. T. 1980. Contemporaneous movement along crossing conjugate normal faults. *Journal of Structural Geology*, **2**, 305–310.

HOSSACK, J. R. 1979. The use of balanced cross-sections in the calculation of orogenic contraction: A review. *Journal of the Geological Society, London*, **136**, 705–711.

HOULAND, M., CROKER, P. F. & MARTIN, M. 1994. Fault-associated seabed mounds (carbonate knolls?) off western Ireland and north-west Australia. *Marine and Petroleum Geology*, **11**, 232–246.

HUGGINS, P., WATTERSON, J., WALSH, J. J. & CHILDS, C. in press. Relay zone geometry and displacement transfer between normal faults recorded in coal-mine plans. *Journal of Structural Geology*.

JACKSON, P. & SANDERSON, D. J. 1992. Scaling of fault displacements from the Badajoz-Córdobe shear zone, SW Spain. *Tectonophysics*, **210**, 179–190.

MARRETT, R. & ALLMENDINGER, R. W. 1991. Estimates of strain due to brittle faulting: sampling of fault populations. *Journal of Structural Geology*, **13**, 735–738.

NICOL, A., WALSH, J. J., WATTERSON, J. & BRETAN, P. 1995. Three dimensional geometry and growth of conjugate normal faults. *Journal of Structural Geology*, **17**, 847–862.

PEACOCK, D. C. P. & SANDERSON, D. J. 1991. Displacements, segment linkage and relay ramps in normal fault zones. *Journal of Structural Geology*, **13**, 721–733.

—— & —— 1994. Geometry and development of relay ramps in normal fault systems. *American Association of Petroleum Geologists Bulletin*, **78**, 147–165.

ROBERTS, A.M., YIELDING, G., KUSZNIR, N.J., WALKER, I. & DORN-LOPEZ, D. 1993. Mesozoic extension in the North Sea: constraints from flexural back-stripping, forward modelling and fault populations. *In*: PARKER, J. R. (ed.) *Petroleum Geology of Northwest Europe: Proceedings of the 4th Conference*. Geological Society, London, **2**, 1123–1136.

ROWAN, M. & KLIGFIELD, R. 1989. Cross section restoration and balancing as aid to seismic interpretation in extensional terranes. *American Association of Petroleum Geologists Bulletin*, **73**, 955–966.

SCHLISCHE, R.W. in press. Geometry and origin of fault-related folds in extensional settings. *American Association of Petroleum Geologists Bulletin*.

VERRALL, P. 1982. *Structural Interpretation with Applications to North Sea Problems*. Course Notes No. **3**. JAPEC, London.

WALSH, J. J. & WATTERSON, J. 1990. New methods of fault projection for coalmine planning. *Proceedings of the Yorkshire Geological Society*, **48**, 209–219.

—— & —— 1991. Geometric and kinematic coherence and scale effects in normal fault systems. *In*: A. M. ROBERTS, YIELDING, G. & FREEMAN, B. (eds) *The Geometry of Normal Faults*. Geological Society, London, Special Publication, **56**, 193–203.

—— & —— 1992. Populations of faults and fault displacements and their effects on estimates of fault-related regional extension. *Journal of Structural Geology*, **14**, 701–712.

——, —— & YIELDING, G. 1991. The importance of small-scale faulting in regional extension. *Nature*, **351**, 391–393.

——, —— & —— 1994. Determination and interpretation of fault size populations: procedures and problems. *In*: AASEN, J. O., BERG, E., BULLER, A. T., HJELMELAND, O., HOLT, R. M., KLEPPE, J. & TORSÆTER, O. (eds) *North Sea Oil and Gas Reservoirs III*. Kluwer Academic Publishers, London, 141–155.

WATTERSON, J., WALSH, J., GILLESPIE, P. A. & EASTON, S. in press. Scaling systematics of fault sizes on a large scale range fault map. *Journal of Structural Geology*.

WHITE, N. J., JACKSON, J. A. & MCKENZIE, D.P . 1986. The relationship between the geometry of normal faults and that of the sedimentary layers in their hangingwall. *Journal of Structural Geology*, **8**, 897–909.

WILLIAMS, G. & VANN, I. 1987. The geometry of listric normal faults and deformation in their hanging-walls. *Journal of Structural Geology*, **9**, 789–795.

WITHJACK, M. O., MEISLING, K. E. & RUSSELL, L. R. 1989. Forced folding and basement-detached normal faulting in the Haltenbanken area, offshore Norway. *In*: TANKARD, A. J. & BALKWILL, H. R. (eds) *Extensional Tectonics and Stratigraphy of the North Atlantic Margins*. AAPG Memoir, **46**, 567–575.

WOODS, E. P. 1992. Vulcan sub-basin fault styles – implications for hydrocarbon migration and entrapment. *APEA Journal*, **32**, 138–158.

YIELDING, G., WALSH, J. J. & WATTERSON, J. 1992. The prediction of small-scale faulting in reservoirs. *First Break*, **10**, 449–460.

The application of cross-section construction and validation within exploration and production: a discussion

JAMES G. BUCHANAN
British Gas Exploration and Production Ltd, Thames Valley Park, Reading, Berkshire RG6 1PT, UK

Abstract: Numerous published studies have shown that cross-section balancing and validation techniques are a powerful method of structural analysis. This paper outlines reasons why cross-section validation also is a valuable methodology within the oil/gas industry.

Cross-section validation is a vehicle of getting analytical rigour into seismic and non-seismic interpretation by testing the section against the rules of structural geology (in particular geometry).

In addition to the analysis of structural traps, cross-section validation can be used in oil or gas sourcing studies, especially in the relative timing of hydrocarbon migration and trap formation/destruction. It is therefore a valuable methodology in both 3D basin analysis and prospect risking/ranking.

Cross-section validation should be an integral part of the petroleum geologist's tool kit to aid further understanding of the 3D evolution of a basin and basin fluids through time. This technique, when used with other methodologies, such as sequence stratigraphy and basin modelling, allows the interpreter to use all the available data sets to constrain geological models on hydrocarbon prospectivity.

This paper discusses the rationale for using cross-section validation (XSV) techniques within hydrocarbon exploration and production companies. In contrast to academia, where XSV is only applied by researchers interested in structural geology or tectonics, explorationists need to use the technique as a way to understand more fully the evolution and prospectivity of sedimentary basins. In many cases, however, users will need to be convinced of the benefits of the technique.

This paper outlines the principal benefits and drawbacks of XSV and some guidelines for the successful use of the technique are presented. This contribution is not intended to be a comprehensive reference on the detailed methodologies and techniques of producing balanced cross-sections. The reader is referred to other sources for that information (De Paor 1988; Woodward *et al.* 1989; Mitra & Nansom 1991).

Within this paper, XSV are used to describe the construction, restoration, analysis and validation of cross-sections. The term cross-section balancing (CSB) is used to describe the process of cross-section analysis where all the rules and assumptions outlined by Elliott (1983) are followed rigorously.

XSV and CSB is now routinely carried in both academia and industry using PC or workstation based computer systems. These systems allow very precise and accurate cross-section construction and restoration and are a major step forward in the application of the technique (Jones 1984; Kligfield *et al.* 1986; Geiser *et al.* 1988; Gibbs & Griffiths 1994; Bishop & Buchanan 1995).

Validation techniques are a major aid in determining sub-surface geometries and can give further understanding of deformation mechanics and dynamics.

Historical perspective

The balanced cross-section technique using the line length methodology was first presented from studies by the oil industry in the Southern Canadian Rocky Mountains (Dahlstrom 1969, 1970). This work built on studies published by Douglas (1950, 1958) where line balancing techniques were used to analyze thrust plane trajectories and thereby make interpretations of structural styles and thrust belt evolution (Fig. 1). These classic studies were not only major contributions both to the study of foreland fold and thrust belts, but also to the exploration efforts within compressional domains (Bally *et al.* 1966, Dahlstrom 1969, 1970) (Fig. 2).

The cross-section balancing methodology was refined by Elliott and co-workers, again mainly from studies in compressional areas (Hossack 1979; Elliott & Johnston 1980; Boyer & Elliott 1982; Elliott 1983). Detailed CSB is a vital part in the geometrical analysis of fault-related fold kinematics and evolution (Suppe 1983; Jamison 1987; Suppe & Medwedeff 1990). The equal area restoration technique was re-discovered at this time

From Buchanan, P. G. & Nieuwland, D. A. (eds), 1996, *Modern Developments in Structural Interpretation,Validation and Modelling*, Geological Society Special Publication No. 99, pp. 41–50.

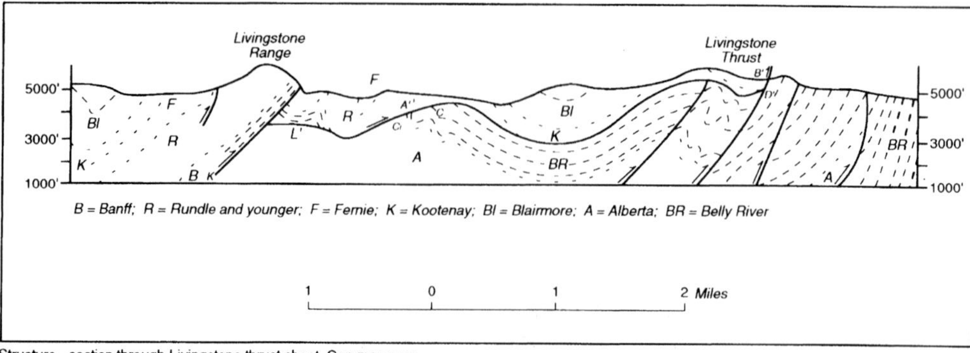

Structure - section through Livingstone thrust sheet, Gap map-area

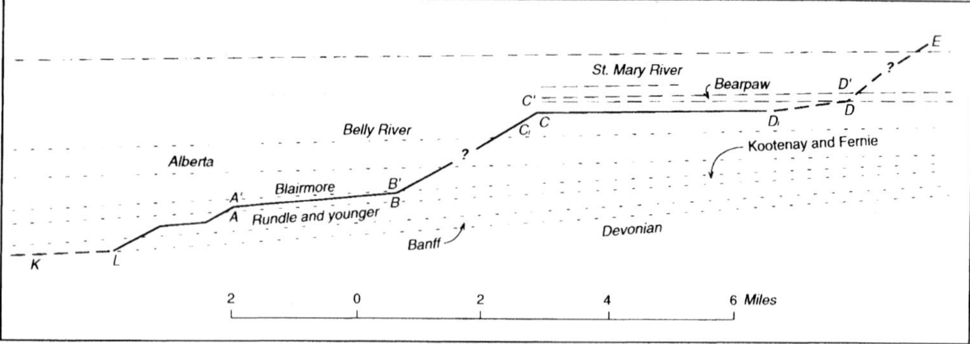

Path (K-E) followed by Livingstone thrust through Upper Paleozoic and Mesozoic strata, as inferred

Fig. 1. One the first examples of detailed cross-section analysis and restoration. The diagram is taken from the classic memoir on the Livingstone Range in SW Alberta Canada by Douglas 1950 (reprinted in 1981). The upper figure shows a cross-section through the Livingstone Range, SW Alberta, in the frontal ranges of the Southern Canadian Rocky Mountains. The lower figure illustrates the trajectory of the Livingstone Thrust plane on a summary stratigraphic template for the area. The distinctive ramp-flat geometry of the fault plane and major detachment or decollement are clearly shown. (Redrawn after Douglas 1950; © Geological Survey of Canada.)

and it also built on 'depth to detachment' studies carried out earlier in the century (Chamberlain 1910, 1919; Goguel 1962; Laubscher 1962; Mitra & Nansen 1989). Many research groups focused on using the balanced cross-section concepts to further understand the evolution of thrust and fold belts (Boyer & Elliott 1982; Butler 1983, 1986; Hossack 1983; Geiser 1988; Boyer 1991) (Fig. 3). At the same time other structural teams moved into the analysis of the North Sea basins and other extensional areas (Gibbs 1983; Davison 1986; Arthur 1993) (Fig. 4). Latterly even salt prone areas (which will have major out-of-plane movements of material) have been studied using the XSV technique (Hossack 1994; Hopper *et al.* 1995; Rowan this volume). Recent studies have focused on the use of the appropriate deformation mechanism in the restoration process i.e. flexural slip or vertical/inclined shear (Hauge & Gray this volume; Verschuren *et al.* this volume).

Cross-section validation rationale

The use of cross-section restoration is a powerful method of section analysis. It helps to ensure the cross-section is strain-compatible with adjoining areas, geometrically valid and internally consistent (Butler 1994). For example, the geometry of the cut-offs and presence of overlaps or gaps in restored sections testify to problems in the initial cross-section interpretation. The technique forces a more focused and detailed analysis as the assumptions made in the construction of the section can be tested. The methodology allows the interpreter to 'get into' the cross-section under analysis and thereby further understand the geometry and evolution of a sedimentary basin. The iterative process of constructing or testing a cross-section is often time consuming but an integral part of the rigorous structural analysis. A study of several validated cross-sections across

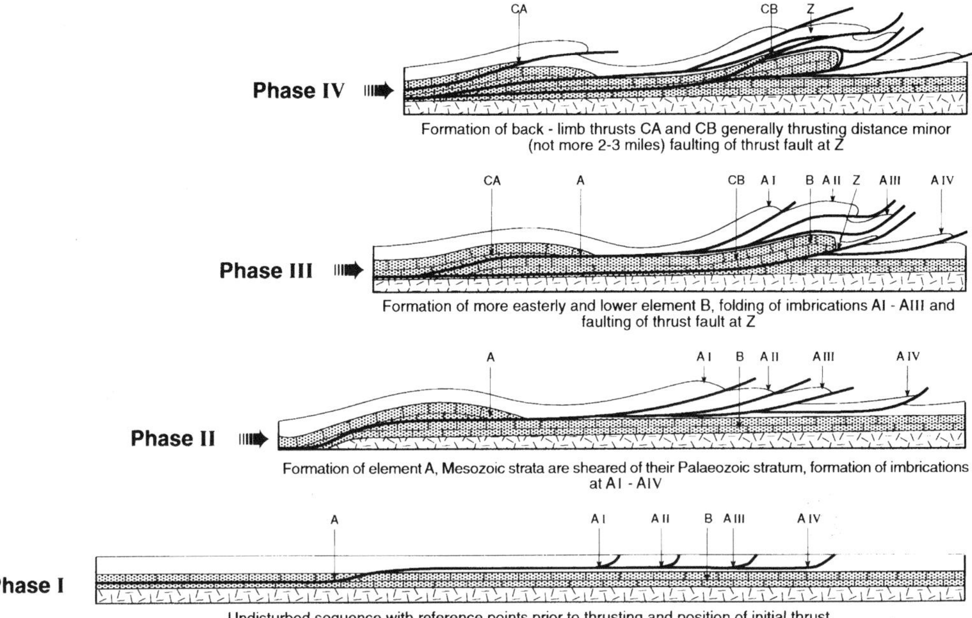

Fig. 2. Sequential cross-sections showing an interpretation of the evolution of the Southern Canadian Rocky Mountains. This highly innovative work published ir 1966 shows initial fault trajectories, the development of hangingwall anticlines and imbricate fans, in addition to documenting the presence of 'out of sequence' thrusts.(Redrawn after Bally *et al.* 1966; © Geological Survey of Candada.)

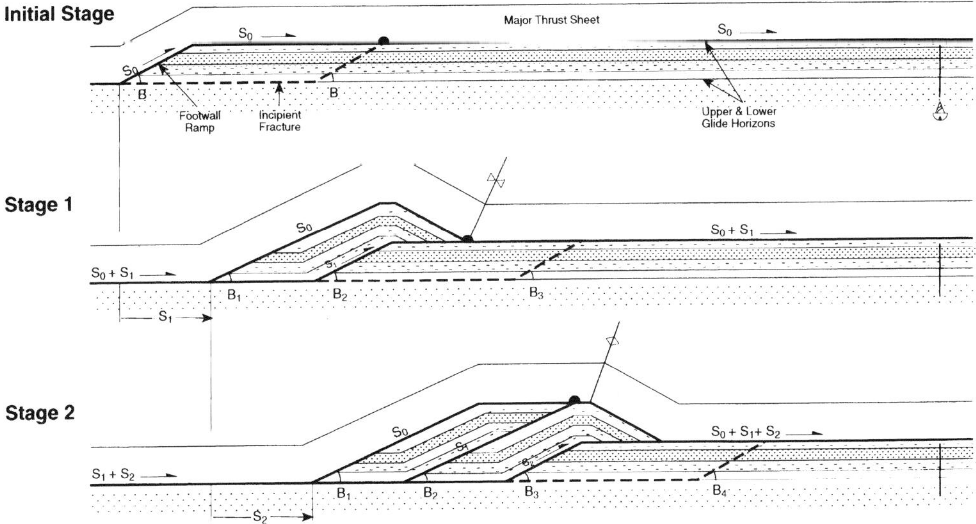

Fig. 3. This figure of the formation of duplexes by progressive footwall collapse is an example of forward modelling using balanced cross sections. In the late 1970s and early 1980s balanced sections were increasing used to analyse and model thrust systems. (Redrawn after Boyer & Elliott 1982; © AAPG.)

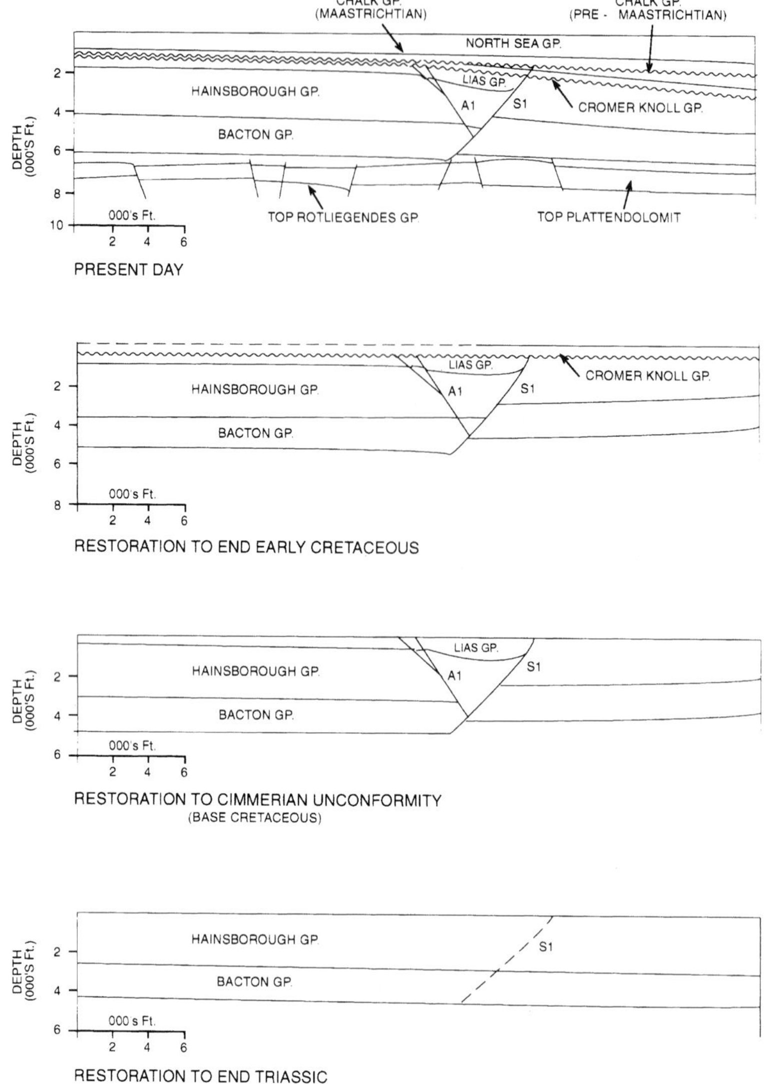

Fig. 4. An example of sequential restored cross-sections across Block 49/28 in the UK sector of the Southern North Sea. Cross section analysis is used to validate the initial interpretation (in this case based in seismic data) and the sequential restorations unable to geoscientist to document and further understand the basin evolution in two dimensions. (Redrawn after Arthur 1993.)

a deformed area can give worthwhile insights into the 3D basin development.

Advantages of cross-section validation within industry

So what if the cross-section balances, will it find more oil/gas in a cost effective manner? The answer has to be yes. The classic papers on the geometry and evolution of the Southern Canadian Rockies (Bally *et al.* 1966; Dahlstrom 1969, 1970) were produced by exploration-focused geologists. Research geologists within the petroleum industry have continued to push forward the XSV technique by testing the application of the methodology in new and different ways such as in fault related folds within frontal thrust belts (Mitra 1990; Mount *et al.* 1990)

As stated above, XSV is a vehicle of getting additional rigour into structural interpretations, thereby giving a better understanding of trap risk on a prospect. The methodology enables analysis of hydrocarbon migration routes as they evolved through basin evolution. XSV tests whether an interpretation is geologically valid and whether it obeys the geometry rules of structural geology. It should be noted that many CSB/XSV techniques are based on geometry and therefore may not be mechanically sound (Hauge & Gray this volume). Validating a trap geometry using XSV is a lot cheaper than testing it with a well. Although obviously there is no alternative to drilling in order to discover, appraise and produce hydrocarbons from a prospect.

In addition to the analysis of the structural trap geometry and formation, the XSV technique can also be used in the study of hydrocarbon migration. Sequentially restored sections can enable changes in geometry of carrier beds through time to be studied thereby testing ideas on the timing of oil/gas migration versus the trap formation. This work helps directly in the problems of timing of the hydrocarbon migration and trap formation especially when the results from a restoration study are integrated with those from basin modelling packages.

The detailed measurement of shortening or extensional strains in the balancing method can allow the interpreter to understand more fully the deformation histories across the basin. These changes in displacement patterns, both spatial and temporally, help document the basin evolution and therefore the controls on hydrocarbon prospectivity. For example, the timing and location of inversion tectonics can often control the generation, expulsion and migration of hydrocarbons. These focused studies on fault displacement and the construction of fault plane maps aid the interpretation or prediction of sealing faults and/or reservoir baffles (Knott 1993).

The XSV methodology enables the interpreter to include both geological field measurements and seismic data into cross-section construction and analysis in addition to the testing of alternative interpretations. The method allows analysis of structural sections in areas with poor quality seismic (unfortunately seismic quality is often inferior in areas of compressional tectonics due to acquisition problems in mountainous terrains).

All technical studies discussed above will improve the ability of the geologist to make decisions on prospect ranking as it can improve the understanding of the risk on the trap, migration and timing aspects of the play. XSV also can be used to make more constrained assessments on structure both before and after drilling.

Limitations or problems of cross-validation within industry

There are limitations to the application of the XSV technique and these should be recognized prior to embarking on a project. In some cases problems can be predicted and therefore kept to a minimum. Some of the major problems are highlighted below and in the following section some solutions are suggested.

Poor planning of the XSV project leads to unfocused and wasted efforts. A major aspect of the planning process ensures the project team is not too narrow, consisting only of structural geologists.

It should be noted that XSV is not a panacea to all structural problems. CSB 'sensu stricto' can only be used in areas of plane strain (Elliott 1983). Care is required therefore in the use of the technique. The regional tectonics need to be reviewed in detail to understand the dominant shortening or extensional directions in the area of interest.

A poorly constrained stratigraphic template or failure to use all the data often results in weak or less rigorous interpretations (this point is discussed more fully in the 'guidelines' section).

As many of the cross-sections requiring validation are based on seismic interpretations, depth conversion methodologies are a major part of the validation process. The change from time to depth domain can alter the angle and length of cut-offs and geometry of fault planes. These changes in geometry need to be noted and, where possible, sensitivity studies on the depth conversion process carried out. Decompaction alogithrims used during restoration studies can also introduce geometrical change within cross-sections. Detailed sensitivity studies can reduce the problems produced by crude or generalised decompaction by giving a better understanding of the compaction process during basin evolution.

XSV studies carried out in isolation can produce spurious results. The results of the validation and/or restoration needs to be integrated with other models or interpretations for the basin.

Guidelines for successful cross-section validation in industry

Planning

Prior to embarking on an XSV project, it is essential to analyze and decide on the aims of the task. Is CSB an appropriate technique to solve the perceived problems? It is important to have focused goals of study (e.g. structural evolution of prospect through time, to test the validity of a migration route).

The successful use of the CSB and XSV technique within exploration and production requires highly skilled, integrated teams of staff drawn from all the relevant geoscience disciplines. It is important to ensure that there is full liaison with regional geological staff and all technical experts.

Tectonic regimes

In dip-slip compressional or extensional areas CSB and XSV has the potential to produce excellent results and therefore is used in a routine manner (Bishop & Buchanan 1995)(Fig. 5). This is especially true where the stratigraphy is homogeneous, 'layer cake' and well understood. Care should be taken in areas of thick-skinned tectonics where structural geometries are often more difficult to interpret due to the lack of deep data and poor understanding of deformation processes within 'basement' rocks.

In areas of oblique-slip or salt tectonics, CSB can not be used as rock material can be moved out of the line of cross-section therefore breaking the plane strain rule (Elliott 1983). However, detailed structural analysis using XSV could still be carried out and will give valuable insight into the tectonic evolution of the area (Rowan, this volume). In many cases problems, initially unrecognised, are highlighted rather than neat solutions identified. In the future, true 3D restoration techniques will enable a complete analysis of basin evolution to be carried out.

Stratigraphy

In all tectonic settings a complete knowledge of the stratigraphy of an area or basin is required to produce a worthwhile validated cross-section. The stratigraphic template is the framework on which the restorations rest. In some areas (e.g. foreland fold/thrust belts) the template is often well constrained and interpretation between data points is relatively simple. Template construction, however, becomes extremely problematic in areas which have active tectonics during sedimentation or where there are major changes in lithofacies across the basin. However, successful restorations in areas of syn-tectonic sedimentation provide valuable

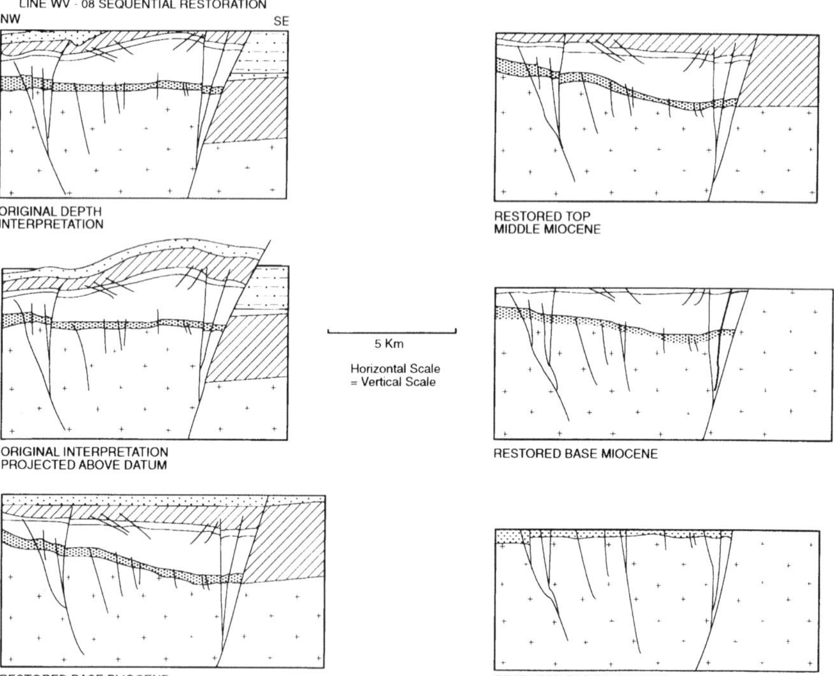

Fig. 5. Sequential cross-section restorations across the Paparoa Basin, west of the South Island, New Zealand. This is one of series of cross-sections validated using modern software presented in this paper. Increasingly CSB and XSV is only being carried out using workstation based computer software which enables projects to be completed more quickly and precisely. (Redrawn after Bishop & Buchanan 1995).

information about the nature and timing of the structural evolution of the basin (Butler *et al.* 1995). The changes in lithostratigraphy, and therefore rheology, should have effects on the structural style interpreted in the cross-sections. The stratigraphic template is usually hung on lithostratigraphic boundaries. However, given sufficient high quality biostratigraphic data, it is possible to construct a template on chronostratigraphic boundaries. Restored cross-sections based on a chronostratigraphic template gives insight into the sedimentation patterns and syntectonic evolution of the basin.

Archive data

Exploration and production companies have a large library of data, much of which cost vast amounts to acquire and process. It is vital, therefore, to use the 'in-house' information and any additional data from outside sources, to make optimal use of this valuable resource.

The analysis of existing literature is an important first task in a XSV project. Even in the remote or frontier areas there will be geological or geophysical data available which will assist in the production of cross-sections. Review all reports from government or academic institutions on the area of interest, not just the standard oil or service company reports. A comprehensive review of the regional tectonics and basin evolution complemented by the most recent plate tectonic reconstruction data will be worthwhile especially if the area of interest is on or near a plate margin. In many cases ancient lineaments control the geometry and evolution of basins and they can be utilized with contrasting displacement directions through the evolution of a basin. The analysis of plate movements can determine the stress vectors for individual plates through the basin history thereby enabling predictions of movement directions on the lineaments and therefore the basin itself.

Geological data

All the surface geological data should be utilized especially if there is limited understanding of basin deformation processes. Use all the topographical and geological maps available, or create maps from satellite data (Landsat or Spot data) to help with the interpretation of surface geological features and styles (Insley this volume). A digital terrain model can be generated from satellite data which will allow the detailed interpretation of regional structural styles. This technique is especially useful in areas where detachments between the surface and sub-surface structural geology are minimal. All satellite maps require ground truth checks to ensure the validity of the interpretations. Problems may occur if too great a reliance is placed on unconstrained interpretations as in many cases surface geology may not reflect the sub-surface,especially in areas with multiple detachments (e.g. in salt prone areas) (Verschuren *et al.* this volume).

In constructing or testing cross-sections it is obviously vital to use all available well data for formation tops and unit thicknesses. Units in offsetting wells should be correlated in order to construct the stratigraphic template for restorations. The correlation study will highlight sediment thickness changes which may result from sedimentary or tectonic processes. The presence of tectonic thickening may highlight the presence of a fault or detachment zone which is unresolved in the seismic data. Data from the well core, dipmeter logs and borehole imaging tools data provides a valuable insight into structural styles. This is especially important where mesoscopic observations are impossible to obtain e.g. in offshore areas. Well test results may also be used to determine the location of possible permeability barriers such as faults, thereby constraining a sub-surface part of the cross-section. Shallow boreholes can be used to understand further subcrop patterns both onshore and offshore. They often have associated stratigraphic, lithological and chronostratigraphical data which may help constrain geological maps. An analysis of the sonic log data can help in the determination of the burial history and therefore correlate with the sequential restoration of the sections (Hillis 1995).

The sedimentology of an area/basin is routinely documented from the analysis of wireline logs and cores from wells. In the onshore domain, well data is supplemented by detailed section logging of the relevant units. These interpretations of depositional environment can be a vital tool in the understanding of modelling in the basin. For example does the validated cross-section predict that water depths will be correct for shallow marine shoreface deposition or do the source rock kerogen facies (lacustrine, marine or mixed) correlate with the restorations of the section?

Geophysical data

The XSV technique can be used to construct a cross-section from scratch using all the available geoscience data sets. The cross-section would be validated during production of the interpretation. Alternatively, the methodology can be employed to test existing cross-section and seismic interpretations. The interpreter should always be aware of the depth conversion methodology used for the seismic data as the change from time to depth domain can alter the geometry of both reflectors and fault planes problems (Kessler & Reshef 1994).

In some offshore areas there is a wealth of shallow data from sparker or other seismic sources. This data may have been acquired for offshore wellsite survey studies or academic research into recent geological processes. This data can help in two ways, firstly, by constraining the near surface velocity model and thereby improving the depth conversion of the seismic section to be restored or validated. Secondly, the data may also help in the interpretation of the structural styles in the higher parts of the section by constraining the position and/or thickness of subcropping units to the sea bed.

Gravity, magnetic or magnetotelluric (MT) data is in many cases cheaper and more readily acquired than seismic data. Many of the technologies and techniques have been borrowed from the mineral exploration industry. The potential field data can be modelled to give insights into the basin geometry especially at depth (Bodard *et al.* 1993; Sparlin & Lewis 1994). A knowledge of basement configuration is often very critical especially when constructing cross-sections in both thin and thick skinned tectonic domains e.g. foreland fold and thrust belts. The use of magnetic and gravity data to determine the basement configuration and the depth to metamorphic or igneous bodies is now common place (Horscroft & Bain this volume). Potential field modelling can identify dykes and sills which will reduce the quality of seismic reflections and thereby aid the interpretation of the sub-surface geology. The MT method is of increasing value at regional, and in some cases, sub-basin scale interpretations especially when integrated with other potential field data sets (Prieto *et al.* 1985). The MT method can be useful in the definition of structure and depth to electrical basement. It is especially valuable where the reflection seismic data is of poor quality but the basin stratigraphy needs to have sufficient contrast in electrical resistance at relevant horizons to make the technique most usable. All potential field datasets have their maximum value in structural analysis when they are integrated and used in conjunction with seismic and surface geological data (Horscroft & Bain this volume).

Other techniques/methodologies

Geometric structural models may be inadequate to fully understand basin tectonics and evolution. Flexural/thermal models should be used in conjunction with validated cross-sections to study basin-forming processes (Wernicke & Burchfiel 1982; Kusznir & Park 1984; Kusznir *et al.* 1987; Egan 1992).

Geochemical/maturity data and burial history modelling can also be a test of sequential restorations. The structural restorations should be consistent with the 1D or 2D basin models. These models are constrained using Vro and other geochemical maturity measurements together with apatite fission track data and they are powerful indicators of basin evolution (Green *et al.* 1995; Hill *et al* 1995; Hill *et al.* this volume). Diagenetic and fluid inclusion studies can also constrain interpretation of basin history (Hardman *et al.* 1993). It is only when structural models are internally consistent with burial history that the XSV technique really helps in the understanding of detailed basin evolution and hydrocarbon prospectivity.

The interpretation of basin evolution should be integrated and tested against those produced by the sequence stratigraphic approach. The validated sections can be used as input to constrain the interpretations for the sequence stratigraphic analysis and vice versa. The sequence stratigraphy may allow predictions of the stratigraphic template to be made and then subsequently tested by structural modelling.

Conclusions

The use of modern structural geology and cross-section construction and validation techniques have been of great value and in many cases will show marked benefits within the oil/gas industry. Unfortunately XSV is perceived to be laborious to use and is therefore thought to be an inaccessible technique. Both the technique itself and the computer programs which carry out restorations can be initially difficult to master. Additional training, more user-friendly software and a greater awareness of the rewards of the technique will enable it to be more widely used.

In order to gain most benefit from the validation/analysis process, it is important to use all the data/datasets not just the available structural data. XSV projects should be performed by multi-discipline teams as a wide range of skills and experience are utilised in producing a well-constrained interpretation. It is the integration of structural studies with other analytical techniques which give the most complete understanding of the basin sedimentary and tectonic history.

XSV and CSB can increase the understanding of the basin evolution and therefore help reduce the risk on structure trap, migration and play dynamics. It is more cost effective to test a structural prospect using XSV techniques rather than drilling techniques. One of the major advantages of XSV lie in the integrated 3D basin analysis methodology hereby obtaining a better understanding of hydrocarbon prospectivity.

The reviews of Dick Nieuwland and Peter Buchanan are acknowledged as they improved both the form and content of this paper. The author would like to thank Peter Buchanan for the invitation to present this paper at the 1994 Structural Validation conference and Dick Nieuwland for his encouragement and patience with the preparation of the paper. The author thanks Andy Carmichael for reviewing an early version of this paper and Ruth Buchanan for her work in the typing and editing of this manuscript.

References

ARTHUR, T. J. 1993. Mesozoic structural evolution of the UK Southern North Sea: insights from analysis of fault systems. *In*: PARKER, J. R. (ed.) *Petroleum Geology of Northwest Europe: Proceedings of 4th Conference*. Geological Society, London, 1269–1280.

BALLY, A. W., GORDY, P. L. & STEWART, G. A. 1966. Structure,seismic data and orogenic evolution of the Southern Canadian Rocky Mountains. *Bulletin of Canadian Petroleum Geology*, **14**, 337–381.

BISHOP, D. & BUCHANAN, P. G. 1995. Development of structurally inverted basins: a case study from the West Coast, South Island, New Zealand. *In*: Buchanan, J. G. & Buchanan, P. G. (eds) *Basin Inversion*. Geological Society, London, Special Publications, **88**, 549–586.

BODARD, J. M., CREER, J. G. & ASTEN, M. W. 1993. Next Generation High Resolution Airborne Gravity Reconnaissance in Oil Field Exploration. *Energy Exploration and Exploitation*, **11**, 198–234.

BOYER, S. E. 1991. Geometric evidence for synchronous thrusting in the southern Alberta and northwest Montana thrust belts. *In*: MCCLAY, K. R. (ed.) *Thrust Tectonics*. Chapman & Hall, London, 377–390.

BOYER, S. E. & ELLIOTT, D. 1982. Thrust Systems. *American Association of Petroleum Geologists Bulletin*, **66**, 1196–1230.

BUTLER, R. W. H. 1983. Balanced cross-sections and their implications for the deep structure of the northwest Alps. *Journal of Structural Geology*, **5**, 125–137.

—— 1985. The restoration of thrust systems and displacement continuity around the Mount Blanc massif, NW external Alpine thrust belt. *Journal of Structural Geology*, **7**, 569–582.

—— 1994 The role of section balancing Understanding the structure and geometric evolution in Thrust Belts. *Abstract Volume: Modern Developments in Structural Interpretation, Validation and Modelling*, London, UK 1994.

——, LICKORISH, W. H., GRASSO, M., PEDLEY, H. M. & RAMBERTI, L. 1995. Tectonics and sequences stratigraphy in the Messinian basins, Sicily: Constraints on the initiation and termination of the Mediterranean salinity crisis. *Geological Society of America Bulletin*, **107**, 425–439.

CHAMBERLAIN, R. T. 1910. The Appalachian folds of Central Pennsylvania. *Journal of Geology*, **18**, 228–251.

—— 1919. The building of the Colorado Rockies. *Journal of Geology*, **27**, 225–251.

DAHLSTROM, C. D. A. 1969. Balanced cross-sections. *Canadian Journal of Sciences*, **6**, 743–757.

—— 1970. Structural geology in the eastern margin of the Canadian Rocky Mountains. *Bulletin of Canadian Petroleum Geology*, **18**, 332–406.

DAVISON, I. 1986. Listric normal fault profiles: Calculation using bed-length balance and fault displacement. *Journal of Structural Geology*, **8**, 209–210.

DE PAOR, D. G. 1988. Balanced Sections in thrust belts, Part 1: Construction. *American Association of Petroleum Geologists Bulletin*, **72**, 73–90.

DIEGEL, F .A. 1986. Topological constraints on imbricate thrust networks, examples from the Mountain City window, Tennessee, USA. *Journal of Structural Geology*, **8**, 269–280.

DOUGLAS, R. J. W. 1950. *Callum Creek, Langford Creek, and Gap Map Areas, Alberta*. Geological Survey Canada Memoirs, **255**.

—— 1958. *Mount Head Map Area, Alberta*. Geological Survey of Canada Memoirs, **291**.

EGAN, S. S. 1992. The flexural isostatic response of the lithosphere to extensional tectonics. *Tectonophysics*, **202**, 291–308.

ELLIOTT, D. 1983. The construction of balanced cross-sections. *Journal of Structural Geology*, **5**, 101.

—— & JOHNSTON, M. R. W. 1980. Structural evolution of the northern part of the Moine thrust belt. *Transactions of the Society of Edinburgh: Earth Sciences*, **71**, 69–96.

GEISER, P. A. 1988. The role of kinematics in the construction and analysis of geological cross-sections in deformed terrains. *In*: MITRA, G. & WOJTAL, S. (eds) *Geometries and mechanisms of thrusting with special reference to the Appalachians*. Geological Society of America Special Papers, **222**.

GEISER, J., GEISER, P. A., KLIGFIELD, R., RATCLIFF, R. & ROWAN, M. 1988. New applications of computer based section construction: strain analysis, local balancing and sub-surface fault prediction. *Mountain Geologist*, **25**, 47–59.

GIBBS, A. D. 1983. Balanced cross-sections from seismic sections in areas of extensional tectonics. *Journal of Structural Geology*, **5**, 153–160.

—— & GRIFFITHS, P. 1994 The third and fourth dimensions in balance and restoration. *Abstract Volume: Modern Developments in Structural Interpretation, Validation and Modelling*, London, UK 1994.

GOGUEL, J. 1962. Tectonics, W. H. Freeman & Co., New York.

GREEN, P. F., DUDDY, I. R. & BRAY, R. J. 1995. Applications of thermal history reconstruction in inverted basins. *In*: BUCHANAN, J. G. & BUCHANAN,

P. G. (eds) *Basin Inversion.* Geological Society, London, **88**, 167–190.

HARDMAN, M., BUCHANAN, J. G., HERRINGTON, P. & CARR, A. 1993. Geochemical modelling of the East Irish Sea Basin: its influence on predicting hydrocarbon type and quality. *In*: PARKER, J. R. (ed.) *Petroleum Geology of Northwest Europe: Proceedings of the 4th Conference.* Geological Society, London, 809–821.

HAUGE, T. A. & GRAY, G. G. 1995. Discrepancy between mechanisms of rock deformation and those chosen for fault prediction and palinspastic restoration. *This volume.*

HILL, K. C. & COOPER, G. T. 1996. A strategy for restoring inversion basins; thermochronology and dip analyses, SE Australia. *This volume.*

——, HILL, K. A., COOPER, G. T., O'SULLIVAN, A. J., O'SULLIVAN, P. B. & RICHARDSON, M. J. 1995. Inversion around the Bass Basin, SE Australia. *In*: BUCHANAN, J. G. & BUCHANAN, P. G. (eds) *Basin Inversion.* Geological Society, London, Special Publications, **88**, 525–547.

HILLIS, R. R. 1995. Regional Tertiary exhumation in and around the United Kingdom. *In*: BUCHANAN, J. G. & BUCHANAN, P. G. (eds) *Basin Inversion.* Geological Society, London, Special Publications, **88**, 167–190.

HOOPER, R. J., GOH, L. S. & DEWEY, F. 1995. The inversion history of the northeastern margin of the Broad Fourteens Basin *In*: BUCHANAN, J. G. & BUCHANAN, P. G. (eds) *Basin Inversion.* Geological Society, London, Special Publications, **88**, 167–190.

HORSCROFT, T.R. & BAIN, J. E. 1996. Validation of seismic data processing and interpretation with integration of gravity and magnetics data. *This volume.*

HOSSACK, J. R. 1979. The use of balanced cross-sections in the calculation of orogenic contraction:- a review. *Geological Society, London*, **136**, 705–711.

—— 1983. A cross-section through the Scandinavian Caledonides constructed with the aid of branch lines. *Journal of Structural Geology*, **5**, 103–122.

—— 1994. Geometric Rules of Section Balancing of Salt Structures in The Gulf of Mexico. *Abstract Volume: Modern Developments in Structural Interpretation, Validation and Modelling*, London, UK 1994.

INSLEY, M. W. 1996. The use of satellite imagery in the validation and verification of structural interpretations for hydrocarbon exploration in Pakistan and Yemen. *This volume.*

JAMISON, W. R. 1987. Geometric analysis of fold development in overthrust terrains. *Journal of Structural Geology*, **9**, 207–219.

JONES, P. B. 1984. Sequence of formation of back limb thrust imbrications: Implications for development of Idaho-Wyoming thrust belt. *American Association of Petroleum Geologists Bulletin*, **68**, 816–818.

KESSLER, D. & RESHEF, M. 1994. Depth processing of structurally complex seismic data. *Abstract Volume: Modern Developments in Structural Interpretation, Validation and Modelling*, London, UK 1994.

KLIGFIELD, R., GEISER, P. & GEISER, J. 1986. Construction of geologic cross-sections using microcomputer systems. *Geobyte*, **1**, 60–66.

KNOTT, I. 1993. Fault seal analysis in the North Sea. *American Association of Petroleum Geologists Bulletin*, **77**, 778–792.

KUSZNIR, N. & PARK, G. 1984. Intraplate lithosphere deformation and the strength of the lithosphere. Geophysical *Journal of the Royal Astronomical Society*, **79**, 513–535.

——, KARNER, G. & EGAN, S. S. 1987. Geometric, thermal and isostatic consequences of detachments in continental lithosphere extension and basin formation. *In*: BEAUMONT, C. & TANKARD, A. J. (eds) *Sedimentary basins and basin forming mechanisms.* Canadian Society of Petroleum Geology Memoirs, **12**, 185–203.

LAUBSCHER, H. P. 1962. Diw Zweiphasenhypothese der Jurafaltung. Ecologae Geologicae Helvetiae, **55**, 1–22.

MITRA, S. 1990. Fault-propagation folds: geometry, kinematic evolution and hydrocarbon traps. *American Association of Petroleum Geology Bulletin*, **74**, 921–945.

—— & NANSOM, J. 1989. Equal area balancing. *American Journal of Sciences*, **289**, 563–599.

—— & —— 1991. *Balanced cross sections in hydrocarbon exploration and production.* AAPG Short Course, Houston, Texas.

MOUNT, V. S., SUPPE, J. & HOOK, S. C. 1990. A forward modelling strategy for balancing cross-sections. *American Association of Petroleum Geologistsa Bulletin*, **74**, 521–531.

PRIETO, C. PERKINS, C. & BERKMAN, E. 1985. Columbia River Basalt Plateau – An integrated approach to interpretation of basalt-covered areas. *Geophysics*, **50**, 2709–2719.

ROWAN, M. G. 1996. Advantages and limitations of section restoration in areas of extensional salt tectonics: an example from offshore Louisiana. *This volume.*

SPARLIN, M. A. & LEWIS, R. D. 1994. Interpretation of the Magnetic Anomaly over the Omaha Oil Field, Gallatin County, Illinois. *Geophysics*, **59**, 1092–1099.

SUPPE, J. 1983. Geometry and kinematics of fault-bend folding. *American Journal of Sciences*, **283**, 684–721.

—— & MEDWEDEFF, D. A. 1990. Geometry of kinematics of fault propagation folding. *Eclogae Geologicae Helvetiae*, **83**, 409–454.

VERSCHUREN, M. A. J., NIEUWLAND, D. A. & GAST, J. 1996. Multiple detachments levels in thrust tectonics: sandbox experiments and palinspastic reconstruction. *This volume.*

WERNICKE, B. & BURCHFIEL, B. C. 1982. Modes of extensional tectonics. *Journal of Structural Geology*, **4**, 105–115.

WOODWARD, N. B., BOYER, S. E. & SUPPE, J. 1989. Balanced geological cross-sections. *American Geophysical Union, Short Courses in Geology*, **6**, 132.

Balancing sections through inverted basins

MIKE P. COWARD
Geology Department, Imperial College, London SW7 2BP, UK

Abstract: Basin inversion and the reactivation of earlier normal faults is now recognized as being widespread in many mountain belts. By ignoring the effects of basin inversion, serious errors in section construction and structural and tectonic interpretations can be made.

Folds produced during thick-skinned basement-involved faulting are generally very different from those produced during thin-skinned thrust tectonics and hence different models must be used for section construction and restoration. In particular, block rotations are important. Rotation about a horizontal axis may lead to shortening of the graben or half-graben, expulsion of material from out of the graben and/or reactivation of the original normal fault. The resultant hanging wall folds are characterized by long gently dipping backlimbs and short hooked forelimbs. Various models for fold development are derived and compared for different inversion geometries and kinematics. Examples are taken from the North Sea.

Many inverted faults involve components of strike-slip or oblique-slip movement, and hence simple 2D methods of section balancing cannot be applied. Basin inversion may have involved rotation of the fault blocks about vertical axes and/or lateral expulsion of material from the basin. Examples are described from South Wales, Syria and the French Alps. Vertical axis rotation appears a common mechanism for intraplate deformation, where the regional shear couple is applied at an angle to earlier fault blocks or crustal lineaments. The analysis of paleomagnetic data together with kinematic data enable the rotation to be determined and amounts of shortening strain estimated.

The presence of early syn-sedimentary faults is now being recognized in many orogenic belts (Dewey 1982; Letouzey 1990) and it is known that much of the intra-continental deformation within the crust is accommodated by the reactivation of pre-existing structures. This is particularly true in areas where compression and uplift affect earlier basins or ocean margins. Some of the better known examples of basin inversion and fault reactivation have been described from commercial seismic data, from areas which are just below sea level or of low relief (Harding 1983; Ziegler 1983, 1987). In collisional mountain belts, such as the Alps or Apennines, where the shortening and uplift are more intense, the reactivation of pre-existing fault systems is extremely important (Gillcrist *et al.* 1987; Hayward & Graham 1989).

In this paper the term inversion is used to describe regions which have experienced a reversal in uplift or subsidence, that is, areas which have changed from being regions of subsidence to regions of uplift or vice-versa. Hence an area which has changed from subsidence to uplift has been affected by positive inversion and an area which has changed from uplift to subsidence has undergone negative inversion (Gillcrist *et al.* 1987, see also Harding 1983; Ziegler 1987). This paper will deal with aspects of positive inversion.

The misinterpretation of folds and thrusts as being related to thin-skinned rather than thick-skinned shortening, involving faults which pass down into the basement resulting in the inversion of sedimentary basins, can have far reaching effects on the structural interpretation of a region, leading to:

(i) use of the wrong method in section construction.
(ii) incorrect calculations of the amount of orogenic shortening.
(iii) incorrect assumptions of the nature of structures at depth, both directly beneath the fold/thrust belt and further back within the hinterland of the mountain belt.

The recognition of inversion structures is even more important in the oil industry as inversion:

(i) will modify the burial history of a sedimentary basin, complicating calculations of timing of maturation and oil generation.
(ii) can uplift sediments above sea level generating a secondary porosity and/or karstification.
(iii) can modify the tilt of a sedimentary package, allowing different directions of fluid migration through time.
(iv) can reactivate older faults, changing their sealing properties and sometimes repumping fluids around the basin.
(v) will form complex structures at depth and it is important to differentiate inversion structures, which reactivate moderate to steeply dipping faults, from thin-skinned thrust structures.

From Buchanan, P. G. & Nieuwland, D. A. (eds), 1996, *Modern Developments in Structural Interpretation, Validation and Modelling,* Geological Society Special Publication No. 99, pp. 51–77.

Positive inversion tectonics occur when basins are uplifted to become positive features, that is, strata are uplifted above their regional level. Note that individual faults may show extension at depth but contraction in their upper portions. Thus in Fig. 1 some of the normal faults show net extension, some show net contraction, while others change from extension at depth to folding and contraction at a higher level.

This paper firstly discusses the geometries of structures associated with thick-skinned fault reactivation and basin inversion and secondly some of the problems which occur when dealing with inversion tectonics.

Geometry of hanging wall folds and section construction through simple inverted basins

The detailed geometries of folds produced during basin inversion will depend on the fault geometry,

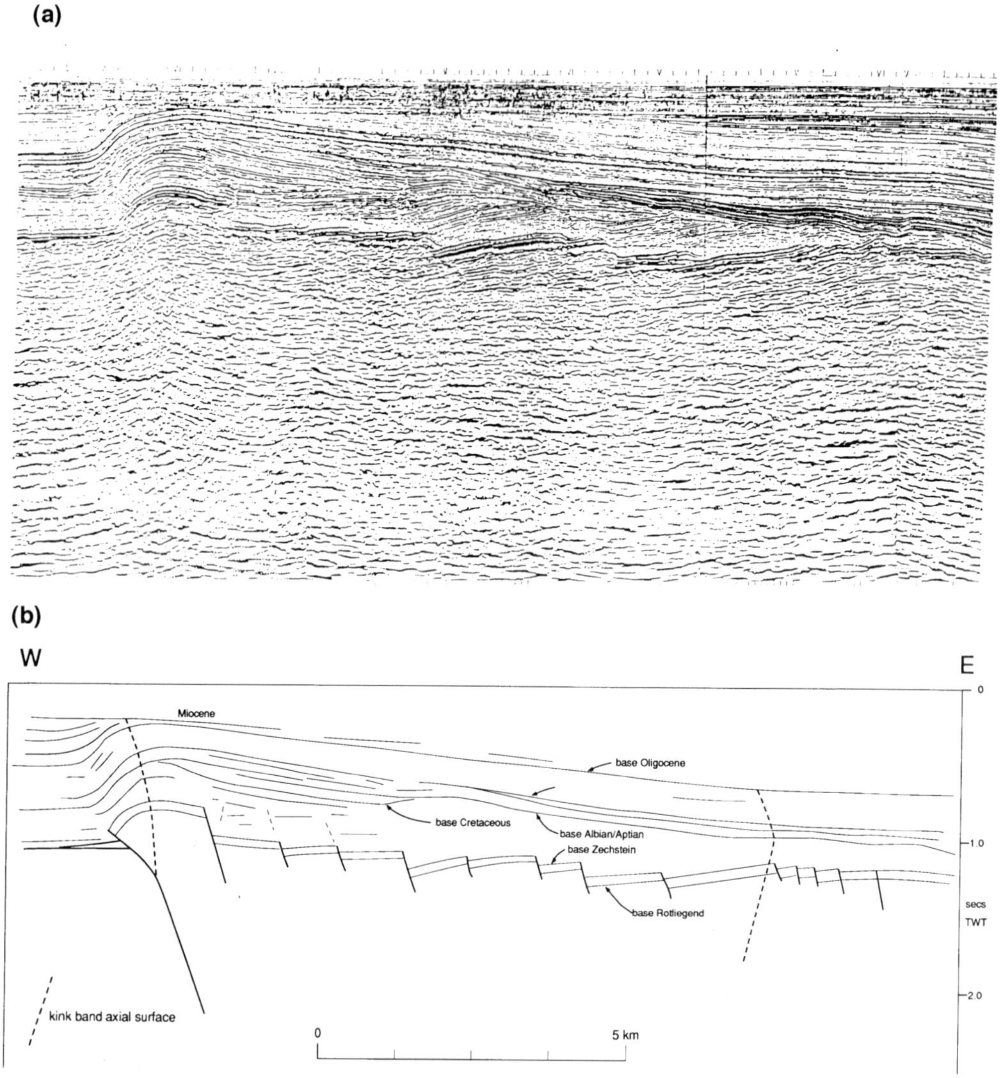

(a)

(b)

Fig. 1. An example of a fold produced by rotational faulting from the southern part of the North Sea. (**a**) Uninterpreted seismic data. (**b**) Line drawing showing the major stratigraphic picks. Note the long, gently-dipping back limb and the short hooked forelimb. (From the South Hewett Fault Zone. Drawn from Badley *et al.* 1989).

that is, if the faults are listric, if they follow a domino rotation pattern, or if they act as single inverted faults.

The folds develop as a result of the decrease in bed length due to reversal in displacement up the fault, and back-rotation of the faults. Several techniques may be applied to cross sections to check interpretations, make predictions about the structures at depth, or help produce balanced sections.

Many of the techniques developed for the analysis of thin-skinned faults may be applied to thick-skinned structures associated with inversion. In particular, the fault-bend fold models derived by Suppe (1983) and Suppe & Medwedeff (1985) can be applied to inversion structures.

Faults associated with multiple block rotation: planar faults

During extension, the development of an array of half-graben has been compared to a collapsing set of dominoes, with associated rotation of the fault blocks (e.g. Morton & Black 1975). Similarly, block faults may back-rotate during inversion causing thrust sense reactivation along the fault plane, until their dip reaches the critical value to stop this reactivation (e.g. Coward 1994; Gillcrist *et al.* 1987). Back rotation will be more pronounced if the displacements are oblique; if they are highly oblique or strike-slip, then the faults could rotate back to vertical. During back rotation the syn-rift fill may be squeezed out into a series of folds and thrusts.

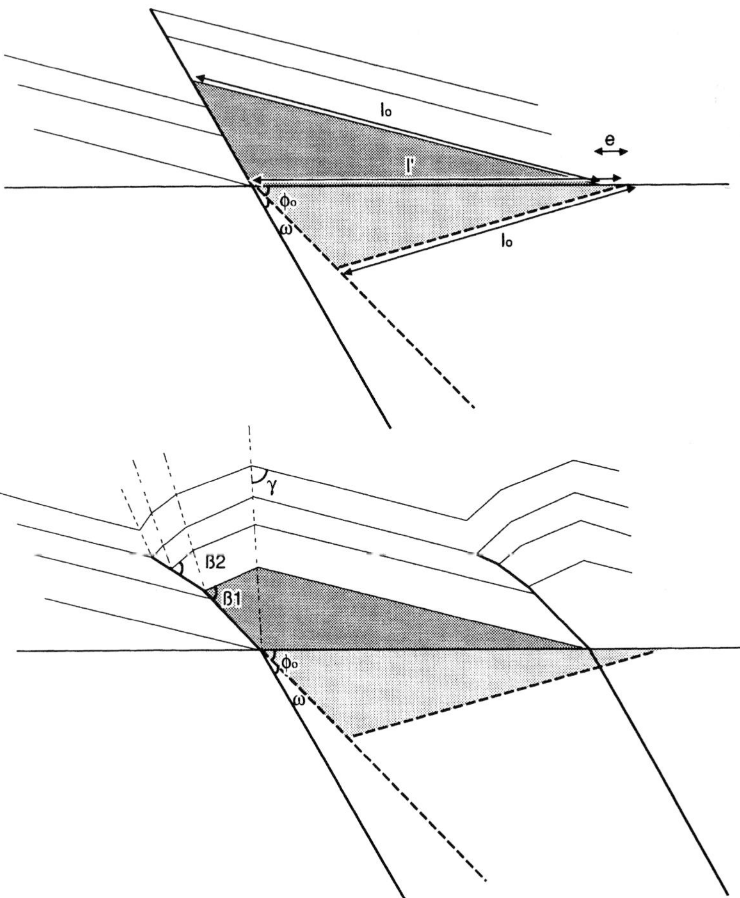

Fig. 2. Model for multiple block rotation leading to basin inversion. Modified from Wang *et al.* (1995). Top: parameters used in estimating shortening from block rotation – note that the extension (e) across the region of inversion is not simply $l' - l_0$. Bottom: block rotation leading to fault bend folds in the post-rift sedimentary cover.

Note that the post-rift cover will also shorten in a series of folds or thrusts, some of which may detach close to the syn-rift/post-rift boundary and, unless deep data are available, may be mistaken for thin-skinned detachment structures.

Examples of folds associated with rotated block faults are shown in Fig. 1. The folds are characterized by relatively planar backlimb dips and short hooked forelimbs. The faults may have extensional geometry at depth, passing through a null point (Williams *et al.* 1989) to a reverse fault at higher levels and then die out in a hooked tip beneath the forelimbs of these structures. A simple geometrical model for their development is shown in Fig. 2. The shortening in this rotational block model is given by

$$\sin \phi_o / \sin \phi = \sin \phi_o / \sin (\phi_o + \omega) \qquad (1)$$

where ϕ_o is the original cut-off angle, ϕ the final cut-off angle and ω is the angle of rotation.

The analysis of kink-band folds by Suppe (1983) shows how the geometry of the hanging wall changes as it moves over a fault with a change in dip. As the hanging wall moves over a curved fault plane, which decreases in dip, the hanging wall cut-off angle will increase. The relationships between the original cut-off angle (ϕ_o), the new cut-off angle (β) (Fig. 2), the change in dip of the fault (θ) and the fold interlimb angle (γ) have been derived by Suppe (1983).

Assuming kink-band geometry, with constant bed thickness around the fold, steeply dipping faults cannot be reactivated to form flat thrusts at high levels. There is no solution, assuming kink-band fold geometry, for a simple ramp-flat fault trajectory where the ramp dip is > 30°. Where the dip of a normal fault is more than 30°, then kink-band fault geometry allows for some decrease in dip of the fault upwards, but always less than the original cut-off angle. Figure 3 shows the limit for θ, the change in dip of a fault, assuming fault bend folding with no change in bed length or bed thickness.

The model of Suppe (1983) for the analysis of kink-band folds can be applied to a limited range of fold structures above steeply dipping ramps. Assuming a model of rotated blocks during inversion (Fig. 2), the backlimb will change dip accompanying the rotation. The forelimb will change dip as the reactivated fault grows into the post-rift cover sequence. For a fault dipping at 55° after 10° rotation (that is, the fault had an initial cut-off angle of 45°), if this fault grows into the post-rift sequence changing angle by 5°, the new hanging wall cut-off angle will be 63° (Fig. 2). The fault could change dip by another 2° to produce a new cut-off angle of 80°. As the fault changes dip its displacment decreases, so that it eventually dies out at a fault tip. This process of up-dip flattening of a fault will generate a steep forelimb with a gradual loss of fault slip (Fig. 2). Note that the resulting fold can be analysed in terms of kink bands or considered as a smooth curve.

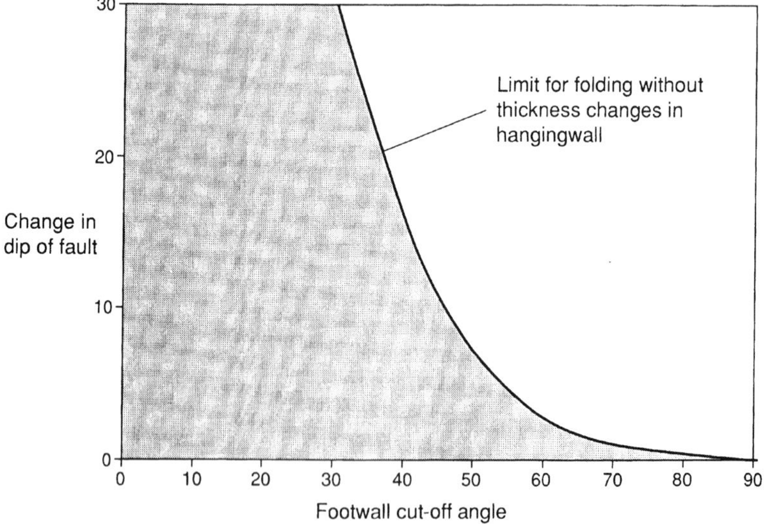

Fig. 3. Graph (shaded) showing the limit for the possible change in dip of a fault producing a fault-bend fold on its hanging wall, assuming different initial cut-off angles.

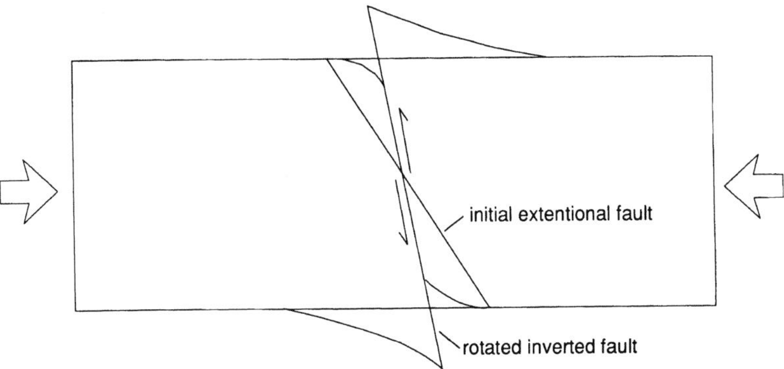

Fig. 4. Model for single fault rotation, leading to inversion: a flexural cantilever model. The fault zone rotates leading to bending of the footwall and hanging wall. Geometrically the model is similar to the flexural cantilever model for extension (Kusznir *et al.* 1991).

Single fault rotation

Single faults, or small groups of faults, which form the boundaries of extensional basins may invert by a process analogous to that of a flexural cantilever (Fig. 4). As the fault rotates the hanging wall flexes upwards while the footwall flexes down. The resultant folds are characterized by long backlimbs and steep forelimbs as shown in the example from the southern North Sea in Fig. 1. The backlimbs can be modelled as fault-bend folds (Fig. 5) where the fault dip increases upwards.

During this type of inversion the shortening is given by:

$$\sin(\phi_o + \omega)/\sin(\phi_o + \omega - \delta) \qquad (2)$$

where ϕ_o is the original cut-off angle, δ the backlimb dip and ω the rotation of the fault.

The final backlimb dip depends on the inital dip of the fault and the angle of rotation. Steeper faults require more rotation and hence more shortening to develop the same backlimb dip. The forced fold may have a strained steep limb or the reactivated fault may curve slightly into the cover and follow a geometry similar to that described for rotated fault blocks in Fig. 2.

Figure 1 shows an interpretation of the South Hewitt Fault structure in the southern North Sea, based on a model assuming cantilever style fault

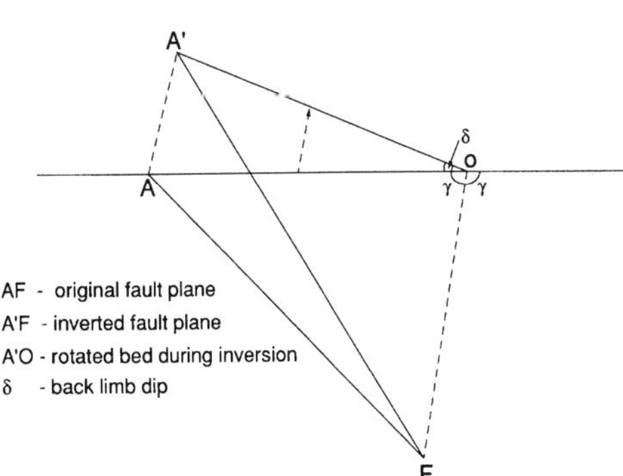

Fig. 5. Model for kink band development for the backlimb of a single rotational fault. This model assumes no change in length of the beds on the hanging wall, i.e. AO = A′O, and also no change in area. A kink band can be defined by interlimb angle (γ). The fault increases in length from AF to A′F.

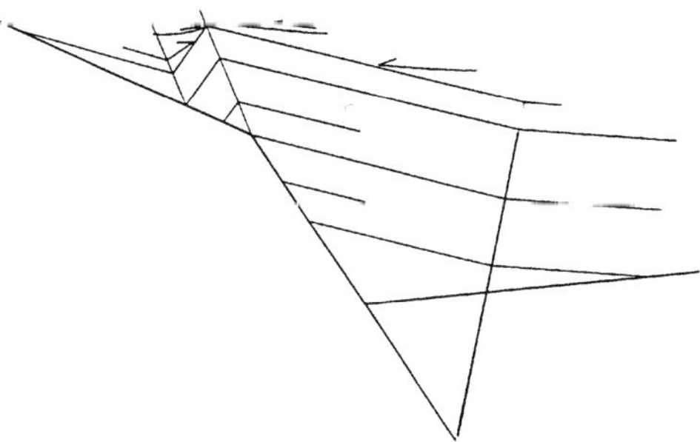

Fig. 6. Cross section through a single rotated inverted fault. The dip of the beds on the footwall will indicate the rotation. Beds are displaced up the steeply dipping original extensional fault and on to a new ramp. A kink-band fault-bend fold originates from the point where the fault changes dip. This example shows a fault- bend fold based on examples in the northern North Sea, where the syn-rift sediments are thrust up a fault plane onto the erosional scarp which acts as a gently dipping reverse fault during inversion. The arrows indicate onlap of sediments deposited during inversion.

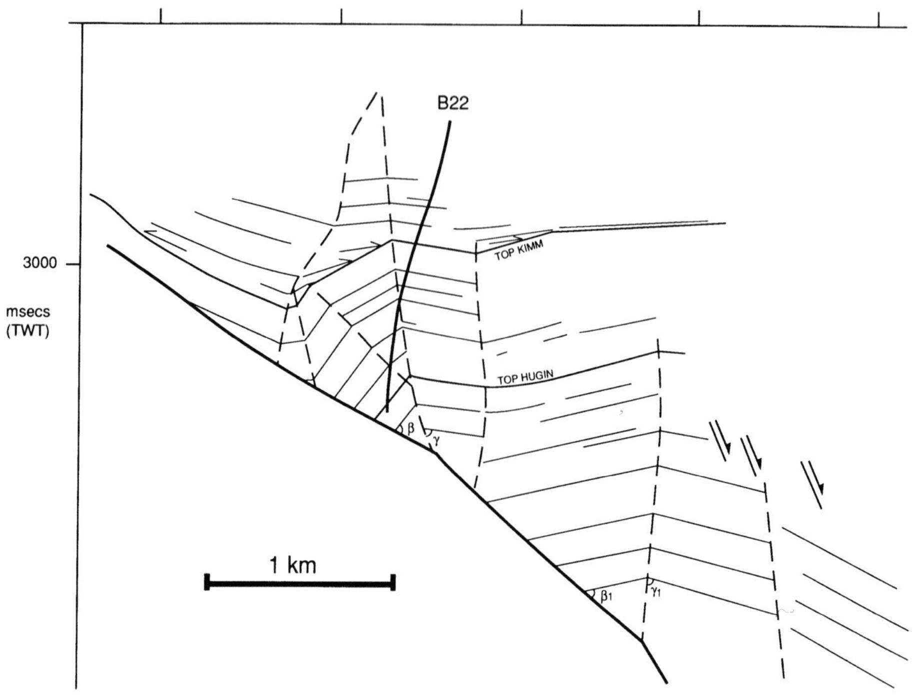

Fig. 7. Interpreted cross section through an inversion structure in the southern part of the Viking Graben, constructed from well data and 3D seismic data.

rotation and kink-band geometry. The shortening is estimated to be 4%. Figure 6 shows a fault bend fold produced above a kink in the fault plane where the fault becomes shallower upwards. The shape of the fault has been derived using fault-bend fold models. Figure 7 shows a similar interpretation based on an inversion fold in the southern Viking Graben. In the example shown the angle of rotation is shown by the dip of the beds on the flatter part of the fault in the W.

The fold shape may vary along strike due to different degrees of inversion. Kink-band panels should be mapped to show these variations.

Rotational faults where shortening is accommodated by strain in the hanging wall block rather than displacement up the fault

During rotation of the fault block, movement may cease up the dip of the fault and hence shortening may occur within the hanging wall of the fault block. Figure 8 shows a simplified model for the rotation of a fault with no displacment along the hanging wall, that is, the length of the fault on the hanging wall remains constant. Kink band geometry is assumed throughout. Assuming no change in cut-off angle (Fig. 8), the change in length $l'/l_o = \sin\alpha/\sin(\alpha + \omega)$. As $\beta = 90° - \omega/2$ (from Fig. 8), then $\alpha = 90° - (\omega/2) - \phi$.

Therefore the shortening is given by

$$l'/l_o = \frac{\sin(90° - \omega/2 - \phi)}{\sin(90° + \omega/2 - \phi)} \qquad (3)$$

where ϕ is the initial cut-off angle and ω the rotation. This relationship is shown graphically in Fig. 9. The intensity of the shortening strain increases with initial dip of the fault and with amount of rotation. Note that for faults with initial dips of 70°, the hanging walls shorten by 10% after only 2.5° rotation. With 5° rotation the shortening is > 20%.

This shortening will result in: (i) Expulsion of the hanging wall in a series of antithetic backfolds and back-thrusts, or (ii) the production of a hanging wall anticline, here modelled as a simple kink-like structure, whose axial surface intersects the fault at its rotational pivot (Fig. 10). The fold interlimb angle and the dip of the backlimb of the fold, are related to the initial dip of the fault and the amount of rotation.

Figure 11 shows a line drawing through weakly inverted half-graben in the southern North Sea.

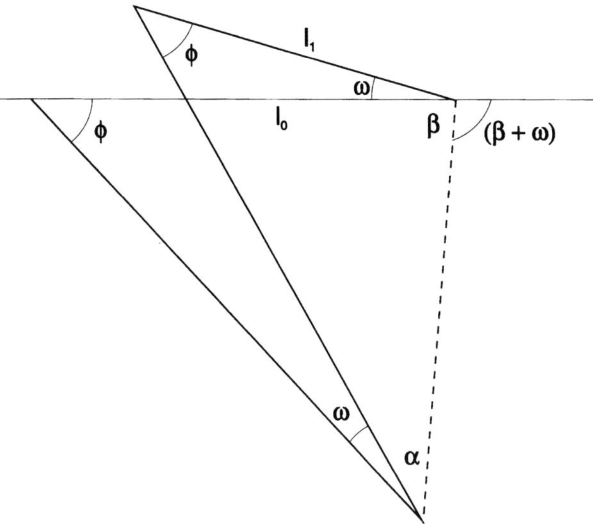

Fig. 8. Model to explain the shortening in the hanging wall of a rotated normal fault, assuming no slip along the fault plane. As there is no slip the cut-off angle ϕ is constant and the bed length changes from l_0 to l'. See text for discussion.

Note the broad anticlinal fold, whose axial surface intersects the normal fault at depth within the basement and compare this structure to the model in Fig. 10.

During rotational inversion there may be both expulsion of material from the hanging wall by slip along the fault plane and also by folding of the hanging wall block, that is, the expulsion may occur by a combination of the mechanisms discussed above and illustrated in Figs 2 and 10. As the fault block rotates it will become less easy for slip to occur along the fault plane and hence

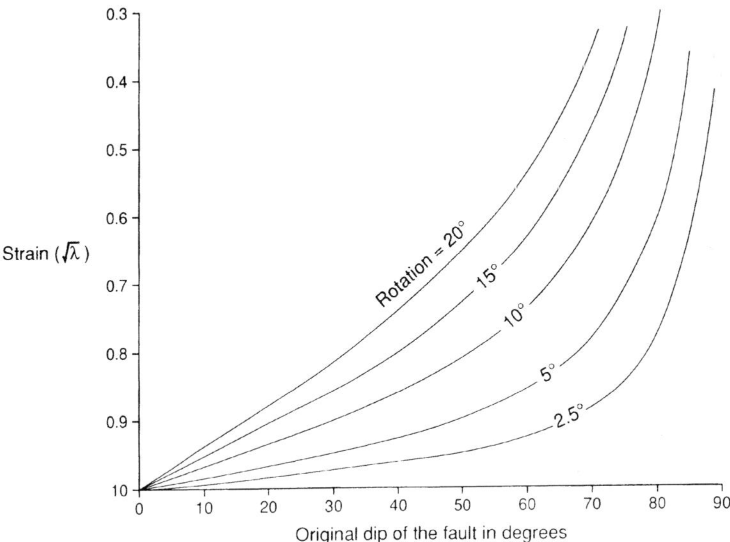

Fig. 9. Graph showing the layer-parallel shortening strain (l'/l_0) in beds on the hanging wall of a normal fault (see Fig. 7 for details), related to the original dip of the fault and amount of rotation (about a horizontal axis), assuming no slip along the fault plane.

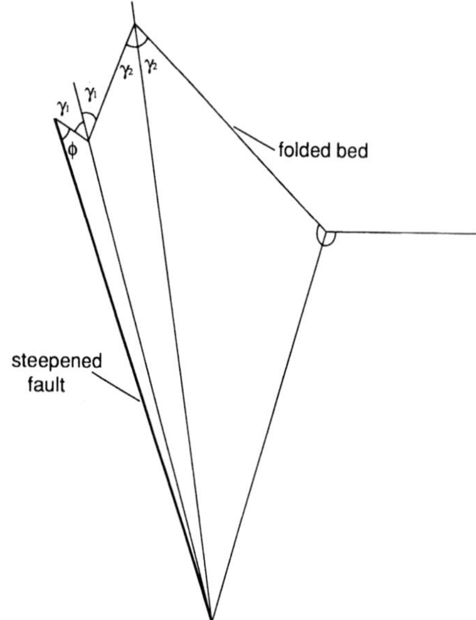

Fig. 10. Kink-band fold produced by shortening of the hanging wall of a rotated fault block, assuming no slip along the fault plane. The hanging wall cut-off angle ϕ is constant.

there will be a tendency for the folding of the hanging wall to occur after displacement up the fault plane. Hence the type of fold produced by the mechanism shown in Fig. 10 will generally form later than that produced by the displacement up the fault, as shown in Fig. 2. The difference in timing of fold growth should be observable from onlap data in the growth sequence. Note that in the example in Fig. 11, the folding of the hanging wall postdates the gentle inversion structure formed by reverse displacement up the fault.

A change in inversion kinematics is sometimes observed from the small-scale structures associated with inversion. Several hanging wall anticlines (e.g. Purbeck Anticline, Dorset (Ameen 1992)) show evidence for low-angle thrusting on the steep fold limb, indicating extra shortening strains developed in the hanging wall fold. Cessation of displacement on the fault may cause the fold to tighten, with the growth of new thrusts. Some of these thrusts may be out-of-syncline structures caused by space problems in the inner arcs of the folds.

In the inverted basins of the Western Alps, Liassic syn-rift sediments in the hanging walls of large Jurassic extensional faults have been tightened into upright chevron folds with locally east verging backfolds and backthrusts. The basement blocks have rotated during Alpine inversion to dips of > 60° (Gillcrist *et al.* 1987). The original normal faults have not been reactivated by reverse movements; instead the half-graben fill has been intensely strained. Shortening strains of > 50% are recorded from tight folds in the inverted half-graben (Gillcrist 1988; Gillcrist *et al.* 1987). These strains are compatible with rotations of 20–30° during Alpine shortening.

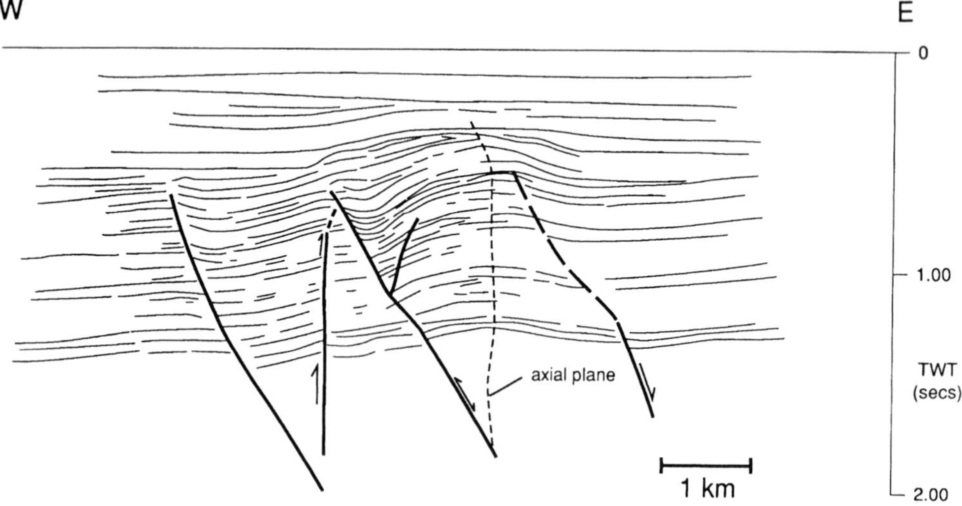

Fig. 11. Line drawing of seismic data through a weakly inverted set of fault blocks, from Quad 53 in the southern North Sea. Note the anticline on the hanging wall of the central fault block and the reverse movement on the antithetic fault on the left-hand (western) fault block.

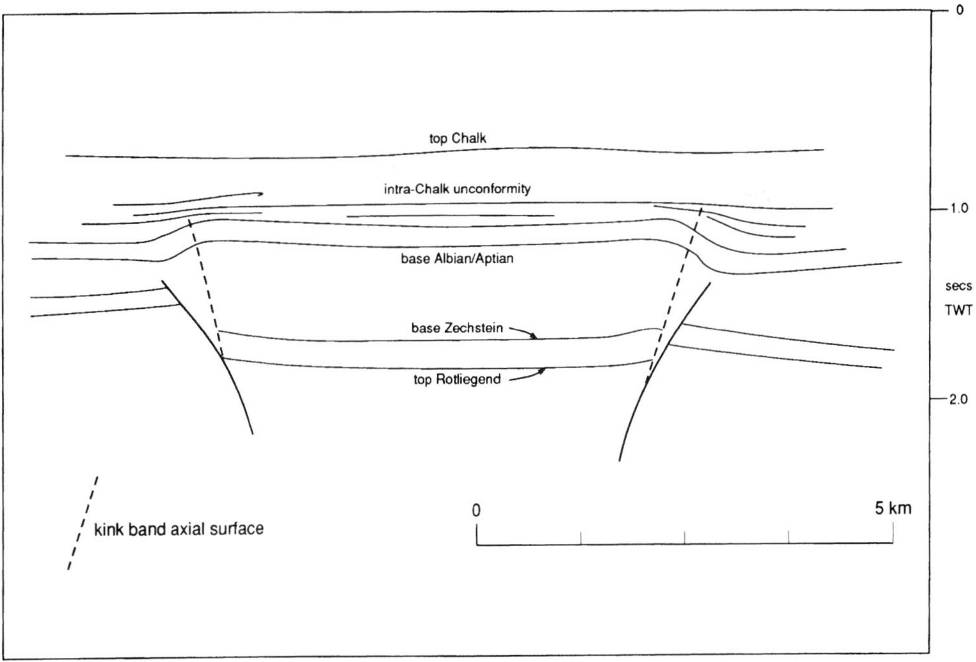

Fig 12. Line drawing of keystone fault block inversion from the southern North Sea. Drawn from data given by Badley *et al.* 1989.

Keystone faults

Figure 12 shows a keystone block from the southern North Sea, uplifted on two faults with opposing dip. This inversion produces no backlimb rotation, although the short hooked forelimbs are similar to those described for rotational faults above. The faults in the basement flatten slightly into the post-rift sequence, causing the flat-topped uplift to be bounded by two monoclines with opposing dips. The geometry of the monoclines can be examined assuming fault-bend fold models. The original normal faults form part of a symmetrical keystone graben; reactivation closes the graben, uplifting the keystone block without change in dip.

Note that faults with opposing dips defining a keystone graben do not necessarily reactivate together to form a keystone uplift. Wang *et al.* (1995). describe a broad anticlinal structure in the East China Sea, where faults with different dips were reactivated at slightly different times giving a complex growth pattern to the anticline.

Rotational listric or bent faults

Many extensional faults are curved or bent in section, in particular, those faults which have several phases of growth will change dip through the syn-rift sequences. During rotational inversion the different sections of the fault blocks will shorten by different amounts, depending on their initial orientation. As shown in Fig. 13, if θ is the change in dip of the fault, ϕ the dip of the fault at depth, ω the rotation causing the inversion, then excess shortening across the upper part of the rotated fault block is given by:

$$\frac{\sin \phi_o}{\sin (\phi_o + \omega)} - \frac{\sin (\phi_o + \theta)}{\sin (\phi_o + \theta + \omega)} \qquad (4)$$

Figure 14a shows the resultant excess shortening in cartoon form. There is excess bed length in the upper part of the rotated fault block. This excess length may result in the following.

(i) Expulsion of the hanging wall in a series of keystone faults (Fig. 14b), possibly reworking earlier antithetic normal faults. The antithetic reverse faults will propagate away from the master synthetic fault. In the example shown in Fig. 11 a steep antithetic fault has been partially reworked during inversion. Note that the keystone uplift shown in Fig. 12 may have been produced by this mechanism.

(ii) Expulsion of the hanging wall in a hanging wall fold. The backlimb dip is produced by antithetic shear.

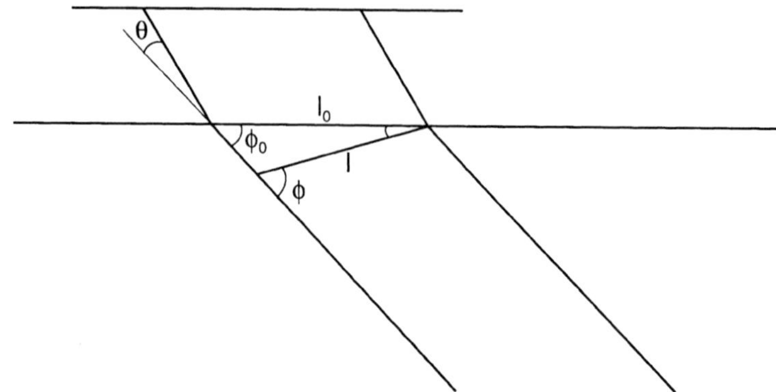

Fig 13. Fault blocks with changing fault dip (θ) with depth. ϕ_0 is the dip of the fault before inversion, ϕ is the dip after inversion. The shortening is given by l/l_0.

(iii) Expulsion of the footwall in a series of steep reverse faults (Fig. 15). The reverse faults will propagate downwards as the zone of strain moves slightly down the hanging wall. These synthetic steep reverse faults have been noted on the models produced by Buchanan & McClay (1992).

(iv) Expulsion of the hanging wall in a series of thin-skinned thrusts (Figs 15 and 16). The thrusts are antithetic to the master inverted fault and may be parallel or sub-parallel to bedding planes in the hanging wall, particularly if weak shales or evaporites are present. Figure 16 shows a simplified diagram illustrating the form of these various structures produced during inversion of curved faults. Backthrusts expelling the rift sequence from out of the half-graben have been reported by many authors (see Bally 1984; Hayward & Graham 1989). Figure 17 shows an example of sediment expelled up the dip of a half-graben in the eastern part of the Brae Province in the northern North Sea. The detachment in this example occurs in Permian salt.

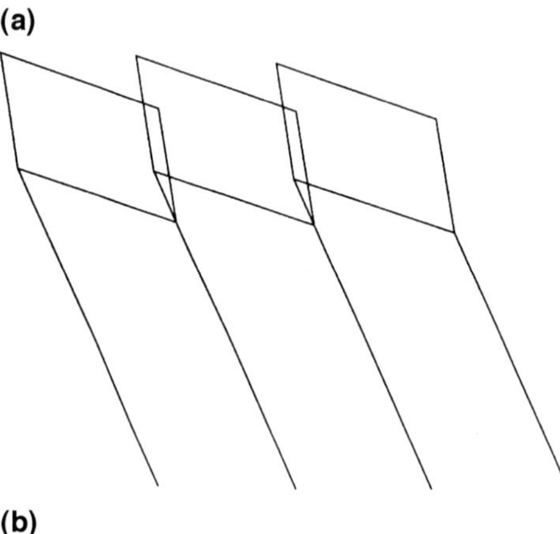

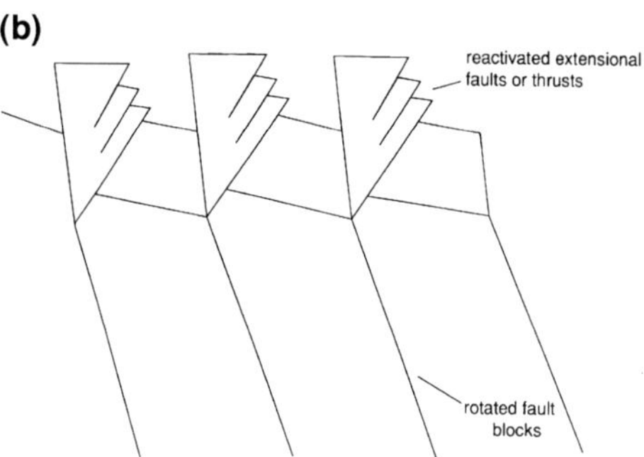

Fig. 14. (**a**) Rotated kinked faults, showing the area loss leading to extra shortening strains in the steeper part of the fault block. (**b**) Expulsion of the hanging-wall of rotated kinked faults, as a series of reworked antithetic faults.

Rotational listric or bent faults with an upward decrease in displacement

Steeply dipping faults are less likely to reactivate than gently dipping faults, particularly as they rotate away from the preferred orientation for failure (see Gillcrist *et al.* 1987). Hence gently dipping faults may reactivate at depth, but may stick at higher levels where they are steeper. The displacement will die out into a fold. Figure 18 shows one simple kink-band model for folding at depth, where the change in angle of the beds allows a decrease in displacement. The tip of the displacement occurs at the change in dip of the fault. As rotation and shortening progress, more beds will be kinked to allow for further strain and the kink-band will migrate away from the change in fault dip. On the hanging wall, adjacent to the steeper dipping portion of the fault, the layers will be similarly kinked but need to develop extra

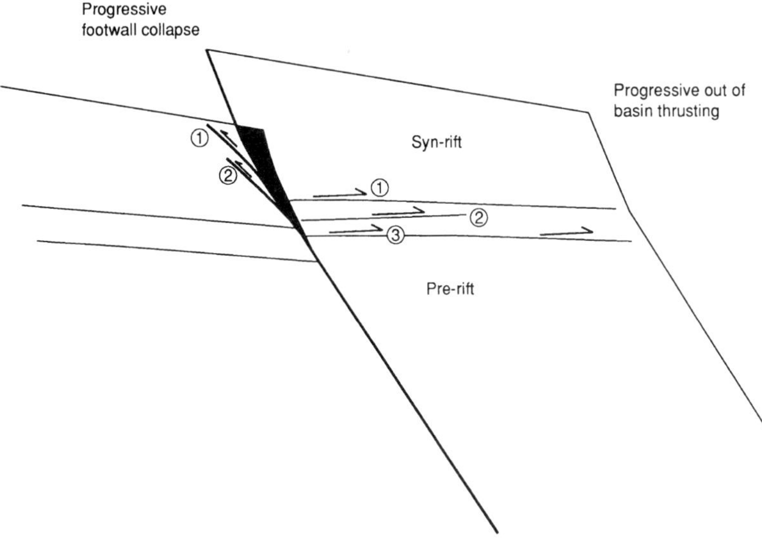

Fig 15. Expulsion of the footwall of rotated kinked faults, in a series of steep reverse faults and/or the hanging wall along antithetic thin-skinned thrusts. The order of fault movement is indicated – note the faults propagate downwards, as the area of strain grows. See text for discussion.

shortening strains, probably in the form of a back-thrust or antithetic shear zone (Fig. 18).

Instead of folding the hanging wall of the fault, a new shortcut thrust may develop. The short-cut structure may be in the footwall or the hanging wall of the listric or kinked thrust. The footwall short-cuts commonly lead to the development of 'floating islands' of pre-rift material, bounded by the original normal fault and by the short-cut. Isolated wedges of footwall rock may be translated onto a thrust hanging wall. At small values of inversion the short-cuts may have only limited displacement

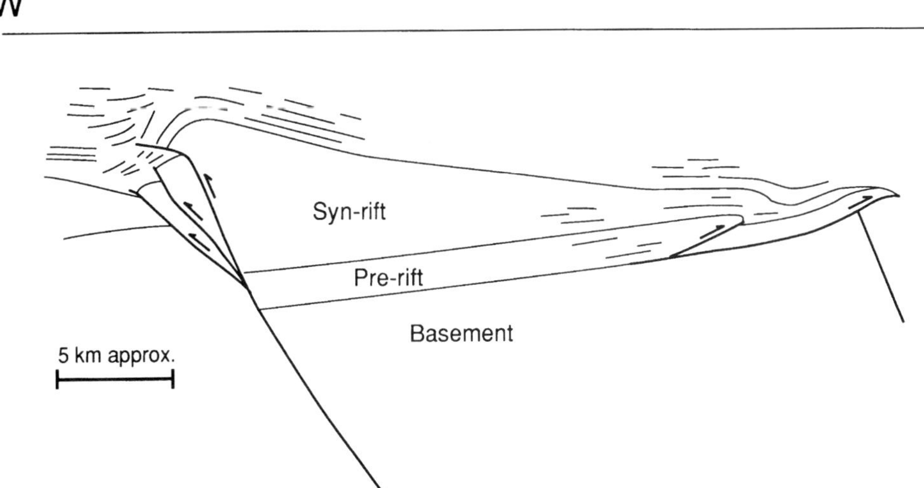

Fig 16. Sketch section through an inverted fault block showing the range of possible structures associated with rotation of a kinked fault.

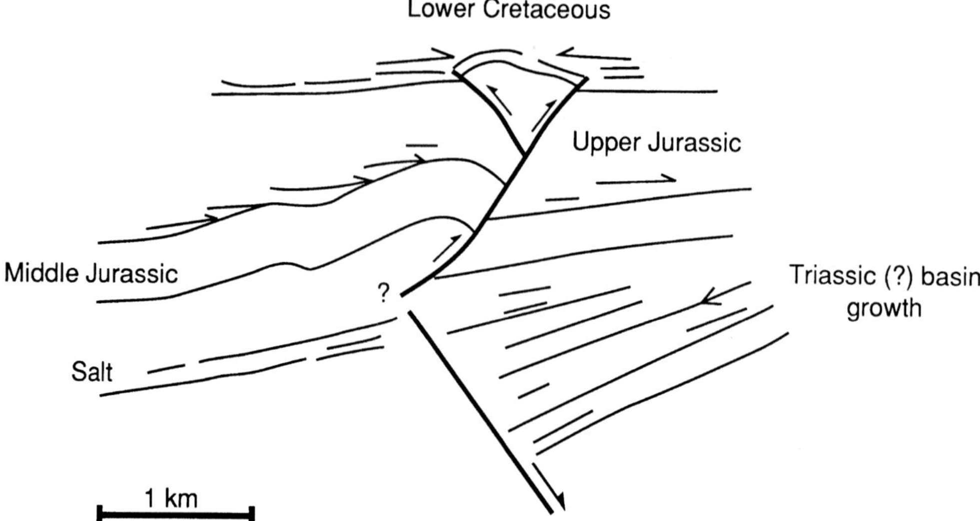

Fig 17. Simplified section through a fold-thrust zone developed near the top of a dip-slope of a half-graben in the southern part of the Viking Graben. The thrusts detach on Permian salt.

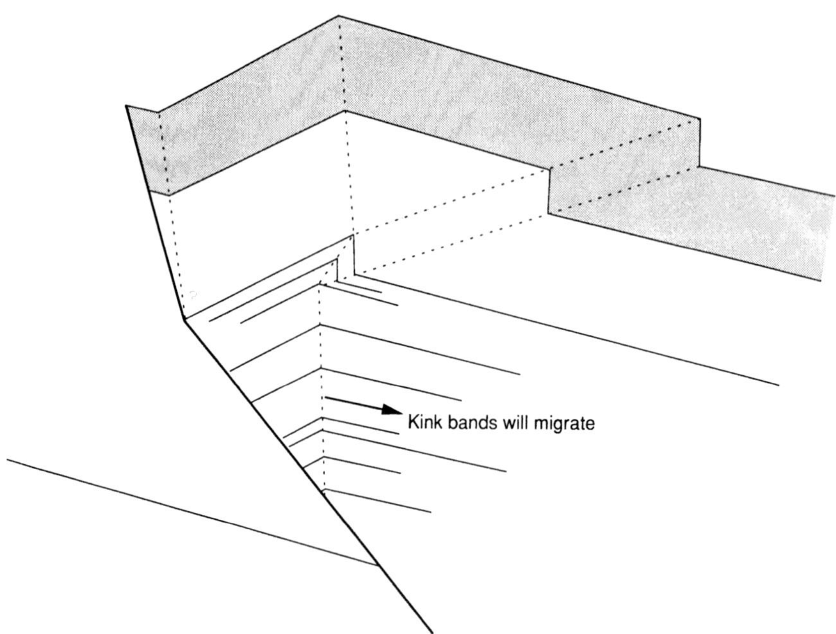

Fig. 18. Kink-band fold produced by variable displacement up the dip of a rotated kinked fault. The structure is analogoues to a fault propagation fold. The slip dies out below the change in dip of the fault plane. No slip is allowed on the steeper part of the fault. A simple kink-band fold develops at depth. This fold passes upwards into an asymmetric back-fold which faces away from the fault plane. Note that from surface data alone, this high-level structure may be mistaken for a fault bend fore-fold and hence a different polarity given to the fault at depth.

and form upward fanning horsetail patterns, similar to those described by Buchanan & McClay (1992). At higher values of contraction the footwall short-cuts may be responsible for generating a lower angle, more smoothly varying thrust trajectory (e.g. Gillcrist *et al.* 1987).

Problems with section balancing in areas in basin inversion: out-of-plane movements

Strike-slip displacements involve out-of section plane movements of material. No cross sections through strike-slip zones can be considered as plane strain sections and hence section balancing should not be attempted through these zones. The simplest pattern of strike-slip displacement involves movement on a vertical fault zone, possibly with some localized push-up of pull-apart deformation at restraining or releasing bends and offsets. Old crustal lineaments or earlier normal faults may be reactivated by intraplate shear stresses. Different styles of strike-slip movement occur depending on the orientation of these crustal lineaments or early normal faults, relative to lithospheric compression or shear.

Related to strike-slip tectonics

Strike-slip related inversion can occur at the restraining bends or offsets along major transcurrent faults. In NW Europe, examples include the following.

(a) The Permo-Carboniferous inversion of the Orcadian Basin, northern Scotland, related to right lateral movement along the Great Glen Fault (Coward *et al.* 1989);

(b) Inversion of the Ronne Graben in the southern Baltic Sea by left lateral movement along the Tornquist Line (Pegrum 1984);

(c) Tertiary inversion in Southern England by right lateral movement along the Bray–Southwest England–St George's Channel lineament, which linked the North Atlantic Rift with the tips of the Bresse–Rhone–Liguren rift systems in France and Germany (Fig. 19) (Coward 1994; Gillcrist *et al.* 1987). Movement on this lineament resulted in inversion at the compressional bends and offsets to produce, for example, the Purbeck Anticline. Where the offset was dilational a new basin formed, for example, in the southern half of Cardigan Bay (Coward 1994). Other major right lateral systems produced the Weald Anticline and the inversion along the Sole Pit system.

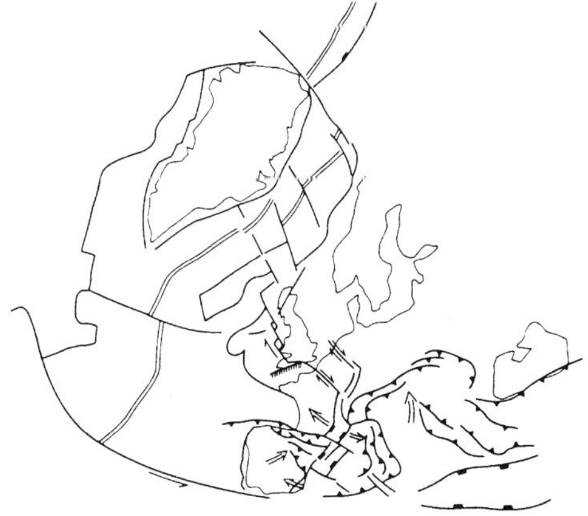

Fig. 19. Map of Northwest Europe showing regional tectonics in the Oligocene and the development of regions of tectonic inversion at compressional jogs on a strike-slip fault. The strike-slip faults were reactivated by intraplate strains caused by variations in the spreading rate along the Atlantic Ocean and the Rhone–Rhine grabens.

The above examples involve a shear couple related to intra-plate deformation, reworking early pre-existing fault zones or basement fabrics. These shear couples can be linked to compressional or extensional plate margins (cf. Ziegler 1987). Collision tectonics are not a pre-requisite of inversion. Alpine-age inversion tectonics occur as far distant from the Alps as the edge of the Rockall Trough, the southern margin of the Porcupine Seabight and the Barents Sea. Inversion structures West of Shetland and in the Voring Basin West of Norway are clearly unrelated to Alpine compression, but originate from stresses derived by movements associated with the major transform faults during Atlantic opening.

Strike-slip tectonics may lead to the development of en-echelon folds and/or secondary faults and shear systems. Zones from different strain regimes may be juxtaposed by the strike-slip movement, for example, extensional faults generated at dilational bends may be juxtaposed next to compressional folds and faults. Cross sections will not balance.

Related to rotational block faulting

Where the shear couple is not parallel to the earlier fault or basement fabric, there may be block rotation about a vertical axis. Shortening or extension can occur across these rotated blocks depending on their initial orientation relative to the shear couple, the sense of shear and the boundary conditions. If the boundary conditions are fixed so that there is no lateral expansion or contraction, and if the original extensional faults rotate so that they eventually lie closer to the orientation of the shear

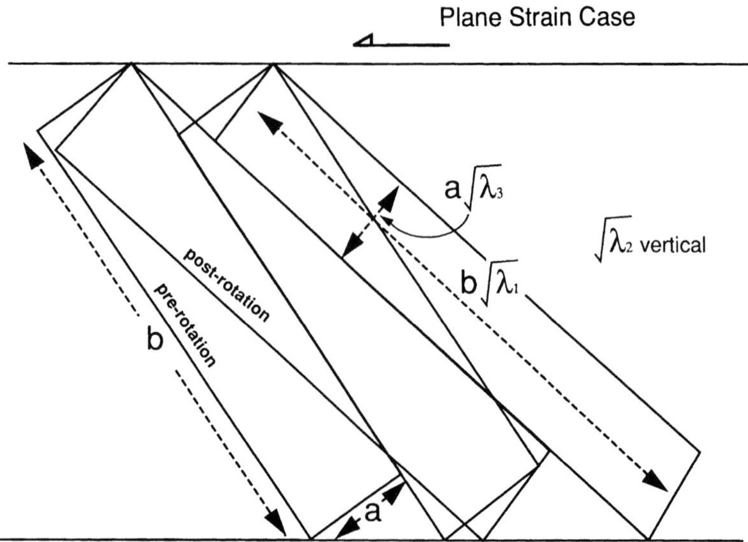

Fig. 20. Fault block rotation about a vertical axis, assuming plane strain with no extension vertically along the rotational axis. The blocks will shorten by $\sqrt{\lambda_1}$ ($\sqrt{\lambda}$is a strain parameter given by original length/final length) and lengthen by $\sqrt{\lambda_3}$. For conditions of no area change $\sqrt{\lambda_1} = 1/\sqrt{\lambda_3}$.

couple, the blocks will need to narrow and lengthen (Fig. 20). If the blocks rotate away from the orientation of the shear couple, the blocks will widen and shorten. The amount of block narrowing or widening can be calculated from the relationship:

$$l_1/l_0 = \sin(\alpha - \omega)/\sin\alpha$$

where α is the original angle made by the faults, relative to the plane of the regional shear and ω is the angle of rotation (Fig. 21). This relationship is summarized in graphical form in Fig. 22.

An example of an area where block rotation is important is the Palmyrides of Syria (Fig. 23, see also Lovelock 1984). The Palmyride Basin is affected by NE–SW-trending folds and thrusts, cut by E–W to ESE–WSW-trending faults (see also Searle 1994). The folds, which are upright to SE verging, appear to be discontinuous and form either en-echelon riedel-type arrays to major ENE trending lineaments, or transfer faults and associated thrusts. The shear sense along the ENE trending folds and faults appears to be right lateral. According to McBride *et al.* (1990) and Chaimov *et al.* (1990, 1992) the structures may be considered as thin-skinned, detaching on Triassic salt, or as thick-skinned structures involving the basement (Fig. 24). As the mountain belts associated with the Bitlis Suture in southern Turkey are *c.* 300 km to the north, the preferred model is one of mainly thick-skinned folds and thrusts associated with displacement on moderately-steeply dipping basement faults. Evidence against a thin-skinned model includes the following.

(i) Area balancing techniques, using the excess area method (Dahlstrom 1969), suggest that if a detachment exists, it is > 10 km deep and well

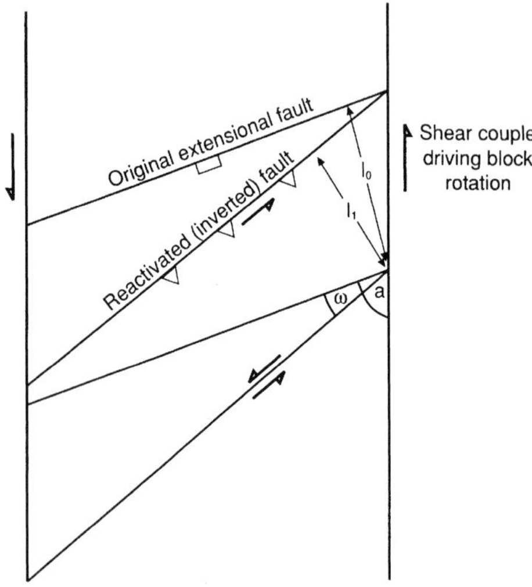

Fig. 21. Relationships used to derive the shortening l_1/l_0 associated with fault block rotation about a vertical axis.

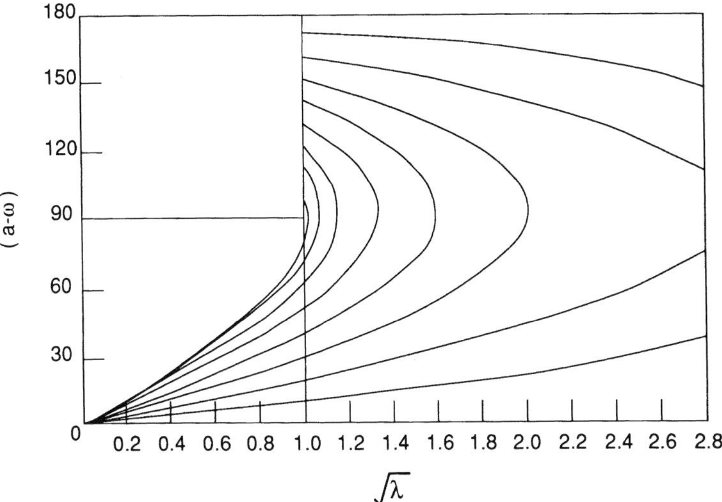

Fig. 22. Graphical relationship for the strain (√λ) across a fault block, related to (1) the initial orientation of the fault block relative to the regional shear couple and (2) the amount of rotation.

below any detachment level in the cover sediments, particularly the Triassic salt;

(ii) Onlaps and truncations in the basins show growth through the Neogene, with no indication of foreland propagation characteristic of thin-skinned mountain belts (e.g. Boyer & Elliott 1982);

(iii) The Upper Cretaceous–Palaeogene sediments thicken southwards towards several major fold-thrust associations, suggesting that these folds lie above basins. Similarly the Triassic-Cretaceous sediments are much thicker within the Palmyrides than beneath the Aleppo and Rutbah plains to the

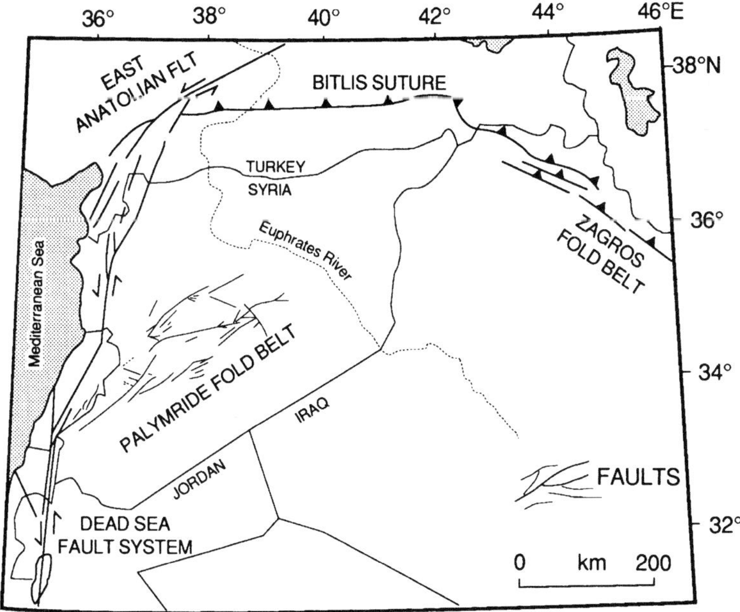

Fig. 23. Map showing the location of the Palmyride belt in Syria, relative to the Bitlis Suture, Zagros Fold Belt and Dead Sea Fault System. Simplified from Chaimov *et al.* (1990).

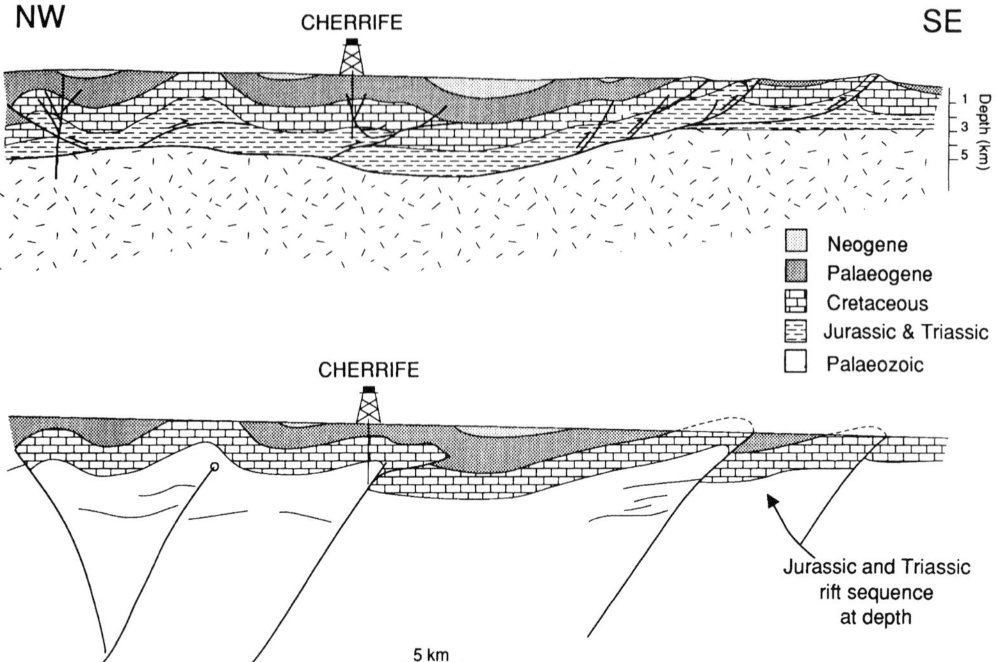

Fig. 24. Simplified cross sections through the Palmyrides. Bottom: thick-skinned interpretation, illustrating the reactivation of basement faults. Top: thin-skinned interpretation with detachment on Triassic salt (modified from Chaimov *et al.* 1990).

north and south, suggesting the presence of an earlier basin (cf. Best *et al.* 1993).

On seismic data the Mesozoic and Cenozoic strata are best considered as a single structural unit (Chaimov *et al.* 1993). Triassic evaporites in some areas decouple these units from lowermost Mesozoic and Palaeozoic rocks below but there is no evidence for a regional decoupling zone.

The top cross sections of Fig. 24 shows the faults as thick-skinned structures, dipping to the north. The Palmyride Ranges are interpreted as the result of inversion of three or four major half-graben. In the northwestern part of the Ranges, a S dipping fault produces a more symmetrical graben. The Palmyride hills or 'jebels' occur as eroded hanging wall anticlines above these faults. They range from broad open folds to steep sided box-like structures. The asymmetry is generally towards the SE. The southern boundaries of the jebels are often mapped as thrusts. Small-scale fault kinematics (author's own unpublished data and also Searle 1994) indicate an oblique thrusting with a component of right lateral shear. Normal faults of Upper Cretaceous–Palaeocene age often occur close to the crests of the jebels; they have not been reworked by the compressional deformation, but uplifted and rotated on the hanging walls of the SE verging thrusts.

The Palmyrides lie north of the Dead Sea left-lateral transform fault zone, which is reported to have a Neogene displacement of *c.* 100 km (Lovelock 1984; Hempton 1987). Palmyride compressional tectonics began at the same time as the opening of the Red Sea and movement on the Dead Sea fault zone, suggesting a causal relationship (Hempton 1987). However the Palmyrides show only 10–15 km total shortening and displacements are associated with right lateral fault movement. Hence movement on the Dead Sea Fault Zone does not simply die out in a zone of compression or left lateral shear in the Palmyrides (cf. Walley 1988).

Chaimov *et al.* (1993) explain the right lateral shear couple along the Palmyrides as due to extrusion of the Aleppo Block to the north. According to their model the northern arm of the Dead Sea Fault Zone would form the left lateral shear zone boundary to this expelled block. However a simpler explanation for Palmyride uplift and inversion, may involve block rotation associated with the regional left lateral shear system, related to the Dead Sea Transform Fault. Palaeo-

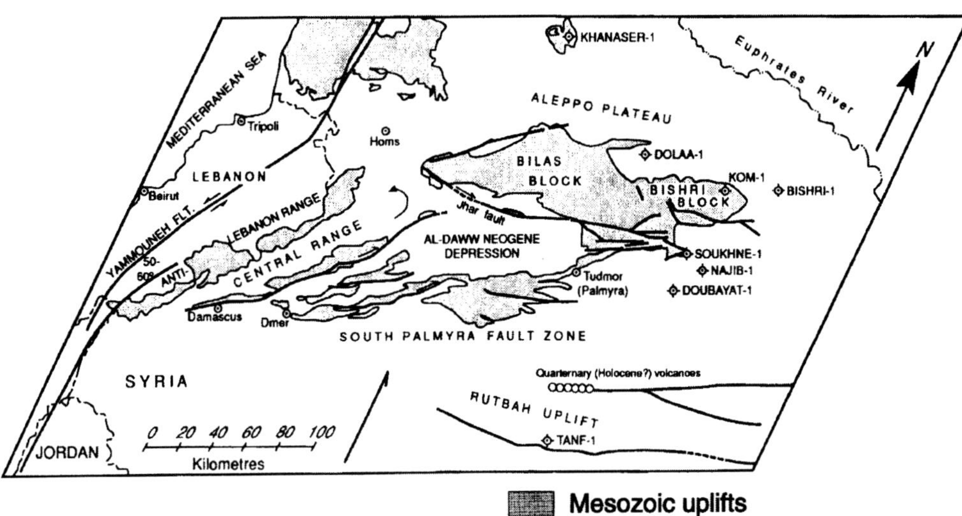

Fig. 25. Map of the Palmyrides showing the left-lateral shear, approximately parallel to the Dead Sea Fault Zone in the west. The amount of rotation and hence the amount of inversion and tightness of the folds increases towards the west. The uplifted blocks are bounded on the southern side by reverse faults with oblique right-lateral movement. The measured anticlockwise rotations in the Lebanon Hills range from 50–60°.

magnetic studies in the western Palmyrides indicate a rotation of > 20°. In the Lebanon Hills to the west, the rotation determined from paleomagnetic data is far greater, > 50° (Gregor *et al.* 1974; Ron 1987; Ron *et al.* 1990). A geometrical relationship between the shear system and faults, similar to that illustrated in Figs 25 and 26, would explain the amount of shortening across the fault blocks. A left-lateral shear couple was developed across western Arabia, associated with the strains developed at the NW tip of the Red Sea. Some of the shear was concentrated along the Dead Sea transform fault, but the remainder was accommodated by block rotation throughout the region.

Thus a diffuse shear system which rotates older crustal blocks may cause extension or rotational-block related inversion. The inversion can occur some distance from the plate collision zone and may not be directly related to continent–continent collision. Inversions attributable to diffuse shear associated with ocean ridge propagation occur along the eastern margin of the Atlantic from west of Shetland to mid-Norway. Other examples include the inverted Carboniferous fault blocks in the southern North Sea (Coward 1993). Coward (1993) suggested that during the Early Carboniferous a NE-trending right lateral shear couple affected the Caledonian structures of southern Britain, causing rotation from a WNW to a NW trend, widening the blocks and allowing stretching in the order of $\beta = 1.1–1.15$. During the Late Carboniferous, the blocks were rotated by a similar trending left lateral shear couple causing tightening across the blocks and inversion of the half-graben.

A model of block rotation can also be used to explain the Variscan inversion of the Palaeozoic basins in South Wales where the original ENE-trending blocks were rotated by up to 30° during

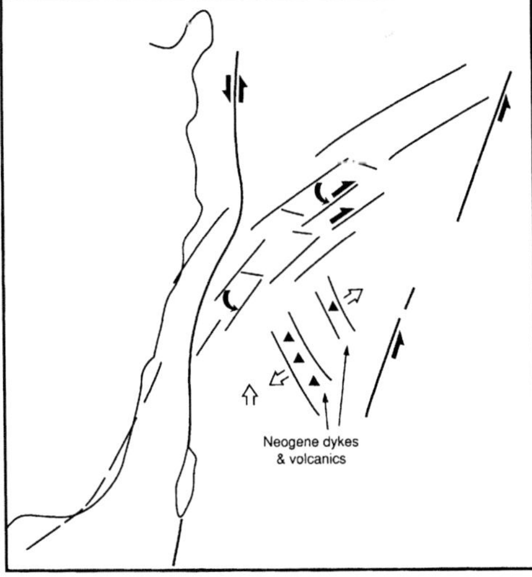

Fig. 26. Simplified model showing the rotation of the Palmyrides explained in terms of rotation of originally ENE-trending extensional fault blocks.

the Variscan Orogeny. This rotation was associated with a NW trending shear couple (Fig 27). Palaeomagnetic data from the folded Devonian sediments indicates *c.* 30° rotation (McClelland-Brown 1983), as does the change in trend of the structures. The NW trend of the shear couple is obtained from regional data on Variscan shortening in SW Britain (cf. Coward 1993). The rotation was associated with *c.* 30% shortening of the half-graben resulting in the expulsion of Devonian and Carboniferous

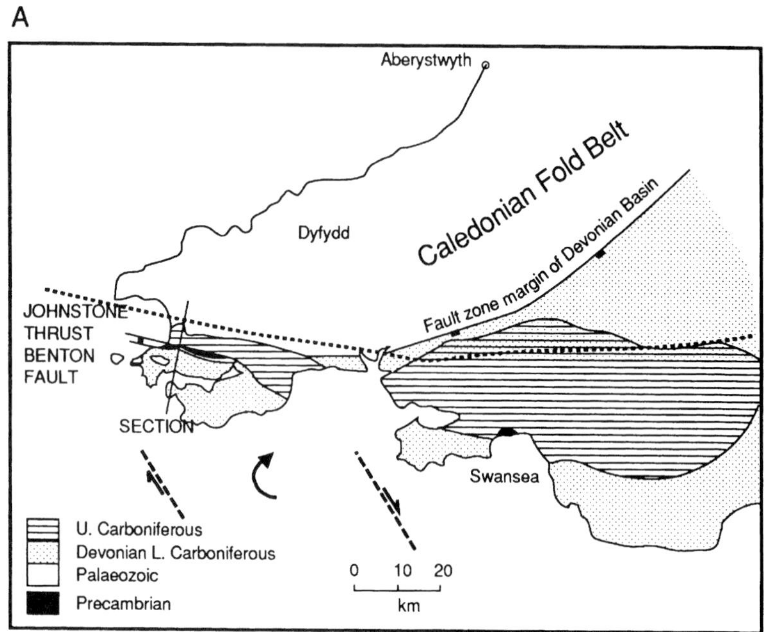

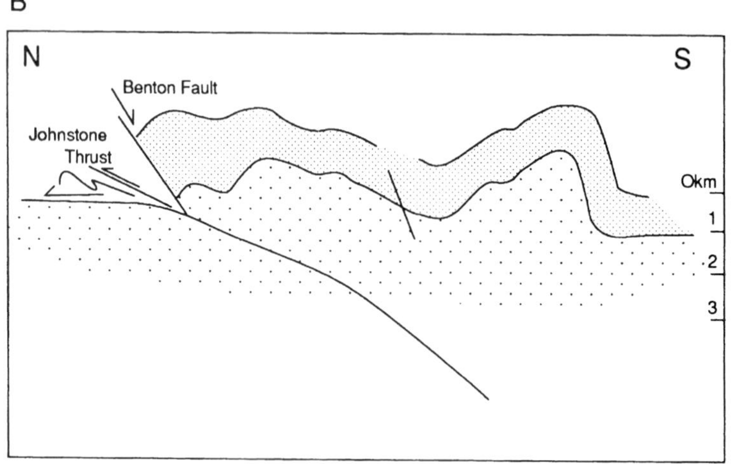

Fig. 27. (**a**) Map showing the change in trend of the Devonian basins in South Wales from the unaffected region in the eastern part of Wales to the strongly inverted region in Pembrokeshire. The sense of the shear couple, parallel to the regional orogenic transport direction, and the sense of rotation, are shown. (**b**) Simplifed cross section showing the inversion geometries of western Pembrokeshire. The stipple indicates folded Devonian sediments preserved on the hanging wall of the inverted normal fault (the Benton Fault). The Johnstone Thrust is a footwall collapse structure on the footwall of this fault. The displacement on the Johnstone Thrust is unknown. From Hayward & Graham (1989).

sediments (Fig. 27b, see also Powell 1989; Hayward & Graham 1989). Synchronous with the folding and thrusting, there was extension along the length of the fold axial surfaces, leading to the development of wide zones of carbonate veining, normal faults and conjugate shear faults.

The inversion associated with block rotation about a vertical axis may be diachronous along strike, as the zone affected by the rotation widens or propagates laterally. Thus in the central North Sea, Permo-Triassic and Late Jurassic rifts were reactivated in a reverse sense during the Late Cretaceous (Cartwright 1989). The orientation of the growth fold axes associated with the inversion are arranged in a segmented pattern, but consistent with net convergence in an approximately NE–SW direction, perpendicular to the strike of the basin. The segmentation not only resulted in variations in amount of shortening, but also in timing of the deformation. There are abrupt changes in the timing of deformation across segment boundaries. These inversion structures may be interpreted as due to Upper Cretaceous clockwise rotation of the earlier extensional fault blocks associated with a NE–SW trending shear couple across the North Sea. The width of the zone affected by this shear couple varied with time, so that the intensity and timing of the inversion vary along the axes of the inversion.

A similar pattern of varying inversion affects basement-involved faults in the western part of the Qaidam Basin in Western China. During the Tertiary the western part of the basin was affected by left lateral movements on the major NE-trending Altyn-Tagh shear zone (Fig. 28) (Wang & Coward 1990; Coward & Ries 1995). NW-trending fault blocks adjacent to the shear zone were first reactivated in the Eocene and Oligocene, forming hanging wall anticlines with long gently dipping backlimbs and short forelimbs. The latter were crossed by N–S-trending normal faults associated with a combination of right lateral shear together with extension along the length of the blocks. Small scale kinematic indicators show right lateral shear associated with reverse sense reactivation of the NW trending normal faults (Wang & Coward 1990; Coward & Ries 1995). During the Neogene the zone of rotational reactivation and inversion widened, the growth sequences related to the folding are diachronous and the depocentre

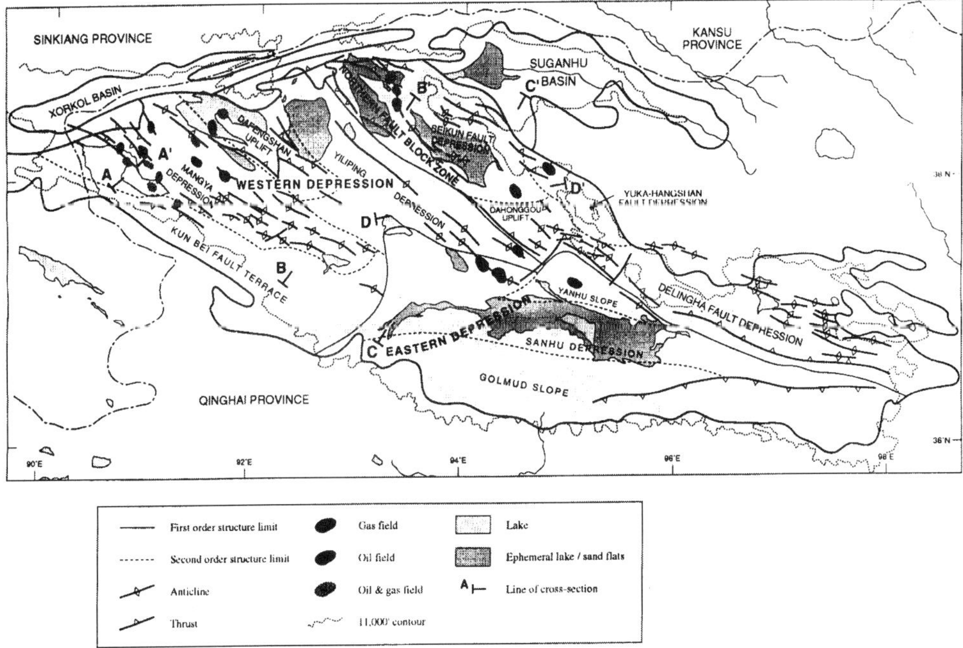

Fig. 28. Map showing inversion anticlines and reverse faults in the western part of the Qaidam Basin in western China produced by rotations associated with a left-lateral shear couple along the Altyn Fault Zone along the NW edge of the basin. The area affected by the rotation increased during the Neogene, so that the folding and basin inversion propagated from west to east across the western part of the basin. From Coward & Ries (1995), published by permission of Petroconsultants.

adjacent to the zone of inversion migrated across the Qaidam Basin from the west to centre of the basin.

Related to the buttressing effect of earlier normal faults

Inversion tectonics involving shortening perpendicular to the original normal faults may lead to strike-slip movements analogous to the development of small scale lateral escape structures. During extension, lateral ramps and transfer zones offset zones of thinned crust forming a tooth-like margin to the basin. Where these lateral ramps are offset, before or during inversion, then the teeth of the stretched crust on one margin may not fit back into the sockets on the opposite margin, leading to local lateral expulsion of material (Fig. 29). Inversion can lead to strain complexities where material does not extrude vertically by crustal thickening but also escapes laterally, so that three dimensional strains and incremental strain histories vary markedly over a small area. Buttressing by basin-bounding faults may lead to local pure shear strains and to lateral expulsion. Thus in the Alps the incremental strain work of Dietrich & Durney (1986), Gourlay (1986) and Spencer (1989) shows a pronounced change in extension direction with time, which can be related to an increase in the rate of lateral expulsion with time (Fig. 30). The most prominent lateral expulsion occurs close to the northwest edge of the Pennine Zone (Fig. 30), an important basin boundary fault during the Cretaceous. Another example of expulsion occurs close to the SE boundary of the Belledonne fault block (Fig. 30).

Quantifying basin shortening

In many basins the exact amount of shortening is difficult to quantify. Overthrusting at the edge of the basin may hide the position and orientation of the original normal fault. Some independant method is required to confirm values of shortening across a basin. This can then be used to help find the shape of the faults at depth.

Palaeomagnetic data have been used in the Palmyrides to determine the amount of rotation of the blocks. As the final orientation of the oblique-slip thrusts is known, the original orientation of the normal faults can be calculated, assuming a model of inversion related to block rotation. From these values the amount of shortening across the fault blocks can be determined (Figs 21 and 22). As the original spacing between the fault blocks is known, then the amount of strike-slip displacement can be calculated for each structure.

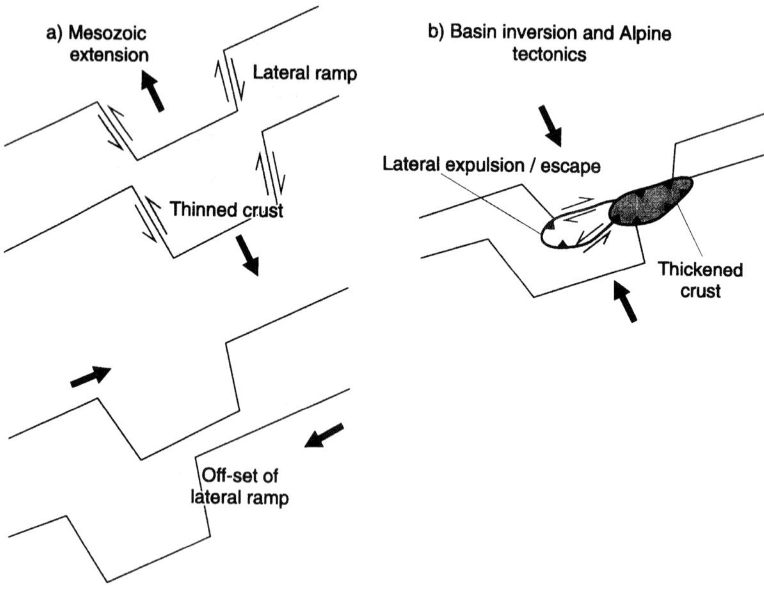

Fig. 29. Tooth and socket model for normal faults and lateral ramps in extended and inverted basins. Inversion causes lateral extrusion of material into weaker parts of the basin. From Coward (1994).

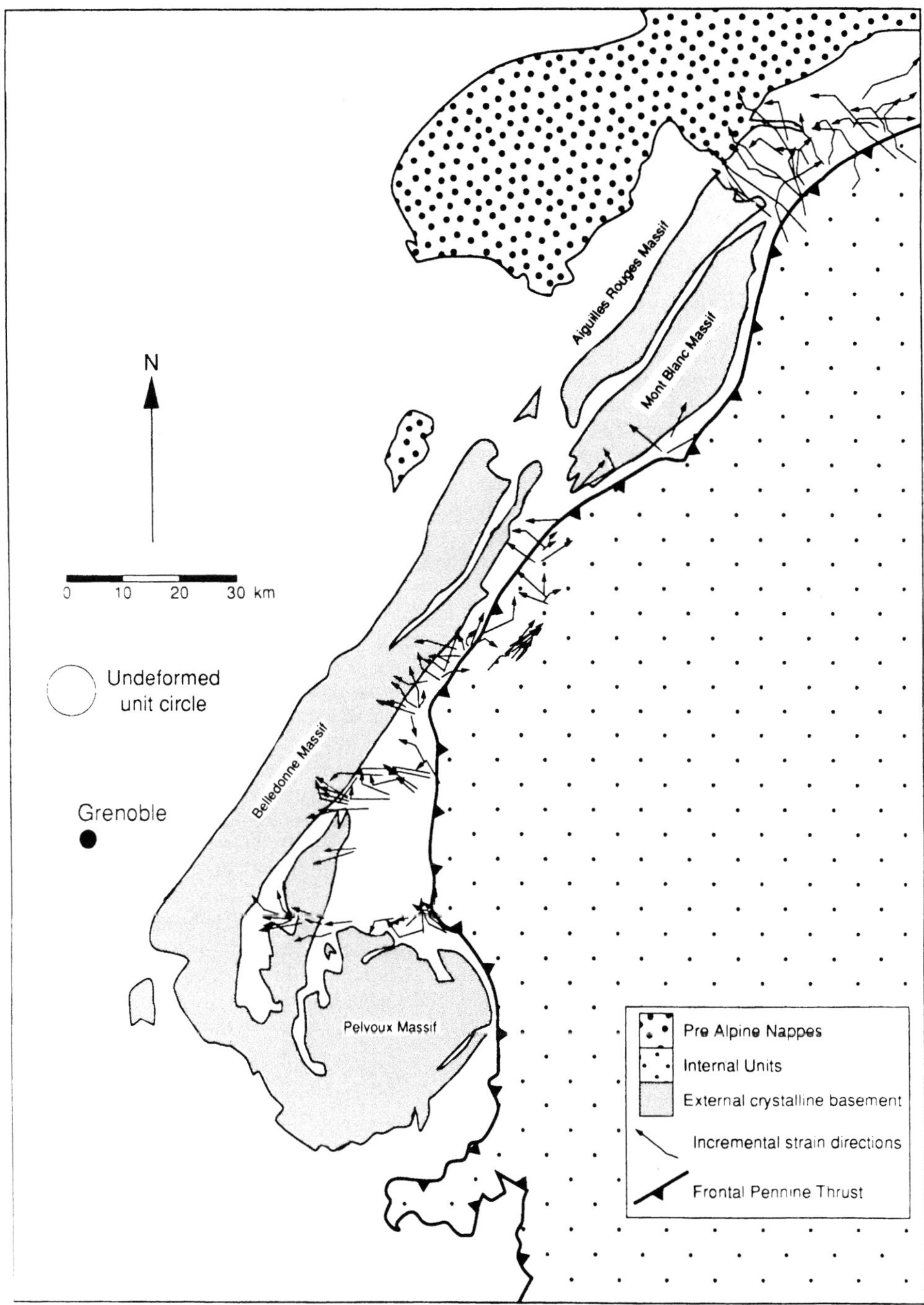

Fig. 30. Map showing incremental strain measurements, from pressure shadow data, from the western part of the Alps, showing variations in incremental stretch direction both laterally and with time. These variations in incremental strain are considered to be due to variations in overthrust direction and the lateral expulsion of the internal part of the Alps away from a high strain zone adjacent to the rigid buttress formed by the outer part of the Southeast France–Dauphinois Basin.

Note that in many regions there may be partitioning of deformation into dip-slip shortening within the blocks and strike-slip or oblique-slip displacement across the block-bounding faults. No one simple cross section on can be constructed and balanced across these deformation zones.

Hence palaeomagnetic data are invaluable in any regional analysis of basement-involved deformation.

Importance of the recognition of inversion

Section balancing techniques which ignore basin inversion can lead to errors in the interpretation of the deep structure of a fold and thrust belt and indeed of the entire mountain belt. Consider for example the data shown in Fig. 31, which represent the edge of a fold-thrust belt. This section is simplified from examples in the Sulaiman and Kohat Ranges, Pakistan (Fig. 32a and b). The section shows the sediments are gently folded, thrusted and uplifted to form a mountain-front monocline against the foreland basin. No thrusts emerge into the foreland basin and all the shortening within the fold-thrust belt has to be transferred onto a backthrust beneath the frontal monocline of the mountain belt (Jadoon *et al.* 1992, 1994). The thin-skinned model (Fig. 33a), highly simplified from Jadoon *et al.* (1992), assumes that the uplift of the thrust-fold belt is a result of thrust imbrication at depth. As the shortening in this deep level imbricate-duplex zone far exceeds the shortening seen in the upper layers of the thrust-fold belt, a passive roof backthrust has to be postulated beneath the upper layer of the fold-thrust zone. As the basement is not imbricated by thin-skinned thrusts, it must continue x km back beneath the hinterland of the fold-thrust belt, where x is the shortening. In the Sulaiman Range, Jadoon *et al.* (1992, 1994) postulate over 250 km shortening in the duplex zone beneath the passive roof backthrust.

The thick-skinned model (Fig. 33b) assumes that much of the uplift beneath the thrust belt is a result of shortening of a sedimentary basin, expulsion of the syn-rift sediments and thick-skinned shortening of the underlying crust. There is no necessity for basement to continue back beneath the hinterland of the thust-fold belt. This model certainly fits best the regional gravity data (see Jadoon *et al.* 1992). Thus it is important to be able to differentiate between folds produced by thin-skinned tectonics and folds produced by inversion.

Assuming thin-skinned thrusting, the shortening across the Sulaiman Ranges is in the order of 250 km (Jadoon *et al.* 1992; 1994). If this interpretation is correct, the shortening must be transferred onto the passive roof backthrust at the mountain front. The Sulaiman Range is an extremely arcuate belt, varying in width from a maximum of 300 km in the centre of the arc, to only a few km in the Sibi re-entrant in the south (Fig. 32a). Thus according to the thin-skinned model, 250 km of displacement need to die out north of the Sibi re-entrant or be lost in some immense backthrust in the re-entrant, for which there is no evidence. Hence a thick-skinned model is preferred. Estimates of the shortening across the folds and thrusts exposed near the surface of the Sulaiman range are in the order of 23 km (Coward *et al.* in preparation), an order of magnitude different from that obtained assuming a thin-skinned model. It is perfectly conceivable that 23 km displacement can die out in the Sibi re-entrant.

The Sulaiman Range is interpreted as part of the inverted margin of the Indian Plate, which has uplifted through the flysch and molassic sediments

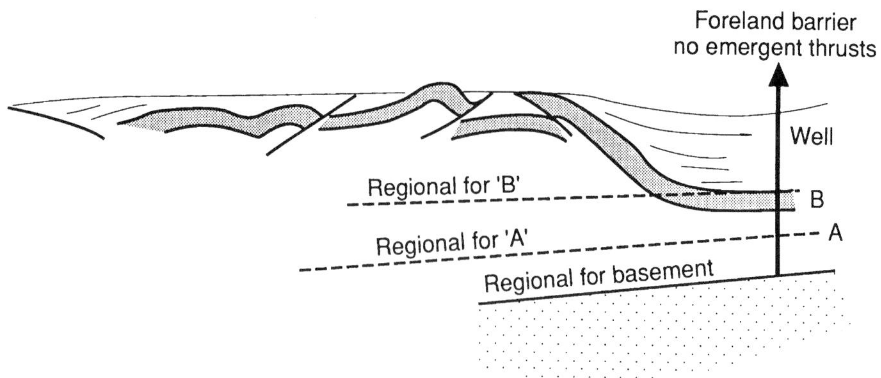

Fig. 31. Highly simplified section illustrating the structure of the frontal part of the eastern Sulaiman Ranges in western Pakistan.

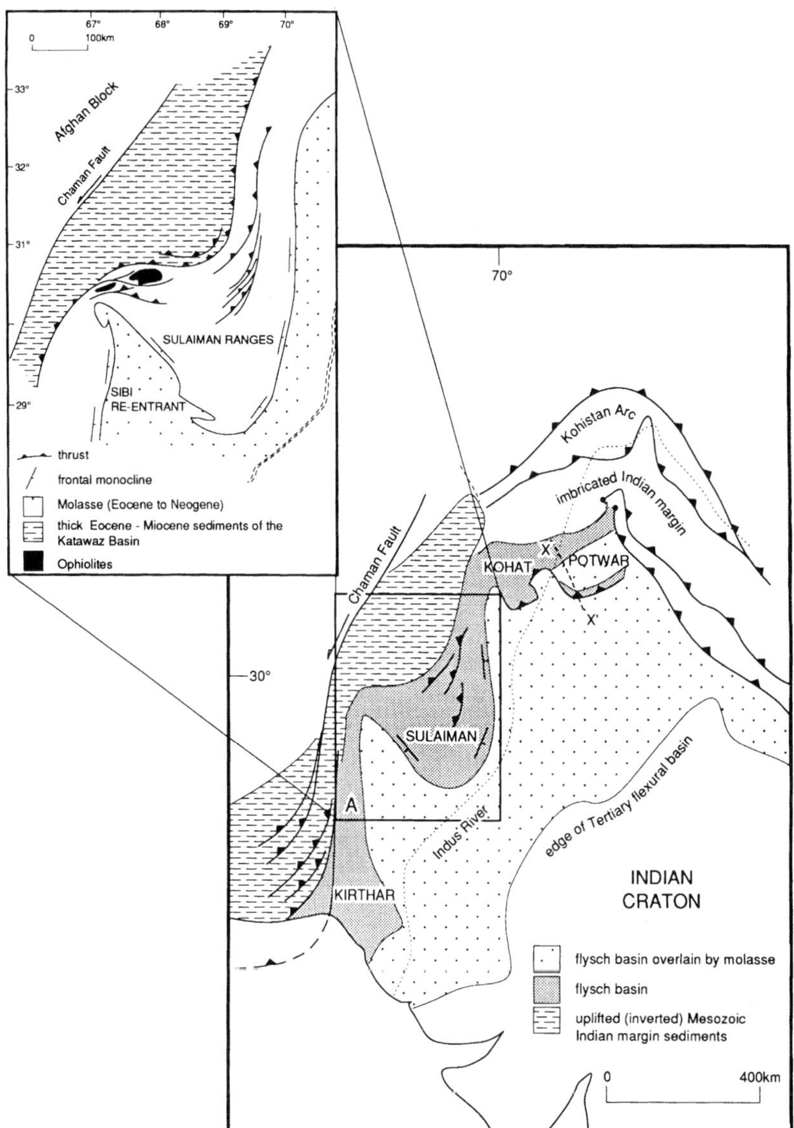

Fig. 32. Simplified map showing the regions of thick-skinned basement uplift in the Kirthar, Sulaiman, Kohat and northern Potwar Ranges. The zone of Himalayan collision tectonics and thin-skinned thrusting lies N and NE of Potwar. Note the width of the flexural basin on the original western edge of the Indian Plate, compared to the much narrower flexural basin produced by Himalayan lithospheric thickening. X–X′ shows the position of the cross section in Fig. 34. *Inset*: Map of the Sulaiman Range, Pakistan showing the uplifted zone relative to the Indus Molasse Basin and the Katawaz Flysch basin. During Paleogene times these basins were part of one system on the western edge of the Indian Plate. The Sulaiman Range was uplifted during the Miocene at the time of collision of western India with the Afghan Block to the NW. The frontal part of the Sulaiman Range is a monocline. No thrust emerges from the Indus Mollasse Basin to the SE of the Sulaiman Range.

of the Indus–Katawaz Basins. The margin was overthrust by ophiolites during the Late Cretaceous-Palaeocene and then uplifted/inverted during the Miocene. The width of the Katawaz–Indus Basin is probably a function of subduction-related loading during the Palaeogene. Collision occurred with the Afghan Block during the Miocene. The Afghan Block was subsequently

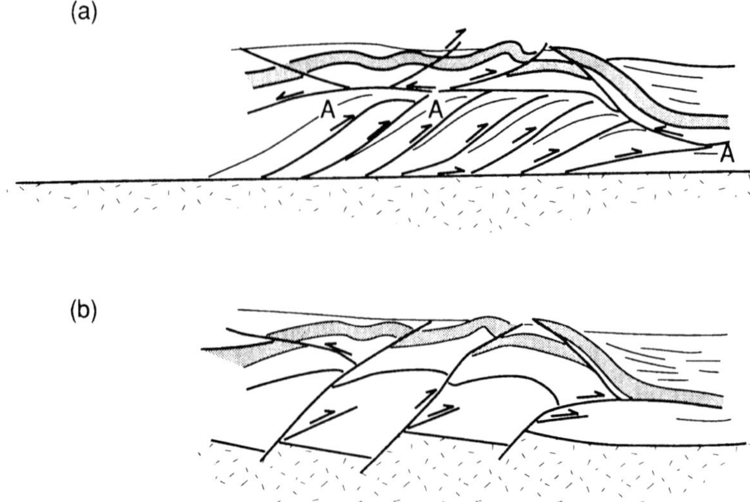

Fig. 33. Interpretations of the structure of the Sulaiman Ranges assuming (**a**) thin-skinned tectonics and (**b**) thick-skinned basin inversion.

expelled further to the SW, over the Makran. The zone of thick-skinned structures, formed during collision, can be traced from the Kirthar Range, near Karachi in the south, to the northern part of the Potwar Plateau in the north (Fig. 31b). The thick-skinned shortening direction is NW–SE, perpendicular to the original plate margin, but parallel to transforms which offset the basin. The shape of the Sulaiman Range probably reflects the original basin shape, bounded by NW-trending transform faults as well as NE-trending rift structures. Earthquake epicentre data from northern Pakistan are aligned on NW-trending zones, with depths down to 40 km (Seeber *et al.* 1981), indicating reactivation of these NW-trending thick-skinned transform faults. Focal mechanisms confirm these movements.

Distinguishing between thin-skinned and thick-skinned tectonics

If one or more of the following criteria hold, then inversion tectonics should be considered viable for the region.

(i) Compressive structures show simultaneous growth. In thin-skinned thrust zones the structures usually propagate towards the foreland, hence increasing the size of the thrust wedge with time, while in thick-skinned or inversion tectonic regimes, the original normal faults often reactivate simultaneously.

(ii) There is a lack of a mountain belt or surface slope which could drive the thin-skinned tectonics.

(iii) There is independent evidence for the presence of an older basin, for example, the presence of null points on sections or maps or the rapid change in thickness or facies of the sediments.

(iv) There is an increase in structural relief up the dip of the fault..

Hidden duplexes are constructed to explain the extra area beneath many thrust belts. Often there is no clear evidence for the nature of this hidden thrust zone.

Examples of reverse faults and folds, which have previously been interpreted as thin-skinned thrusts, but which may warrant re-interpretation as thick-skinned inversion structures include parts of the Apennines of Italy, the Palmyrides of Syria, the Zagros Ranges of Iraq and western Iran and the Kirthar, Sulaiman and Kohat Ranges of Western Pakistan. In addition many overthrust belts, such as most of the South American cordilleras, include components of basin inversion. Figure 34 shows a simplified section through the Salt Ranges and Potwar region of Pakistan. This area has previously been interpreted as totally thin-skinned. The revised interpretation of the Sulaiman and Kohat regions suggests that the thin-skinned Salt Ranges Thrust Zone may be underthrust by thick-skinned faults.

Similarly in the Western Alps the initial thin-skinned overthrusts are considerably modified by extreme basin inversion uplifting the external basement blocks. This inversion thickened the crust ahead of the main thrust zone; later thrusts formed

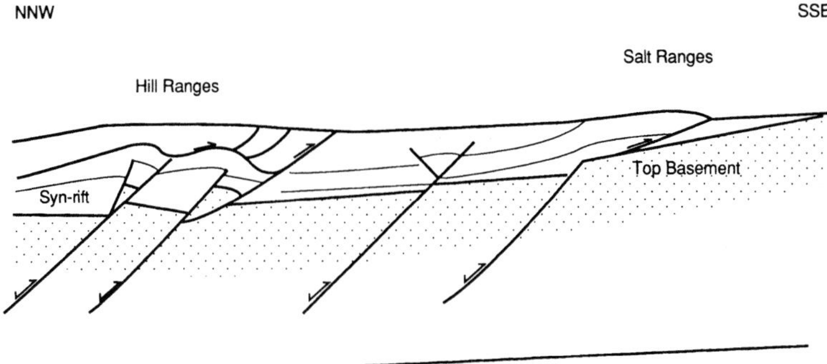

Fig. 34. Simplified section through the frontal ranges of the Himalayas, from the Kohat/western Potwar region in the north to the Salt Ranges. The position of the section line is shown on Fig. 32.

as a break-back sequence on the hinterland of these inverted grabens. The problem in identifying thick-skinned tectonics is often aggravated by the development of new thin-skinned detachments driven by gravity gliding from the uplift zones caused by basin inversion. In the Argentinian Andes there has been uplift of Triassic–Jurassic basins during the Tertiary. In the Neuquen Basin the Mesozoic post-rift sequence contains three evaporite horizons which form detachments during inversion. Inversion anticlines in the frontal regions of the Andes uplift and fold earlier detachments in the evaporites formed partially by gravitational collapse off earlier zones of uplift in the W.

Hence the thin-skinned detachments at the frontal regions of the mountain belts needs to be balanced against:

(i) regional plate collisional events in the interior of the mountain belt;
(ii) gravitational gliding off a topographic high formed by earlier uplift events; and
(iii) basement shortening associated with the frontal uplift zones.

In areas such as the Western Himalayas or the Andes, it is important to consider all these kinematic mechanisms and estimate the relative components of each. Previous calculations for the shortening across the frontal Himalayas (e.g. Coward & Butler 1985) considered only one component, the thin-skinned shortening due to plate collision, and hence the results are probably incorrect.

Conclusions

(1) Rotation is important in fault reactivation during basin inversion, involving both rotation about horizontal and vertical axes.
(2) Rotation about a horizontal axis causes shortening of the half-graben, leading to renewed displacement up the fault and/or expulsion of material from out of the half-graben. The resultant hanging wall fold is characterized by long gently dipping back-limbs and short hooked forelimbs. The fore-limbs can be modelled as fault-bend folds, where displacement decreases as the fault dip decreases.
(3) The rotation may be modelled as domino-style fault block rotation or due to rotation of a single fault as a rotational cantilever.
(4) Rotation of listric faults leads to excess shortening in the upper part of the half-graben, resulting in the development of reverse faults or thrusts in the footwall of the reactivated fault, or antithetic thrusts or reverse faults in the hanging
(5) Rotation may occur about a vertical axis associated with a regional shear couple oblique to earlier normal faults or basement lineaments. This rotation causes further extension across the fault blocks or shortening and basin inversion across the blocks. The strain depends on the initial orientation of the fault blocks to the regional shear couple and the sense and amount of shear. Examples of fault block rotation are described from the Palmyrides of Syria. However the mechanism is common, explaining many regions of basin inversion including parts of the North Sea, Mediterranean and Central Asia (e.g. Tarim and Qaidam Basins). Hence paleomagnetic data are invaluable in any regional analysis.
(6) Rotation about a vertical axis leads to non-plane strain inversion. However inversion structures are often associated with even more extreme out-of section movements, related to

strike- slip faulting and lateral explusion of material into weaker parts of the basin.

(7) Basin inversion involving reactivation of earlier faults or basement lineaments is a common intraplate deformation mechanism. Inversion geometries have been recognized in the frontal regions of many orogenic belts, from the Andes to the Himalayas and from the Caledonides to the Alps. Often thin-skinned thrust tectonics may hide inversion geometries at depth. However it is very important to recognize or test for basin inversion and thick-skinned thrust tectonics rather than assume a thin-skinned thrust model; the assumption of thin-skinned thrusting may lead to gross errors in structural and tectonic interpretation.

References

AMEEN, M. S. 1992. Strain pattern in the Purbeck–Isle of Wight Monocline: a case study of folding due to dip-slip faulting in the basement. *In*: BARTHOLOMEW, M. J. *et al.* (eds) *Basement tectonics*, **8**, Kluwer Academic Publishers, 559–78.

BADLEY, M. E., PRICE, J. D. & BACKSHALL, L. C. 1989. Inversion, reactivated faults and related structures: seismic examples from the southern North Sea. *In*: COOPER, M. A. & WILLIAMS, G. D. (eds) *Inversion Tectonics*, Geological Society, London, Special Publications, **44**, 201–219.

BALLY, A. W. 1984. Tectonogenese et sismique reflection. *Bulletin de la Societe Geologique de France*, **26**, 279–286.

BEST, J. A., BARAZANGi, M., AL SAAD, D., SAWAF, T. & GEBRAN, A. 1993. Continental margin evolution of the northern Arabian Platform in Syria. *American Association of Petroleum Geologists Bulletin*, **77**, 173–193.

BOYER, S. E. & ELLIOTT, D. 1982. Thrust systems.*American Association of Petroleum Geologists Bulletin*, **66**, 1196–1230.

BUCHANAN, P. G. & MCCLAY, K. R. 1992. Sandbox experiments of inverted listric and planar fault systems. *Tectonophysics*, **188**, 97–115.

CARTWRIGHT, J. A. 1989. The kinematics of inversion in the Danish Central graben. *In*: COOPER, M. A. & WILLIAMS, G. D. (eds) *Inversion Tectonics*. Geological Society, London, Special Publications, **44**, 153–75.

CHAIMOV, T. A., BARAZANGI, M., AL SAAD, D., SAWAF, T. & GEBRAN, A. 1990. Balanced cross sections and shortening in the Palmyride fold belt of Syria and implications for movement along the Dead Sea fault system. *Tectonics*, **9**. 1369–86.

——, ——, ——, —— & —— 1992. Mesozoic and Cenozoic deformation inferred from seismic stratigraphy in the southwestern intracontinental fold-thrust belt, Syria. *Geological Society of America Bulletin*, **104**, 704–15.

——, ——, ——, —— & KHADDOUR, M. 1993. Seismic fabric and 3D structure of the southwestern intra-continental Palmyride fold belt, Syria. *American Association of Petroleum Geologists Bulletin*, **77**, 2032–47.

COWARD, M. P. 1993. The effects of Late Caledonian and Variscan continental escape tectonics in basement structure, Paleozoic basin kinematics and subsequent Mesozoic basin development in NW Europe. *In*: PARKER, J. R. (ed.) *Petroleum Geology of Northwest Europe. Proceedings of the 4th Conference.* Geological Society, London, 1095–1108.

—— 1994. Inversion Tectonics. *In*: HANCOCK, P. R. (ed.) *Continental Deformation*, Pergamon Press, 289–304.

—— & BUTLER, R. W. H. 1985. Thrust tectonics and the deep structure of the Pakistan Himalaya. *Geology*, **13**, 417–20.

—— & RIES, A. C. 1995. *The Qaidam Basin.* Basin Monitor Series, Petroconsultants SA Geneva.

——, ENFIELD, M. A & FISCHER, M. W. 1989. Devonian basins of northern Scotland: extension and inversion related to Late Caledonian-Variscan tectonics. *In*: COOPER, M. A. & WILLIAMS, G. D. (eds) *Inversion Tectonics.* Geological Society, London, Special Publications, **44**, 275–308.

——, GILLCRIST, R. & TRUDGILL, B. 1991. Extensional structures and their tectonic inversion in the Western Alps. *In*: ROBERTS, A. M., YIELDING, G. & FREEMAN, B. (eds) *The Geometry of Normal Faults.* Geological Society, London, Special Publications, **56**, 93–112.

DAHLSTROM, C. D. A. 1969. Balanced cross sections. *Canadian Journal of Earth Sciences*, **6**, 743–57.

DEWEY, J. F. 1982. Plate tectonics and the evolution of the British Isles. *Journal of the Geological Society, London*, **139**, 371–414.

DIETRICH, D. & DURNEY, D. W. 1986. Change in direction of overthrust shear in the Helvetic nappes of western Switzerland. *Journal of Structural Geology*, **8**, 373–381.

GILLCRIST, R. 1988. *Mesozoic basin development and structural inversion in the external French Alps.* PhD thesis, University of London.

GILLCRIST, R., COWARD, M. P. & MUGNIER, J. L. 1987. Structural inversion, examples from the Alpine Foreland and the French Alps. *Geodinimica Acta*, **1**, 5–34.

GOURLAY, P. 1986. La deformation de socle et des courverture delphino helvetiques dan la region du Mont Blanc (Alpes occidentales). *Bulletin of the Geological Society of France*, **8**, 159–69.

GREGOR, C. B., MERTZMAN, S., NAIRN, A. E. M. & NGENDANK, J. 1974. Paleomagnetism of some Mesozoic and Cenozoic volcanic rocks from the Lebanon. *Tectonophysics*, **21**, 375–95.

HARDING, T. P. 1983. Seismic characteristics and identification of negative flower structures, positive flower structures and positive structural inversion. *American Association of Petroleum Geologists Bulletin*, **69**, 582–600.

HAYWARD, A. B. & GRAHAM, R. H. 1989. Some geometrical characteristics of inversion. *In*: COOPER, M. A. & WILLIAMS, G. D. (eds) Inversion Tectonics. Geological Society, London, Special Publications, **44**, 17–40.

HEMPTON, M. 1987. Constraints on Arabian plate motion and extensional history of the Red Sea. *Tectonics*, **6**, 687–705.

JADOON, I. A. K., LAWRENCE, R. D & LILLIE, R. J. 1992. Balanced and retrodeformed geological cross sections from the frontal Sulaiman lobe, Pakistan: duplex development in thick strata along the western margin of the Indian Plate. *In*: MCCLAY, K. (ed.) *Thrust Tectonics*. Chapman & Hall, London, 343–56.

——, LAWRENCE, R. D & LILLIE, R. J. 1994. Seismic data, geometry, evolution and shortening in the active Sulaiman fold-and thrust-belt of Pakistan, southwest of the Himalayas. *American Association of Petroleum Geologists Bulletin*, 78, 758–74.

KUSZNIR, N. J., MARSDEN, G. & EGAN, S. S. 1991. A flexural-cantilever simple-shear/pure shear model of continental lithospheric extension: applications to the Jeanne d'Arc Basin, Grand Banks and Viking Graben, North Sea. *In*: ROBERTS, A. M., YIELDING, G. & FREEMAN, B. (eds) *The Geometry of Normal Faults*. Geological Society, London, Special Publications, **56**, 41–60.

LETOUZEY, J. 1990. Fault reactivation, inversion and fold-thrust belt. *In*: LETOUZEY, J. (ed.) *Petroleum and Tectonics in Mobile Belts*. Editions Technip, Paris, 101–28.

LOVELOCK, P. 1984. A review of the tectonics of the northern Middle East region. *Geological Magazine*, **121**, 577–87.

MCBRIDE, J. H., BARAZANGI, M., BEST, J., AL SAAD, D., SAWAF, T., AL OTRI, M. & GEBRAN, A. 1990. Seismic reflection structure of the intracratonic Palmyride fold-thrust belt and surrounding Arabian platform, Syria. *American Association of Petroleum Geologists Bulletin*, **74**, 238–59.

MCCLAY, K. R. 1989. Analogue models of tectonic inversion. *In*: COOPER, M. A. & WILLIAMS, G. D. (eds) *Inversion Tectonics*. Geological Society, London, Special Publications, **44**, 41–62.

—— & BUCHANAN, P. G. 1992. Thrust faults in inverted extensional basins. *In*. MCCLAY, K. R. (ed.) *Thrust Tectonics*. Chapman & Hall, 93–104.

MCCLELLAND-BROWN, E. 1983. Paleomagnetic studies of fold development and propagation in the Pembrokeshire Old Red Sandstone. *Tectonophysics*, **98**, 131–149.

MORTON, W. H. & BLACK, R. 1975. Crustal attenuation in Afar. *In*: PILGER, A & ROSLER, A. (eds) *Afar depression of Ethiopia*. Interunion Committee Geodynamics Science Report no. **14**, 55–65. E Schwiezerbart'sche Verlagsbuchhandlung, Stuttgart.

PEGRUM, R. M. 1984. The extension of the Tornquist Zone in the Norwegian North Sea. *Norsk Geologisk Tidsskrift*, **64**, 39–68.

POWELL, C. M. 1989. Structural controls on Paleozoic basin evolution and inversion in southwest Wales. *Journal of the Geological Society of London*, **146**, 439–46.

ROBERTS, D. G. 1989. Basin inversion in and around the British Isles. *In*: Cooper, M. A. & Williams, G. D. (eds) Inversion Tectonics. Geological Society, London, Special Publications, **44**, 131–50.

RON, H. 1987. Deformation along the Yammuneh, the restraining bend of the Dead Sea Transform: paleomagnetic and kinematic implications. *Tectonics*, **6**, 653–66.

——, NUR, A. & EYAL, Y. 1990. Multiple strike-slip fault sets: a case study from the Dead Sea Transform. *Tectonics*, **9**, 1421–31.

SEARLE, M. P. 1994. Structure of the intraplate eastern Palmyride Fold Belt, Syria. *Geological Society of America Bulletin*, **106**, 1132–50.

SEEBER, L., ARMBRUSTER, J. G. & QUITTMEYER, R. C. 1981. Seismicity and continental subduction in the Himalayan Arc. *In*: GUPTA, H. K. & DELANY, F. M. (eds) *Zagros, Hindu Kush, Himalaya, Geodynamic Evolution*. American Geophysical Union, Geodynamiocs Series, **3**, 215–242.

SPENCER, S. 1989. *The nature of the North Pennine Front, French Alps*. PhD thesis, University of London.

SUPPE, J. 1983. Geometry and kinematics of fault-bend folding. *American Journal of Science*, **283**, 684–721.

—— & MEDWEDEFF, D. A. 1985. Fault-propagation folding. *Geological Society of America Abstracts with programs*, **16**, 670.

WALLEY, C. D. 1988. A braided strike-slip model for the northern continuation of the Dead Sea Fault and its implications to Levantine tectonics. *Tectonophysics*, **145**, 63–72.

WANG, Q. M., COWARD, M. P., YUEN, W., ZHAO, Z., LUI, S. & WANG, W. 1995. Fold growth during basin inversion – example from the East China Sea. *In*: BUCHANAN, J. & BUCHANAN, P. (eds) *Basin Inversion*. Geological Society, London, Special Publications, **88**, 493–522.

——. & —— 1990. The Chaidam Basin (NW China): formation and hydrocarbon potential. *Journal of Petroleum Geology*, **13**, 93–112.

WILLIAMS, G. D., POWELL, C. M. & COOPER, M. A. 1989. Geometry and kinematics of inversion tectonics. *In*: COOPER, M. A. & WILLIAMS, G. D. (eds) *Inversion Tectonics*. Geological Society, London, Special Publications, **44**, 3–16.

ZIEGLER, P. A. 1983. Inverted basins in the Alpine Foreland. *In*: Bally, A. W. (ed.) *Seismic expression of structural styles*. American Association of Petroleum Geologists, Studies in Geology, **15**, 3.3.3–3.3.12.

—— 1987. Compressional intra-plate tectonics in the Alpine Foreland. *Tectonophysics*, **137**, 389–420.

—— 1989. A geodynamic model for Alpine intra-plate compressional deformation in Western and Central Europe. *In*: COOPER, M.A. & WILLIAMS, G.D. (eds) *Inversion Tectonics*. Geological Society, London, Special Publications, **44**, 63–86.

Construction and validation of extensional cross sections using lost area and strain, with application to the Rhine Graben

RICHARD H. GROSHONG, JR
Department of Geology, The University of Alabama, Tuscaloosa, AL 35487-0338, USA

Abstract: The simultaneous area balance of multiple horizons in a graben system formed above a detachment is the basis for the lost-area section construction and validation technique. Area balance requires that the area displaced below the original regional level of each bed in a graben (the lost area) be balanced by the displaced area at the boundary of the system, given by the product of the displacement times the depth to detachment. For multiple horizons, this relationship is a straight line on a plot of lost area versus depth. The slope of the line is the displacement and the depth intercept is the depth of the detachment. In general, beds within the graben system undergo both visible and homogeneous layer-parallel extension. The homogeneous component can be calculated from the width of the graben system, the lost area, and the depth to detachment. Conversely, the detachment depth may be calculated from the total layer-parallel extension and the lost area. The technique is illustrated by application to the Rhine graben. The Rhine graben is inferred to have formed with its detachment at the base of a normal thickness (31 km) of crust and to have been uplifted. Crustal necking with a detachment at 15 km depth is a viable alternative but requires three times the layer-parallel extension of 6.3% measured at the top of the Permian and older basement. The large difference between the layer-parallel extension (6.3%) and the crustal extension (29%) is in close agreement with an area-balanced full-graben model having boundaries that slope at 60° toward the centrer of the graben.

Most cross-section balancing and validating techniques are based on specific kinematic models for the deformation, for example, constant bed length and bed thickness (Dahlstrom, 1969) or oblique simple shear (White *et al.* 1986). If the kinematic model is not appropriate for the cross section, a correct cross section may be considered to be invalid and inferences made from incomplete data may be incorrect. Lost-area balancing (Groshong, 1994) allows a cross section to be tested for balance and internal consistency, regardless of the kinematic model, on the basis of the relationships displayed on a graph of lost area versus depth. The displacement required to produce a structure is the slope of the area–depth line. This is a bed-length independent measure of the displacement and provides the information needed to calculate any layer-parallel strains that may be present in the structure. The 'lost area' is the area that drops below the original depositional level (the regional, McClay 1992) as a result of extensional displacement. The method is scale independent, as long as the complete structural system is considered. A structural system is the entire region that, due to deformation, departs from the regional elevation. The fundamental assumptions of the technique are that the cross section is area balanced and that the structure terminates at a detachment.

The discussion begins with a review of the area-balance relationships and the lost-area diagram, followed by the equations for determining strain from the area balance. A full-graben model, with and without necking is described and applied to a profile across the central Rhine graben. It is shown that the depth to detachment for the Rhine Graben can be determined from the lost area of one horizon, together with the layer-parallel extension of the horizon as seen on the cross section. The layer-parallel extension in the full graben model is similar to the values determined from the sedimentary cover in examples like the Rhine Graben and the North Sea and is much less than the amount of crustal extension.

Theory

Area balance, depth to detachment, and displacement

If the area of a cross section is constant, then the area displaced along the lower detachment is equal to the area dropped below the regional in extensional structures (Fig. 1):

$$S = DH, \qquad (1)$$

where S = displaced area, D = displacement of the block on the lower detachment and H = depth

From Buchanan, P. G. & Nieuwland, D. A. (eds), 1996, *Modern Developments in Structural Interpretation,Validation and Modelling*, Geological Society Special Publication No. 99, pp. 79–87.

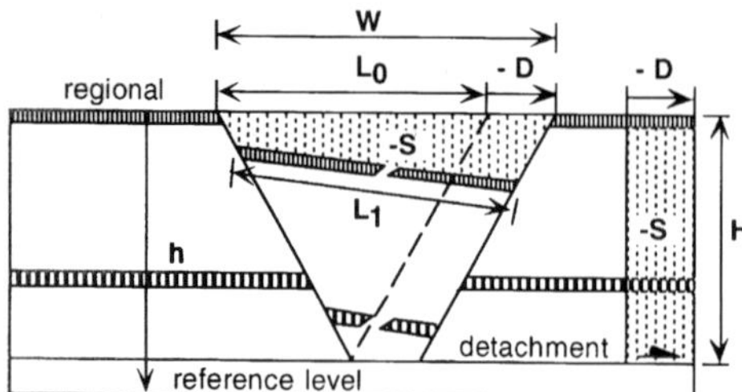

Fig. 1. Area-balanced extension. The depth to detachment (H) and the depth to the reference horizon (h) are always measured from the original position of the horizon (the regional). The displacement on the lower detachment (D) and the displaced or lost area (S) are always negative in extension. The original bed length (L_0) is extended in the formation of the graben and so the final bed length (L_1) is shown schematically as having been pulled apart.

to detachment (Chamberlin 1910; Hansen 1965; Voight 1974; Gibbs 1983). The sign conventions are that area displaced above the regional (excess area) and contractional displacement are positive (Epard & Groshong 1993) and area dropped below regional (lost area) and extensional displacement are negative. The cross section must obey this relationship at every structural horizon at depth, h, above a common reference level (Fig. 1). The depth h is positive above and negative below the reference level. Each horizon has its own regional but all horizons have the same reference level. Plotted on an area–depth graph (Fig. 2), the relationship is the straight line

$$S = Dh + S_a, \tag{2}$$

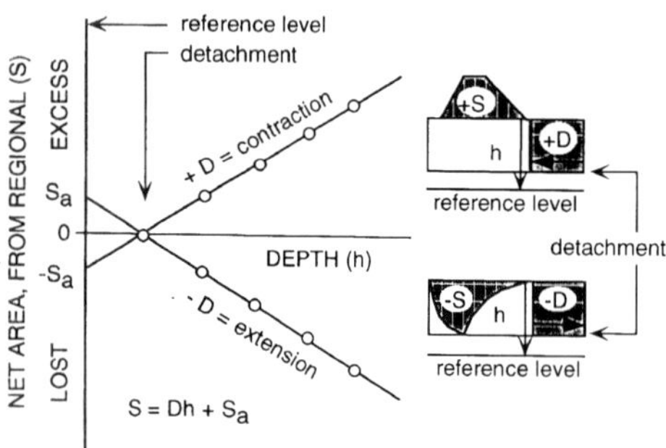

Fig. 2. Area–depth graph for extension and contraction. The lines represent the area–depth relationships for balanced cross sections (equation (2)). Lost areas are negative and the slope of the line for an extensional structure is negative; excess areas are positive and the slope of the line for a contractional structure is positive. D, h, and S are defined in Fig. 1.

where S_a is the area intercept of the line. The slope of this line is the displacement on the lower detachment. The detachment itself is located at the point at which the area S goes to zero:

$$H = -S_a/D \tag{3}$$

and may be above or below the arbitrary reference level. In a typical example (Fig. 3), the displaced areas from multiple horizons are plotted to define the area–depth line from which the depth to detachment is determined. If units are displaced both upward and downward from regional, then the net displaced area is plotted on the area–depth diagram to obtain the true detachment (Groshong 1994). The slope of the area–depth line is the displacement on the lower detachment required to form the structure. The slope can be determined from the relationship

$$D = \Delta S/\Delta h, \tag{4}$$

or by solving equation (2) for D as

$$D = (S - S_a)/h. \tag{5}$$

Vertical exaggeration does not alter the straight-line area–depth correlation but does change the slope of the line. Thus the prediction of the detachment depth is unchanged but the implied amount of

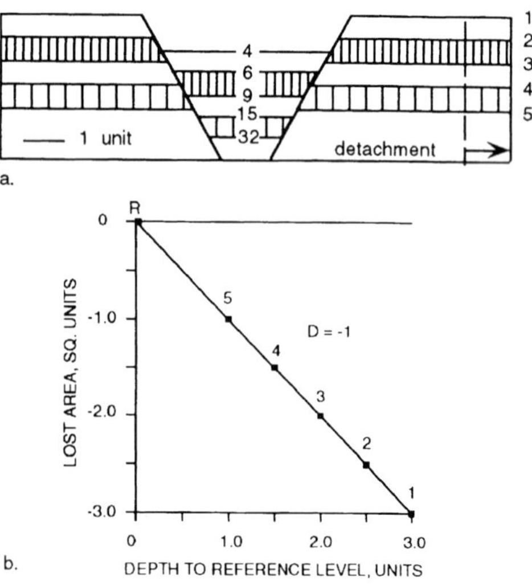

Fig. 3. Area balance of multiple horizons in a full graben. The model is analytically area-balanced (Groshong, 1994). The detachment is the reference level, R. (**a**) Cross section. The numbers in the graben are the requisite layer-parallel strains (equation (8)) in percent for horizons 1–5. (**b**) Lost-area diagram in which the lost areas of individual horizons in the graben are plotted against the depth to the reference level.

displacement is changed by vertical exaggeration. Vertical exaggeration will also affect the inferred strain because of its effect on the displacement. Growth beds do not fall on the area–depth line defined by the pre-growth beds because each growth unit has its own value of displacement. Growth beds may fall on a straight area–depth line if the rate of deposition is proportional to the rate of displacement. In this case the slope of the line is not the displacement. A non-linear area–depth curve will result if the vertical displacement profile is non-linear (cf. appendix in Epard & Groshong 1993).

Strain and depth to detachment

The bed length on the cross section (L_1) is not necessarily, or even usually, equal to the original bed length (L_0). The original bed length (Fig. 1) is:

$$L_0 = W + D, \qquad (6)$$

where W is the width of the structure at regional for the horizon in question and D is negative in extension. A change in bed length means that there is layer-parallel strain (e) which can be calculated from

$$e = (L_1 - L_0)/L_0. \qquad (7)$$

Substituting equations (1) and (6) into (7) gives

$$e = (L_1 H/(HW + S)) - 1. \qquad (8)$$

The value of e calculated from this equation is the layer-parallel strain required for the cross section to be balanced and so is termed the requisite strain (Groshong & Epard 1994). This value of e is not necessarily a principal strain. Alternatively, the shortening (equations (6) and (7)) can be expressed as the length change ΔL, where,

$$\Delta L = L_1 - L_0 = L_1 - (W + D). \qquad (9)$$

This value of ΔL is the requisite length change, much of which may occur on second-order faults (Fig.4). If bed length is constant, then $L_1 - L_0 = 0$, and $\Delta L = e = 0$.

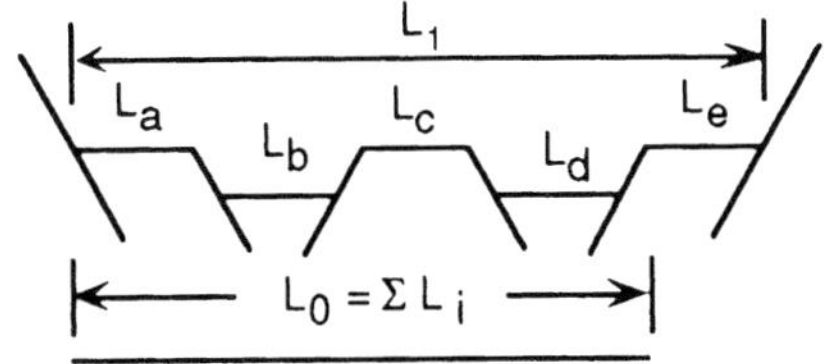

Fig. 4. Layer-parallel extension along visible faults. L_a, L_b, etc. represent individual bed segments. The strain is defined by equation (7).

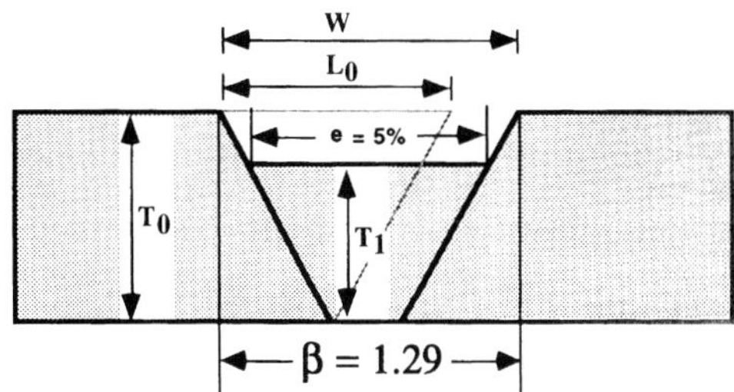

Fig. 5. Crustal extension β (equation (10)), of the full graben in Fig. 3 contrasted to the layer-parallel strain e (equation (8)) of the uppermost bed in the graben.

The requisite layer-parallel strain is very different from the system-scale stretching factor β (McKenzie 1978), which is given by the equations:

$$\beta = T_0/T_1, \text{ or} \qquad (10)$$

$$\beta = W/L_0. \qquad (11)$$

where T_0 is the original total thickness and T_1 is the final total thickness (Fig. 5). As long as the thinning is uniform over the region of interest, these equations give the correct system-scale extension, regardless of the kinematic model for the deformation. The use of equation (10) in a symmetrical full graben is relatively straightforward wherever it is measured, but the use of equation (11) contains pitfalls. Clearly the layer-parallel extension, $e = 5\%$, of the top horizon in the area-balanced graben (Fig. 5) is much less than β for the whole graben system which is 29% (β = 1.29, where e in $\% = (\beta - 1) \times 100$). This difference is due to the fact that the strains relate to different aspects of the system. Many extension values calculated using equation (11) have incorrectly used the present-day length, L_1, instead of L_0. The value of β can be correctly calculated from equation (11) but requires restoration of the upper horizon to its original length to determine L_0. The length of a bed in the graben may include a component of homogeneous strain that cannot be restored without additional knowledge of the behaviour of the system, as in Fig. 5. The anticipated result of this pitfall is that the length of L_0 will be over-estimated, even if visible faults are restored, and therefore β will be underestimated if computed from equation (11). These pitfalls provide an explanation for the common observation (Ziegler 1992*b*; Bois 1993) that β from equation (10) is greater than the value from (11).

An alternative method for finding the detachment depth is to solve equations (1) or (8) for H:

$$H = S/D, \qquad (12)$$

$$H = S(e + 1)/(L_1 - W - eW). \qquad (13)$$

Equation (12) is appropriate where all of the bed-scale extension is visible or is truly zero; in this case D is known from equation (4), and the method is that of Chamberlin (1910). Equation 13 must be used if there is a known component of homogeneous strain that is not directly visible on the cross section. If $e = 0$, then $L_1 = L_0$ and equation (13) reduces to equation (12). The equations apply to any structural system that returns to regional, regardless of the kinematic model. Both equations apply to any bed, whether or not it is part of a growth sequence, and thus provides a method for calculating the depth to detachment in growth intervals.

Necking

Necking is the extensional thinning of a region from both the top and bottom (Fig. 6). Each side of the neck is separately area balanced and the total extension is also area balanced:

$$S_1 = DH_1, \quad (14)$$

$$S_2 = DH_2, \quad (15)$$

$$S_1 + S_2 = DT. \quad (16)$$

Each side of the neck will have an area–depth relationship analogous to a structure detached at the effective detachment horizon (Fig. 6). The effective detachment horizon is a surface of structural disharmony with independent deformation on opposite sides, not necessarily a displacement surface. It is not necessary to know the location of the effective detachment horizon *a priori*; area-depth values from one side of the neck will go to zero area at this level. The effective detachment horizon need not be in the middle of the extended region and the necks may be offset from one another (Okaya & Thompson 1986; Klemperer & White 1989; Kusznir & Egan 1989; Reston 1990; Kusznir & Ziegler 1992) without affecting the area balance. If the necks are offset, then the detachment horizon may be a displacement surface (Okaya & Thompson 1986; Reston 1990; Harry & Sawyer 1992) that transfers the extension from one side of the neck to the other.

The depth to the effective detachment horizon is not controlled by the area balance in a neck with vertical boundaries. This is the uniform stretching model of McKenzie (1978) for which the elevation of the beds within the neck is controlled only by isostasy and thermal expansion. Isostasy and thermal expansion should play a role in necking by the model of Fig. 6 but the effects may be limited by the geometric boundary conditions and the flexural rigidity of the crust (Buck *et al.* 1988; Kooi *et al.* 1992; Kusznir & Ziegler 1992). The possible effects of isostasy and thermal expansion on a crustal-scale graben of the style of Figs 3 & 5 will be discussed below in the context of the Rhine Graben.

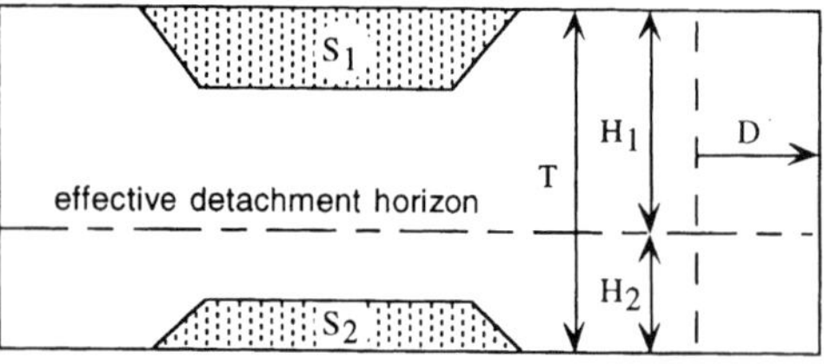

Fig. 6. Extension by necking. The lost areas (shaded) on both sides of the neck are area balanced with respect to the effective detachment horizon and the total deformation is area balanced with respect to the total thickness T (equations.(14–16)).

Depth to detachment in the central Rhine graben

The Rhine graben is a classic small-displacement crustal-scale rift. Regional overviews of the geology have been given by Illies (1977), Meier & Eisbacher (1991), Brun *et al.* (1992) and Ziegler (1992*a*). Rifting began in the Late Eocene and regional doming commenced in the mid-Late Miocene (Ziegler 1992*a*). The near-surface structure (Fig. 7) is from a very detailed cross section which Doebl & Teichmüller (1979) constructed by using surface, well-log and seismic reflection data. The Doebl & Teichmüller section is balanced and restorable. The length and area measurements reported below were made on the original Doebl & Teichmüller cross section. The base of the crust is shown as the seismic refraction Moho (Illies 1977; Ziegler 1992*a*; Rousset *et al.*, 1993). The profile is from the central area of the Rhine graben where the graben is relatively symmetrical. The graben is asymmetrical to the north and south of the profile selected here (Brun *et al.*, 1992; Bois 1993; Rousset *et al.*, 1993) and the symmetry reverses across this area. The profile selected is the only one suitable for interpretation of the four presented by Doebl & Teichmüller because it is the only profile on which the regional can be accurately determined from the same horizon preserved on both rift shoulders. The preserved contact (shown in Fig. 7) is between the Permian and older basement and the Triassic Buntsandstein. Preservation of this contact outside the graben demonstrates that footwall uplift across the border faults is minimal in this vicinity. The absence of other horizons for which the regional

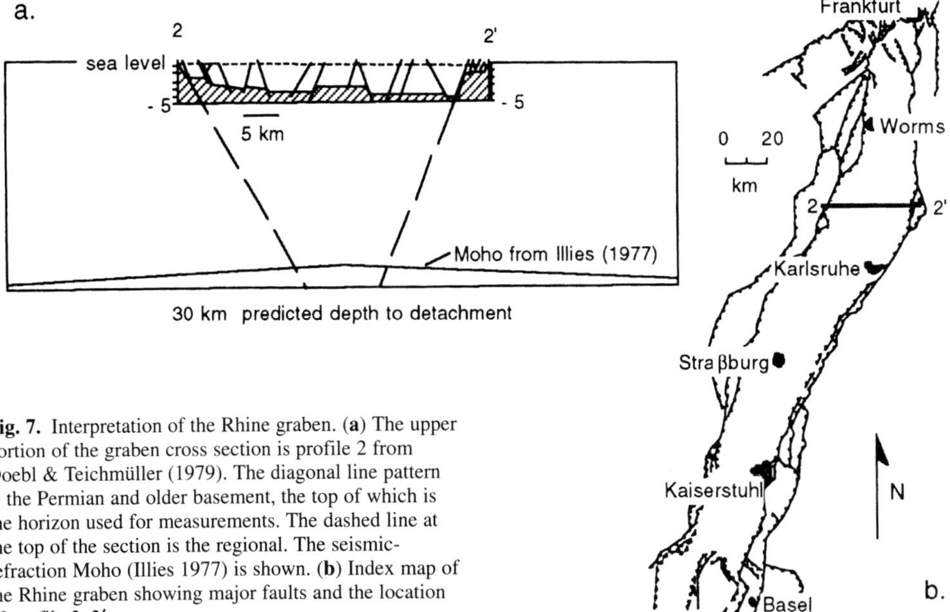

Fig. 7. Interpretation of the Rhine graben. (**a**) The upper portion of the graben cross section is profile 2 from Doebl & Teichmüller (1979). The diagonal line pattern is the Permian and older basement, the top of which is the horizon used for measurements. The dashed line at the top of the section is the regional. The seismic-refraction Moho (Illies 1977) is shown. (**b**) Index map of the Rhine graben showing major faults and the location of profile 2–2′.

can be established means that the detachment depth cannot be predicted using a lost-area diagram. The interpretation will be developed from the extension of the top of the pre-rift basement and represents the net result of all Triassic and younger deformation. The extension and lost area will be used with equation (12) to determine detachment depth. Because shallowly buried rocks are both brittle and weak, it is mechanically reasonable, although not certain, that nearly all of the layer-parallel extension at this shallow level is by faulting. The first detachment-depth calculation is based on the assumption that all the extension is by the faulting visible on the original cross section by Doebl & Teichmüller. The necessary measurements are $S = 118.7\ \text{km}^2$, $W = 42.1$ km, $L_0 = 38.3$ km, $L_1 = 40.7$ km. The displacement is 3.8 km (equation (6)) and the extension of the top-basement horizon is 6.3% (equation (7)). The detachment depth from equation 12 is 31.2 km, which matches the depth to the Moho (30–31 km) away from the Rhine graben (Illies 1977; Meier & Eisbacher 1991). The implication of this result is that the detachment for the Rhine graben formed at the base of the crust. This is reasonable based on lithospheric strength profiles (Carter & Tsenn 1987; Ranalli & Murphy 1987; Molnar 1988) which show a drastic strength minimum at the Moho and a strength maximum just below the Moho, making the Moho a likely position for a detachment.

The crust below the Rhine graben does not have the full thickness of 30–31 km, however, and so the graben is not an exact analogue to the full-graben model of Figs 3 & 5. The Moho is at a depth of 27 km directly below the cross section (Fig. 7). The detachment might be located at a depth of 27 km (compare Echtler *et al.*, 1994). This interpretation would be consistent with the cross-section data if there were an additional homogeneous and therefore unmeasured component of layer parallel strain. If the detachment depth is 27 km, then from equation (8), a requisite layer-parallel strain of 1.6% is required at the level of the top of the Permian in addition to the 6.3% extension on the visible faults. A strain of 1.6% could easily occur in a rock deformed at low temperature and just as easily be overlooked without careful thin-section scale measurements, such as of twinned-calcite strain (Groshong 1988). While this is quantitatively reasonable, displacement on a regional detachment at 27 km, just above the base of the crust, seems mechanically less likely than a detachment right at the base of the crust. This hypothesis also fails to explain the crustal thinning below the detachment.

Necking can explain simultaneous thinning at both the top and base of the crust. According to the model of Fig. 6, measurements taken above the neck should be consistent with the effective detachment level. The effective detachment level might be in the middle of the crust at a depth of, say, 15 km, the location of the strength minimum predicted by the extensional crustal model of Ranalli & Murphy (1987). From equation (8), an effective detachment at 15 km depth requires a requisite layer-parallel strain of 12.8% in addition to the visible 6.3%. A requisite strain greater than 4–6% is probably too much for crystal plastic strains in shallowly buried rocks, but is reasonable if accommodated by second-order faults. Thus necking is a viable hypothesis but requires a significant amount of additional small-scale strain.

Physical uplift of the Moho by a process linked to extension is another possible explanation. Several mechanisms have been suggested to explain the uplift of the Moho below a rift. Isostatic uplift of some form is required due to tectonic thinning. Other mechanisms could not only lift the Moho but the entire crust as well, including uplift by intrusions at the base of the crust (Illies 1977; Ziegler 1992*b*), phase transformation (Prodehl *et al.* 1992), and thermal expansion (McKenzie 1978; Kusznir & Ziegler 1992). Mechanical models of lithospheric necking indicate that extension above a detachment greater than 20 km deep should lead to regional doming of the rift (Kooi *et al.* 1992; Ziegler 1992*b*). The effect of uplift can be represented by a simple model in which the graben is formed by stretching of the entire crust and then uplifted (Fig. 8). The block geometry of the model is identical to that of the previous full graben model (Fig. 5), except that the graben, instead of remaining fixed to the lower detachment, is uplifted by the amount shown. To maintain continuity across the boundary faults, the displacement on the lower detachment must be less than for the zero-uplift model. Both the lost area and the displacement are reduced, but in proportion, such that the calculated depth to detachment turns out to be the same as for the zero-uplift model. Thus an externally supported uplift of the graben need not alter the depth to detachment calculation in the model (Fig. 8). Applied to the Rhine graben, the model suggests that the graben could have been vertically uplifted from an original full-thickness crust that was extended above a Moho detachment. The refraction Moho (Fig. 7) does not have sharp corners at the base of the graben like the model in Fig. 8, but seismic reflection profiles of the Moho (Brun *et al.* 1992; Rousset *et al.* 1993) show definite graben-boundary offsets. A version of the model with a smoothed lower detachment would resemble the mechanical models of the Rhine graben presented by Villemin *et al.* (1986) and Ricke & Mechie (1989).

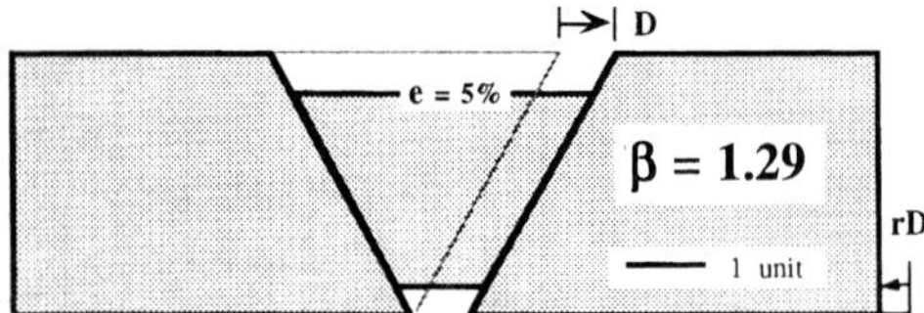

Fig. 8. Graphical area-balanced model of graben uplift. Block geometries are as in Figs 3 & 5 except that the graben is uplifted and the displacement on the lower detachment is reduced by the amount *rD*, indicated by the arrow. Model measurements are $D = 0.58$ units and $S = 1.73$ units2, giving $H = 3.0$ units, the true depth to the lower detachment from equation (12).

Taking the original thickness of the crust in the Rhine graben as 31 km and the final thickness as 24 km (27 km minus 3 km of post-Permian sediments in the graben), the β value is 1.29 (29%) from equation (10). This is much larger than the extension of the top Permian horizon (6.3%) but is not a discrepancy; it is expected from the model. The full-graben models of Figs 5 & 8 have the same β value and nearly the same *e* value as the Rhine graben. The difference between β and *e* is expected in a graben that does not have vertical sides. The total crustal extension is reduced by the uplift of the graben block from the base, even though the thinning remains the same. The crustal extension calculated from the surface width of the Rhine graben using equation (11) is $\beta = 1.10$, significantly less than the value from equation (10). In the model of Fig. 8, $W = 4.09$ units and $L_0 = 3.51$ units, giving an extension of $\beta = 1.17$ from equation (11). The different β values and the layer-parallel extension in the model (Fig. 8) are very similar to the corresponding values for the Rhine graben on the profile of Fig. 7.

Previous investigators have obtained shallower detachment depths for the Rhine graben by essentially the same method used here to obtain the depth of 31.2 km. Voight (1974) determined a depth of 24.5 km from a cross section about 20 km to the south of the line interpreted here. Near the northern end of the graben, north of Worms, much shallower depths of 16.6 and 18.3 km have been inferred by Meier & Eisbacher (1991). The differences may be due to a change in detachment level along the axis of the graben. The difference should not be the result of a change in structural style from full graben to half graben because model studies (Groshong 1994) indicate no difference is required. Brun *et al.* (1992) interpreted the master fault of the Rhine graben to extend to the Moho at both the north and south ends of the graben, and the faults in the center portion of the graben probably do so also.

Discussion

There are different ways to measure the strain in a graben system and they do not give equivalent results. Layer-parallel strain is the strain of a bed within the graben. In upper crustal and sedimentary rocks, this strain is usually heterogeneous, much of it occurring on faults. This is the component typically measured from a cross section or seismic line across a graben. The total layer-parallel strain may also include a 'sub-seismic' or homogeneous strain component, which is strain not visible at the scale of observation. Such smaller-order structures can be very important as barriers to or conduits for fluid flow. If the depth to detachment is known, either from the lost-area diagram or independently, then the strain required for area balance, the requisite strain, may be determined from equation (8) and is the missing homogeneous component. The computed requisite length changes or strains should be interpreted as the values required if the cross section is exactly as drawn. In the full graben model (Fig. 3a) the requisite strains are small at the upper levels but increase downward in the graben and near the lower detachment are potentially very large. A bed halfway between layer 5 and the detachment is extended 100% and the strain approaches infinity at the detachment for the particular model illustrated (Figs 3 & 5) in which, before deformation, the master faults joined at the detachment. The calculated requisite strain is an average value along the entire length of the horizon being measured and so true strain values both higher and lower may occur along the bed. For example, the strain may be large in a fold hinge where the bed curvature is the greatest and be zero elsewhere along the layer.

The β value from equation (10) is the whole-system extension and does not change with elevation above the detachment. Only for a vertical-sided region is β equal to the layer-parallel extension. The small near-surface strain in a full graben with sloping boundaries may explain the common observation that layer-parallel, fault-related extensions are usually substantially less than computed crustal extensions (Wood & Barton, 1983; Ziegler 1983; Badley *et al.* 1988). It has been argued that fractal models of fault distributions imply that a substantial amount of strain occurs at scales below that of the observations (e.g. Marrett & Allmendinger 1991, 1992; Walsh *et al.* 1991; Walsh & Watterson 1992) and it has also been argued that small-scale faults account for little additional strain (Scholz & Cowie 1990; Scholz *et al.* 1993) or that the strains at different scales might not represent the same strain ellipsoid (Peacock & Sanderson 1994). The validity of such models can be tested by comparison to the calculated requisite strain on cross sections for which the detachment is known

Conclusions

A plot of lost area versus depth, called a lost-area diagram, allows the depth to detachment, displacement on the lower detachment and the internal consistency of an extensional cross section to be determined. Departures from internal consistency, including growth beds, can be recognized as the failure of the points to fall on the same line. The method applies to any structure for which the lost area is equal to the displacement times the detachment depth, including full grabens, listric and ramp-flat half grabens, graben systems with footwall uplifts and combinations of these styles.

For a given cross section to be area balanced, layer-parallel extension may be required. If the extension is homogeneous, that is, not visible at the scale of the cross section, it can be calculated from the area balance and is termed the requisite strain. The requisite strain may be accommodated along faults that are too small to be resolved on the cross section but that may be important for fluid migration.

The graben-system-scale strain, β, as calculated from the change in vertical thickness of the interval of interest (i.e. the crust) is not the same as the layer-parallel extension of a horizon in the graben. The model of the full graben shows that the strain within the upper part of the graben should be much less than β value, as is commonly observed in crustal-scale grabens.

The interpretation of the Rhine graben is based on a single detailed cross section which is consistent with the interpretations that (1) there is no strain below the scale of the cross section, (2) the graben formed with its basal detachment at the base of the crust, and (3) the graben has been uplifted a few km by mechanism(s) independent of (but presumably related to) the extension that formed the graben. Necking of the crust is an equally reasonable hypothesis. Necking with a 15 km deep effective detachment requires a requisite strain of nearly 13% compared to zero requisite strain for a detachment at the base of the crust. This difference should be readily detectable in outcrop and thin-section studies.

An early version of this manuscript benefited from the helpful comments of Claus von Winterfeld and two anonymous reviewers. The clarity of the present version has been improved because of suggestions by Dennis Harry.

References

BADLEY, M. E., PRICE, J. D., RAMBECH DAHL, C & AGDESTEIN, T. 1988. The structural evolution of the northern Viking Graben and its bearing on extensional modes of basin formation. *Journal of the Geological Society, London,* **145**, 455–472.

BOIS, C. 1993. Initiation and evolution of the Oligo-Miocene rift basins of southwestern Europe: contribution of deep seismic profiling. *Tectonophysics*, **226**, 227–252.

BRUN, J. P., GUTSCHER, M. A. & DEKORP-ECORS team. 1992. Deep crustal structure of the Rhine Graben from DEKORP-ECORS seismic reflection data: a summary. *Tectonophysics*, **208**, 139–147.

BUCK, W. R., MARTINEZ, F., STECKLER, M. S. & COCHRAN, J. R. 1988. Thermal consequences of lithospheric extension: pure and simple. *Tectonics*, **7**, 213–234.

CARTER, N. L. & TSENN, M. C. 1987. Flow properties of continental lithosphere. *Tectonophysics*, **136**, 27–63.

CHAMBERLIN, R. T. 1910. The Appalachian folds of central Pennsylvania. *Journal of Geology*, **18**, 228–251.

DAHLSTROM, C. D. A. 1969. Balanced cross sections. *Canadian Journal of Earth Sciences,* **6**, 743–757.

DOEBL, F. & TEICHMÜLLER, R. 1979. Zur Geologie und heutigen Geothermik im mittleren Oberrhein-Graben. *Fortschritte Geologie von Rheinland und Westfalen*, **27**, 1–17.

ECHTLER, H. P., LÜSCHEN, E. & MAYER, G. 1994. Lower crustal thinning in the Rhinegraben: Implications for recent rifting. *Tectonics*, **13**, 342–353.

EPARD, J.-L. & GROSHONG, R. H., Jr. 1993. Excess area and depth to detachment. *American Association of Petroleum Geologists Bulletin*, **77**, 1291–1302.

GIBBS, A. D. 1983. Balanced cross section construction from seismic sections in areas of extensional tectonics. *Journal of Structural Geology*, **5**, 153–160.

GROSHONG, R. H., Jr. 1988. Low-temperature deformation mechanisms and their interpretation. *Geological Society of America Bulletin*, **100**, 1329–1360.

—— 1994. Area balance, depth to detachment and strain in extension. *Tectonics*, **13**, 1488–1497.

—— & EPARD, J.-L. 1994. The role of strain in area-constant detachment folding. *Journal of Structural Geology*, **16**, 613–618.

HANSEN, W. R. 1965. Effects of the earthquake of March 27, 1964, at Anchorage. Alaska. *US Geological Survey Professional Paper,* **542-A**, 1–68.

HARRY, D. L. & SAWYER, D. S. 1992, A dynamic model of extension in the Baltimore Canyon trough region. *Tectonics*, **11**, 420–436.

ILLIES, J. H. 1977. Ancient and recent rifting in the Rhinegraben. *Geologie en Mijnbouw,* **56**, 329–350.

KLEMPERER, S. L. & WHITE, N. 1989. Coaxial stretching or lithospheric simple shear in the North Sea? Evidence from deep seismic profiling and subsidence. *In*: TANKARD, A. J. & BALKWILL, H. R. (eds) *Extensional Tectonics and Stratigraphy of the North Atlantic Margins.* American Association of Petroleum Geologists Memoir **46**, 511–522.

KOOI, H., CLOETINGH, S. & BURRIS, J. 1992. Lithospheric necking and regional isostasy at extensional basins 1. Subsidence and gravity modelling with an application to the Gulf of Lions margin (SE France). *Journal of Geophysical Research*, **97**, 17553–17571.

KUSZNIR, N. J. & EGAN, S. S. 1989. Simple-shear and pure shear models of extensional sedimentary basin formation: Application to the Jeanne d'Arc Basin, Grand Banks of Newfoundland. *In*: TANKARD, A. J. & BALKWILL, H. R. (eds) *Extensional Tectonics and Stratigraphy of the North Atlantic Margins.* American Association of Petroleum Geologists Memoir **46**, 305–322.

KUSZNIR, N. J. & ZIEGLER, P. A. 1992. The mechanics of continental extension and sedimentary basin formation: A simple shear/pure shear flexural cantilever model. *Tectonophysics*, **215,** 117–131.

MARRETT, R. & ALLMENDINGER, R. W. 1991. Estimates of strain due to brittle faulting: sampling of fault populations. *Journal of Structural Geology*, **13**, 735–738.

—— & —— 1992. Amount of extension on 'small' faults: An example from the Viking graben. *Geology*, **20**, 47–50.

MCCLAY, K. R. 1992. Glossary of thrust tectonics terms. *In*: MCCLAY, K. R. (ed.) *Thrust Tectonics*, Chapman & Hall, London, 419–433.

MCKENZIE, D. 1978. Some remarks on the development of sedimentary basins. *Earth and Planetary Science Let*ters, **40,** 25–32.

MEIER, L. & EISBACHER, G. H. 1991. Crustal kinematics and deep structure of the northern Rhine graben, Germany. *Tectonics,* **10**, 621–630.

MOLNAR, P. 1988. Continental tectonics in the aftermath of plate tectonics. *Nature*, **335,** 131–137.

OKAYA, D. A. & THOMPSON, G. A. 1986. Involvement of deep crust in extension of Basin and Range province. *Geological Society of America Special Paper* **208**, 15–22.

PEACOCK, D. C. P. & SANDERSON, D. J. 1994. Strain and scaling of faults in the chalk at Flamborough Head, U. K. *Journal of Structural Geology*, **16**, 97–107.

PRODEHL, C., MUELLER, St., GLAHN, A., GUTSCHER, M. & HAAK, V. 1992. Lithospheric cross sections of the European Cenozoïc rift system. *Tectonophysics*, **208**, 113–138.

RANALLI, G. & MURPHY, D. C. 1987, Rheological stratification of the lithosphere. *Tectonophysics*, **132**, 281–295.

RESTON, T. J. 1990. The lower crust and the extension of the continental lithosphere: kinematic analysis of BIRPS deep seismic data. *Tectonics*, **9**, 1235–1248.

RICKE, M. & MECHIE, J. 1989. Finite element modelling of a continental rift structure (Rhinegraben) with a large-deformations algorithm. *Tectonophysics*, **165**, 81–91.

ROUSSET, D., BAYER, R., GUILLON, D. & EDEL, B. 1993. Structure of the southern Rhine Graben from gravity and reflection seismic data (ECORS-DEKORP program). *Tectonophysics*, **221**, 135–153.

SCHOLZ, C. H. & COWIE, P. A. 1990. Determination of total strain from faulting using slip measurements. *Nature*, **346**, 837–838.

——, DAWERS, N. H., YU, J.-Z., ANDERS, M. H. & COWIE, P. A. 1993. Fault growth and fault scaling laws: preliminary results. *Journal of Geophysical Research*, **98**, 21,951–21,961.

VILLEMIN, T., ALVAREZ, F. & ANGELIER, J. 1986. The Rhinegraben: extension, subsidence and shoulder uplift. *Tectonophysics*, **128**, 47–59.

VOIGHT, B. 1974. Thin-skinned graben, plastic wedges, and deformable-plate tectonics. *In*: ILLIES, J. H. & FUCHS, K. (eds) *Approaches to Taphrogenesis, Inter-Union Commission on Geodynamics Scientific Report No. 8*, E. Schweizerbart'sche Verlagsbuchhandlung, Stuttgart, 396–419.

WALSH, J. J. & WATTERSON, J. 1992. Populations of faults and fault displacements and their effects on estimates of fault-related regional extension. *Journal of Structural Geology*, **14**, 701–712.

——, —— & YIELDING, G. 1991. The importance of small-scale faulting in regional extension. *Nature*, **351**, 391–393.

WHITE, N. J., JACKSON, J. A. & MCKENZIE, D. P. 1986. The relationship between the geometry of normal faults and that of the sedimentary layers in their hanging walls. *Journal of Structural Geology*, **8**, 897–909.

WOOD, R. & BARTON, P. 1983. Crustal thinning and subsidence in the North Sea. *Nature*, **304**, 134–136.

ZIEGLER, P. A. 1983. Crustal thinning and subsidence in the North Sea; matters arising. *Nature*, **304**, 561.

—— 1992*a*. European Cenozoic rift system. *Tectonophysics*, **208**, 91–111.

—— 1992*b*. Geodynamics of rifting and implications for hydrocarbon habitat. *Tectonophysics*, **215**, 221–253.

A critique of techniques for modelling normal-fault and rollover geometries

T. A. HAUGE & G. G. GRAY

Exxon Production Research Company, PO Box 2189, Houston, TX 77252-2189, USA

Abstract: A review of the published literature indicates that antithetic oblique shear, at an angle of 30 ± 10° measured from the vertical, is the most appropriate algorithm for quantitative modelling of geometric relationships between downward-flattening normal faults and the strata in their hanging walls. This conclusion is based on 2D analysis of 11 natural examples (listric normal faults without apparent salt involvement, shale mobility, or footwall deformation) and selected laboratory physical models. It may not apply where mobile salt, mobile shale, or significant footwall tilting or deformation is present. Antithetic oblique shear, like all modelling algorithms, does not accurately simulate natural deformation processes, which are significantly more complex.

Exploration for and production of hydrocarbons in extensional settings requires accurate interpretation of faults and the strata they offset. This paper addresses the problem of how to select a modelling algorithm[1] for use in quantitative structural analysis of a normal fault and the strata in its hanging wall. In particular, this paper is concerned with downward-flattening (listric) normal faults, a common normal-fault geometry. Analysis of this sort requires that cross sections are in the plane of movement for the fault being examined, that compaction is accounted for, and that the footwall remains essentially undeformed. For examples cited in this paper, which are drawn from the published literature, we assume that this is the case (e.g. White *et al.* 1986, Xiao and Suppe 1989; White 1992).

Many normal faults are non-planar. Consequently, strata in their hanging walls typically deform during displacement (Fig. 1). The geometry of this deformation is controlled by the depositional geometry of the strata, the geometry of the fault, compaction of the strata, and the deformation mechanism[1] of the hanging-wall strata (see, for example, White, 1992; Withjack & Peterson 1993). Validation of interpretations of normal faults and the strata in their hanging walls typically includes testing whether the hanging-wall strata are geometrically compatible with the fault geometry. Despite the fact that natural deformation can be considerably more complex than the modelling algorithms, simple mechanisms are routinely employed in quantitative structural analysis because they are relatively easy to implement, and they conserve the area of the section in the restored state. The choice of the most reasonable modelling algorithm is important, because different modelling algorithms produce markedly different results (Figs 2 & 3). The problem addressed in this paper is how to choose the best (although admittedly oversimplified) modelling algorithm with which to address such problems.

[1] For clarity, the term 'deformation mechanism' is used to describe the actual deformation of rocks (or, rather, our incomplete understanding of that deformation), and the term 'modelling algorithm' is used to describe techniques used in quantitative structural analysis to simulate that deformation.

Quantitative analysis of normal faults and their hanging walls

Choice of oblique shear

Shear oblique to bedding is becoming the generally preferred modelling algorithm for listric normal fault structures (e.g. White *et al.* 1986; Xiao & Suppe 1989; Groshong 1990; Dula 1991; Roberts *et al.* 1991; White 1992; Xiao & Suppe 1992; Withjack & Peterson 1993) although this opinion is not unanimous (Williams & Vann 1987; Keller 1990; Dula 1991). Among the proponents of oblique shear, there remains uncertainty as to the most appropriate angle of shear to be applied, in part because published analyses of physical models and natural examples (White *et al.* 1986; Dula 1991; White 1992; Xiao & Suppe 1992; Kerr & White 1993) suggest a broad range of acceptable shear angles.

Data selection and analysis

The published studies cited in the previous paragraph evaluated modelling algorithms by applying

From Buchanan, P. G. & Nieuwland, D. A. (eds), 1996, *Modern Developments in Structural Interpretation, Validation and Modelling*, Geological Society Special Publication No. 99, pp. 89–97.

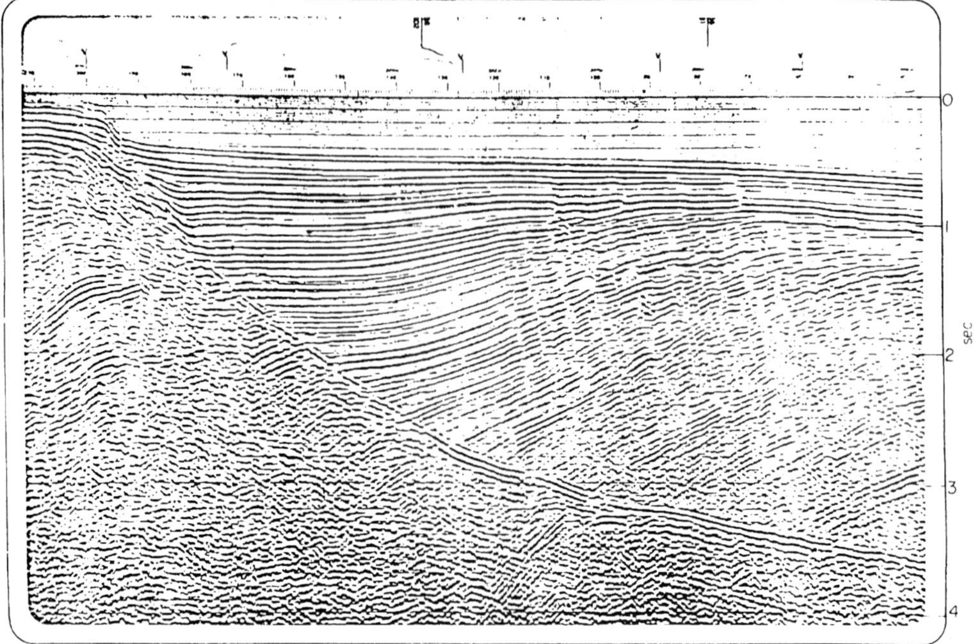

Fig. 1. Seismic example of listric normal fault and folded hanging-wall beds, from the Gulf of Oman (Reprinted from Wernicke & Burchfiel 1982 with kind permission of Elsevier Science Ltd).

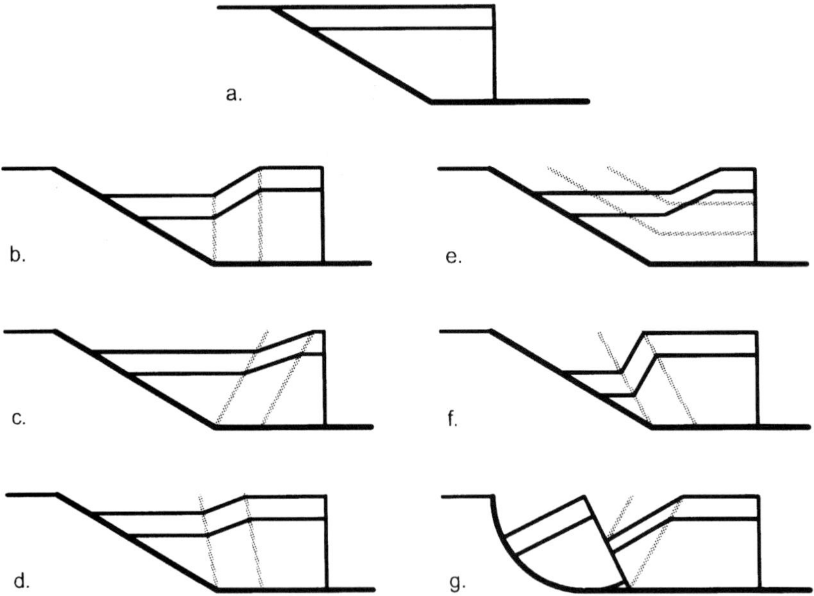

Fig. 2. Deformation mechanisms commonly used for modelling geologic structures. All except rigid-body rotation (**g**) employ simple shear, but they vary according to the orientation of the shear planes. Fault slip measured along the horizontal segment of the faults is equal in all examples, and the dip of the ramp (except for g) is 30°: (**a**) starting configuration for **b-f**; (**b**) vertical shear; (**c**) antithetic oblique shear; (**d**) synthetic oblique shear; (**e**) fault parallel shear (constant slip); (**f**) bedding-parallel shear (flexural slip); (**g**) rigid body rotation with adjacent antithetic oblique shear. In oblique and vertical shear, shear surfaces are planar and maintain their orientation with respect to the vertical during (retro)-deformation. In bedding-parallel shear and fault-parallel shear, shear surfaces are curviplanar; in bedding-parallel shear the shear surfaces change shape during (retro)-deformation. Stippled lines are representative shear surfaces, except in (**f**), where they are axial surfaces to symmetric flexural-slip folds.

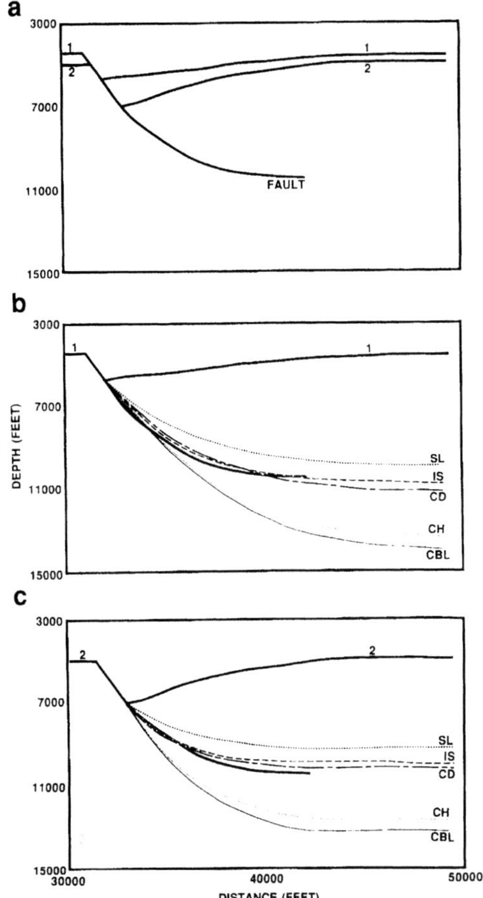

Fig. 3. Results of fault-plane prediction, based on a depth section of an interpreted onshore growth fault from the Gulf of Mexico area (Dula 1991). (**a**) Depth section. (**b**) Model fault geometries constructed from the shape of marker bed 1. (**c**) Model fault geometries constructed from the shape of marker bed 2. SL, fault-parallel shear; IS, inclined shear with 20° antithetic shear, measured from vertical; CD, constant displacement; CH, vertical shear; CBL, flexural slip. Note that the best-fit solution was provided by antithetic oblique shear or constant displacement modelling algorithms.

them to natural examples (outcrops and seismic profiles) and physical models. When inappropriately scaled modelling experiments are deleted from this group of examples, the resulting data set consists of eleven natural examples and five model examples of listric normal faults. The basis for our selection of appropriate data is discussed in Appendix 1.

The best-fit shear angles for the chosen examples of listric normal faults are shown in Fig. 4. The natural examples include basement-involved and basement-detached extensional structures, and most involve rocks of Tertiary age. The best-fit shear angles have been determined through a variety of methods, which are described in Appendix 2. These data suggest a 'rule of thumb': an antithetic shear angle of 30° ± 10° should provide a reasonable approximation for modelling gross geometric relationships between listric normal faults and the strata in their hanging walls. This rule of thumb would be useful in the absence of explicit models of individual examples, such as may be provided by inversion techniques (Kerr & White 1992). These data indicate no correlation between shear angle and average dip of the fault, a correlation which has been suggested (Kerr & White 1992) based largely on studies of laboratory experiments (see Appendix 1).

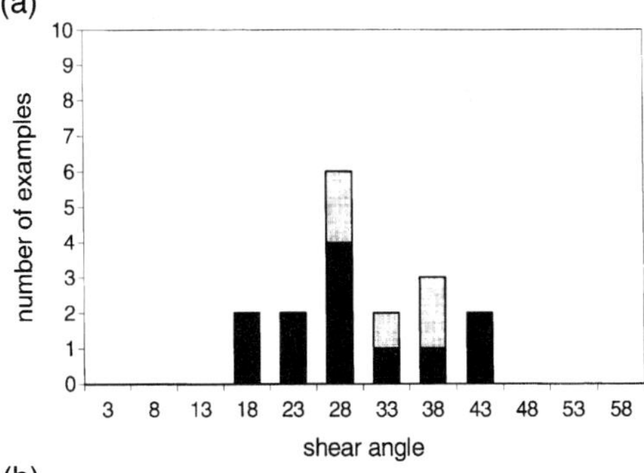

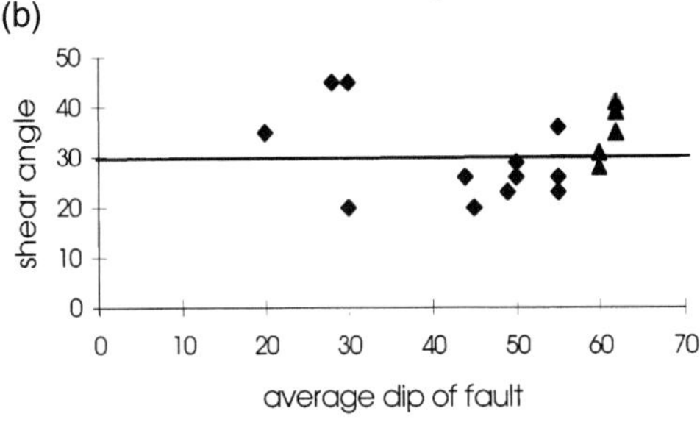

Fig. 4. (**a**) Histogram of best-fit oblique-shear angles (measured from the vertical) for twelve natural examples (black) and five experimental examples (grey) of listric normal faults. Note the median value of 26–30°.
(**b**) Plot of the best-fit angle of shear versus average dip of faults, after Kerr & White 1992 (natural examples are diamonds; laboratory experiments are triangles). 'Average dip' refers to the average of the near-surface fault dip and the dip of the detachment at depth (see Kerr & White 1992). The laboratory data are from experiments which have not used a constant-slip boundary condition (see Appendix 1). Shear angle data from Dula (1991), Kerr & White (1992, 1993), White (1992) and Xiao & Suppe (1992).

Extensional deformation observed in nature

Observations of natural deformation

Several deformation mechanisms have been observed to accommodate extensional deformation of the hanging walls of listric (downward-flattening) normal faults, including bedding parallel slip (flexural slip), antithetic normal faulting, synthetic normal faulting, and apparent rigid-body rotation[2]. Seismic reflection profiles (Figs 5 & 6; e.g. Wernicke & Burchfiel 1982; White *et al.* 1986; Dula 1991; Xiao & Suppe 1992; White 1992) and outcrops (Fig. 7; Higgs *et al.* 1991) indicate that, in nature, deformation is typically accommodated by more than one of these mechanisms. In our experience, antithetic oblique shear and apparent rigid-body rotation are the most commonly observed natural deformation mechanisms in the hanging walls of listric normal faults (Fig. 6). Less commonly, synthetic oblique shear is the dominant observable deformation mechanism (Fig. 5).

[2] 'Rigid-body rotation' implies no internal deformation of a fault-bounded mass of rock. However, this condition typically is assumed, rather than documented, especially on seismic sections. Bedding-parallel slip may be difficult to detect either in outcrop or on seismic data, and bedding-oblique faulting at a subseismic scale would be undetected on seismic data. As a result, in this paper the term is qualified as 'apparent.'

Natural deformation compared with modelling algorithms

Although natural deformation appears to be more kinematically complex than the modelling algorithms used in most restorations, some authors have raised the possibility of using the shear angles derived from restorations to infer which mechanisms may have been dominant in the extensional system during deformation (e.g. Rowan & Kligfield 1989; Dula 1991). If this relationship were true, it might provide valuable input for studies of the finer-scale deformation of normal fault hanging

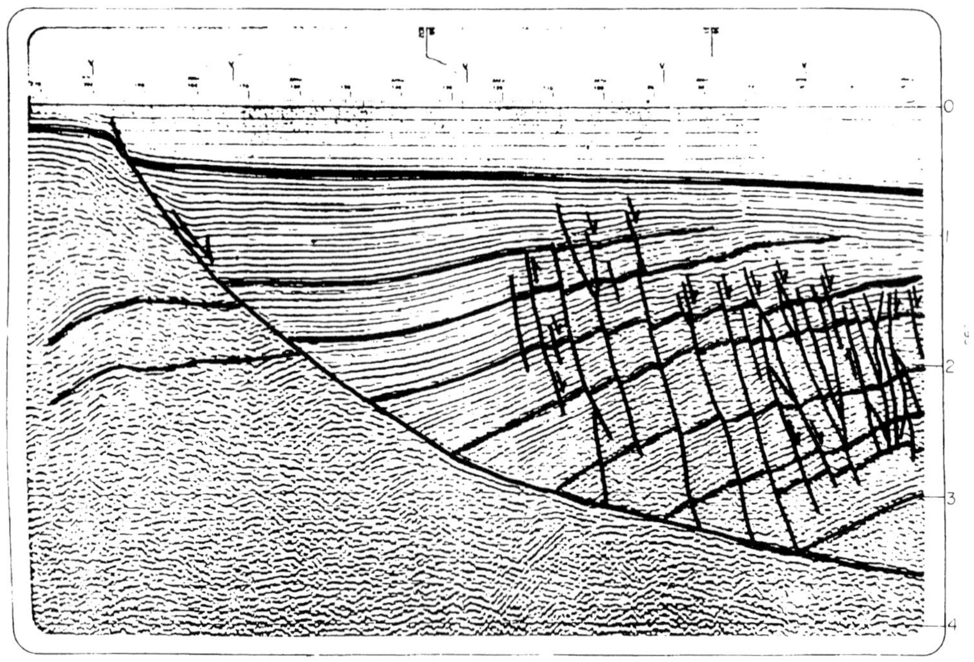

Fig. 5. Interpreted version of Fig. 1 (Wernicke & Burchfiel 1982). Extensional fault-bend folding of the hanging wall in this example appears to have been accommodated by synthetic oblique shear, indicated by numerous secondary synthetic normal faults interpreted by Wernicke & Burchfiel. The slivers of rock between these synthetic faults appear to have tilted domino-style during slip on the synthetic faults. Hanging-wall beds dip into the master fault but are apparently undeformed. Rigid-body rotation, bedding-parallel slip and oblique antithetic shear may have accommodated the tilting of these beds. Reprinted with kind permission of Elsevier Science Ltd.

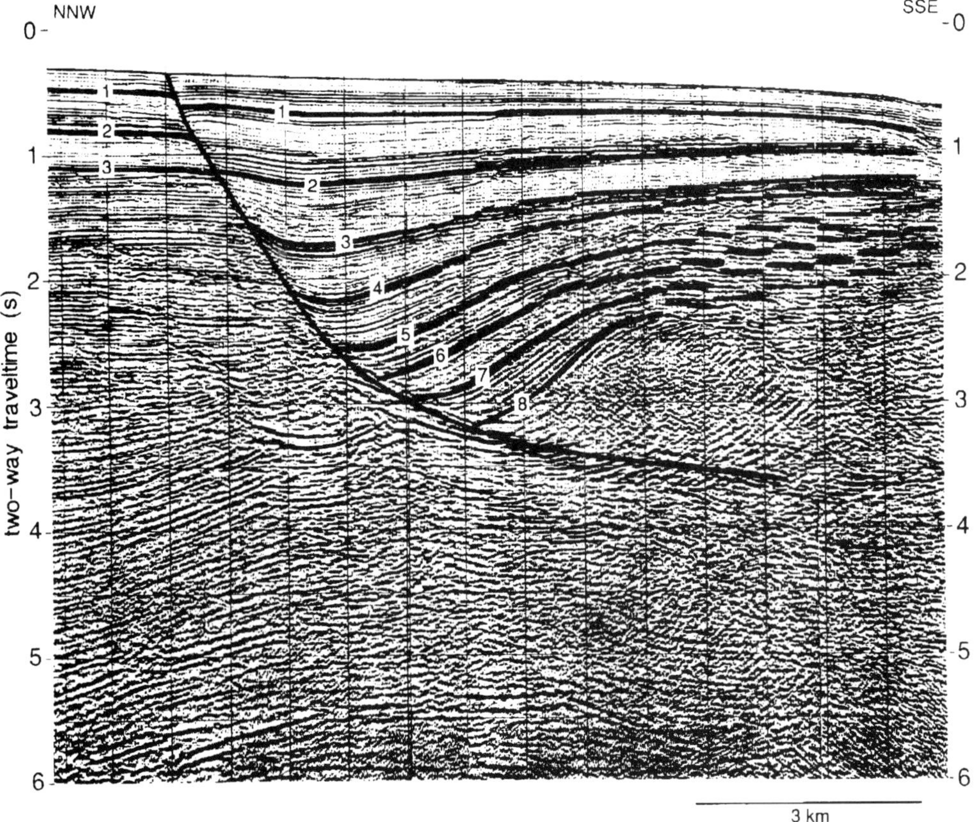

Fig. 6. Interpreted seismic reflection image of listric normal fault from the Gulf of Mexico, from Xiao & Suppe (1992). Extensional fault-bend folding of the hanging wall, which is required by the listric geometry of the fault, appears to have been accommodated by antithetic oblique shear, indicated by numerous secondary antithetic fault offsets interpreted by Xiao & Suppe. Several secondary synthetic fault offsets are also interpreted, such as in the area of greatest curvature of the hanging-wall beds, where a keystone graben has developed. Rigid-body rotation appears to have occurred in the region of labels 6 and 7. Note that folding of hanging-wall beds during faulting accommodated significant expansion of strata into the fault. Reproduced with permission of AAPG.

walls, and for the attempts to use self-similarity principles to predict reservoir compartmentalization. White (1992) has determined the best-fit shear angle for a natural example published by Wernicke & Burchfiel (1982), shown in Figs 1 (uninterpreted) and 5 (interpreted). White's results are shown in Fig. 8. In this example, antithetic shear 34° from the vertical, in conjunction with 50% compaction, predicted the fault shape to a very good approximation. In contrast, observed folding of the hanging wall was accommodated by faults with synthetic slip. This result suggests that a shear angle determined through inversion or restoration techniques does not necessarily correspond to the dominant mechanism operating within a deforming hanging wall. Instead, the best-fit angle of shear may simply approximate the overall lengthening of the hanging wall that occurs as part of the deformation process, rather than revealing the kinematics of natural deformational processes.

Conclusions

Several recommendations for modelling normal-fault and rollover geometries arise from the data discussed here.

- Modelling of hanging-wall deformation associated with naturally-occurring, downward-flattening listric normal faults is likely to be well approximated by a shear angle of 30° ± 10° antithetic, measured from the vertical. The shear angle used for modeling is independent of the dip of the fault.

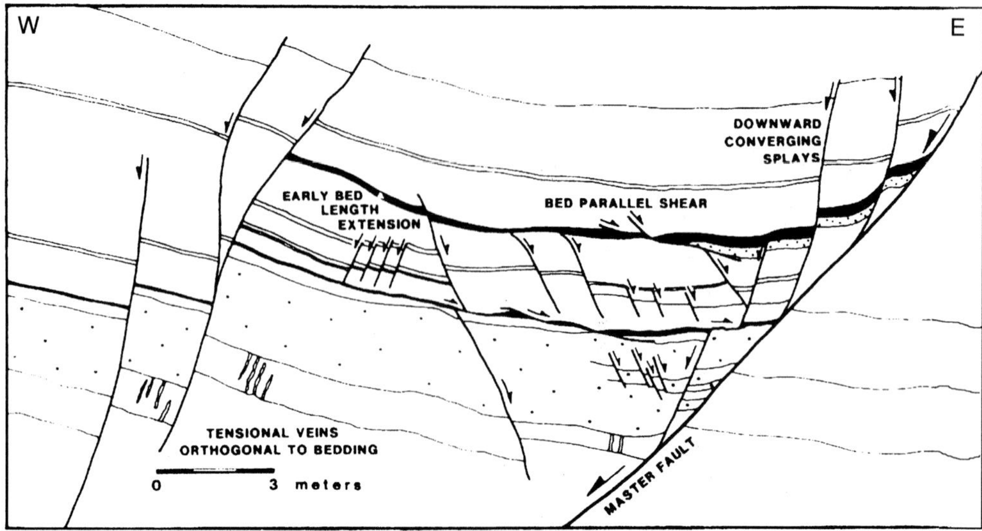

Fig. 7. Drawing of an outcrop of a normal fault and its hanging wall, from Higgs *et al.* (1991). The hanging-wall beds are tilted and warped, whereas the footwall beds are gently dipping. The hanging-wall beds presumably deformed as a result of fault-bend folding in response to slip along the master fault, although the geometry of the master fault at depth is not known. The mechanism of tilting of the footwall beds is not known. The dominant deformation mechanism in this example appears to be synthetic secondary faulting and domino-style tilting, accompanied by minor synthetic and bedding-parallel faulting. Reproduced by permission of GSA.

- This modelling approach is intended for application to a single master normal fault and its hanging wall, in the absence of mobile salt, mobile shale, and footwall deformation. This rule may be relaxed for palinspastic restorations of multi-fault systems.
- There appears to be no direct correspondence between modelling algorithms and natural deformation mechanisms of extensional structures.

The authors appreciate the thoughtful reviews of Peter Buchanan, Jay Jackson, Steve Lingrey, Steve May, John Sumner, Al Tuminas and Scott Wilkerson. Many of the ideas presented here are the outgrowth of conversations with Nicky White, and his critiques are appreciated. The authors would also like to thank Cliff Ando, Rick Gottschalk and Chris Hedlund, with whom these and other ideas related to fault-related folding have been discussed.

Appendix 1: Limitations of physical modelling studies of normal faulting

Certain physical modelling techniques are inappropriate for quantitative analysis of the type discussed in this paper. Specifically, a nonstretching flexible plastic sheet (e.g. mylar) at the base of the hanging wall (Fig. A1) imposes a constant-slip constraint on hanging-wall deformation, inhibiting extension of the hanging wall. In areas of downward-steepening faults, this typically causes contractional structures to form (Fig. A2). To our knowledge, such contractional structures are not observed in nature. Many published physical modelling experiments employed this technique (e.g. McClay & Ellis, 1987*a*, 1987*b*; Ellis & McClay 1988; Kerr & White 1992; Kerr *et al.* 1993). The reader is referred to McClay & Ellis (1987*b*) and Ellis & McClay (1988) for a description of various physical modelling techniques.

Because of the unnatural artifacts contained in these models, quantitative analysis of fault–fold relationships based on these examples produces misleading results. This is evident in the work of Kerr & White (1992), who calculate best-fit shear angles for both model experiments and natural examples of listric normal faults. Their model examples include many which were created using a plastic sheet at the base of the hanging wall. Their study infers a relationship between the average dip of a fault surface and the resulting best-fit angle of shear of its hanging wall (Fig. A3). Natural examples alone, and experimental examples not using a plastic sheet, suggest no systematic relationship between shear angle and the average dip of a fault (Fig. 5).

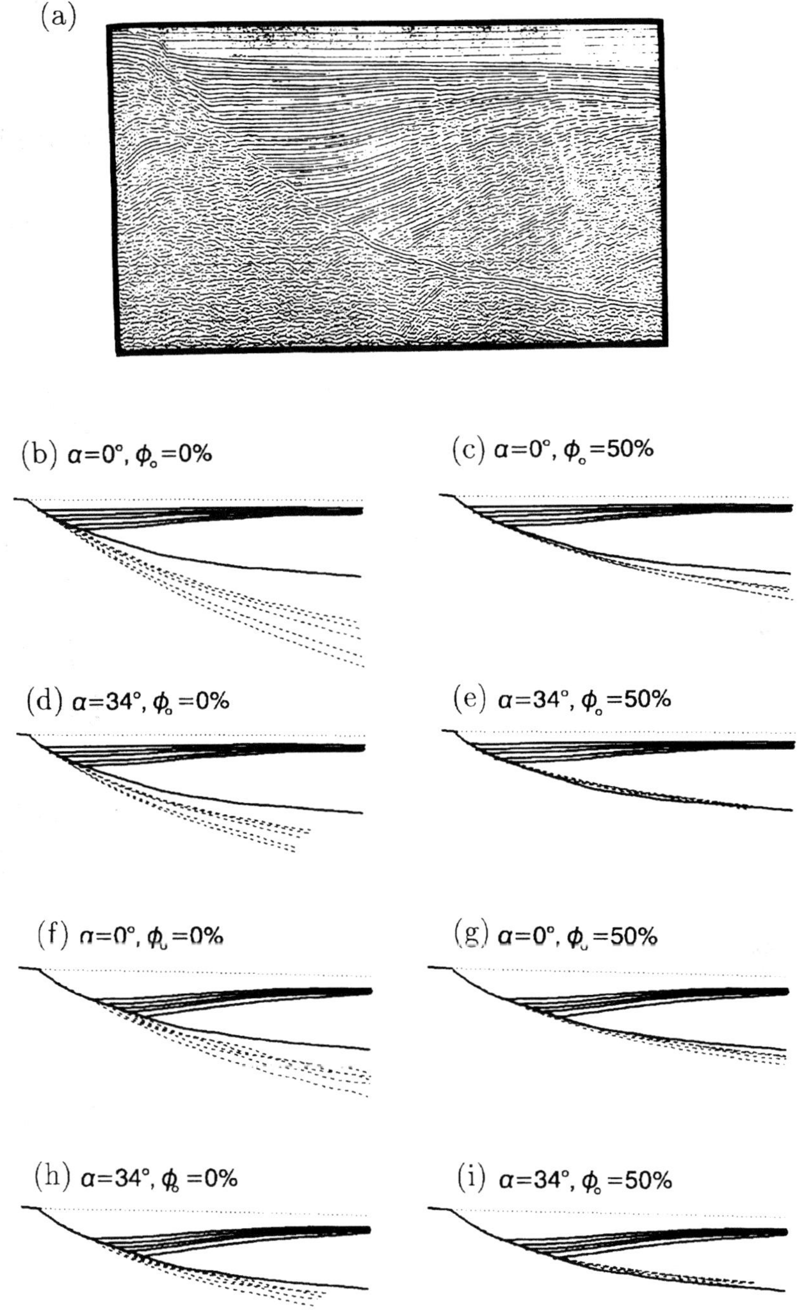

Fig. 8. Inversion of fault geometry from depth-conversion of hanging-wall beds shown in Figs 1 & 5. This analysis shows the observed (solid line) and predicted (dotted lines) fault trajectories for five shallow hanging-wall beds (**b–e**) and five deeper hanging-wall beds (**f–i**). Note that a shear angle of 34° antithetic and an initial porosity of 50% resulted in both convergence of the solutions of the five modelled beds and correspondence of the model solution with the observed fault trajectory. Note how closely the antithetic shear angle predicts the fault trajectory, whereas the dominant macroscopic deformation mechanism from the seismic interpretation (Fig. 5) is rigid-body rotation and synthetic shear. From White (1992).

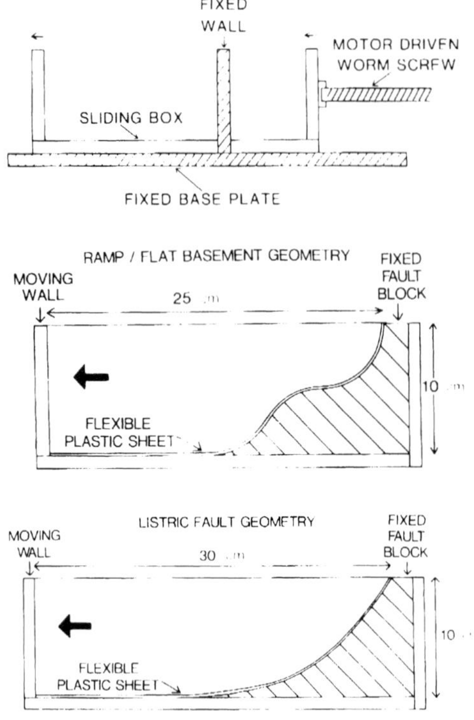

Fig. A1. Modelling apparatus employing flexible plastic sheet. The plastic sheet is attached to the moving wall of the model and is dragged across the precut fault surface of the rigid footwall. From Ellis & McClay (1988). Reproduced by permission of Blackwell Science.

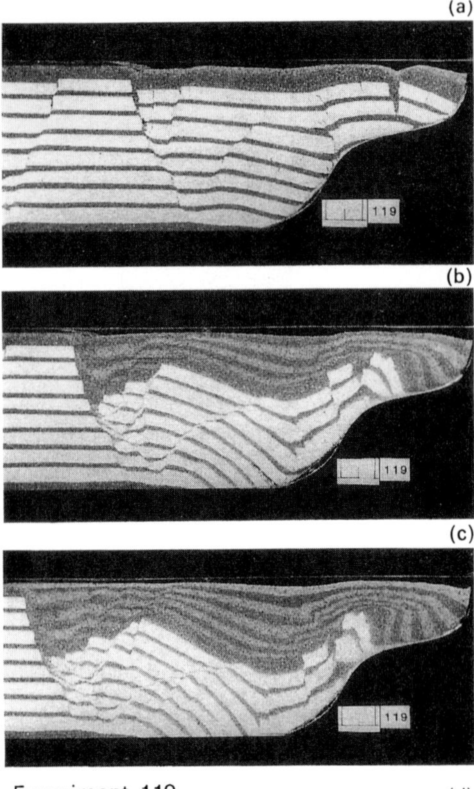

Fig. A2. Physical model employing a flexible plastic sheet at the base of the hanging wall, from Ellis & McClay (1988). The footwall of the fault model is a rigid block which defines downward-steepening and downward-flattening bends in the fault surface. The plastic sheet, which is attached to and moves with the hanging wall, imposes a constant-slip constraint on the hanging wall. Note the reverse faults that formed in the area of the downward-steepening bend of the fault. The plastic sheet may be thought of as a neutral surface, above which extension occurs where the fault is concave upward, contraction occurs where the fault is concave downward, and no extension occurs where the fault is planar. In contrast, the hanging walls of normal faults in nature collapse extensionally under their own weight, thereby conforming to the shape of the underlying master fault without the formation of secondary reverse faults. Reproduced by permission of Blackwell Science.

Appendix 2: Techniques for explicit determination of best-fit shear angles for modelling of normal-fault structures

Two types of techniques for determining best-fit shear angles for modelling hanging-wall deformation above listric normal faults have been described.

The inversion technique of White (1992) is particularly robust, as it presumes no knowledge of the geometry of the fault beyond the point at which it tips out upward. For each of several hanging-wall beds, the technique makes calculations of fault trajectory for a prescribed range of values of alpha (shear angle) and phi (initial porosity). It then seeks the values of alpha and phi for which the solutions for all the modelled beds are most similar. Applying this technique to examples for which the fault trajectory is known, White has found that the solution upon which calculations for the beds converge corresponds with the known fault trajectory. (See Kerr and White, this volume.)

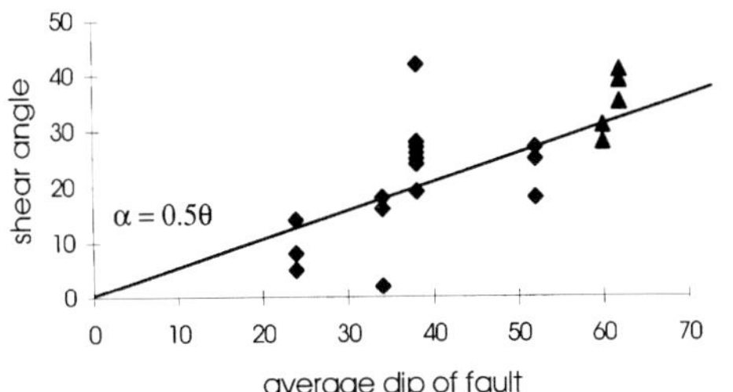

Fig. A3. Plot of the best fit angle of shear (measured from the vertical) versus average dip of faults for 20 sandbox experiments. Fifteen of these experiments were constrained to constant-slip by a basal plastic sheet (diamonds), whereas five were not so constrained (triangles). These data indicate a relationship between shear angle and average fault dip. Plotted on this graph is a line of $\alpha = 0.5\theta$ average fault dip, an expected geometric consequence of using the plastic sheet in the experiment. Data from Kerr & White (1992, this volume).

The techniques of Xiao & Suppe (1992) and Dula (1991) are in most applications less robust than that of White, because they require a degree of knowledge of the shape of the fault. Dula (1991) inferred shear angle from observation of the orientation of antithetic shears or by iteratively maximizing agreement between fault shapes modelled for two horizons (calculated individually for each horizon) and the observed, near surface segment of the fault. The technique of Xiao & Suppe (1992) calculates the shear angle from interpreted values of axial surface orientation, average dip of folded beds, and footwall cutoff angle.

References

DULA, W. F., Jr. 1991. Geometric models of listric normal faults and rollover folds. *American Association of Petroleum Geologists Bulletin*, **75**, 1609–1625.

ELLIS, P. G. & MCCLAY, K. R., 1988, Listric extensional fault systems – results of analogue model experiments: *Basin Research*, **1**, 55–70.

GROSHONG, R. H. 1990. Unique determination of normal fault shape from hanging-wall bed geometry in detached half grabens. *Eclogae Geologicae Helvetiae*, **83**, 455–471.

HIGGS, W. H., WILLIAMS, G. D. & POWELL, C. M., 1991, Evidence for flexural shear folding associated with extensional faults. *Geological Society of America Bulletin*, **103**, 710–717.

KELLER, P. 1990. Geometric and kinematic model of bed length balanced graben structures. *Eclogae Geologicae Helvetiae*,. **83**, 473–493.

KERR, H. G. & WHITE, N. 1992. Laboratory testing of an automatic method for determining normal fault geometry at depth. *Journal of Structural Geology*, **14**, 873–885.

—— & —— 1993. The application of an automatic method for determining normal fault geometries. *Journal of Structural Geology*, **14**, 7:873–885.

—— & —— 1996. Kinematic modelling of normal fault geometries using inverse theory. *This volume.*

——, ——, & BRUN, J. 1993. An automatic method for determining three-dimensional normal fault geometries. *Journal of Geophysical Research*, **98**, 17837-17857.

MCCLAY, K. R. & ELLIS, P. G. 1987*a*. Geometries of extensional fault systems developed in model experiments. *Geology*, **15**, 341–344.

—— & ELLIS, P. G. 1987*b*. Analogue models of extensional fault geometries. *In*: COWARD, M. P., DEWEY, J. F. & HANCOCK, P. L. (eds) *Continental Extensional Tectonics.* Geological Society, London, Special Publications, **28**, 109–125.

ROBERTS, A. G., YIELDING, G. & FREEMAN, B. (eds) 1991. *The Geometry of Normal Faults.* Geological Society, London, Special Publications, **56**.

ROWAN, M. G. & KLIGFIELD, R., 1989, Cross-section restoration and balancing as aid to seismic interpretation in extensional terranes: *Bull. Am. Assoc. Petrol. Geol.* **73**, 955–966.

WERNICKE, B. P. & BURCHFIEL, B. C. 1982. Modes of extensional tectonics: *Journal of Structural Geology*, **4**, 105–115.

WHITE, N. 1992. A method for automatically determining normal fault geometry at depth. *Journal of Geophysical Research*, **97**, 1715–1733.

——, JACKSON, J. A. & MCKENZIE, D. P. 1986. The relationship between the geometry of normal faults and that of the sedimentary layers in their hanging walls. *Journal of Structural Geology*, **8**, 897–909.

WILLIAMS, G. & VANN, I. 1987, The geometry of listric normal faults and deformation in their hanging walls. *Journal of Structural Geology*, **9**, 789–795.

WITHJACK, M. O. & PETERSON, E. T. 1993. Prediction of normal fault geometries – a sensitivity analysis. American Association of Petroleum Geologists Bulletin, **77**, 1860–1873.

XIAO, H. & SUPPE, J. 1989. Role of compaction in listric shape of growth normal faults. American *Association of Petroleum Geologists Bulletin*, **73**, 777–786.

—— & —— 1992. Origin of rollover. American Association of Petroleum Geologists Bulletin, **76**, 509–529.

A strategy for palinspastic restoration of inverted basins: thermal and structural analyses in SE Australia

KEVIN C. HILL & GARETH T. COOPER

VIEPS School of Earth Sciences, La Trobe University, Melbourne, Australia 3083

Abstract: Constructing balanced sections across inverted basins is difficult because of the different directions of tectonic transport during extension and compression and the imprecise estimates of the magnitude and timing of inversion events. Sections oriented within 25° of all transport directions can be restored and result in ≤10% error in the amount of extension-compression. Extensional faults are commonly oblique to the transport direction, so the inferred extension direction may be verified by dip analysis of the pre-rift section. Where oblique extensional faults are reactivated during compression, the compressional vector is best determined from the orientation and magnitude of large, curved inversion anticlines. For strongly inverted and eroded anticlines, where T_{max} is not at the present day, palaeotemperature gradients determined regionally from borehole vitrinite reflectance and fission track profiles constrain the amount of denudation allowing reconstruction of the original anticline morphology. The timing of inversion and erosion determined by apatite fission track analysis allow the eroded strata to be replaced on the section and the inversion restored to reveal the amounts of extension and compression and the syn-rift basin morphology, which may constrain hydrocarbon migration paths. For the Otway Ranges in SE Australia 45% Early Cretaceous extension followed by 10% mid-Cretaceous compression are inferred.

The aim of this paper is to outline a method for constructing two-dimensional balanced and restored cross sections across an inverted basin, in order to understand the magnitude, timing and kinematics of inversion. The term 'inversion' is used here to describe the compressional reactivation of previously extensional faults and basins, rather than the regional uplift of whole basins and/or their margins.

Firstly, some of the problems inherent with inversion are discussed, particularly multiple deformations each with a different stress orientation, and the poor control on the amount and timing of uplift. We then propose a strategy to interpret and restore inversion structures, emphasising the inclusion of thermochronological data and the basinwide linkage, or otherwise, of faults at depth. Thermochronological data are particularly important in strongly inverted basins in that they constrain estimates of the original thickness of syn-rift section, subsequently eroded. Correctly restoring the syn-rift section helps constrain the interpretation of faults at depth, commonly to mid-crustal levels, which will lead to better understanding of the kinematics of inversion.

This study uses the Cretaceous–Tertiary Otway Basin of Bass Strait in SE Australia as an example (Fig. 1). The details of the Bass Strait study have been published elsewhere (Cooper 1995; K. A. Hill *et al.* 1995; K. C. Hill *et al.* 1995) and the results are only summarized here to illustrate the method. Both inversion and regional uplift are recorded in the Bass Strait failed rift between Australia and Tasmania, but the example in this paper is confined to the compressional reactivation of the Otway Ranges area along the NW margin of Bass Strait (Figs 1 & 2). Early Cretaceous rifting along Australia's southern margin passed through Bass Strait, but mid-Cretaceous breakup passed south of Tasmania along the Sorell Fault and Tasman Fracture zone (e.g. Veevers *et al.* 1991). Bass Strait was left as a failed rift, whilst the Otway Basin to the west of the Sorell Fault (Fig. 2) underwent Late Cretaceous rifting and slow spreading with the development of a passive margin in the Tertiary. Several inversion episodes have been recorded around the Bass Strait failed rift since the mid-Cretaceous, particularly uplift and erosion of the Otway Ranges in the mid-Cretaceous and Pliocene (Fig. 2; K. C. Hill *et al.* 1995). The main events in the Otway Basin are summarized in Table 1 and the stratigraphic units and principal unconformities shown in Fig. 3.

Problems restoring inversion terranes

Two of the issues that arise when restoring an inversion terrane as opposed to extensional or

From Buchanan, P. G. & Nieuwland, D. A. (eds), 1996, *Modern Developments in Structural Interpretation, Validation and Modelling*, Geological Society Special Publication No. 99, pp. 99–115.

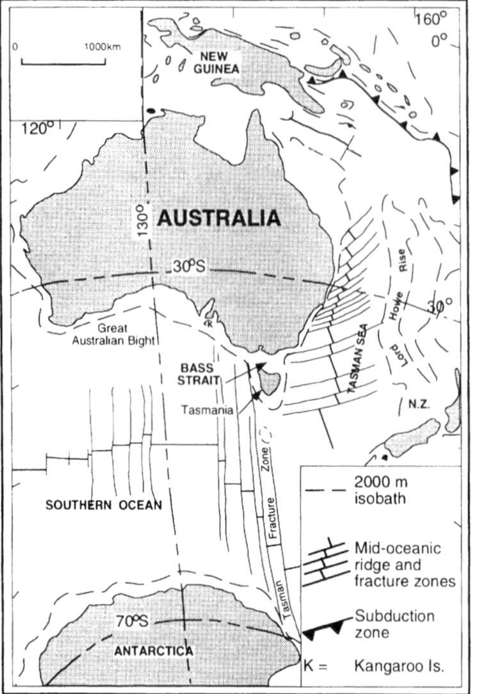

Fig. 1. Plate tectonic setting of SE Australia showing the location of the Bass Strait failed rift, the margins of which have been inverted (K. C. Hill *et al.* 1995). Early Cretaceous rifting through Bass Strait gave way to Late Cretaceous slow spreading south of Tasmania and Tertiary fast spreading to open the Southern Ocean. Tasman Sea rifting and spreading occurred in the Late Cretaceous–Palaeocene. Island arc collision with New Guinea placed the continent into compression from *c.* 10 Ma.

compressional deformation individually, are:

(1) inversion involves two or more deformation events which only rarely have the same direction of tectonic transport and
(2) the timing and amount of inversion and denudation are difficult to determine, particularly at major unconformities.

A balanced section is only valid if constructed in the movement direction, which may not be possible for inverted structures. Diagrams 1a and 2a in Fig. 4 illustrate compressional reactivation perpendicular to the original extension along a curved extensional fault. Such inversion cannot be represented on a two-dimensional cross section and requires three-dimensional balancing, applying geometrical and/or flexural-isostatic 3D balancing techniques, currently being developed (e.g. Gibbs & Griffiths 1994; Ma & Kusznir 1994). Diagrams 1b and 2b (Fig. 4) show sedimentation followed by compressional reactivation at *c.* 40° to the original extension direction, which can be represented on a cross section bisecting the two movement directions, as shown schematically on 2c (Fig. 4; see below for further discussion). To construct such a section requires knowledge of the extension and compression directions, which is difficult to obtain from anticline trends or seismic profiles, as illustrated from the curved anticline and the orthogonal sections on 2b (Fig. 4).

When restoring an inverted section it is important to know the amount of denudation and when it occurred, such that the missing section can be replaced at the appropriate time allowing section restoration. This is especially important for the eroded syn-rift section adjacent to large, basin-bounding faults, as this constrains estimates of the amount of extension and the syn-rift basin morphology. Immediately prior to inversion is often the time of maximum burial of the syn-rift section and the time of hydrocarbon generation. The hydrocarbon migration paths are largely controlled by the basin geometry at that time, so it is important to determine the pre-inversion basin morphology, taking into account structural restoration and decompaction following removal of the overburden.

To overcome some of these problems, a strategy for constructing restorable 2D sections across an inverted terrane is proposed, illustrated in Fig. 5. The individual elements of this strategy are discussed below, using the eastern Otway Basin as an example (Fig. 2), particularly focusing on the Otway Ranges area (for details see Cooper *et al.* 1993; Hill *et al.* 1994*a* & *b*; Cooper 1995 and K. C. Hill *et al.* 1995).

Modelling techniques

A general problem in basin scale restoration is the method applied, geometrical balancing and/or flexural-isostatic modelling (e.g. Dula 1991; Kusznir & Ziegler 1992). Both methods include backstripping and decompaction and both have their application, depending upon the structural interpretation and the scale of restoration. Geometrical balancing assumes constant displacement along faults and that horizons were continuous and roughly horizontal at the time of deposition, although palaeo-surface topography is easily incorporated if known. Many geometrical restorations involve listric faults (e.g. Williams & Vann 1987; Dula 1991), but it is equally easy to model planar faults passing down into a mid-crustal ductile shear zone as suggested by Kusznir & Ziegler (1992) and many others. In the latter case a hypothetical detachment deep in the lower crust

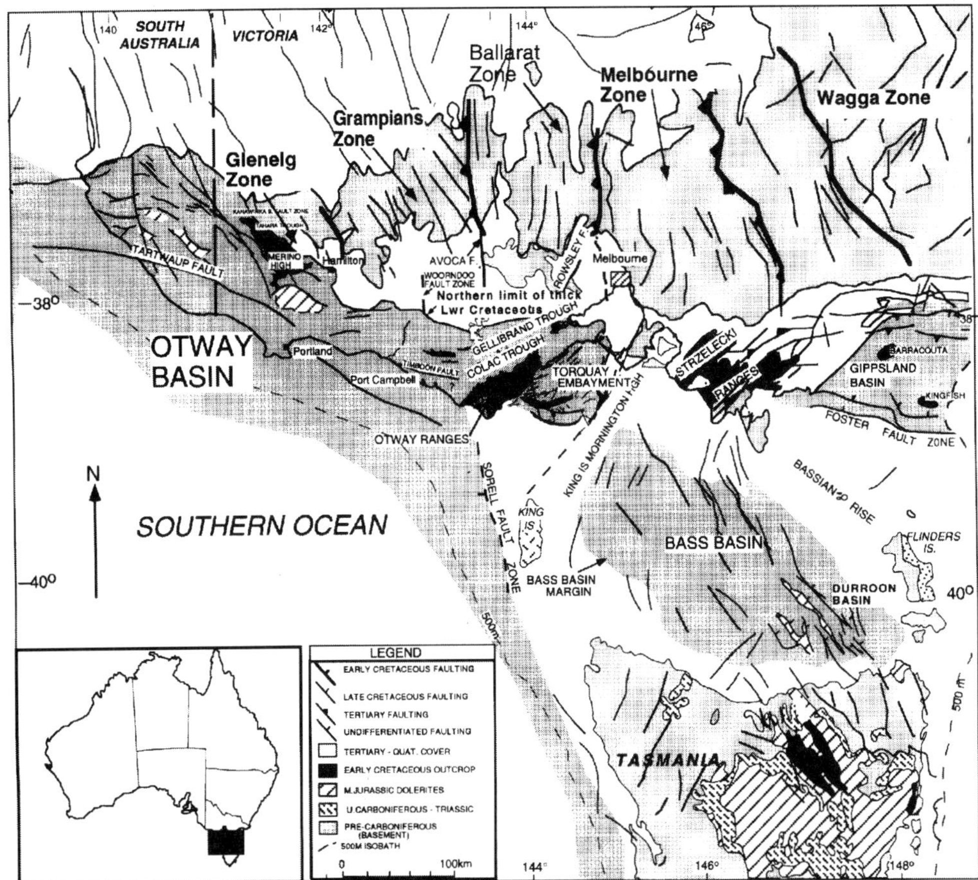

Fig. 2. Location of the inverted Otway Ranges and Colac Trough and the adjacent Torquay Embayment, part of the Cretaceous-Tertiary Otway, Bass and Gippsland sedimentary basins (shaded). The offshore Sorrel fault zone and onshore Woorndoo fault zone link into the Tasman Fracture Zone to the south (Fig. 1), the position of which was probably controlled by the Palaeozoic and Proterozoic Grampians–Glenelg structural grain (K. A. Hill *et al.* 1995). Together the fault zones separate the Otway Basin passive margin from the Bass Basin failed rift with inverted margins (K. C. Hill *et al.* 1995).

may be used as a convenient tool to restore upper crustal fault blocks along planar steep-dipping faults (see below).

The advantages of geometrical restoration are that the entire section including the syn-rift units can be restored, it is not model driven and it does not require input of a beta value (McKenzie 1978) and a value for the effective elastic strength of the lithosphere. In contrast, flexural-isostatic (lithospheric) modelling does assume a mechanism for deformation of the lithosphere and usually does need such input, but has the advantages of predicting footwall uplift, basin subsidence, rollover of the hangingwall and thermal effects (e.g. Kusznir & Ziegler 1992; Kusznir *et al.* 1994). These alternative methods are not reviewed further here, but geometrical balancing is used to illustrate the inclusion of dip analysis and thermochronological data in restoring inverted basins.

Extension and compression directions

An estimate of the tectonic transport direction during the compression (inversion) stage can be obtained from field and/or seismic mapping of the major anticlinal structures. Kinematic indicators, such as slickensides, are often used, but usually show only the most recent direction of movement along the fault which may not be relevant to older

Table 1. *The main tectonic events in the Otway Basin*

AGE		TECTONIC EVENT
NEOGENE	PLIO	EAST OTWAY INVERSION + NEWER VOLCANICS
	MIO	REGIONAL SUBSIDENCE; OLDER VOLCANICS
PALEOGENE	OLIG	OLDER VOLCANICS
	EOC	COMMENCE RAPID SPREADING OF SOUTHERN OCEAN
	PAL	
CRETACEOUS	U	WEST OTWAY SLOW DRIFT-SUBSIDENCE
	97.5 Ma	INITIAL BREAKUP + EAST OTWAY INVERSION
	L	REGIONAL SUBSIDENCE
		RIFTING
JURASSIC	U	
	M	TASMANIA + OTWAY? DOLERITES
	L	

After Veevers *et al.* (1991) and K. C. Hill *et al.* (1995)

inversion events. Faults with linear traces result in linear anticlines which may not indicate the compression direction as they may be reactivated extensional faults oblique to the compression. However, the relative amplitude of a curved anticline developed above a curved fault may be used to indicate the compression direction (Fig. 4, 2b). Although now denuded, the morphology of large inversion anticlines may be estimated by restoring the eroded section, determined from thermochronological analysis (see 'thermochronology' below).

The tectonic transport direction during extension is more difficult to constrain as rocks have relatively little strength during extension and may break (fault) in any orientation, often controlled by the pre-existing basement fabric. The strike of extensional faults is often assumed to be perpendicular to the extension direction, but such an interpretation may be untenable as shown by the coeval NW–SE, E–W and NE–SW striking Early Cretaceous faults in the Otway Basin (Hill *et al.* 1994*a* & *b*). Scott *et al.* (1994) demonstrated

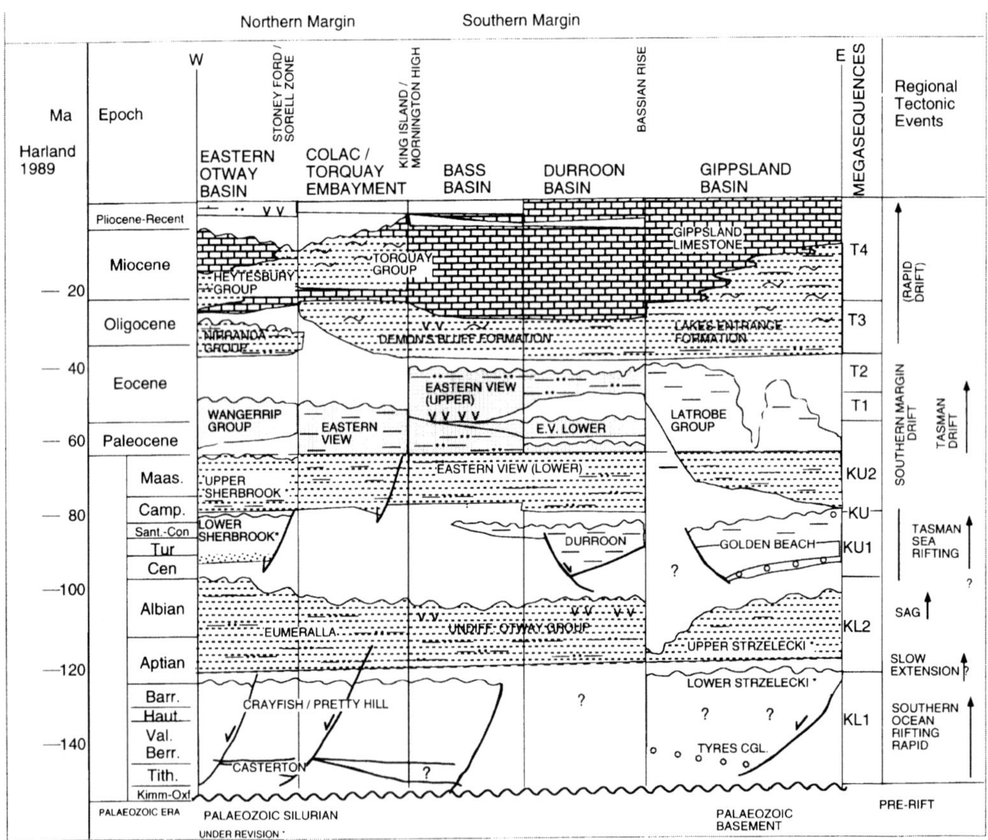

Fig. 3. The main stratigraphic sequences in the Bass Strait region tied into regional tectonic events, after K. C. Hill *et al.* (1995) and Featherstone *et al.* (1991)

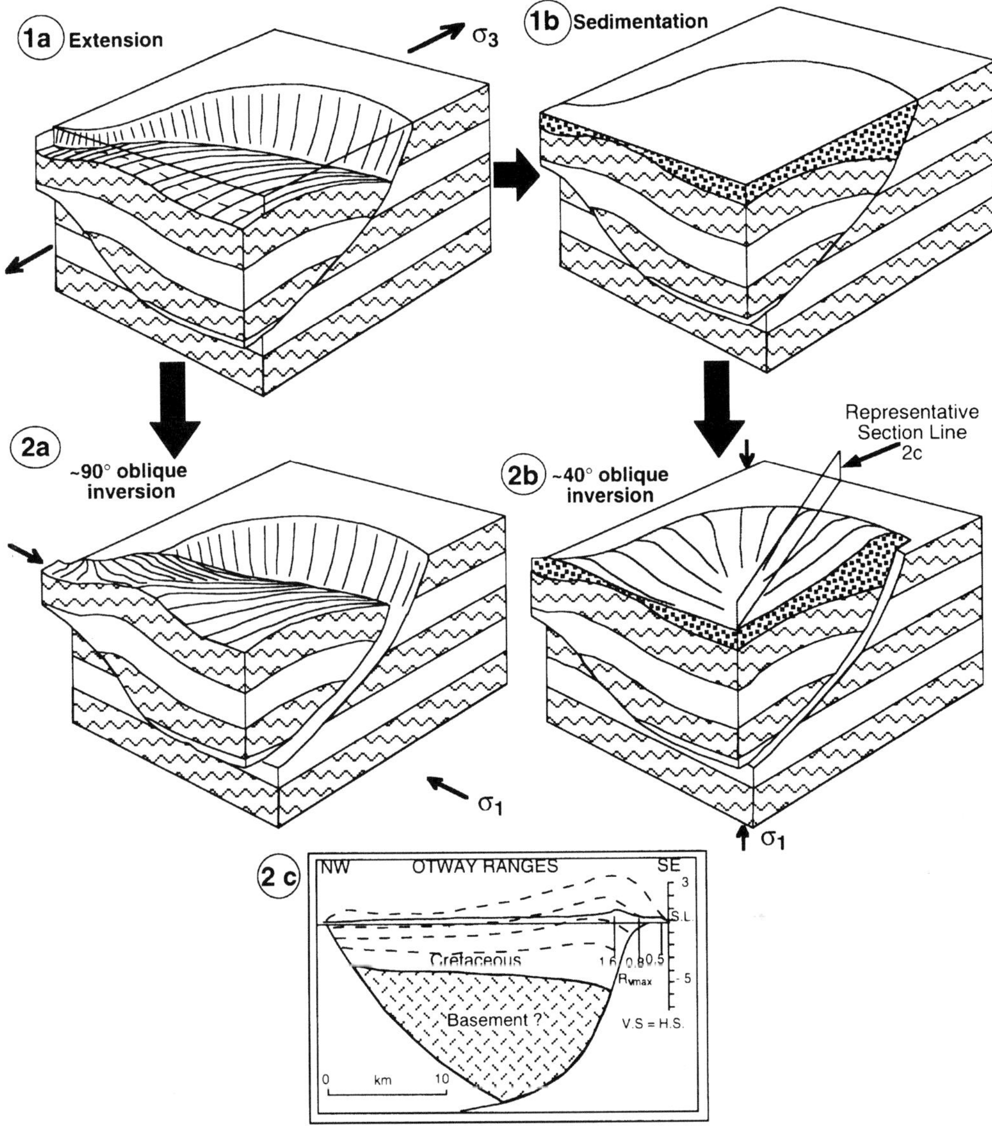

Fig. 4. Restoring inversion terranes that have variable tectonic transport directions. (**1a**) A curved extensional fault with frontal and lateral ramp (transfer zone) components, illustrating oblique extension along a large part of the fault; (**2a**) inversion of the same fault, but with compression at 90° to the original extension direction. This cannot be restored without rigorous 3-dimensional balancing: (**1b**) sediments fill in the hole created by extension in (**1a**); (**2b**) inversion of the same fault at ~40° to the original extension direction. Although not rigorously balanced, a good representation of the restored structure can be obtained along a section line bisecting the extension and inversion directions, as shown in the schematic example from the Otway Ranges in SE Australia (**2c**). With care, representative restored sections can be constructed for extension and inversion directions up to 50° apart.

the utility of dip analysis of pre-rift beds in the hangingwall of faults to determine the extension direction directly. From seismic line intersections the dip directions of the hangingwall beds can be determined and, for a large population, the dip directions are parallel to the extension direction, regardless of fault orientation. However, care must be taken to analyse the dips in areas that are not inverted and to take account of possible rotation due to regional basin sag.

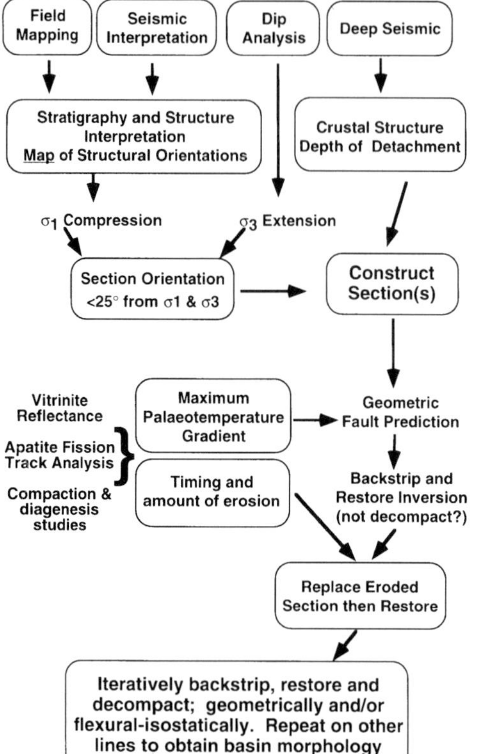

Fig. 5. A strategy for constructing a balanced and restored cross section across an inverted terrane. Key elements include: confirming the directions of extension and compression from mapping and dip analysis to define the section orientation and determining the timing and amount of erosion by thermochronological analysis.

For the Otway Ranges, in the absence of other data, the compression direction is taken to be *c*. 135°, which is roughly perpendicular to the strike of the large mid-Cretaceous anticline defined thermochronologically (see 'thermochronology' below). To determine the extension direction, Cooper (1995) carried out dip analysis of the Torquay Embayment (Fig. 6) placing most emphasis on the vectors in the uninverted parts of the basin, rather than the inverted portions in the NW (Fig. 7). Cooper (1995) determined a resultant mean extension direction of 166°, but interpreted a Barremian extension direction of 193° in the south and an Aptian–Albian extension direction of 150° in the north and east (Fig. 6).

Section orientation

A balanced cross section should be constructed in the direction of tectonic transport, but for inversion structures there are two or more such directions. How divergent can those directions be whilst still allowing a valid balanced cross section?

Figure 8 shows a profile of a listric fault and the corresponding hangingwall assuming 90° shear, drawn using the Chevron construction (e.g. Williams & Vann 1987; Dula 1991). Also shown are profiles of the same fault constructed at 25° and 60° to the direction of extension. It can be seen from Fig. 8 that the 60° oblique section produces a substantially different profile, but that the 25° oblique section is similar to the true section and results in only a 10% overestimate of the amount of extension. In both cases the diagram shows that the same detachment depth is predicted geometrically from the hanging-wall profile regardless of section obliquity, if 90° shear is used. Even if 70° antithetic shear is used, as suggested by Dula (1991), the detachment is only 5% deeper for the 25° oblique section. It is suggested therefore, that a reasonable limit to the divergence of the section from the tectonic transport direction is 25°, assuming the section does not cross lateral ramps or tear faults. Applying this rule to both compression and extension directions, a representative cross section could be constructed for directions up to a maximum of 50° apart, if the section bisected the two transport directions (Fig. 4, 2b).

In the Otway Ranges area, the balanced section orientation was largely dictated by the availability of data, particularly the orientation of seismic data and of creeks for field measurements. The balanced section trends towards 150°, within 16° of the mean extension direction and 15° of the inferred compression direction (Fig. 6). However, the section is oblique by 43° to Cooper's (1995) Barremian extension direction for the Snail Terrace which may distort restoration of the southern portion (Fig. 6).

Section construction – detachments

Having determined the section orientation and position, intersecting the most data and avoiding tear faults, the data from field mapping, seismic interpretation, wells, deep seismic data and other sources are entered on the depth section with no vertical exaggeration. Field mapping and interpretation of industry reflection seismic data (e.g. Fig. 7) constrain the fault shape down to depths of *c*. 5 km, but often provide few constraints on the deeper fault orientation and possible detachment depth. Some indication of the deep structure can be obtained from deep reflection seismic data, as shown by Finlayson *et al.*'s (1994) interpretation across the Otway Ranges. They imply detachments at *c*. 6 s TWT, although the obliquity and tortuosity

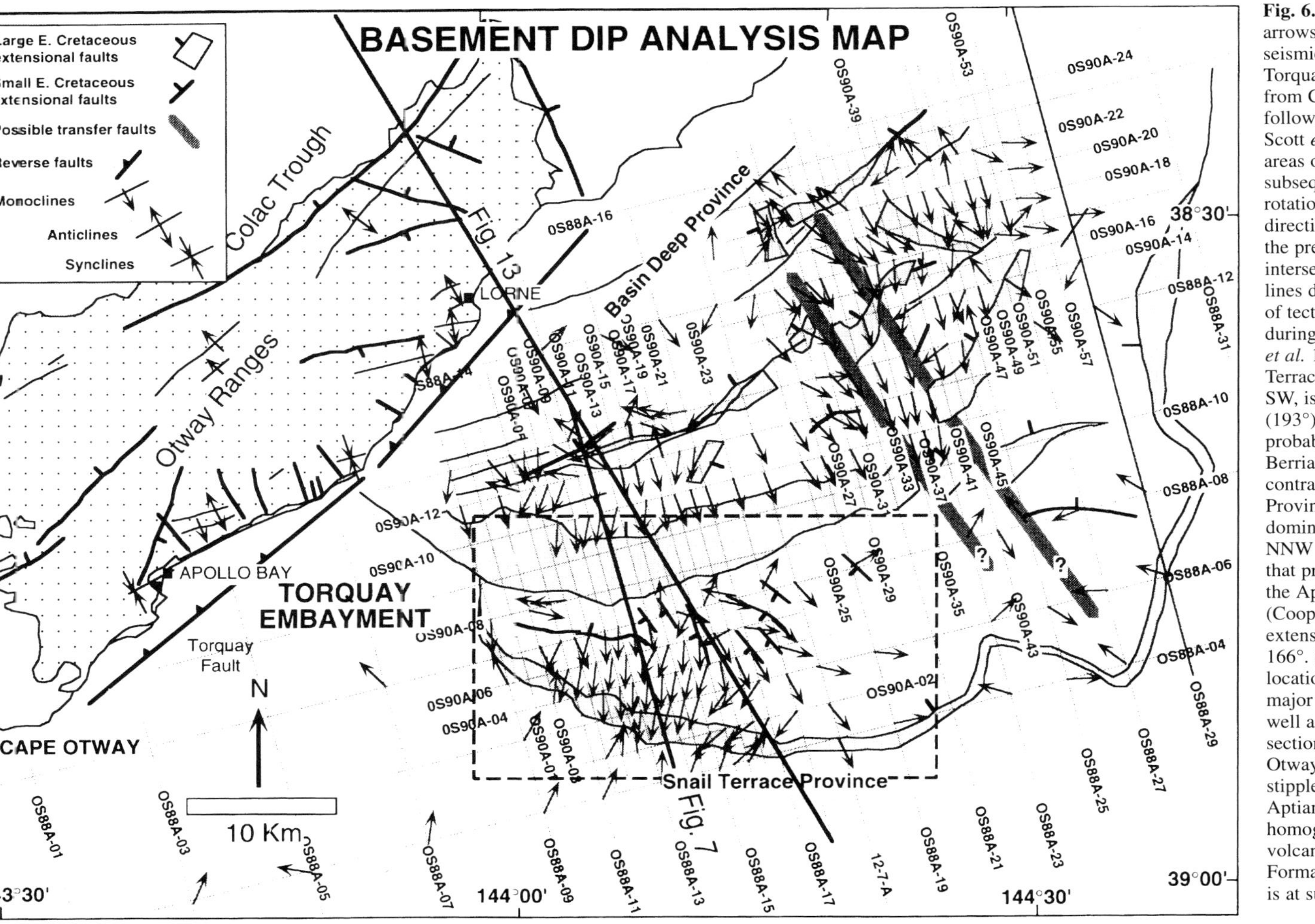

Fig. 6. Dip direction arrows determined from seismic basement for the Torquay Embayment, from Cooper (1995) following the method of Scott *et al.* (1994). In areas of minimal subsequent inversion and rotation, the dip directions measured from the pre-rift section at the intersections of seismic lines define the direction of tectonic transport during extension (Scott *et al.* 1994). The Snail Terrace Province in the SW, is dominated by N–S (193°) extension, that probably occurred in the Berriasian-Barremian. In contrast, the Basin Deep Province in the NE, is dominated by SSE–NNW (150°) extension, that probably occurred in the Aptian–Albian (Cooper 1995). The mean extension direction is to 166°. Seismic line locations and some of the major faults are shown as well as the balanced cross section (Fig. 13). The Otway Ranges stipple shows where Aptian–Albian homogeneous, volcanolithic, Eumeralla Formation (KL2, Fig. 3) is at surface.

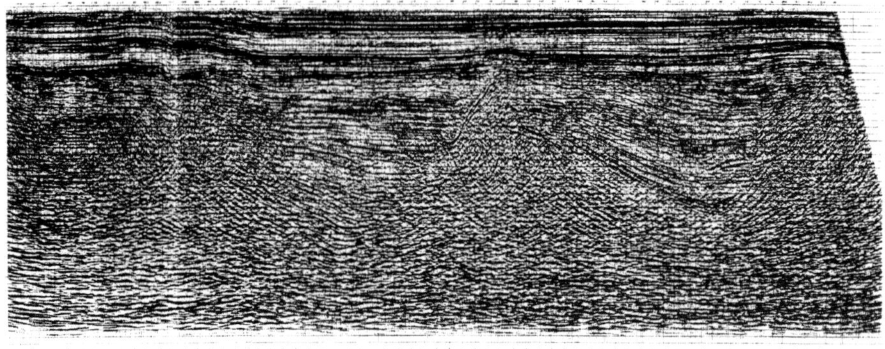

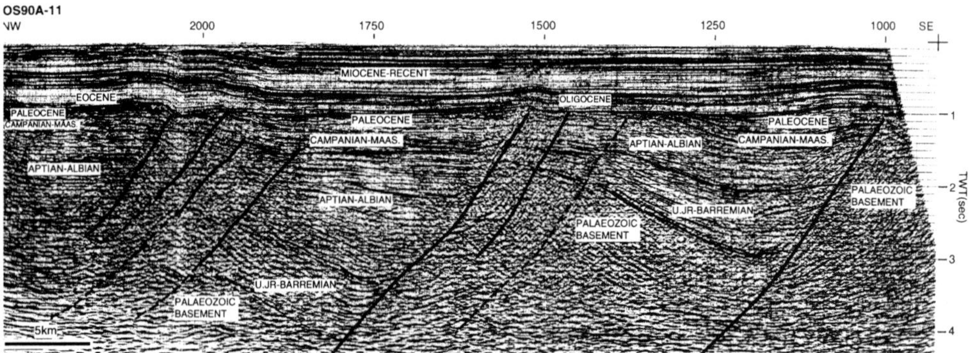

Fig. 7. A reflection seismic line from the Torquay Embayment (see Figs 6 & 10 for location) illustrating Early Cretaceous extension on NW-dipping faults and significant pre-Eocene truncation. Pliocene inversion at the NW end of the line is evident from the thickened but uplifted Eocene section and is interpreted to have reactivated Early Cretaceous extensional faults. The structural style there is similar to that of the adjacent onshore Otway Ranges (e.g. Fig. 4 (2c)). From K. C. Hill *et al.* (1995 fig. 4a).

of the line, particularly at the southern end (Figs 9 & 10) distort the geometries.

If the faults are listric, the detachment depth can be estimated by applying the Chevron construction (e.g. Williams & Vann 1987; Dula 1991), using the heave of major faults, their hanging-wall profile and an estimate of the regional elevation. This method produces variable results depending on the shear angle used in the Chevron Construction (Williams & Vann 1987; Dula 1991). In the Torquay Embayment, modelling of the steep, basin-forming, Early Cretaceous faults (Fig. 7) suggests detachments at > 20 km depth which is inconsistent with the high temperature gradients in the Early Cretaceous (see 'thermochronology', below) which suggest a shallow brittle-ductile transition. However, such a hypothetical deep detachment can be used to restore upper crustal fault blocks.

Thermochronology

Thermochronological analysis is important in inversion terranes to determine the timing of cooling, the amount of cooling and the palaeo and present temperature gradients to determine the amount of uplift and erosion (Green *et al.* 1995). The amount of burial and denudation at unconformities can be estimated from compaction-porosity reduction curves for standard lithologies which can be determined from electric logs if boreholes exist. However, this method does not give the temperature data important for hydrocarbon exploration. Alternatively, the time–temperature and denudation parameters can be determined from a combination of vitrinite reflectance and apatite fission track analyses (e.g. Bray *et al.* 1992).

If the strata beneath an inversion unconformity are not at their maximum temperature now, borehole vitrinite reflectance and apatite fission track

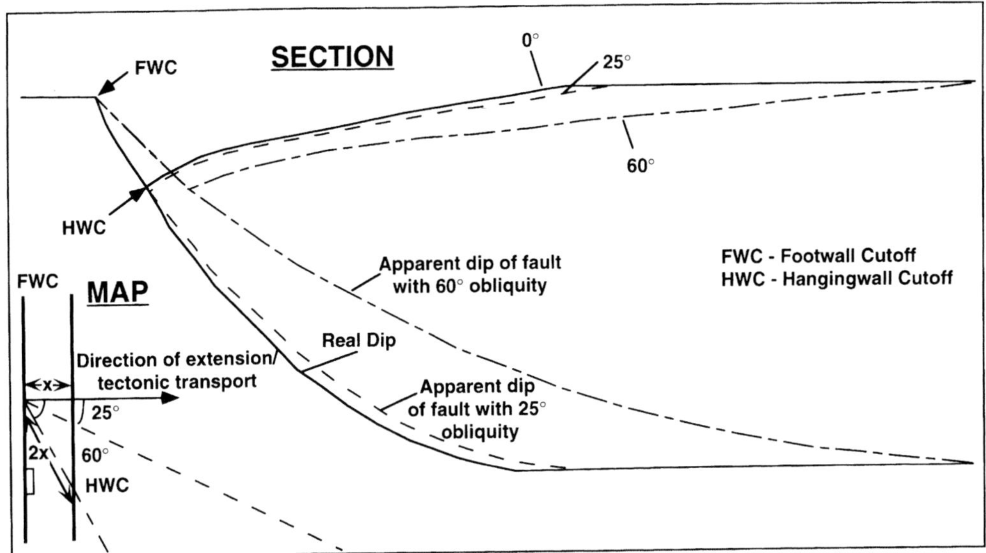

Fig. 8. The prediction of fault shape and detachment depth from the mapped hanging-wall profile on a section line drawn oblique to the direction of tectonic transport; using the Chevron construction with 90° shear (e.g. Williams & Vann 1987; Dula 1991). Notice that with 90° shear, the detachment depth predicted is constant regardless of section obliquity. In addition, a section that is 25° from the direction of extension has only a 10% error in the calculated amount of extension and the fault and hanging-wall profiles are almost the same as the true profiles.

values from those strata allow determination of the maximum palaeotemperature gradient. Knowing the temperature gradient, the vitrinite reflectance and apatite fission track values at the unconformity can be used to determine the amount of erosion, due to inversion, with typical two standard deviation errors of ± 50% (Green *et al.* 1995; Bray *et al.* 1992; Sweeney & Burnham 1990: see 'limitations'

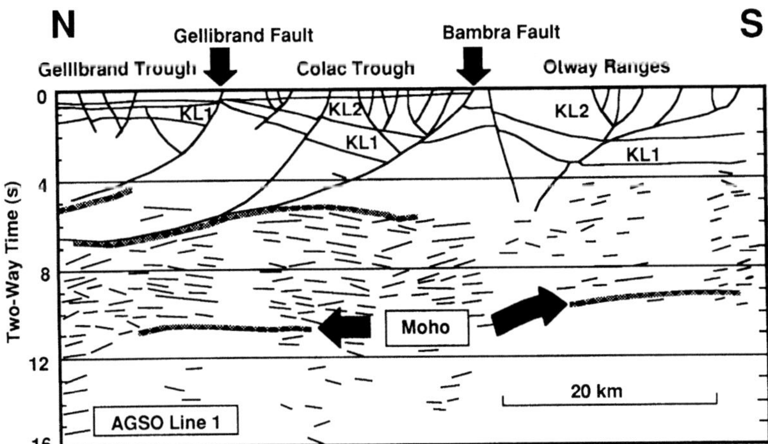

Fig. 9. Line diagram interpretation of an Australian Geological Survey Organisation deep reflection seismic profile across the Otway Ranges and Colac Trough, from Finlayson *et al.* (1994). See Fig. 10 for location and note the irregular and oblique orientation of the southern end of the line, where it follows a road across the Ranges, creating irregular geometries on the seismic line. The interpretation shows relatively steep, north-dipping faults soling at mid-crustal or deeper levels and bounding half graben formed in the Early Cretaceous. The Otway Ranges is the deepest graben, downfaulted on both sides and overlies thinned crust. It is also the area of greatest inversion. See Fig. 7 for an offshore continuation of this line.

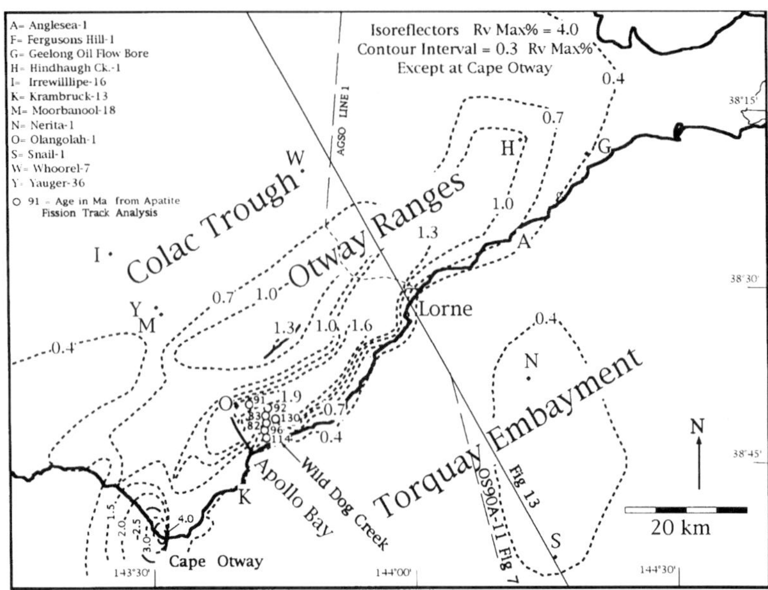

Fig. 10. Vitrinite reflectance ($R_{v\ max}$%) isoreflectors for the Otway Ranges (after Cooper *et al.* 1993 and Struckmeyer & Felton 1990) and apatite fission track analysis ages along Wild Dog Creek, from K. C. Hill *et al.* (1995). See Fig. 4 for location of Otway Ranges. 105 vitrinite reflectance datapoints were contoured, with reflectance measured on telovitrinite and occasionally on detrovitrinite, in accordance with Australian Standards 2468 and 2061 and the guidelines of Cook & Kantsler (1982). The vitrinite reflectance data outline a broad structural high and demonstrate considerable uplift and erosion as they indicate that the sediments now at surface in the core of the Ranges have previously been heated to temperatures ≫150°C. The apatite fission track ages of *c.* 95–90 Ma show that the sediments cooled from temperatures of > 120°C to < 60°C around that time (K. C. Hill *et al.* 1995).

below). If done regionally the magnitude and morphology of the inversion anticline can be reconstructed. Apatite fission track analysis can determine maximum palaeotemperatures of samples within the range of *c.* 60–130°C, but also records the time and rate of cooling below 100–130°C, depending on fluorapatite-chlorapatite composition (Green *et al.* 1989, Dumitru *et al.* 1991, Laslett *et al.* 1987). Combining the two techniques reveals the time-temperature history of rocks below *c.* 150°C (Bray *et al.* 1992; Green *et al.* 1995).

Thermochronological analyses of the Otway Ranges are discussed below to illustrate the application of the techniques and their limitations.

Thermochronology of the Otway Ranges

The Otway Ranges comprise relatively steep-sloped, forested hills up to 500 m high and expose 2–5 km thick, monotonous, incompetent volcano-lithic Aptian–Albian clastic sediments throughout much of their length (stipple on Fig. 6). They were uplifted in the Plio-Pleistocene as shown by the deformation of Pliocene sediments observed on Torquay Embayment seismic data (Fig. 7; K. C. Hill *et al.* 1995). However, surface vitrinite reflectance measurements range from $R_{v\ max}$ 0.4% along the coast to > 2.0% in the core of the Ranges (Fig. 10) which correspond to maximum palaeotemperatures of *c.* 80°C and 200°C respectively (Fig. 11; using the methods of Sweeney & Burnham 1990). Analysis of vitrinite reflectance profiles of the Lower Cretaceous sequence from wells (well locations on Fig. 10) indicates Early Cretaceous temperature gradients of *c.* 60°C km^{-1} (Cooper *et al.* 1993; Duddy 1994) allowing the palaeotemperature estimates to be converted to denudation for the Otway Ranges (Fig. 11). The high temperature gradients also suggest a shallow brittle-ductile transition during the Early Cretaceous.

Apatite fission track analysis reveals the timing of erosion and cooling to present temperature gradients of *c.* 30°C km^{-1}. Apatite grains heated to > 130°C are usually completely annealed such that tracks created after cooling reveal the time of cooling. Samples from the core of the Ranges along Wild Dog Creek record cooling ages of *c.* 95–90 Ma (Figs 10 & 11; K. C. Hill *et al.* 1995)

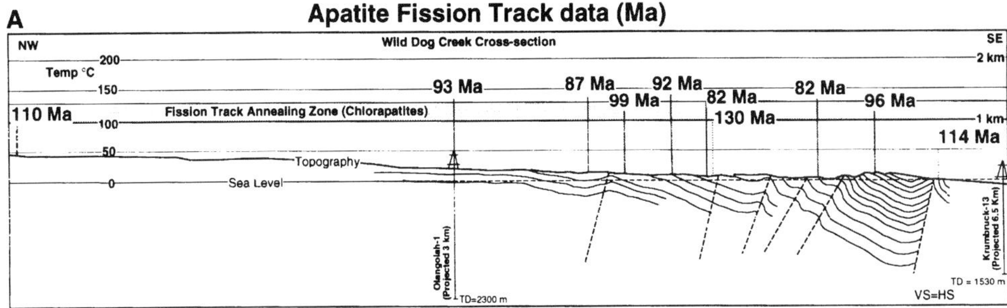

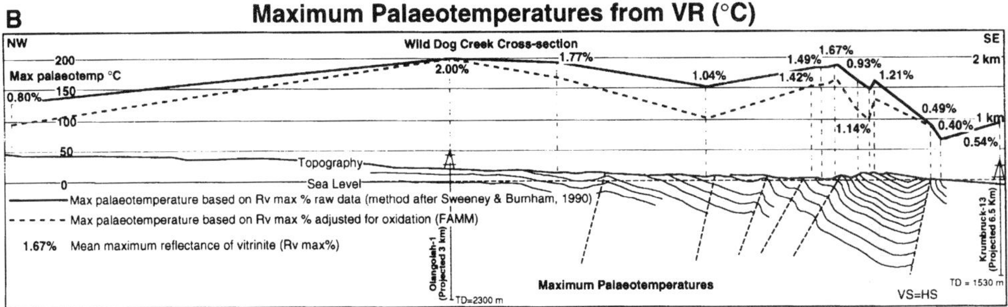

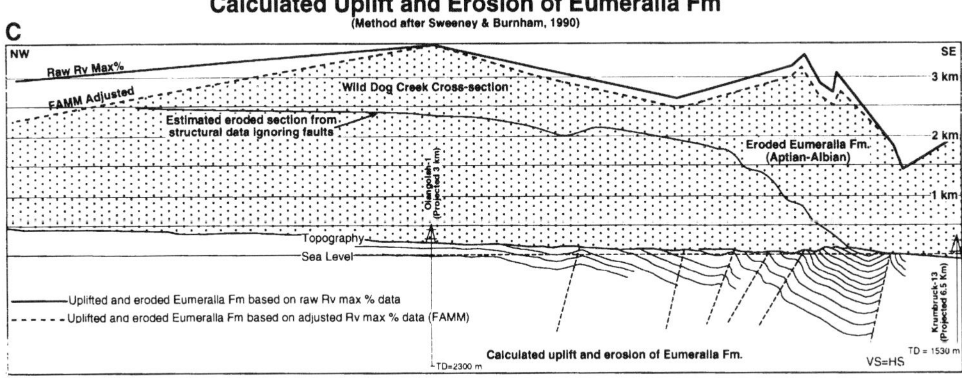

Fig. 11. Transect along Wild Dog creek (Fig. 10) showing the geological structure (simplified) of the homogeneous Eumeralla Formation (KL2) interpreted from field dip measurements and the thermochronological data used to calculate the amount of uplift and erosion during inversion. (**a**) The apatite fission track ages are shown relative to the chlorapatite annealing zone (e.g. Duddy 1994; Laslett *et al.* 1987). Totally reset ages (above the annealing zone) record the time of cooling from temperatures ≥130°C (for chlorapatite) to less than *c.* 60°C. Partially reset ages are within the annealing zone, whilst the 114 Ma age plotted below the zone records the stratigraphic age (volcanogenic apatites). The juxtaposition of totally and partially annealed samples suggests local faults. (**b**) The solid line records maximum temperature estimated from vitrinite reflectance analysis, whilst the dashed line is corrected for oxidation (FAMM). (**c**) The lower diagram records the estimated amount of erosion using temperature gradients of *c.* 60°C km^{-1} inferred from vitrinite reflectance profiles from the Lower Cretaceous section in wells (Cooper *et al.* 1993; Duddy 1994). The amount of material eroded varies considerably from that estimated from simple structural projections, which are unable to account for the affects of faulting.

which can be converted to a geohistory profile using palaeotemperature gradients (Fig. 12). From Fig. 12, it can be seen that although there was inversion in the Plio-Pleistocene, the major inversion of the Otway Ranges was in the mid-Cretaceous. Furthermore, the amount of section eroded can be estimated, such that it can be restored onto a mid-Cretaceous balanced cross section (Fig. 5). This has important implications for decompaction as maximum burial in the Otway Ranges and Colac Trough was in the Albian, so Early Cretaceous units in those areas should not be decompacted after stripping off the Tertiary and Upper Cretaceous section.

Limitations of thermochronological data

The thermochronological analysis described above is largely limited to strongly inverted areas where the strata beneath the inversion unconformity have been hotter in the past than at the present day. Using only samples below the unconformity, the palaeotemperature gradient can be estimated and hence the amount of erosion, assuming that the eroded material has the same thermal conductivity as the strata analysed. Vitrinite reflectance analyses of samples above the unconformity can yield information about subsequent temperature gradients. If these subsequent gradients are higher than the pre-inversion gradients, then both vitrinite reflectance and fission track data are likely to be overprinted so the amount of denudation cannot be determined. If the temperature gradient decreases substantially following inversion, then the strata may be buried deeper than pre-inversion and still remain at lower temperatures so that, with care, the amount of erosion resulting from inversion can still be determined.

In areas with multiple superimposed inversion events, only the most recent event is likely to be recorded if they are all of similar (large) magnitude. However, the largest event will be recorded if it is considerably bigger than the subsequent events, as is the case for the Otway Ranges.

Considerable care is necessary when making temperature estimates from the vitrinite reflectance and apatite fission track data using the available kinetic models (e.g. Sweeney & Burnham 1990; Laslett *et al.* 1987). Two factors are particularly significant, apatite composition and vitrinite oxidation. Although the temperature range of annealing for the more common fluorapatite is *c.* 60–110°C, for chlorine-rich apatite grains it can be as high as 100–150°C (Duddy 1994). This is illustrated by the sample with 130 Ma apatite fission track age from the core of the Otway Ranges, where the vitrinite reflectance data suggest maximum palaeotemperatures of > 150°C and all the nearby samples are consistent with 95–90 Ma cooling from such temperatures (Fig. 11; K. C. Hill *et al.* 1995). We infer that the sample includes some chlorine-rich apatite grains which have resisted annealing even at high temperatures retaining their ?Palaeozoic crystallisation age, giving a mean age of 130 Ma for all the grains.

However, even chlorine-rich apatite grains should be annealed at 150°C, suggesting that the temperature estimate from the vitrinite data may also be misleading, perhaps due to oxidation. Assessing the effects of oxidation on surface samples is relatively difficult, especially for low level coal deterioration (McHugh *et al.* 1991), but can be determined using Fluorescence Alteration of Multiple Macerals (FAMM) analysis (Wilkins *et al.* 1992; Ellacott *et al.* 1994). The FAMM technique has been shown to be comparable with traditional oxidation assessment techniques such as dilato-

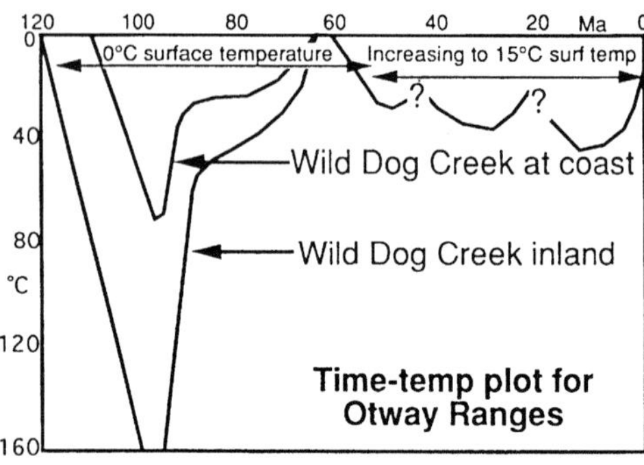

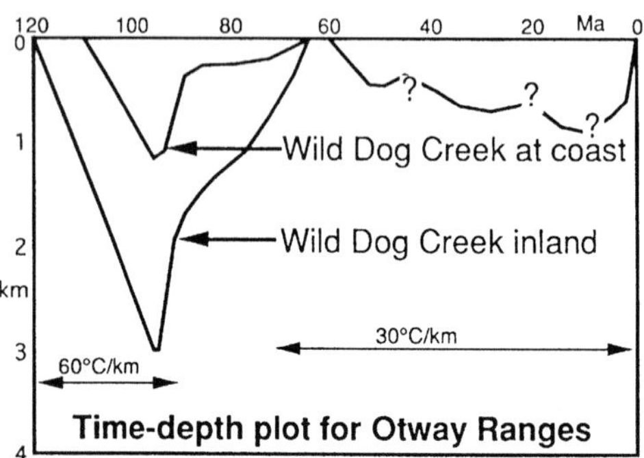

Fig. 12. Wild Dog Creek time–temperature and –depth plots (from K. C. Hill *et al.* 1995) showing the timing and amount of uplift and erosion from apatite fission track and vitrinite reflectance analysis (Fig. 11). The time of cooling is constrained by apatite fission track analysis and the conversion to depth by vitrinite reflectance analysis, which indicates high temperature gradients during Early Cretaceous extension. Surface temperatures reflect the high latitude of the Otway Ranges in the Cretaceous.

metry and Gieseler plastometry (McHugh *et al.* 1991). FAMM analysis showed no significant oxidation of Otway Ranges coastal samples, but a sample from Wild Dog Creek had Equivalent R_0% (FAMM) *c.* 0.30 R_0% lower than measured R_0% probably due to oxidation during early burial diagenesis (Fig. 11). The effect of such oxidation is more pronounced in samples where R_0% = 1.0–1.2 (McHugh *et al.* 1991). Thus the FAMM analysis suggests that the maximum temperature estimate from vitrinite reflectance data for parts of Wild Dog Creek may be 110–120°C rather than 150°C, therefore consistent with the apatite fission track analysis (Fig. 11).

Structural analysis of thermochronological data

The map of vitrinite reflectance data and particularly the derived denudation map at the surface or at an unconformity may greatly aid structural interpretation. In the case of the Otway Ranges, the map of vitrinite reflectance values reveals two large elongate anticline structures generated by mid-Cretaceous inversion (Fig. 10). These are interpreted to be perpendicular to the direction of mid-Cretaceous compression, as opposed to the smaller oblique anticlines near Lorne, shown on Fig. 6. In addition, the thick homogeneous Aptian–Albian section exposed in the Ranges has no internal markers, so that only form lines are obtained from geological mapping (Fig. 11). The vitrinite data therefore place considerable constraints on the shapes of the mid-Cretaceous structures and the closely spaced vitrinite reflectance and apatite fission track analyses can be used to indicate the likely presence of faults.

Balancing and restoring the section

The restoration of the inverted section in the Otway Ranges is illustrated by geometrical area balancing using Geosec^{TM}. Planar, steep-dipping faults in the upper crust are restored using a hypothetical *c.* 20 km deep detachment, as indicated by fault prediction from hangingwall shape. However, this detachment is a tool for section restoration purposes only and not intended to indicate a real detachment fault. Rather, the upper crustal fault blocks probably pass into a mid-crustal ductile shear zone at *c.* 10–15 km. Such an architecture could be modelled lithospherically applying the beta factors obtained from geometrical restoration.

The Otway Ranges section was constructed by entering all available data, then using the hanging wall shape and fault offset of basement to predict a > 20 km deep detachment. The estimated eroded Mio-Pliocene section was restored over the Ranges (Fig. 13, present day) then the Pliocene inversion structures restored using oblique shear and area balancing. The latter was necessary as the observable faults do not propagate to surface, where the deformation is manifested as folds. This ductile or distributed deformation was restored by area balancing. The Tertiary section was then stripped off and the underlying section decompacted prior to structural restoration of hanging-wall and footwall cut-offs.

In restoring inverted sections a key issue is the timing of maximum compaction, which may occur prior to inversion rather than at the present day. For instance, in the case of the Otway Ranges and Colac Trough (Fig. 10) the thermochronological data indicate that maximum temperatures in the Lower Cretaceous section were reached just prior to mid-Cretaceous uplift and denudation. A substantial drop in heat flow is then inferred so that the section could theoretically have been buried deeper during the Tertiary. However, only a thin Tertiary section onlaps the Ranges, so the mid-Cretaceous is interpreted as the time of maximum burial and therefore maximum compaction of the underlying sediments. However, maximum burial of the Torquay Embayment is probably now. Thus, when the Tertiary section is stripped off, the Torquay Embayment should be decompacted, but the Otway Ranges and Colac Trough should not, as they underwent maximum compaction in the Early Cretaceous. This results in the abrupt offsets in the base of the mid-Cretaceous section (Fig. 13), shown by transmitting the decompaction downwards.

The mid-Cretaceous section was restored to a horizontal surface, but this does not take into account the amount of material eroded. Thermochronological analyses indicate that from 95–90 Ma, 1–3 km of sediment was eroded over the Otway Ranges and Colac Trough, so this was replaced onto the section (Fig. 13, mid-Cretaceous). The section was then restored structurally to illustrate the amount of mid-Cretaceous compression (Fig. 13, 100 Ma). Finally the Aptian–Albian (KL2) sequence was stripped off and the KL1 sequence decompacted across the whole section, prior to restoration of the Early Cretaceous extension, allowing an estimate of the amount of extension (Fig. 13).

Discussion

The strategy outlined here illustrates how a multidisciplinary approach can place considerable constraints upon the interpretation of inverted

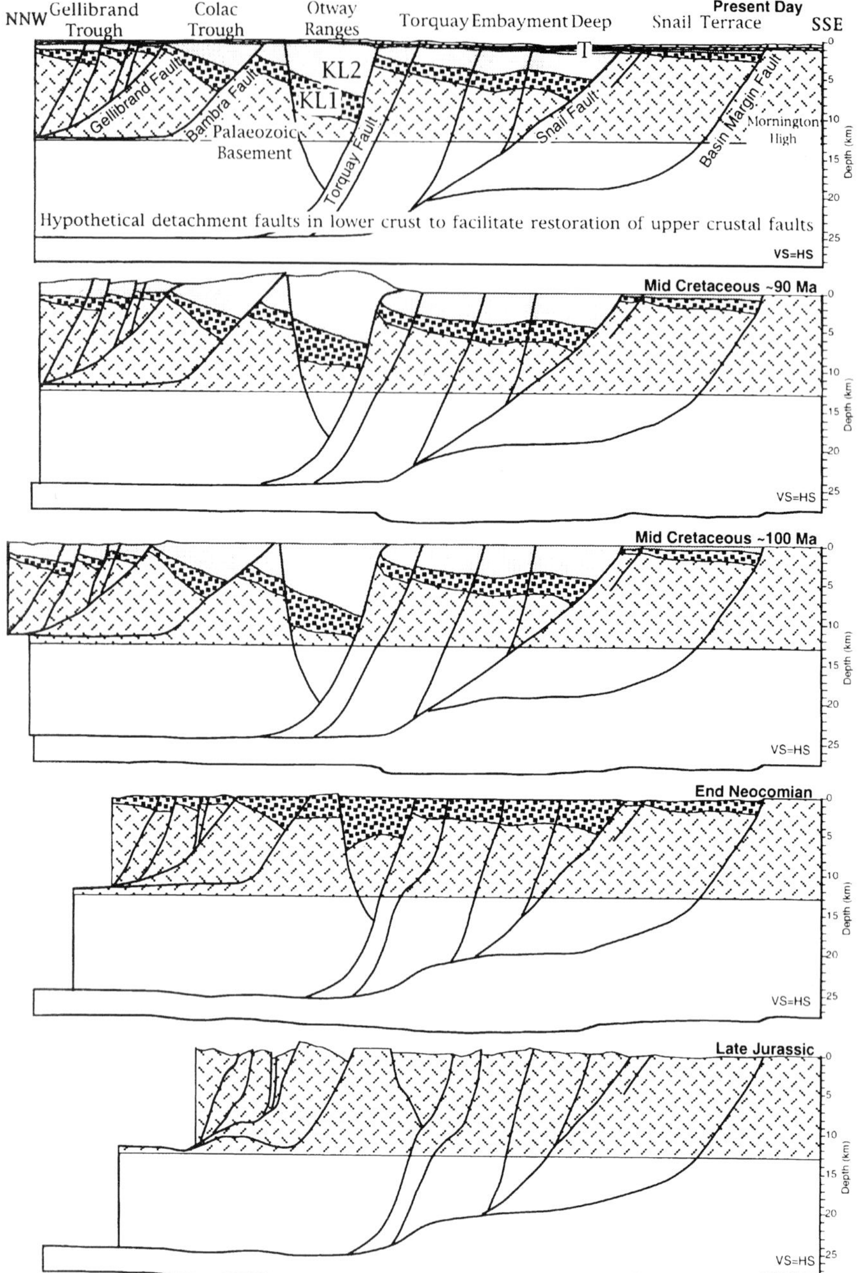

Fig. 13. Preliminary balanced, restored and decompacted section across the inverted Otway Ranges using Geosec^{TM}. The section trends approximately NNW–SSE, parallel to the extension direction for the Torquay Basin Deep (Fig. 7) and within 16° of the average extension direction for the basin and 15° of the estimated compression direction. The *c*. 20–25 km deep detachments, predicted from the steep faults observed on seismic data, are hypothetical and are used as a tool to restore the upper crustal fault blocks along steep-dipping planar faults to depths of 10–15 km (shaded on the section). To construct the 90 Ma section, the Tertiary horizons were stripped and the faults restored, then the amount of material eroded was replaced, consistent with the apatite fission track and vitrinite reflectance analyses (Figs 11 & 12) illustrating the amount of inversion. The faults were then restored to create the 100 Ma, pre-inversion section. The preliminary restoration of basement indicates the amount of crustal extension.

basins and particularly on their syn-rift morphology. For a strong inversion event resulting in a major unconformity, thermochronological analysis can reveal the amount of section denuded and the timing of denudation. This is particularly important in restoring the eroded syn-rift section adjacent to major faults, where the amount of growth due to extension may be understimated. When serial sections are restored to the pre-inversion state, they will define the basin morphology at the probable time of maximum heating and of hydrocarbon generation and migration, helping to delineate migration paths.

Correctly restoring the syn-rift section eroded during inversion also allows the true amount of extension to be estimated and helps constrain the fault interpretation at depth, often to mid-crustal levels. Regional cross sections can then be geometrically balanced, restored and decompacted, without assumptions regarding beta factor and elastic strength of the lithosphere inherent in lithospheric modelling. The use of a hypothetical detachment within the ductile lower crust allows geometrical restoration of upper crustal fault blocks along relatively steep planar faults or listric faults as necessary. The syn-rift section can be balanced, restored and decompacted giving an estimate of the beta factor for input into flexural-isostatic models.

The apatite fission track and vitrinite reflectance analyses techniques briefly reviewed here have most application in strongly inverted basins, as both datasets are overprinted if maximum temperatures are at the present day or if post inversion temperature gradients increase. The techniques are particularly applicable to areas where the entire post-rift section is denuded or even areas of total basin inversion (Bally 1984). In such areas apatite fission track analysis of basement samples can be used to determine the amount of denudation, albeit with less precision than when vitrinite reflectance data are also available. In areas with multiple superimposed inversion events, only the most recent event is likely to be recorded, but the largest event will be recorded if it is considerably bigger than the subsequent events, as in the Otway Ranges.

The preliminary geometrically balanced section for the Otway Ranges illustrates Early Cretaceous extension on steep NW-dipping faults, that are relatively planar to mid-crustal depths followed by mid-Cretaceous inversion, interpreted to reactivate the same faults, as observed on seismic data from the Colac Trough (K. C. Hill *et al.* 1995). Compressional reactivation of such steep faults is kinematically difficult, but may be explained by fluid lubrication of the faults following the release of elevated hydrostatic pressure at depth (Sibson, 1995). The magnitude and effects of mid-Cretaceous inversion are shown by incorporating the thermochronological data and indicate < 5 km shortening over 50 km (excluding the uninverted Torquay Embayment), implying < 10% shortening. Early Cretaceous extension over the same area was *c.* 15 km or *c.* 45%. These estimates of extension and inversion are only possible by balancing the section in the optimal orientation and by incorporating the thermochronological data. The importance of the latter is emphasized by the relative scale of the mid-Cretaceous structures as opposed to the relief created by Plio-Pleistocene inversion (Fig. 13, mid-Cretaceous v. present day).

Conclusions

(1) To construct 2D sections over inverted terranes, the cross section should be oriented within 25° of both the compressional and extensional directions of tectonic transport.

(2) The compressional direction of transport can be estimated by the orientation of curved anticlines formed by inversion. The shape and orientation of these may be further constrained by regional thermochronological mapping.

(3) The extensional direction of tectonic transport often estimated from fault orientations can be determined directly from dip analysis of the pre-rift section (Scott *et al.* 1994).

(4) The relative magnitude and timing of strong inversion events can be determined by vitrinite reflectance and apatite fission track analyses of borehole samples and samples from the inversion unconformity, particularly if it is a present day land surface. This complements structural, diagenetic and porosity-compaction studies.

(5) Inclusion of such thermochronological data into geometrically balanced cross sections allows estimates of the amount and % of extension and compression during inversion and palinspastic restoration of the syn-rift basin morphology, which may constrain hydrocarbon migration paths.

Research was sponsored by BHP Petroleum, the Shell Company of Australia, SAGASCO, Bridge Oil NL, Gas & Fuel Exploration, Esso Australia and Geotrack International. Cogniseis, Petrosys NL and Schlumberger-Seaco have made available discounted software for data analysis. The research was also funded through the Australian Research Council and Australian Institute of Nuclear Science and Engineering grants. Part of this work was conducted within the the National Geoscience Mapping_Accord (NGMA) Otway Basin Project. The authors would also like to thank the VIEPS students who have contributed to mapping of the Bass Region: Chris Durrand, Ildy Horvath, Mike Gilbert, Ciaran Lavin, Andrea O'Sullivan and Jane Richardson.

References

BALLY, A. W. 1984. Tectongénèse et sismique réflexion. *Bulletin Société Géologique de France*, **26**, 279–285.

BRAY, R. J., GREEN, P. F. & DUDDY, I.R. 1992. Thermal history reconstruction using apatite fission track analysis and vitrinite reflectance: a case study from the UK East Midlands and Southern North Sea. *In*: HARDMAN, R. F. P. (ed) *Exploration Britain: Geological Insights for the Next Decade*, Geological Society, London, Special Publications, **67**, 3–25.

COOK, A. C. & KANTSLER, A. J. 1982. *The origin and petrology of organic matter in coals, oil shales and petroleum source rocks,* The University of Wollongong, Wollongong.

COOPER, G. T. 1995. Seismic structure and extensional development of the eastern Otway Basin–Torquay Embayment. *The Australian Petroleum Exploration Association Journal*, **35**, 241–256.

——, HILL, K. C. & WLASENKO, M. 1993. Thermal modelling in the eastern Otway Basin. *The Australian Petroleum Exploration Association Journal*, **33**, 205–213.

DUDDY, I. R. 1994. The Otway Basin: Thermal, structural, tectonic and hydrocarbon generation histories. *In*: FINLAYSON, D. M. (ed.) NGMA/PESA Otway Basin Symposium Abstracts, *AGSO Record* **1994/14**, 35–42.

DULA, W. F., JR. 1991. Geometric models of Listric Normal Faults and Rollover Folds. *American Association of Petroleum Geologists Bulletin*, **75**, 1609–1625.

DUMITRU, T. A., HILL, K. C., COYLE, D. A., DUDDY, I. R., FOSTER, D. A., GLEADOW, A. J. W., GREEN, P. F., LASLETT, G. M., KOHN, B. P. & O'SULLIVAN, A. B. 1991. Fission Track Thermochronology: Application to Continental Rifting of Southeastern Australia. *The Australian Petroleum Exploration Association Journal*, **31**, 131–142.

ELLACOTT, M. V., RUSSELL, N. J. & WILKINS, R. W. T. 1994. Troubleshooting vitrinite reflectance problems using FAMM: A Gippsland and Otway Basin case study. *The Australian Petroleum Exploration Association Journal*, **34**, 216–230.

FEATHERSTONE, P., AIGNER, T., BROWN, L., KING, M. & LEU, W. 1991. Stratigraphic Modelling of the Gippsland Basin. *The Australian Petroleum Exploration Association Journal*, **31**, 105–115.

FINLAYSON, D. M., JOHNSTONE, D. W., OWEN, A. J. & WAKE-DYSTER, K. D. 1994. Deep seismic profiling: basement controls on Otway Basin development. *In*: FINLAYSON, D. M. (ed.) NGMA/PESA Otway Basin Symposium Abstracts, *AGSO Record* **1994/14**, 13–18.

GIBBS, A. D. & GRIFFITHS, P. A. 1994. The third and fourth dimensions in balance and restoration. Abstract in 'Modern Developments in Structural Interpretation, Validation and Modelling', Geological Society of London, February 1994.

GREEN, P. F., DUDDY, I. R. & BRAYY, R. J. 1995. Applications of Thermal History Reconstruction in inverted basins. *In*: BUCHANAN, J. G. & BUCHANAN, P. G. (eds) *Basin Inversion.* Geological Society, London, Special Publications **88**, 149–165.

——, ——, GLEADOW, A. J. W. & LOVERING, J. F. 1989. Apatite fission track analysis as a palaeotemperature indicator for hydrocarbon exploration. *In*: NAESER, N. D. (ed.) *Thermal history analysis in sedimentary basins.* Springer Verlag, 181–195.

HILL, K. A., FINLAYSON, D. M., HILL, K. C. & COOPER, G. T. 1995. Mesozoic tectonics of the Otway Basin Region: the legacy of Gondwana and the Active Pacific margin; – a review and ongoing research. *The Australian Petroleum Exploration Association Journal*, **35**, 332–358.

——, ——, ——, PERINCEK, D. & FINLAYSON, B. 1994*a*. The Otway Basin: pre-drift tectonics. *In*: FINLAYSON, D. M. (ed.) NGMA/PESA Otway Basin Symposium Abstracts. *AGSO Record* **1994/14**, 43–48.

—, COOPER, G. T., RICHARDSON, M. J. & LAVIN, C. J. 1994b. Structural framework of the Eastern Otway Basin: inversion and interaction between two major structural provinces. *Exploration Geophysics*, **25**, 79–87.

HILL, K. C., HILL, K. A., COOPER, G. T., O'SULLIVAN, A. J., O'SULLIVAN, P. B. & RICHARDSON, M. J. 1995. Inversion around the Bass Basin, SE Australia. *In*: BUCHANAN, J. G. & BUCHANAN, P. G. (eds) *Basin Inversion*. Geological Society, London, Special Publications, **88**, 525–547.

KUSZNIR, N. J. & ZIEGLER, P. A. 1992. The mechanics of continental extension and sedimentary basin formation: A simple-shear/pure-shear flexural cantilever model. *Tectonophysics*, **215**, 117–131.

—, ROBERTS, A. & MORLEY, C. 1994. Forward and reverse modelling of rift basin formation. *In*: LAMBIASE, J. J. (ed) *Hydrocarbon Habitat in Rift Basins.* Geological Society, London, Special Publications, **80**, 33–56.

LASLETT, G. M., GREEN, P. F., DUDDY, I. R. & GLEADOW, A. J. W. 1987. Thermal annealing of fission tracks in apatite, 2. A quantitative analysis. *Chemical Geology*, **65**, 1–13.

MA, X. Q. & KUSZNIR, N. J. 1994. Effects of rigidity layering, gravity and stress relaxation on 3-D subsurface fault displacement fields. *Geophysical Journal International*, **118**, 201–220.

MCHUGH, E. A., DIESSEL, C. F. K. & KUTZNER, R. 1991. Use of fluorescence microscopy in the detection of low level oxidation in bituminous coals. *Fuel*, **70**, 647–653.

MCKENZIE, D. P. 1978. Some remarks on the development of sedimentary basins. *Earth and Planetary Science Letters*, **40**, 25–32.

SCOTT, D. L., BRAUN, J. & ETHERIDGE, M. A. 1994. Dip analysis as a tool for estimating regional kinematics in extensional terranes. *Journal of Structural Geology*, **16**, 393–401.

SIBSON, R. H. 1995. Selective reactivation of faults during basin inversion: Implications for fluid redistribution. *In*: BUCHANAN, J. G. & BUCHANAN, P. G. (eds) *Basin Inversion.* Geological Society, London, Special Publications, **88**, 3–19.

STRUCKMEYER, H. I. M. & FELTON, E. A. 1990. The use of organic facies for refining palaeoenvironmental interpretations: A case study from the Otway Basin, Australia. *Australian Journal of Earth Sciences*, **37**, 351–365.

SWEENEY, J. J. & BURNHAM, A. K. 1990. Evaluation of a simple model of vitrinite reflectance based on chemical kinetics. *American Association of Petroleum Geologists Bulletin*, **74**, 1559–1570.

VEEVERS, J. J., POWELL, C. M. & ROOTS, S. R. 1991. Review of Sea floor spreading around Australia I. Synthesis of the patterns of spreading. *Australian Journal of Earth Sciences*, **38**, 373–98.

WILKINS, R. W. T., WILMHURST, J. R., RUSSELL, N. J., HLADKY, G., ELLACOTT, M. V. & BUCKINGHAM, C. 1992. Fluorescence alteration and the suppression of vitrinite reflectance. *Organic Geochemistry*, **18**, 629–640.

WILLIAMS, G. D. & VANN, I. R. 1987. The geometry of listric normal faults and deformation in their hangingwalls. *Journal of Structural Geology*, **9** (7), 789–795.

Discussion of potential errors in fault heave methods for extension estimates in rifts, with particular reference to fractal fault populations and inherited fabrics

C. K. MORLEY

Department of Petroleum Geoscience, University of Brunei Darussalam, Gadong 3186, Brunei

Abstract: Fault heave methods for estimating extension (including balanced cross-sections and the flexural cantilever model) are important because they are the only methods that are universally applicable to estimating extension from the syn-rift stratigraphic section. One of the main criticisms of the methods when based on reflection seismic data is the failure to account for crustal extension along sesimically invisible faults (displacements up to a few tens of metres). The fractal distribution of several fault populations has been cited as evidence for the existance of such minor faults. It has been suggested by several workers that the discrepancies between extension estimates for the North Sea rifts calculated from the syn-rift section and by back-stripping (using the McKenzie model) can be explained by a substantial amount of crustal extension having taken place on seismically invisible faults. This paper examines whether it is reasonable to extrapolate such conclusions to other rifts.

Estimates for the contribution of minor faults to crustal extension based on fractal distributions of faults need to be treated with caution for several reasons: (1) in some rifts large boundary faults (several kilometres displacement) may account for 80–95% of the total extension amount. In such rifts minor faults may not contribute significantly to crustal extension (less than 5%). (2) The different distribution of fault size and spacing between different zones within rifts suggests that a single fractal distribution along a regional rift traverse is commonly unrealistic. (3) Fractal fault populations measured in syn-rift sedimentary sequences may be measuring faults involved in gravity tectonics and compaction, not just those faults that contribute to crustal extension.

Regions with very complex pre-rift structural histories (e.g. North Sea) appear to have significant amounts of extension on minor faults. This contrasts with rifts developed on crusts with a relatively stable history (e.g. East African rift), where much of the extension is accommodated on a few large faults. The differences between these two rifts (despite both rifts having undergone similar amounts of extension) illustrate that pre-existing fabrics are extremely significant for understanding fault populations, fault orientations, partitioning of extension between different sizes of fault, the types of linkage between faults (transfer zones), syn-rift basin subsidence and fault lengths. An appreciation of this variability should lead to more accurate prediction of the contribution of seismically invisible faults to extension on balanced cross-sections.

There are six main methods for estimating extension in rifts: (1) the McKenzie model (pure shear, Airy isostacy assumption for syn-rift geometry, McKenzie 1978); (2) summing fault heaves from seismic data, or from surface data (balanced sections); (3) tables of extension for bedding and fault angle (Wernicke & Burchfiel 1982); (4) the flexural cantilever model (combined simple shear-pure shear model, flexural isostatic assumption for syn-rift section, Kusznir & Egan 1990); (5) back-stripping of the post-rift sag basin (using Airy or flexural isostacy, e..g. McKenzie 1978; Steckler & Watts 1978; Royden & Keen 1980; Roberts *et al.* 1993); and (6) Moho geometry.

Much work on estimating extension has been conducted in the North Sea, where a wide range of techniques has been applied. Initially due to data quality and availability constraints, extension in the North Sea was measured by relating the thickness of the syn-rift section, or the post rift section, in wells to extension using the McKenzie model (McKenzie 1978). This model had the benefit of not requiring detailed regional cross-sections to estimate extension. As more data became available, in particular good quality regional seismic lines (industry and deep reflection seismic lines), other techniques were used to estimate extension including the Moho geometry (e.g. Klemperer & Hurich 1990), and fault heave methods (e.g. Roberts *et al.* 1993). The different methods have resulted in considerable variance in extension estimates and doubt has been raised about the accuracy of several

From Buchanan, P. G. & Nieuwland, D. A. (eds), 1996, *Modern Developments in Structural Interpretation, Validation and Modelling,* Geological Society Special Publication No. 99, pp. 117–135.

techniques, particularly those involving direct measurement of extension from syn-rift structures (e.g. Ziegler 1982; Barton & Wood 1984; Ziegler & Van Hoorn 1990; White 1990; see discussion in Marrett & Allmendinger 1991).

Extension estimates for other rifts may have to be made using a more restricted range of techniques (in particular fault heaves), due either to the nature of the data base, or because only syn-rift basins are present (the post-rift sequence was either eroded or has not yet developed). Yet experience from the North Sea suggests that such estimates of extension from fault heaves can be significantly inaccurate. This paper examines various problems and errors associated with estimating extension based on fault geometries and in particular concentrates on the problems of predicting seismically invisible minor faults and their contribution to crustal extension. Reasons for variations in fault populations, orientations and displacement amounts are then examined.

Problems estimating extension from a syn-rift section

Measuring fault heaves from cross-sections

The most comprehensive way to measure extension in the upper crust is to make a balanced section to ensure the geometric viability of the section, and measure the extension by comparing the line lengths of the restored and present day cross-sections (e.g. Dahlstrom 1969; Hossack 1979; Gibbs 1983; Rowan & Kligfield 1989). The difference in line length will approximately equal the sum of heaves on the faults since there is little folding in rifts. Balanced cross-sections can be constructed from surface, seismic and well data or a combination of all three. In principal if sufficient information were available then the methods based on fault geometry would be the best way of estimating extension in the upper crust. However there are usually practical limitations to the method as discussed below.

For balanced cross-sections in a syn-rift section, the assumptions of consistent bed lengths for multiple units cannot be made. The bed length of a horizon near the top of the syn-rift sequence should be longer than the bed lengths of a unit deposited near the base. Balancing of extensional fault sections therefore tends to be a series of kinematic restorations where units, whose shape tends to be controlled by the fault geometry, are progressively removed. This can permit a greater degree of variability in balancing extensional systems than exists for sections where constant bed length assumptions can be made (such as some regions of fold and thrust belts).

Extension estimates from surface data alone have several potential problems. Often outcrops are discontinuous, consequently fault and layer dip data and fault traces have to be projected. Fault dip angle is very important for extension estimates (see Wernicke & Burchfiel 1992, for example), yet difficult to measure at a surface. Whether a fault should remain planar or become listric or curved in some other way in the subsurface can be difficult to determine. Although from surface data alone it may be very difficult to accurately reconstruct the syn-rift geometry, provided that the exposure is good, analysis of surface data has the advantage of allowing extension at all scales to be observed.

Extension estimates from balancing cross-sections requires that the direction of extension is known. This can be a significant problem in many rifts since poor or non-existant exposure of fault planes may make direct measurement of striations impossible. In the western branch of the East African rift system, for example there has been considerable variance in the estimated extension direction by up to 60° (e.g. Rosendahl *et al.* 1986; Morley 1988; Ebinger 1989; Wheeler & Karson 1989; Kilembe & Rosendahl 1992).

Extension estimates from seismic reflection data

Seismic reflection data, if it is of good quality, will provide excellent images of the basin fill and fault geometry and enable accurate estimates to be made of extension on faults with more than a few tens of metres displacement. To accurately estimate extension, the sections need to be depth converted accurately. Several other potential problems with estimates from seismic reflection data are as follows.

(1) Estimating the eroded geometries of large footwall uplifts (Fig. 1). Commonly approaching rift margins the large boundary faults are associated with significant amounts of footwall uplift. In many cases no markers can be found to correlate offsets of the footwall and hanging walls, because erosion has removed the units in the footwall or they were never deposited. Some kind of flexural-isostatic modelling can help estimate the height of the eroded footwall (e.g. Kusznir & Egan 1990; Kusznir *et al.* 1991). The model balances lower crustal ductile extension and upper crust brittle extension, hence the actual surface geometry and Moho profile can be cross-checked with the model. Application of the flexural cantilever model to Lake Tanganyika, which estimates the amount of footwall erosion, suggests that the extension on the boundary faults using sesimic data and the eroded fault block geometries (Morley 1988) appears to

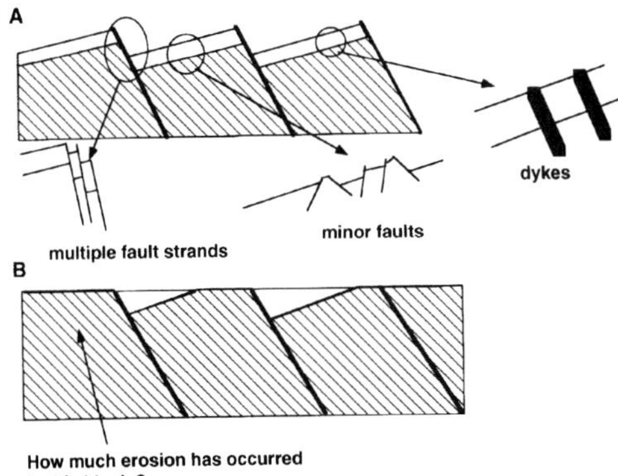

Fig. 1. Common causes of extension estimate error in cross-sections through rifts. (**a**) Extension that is commonly difficult to quantify, over-estimation of extension due to interpretation of single boundary fault in place of multiple faults; extension on minor faults and extension on intrusions. (**b**) Problem of reconstructing eroded fault blocks. In two blocks projection of horizons and faults is a relatively simple procedure. On the left-hand footwall block erosion has removed all the marker horizon so it is not possible to determine how much erosion of the block has occurred. The observed displacement is a minimum estimate only.

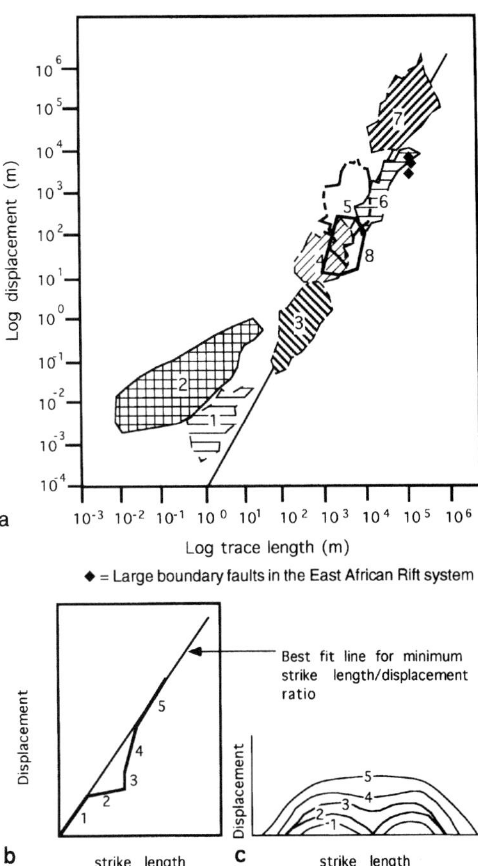

Fig. 2. Log displacement–log fault strike length relationship for faults (compiled by Davidson 1994; 1, Peacock & Sanderson 1991; 2, Davison 1994; 3, Walsh & Watterson 1991; 4, 5, 6 and 7, Marrett & Allmendinger 1991; 8, Central Kenya rift, Grimaud *et al.* 1994). The large boundary faults in the East African rift system are located in the Rukwa, Tanganyika and northern Kenya rift segments. (**b**) and (**c**) (c) is modified from Davison 1994) show how linkage of faults can create significant variations in the length–displacement relationships. A fault that grows in isolation might be expected to increase stike length and displacement in proportion, which corresponds to the straight line in (b). However if the fault links with other faults during its growth then it will depart from that linear progression. The initial linkage will produce an abrupt decrease in the displacement: strike length ratio (2 in (b) and (c)). The fault will then adjust by the displacement increasing with no (3) or little (4) increase in the strike length, until it reaches the proportions associated with the linear displacement-strike length relationship.

have considerably underestimated the extension by 30–40% (Kusznir *et al.* 1995). It should be noted that much attention on missing extension has focused on small faults, whilst the flexural cantilever method helps resolve missing extension on large faults.

(2) Imaging of fault zones as a single fault. In complex fault zones several closely spaced faults may actually be present (Fig. 1). However the seismic reflection data may simply show a no data zone across which a single fault is interpreted. Consequently the line-lengths of reflectors within the fault zone are ignored and the amount of extension is over-estimated (Fig. 1).

(3) The presence of seismically invisible faults that contribute significantly towards crustal extension. This is a significant topic of discussion and is treated separately in the following section.

Contribution of seismically invisible faults

Statistical studies of extensional faults suggests that many brittle fault populations may follow fractal size distributions (power-law exponent on a plot of log size versus log cumulative number (Kakimi 1980; Villemin & Sunwood 1987; Scholz & Cowie 1990; Walsh *et al.* 1991; Marrett & Allmendinger 1991)). Within the range of scale invariance, the

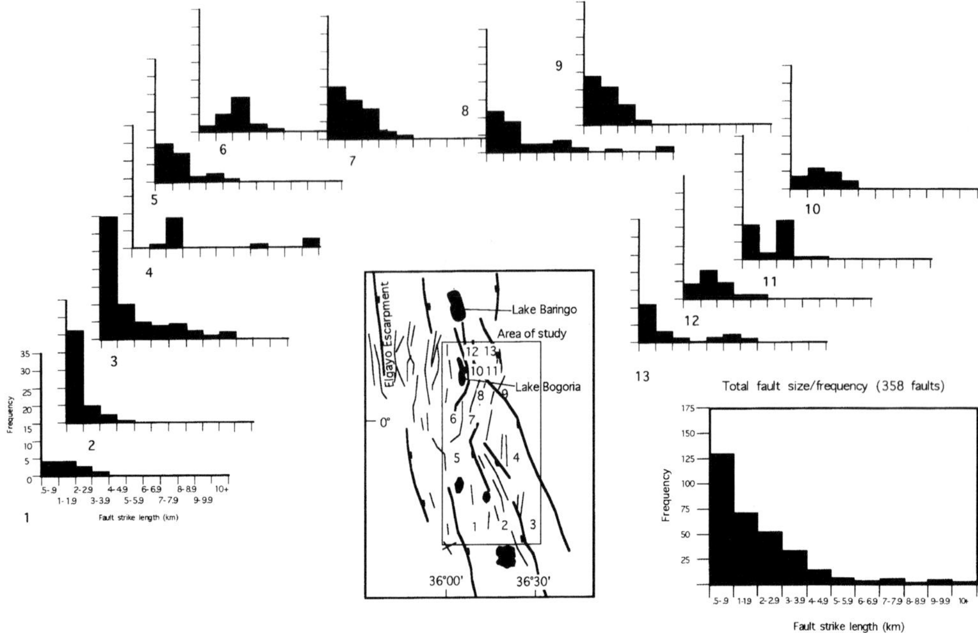

Fig. 3. Graph of fault length plotted against frequency for extensional faults measured from field maps, landsat and Spot satellite imagery over the central Kenaya rift (unpublished data and partially after Grimaud *et al.* 1994). The data illustrate considerable variation in fault characteristics between areas. Resolution of fault size below 1 km is not good, so the fault frequencies determined for the first column must be regarded as incomplete data (see Fig. 9 for location).

largest faults in a population may be used to estimate the relative numbers and sizes of smaller faults (Figs 2 & 3). Hence the total extension on small faults may be estimated (Kakimi 1980; Marrett & Allmendinger 1991, 1992).

Differences exist between extension estimates for the Viking Graben using restorations of faults derived from seismic profiles (Ziegler & Van Hoorn 1990; Marsden *et al.* 1991), and those derived from area balancing of the entire crust using deep crustal reflection profiles and gravity modelling (e.g. Klemperer & Hurich 1990). Marrett & Allmendinger (1992) have cited the Viking Graben as an example where extension on small faults can help explain the discrepancy, suggesting that extension estimates from seismic data are actually underestimates. This reasoning implies that the other methods are more accurate than the fault summing method, this is not necessarily so.

Fractal fault analysis for a region can be done in 2D map or 3D volumetric analysis or more specific analysis can be made along a one dimensional traverse (Marrett & Allmendinger 1992). All methods have significant shortcomings in practical application for estimating the extension amount on sub-resolution (seismic or other) faults.

Fractal analysis of 2D and 3D fault populations

In order to make use of fractal analysis involving large populations it is usually necessary to use samples from a rock volume or area and not from a specific single line traverse. The regional data plotted on a fault size-cumulative frequency log–log graph commonly closely corresponds to a straight line (Fig. 2). When it does the data can be assumed to be fractal within that range and used to infer the amount of extension that is attributable to a certain size of fault (e.g. Walsh *et al.* 1991). This method can work very well for relatively local studies (e.g. individual oil fields (e.g. Gauthier & Lake 1993)), but it reaches a point where vertical and areal differences in fault geometry make the method inaccurate and invalid.

There is also another major inaccuracy involved with the sampling of faults from 2D or 3D data in order to estimate extension amount. In the analysis each fault is only counted once hence a large fault that crosses the entire map will be counted equally with a minor fault that may cross only a small portion of the map. Figure 3 illustrates the problem. There are one 500 m fault, five 100 m faults, fifteen

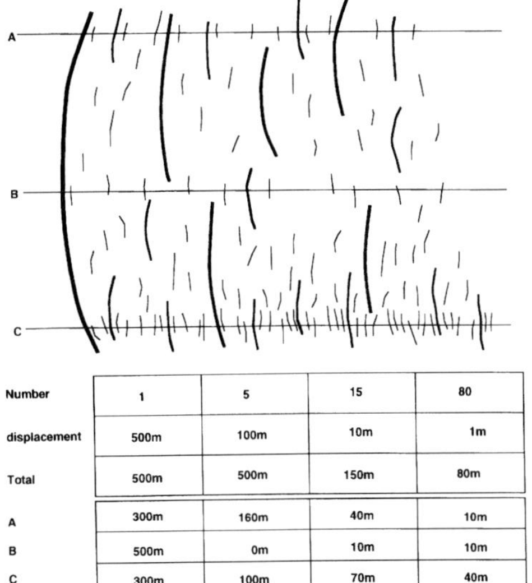

Number	1	5	15	80
displacement	500m	100m	10m	1m
Total	500m	500m	150m	80m
A	300m	160m	40m	10m
B	500m	0m	10m	10m
C	300m	100m	70m	40m

Fig. 4. Changes in minor fault population caused by variations in the extension amount of the largest fault. Assuming that the extension amount is constant throughout the area, then as the largest fault changes displacement along strike, so the extension amount on the minor fault population must change also to compensate. Measuring faults from an area would not enable accurate prediction of the fault population likely to be encountered on a traverse (see difference between the total fault population displacement and the displacements on individual lines A, B, and C).

10 m faults and eighty 1 m faults. If extension was (incorrectly) calculated using the whole fault population the results would show it was distributed amongst the faults 40.5%, 40.5%, 12% and 7% respectively. Extension amount is not however an areal quantity: it should be measured along a line. Three traverses are made across the map to measure extension. It is apparent that the partitioning of extension measured along the traverses differs significantly from that derived from the map in two ways. (1) The largest fault is counted on all three traverses whilst only a fraction of the total smaller fault population is encountered on each traverse. Consequently the contribution of the larger faults to the total extension amount (even though the extension varies along strike) is increased in comparison with the area estimate. (2) The largest fault is shown to change extension amount along strike, yet the overall extension amount is kept approximately constant. As extension amount decreases on the larger fault the contribution of minor faults to the total extension amount becomes more important (this is seen in the Lake Tanganyika and Gulf of Suez example given below). Consequently the distribution of fault sizes along a specific traverse will often differ considerably from those of the overall fault population.

To illustrate the same problem with natural data three seismic lines from the same rift segment in Lake Tanganyika are considered (Fig. 5). Line 24 is a half graben, as such it is bounded by a single major fault. This fault has a throw of 5 km. There is then a jump down to the next fault size (tens to hundreds of metres throw). There is no intermediate fault population (500 m–5 km throw). The partitioning of the throw between the single boundary fault (5 km throw) and the remaining seismically visible fault population (2.1 km total throw) at the Precambrian basement level is 75% of the throw is attributable to the boundary fault and 25% throw on the minor faults. Line 210 is a full graben bounded by two major faults (Fig. 5, faults 1 and 2). At the top Precambrian basement level the throw on the two boundary faults (from flexural cantilever modelling) is 12.5 km, whilst the cumulative throw on the minor faults is 1.4 km. Thus 90% throw is attributable to the two major faults whilst 10% throw is attributable to seismically visible minor faults. Line 38 is a full graben where only one minor fault is visible between the two major boundary faults (faults 1 and 2, Fig. 5c), virtually all the throw in the main part of the rift is concentrated on the boundary faults. With rifts of the type described above the contribution of seismically invisible faults to total extension is likely to be small, but varying, being probably more significant for line 24 than the other two lines. If the rift does not contain such large boundary faults then the influence of minor faults will become much more important. This relationship has been discussed by Marrett & Allmendinger (1991).

The Gulf of Suez also illustrates a systematic change in fault populations, apparently related to large transfer zones. Cross-sections spaced every 20 km have been constructed (Figs 6 & 7) to illustrate the location of faults and the amount of throw on the fault (for seismically visible faults). The proportions of extension between the largest faults in the section and the other visible faults can be seen to vary along the rift. In the centres of large-scale dip provinces the large faults contribute highly to the extension amount. At transfer zones their importance diminishes and the contribution of minor faults to the total extension amount becomes greater. Plotting extension amounts on each section from NW to SE along the Gulf (Fig. 8) it can be seen that the total extension increases to the south. If a smooth line is fitted to the total extension then the difference between the smoothed and observed lines gives some crude impression of

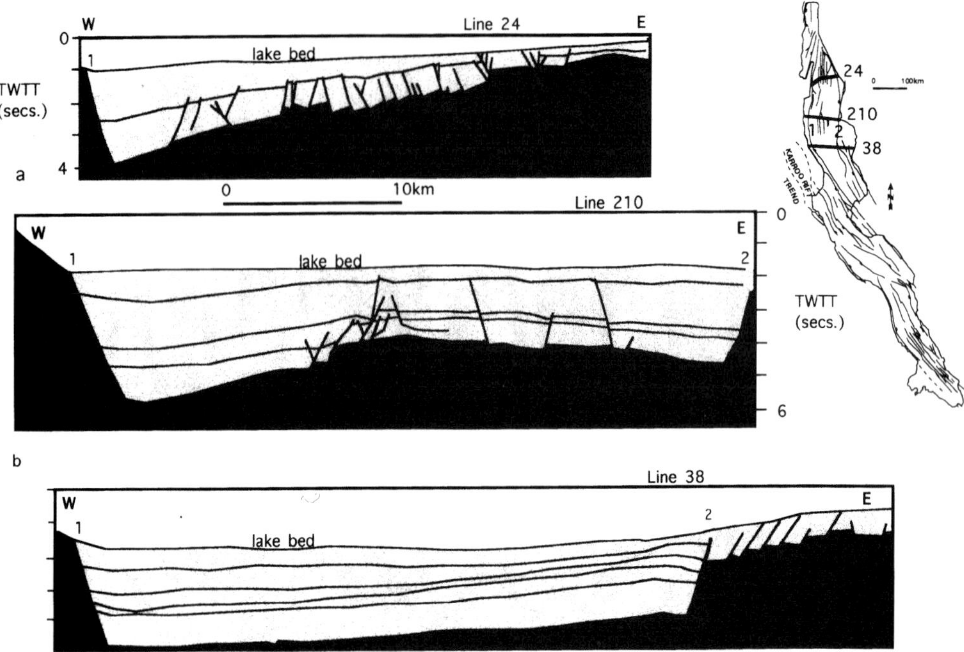

Fig. 5. Variations in the seismically visible fault population within a rift segment, sections taken from seismic lines across Lake Tanganyika. The profiles illustrate how minor faults are more frequent and contribute more to the total extension than the visible minor faults associated with full grabens. Note also the change from numerous minor faults to virtually no minor faults passing from the eastern footwall to the full graben floor of Line 38, thus indicating that it can be dangerous to predict regional fault population patterns from data originating from a few limited areas.

the variations in importance of the invisible minor faults to the extension amount. The data suggests that invisible faults contribute more to crustal extension in the regions of major transfer zones than at the centres of dip provinces. This is in agreement with the observation that the density of detectable visible minor faults increases near transfer zones (Figs 6 & 8).

The central Kenya rift displays considerable variation of fault sizes and frequencies in different areas (Fig. 3). The total fault population (of faults mappable at a scale of 1:50 000) for an area of 2100 km^2 displays the classic distribution of decreasing frequency of occurrence with increasing fault length. However different subsets of the total population (measured as faults that lie within or cut the boundary of a 5 km^2 area) display considerable departures in length–frequency relationships from that of the total fault population (Fig. 3). These changes in distribution, unlike the examples above, are for an area which is not dominated by large boundary faultts. They demonstrate that minor fault populations can also display considerable lateral variations in their characteristics.

Fractal analysis of faults along a 1D traverse

From the discussion above it is apparent that to obtain meaningful estimates of extension from fractal populations the data must be collected from a traverse in close proximity to the section along which extension is being estimated. Unfortunately for many rifts this may be impossible to do in practice because it requires: (1) that a high number of measurements can be made in a relatively small area, and (2) that the measurements correspond to the structural level at which extension is being estimated. In order to estimate the maximum extension for example, measurements must be made at the base of the syn-rift sequence.

Even along a specific traverse it is necessary to question the fractal assumptions, because fault spacing and displacement varies considerably within rifts. Some blocks may be intensely broken up by minor faults, other blocks may only be bounded by larger faults and internally little deformed (Fig. 5c). The assumption that a single fractal fault distribution can be applied along a regional traverse therefore may break down. The

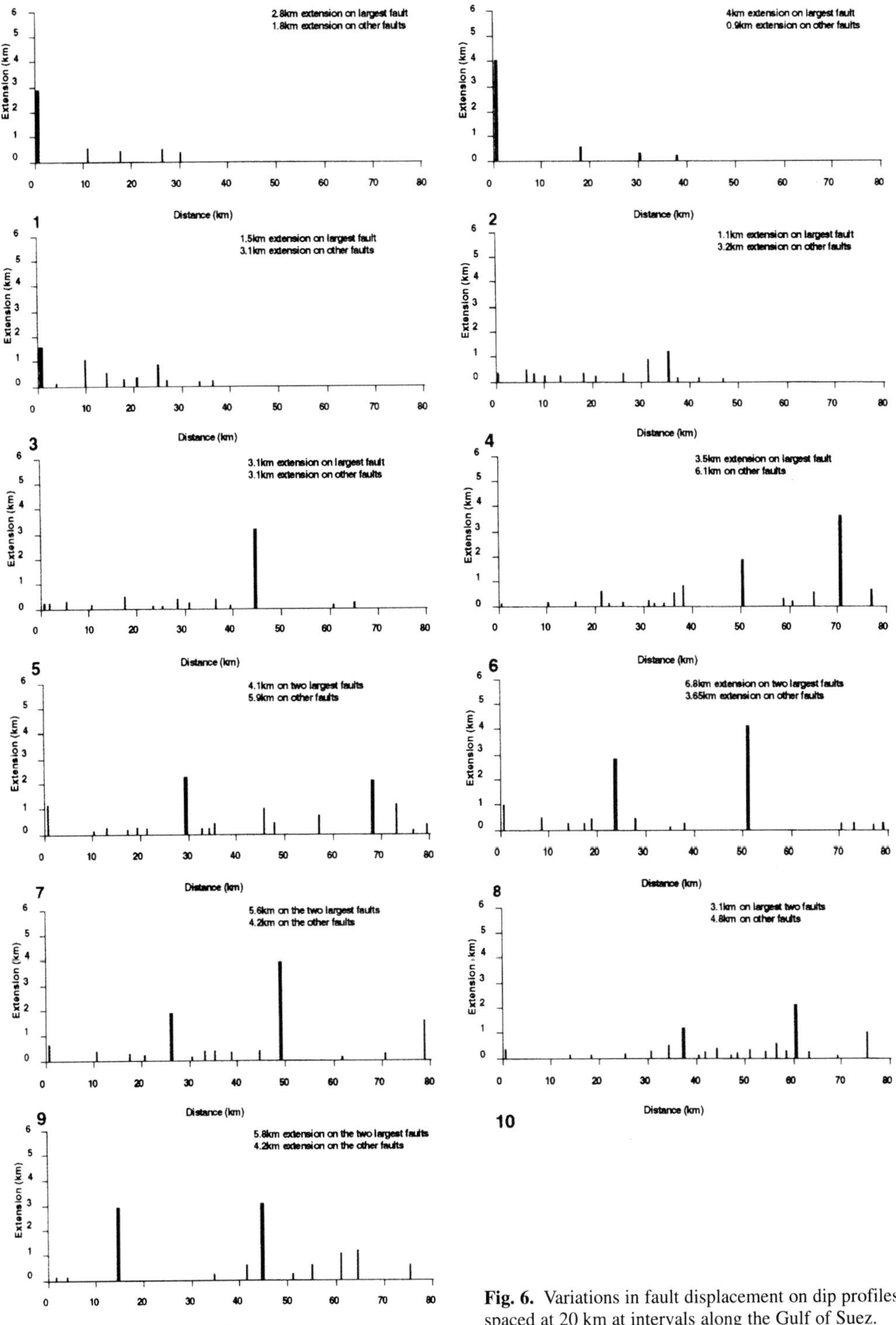

Fig. 6. Variations in fault displacement on dip profiles spaced at 20 km at intervals along the Gulf of Suez.

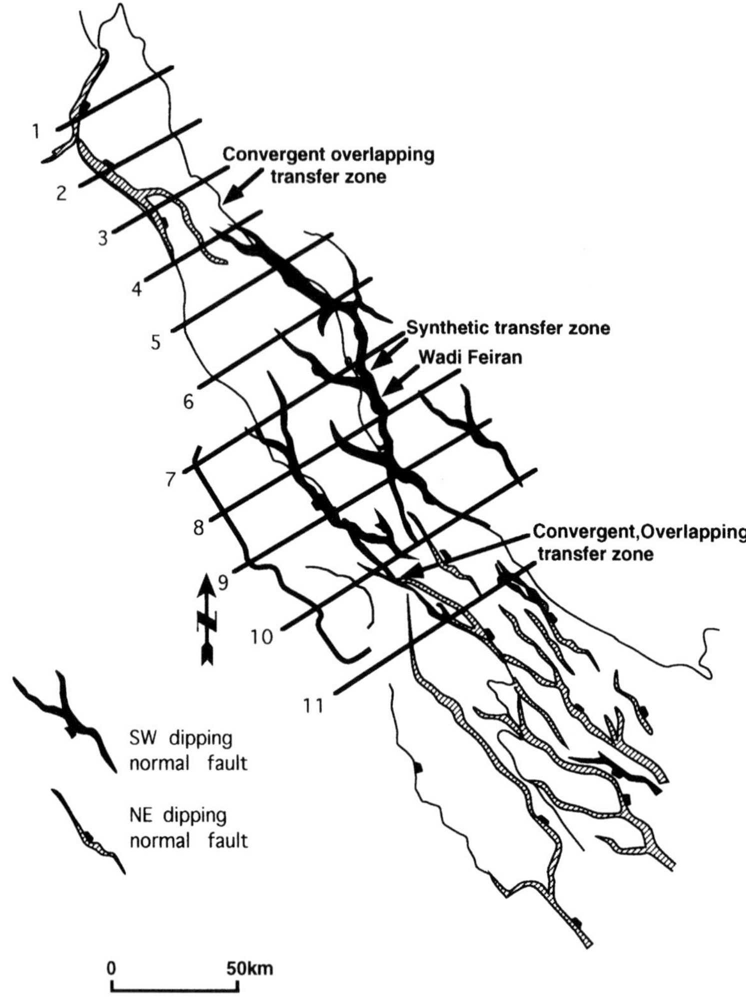

Fig. 7. Location map of the Gulf of Suez (for Fig. 6, fault pattern after Patton *et al.* 1994).

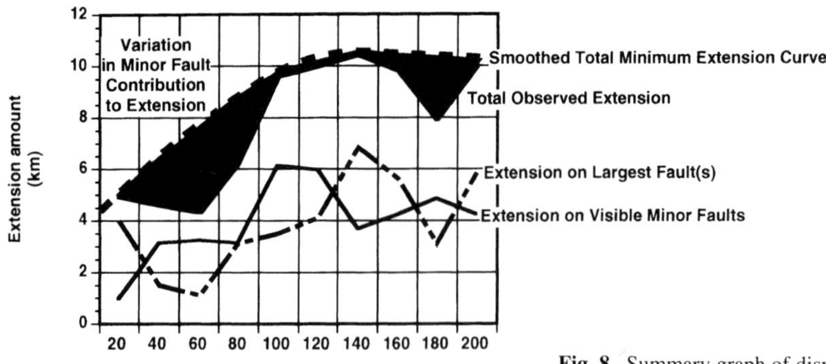

Fig. 8. Summary graph of displacement data for the Gulf of Suez (Fig. 6).

fault distribution analysis needs to be targeted for specific regions within a rift. One further problem with measuring fault populations in the syn-rift section is the origin of the faults. The whole fault population in a rift can be partitioned into faults that contribute to crustal extension and faults that are only related to shallow deformation and do not link to crustal extension, for example gravity sliding, local stresses (e.g. outer arc extension on a fold) or compaction. The tops of tilted fault blocks are the regions where shallow detachment faults, gravity sliding and slumping have been extensively recorded. These are the very areas where most cores have been taken by industry, and that may have subsequently been used in fractal analysis of fault populations (Walsh *et al.* 1991).

Comparison of fault patterns in the East African rift and North Sea

From the discussion so far it is apparent that fault population properties change significantly within and between rift systems. In order to use fault population statistics to predict extension amount on faults beyond the range of observable data (for example satellite or seismic data) it is necessary to understand some of the potential causes for the variations (e.g. areal variations in deformation intensity, changes in lithospheric conditions such as temperature, elastic thickness, mineral composition and pre-esisting fabrics). Here it is suggested that the pre-existing fabric pattern exerts a considerable influence on the fault patterns and populations that are developed in rifts. To demonstrate this point the East African rift system and the North Sea late Jurassic rifts are compared below.

Nature of the pre-existing fabrics

The East African rift system is a late Tertiary age rift system that is split into two branches. Basement control at on rift geometry at the largest scale is seen by the rift branches, which avoid the Archaean cratonic areas and lie within the Proterozoic mobile belts (McConnell 1972). The rift sediments and volcanics commonly lie directly on crystalline Precambrian basement, although in places the rift cross-cuts or follows segments of late Palaeozoic (Karroo) and Cretaceous–Palaeogene rift systems (Fig. 9). The effects of pre-existing structure on rift geometry are largely controlled by ductile fabrics in Precambrian basement (McConnell 1972; Daley *et al.* 1989; Smith & Mosley 1994). In places the Karroo and Cretaceous–Palaeogene rift basins structural trends appear to exert an influence on the East African rift system. These earlier rifts themselves tend to follow the ductile Precambrian basement fabrics, hence basement fabrics commonly affect superimposed rift systems.

The Western branch of the rift system appears to have been avtive from the Upper Miocene and is still seismically active (e.g. Ebinger 1989), i.e. a total rift duration of about 10 Ma. Heat flow data and the absence of extensive volcanism indicate that the rift branch is relatively cold.

The North Sea is a large rift province that displays a variety of rift geometries, some of which are controlled by pre-existing fabric. Perhaps one of the most extreme examples of the control of pre-existing fabrics is the Witch Ground Graben. It is developed on a thick sedimentary sequence whose ages spans the Palaeozoic and Mesozoic. A considerable number of tectonic events have affected these sedimentary rocks including Caledonian compressive and strike-slip deformation, Hercynian compression, strike-slip and extension and Triassic extension, although not all the tectonic events are necessarily present in one area (e.g. Coward 1993). Thus many brittle fabrics were present in the sedimentary section prior to rifting during the Late Jurassic (e.g. Batholomew *et al.* 1993). In the Witch Ground Graben minor extension affected the Sgiath and Piper Formations, the main rift phase occurred during Kimmeridge Clay Formation times, and died out in the Early Cretaceous. The duration of rifting was about 15 Ma (Boldy & Brealey 1990; O'Driscoll *et al.* 1990; Hibbert & MacKertich 1993).

Fault density and displacement

Seismic reflection data acquired over Lakes Malawi, Rukwa and Tanganyika, and the northern Kenya rift have shown that the extension and subsidence is dominated by large boundary faults (e.g. Ebinger *et al.* 1984; Burgess *et al.* 1988; Morley 1988; Morley *et al.* 1992*a*, *b*). Some faults are 100–250 km long, display up to 7–10 km of extension (heave) and up to 7 km of syn-rift fill (throw) in the hanging wall (Fig. 10). As previously discussed boundary faults appear to be responsible for a large percentage of the total extension (75–95%), with minor seismically visible faults playing a much less important role. With rifts of the type described above the contribution of seismically invisible faults to total extension at the top basement level is likely to be small. It should be noted that the Kenya rift displays mixed deformation styles with an early (Palaeogene–Mid-Miocene) half graben style dominated by large boundary faults, and a later (Late Miocene–Pleistocene) minor fault swarms style, that may be related to the development of large intrusive bodies shallow in the crust.

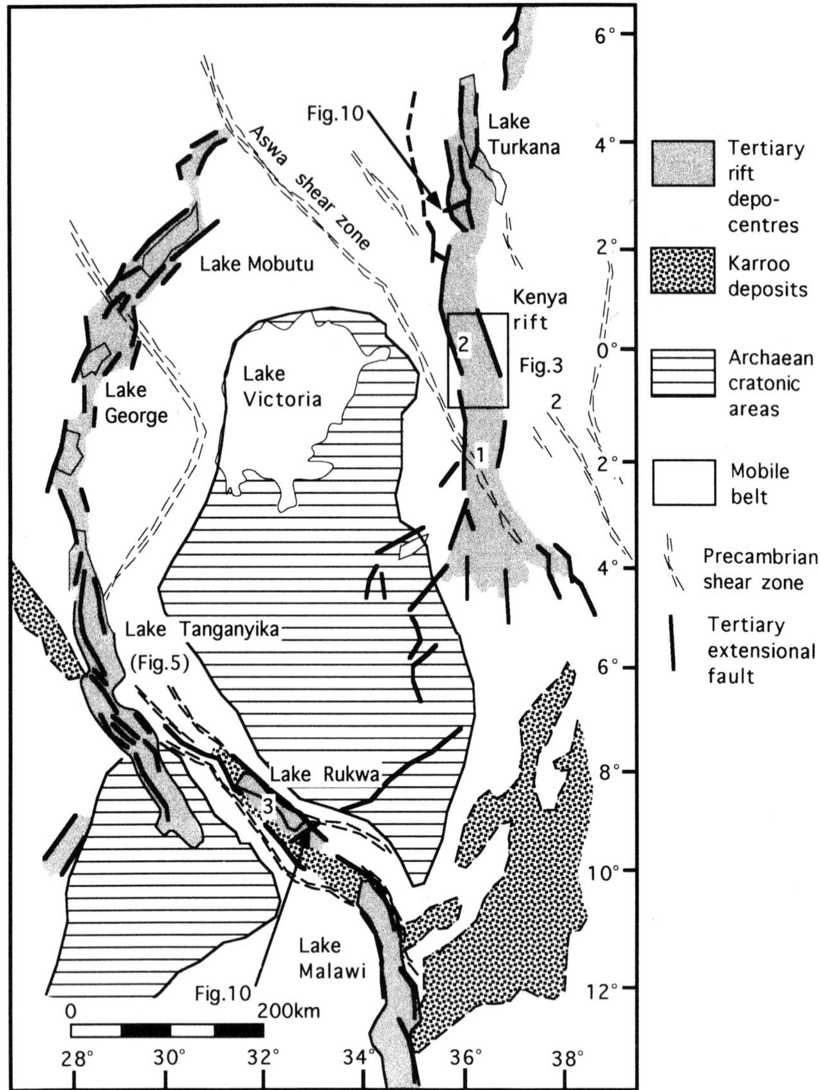

Fig. 9. Location map for the East African rift system. Modified from Morley (1995), and compiled from Daley *et al.* (1989), Morley *et al.* (1992*b*) and Smith & Mosley 1994. Number 1, 2 and 3 denote location of areas for fault orientation data (Fig. 11).

The syn-rift section in the Witch Ground Graben even near major boundary faults rarely exceeds 1 km. However in places thicknesses may reach 2 km when rifting continued into the Lower Cretaceous. Extension on the main faults zones reached a maximum of about 3 km, and in general extension is more evenly distributed across numerous minor faults than in the East African rift system (Table 1).

Despite the similar duration of rift activity and similar Beta factors significant amounts of extension are distributed across many faults of different sizes in the Witch Ground Graben. On the contrary extension is concentrated on a few large faults in much of the East African rift system including Lakes Tanganyika (Fig. 5), Malawi and Rukwa (Fig. 9).

It should be noted that outside of the Witch Ground Graben but within the North Sea area there is a great range in the size of boundary faults. For example the south Viking Graben appears to be dominated by large boundary faults (e.g. Rattey & Hayward 1993). Even between the northern and southern areas of the Viking Graben there are

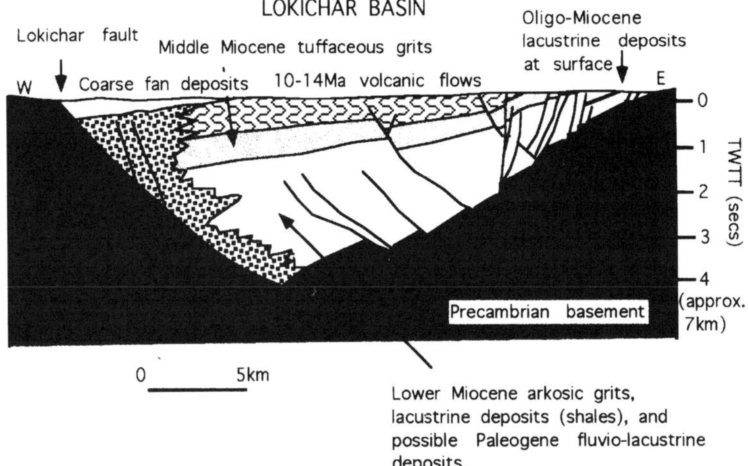

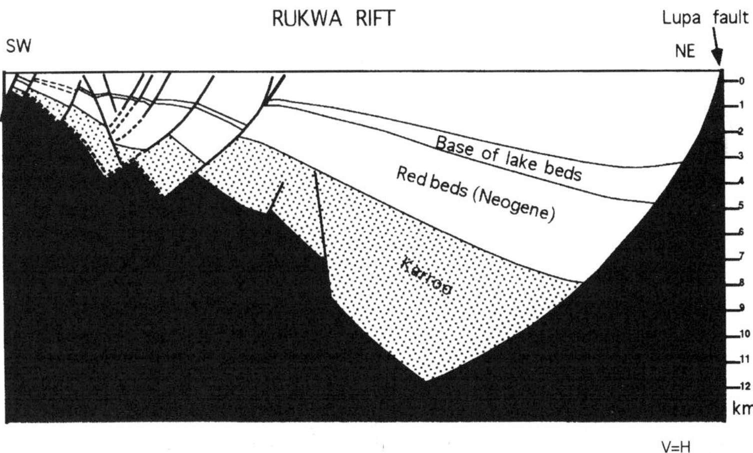

Fig. 10. Cross-sections through a quasi-pure extension portion of the East African rift, the Lokochar basin, northern Kenya rift, and an oblique extension area, Lake Rukwa (see Fig. 9 for location). The basin faults and geometry are based upon seismic reflection data (see Morley *et al.* 1992*b*). For the Lokichar basin cross-section lithologies and ages are speculative, projected from the surface geology, inferred from seismic reflector geometries and generalized modes for rift basin fill.

differences, in the north Viking Graben minor seismically visible faulting appears to be less important (e.g. Rattey & Hayward 1993, figs 11 & 12). In those parts of the North Sea where boundary faults are dominant the syn-rift section rarely exceeds 3 km thickness (i.e. only half the thickness of the section commonly found in the East African rift system). The presence of larger faults may indicate areas of fewer pre-existing fabrics or areas where fabrics of a more consistent orientation are present, or areas where the fabrics do not have low shear strengths.

Table 1. *Comparison of fault size and spacing between Lake Tanganyika and the Witch Ground Graben*

	Av. spacing	Range	Throw on major faults	β factor
Witch Ground Graben (337 km)	1 fault every 2 km	1.5–2.25 km	< 3 km	c. 1.3 +
Lake Tanganyika (775 km)	1 fault every 5 km	2.4–10 km	6–7 km	c. 1.1–1.2

Fault orientations

The marked difference in fault displacement and fault density distribution between the Witch Ground Graben and the Western Branch of the East African rift is accompanied by a significant difference between the pattern of fault orientations in the two areas. These differences point to the influence of pre-existing fabrics on the rift geometry.

The regional extension direction in the East African rift system is thought to be E–W (Morley 1988; Strecker *et al.* 1990; Ring *et al.* 1993), consequently N–S-oriented rift segments have undergone approximately pure extension. The important fault orientations for individual N–S segments in the East African rift tend to be restricted to a narrow range of angles. One of the simplest patterns is found in the southern Kenya rift where the fault strike orientations are spread symmetrically about an orientation parallel to the rift axis (Fig. 11). The faults orientations occupy a span of about 20–25° on either side of the (N–S) symmetry axis, this is similar to physical models where the material is isotropic (e.g. Freund & Merzer 1976). The spread of frequently used orientations is commonly slightly larger in other rifts (e.g. Gulf of Suez, 35°, Robson 1971, Viking Graben, 25–35°), where the influence of a wider range of pre-existing fabric is manifest. Where the East African rift is deflected from a N–S orientation, there is correspondence between the basement fabric orientation and the strike of the deflected rift axis (McConnell 1972; Daly *et al.* 1989). The obliquely oriented Lake Rukwa displays a mixture of the N–S oriented faults, formed perpendicular to the regional extension direction and the dominant NW–SE trend formed parallel to the basement fabric (Fig. 11). Thus the fault orientations in the East African rift display a restricted range.

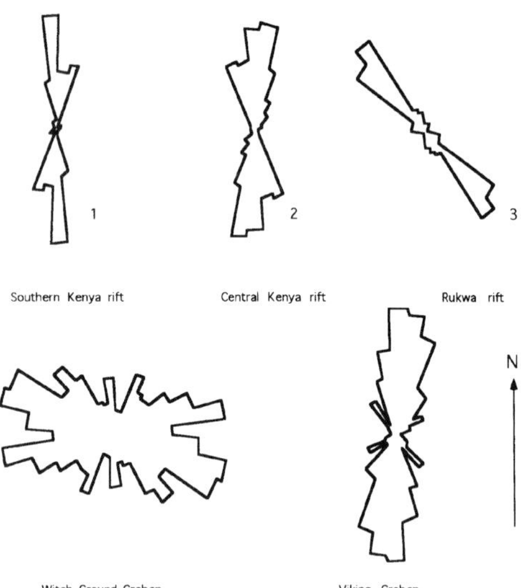

Fig. 11. Variations in fault orientation in the East African rift system (see Fig. 9 for location), the Witch Ground Graben and the Viking Graben. The diagrams represent the average orentations of 1 km long fault segments and their frequency of occurrence.

Within the Viking Graben the fault orientations are intermediate between the simplest parts of the Kenya rift (Fig. 11) and the complexity of the Witch Ground Graben. The fault frequency versus orientation and length versus orientation distributions are only slightly asymmetrical, but show a wider spread of fault orientations (25–35°) either side of the symmetry axis than the simple parts of the southern Kenya rift. On the Shetland Platform the influence of pre-existing trends on Jurassic faults is very apparent. In particular the north to northwest and northeast Caledonian trends and NE–SW-trending Triassic faults (Roberts *et al.* 1993; Platt 1995). The overall N–S trend to the platform is frequently interrupted by faults or fault segments with NE or NW trends. The Tern–Eider ridge for example forms an important NE–SW-trending oblique element (Panhuys-Sigler *et al.* 1991), whilst smaller oblique faults bound many tilted fault blocks (e.g. Harding 1984; Spencer & Larsen 1990).

The fault orientation pattern for the Witch Ground Graben displays multiple important fault orientations where there is no simple symmetry as that described above for the Kenya rift. E–W, WNW–ESE, N–S and NE–SW trends are all important (Fig. 11). This reflects the influence of pre-existing fabrics where the E–W orientation may reflect Triassic rift trends and the NE–SW orientation reflects Caledonian and Carboniferous trends.

The Witch Ground Graben probably formed under oblique extension. The most similar area in the East African Rift for comparison is Lake Rukwa, which also formed under oblique extension (Morley *et al.* 1992*a*). Both areas are oriented oblique to the regional extension direction because it was easier for faults to exploit the pre-existing fabric, than to create new faults that would form a fault system sub-perpendicular to the regional extension direction. Despite the similar tectonic position, the fault pattern produced by the two rift segments is very different (Fig. 11). This points to differences in the type, frequency and variety of orientations of pre-existing fabrics affecting the two rift segments.

The increased influence of pre-existing fabrics in oblique zones suggests that in comparing rift segments within and between rifts the extension-perpendicular rift segments might exhibit more similarities than the oblique segments. This is true of the comparison of fault orientations between the Rukwa rift and the Witch Ground Graben and the southern Kenya rift and the Viking Graben. Despite the different fabric pattern the Rukwa rift still exhibits one of the main characteristics of the East African rift system, that is the dominance of the boundary faults, the south eastern part of the lake displays up to 7.5 km thickness of Upper Miocene to recent rift-fill sediments (Figs 5 & 10). This is not matched anywhere in the North Sea.

Boundary faults in the East African rift system may zig-zag and gently curve, but in general do not undergo abrupt changes in orientation. The range of variation in orientation of major fault systems in the East African rift system are given in Figs 9 & 11. They tend to display a limited range of orientations within about 30–40°. On the other hand the Witch Ground Graben displays boundary fault systems that undergo very pronounced changes in orientation (such as the Halibut Horst boundary fault zone, Fig. 12), and in general the largest faults in the Witch Ground Graben tend to be shorter than those in the East African rift system (Table 1).

Transfer zone geometries

Transfer zones or accommodation zones occur where faults die out and transfer their displacement onto other fault systems (e.g. Bosworth 1985; Morley *et al.* 1990). With time faults in rifts may propagate along strike, amalgamate, deactivate and reactivate, and new faults may be initiated. Consequently the geometry of transfer zones will also change. It is also to be expected that pre-existing fabric may exert a strong control on the location and geometry of transfer zones. Laboratory experiments have shown that pre-existing discontinuities may inhibit fracture propagation if the frictional shear strength of the discontinuity is sufficiently low relative to the tensile strength of the surrounding material (Teufel 1979; Teufel & Clark 1984). Fractures may also be diverted by pre-existing fabrics. In low mean stress environments the angle between the propagation direction and cross-trends is important. Fractures preferentially turn into discontinuities at low angles (30–60°), whilst propagating across them at high angles (Blanton 1982). The relationship between fault length and displacement can be highly variable depending upon the maturity of the fault linkage (i.e. high length: displacement ratios when two or more faults have just linked, to low length: displacement ratios after considerable displacement on the linked system). Plots of fault length against displacement show a scattered but linear relationship for many fault populations and commonly display minimum ratios of about 10:1 to 20:1 (e.g. Gillespie *et al.* 1993; Davison 1994, fig. 2). Hence if pre-existing fabrics limit the strike length of faults then it means that the maximum displacement on those faults will also be limited. If two fault systems are compared, one where the lateral propagation of faults is inhibited by discontinuities, the other where there are no discontinuities, then it is to be expected that more,

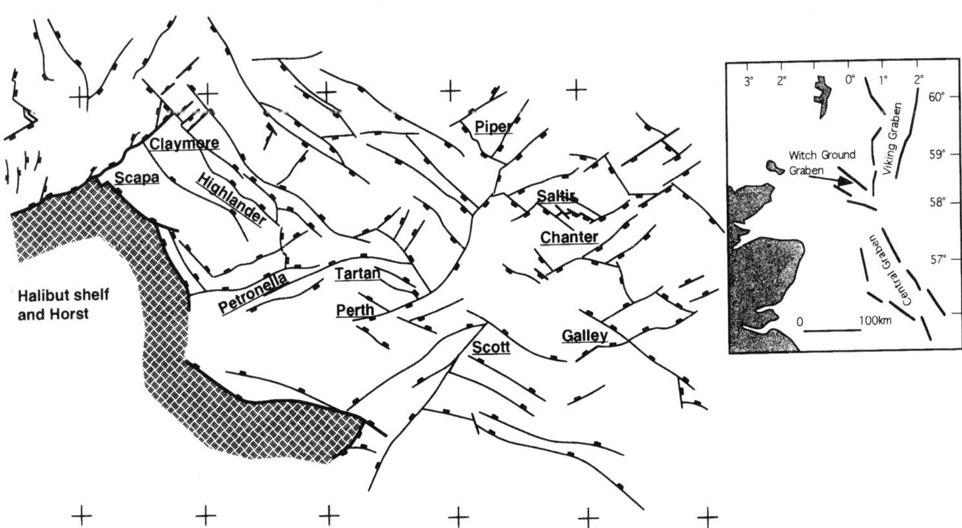

Fig. 12. Regional fault map, Witch Ground Graben.

shorter faults will be required to accomplish the same amount of extension in the region where the most numerous oblique discontinuities are present. Such considerations suggest that pre-existing fabric type, relative strength, frequency and orientation will play an important role in controlling the fault size, distribution and fault population sizes.

In the East African rift system the transfer zones between boundary faults have relatively simple geometries (e.g. Rosendahl *et al.* 1986; Morley *et al.* 1990). Large faults are sub-parallel and commonly involve overlapping geometries. Only rarely do faults markedly change orientation passing into transfer zones, although some transfer zones are associated with basement fabrics trending oblique to the rift axis (Fig. 9) (also see Milani & Davison 1988 for examples in Brazil). The absence of numerous fabrics trending at a high angle to the rift suggest that fault propagation and linkage developed in a manner that permitted the development of large boundary faults. The complete dominance of major faults in some parts of the rift system (Fig. 5c) and absence of minor faults indicates that the accommodation of strain by the boundary fault systems commenced very early on in the rift history.

The transfer zone pattern is a mixture of simple and complex geometries in the Witch Ground Graben. Simple transfer zone patterns can be recognized, at the largest scale, between faults that are relatively short (a maximum of 10–20 km long). Since the fault segments of a fixed orientation are short, it is common that transfer zones between fault sets of different orientation lie very close to each other or become intermingled (Fig. 12). Synthetic transfer zones (Morley *et al.* 1990), are common along long fault zones and in some parts ofd the Witch Ground Graben add a further complication within conjugate transfer zones generated between fault sets of different orientation (Fig. 12).

The Saltire field illustrates the influence of pre-existing fabrics on transfer zone geometry. In general terms the field is formed on an WNW–ESE-oriented horst block at the Zechstein level (Casey *et al.* 1993). A Carboniferous normal fault trend crosses the E–W-oriented Saltire field in a NE–SW direction (Casey *et al.* 1993). An E–W north-dipping normal fault forms the northern boundary to the horst. This fault splays where it meets the NE–SW trend and one of the splay trends NE–SW north of the field (Fig. 13). The WNW–ESE-striking, SSE-dipping southern boundary fault to the horst is the main fault and was active during the Late Jurassic extension. A fault of similar orientation, active during the Late Jurassic is present at the north end of horst block and it cuts the older fault that defines the horst block. Passing

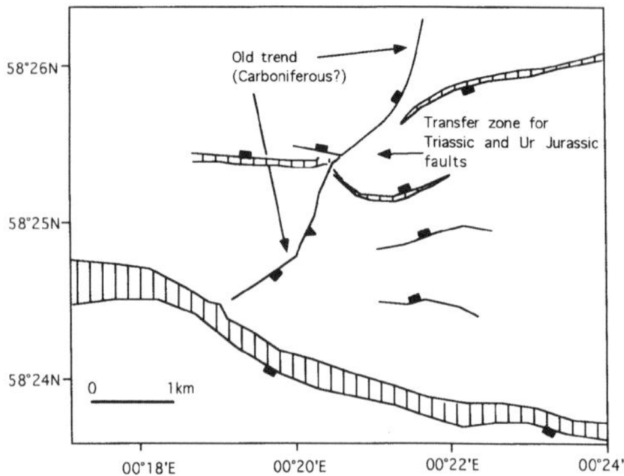

Fig. 13. Fault map of Saltire field, Witch Ground Graben. Modified from Casey *et al.* (1993).

westwards this fault swings around to a NE–SW orientation and looses displacement when it encounters the Carboniferous trend. Thus there is a complex zone in the northern central part of the field where several different fault trends of different ages meet and interact. It demonstrates how a pre-existing fabric can interrupt later faults. The presence of many such trends will result in numerous short fault segments of a certain orientation or short faults.

Lithospheric controls on fault size

Causes other than pre-existing fabrics for the presence or absence of large boundary faults are possible. The primary variations in extensional geometry between wide rifts, narrow rifts and core complexes can be explained by variations in certain lithospheric parameters as demonstrated by theoretical models (e.g. Kusznir & Park 1986; Buck 1991). Such models are based on simplified assumpions about the physical behaviour of the lithosphere including: vertical variations in the strength and density of layers many kilometres thick, strain rate, the elastic thickness of the crust and large-scale temperature variations. It has also been suggested that variations, between rifts, in fault spacing and syn-rift basin width might be related to the elastic thicknesses of the crust (Jackson & White 1989; Jackson & Blenkinsop 1993). Whilst this idea is reasonable, the evidence for widely varying elastic thicknesses between rifts is questionable. Within the flexural cantilever model the elastic thickness of the crust is a very important parameter and different studies have accorded very different elastic thicknesses to rifts (Watts *et al.* 1982; Bechtel *et al.* 1987; Ebinger *et al.* 1989; Kusznir *et al.* 1991). Regional gravity studies of the East African rift have suggested elastic thicknesses in the order of 17–38 km (Bechtel *et al.* 1987; Ebinger *et al.* 1989). However when such thicknesses are built into a flexural cantilever model the basin geometry becomes unrealistic and the maximum bending stresses predicted by the model become extremely high (Hendrie et al. 1994). For most rifts an elastic strength of about 3–5 km seems to work best in the models, producing low bending stresses and a good match to the actual rift geometry.

Both the North Sea and the East African rift system can be modelled using similar lithospheric physical properties including similar elastic thicknesses (e.g. Roberts *et al.* 1993; Hendrie *et al.* 1994; Kusznir *et al.* 1995). Consequently the differences in geometry between the rifts do not appear to be linked with differences in basic lithospheric physical properties. Hence another cause for the differences in fault size and spacing must be sought, such as pre-existing fabrics.

The discussion above centred on the large half graben systems in the East African rift. It is necessary to point out that in the Kenya rift there is a late stage (approximately last 5–6 Ma) system of minor faults that has been superimposed on the older fault systems. This new system of faults has completely different population characteristics to the old half graben systems, and this change probably does reflect varying lithospheric conditions (such as temperature, strain rate and the amount of intrusive material within the crust) (see Morley 1994 for a review).

Conclusions

Estimates of extension from a syn-rift section alone requires using some type of fault-heave summing method or the Moho geometry. Fault-heave summing methods require detailed information. To correctly estimate fault angle, expansion of the syn-rift section into major faults and to locate blind faults, cross-sections must be constructed using seismic reflection data or exceptional well control (e.g. Gulf of Suez). In the centre of most rifts there is no deep erosion of the syn-rift section, but approaching the rift flanks the tilted fault blocks may be eroded down to basement. Flexural–isostatic modelling helps to reconstruct the geometry of, and the extension required to generate such uplifts. Most fault heave approaches assume plane (2D) strain, which requires the extension direction to be known.

Even well constrained sections may be unreliable because usually they fail to account for crustal extension along seismically invisible faults (displacements up to tens of metres). The fractal distribution of several fault populations has been cited as evcidence for this error (e.g. Walsh *et al.* 1991; Marrett & Allmendinger 1991, 1992). However the fractal approach also needs to be treated with caution for several reasons. (1) For rifts with large boundary faults (several kilometres displacement) the distribution of extension is concentrated on those large faults (which contribute perhasps 80–95% of the total extension amount). Hence minor faults may not contribute significantly to crustal extension (less then 5%). (2) The different distribution of fault size and spacing between different zones within rifts suggests that extrapolation of a single fractal distribution along a regional rift traverse is unrealistic. This change in distribution can occur both in the dip direction between different tilted fault blocks, and along strike as displacement varies on major fault zones. (3) Fractal fault populations measured in syn-rift sedimentary sequences may be measuring faults created by local stresses, gravity tectonics and compaction, not just those faults that contribute to crustal extension.

In order to use fault population statistics to predict invisible or pseudo-ductile strains accomplished by minor faults it is necessary to recognize areas where important changes in fault population distributions might occur. Two controlling factors identified in this study are: (1) at large transfer zones the contribution of minor faults to the total extension amount becomes more important relative to the dip provinces between large transfer zones and (2) in of rifts with numerous pre-existing fabrics (particularly brittle structures) of widely ranging origin and relatively low shear strengths minor faults may account for several tens of percent of the total extension. Rifts with only a limited range of pre-existing fabrics orientations (particularly ductile fabrics) or with brittle fabrics of relatively high shear strength, tend to develop large boundary faults that are the dominant contributors to the total extension (70–95%), at least at the early stages of rifting.

Pre-existing fabrics control not only fault orientation and dip, but can also affect fault length, displacement amount, fault spacing and other fault population characteristics. With increasing variety of orientation and frequency of occurrence pre-existing fabrics will tend to favour the development of a wide range of fault sizes in rifts that will have the appearance of a single fractal population. If this is true the fractal pattern should be regarded as a pseudo pattern.

The authors would like to thank the two reviewers for very helpful comments that improved the manuscript. Discussions with colleagues at Amoco and Elf Aquitaine helped considerably with the ideas expressed in this paper.

References

BARTHOLOMEW, I. D., PETERS, J. M. & POWELL, C. M. 1993. Regional structural evolution of the North Sea: oblique slip and reactivation of basement lineaments. *In*: PARKER, J. R. (ed.) *Petroleum Geology of Northwest Europe: Proceedings of the 4th Conference*. Geological Society, London, 1109–1122.

BARTON, P. & WOOD, R. 1984. Tectonic evolution of the North Sea Basin: crustal stretching and subsidence. *Geophysical Journal of the Royal Astronomical Society*, **79**, 987–1022.

BECHTEL, T. T., FORSYTH, D. W. & SWAIN, C. J. 1987. Mechanisms of isostatic compensation in the vicinity of the East African Rift, Kenya. *Geophysical Journal of the Royal Astronomical Society*, **90**, 445–465.

BLANTON, T. L. 1982. An Experimental Study of Interaction Between Hydraulically-Induced and Pre-existing Fractures. SPE/DOE preprint 10847, *Proceedinsgs of the Unconventional Gas Recovery Symposium, Pittsburgh*, 559–571.

BOLDY, S. A. R. & BREALEY, S. 1990. Timing, nature, and sedimentary result of Jurassic tectonism in the Outer Moray Firth. *In*: HARDMAN, R. F. P. & BROOKS, J. (eds) *Tectonic Events Responsible for Britain's Oil and Gas Reserves*. Geological Society, London, Special Publications, **55**, 254–279.

BOSWORTH, W. 1985. Geometry of propagating continental rifts. *Nature*, **316**, 625–627.

BUCK, W. R. 1991. Modes of continental lithosphere extension. *Journal of Geophysical Research*, **96**, 20,161–20,178.

BURGESS, C. F., ROSENDAHL, B. R., SANDER, S., BURGESS, C. A., LAMBIASE, J., DERSKEN, S. & MEADER, N. 1988. The structure and stratigraphic evolution of Lake Tanganyika. *In*: MANSPREIZER, W. (ed.) *A case study of continental rifting, Triassic/Jurassic rifting*. Developments in Geotectonics, **27**, Elsevier, Amsterdam, 859–881.

CASEY, B. J., ROMANI, R. S. & SCHMITT, R. H. 1993. Appraisal geology of the Saltire field, Witch Ground Graben, North Sea. *In*: PARKER, J. R. (ed.) *Petroleum Geology of Northwest Europe: Proceedings of the 4th Conference*. Geological Society, London, 507–518.

COWARD, M. P. 1993. The effect of Late Caledonian and Variscan continental escape tectonics on basement structure, Paleozoic basin kinematics and subsequent Mesozoic basin development in NW Europe. *In*: PARKER, J. R. (ed.) *Petroleum Geology of Northwest Europe: Proceedings of the 4th Conference*. Geological Society, London, 1095–1108.

DAHLSTROM, C. D. A. 1969. Balanced cross-sections. *Canadian Journal of Earth Sciences*, **6**, 743–757.

DALY, M. C., CHOROWICZ, J. & FAIRHEAD, D. 1989. Rift basin evolution in Africa: the influence of reactivated steep basement shear zones. *In*: COOPER, M. A. & WILLIAMS, G. D. (eds) *Inversion Tectonics*. Geological Society, London, Special Publications, 44, 309–334.

DAVISON, I. 1994. Linked fault systems; extensional, strike-slip and contractional. *In*: HANCOCK, P. L. (ed.) *Continental Deformation*. Pergamon Press, Oxford, 121–142.

EBINGER, C. J. 1989. Tectonic development of the western branch of the East African rift system. *Geological Society of America Bulletin*, **101**, 885–903.

——, BECHTEL, T. D., FORSYTH, D. W. & BOWIN, C. O. 1989. Effective elastic thickness beneath the East African and Afar Plateaux and dynamic compensation of the uplifts. *Journal of Geophysical Research*, **94**, 2883–2901.

——, CROW, M. J., ROSENDAHL, B. R., LIVINGSTONE, D. A. & LEFOURNIER, J. 1984. Structural evolution of Lake Malawi, Africa. *Nature*, **308**, 627–629.

FREUND, R. & MERZER, A. M. 1976. The formation of rift valley and their zigzag fault patterns. *Geological Magazine*, **113**, 561–568.

GAUTHIER, B. D. M. & LAKE, S. D. 1993. Probabilistic modeling of faults below the limit of seismic resolution in Pelican Field, North Sea, Offshore United Kingdom. *American Association of Petroleum Geologists Bulletin*, **77**, 761–777.

GIBBS, A. D. 1983. Balanced cross-section construction from seismic sections in areas of extensional tectonics. *Journal of Structural Geology*, **5**, 153–160.

GILLESPIE, P. A., HOWARD, C. B., WALSH, J. J. & WATTERSON, J. 1993. Measurement and characterisation of spatial distributions of fractures. *Tectonophysics*, **226**, 113–141.

GRIMAUD, P., RICHERT, J-P., ROLET, J., TIERCELIN, J-J., XAVIER, J-P., MORLEY, C. K., COUSSEMENT, C., KARANJA, S. W., RENAUT, R., GUERIN, G., LE TURDU, C. & MICHEL-NOEL, G. 1994. Fault geometry and extensional mechanisms in the central Kenya Rift, East Africa. A 3D remote sensing approach. *Bulletin du Centres Recherches Exploration-Production Elf Aquitaine*, **18**, 1, 59–92.

HARDING, T. P. 1984. Graben hydrocarbon occurrences and structural style. *American Association of Petroleum Geologists Bulletin*, **68**, 33–362.

HENDRIE, D. B., KUSZNIR, N. J., MORLEY, C. K. & EBINGER, C. J. 1994. A quantitative model of Cenozoic rift basin development in Northern Kenya. *Tectonophysics*, **236**, 409–438.

HIBBERT, M. J. & MACKERTICH, D. S. 1993. The structural evolution of the eastern end of the Halibut Horst, Block 15/21, Outer Moray Firth, UK North Sea. *In*: PARKER, J. R. (ed.) *Petroleum Geology of Northwest Europe: Proceedings of the 4th Conference*. Geological Society, London, 1179–1188.

HOSSACK, J. R. 1979. The use of balanced cross-section in the calculations of orogenic contraction. *Journal of the Geological Society, London*, **136**, 705–711.

JACKSON, J. & BLENKINSOP, T. 1993. The Malawi earthquake of March 10, 1989: deep faulting within the East African rift system. *Tectonics*, **12**, 1131–1139.

—— & WHITE, N. J. 1989. Normal faulting in the upper continental crust, observations from regions of active extension. *Journal of Structural Geology*, **11**, 15–36.

KAKIMI, T. 1980. Magnitude-frequency relation for displacement of minor faults and its significance in crustal deformation. *Geological Survey of Japan Bulletin*, **31**, 467–487.

KILEMBE, E. A. & ROSENDAHL, B. R. 1992. Structure and stratigraphy of the Rukwa Rift. *Tectonophysics*, **209**, 143–158.

KLEMPERER, S. L. & HURICH, C. A. 1990. Lithospheric structure of the North Sea from Deep Seismic Refraction Profiling. *In*: BLUNDELL, D. J. & GIBBS, A. D. (eds) *Tectonic Evolution of the North Sea Rifts*. Clarendon Press, Oxford, 37–63.

KUSZNIR, N. J. & EGAN, S. S. 1990. Simple-shear and pure-shear models of extensional sedimentary basin formation: application to the Jeanne D'Arc basin, Grand Banks of Newfoundland. *In*: TANKARD, A. J. & BALKWILL, H. R. (eds) *Extensional tectonics and stratigraphy of the North Atlantic margins*. American Association of Petroleum Geologists Memoir, **46**, 305–322.

—— & PARK, R. G. 1987. The Extensional Strength of the Continental Lithosphere: Its Dependence on Geothermal Gradient and Crustal Composition and Thickness. *In*: COWARD, M. P., DEWEY, J. F. & HANCOCK, P. L. (eds) *Continental Extensional Tectonics*. Geological Society, London, Special Publications, **28**, 35–52.

——, MARSDEN, G. & EGAN, S. S. 1991. A Flexural-Cantilever Simple-Shear/Pure-Shear Model of Continental Lithosphere Extension: Applications to the Jeanne d'Arc Basin, Grand Banks and Viking Graben, North Sea. *In*: ROBERTS, A. M., YIELDING, G. & FREEMAN, B. (eds) *The Geometry of Normal Faults*. Geological Society, London, Special Publications, **56**, 41–60.

——, ROBERTS, A. & MORLEY, C. 1995. Forward and reverse modelling of rift basin formation. *In*: LAMBIASE, J. J. (ed.) *Hydrocarbon Habitat in Rift Basins*. Geological Society, London, Special Publications, **80**, 33–56.

MARRETT, R. & ALLMENDINGER, R. W. 1991. Estimates of strain due to brittle faulting: Sampling of fault populations. *Journal of Structural Geology*, **13**, 735–738.

—— & —— 1992. Amount of extension on 'small' faults: An example from the Viking Graben. *Geology*, **20**, 47–50.

MARSDEN, G., YIELDING, G., ROBERTS, A. M. & KUSZNIR, N. J. 1991. Application of a flexural cantilever simple-shear/pure-shear model of continental lithosphere to the formation of the northern North Sea Basin. *In*: BLUNDELL, D. J. & GIBBS, A. (eds) *Tectonic Evolution of the North Sea rifts*. Oxford University Press, 236–257.

MCCONNELL, R. B. 1972. Geological development of the rift system of eastern Africa. *Geological Society of America Bulletin*, **83**, 2549–2572.

MCKENZIE, D. P. 1978. Some remarks on the development of sedimentary basins. *Earth and Planetary Science Letters*, **40**, 25–32.

MILANI, E. J. & DAVISON, I. 1988. Basement control and transfer tectonics in the Reconcavo-Tucano-Jatoba rift, Northeast Brazil. *Tectonophysics*, **154**, 41–70.

MORLEY, C. K. 1988. Variable extension in Lake Tanganyika. *Tectonics*, **7**, 785–801.

—— 1994. Interaction of deep and shallow processes in the evolution of the Kenya rift. *Tectonophysics*, **236**, 81–91.

—— 1995. Developments in the structural geology of rifts over the last decade and their impact on hydrocarbon exploration. *In*: LANBIASE, J. J. (ed.) *Hydrocarbon Habitat in Rift Basins*. Geological Society, London, Special Publications, **80**, 1–32.

——, CUNNINGHAM, S. M., HARPER, R. M. & WESTCOTT, W. A. 1992*a*. Geology and Geophysics of the Rukwa Rift, East Africa. *Tectonics*, **11**, 69–81.

——, NELSON, R. A., PATTON, T. L. & MUNN, S. G. 1990. Transfer zones in the East African Rift System and Their Relevance to Hydrocarbon Exploration in Rifts. *American Association of Petroleum Geologists Bulletin*, **74**, 1234–1253.

——, WESCOTT, W. A., STONE, D., HARPER, R. M., WIGGER, S. T. & KARANJA, F. M. 1992*b*. Tectonic evolution of the northern Kenya Rift. *Journal of the Geological Society, London*, **149**, 333–348.

O'DRISCOLL, D., HINDLE, A. D. & LONG, D. C. 1990. The structural controls on Upper Jurassic and Lower Cretaceous reservoir sandstones in the Witch Ground Graben, UK North Sea. *In*: HARDMANN, R. F. P. & BROOKS, J. (eds) *Tectonic Events Responsible for Britain's Oil and Gas Reserves*, Geological Society, London, Special Publications, **55**, 299–323.

PANHUYS-SIGLER, VAN, M., BAUMANN, A. & HOLLAN, T. C. 1991. The Tern Field, Block 210/25a. UK. North Sea. *In*: ABBOTTS, I. L. (ed.) *United Kingdom Oil and Gas Fields, 25 Years Commemorative Volume*. Geological Socety, London, Memoir, **14**, 191–198.

PATTON, T. L., MOUSTAFA, A. R., NELSON, R. A. & ABDINE, A. S. 1994. Tectonic evolution and structural setting of the Suez Rift. *In*: LANDON, S. M. (ed.) *Interior Rift Basins*. American Association of Petroleum Geologists, Memoirs, **59**, 9–55.

PEACOCK, D. C. P. & SANDERSON, D. 1991. Displacements, segment linkage and relay ramps in normal fault zones. *Journal of Structural Geology*, **13**, 721–733.

PLATT, N. H. 1995. Structure and tectonics of the northern North Sea: new insights from deep penetration regional sesimic data. *In*: LAMBIASE, J. J. (ed.) *Hydrocarbon Habitat in Rift Basins*. Geological Society, London, Special Publications, **80**, 103–113.

RATTEY, R. P. & HAYWARD, A. B. 1993. Sequence stratigraphy of a failed rift system: the Middle Jurassic to Early Cretaceous basin evolution of the Central and Northern North Sea. *In*: PARKER, J. R. (ed.) *Petroleum Geology of Northwest Europe: Proceedings of the 4th Conference*. Geological Society, London, 215–250.

RING, U., BETZLER, C. & DELAVAUX, D. 1993. Normal vs. Strike-slip faulting during rift development in East Africa: The Malawi rift. *Geology*, **20**, 1015–1018.

ROBERTS, A. M., YIELDING, G., KUSZNIR, N. J., WALKER, I. & DORN-LOPEZ, D. 1993. Mesozoic extension in the North Sea: constraints from flexural backstripping, foreward modelling and fault populations.

In: PARKER, J. R. (ed.) *Petroleum Geology of Northwest Europe: Proceedings of the 4th Conference.* Geological Society, London, 1123–1136.

ROBSON, D. A. 1971. Structure of the Gulf of Suez (Clysmic Rift), with Special Reference to the Eastern Side. *Quarterly Journal of the Geological Society of London*, **115**, 247–276.

ROSENDAHL, B. R., REYNOLDS, D. J., LORBER, P. M., BURGESS, C. F., MCGILL, J. SCOTT, D., LAMBIASE, J. J. & DERKSEN, S. J. 1986. Structural expressions of rifting: lessons from Lake Tanganyika, Africa. *In*: FROSTICK, L. E., RENAUT, R. W., Reid, I. & TIERCELIN, J-J. (eds) *Sedimentation in the East African Rifts.* Geological Society, London, Special Publications, **25**, 29–43.

ROWAN, M. G. & KLIGFIELD, R. 1989. Cross section restoration and balancing as aid to seismic interpretation in extensional terranes. *American Association of Petroleum Geologists Bulletin*, **73**, 955–966.

ROYDEN, L. & KEEN, C. E. 1980. Rifting process and thermal evolution of the continental margin of Eastern Canada determined from subsidence curves. *Earth and Planetary Science Letters*, **51**, 343–361.

SCHOLZ, C. H. & COWIE, P. A. 1990. Determination of total strain from faulting. *Nature*, **346**, 837–839.

SMITH, M. & MOSLEY, P. 1994. Crustal heterogeneity and basement influence on the development of the Kenya Rift, East Africa. *Tectonics*, **12**, 591–606.

SPENCER, A. M. & LARSEN, V. B. 1990. Fault traps in the Northern North Sea. *In*: HARDMAN, R. F. P. & BROOKS, J. (eds) *Tectonic Events Responsible for Britain's Oil and Gas Reserves.* Geological Society, London, Special Publications, **55**, 281–298.

STECKLER, M. S. & WATTS, A. B. 1978. Subsidence of the Atlantic-type margin off New York. *Earth and Planetary Science Letters*, **41**, 1–13.

STRECKER, M. R., BLISNIUK, & EISBACHER, G. H. 1990. Rotation of extension direction in the Central Kenya Rift. *Geology*, **18**, 299–302.

TEUFEL, L. W. 1979. *An Experimental Study of Hydraulic Fracture propagation in Layered Rock.* PhD thesis, Department of Geology, Texas A & M University, College Station, 99.

—— & CLARK, J. A. 1984. Hydraulic fracture propagation in layered rock: experimental studies of fracture containment. *Society of Petroleum Engineers Journal*, **24**, 19–32.

VILLEMIN, T. & SUNWOOD, C. 1987. Distribution logarithmique self-similaire des rejets et longuers de failles: Exemple du Bassin Houiller Lorrain. *Compte Rendus* ser. II, **305**, 1309–1312.

WALSH, J. & WATTERSON, J. 1991. New methods of fault projection for coalmine planning. *Proceedings of the Yorkshire Geological Society*, **48**, 209–219.

——, —— & YIELDING, G. 1991. The importance of small-scale faulting in regional extension. *Nature*, **351**, 391–393.

WATTS, A. B., KARNER, G. D. & STECKLER, M. S. 1982. Lithosphere flexure and the evolution of sedimentary basins. *In*: KENT, P., BOTT, M. H. P., MCKENZIE, D. P. & WILLIAMS, C. A. (eds) *Philosophical Transactions of the Royal Society of London*, **305**, 149–281.

WERNICKE, B. & BURCHFIEL, B. C. 1982. Modes of extensional tectonics. *Journal of Structural Geology*, **4**, 105–115.

WHEELER, W. H. & KARSON, J. A. 1989. Structures and kinematics of the Livingstone Mountains border fault zone, Nyasa (Malawi) Rift southwestern Tanzania. *Journal of African Earth Sciences*, **8**, 393–413.

WHITE, N. 1990. Does the uniform stretching model work in the North Sea? *In*: BLUNDELL, D. J. & GIBBS, A. D. (eds) Tectonic *Evolution of the North Sea Rifts.* Clarendon Press, Oxford, 217–239.

ZEIGLER, P. A. 1982. *Geological Atlas of Western and Central Europe.* Shell Internationale Maatschappij, B.V., The Hague.

—— & VAN HOORN, B. 1990. Evolution of North Sea rift system. *In*: TANKARD, A. J. & BALKWILL, H. R. (eds) *Extensional tectonics and stratigraphy of the North Atlantic margins.* American Association of Petroleum Geologists Menoir, **46**, 471–500.

Forward modelling of compaction above normal faults: an example from the Sirte Basin, Libya.

A. G. SKUCE,

Husky Oil Operations Ltd,Box 6525, Station D, Calgary, Alberta T2P 3G7, Canada

Abstract: Seismic data reveal the structure of a large normal fault that forms the western boundary of the Agedabia Trough in the central Sirte Basin, Libya. A significant monoclinal flexure in the sediments forming the hanging wall of this fault is interpreted to have been generated by their compaction. A computer forward model supports this interpretation. Characteristics of this kind of structure are a vertical synclinal axial plane above the hanging-wall cutoff of the basement and an inclined but less distinct anticlinal axial plane above the footwall cutoff of the basement. Dips increase with depth. The model also shows how compaction can cause sedimentary growth within the hanging wall rather than just at the fault plane. In some circumstances, compaction can produce anticlinal structures in normal fault hanging walls that could easily be confused with inversion structures. Structures can be produced by compaction in normal fault hanging walls even where no lithology contrast occurs across the fault. Compaction structures generally will be most obvious over large planar faults that involve basement. Nevertheless, compaction structures are also important in the case of detached listric faults: in this case the effects are often obscured by the superimposition of fault bend folds that will have similar or greater magnitudes but are typically anticlinal, whereas the compaction structures are usually synclinal. Compaction effects compound the difficulties of geometric inverse modelling.

Many recent publications have concerned themselves with the detailed geometric modelling of oilfield scale rollover structures in the hanging walls of listric normal faults (e.g. Withjack & Peterson 1993; Xiao & Suppe 1992): examples of these kinds of structures are often taken from deltaic areas. In contrast, other recent publications deal exclusively with planar normal faults and the modelling of the basin-scale structures that result from the isostatic and thermal consequences of extension in rifts and continental margins (e.g. Kusznir *et al.* 1991; Westaway & Kusznir 1993). A common feature of both types of model is that the beds in the hanging walls predominantly have dips counter to that of the fault plane itself (i.e. antithetic dips).

In practice, however, it is observed that a large proportion of normal faults have beds in their hanging walls with significant basinward (synthetic) dips. The structure from the Sirte Basin of Libya discussed in this paper is an example of this. There are several possible causes for synthetic hanging wall dips, some of which are discussed later. The most important of these, compaction, the subject of this paper, will occur to some extent in nearly all normal faulting environments. Compaction structures in the hanging walls of normal faults, the geometry of which is sometimes termed 'lick-up' (e.g. Graham 1992), have previously been recognized in the North Sea (Evans & Parkinson 1983; White *et al.* 1986; Badley *et al.* 1988; Milton *et al.* 1990). Nevertheless, compaction structures have frequently been either ignored or have been misinterpreted as having arisen from other mechanisms. Two misconceptions are behind these oversights: first, it is often assumed that significant structures only form by differential compaction when basement rocks are present in the footwall of the fault; secondly, it is also assumed that the amount of compaction that takes place after the first few hundred metres of burial is negligible.

The effects of compaction have often been neglected in geometric modelling studies. Notable exceptions to this are the articles by N. J. White and his co-workers (White *et al.* 1986; White & Yielding 1991; Kerr & White 1992) which attempt inverse modelling (i.e. the estimation of an unknown fault geometry from a known geometry of hanging-wall horizons) of structures resulting from compaction and fault bend folding. Their work necessitated making several important simplifying assumptions that have serious consequences in some situations. The simple forward models described in this paper illustrate the complexity and scale of compaction structures produced by permutations of fault movement, subsidence and the deposition of different rock types. The structural consequences of compaction have economic significance because hanging wall fault traps formed by this process can be a significant kind of hydrocarbon prospect, especially in mature

From Buchanan, P. G. & Nieuwland, D. A. (eds), 1996, *Modern Developments in Structural Interpretation, Validation and Modelling,* Geological Society Special Publication No. 99, pp. 135–146.

exploration areas where the more obvious anticlines and footwall fault traps have previously been drilled.

An example from the Sirte Basin

The Sirte Basin of Libya (Fig. 1), comprises a number of platforms and sub-basins formed by lithospheric extension during the Late Cretaceous (Gumati & Nairn 1991). The ensuing thermal subsidence was interrupted by a minor rifting phase in the Eocene which reactivated the faults formed during the earlier phase of rifting. The deepest of the sub-basins is the Agedabia (or Sirte) Trough that contains Upper Cretaceous and Tertiary sediments more than 6 km thick. At the time of writing, no well has yet penetrated the complete sedimentary section within the trough.

Figure 2 summarizes the formation names and lithologies of the major units in the Agedabia Trough. The lower part of the Rakb and the Etel Formations are not present on the Zelten Platform. The depth to basement and the presence of the Etel and Sarir Formations within the Agedabia Trough are uncertain in the study area. The challenge of seismic interpretation and structural modelling is to predict the nature and geometry of the deeper formations that have not yet been penetrated by drilling.

Figure 3 shows an interpreted time-migrated seismic line over the normal fault that separates the Agedabia Trough from the Zelten Platform. Two wells have been projected a few hundred metres on to the line and control the identification of seismic markers down to the top Kalash Horizon within the trough and just above the top basement horizon on the platform. Other, more distant wells in the area reach the basement on the platform and the Orange reflector within the trough. A depth converted line drawing of the seismic section is shown in Fig. 4.

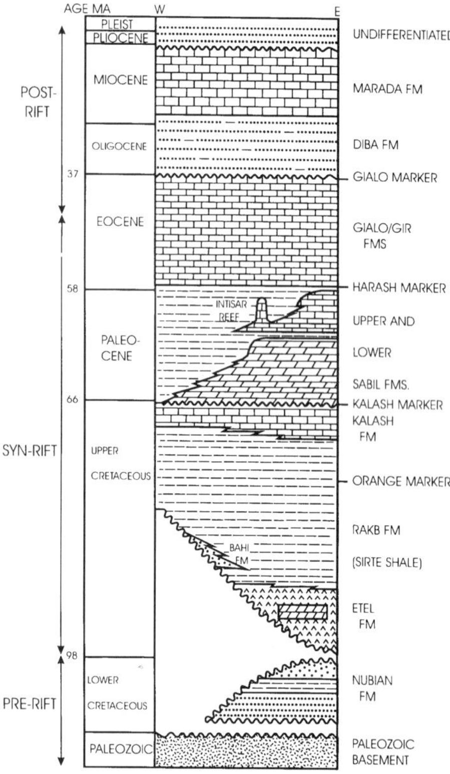

Fig. 2. Generalized stratigraphic column for the study area. The lower part of the Rakb Formation and the Etel Formation are not present on the Zelten Platform. No wells have penetrated units older than the Orange horizon in the Agedabia Trough in the immediate area, so the deep geology is uncertain.

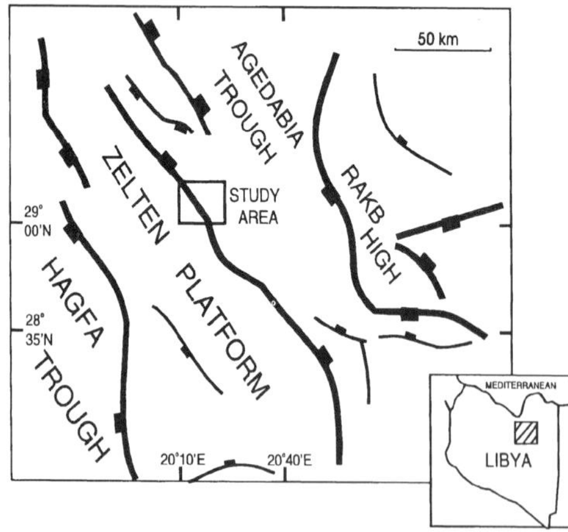

Fig. 1. Location of the Agedabia Trough and the Zelten Platform in the Sirte Basin of Libya.

The position of the normal fault is defined in the upper and middle parts of the section (i.e. above 2.5 s) by reflection terminations but its deeper geometry is not completely clear. The NE-dipping reflectors at depth are thought to mostly represent real structural dips but, in places (below the Orange Horizon and in the footwall), they may be wavefront artifacts produced by the migration. Apart from the normal fault itself, the most significant structure on the section is the monocline involving all non-basement reflectors that is located above the fault. The structure dips to the NE, away from the fault and is bounded on its SW and NE sides respectively by anticlinal and synclinal fold axes. Two characteristics of this monocline are important: (a) the steepness of the dips is greater

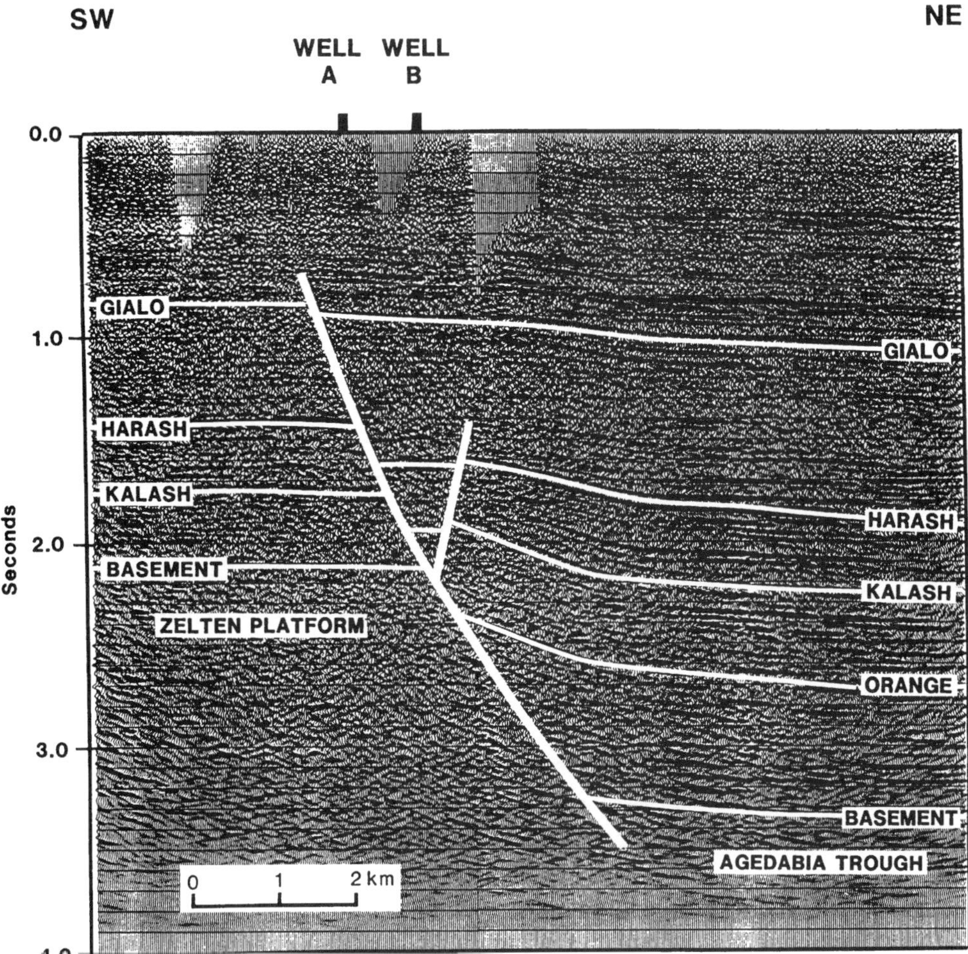

Fig. 3. Migrated Vibroseis line across the fault that separates the Zelten Platform from the Agedabia Trough. The line is located in the southern part of the study area outlined in Fig. 1.

on deeper horizons, i.e. 4° at the Gialo horizon, 10°at the Harash, 18° at the Kalash and 21° at the Orange; (b) the synclinal axial plane that defines the NE limit of the monocline is approximately vertical.

Consideration of these two factors led to the interpretation of the monocline as a compaction effect over the basement fault block. Generally, one would expect compaction effects to increase with depth, and since the process is driven by gravity, displacements should be vertical in the absence of other forces. Any resulting deformation can be modelled by distributed vertical simple shear, therefore any folding induced by compaction would be expected to produce vertical axial planes. In contrast, Waltham (1990) proposed that displacements due to compaction are parallel to the fault plane, whereas White & Yielding (1991) assumed that compaction displacements are parallel to the simple shear direction used in the modelling of the fault bend folding, i.e. usually antithetically inclined. In both these cases, the direction of compaction driven shear deformation was chosen for computational expediency rather than on geological grounds. Such assumptions would produce different structural geometries, including inclined fold axes.

Seismological studies of currently active normal faults in extensional areas show that fault planes typically have dips in the range 45–65° (Jackson 1987; Westaway & Kusznir 1993). The interpretation presented here (Fig. 4) shows a fault with a slight bend at 3.5 km depth. The fault has dips

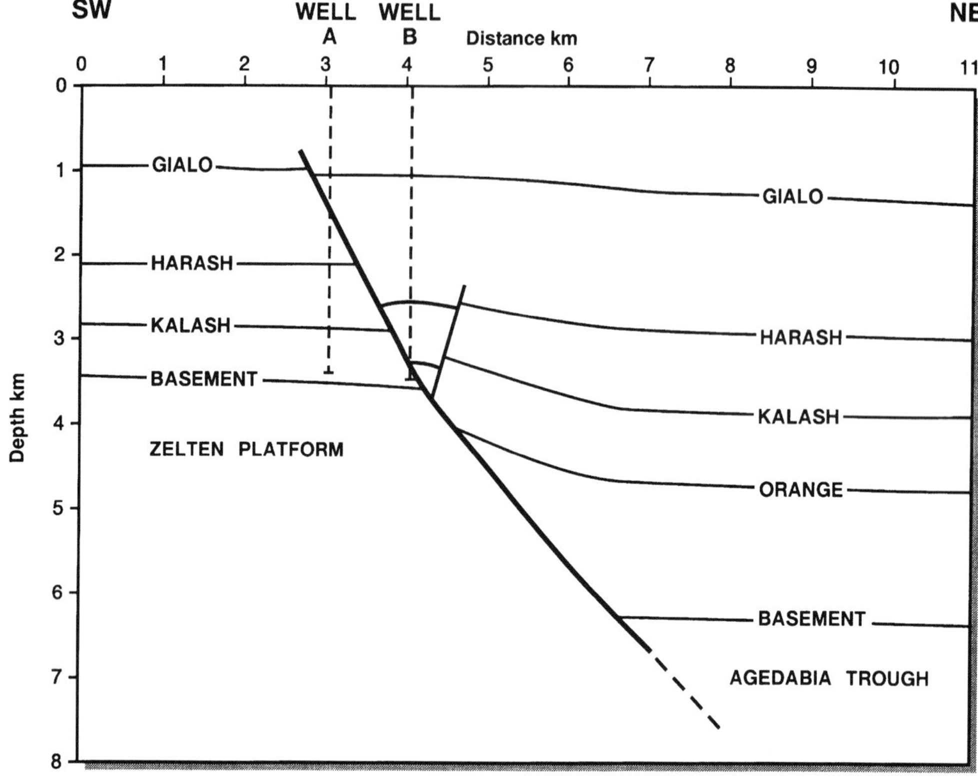

Fig. 4. Depth converted interpretation of data in Fig. 3.

in the range of 50–60°. The fault is considered essentially planar, and is unlike the frequently illustrated listric faults which become sub-horizontal at depth (Morley *et al.* 1990).

If a planar fault geometry is assumed and, thus, fault bend folding effects can be neglected, then the top basement depth within the trough can be quickly, but roughly, estimated as lying at the downward projection of the fault plane and the synclinal axial plane of the compaction fold. 'Basement' for these purposes would include any non-compactable or pre-compacted sediments such as the pre-rift Sarir Sandstones of the Nubian Formation (Fig. 2).

When the interpretation described above was first made, it was unclear whether or not it was quantitatively reasonable. Specifically, it was not known whether compaction alone was sufficient to produce features of the size and shape of those observed in the data and, if not, what additional mechanisms would have to be invoked to explain the origin of the structures. A forward model was produced using the GEOSEC software to determine the geometry and the order of magnitude of compaction structures, given a set of simplified quantitative assumptions about the structural and sedimentary history of the part of the Sirte Basin examined in this paper. No attempt was made to produce an exact match between model and example.

Figure 5 shows the development of the model, stage by stage. The top of the model is assumed to lie at sea level. At each stage in which sediments are added, the sediments are assumed to completely fill any voids up to sea level. In order to simplify the model and to isolate the effects of compaction, the fault is assumed to be completely planar with a dip of 55°, thus no fault bend folding occurs. Subsidence occurs in two ways, either a generalized vertical subsidence of both hanging and footwalls, or by differential subsidence of the hanging wall alone by means of movements across the fault plane.

The various lithologies assumed for each layer of the model are shown in Fig. 5 and were chosen to roughly reflect the geology of the study area. The rock properties determine the compaction response of the layers as defined within the GEOSEC software. The compaction profile selected for the shale units was the Baldwin–Butler (1985) relationship. The maximum permitted porosity of shales within the program (i.e. at zero depth), is limited to 45%, which is equivalent to the porosity at 135 m in the Baldwin–Butler function. This means that a fully compacted shale at depth will have bed thicknesses equal to 55% of the bed thicknesses at the time of deposition. The mixed lithologies compact at rates proportionally between the rates for the pure lithologies. Sands compact according to the Sclater–Christie (1980) curve and carbonates according to the Schmoker–Halley (1982) relationship. Ideally, compaction curves derived from the Sirte Basin for each of the rock types should have been used. However, due to the nature of the stratigraphy (Fig. 2), few inferences could be drawn, for example, about the compaction of the shales in the Rakb Formation over the important first 2.5 km of burial because the rocks presently found at those depths are predominantly sandstones and carbonates (Fig. 4).

Within the program, compaction occurs according to the thickness of sediments directly above the layer at a particular point. Porosities do not vary with depth for any particular vertical column within a layer and a layer does not compact under its own weight. These assumptions tend to cause compaction to occur later in the model than it would naturally. Such effects can be reduced by modelling with thinner layers (see below).

The first compaction structures in the model appear at stage 4. In stage 5, unit c is deposited as a thicker unit in the hanging wall than the footwall due to compaction subsidence of unit a. The thickening of unit c does not occur due to growth across the fault plane. Stages 6 and 7 show the effects of further compaction, subsidence and deposition. For the first time, in stage 8, compaction occurs in the footwall. The fault moves again in stage 9, cutting through and displacing units c and d, followed by the deposition of unit e.

Note that unit e thickens both at the fault plane and within the hanging wall due to compaction of lower units. The fault in the shallower section in stage 9 is assumed to have the same dip as its continuation below where it offsets the basement. At stage 10, compaction has resulted in the upper part of the fault rotating with a bend at the footwall cutoff of the basement horizon. Continued movement on the fault (not shown), would result in fault bend folding in the units above the bend.

The top of unit e in stage 10 has formed an anticline. This feature is, in part, an artifact resulting from the thinning by faulting of unit d before compaction. The use of thinner intervals with intermediate compaction stages for unit e would diminish this effect. Nevertheless, there are geological circumstances where such an anticline could develop as a result of compaction in the hanging-wall of a fault (see below). The final two stages of the model show the deposition of unit f and the continued compaction of the deeper units. The main purpose of modelling units e and f was to compact the units beneath them: the structures formed in the horizons that form the tops of these units in the model are not realistic.

Comparison of the model with the actual depth interpretation (Fig. 4), shows that many of the features of the model match the interpretation of the data quite well, especially the vertical nature of the synclinal fold axis and its location immediately above the hanging-wall cutoff of the basement. The anticlinal axial plane of the fold is not vertical and is not related in a simple way to the footwall cutoff of the basement. The units c, d, e and f are all thicker in the trough than over the platform, but only unit e thickens at the fault plane; the rest of the thickening occurs in the hanging wall of the fault above the basement fault heave. Such effects are also clearly seen in the data. Thus, sedimentary growth across a fault need not occur in its entirety at the fault plane, as is often shown in models of growth faults. There is very little change in thickness of the Harash–Kalash interval observed between wells A and B, despite the fact that they are on different sides of the fault, but the thickness of this interval in the trough more distant from the fault is known to be greater.

The dips of the modelled compaction structures (around 30°) are greater than is observed in the data (up to 21°) in the deeper part of the section. There are several possible reasons for this.

(1) The assumption that the porosity in shales at zero depth is 45%. Choosing lower values for this parameter would lead to less compaction at depth.

(2) The Baldwin–Butler curve predicts very low porosities (i.e. less than 2%), at depths greater than 5 km. Other functions, such as Dickinson's (13% porosity at 5 km) (Baldwin & Butler 1985), might be more appropriate for the thick shale packages used in the model. Seismic interval velocities suggest that the deeper parts of the Rakb Formation could be undercompacted, but these data are too imprecise to be used to constrain the model.

(3) The lithology of the deep section is unknown. It is possible that there are considerable thicknesses of evaporites, carbonates and coarse clastics in this sequence, all of which would compact less than the pure shale units assumed in the model. Clearly, more deep geological control on lithologies and porosities is required to resolve questions over this point and point 2.

(4) The use of thinner layers would tend to reduce the magnitude of the compaction structures since more compaction would occur at an earlier

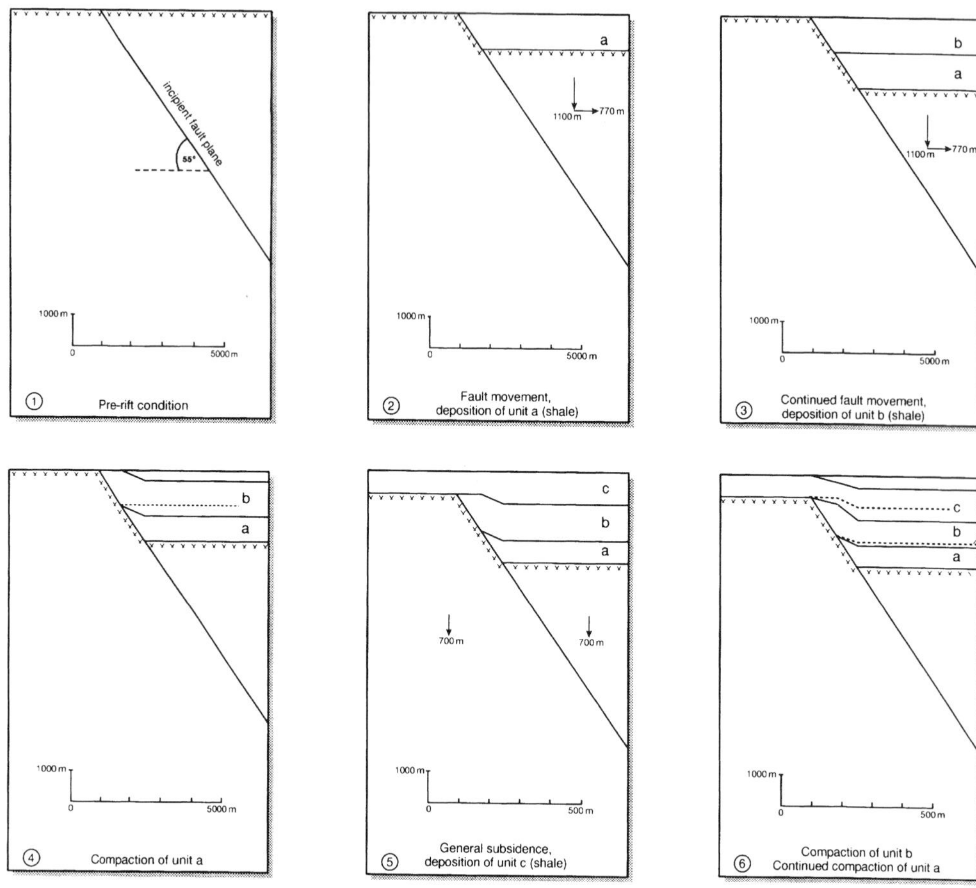

stage. Figure 6 illustrates the effect of modelling with 100 m thick layers compared with the 1 km thick layers used in Fig. 5. Using the Baldwin–Butler (1985) shale compaction function, the thin layer model produces hanging wall dips that are less than the thick layer model by about 6° and produces 80 m less compaction overall. Using instead the Sclater–Christie (1980) shale compaction function in the 100 m thick layer case (not shown) would produce greater hanging-wall compaction than even the thick layer Baldwin–Butler model, with dips in the hanging-wall horizons of around 37°. Thus, although using thick layer models exaggerates compaction effects somewhat, bigger uncertainties are present in the choice of the compaction function.

A further mismatch between the model and the data can be seen by comparing the shape of the Harash reflection in Fig. 4 with the top of unit d in stage 12 of Fig. 5, in the immediate hanging wall of the fault. In the data, the horizon rolls over, dipping very slightly to the SW, whereas in the model, the horizon dips to the NE. A related minor antithetic fault is observed in the hanging wall of the fault (Fig. 3). The rollover and antithetic faulting could be a result of minor folding above a slight concave bend observed in the fault but, for the sake of simplicity, this effect was not modelled. Combining fault bend folding and compaction in a single model is likely to be difficult since more than one deformation mechanism is likely to be active simultaneously. Here, compaction is assumed to be accommodated by vertical simple shear, whereas rollover structures are apparently best modelled by using antithetically inclined simple shear (White *et al.* 1986; Xiao & Suppe 1992).

The reduced dip of the upper part of the fault plane predicted in the model is not observed in the data. This could be a consequence of overestimating the amount of compaction in unit d. This unit corresponds to the Upper and Lower Sabil Formations which are mostly made up of carbonates. In carbonates, petrographic work is required to distinguish between loss of porosity by

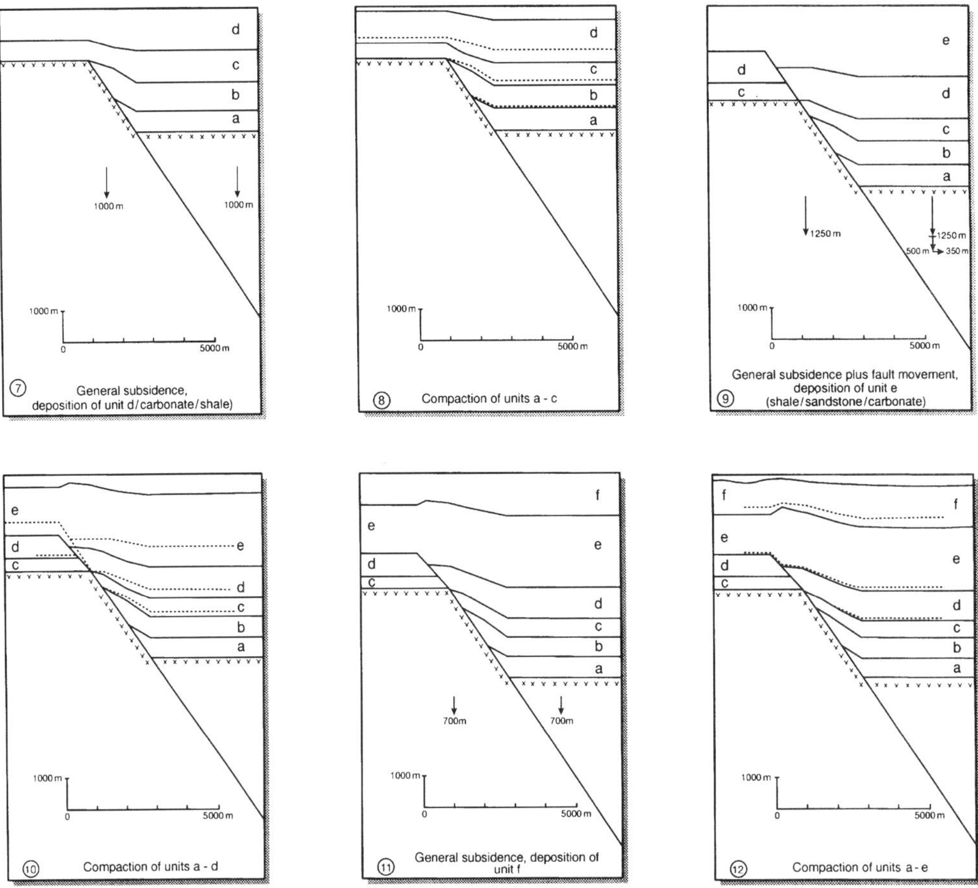

Fig. 5. Forward model of fault movement, sedimentation and compaction over a planar normal fault. Units a, b and c correspond approximately to the Rakb and Kalash Formations; unit d to the Upper and Lower Sabil Formations; unit e to the Gialo and Gir Formations; and unit f to Oligocene and younger units. The fault movements in stages 2 and 5 correspond to the main period of late Cretaceous rifting in the Sirte Basin; the fault movement in stage 9 represents a minor Eocene faulting episode (Gumati & Nairn 1991).

compaction or through cementation: data is lacking to make this distinction in this case. As noted above, the assumption in the model was that the original dip of the fault plane in the upper section before compaction (stage 9) was the same as that in the deeper section. If it had been assumed that the original fault dip in the shallow section was around 62°, then after compaction the dip of the fault would have been about the same as in the deeper section. Alternatively, if the simple shear accommodating the compaction was assumed to be inclined in the direction of the fault plane (instead of being vertical) then the fault plane dip would not change after compaction. Of course, changing this parameter would also alter many of the other compaction structures in the model. Waltham's (1990) finite difference models of compaction assume displacements parallel to the fault plane, this results in no syncline being generated in the hanging wall.

Illustrated in Fig. 7 is one way in which a compaction anticline could form in a simple sequence of sandstones and shales. The shale units in the model compact more than the overlying sandstones; however, the shales are tectonically thinned by the fault, allowing an anticline to form as a result of compaction. The limits of the anticline are between lines directly above the footwall cutoff of the top of the shale sequence and the hanging wall cutoff of the base of the shales. Other circumstances in which anticlines might form in this way are where a fault thinned unit compacts at a later than normal stage, for example, in carbonates or undercompacted shales. A compaction anticline is not observed in the example shown here (Fig. 3),

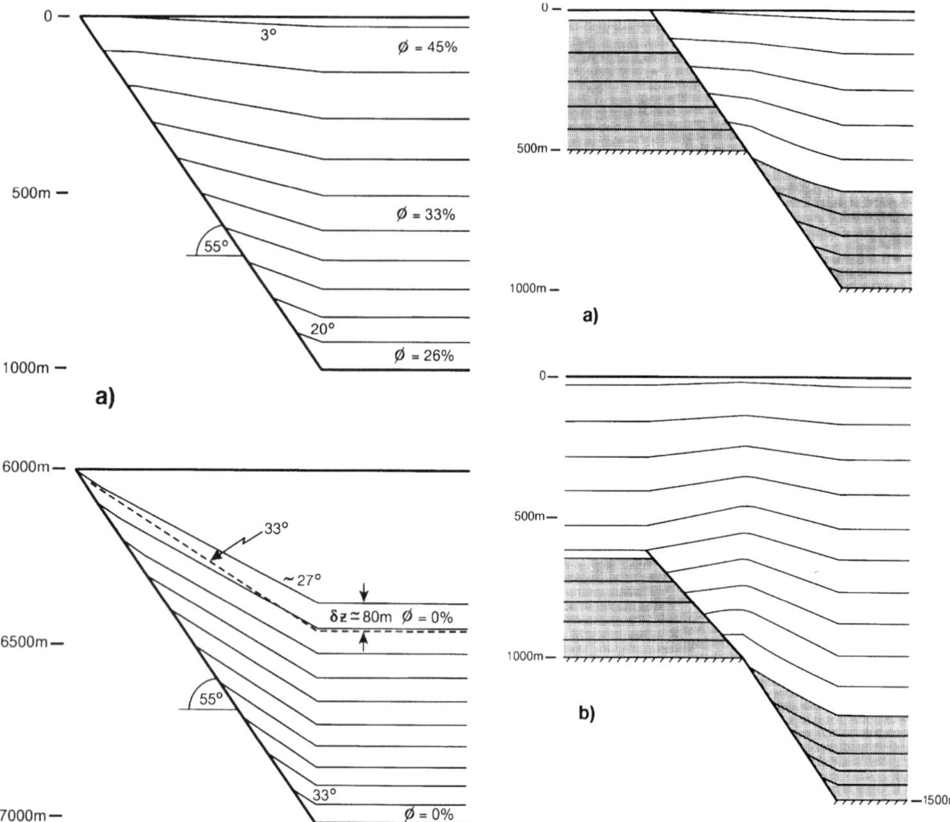

Fig. 6. Comparison of compaction modelling using thin and thick layers. (**a**) The hanging wall subsides in steps of 100 m; at each stage shale is deposited and subsequently compacted. Note how the upper layers are thicker than the lower layers and that the compaction induced dips (shown on the figure) increase with depth. The Baldwin-Butler (1985) shale compaction function is used. The porosities (ϕ) are shown for selected layers. The average porosity of the section is around 34%. (**b**) The previous stage of the model is taken to depths of 6–7 km and compacted. Porosities are zero at these depths. The dashed line shows the result, for the top horizon only, that would be obtained if the ten 100 m stages in (a) were modelled with a single 1 km thick unit. The figure shows that using thinner layers produces 80 m less compaction overall (δz) and less steep hanging wall dips (27° versus 33°).

Fig. 7. Formation of a compaction anticline. (**a**) Before faulting, the basement subsides in five stages of 100 m, depositing five layers of shale (shaded) which compact at each stage. Subsequent faulting in five stages of 100 m allows the deposition of five layers of sandstone (unshaded) which compact themselves and the underlying shales in the hanging wall at each stage. The compaction functions used are the exponential functions for shales and sandstones derived from the North Sea by Sclater & Christie (1980). (**b**) Faulting ceases and the entire model subsides in five 100 m stages with deposition of the five top layers of sandstone. Compaction of the sediments in the footwall results in rotation of the fault plane and the formation of the left flank of the anticline; compaction of the hanging wall steepens the right flank of the anticline. The amplitude of the anticline has a maximum of about 70 m at 500 m depth.

but possible instances of such features have been observed in the Niger Delta (P. B. Buchanan, pers. comm.) and in the Fort St John Graben in NE British Columbia (L. Brady, pers. comm.). A feature of compaction anticlines is that some horizons in the hanging wall of the fault are elevated above corresponding beds in the footwall: this is an observation that would often be regarded as evidence of structural inversion.

Alternative models

There are several other mechanisms capable of producing basinward (synthetic) dips in the hanging walls of normal faults. A full treatment of each of the mechanisms is beyond the scope of this paper, but a brief discussion of four of them follows. Mechanisms such as salt structures in normal fault hanging walls (Enachescu 1988) and original sedimentary structures (Xiao & Suppe 1992, Waltham *et al.* 1993) are additional possibilities for producing synthetic hanging wall dips but are not discussed further here.

Drag

'Normal-drag' and 'reverse-drag' are commonly, if informally used terms to describe, respectively, structures with geometries such as those discussed in this paper (lick-up) and the well-known rollover structures formed above listric faults. Waltham (1990) proposed that drag in the vicinity of the fault is a more likely cause of hanging wall synclines than compaction. Drag effects are observed in dipmeters from wells in the Norwegian North Sea (J. Hesthammer, pers. comm.) and are used to locate and identify normal faults. Typically, such drag zones are tens of metres wide and have synthetic dips in the footwall and hanging wall. It is unclear whether such features form before or during faulting. Effects on such a scale would not usually be observable seismically. In the Sirte Basin example shown in this paper, the zone of basinward dips increases in width at shallow depths where the fault displacements are smallest; counter to the effect one might expect to see if the structures were caused by drag, where the width of the drag zone ought to increase with fault displacement, or at least remain constant. Furthermore, it is difficult to imagine how drag could have caused significant effects kilometres away from the fault in the hanging wall, while at the same time producing no observable drag effects in the footwall.

Folding over convex fault bends

Xiao & Suppe (1992) have proposed that some 'normal drag' features may arise from folding over convex bends of normal faults. This is a geometrically plausible mechanism that undoubtedly applies in some situations. Such fault bends can be estimated by inverse modelling the hanging wall horizon shapes. When attempted with the example in this paper, faults are predicted that steepen to near vertical at great depths in the crust, a solution that would suggest that the faults bounding the Agedabia Trough are quite different from other large normal faults observed in rifts elsewhere (Westaway & Kusznir 1993): an unnecessary and unacceptable conclusion. In addition, different deep fault geometries are predicted by inverse modelling on different horizons when, of course, the same fault prediction should be common to all horizons. In principle, closer convergence of results might be achievable by adjusting certain parameters, such as the shear angle, in the manner of White & Yielding (1991).

Extensional forced folding

Withjack *et al.* (1989) described a number of models and examples of extensional forced folding, a process that occurs when a ductile detachment layer such as salt is present above an active normal fault. Such a mechanism might, in principle, account for some of the characteristics of the structures observed in the example shown here. However, the synthetically dipping monocline produced by a simple extensional forced fold should have homogeneous dips, contrary to observations in the example described in this paper. Figure 7 in Withjack *et al.* (1989) is an interpreted seismic line across a major normal fault that has a basically similar hanging wall geometry to the example cited here in my Figure 3: however, they, perhaps wrongly, did not consider compaction as a possible mechanism for forming their structure.

Mitra (1993) proposed a type of structure that he termed 'extensional fault propagation folding'. No geological examples of this kind of structure were described but the clay modelling examples correspond in some respects to models of extensional forced folds.

Inversion, compression

Any interpretation of the structures seen in Fig. 3 as involving structural inversion or compressional tectonics can be rejected here since: (a) no such compressional event is known in this part of Libya; (b) no reverse faulting is observed in the vicinity; (c) inversion, by itself, could not account for the observed thinning of several of the units in the hanging wall as they approach the normal fault.

In map view (Skuce 1994; Morley *et al.* 1990), the normal fault forming the W margin of the Agedabia Trough is not linear but is displaced in numerous minor and major transfer structures. The fault is most unlikely therefore, to be a strike-slip fault and the structures that are confined to the NE of the fault cannot be attributed to any kind of 'flower-structure'.

Discussion

The compaction of shales, despite a great deal of study, is still poorly understood and published relationships between shale porosities and depth show wide variance. Shale porosity is not only dependant on depth: time, temperature and chemistry are also important factors (Issler 1992). Shale porosites at a depth of 1 km are typically in the range of 20–32%, although higher values are sometimes observed in areas of rapid deposition (Issler 1992; Baldwin & Butler 1985). Reported low values of porosity at shallow depths (e.g. 10% at 1 km) are anomalous and are often due to miscalculation of shale porosities from well logs or underestimation of the amount of post-compaction uplift and erosion (D. Issler, pers. comm.). Thus, compaction of normal shales, even below burial depths of 1 km, is likely to be considerable and cannot safely be neglected.

There is a commonly made but erroneous assumption that compaction cannot produce significant dips in the hanging walls of normal faults unless basement is involved in the footwall. In Fig. 6a and in Fig. 5 (stage 4) structures have formed in the hanging walls of the faults which are independent of the lithology of the footwall. Since the footwall had not been additionally loaded in either case, no compaction of the footwall could have occurred, regardless of its composition. In a case where the hanging wall and footwall are composed of the same rocks, continued burial of both sides of the fault would result in compaction of the whole section, causing some flattening of the previously formed hanging wall compaction structures, but not their removal. Thus, compaction structures do not require contrasting lithologies across the fault in order to form.

Compaction is often neglected as a first order effect in structural modelling in areas of detached faulting (e.g. Xiao & Suppe 1992). In fact, compaction structures are likely to be considerable above listric faults. One reason such features are sometimes not easily recognized is because such faults produce anticlinal (rollover) structures that usually have a magnitude equal to or greater than the normally synclinal structures produced by compaction. The two kinds of structures will interact, sometimes destructively. Differential subsidence will be largest close to the fault and will reduce to zero above the point where the listric fault flattens out (see Withjack & Peterson 1984, fig. 13b). Thus, compaction structuring, a consequence of subsidence, will be strongest close to the fault. A typical product of compaction and rollover might, therefore, be a syncline near the fault with a larger anticline further away (White & Yielding 1991). Published examples of such features can be seen in McClay *et al.* (1991), fig. 6 (Gulf of Mexico); Xiao & Suppe (1992), fig. 21 (Mississippi Canyon) and Beach & Trayner (1991) figs 4 and 8 (Nile Delta). In none of these cases was compaction considered the cause of the synclinal structures.

Sandbox models of extensional listric faults (McClay *et al.* 1991) show hanging wall horizons which dip increasingly steeply as they approach the fault plane. They found that these steep dips could not be imaged in seismic models based on the sandbox geometries: this was attributed to spatial aliasing problems. On a real data example (McClay *et al.* 1991; their fig. 6), while there are some genuine imaging problems around 3.5 s and deeper; the shallow section, above 3.0 s, shows synclinal structures in the hanging wall, adjacent to the fault plane, that are most probably real. Sandbox models do not take compaction into account, which process is a more likely source of these synclines in the real data example than any kind of seismic imaging problem.

Geometric modelling of compaction and fault bend folding (rollover) together is likely to be a complex practical problem and has not been attempted here. There are problems with the relative timing of the deformation (i.e. does compaction occur immediately upon burial) and a further problem arising from the likelihood that the two deformation processes are best modelled by different mechanisms, for example, vertical and inclined simple shear. Although the displacement of any particular point could be expressed as the vector sum of the two mechanisms, the direction as well as the magnitude of this vector would vary with position within the model.

White & Yielding (1991) and Kerr & White (1992) described methods for inverse structural modelling that consider compaction as well as deformation of the hanging wall horizons over bends in the fault plane. However, for computational expediency they assumed that compaction and the fault bend folding both occur by simple shear oriented in the same direction, while recognizing that compaction should ideally be modelled as occurring vertically. Deformation is also assumed in their models to occur in one phase, the whole section has uniform lithology, and the footwall is assumed rigid. In many practical cases, one or more of these assumptions will not apply.

Withjack & Peterson (1993) showed that geometric inverse models are very sensitive to small errors in four input variables, as follows: (i) fault dip at shallow levels; (ii) the geometry of the hanging wall horizon close to the fault; (iii) the inclination of the simple shear deformation used to model the hanging wall; (iv) the correlation of horizons across the fault plane. Small uncertainties in any one of these factors lead to large uncertainties in the prediction of the fault plane geometry at depth. All of these factors are dependent on the amount of compaction of either the footwall (i), the hanging wall (ii) and (iii), or any part of the section (iv) (Figs 5 and 7). Thus, any attempt at inverse modelling must have, as a starting point, an accurate knowledge of the amount and timing of compaction. Since this requires the deep geology of the section to be known in some detail, including the fault geometry, it might reasonably be asked what purpose inverse modelling would then serve.

The forward models described in this paper show that the processes of fault movement, sedimentation and compaction are non-commutative, i.e. the order in which they happen matters; similar observations were made by White *et al.* (1986). Obviously, the relative timing of structural processes will be equally important in inverse modelling. An unsuccessful attempt was made to invert the forward model produced in Fig. 5, considering decompaction and using the same software. The program (GEOSEC) predicted faults steepening with depth, a different one for each hanging wall horizon, even though the rock properties used to produce the forward model and the final geometries of the horizons were known precisely. White & Yielding's (1991) method for making the fault plane solutions for different horizons converge, by adjusting the simple shear direction and the compaction parameters, could not work in this case because these parameters are fixed, having been used in building the forward model. The inverse modelling failed because it assumed a single deformation step, in which decompaction and fault bend folding occur simultaneously, whereas the forward model was produced in multiple steps.

Conclusions

Compaction can be the cause of large fold structures in the hanging walls of normal faults. Such structures can have oilfield scale dimensions and dips up to 20° or more. The magnitude and geometry of compaction structures is, in part, dependent on the compaction characteristics of the hanging wall and footwall rocks but there does not have to be a lithology or porosity contrast across the fault in order for such structures to form. Forward models show that the relative timing of subsidence, fault movement and compaction are critical in producing a certain geometry. The forward models presented in this study make the simplifying assumption that the faults are completely planar, which is not true in most cases. Forward modelling of structures that involve compaction, sedimentary growth and folding over non-planar faults will produce a wide variety of complex structures, including compaction anticlines.

Inverse geometric modelling that considers compaction as well as fault bend folding is as complex as the forward modelling process, but with the added difficulty that the output of the process is highly sensitive to small uncertainties in some of the input parameters (Withjack & Peterson 1993), that are, in turn, dependent on the magnitude and timing of compaction. Existing processes that attempt inverse modelling make important simplifying assumptions that limit their applicability.

The managements of the National Oil Company of Libya, ÖMV AG and Husky Oil International are thanked for granting permission to publish this paper. Peter Buchanan performed the GEOSEC modelling and made many useful suggestions. I. Davison and J. D. van Wees are thanked for their careful reviews and constructive criticism. Maria Nemethy and Shannon Wahler prepared the figures.

Appendix

The compaction functions used in the modelling are as follows.

ϕ = porosity; z = depth in km

1. Baldwin–Butler (1985) (shales)

$$\phi = 1 - (0.166 \times z)^{0.1575}$$

ϕ is restricted to the range 0 0.45

2. Sclater–Christie (1980) (shales)

$$\phi = 0.63\ e^{(-z \times 0.520)}$$

3. Sclater–Christie (1980) (sandstones)

$$\phi = 0.49\ e^{(-z \times 0.270)}$$

4. Schmoker–Halley (1982) (carbonates)

$$\phi = 0.417\ e^{(-z \times 0.4)}$$

References

BADLEY, M. E., RAMBECH DAHL, C. & AGDESTEIN, T. 1988. The structural evolution of the Northern Viking Graben and its bearing on extensional modes of basin formation. *Journal of the Geological Society, London*, **145**, 455–472.

BALDWIN, B. & BUTLER, C. O. 1985. Compaction curves. *American Association of Petroleum Geologists Bulletin*, **69**, 622–626.

BEACH, A. & TRAYNER, P. 1991. The geometry of normal faults in a sector of the offshore Nile Delta, Egypt. *In*: ROBERTS, A. M., YIELDING, G. & FREEMAN, B. (eds) *The Geometry of Normal Faults*. Geological Society, London, Special Publications, **56**, 173–182.

ENACHESCU, M. E. 1988. Extended basement beneath the intracratonic rifted basins of the Grand Banks of

Newfoundland. *Canadian Journal of Exploration Geophysics*, **24**, 48–65.

EVANS, A. C. & PARKINSON, D. N. 1983. A half graben and tilted fault block structure in the Northern North Sea. *In*: BALLY, A.W. (ed). *Seismic Expressions of Structural Styles*. American Association of Petroleum Geologists Studies in Geology Series, **15**, 2.2.2-7–2.2.2-11.

GRAHAM, R. H. 1992. Comparison of the Tethyan margin of the SW Alps with the North Sea. American Association of Petroleum Geologists field trip handbook, *Alpine Mesozoic Basin in the SE of France*, unpublished.

GUMATI, Y. D. & NAIRN, A. E. M. 1991. Tectonic subsidence of the Sirte Basin, Libya. *Journal of Petroleum Geology*, **14(1)**, 93–102.

ISSLER, D. R. 1992. A new approach to shale compaction and stratigraphic restoration, Beaufort–Mackenzie Basin and Mackenzie Corridor, Northern Canada. *American Association of Petroleum Geologists Bulletin*, **76**, 1170–1189.

JACKSON, J. A. 1987. Active normal faulting and crustal extension. *In:* COWARD, M. P., DEWEY, J. F. & HANCOCK, P. L. (eds) *Continental Extensional Tectonics*. Geological Society, London, Special Publications, **28**, 3–17.

KERR, H. G. & White, N. 1992. Laboratory testing of an automatic method for determining normal fault geometry at depth. *Journal of Structural Geology*, **14**, 873–885.

KUSZNIR, N. J., MARSDEN, G. & EGAN, S. S. 1991. A flexural-cantilever, simple-shear/pure-shear model of continental lithosphere extension: applications to the Jeanne d'Arc Basin, Grand Banks and Viking Graben, North Sea. *In*: ROBERTS, A. M., YIELDING, G. & FREEMAN, B. (eds) *The Geometry of Normal Faults*, Geological Society, London, Special Publications, **56**, 41–60.

MCCLAY, K. R., WALTHAM, D. A., SCOTT, A. D. & ABOUSETTA, A. 1991. Physical and seismic modelling of listric normal fault geometries. *In:* ROBERTS, A. M., YIELDING, G. & FREEMAN, B. (eds) *The Geometry of Normal Faults*. Geological Society, London, Special Publications, **56**, 231–239.

MILTON, N. J., BERTRAM, G. T. & VANN, I. R. 1990. Early Palaeogene tectonics and sedimentation in the central North Sea. *In*: HARDMAN, R. F. P. & BROOKS, J. (eds) *Tectonic Events Responsible for Britain's Oil and Gas Reserves*. Geological Society, London, Special Publications, **55**, 339–351.

MITRA, S. 1993. Geometry and kinematic evolution of inversion structures. *American Association of Petroleum Geologists Bulletin*, **77**, 1159–1191.

MORLEY, C. K., NELSON, R. A., PATTON, T. L. & MUNN, S. G. 1990. Transfer zones in the East African Rift system and their relevance to hydrocarbon exploration in rifts. *American Association of Petroleum Geologists Bulletin*, **74**, 1234–1253.

SCHMOKER, J. W. & HALLEY, R. B. 1982. Carbonate porosity versus depth: a predictable relationship for south Florida. *American Association of Petroleum Geologists Bulletin*, **66**, 2561–2570.

SCLATER, J. G. & CHRISTIE, P. A. F. 1980. Continental stretching: an explanation of the post-mid-Cretaceous subsidence of the central North Sea basin. *Journal of Geophysical Research*, **85**, 3711–3739.

SKUCE, A. G. 1994. A structural models of a graben boundary fault system, Sirte Basin, Libya: compaction structures and transfer zones. *Canadian Journal of Exploration Geophysics*, **30**, 73–83.

WALTHAM, D. 1990. Finite difference modelling of sandbox analogues, compaction and detachment free deformation. *Journal of Structural Geology*, **12**, 375–381.

——, HARDY, S. & ABOUSETTA, A. 1993. Sediment geometries and domino faulting. *In*: WILLIAMS, G. D. & DOBB, A. (eds) *Tectonics and Seismic Sequence Stratigraphy*. Geological Society, London, Special Publications, **71**, 67–81.

WESTAWAY, R. & KUSZNIR, N. J. 1993. Fault and bed rotation during continental extension: block rotation or vertical shear? *Journal of Structural Geology*, **15**, 753–770.

WHITE, N. J. & YIELDING, G. 1991. Calculating normal fault geometries at depth: theory and examples. *In*: ROBERTS, A.M., YIELDING, G. & FREEMAN, B. (eds) *The Geometry of Normal Faults*, Geological Society, London, Special Publications, **56**, 251–260.

——, JACKSON, J. A. & MCKENZIE, D. P. 1986. The relationship between the geometry of normal faults and that of the sedimentary layers in their hanging walls. *Journal of Structural Geology*, **8**, 879–909.

WITHJACK, M. O. & PETERSON, E. T. 1993. Prediction of normal fault geometries – a sensitivity analysis. *American Association of Petroleum Geologists Bulletin*, **77**, 1860–1873.

——, MEISLING, K. E. & RUSSELL, L. R. 1989. Forced folding and basement-detached normal faulting in the Haltenbanken area, offshore Norway. *In*: TANKARD, A. J. & BALKWILL, H. R. (eds). *Extensional Tectonics and Stratigraphy of the North Atlantic Margins*, American Association of Petroleum Geologists Memoir **46**, 567–575.

XIAO, H. & SUPPE, J. 1992. Origin of rollover. *American Association of Petroleum Geologists Bulletin*, **76**, 509–529.

Benefits and limitations of section restoration in areas of extensional salt tectonics: an example from offshore Louisiana

MARK G. ROWAN

Department of Geological Sciences and Energy & Minerals Applied Research Center, Campus Box 250, University of Colorado, Boulder, CO 80309, USA

Abstract: A composite, multi-level salt system from offshore Louisiana is used to illustrate some benefits and limitations of cross section restoration. The system has a complex three-dimensional geometry in which both salt and overburden have apparently moved in multiple directions during Miocene-Pleistocene deformation. Sequential restorations are beneficial in that they provide a framework for: (1) analysis of the structural evolution of salt and overburden; (2) determination of rates of various processes; (3) prediction of changes in the sea-floor paleotopography; (4) evaluation of evolving sediment transport and deposition systems; and (5) analysis of the timing and pathways of hydrocarbon migration. The limitations of section restoration include possible inadequacies of restoration algorithms to model correctly actual rock deformation. More importantly, the assumption of plane-strain deformation is shown to be invalid in this example. Local pin lines are used to demonstrate that there is significant out-of-plane movement. Thus, cross-sectional area for both salt and overburden may need to be added or removed from the section during restoration if the goals are to illustrate the geometric evolution. If, on the other hand, the goal of restoration is to validate and, if necessary, modify the interpretation, then 2D restoration is not appropriate in this case. Ultimately, 3D restoration is needed to restore accurately complex structural geometries.

Cross section restoration was originally practiced in fold-and-thrust belts in an attempt to generate reasonable interpretations of field and subsurface data (e.g. Bally *et al.* 1966, Dahlstrom 1969, Royse *et al.* 1975). The idea is that the rules of balancing (Dahlstrom 1969) can place constraints on subsurface geometry. There is always room for interpretational license in the analysis of structural data, and restoration and balancing allow a given interpretation to be evaluated and, if necessary, modified. Errors may be identified in the restorations because of a lack of balance, and the interpretation can then be corrected to produce a viable and admissible geometry (Elliott 1983).

Section balancing in fold-and-thrust belts can also be used to help understand original (undeformed) facies relationships, to evaluate hydrocarbon generation and migration, and to determine the structural evolution of folds and faults (Woodward *et al.* 1985). In the latter applications, it is necessary to sequentially restore the section to different stages of the progressive deformation. Unfortunately, however, syntectonic sediments are usually lacking or sparse due to uplift and erosion of mountain belts. The result is that restoration is usually limited to reconstructing the original, undeformed configuration. The main focus, therefore, of restoration and balancing in fold-and-thrust belts, within both industry and academia, remains section validation and modification.

Section restoration and balancing was first applied to extensional terranes by Gibbs (1983). Because many extensional basins have relatively complete records of syntectonic deposition, sequential restoration has become a more powerful tool and has been used to determine and illustrate the structural evolution of a variety of extensional terranes (e.g. Rowan & Kligfield 1989; Worrall & Snelson 1989; Nunns 1991; Beach & Rowan 1992; Schultz-Ela 1992; Rowan 1993; Bishop *et al.* 1995; Diegel *et al.* 1995; Schuster 1995; Peel *et al.* 1995; Rowan 1995). A series of sequential restorations can also serve as the framework for analyses of facies development, the changing sea-floor topography through time, hydrocarbon generation and migration, and the rates of various processes such as sedimentation, subsidence, extension and salt flow (Rowan 1993). Nevertheless, section validation remains one of the primary applications of restoration and balancing in extensional terranes (e.g. Rowan & Kligfield 1989; Nunns 1991; Beach & Rowan 1992).

There are a number of assumptions and limitations inherent in section restoration. Probably the most critical of these is that restoration is a 2D process that assumes plane-strain deformation, i.e. that there is no movement of material into or out of the plane of the cross section. Thus, cross-sectional area is maintained during restoration, such that individual stratigraphic units, though they

From Buchanan, P. G. & Nieuwland, D. A. (eds), 1996, *Modern Developments in Structural Interpretation, Validation and Modelling,* Geological Society Special Publication No. 99, pp. 147–161.

may have different shapes at various stages of a sequential restoration, always have the same area. This is generally considered to be a safe assumption as long as the section is oriented in the direction of material transport (e.g. Woodward *et al.* 1985).

The assumption of plane-strain deformation may be invalid, however, in the case of salt tectonics because of the 3D nature of salt flow and the possible role of dissolution. Numerous authors have acknowledged this issue, either explicitly or implicitly, by allowing the cross-sectional area of salt to vary during restoration (Rowan & Kligfield 1989; Worrall & Snelson 1989; Diegel & Cook 1990; Hossack & McGuinness 1990; Wu *et al.* 1990; Moretti *et al.* 1990; Seni 1992, Schultz-Ela 1992; Rowan 1993; McGuinness & Hossack 1993; Bishop *et al.* 1994; Koyi 1994; Diegel *et al.* 1995; Schuster 1995; Peel *et al.* 1995; Rowan 1995). However, the assumption of plane-strain deformation is invariably maintained for the supra- and sub-salt stratigraphy. Hossack (1995) nicely summarizes the prevailing opinion:

> Geologists have long suspected that section balancing is more difficult in salt structures because of the ability of the salt to flow in and out of the plane of section and also to dissolve and thereby violate constant volume [area] considerations. However, the surrounding sediments generally deform by brittle plastic processes and are less able to flow out of the plane of a properly chosen section. The pragmatic approach is to restore sections by assuming constant volume [area] conditions for the sediment structures alone and to leave the salt area as gaps which may change in area through time."

This approach is probably valid in many instances of salt deformation, especially where relatively linear structures are predominant. In this paper, however, the 3D geometry of a complex salt system in the Louisiana shelf of the Gulf of Mexico is used to demonstrate that, at least in some areas of salt tectonics, it is impossible to choose a properly-oriented regional section. It will be argued, therefore, that the assumption of plane-strain deformation is invalid not only for the salt, but locally also for the sediments. Thus, restorations must be 3D if they are to be accurate. An important implication is that 2D restoration in this type of terrane generally cannot be used to validate or modify an interpretation. Even relatively large errors in the restored geometries may reflect limitations inherent in 2D restoration rather than real mistakes that need to be corrected. If, however, the goal of section restoration is to determine and illustrate the evolution of salt-related structures, then incorporating area changes in the overburden stratigraphy may be warranted. This statement contradicts the standard

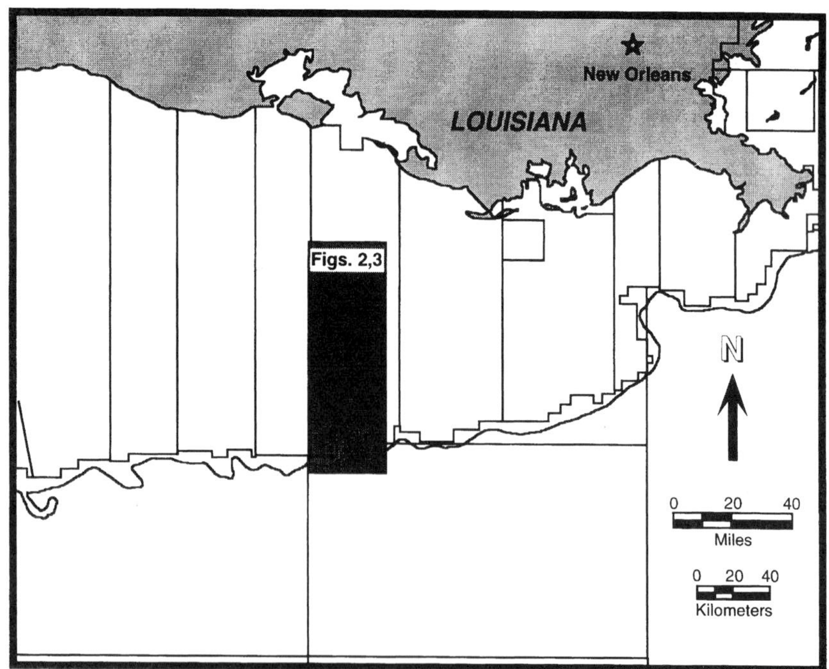

Fig. 1. Location map showing study area, offshore Louisiana lease areas and shelf-slope break (smooth curve).

'rules' of cross section restoration, but it will be justified below.

Data base

The study area is located in the Eugene Island lease area of offshore Louisiana (Figs 1 & 2). It consists of a north–south corridor extending over much of the continental shelf to the shelf-slope break, with a maximum north–south extent of 150 km and a maximum east–west extent of 60 km. Approximately 2500 km of 2D migrated, multifold seismic data, acquired between 1982 and 1988, were provided by industry sources to the University of Colorado (Fig. 2). In addition, well logs and biostratigraphic data from 19 wells (Fig. 2) were also provided which were used to generate synthetic seismograms and constrain the seismic interpretation.

Eight horizons were interpreted in the area. These generally correspond to condensed sections; the deeper ones are also sequence boundaries because of the absence of transgressive and highstand systems tracts (Rowan *et al.* 1994*b*; Alexander & Flemings 1995). Although some of the wells penetrate strata as old as 5.8 Ma, most of the wells are shallower, and the oldest interpreted horizon is dated at 4.0 Ma. These limitations affect the restoration procedure: although restorations can be generated with relative confidence back to 4.0 Ma, older stages are schematic and based only on the observed geometry of salt and bedding rather than actual restorations of correlated strata.

Salt system geometry

The southern portion of Eugene Island is dominated by a complex, multi-level salt system consisting of a series of salt welds (Jackson & Cramez 1989) and diapirs. The geometry of the salt system is summarized below (see also Rowan *et al.* 1994*c*; Rowan 1995).

Figure 3 is a structure contour map in two-way travel time (TWT) of the salt system, and Fig. 4 shows two interpreted north–south seismic reflection profiles. The most prominent feature of the three-dimensional geometry is the structural relief on the salt system. Depths to the top salt or salt weld range from less than 1 s TWT over the tallest diapir to greater than 7 s. This topography is highly irregular, consisting of numerous lows bounded by ramps and separated from each other by saddles at various levels. There are two types of lows (Fig. 3). To the north (right end of Fig. 4b) is a laterally extensive low that is separated from lows to the south by a continuous, horizontal portion of the salt weld that is above 4.0 s and that trends NNW–SSE.

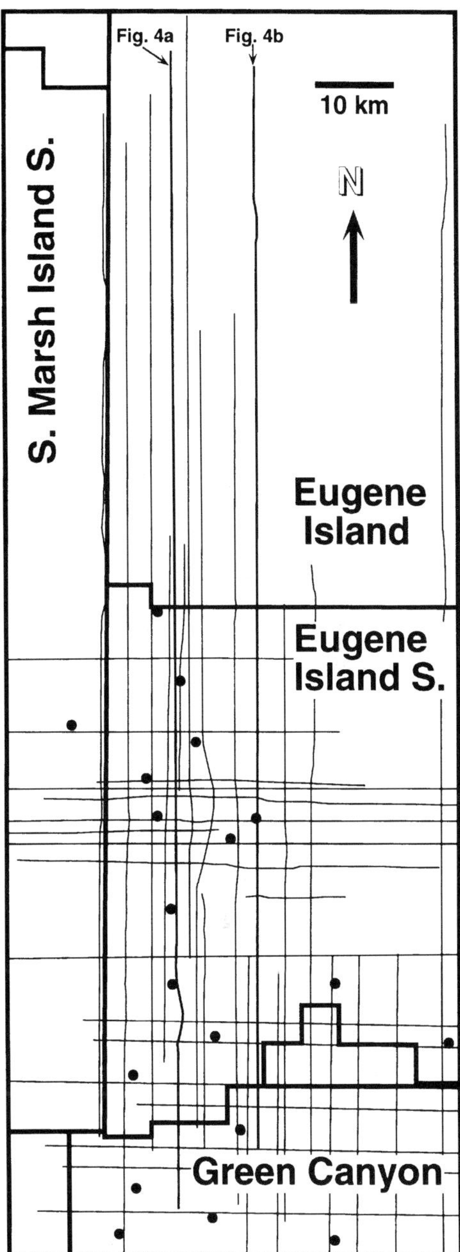

Fig. 2. Base map showing offshore Louisiana lease areas, seismic data and well locations. Seismic profiles illustrated in Fig. 4 are indicated.

In contrast, the southern lows are isolated, circular to elliptical depressions.

The various levels of salt welds are connected rather than discrete. For example, there are three different levels of subhorizontal welds discernible

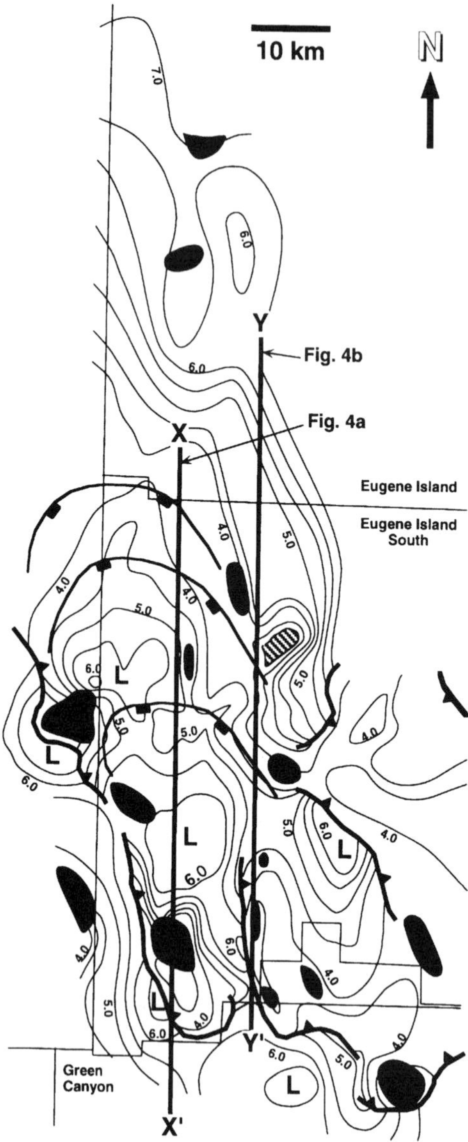

Fig. 3. Structure contour map (in TWT) of salt system in southwestern Eugene Island (C.I. = 0.5 s TWT). Only major diapirs and the salt weld are mapped; small salt rollers and other minor salt bodies are ignored. Diapirs are in black, diagonal ruling indicates no shallow salt, elliptical lows are marked by L, thick lines with triangles are salt overhangs, and thick lines with rectangles are growth faults. Locations of profiles illustrated in Fig. 4 are shown.

in Fig. 4 at around 6.5, 5.0 and 4.0 s TWT. These welds are not separate and distinct salt sheets, but are connected by a variety of ramps with different orientations (Fig. 3). Although there is a prominent trend created by the NNW-striking, level high in the weld in the north-central portion of the map, inspection of Fig. 3 will show that there are major ramps that dip in all directions.

The topography on the weld system could be secondary, i.e., caused by deformation of a sub-horizontal weld above deeper faults or salt bodies. If this were the case, reflections underlying the ramps would have approximately the same orientation as the ramps themselves because of folding. Instead, the pattern observed in Eugene Island and in much of northern Green Canyon and Ewing Bank is that underlying reflections are truncated by the ramps (for example, on either side of the shallow portion of the weld in the middle of the section; Fig. 4b). These cutoffs are interpreted as marking the original overhanging interface between roughly horizontal bedding and a pre-existing salt diapir that has since been evacuated (see Rowan 1995). The salt weld topography, therefore, is thought to be primary (although this does not preclude some possible modification by subsequent, deeper deformation).

The implication is that the lows in the weld surface mark the locations of early salt bodies that fed salt to higher levels during sedimentary loading. This interpretation is similar to that of Sumner *et al.* (1991), in which they describe a series of connected horizontal sills. It is also supported by the geometry of an existing salt body presented by McGuinness & Hossack (1993, fig. 18). They mapped the base of a shallow salt sheet and determined that it had significant topography with three lows. The sheet was interpreted as an amalgamated salt body sourced from depth by vertical diapirs centred beneath the three lows. If the salt were to be evacuated from this salt body, the resulting salt weld would have a geometry very similar to that mapped in southwestern Eugene Island.

Another characteristic feature of the structural geometry is that the dominant listric normal faults tend to extend up from the ramps in the weld surface (Fig. 4). In map view, these curve around the updip and lateral margins of the elliptical lows (Fig. 3). Thus, faults have variable strikes that often mimic the strikes of the associated ramps in the salt weld. The strong association between faults and the salt weld geometry suggests that the topography on the weld is controlling both the orientation and location of faults, and thereby the location and size of supra-salt sedimentary basins.

The normal faults accommodate extension on a detachment formed by the salt weld surface. The amount of extension, as measured by the heaves on the faults, is relatively minor and probably limited by the southern termination of the detachment and by the relief of the detachment surface. The extension must be accommodated somehow by shortening; this can be observed in the contractional folds near the southern end of the salt system (Fig. 4b) and by the narrowing of diapirs through time (see below). However, the amount of extension and contraction observed on any given profile should not necessarily be equal because balance may take place out of the plane of the cross section.

The thick lines on Fig. 3 mark the edges of salt overhangs. These do not separate distinct shallow sheets from deeper ones, but instead separate shallower portions of the same salt system from deeper portions. The overhanging part can always be traced around and found to be in continuity with the part that is overhung. The three-dimensional geometry is thus very complex.

Although most of the area is underlain by a salt weld, scattered diapirs are present, and they are not randomly located. The majority are located near the frontal or lateral edges of the shallow, overhanging parts of the composite salt system, and most of the others are located over the saddles between lows (Fig. 3). Thus, very few of the diapirs have deep roots. Furthermore, they are often located in the footwalls of major faults. Whether faults served to localize salt diapirs or diapirs served to localize faults is unclear, but modeling by Vendeville & Jackson (1992) suggests that the former may be the case.

Restoration method

Technique

Restorations of two north–south profiles were carried out using the software package GEOSEC, developed and marketed by CogniSeis Development Co. A technique was used that was designed specifically for restoration in this type of salt province (Rowan 1993); for further information, this detailed paper should be consulted.

The method attempts to isolate and remove the effects of the various processes that have affected the changing geometry of a cross section. These include sedimentation, compaction, loading subsidence, salt movement and folding and faulting. The first step, however, is to convert the time interpretation to depth. Velocity surveys from the southwestern Eugene Island area were compiled to generate a depth-dependent velocity function that was applied to the time data. The data define a tightly-constrained velocity gradient down to the maximum depths penetrated by wells (Rowan *et al.* 1994*b*), and this gradient was simply projected in order to depth convert the deeper part of the section.

The depth section is restored sequentially back through time. For any given stage in this process, the top sedimentary layer is first stripped off and the underlying units are decompacted using appropriate standard porosity versus depth functions. The sedimentary load associated with the removed over-burden is then calculated by applying an Airy (one-dimensional) isostatic model to places along the section where there has been no change in salt thickness during the time interval being restored. Identification of these locations is based on a subjective analysis of the deformed geometry prior to restoration, and a reiterative process may be necessary after the initial restoration.

The same locations along the section where the Airy correction was calculated can now be used to determine the change in paleobathymetry, because this is part of the isostatic calculation, and the resulting water depths are connected to generate a new sea-floor template for structural restoration. The supra-salt sedimentary section is then restored to this template by removing fault offsets and using an appropriate algorithm to remove bedding deformation, thereby generating a restored top salt horizon. The base salt and sub-salt section are ignored during this phase because they have already been restored using the combination of decompaction and isostatic adjustment. Thus, the space between the restored top salt and the adjusted base salt defines the preexisting salt geometry.

The result is an intermediate restoration with a new paleobathymetric profile, bedding and fault geometries, and salt geometry. The next stage of restoration is then carried out in the same manner, starting with removal of the top layer, and this process proceeds back to the oldest interpreted horizons. Note that no assumptions are made about the kinematics or 'balance' of salt deformation, so that the cross-sectional area of salt is allowed to vary with time, whereas the cross-sectional areas of the supra- and sub-salt sequences are maintained.

Assumptions and limitations

There are several assumptions and limitations inherent in section restoration that must be understood before the results can be evaluated with any confidence. First, a restoration algorithm is applied that may or may not accurately model the actual rock deformation. Typical algorithms in extensional terranes include vertical simple shear (Verrall 1981), antithetic simple shear (White *et al.* 1986) and bed-length restoration (Rowan & Kligfield 1989), and each will produce a different restored geometry. These differences could be considered critical if the restorations are being used to evaluate and modify details of the interpretation. However, the uncertainty in knowing which restoration algorithm most accurately represents the actual rock deformation suggests that such an application

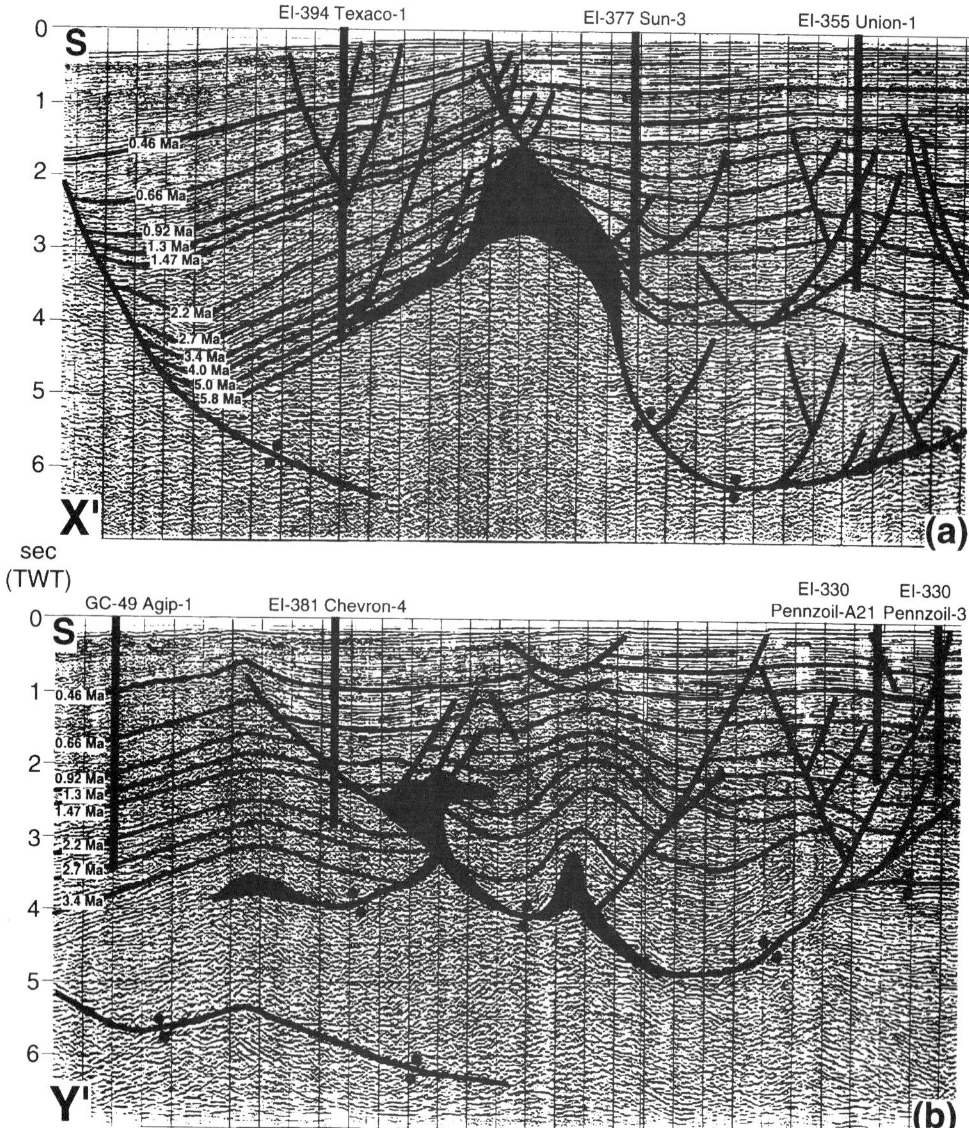

is unwarranted (Rowan 1993; Hauge & Gray 1995). If, on the other hand, the goal is to determine the overall geometric evolution, the variations in restored geometry produced by using different algorithms can be neglected at the scale of the sections being restored here.

A second limitation concerns the isostatic adjustment. Errors may be introduced for several reasons. First, because the crust has a flexural rigidity determined in part by its effective elastic thickness, a flexural isostatic model should ideally be used. However, the Airy model provides a satisfactory approximation in the Tertiary of the Gulf of Mexico (Rowan 1993). Second, measuring the net load added during any given time interval may be complicated because of the uncertainty in determining the net salt flux during the same interval. This becomes especially difficult when restoring the older part of the section because of the relative abundance of salt, the greater degree of salt deformation and the scarcity of deep well control.

A third limitation inherent in restoration is much more critical: this is the plane-strain assumption already discussed. The restoration method

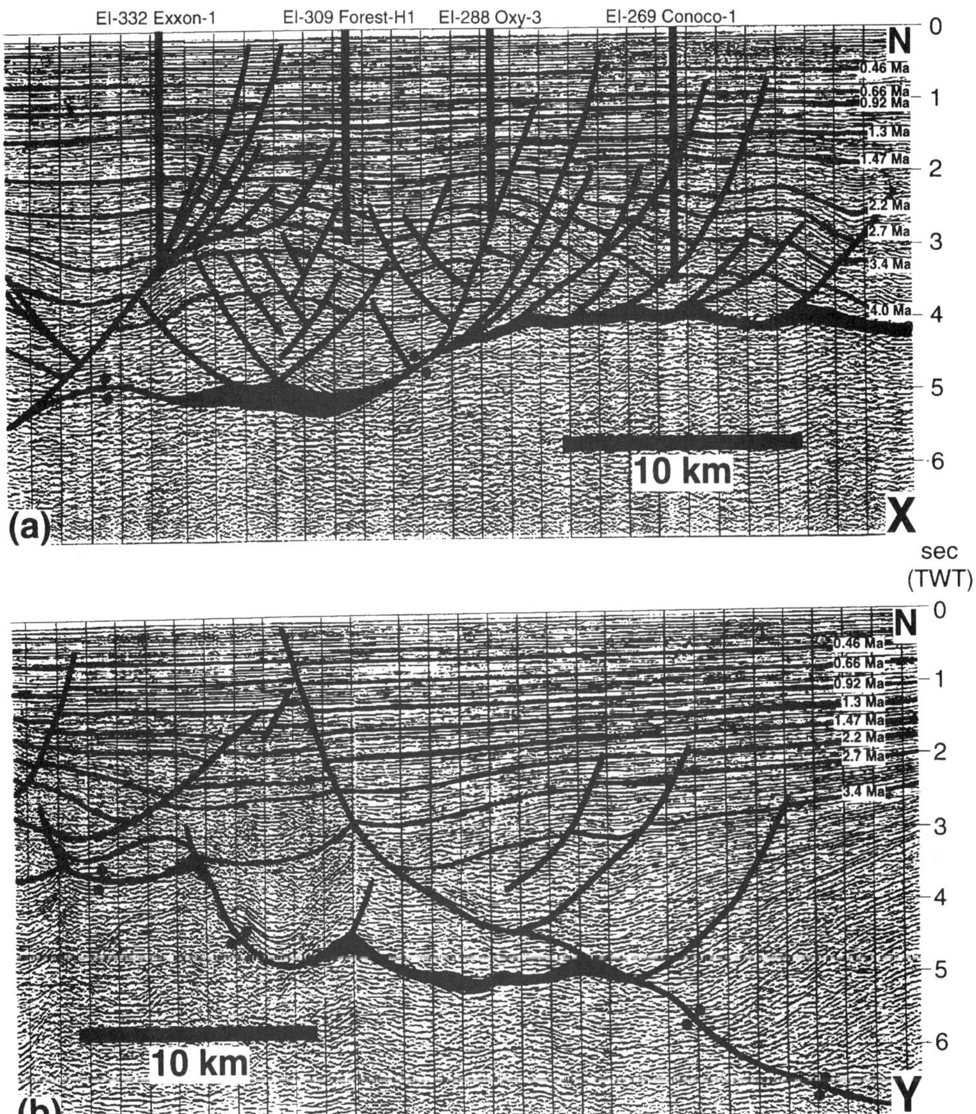

Fig. 4. Interpreted N–S seismic profiles through southwestern Eugene Island: (**a**) profile X–X′; (**b**) profile Y–Y′. The interpretations show well control (vertical grey bars), salt (black), salt welds (indicated by dot pairs), faults and stratigraphy. Profile locations are shown in Figs 2 & 3. Data provided courtesy of Halliburton Geophysical Services.

described above does not impose this condition for salt, but it does hold for the overlying strata. Yet deformation of the entire section is very much a three-dimensional process. It has already been shown that fault orientations mimic the topography on the salt weld surface; sediments, therefore, may move in many different directions as salt is withdrawn. This restriction imposes some very real limitations on restoration in the Eugene Island region, as is discussed more thoroughly in the next section.

A final limitation is that any given cross section cannot accurately represent the real three-dimensional geometry or evolution of the area. Along-strike variations in geometry are the rule, rather than the exception; in fact, it is virtually impossible to define a regional strike direction. Furthermore, deformation seen in one part of a profile may affect or be affected by events in areas not traversed by the section. For example, extension measured on a profile may be balanced by contraction on another profile. By carefully examining restorations of serial, two-dimensional, profiles while bearing in mind the three-dimensional geometry, much useful information can be obtained. However, a true three-dimensional evolutionary model can be generated only by conducting three-dimensional restorations.

Area balance during restoration

It is important in section restoration to define the boundary conditions (e.g. Dahlstrom 1969; Woodward *et al.* 1985; Geiser 1988). These are termed pin lines, and in the strictest sense are lines, placed on the section, that have undergone no net angular shear strain (Geiser 1988). Figure 5a shows two regional pins for a north-south profile, one placed to the north of the major north-dipping ramp, where there has been no lateral movement, and the other placed just south of the southern termination of the salt system. Although the latter one is probably not a pin line *sensu stricto* because of the local tilting, and therefore possible bedding-parallel shear, all extension and contraction on the salt system should be confined to the area between the two pins.

It cannot be a coincidence that major faults in this area are invariably located over ramps in the salt weld surface (Figs 3, 4 & 5). Yet restoration may move these faults a significant distance in the direction of their footwalls (see below and Fig. 6), suggesting that they just accidentally ended up over the ramps at the present stage in the deformation history. Although there is probably some movement of footwalls over the ramps and some associated development of footwall splays, it is relatively minor. It is likely that the main faults initiated at the ramps because of the geometry of the underlying salt weld surface and that they have moved laterally only small distances during subsequent deformation. Note that this is not necessarily true for other faults such as those merging with a horizontal portion of the salt weld; they may well be displaced by movement on the underlying detachment.

If one accepts that the major normal faults have moved only insignificantly during the progressive deformation, then they can be used as local pin lines (Woodward *et al.* 1985). However, they should be used as such in a flexible manner, meaning that, although they should not move from the vicinity of the ramp during restoration, there can be some minor deformation and translation (the magnitude of such movement has not been determined). If restoration does result in significant movement, so that a fault currently located over a ramp restores to a position away from the ramp, then this restored geometry is probably in error.

Figure 6a shows a vertical-simple-shear restoration of the profile in which such local pins are ignored. There are minor overlaps of the northern fault blocks that may be significant but that may

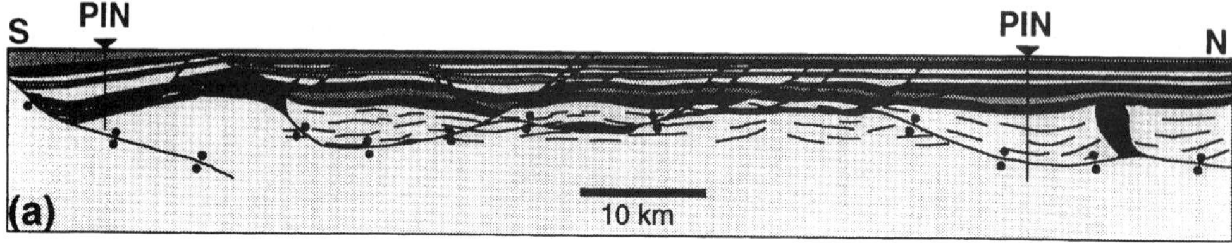

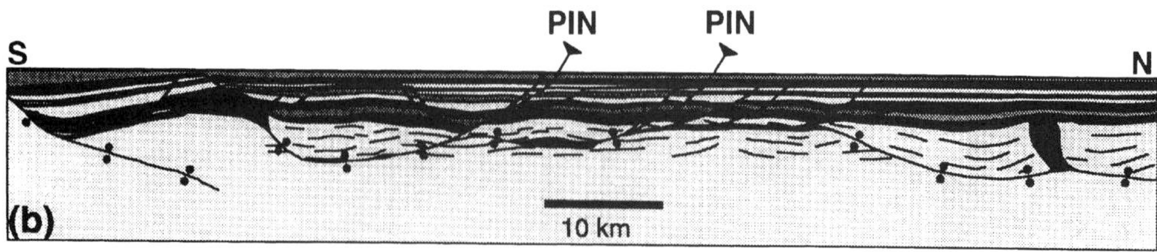

Fig. 5. Boundary conditions for restoration of seismic profile X–X′ (Fig. 4a): (**a**) regional pins; (**b**) local pins. Note that these are not necessarily pin lines *sensu stricto* (see text for discussion). Illustrations are 1:1 depth sections with salt in black and salt welds indicated by dot pairs.

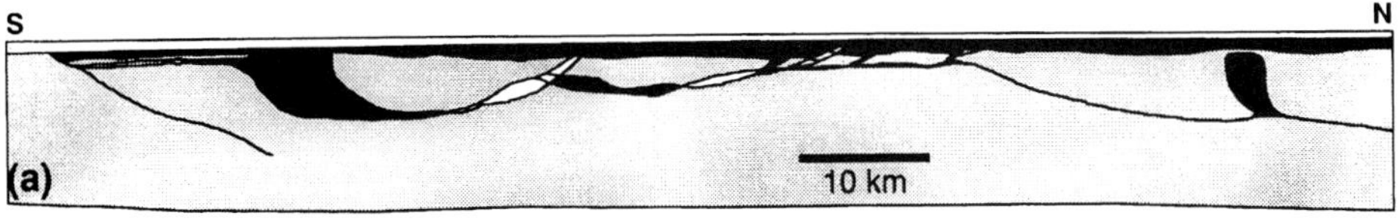

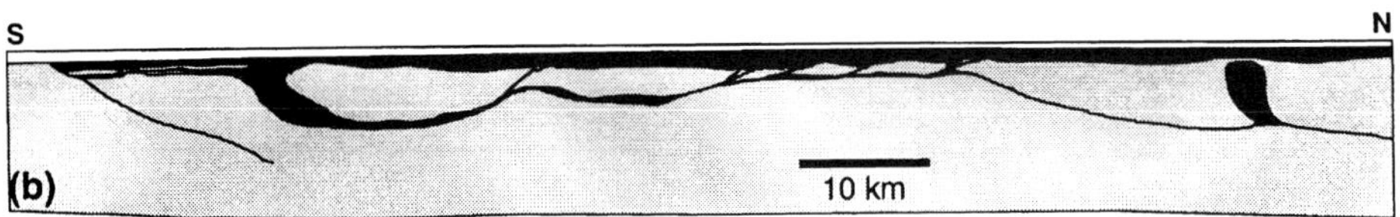

Fig. 6. 1:1 restorations of seismic profile X–X′ (Fig. 4a) to 3.4 Ma: (**a**) vertical simple shear restoration (overlaps are shown in white); (**b**) bed-length restoration (note shear of faults and salt diapir). See text for discussion.

also be artifacts beyond the resolution of the restoration technique. More important, however, are the prominent overlaps of the overburden and the sub-salt section in the areas of the two south-dipping ramps. The same problem is seen in a different way in a bed-length restoration (Fig. 6b). There are no overlaps because of the way the algorithm was applied, but there is significant layer-parallel shear indicated by the bends in both the restored fault trajectories and the edge of the southern salt diapir. In essence, the upper part of the overburden has been restored farther to the right because of closing up extension on the faults, whereas the lower part has hardly moved at all because it is attached to the ramps. There are other ways of applying both algorithms (for example, so that fault overlaps are created during bed-length restoration), but similar problems will result irrespective of the method used.

The problems that arise when standard section restoration techniques are applied to the interpreted profile (Fig. 6) may have several explanations. First, they could represent limitations in the applied restoration algorithms. However, it has already been shown that problems are generated irrespective of the method used. Second, they could indicate that there are mistakes in the interpretation; the restorations appear to invalidate the interpretation and suggest that modifications should be made. Whilst this is certainly possible, the well control provides good correlation from footwall to hanging wall for sequences younger than 3.4 Ma. It is unlikely that minor mis-ties could account for the 3 km of horizontal overlap shown in the 3.4 Ma restoration (Fig. 6a).

A more probable explanation lies in the complex three-dimensional geometry of the region, such that the problems are probably caused by the orientation of the profile. There are two major ramps, and associated faults, in the detachment system in the central part of the section (where the prominent overlaps are shown in Fig. 6a). The southern ramp and its fault strike E–W where crossed by the profile (Fig. 3), so that the profile has the correct orientation for restoration. The northern faults, however, are subparallel to the NNW–SSE trending, elongate high in the salt weld surface (shallower than 4.0 s) just north of the centre of the area (Fig. 3), and are thus highly oblique to the profile. Much of the displacement on the northern fault system may be out of the plane of the section and into the adjacent basin to the SW.

The problems apparent in the restorations (Fig. 6) result, in different ways, from moving the major faults away from their positions above the ramps by restoring the extension on the faults farther north. If (and only if) the goal of the restorations is to show the large-scale evolution through time, one solution is to keep the ramp faults fixed during restoration (as local pins) and add cross-sectional area to the footwalls as needed. In essence, this is an attempt to restore material that has moved out of the plane of the section. Admittedly, it is not rigorous and it disobeys the standard rules of section restoration. However, it is a reasonable approach given the complex three-dimensional deformation kinematics, and it has been implemented in the restorations shown in Fig. 7.

It is sometimes possible that this type of problem can be avoided by choosing properly oriented sections. Whereas this is certainly possible where structures are more linear and tectonic transport is close to unidirectional, it is impossible on regional profiles through an area such as southern Eugene Island. Analysis of the structural map (Fig. 3) shows that there is no consistent fault orientation or dip direction that would indicate the appropriate orientation for section restoration. The problem is

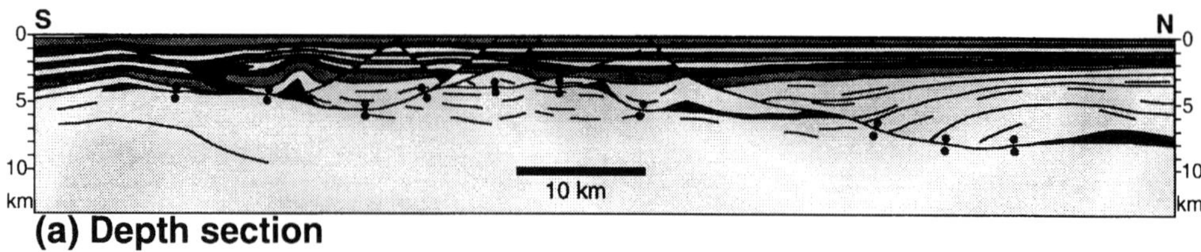

(a) Depth section

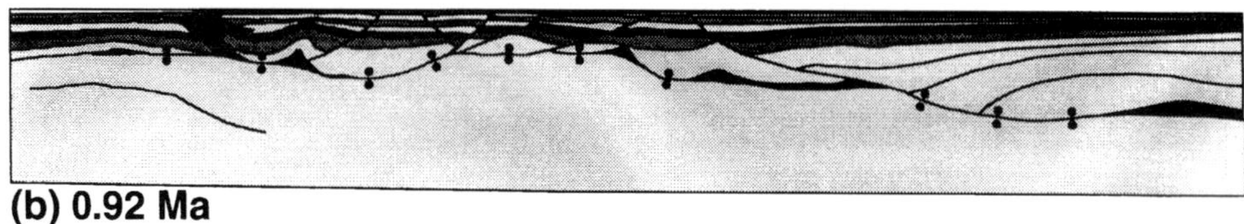
(b) 0.92 Ma

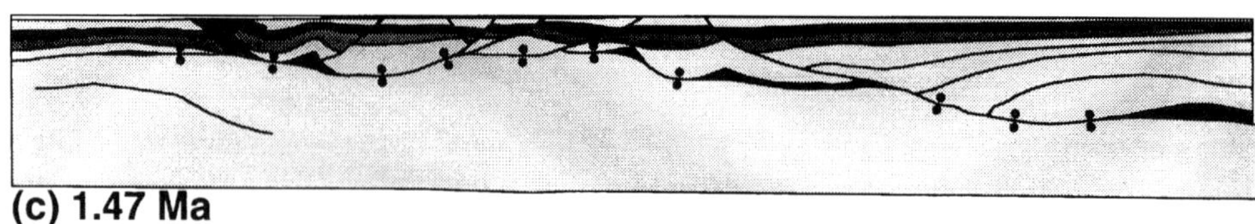
(c) 1.47 Ma

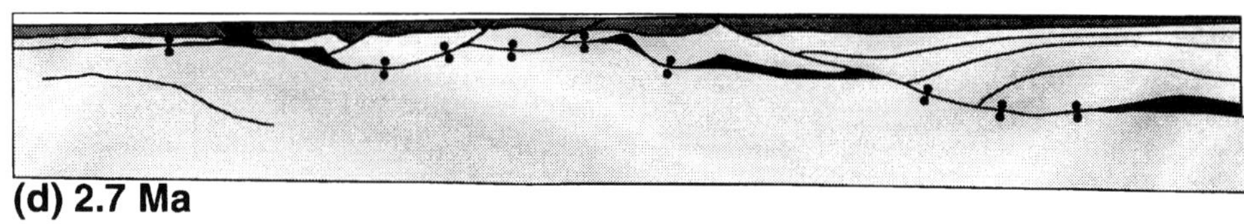
(d) 2.7 Ma

not unique to southern Eugene Island, as mapping in a large area to the southeast of the study area (in Ewing Bank and northern Green Canyon) shows similar structural geometries with highly variable fault and salt body orientations (Rowan *et al.* 1994*a*).

Variable fault strikes do not necessarily require variable displacement directions. For example, an arcuate ramp fault may have dip-slip movement in the centre where it is south-dipping, but oblique to strike-slip movement where the fault curves around to more N–S strikes. If this were the case, properly-oriented cross sections (roughly N–S) could be restored under plane-strain conditions. In the study area, the particle displacement vector patterns are unknown: they could be consistent or variable. However, it has been shown that restorations of profiles oriented parallel to the regional down-slope sediment and tectonic transport directions (roughly N–S) do result in errors in the reconstructed locations of the ramp faults. The implications are that displacement vectors are not consistent and that there must be a significant component of out-of-the-plane movement. Ultimately, only three-dimensional restoration will resolve this issue.

Restoration results

The profile illustrated in Fig. 4 has been converted to depth and sequentially restored (Fig. 7). The restorations are well-constrained only back to 3.4 Ma; earlier restorations are schematic best guesses based on observed bedding and salt geometries and rough estimates of the ages of the deeper sequences. Although the restorations were generated and are illustrated going back in time, the results will be described in the opposite sense, consistent with the way in which the structures evolved.

The ages of the first few restorations (Fig. 7g, h) are unconstrained, and it is postulated that they may represent the Miocene history of the area. In any

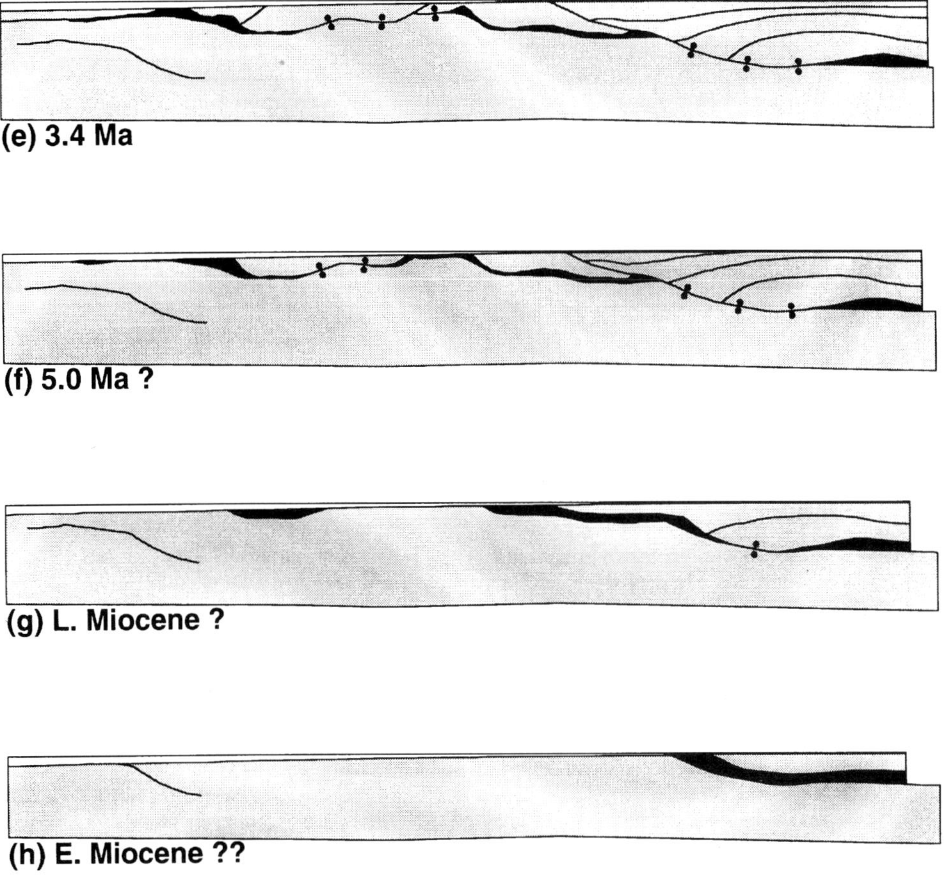

Fig. 7. Sequential restorations of seismic profile Y–Y′ (Fig. 4b) using vertical simple shear. Salt is in black, salt welds are indicated by dot pairs and water is in white. The three oldest restorations (**f-h**) are schematic and unconstrained by well control. See text for description of evolution.

case, they show that a deep-level salt sheet (Fig. 7h) was loaded by sediment progradation from the north such that the salt moved up and basinward at the sea floor to a new level (Fig. 7g) and then again to a yet higher level by the end of the Miocene (Fig. 7f). In the meantime, salt moved from out of the plane of the section into the southern part of the profile (Fig. 7g), and this too was displaced up and laterally by sedimentary loading by 5.0 (?) Ma (Fig. 7f). By the end of the Miocene, salt from the two sources (north and south) had merged at the sea floor in the centre of the profile (Fig. 7f).

By 3.4 Ma, the oldest constrained restoration (Fig. 7e), the situation was broadly similar to the present day geometry. The same salt system was present, with the only major differences found near the southern end of the profile. At 3.4 Ma, this area was dominated by a broad salt body with a thin tongue extending to the south. Between 3.4 and 2.7 Ma (Fig. 7d), salt was withdrawn from part of this body into a growing diapir whose crest remained at the sea floor (local sea-floor topography is not shown). At the same time, extension on faults to the north was accommodated, at least in part, by contractional folding over the northern part of the broad southern salt body.

By 1.47 Ma (Fig. 7c), extension on the south-dipping ramp fault left of the centre of the profile had accelerated, probably in response to complete weld formation in the hanging wall basin, although this cannot be seen on this profile (see Rowan 1995). There was further growth of the contractional anticline to the south, but this appears to have been unable to accommodate all the shortening. The excess shortening was taken up by narrowing of the southern salt diapir and thrusting of the north flank higher than the south flank. This squeezing of the diapir may explain why it continued to grow and remain at the sea floor even though further salt supply was apparently cut off by surrounding salt welds. It also appears to have forced some of the salt out, probably as a lateral extrusion at the sea floor (see Fletcher *et al.* 1995), thereby creating the northern overhang observed today.

Extension slowed after 1.47 Ma, and so did the associated contractional folding (Rowan 1995). Likewise, without lateral compression to squeeze salt up in the southern diapir, it ceased growing and became buried by 0.92 Ma (Fig. 7b). Since that time, there has been only minor faulting and other deformation along the profile, leading to the geometry observed today (Fig. 7a).

The results of restoration of a single profile may be misleading and cannot adequately depict the true three-dimensional evolution of the area. At a minimum, multiple profiles need to be restored and combined with analysis of the mapped geometry. Ideally, restoration should be three-dimensional, but the technologies for accomplishing this are still in their infancy, and such work is still in its preliminary stages (e.g. McBride *et al.* 1995).

The three-dimensional evolution of the study area is beyond the scope of this paper, but the reader is referred to other works that present a model based on two-dimensional restorations and the mapped geometry and distribution of salt bodies and salt welds (Rowan *et al.* 1994*b*, c; Rowan 1995). In essence, the elliptical, ramp-bounded lows in southwestern Eugene Island are thought to represent the evacuated remnants of a salt-stock canopy, whereas the geometries to the north are characteristic of a stepped counter-regional system (Schuster 1995) of allochthonous salt sheets. The two systems merged near the end of the Miocene and were subsequently modified by salt evacuation, extensional faulting, and associated deformation into the present geometry.

Discussion

Cross section restoration can be an invaluable tool in the analysis of any deformed terrane. Probably its most important benefit is that going through the process forces a geologist or geophysicist to examine carefully the consequences of a given interpretation. The interpretation, and the evolutionary history depicted by the restorations, can be weighed against other information that may be available, such as the regional tectonic setting, the changes in facies development through time, the evolving paleoecological environments, and the distribution of hydrocarbon accumulations. By integrating all such data within the temporal and geometric framework provided by the sequential restorations, the large-scale features of an interpretation can be evaluated and understood more clearly. For example, are the restoration results compatible with known changes in depositional environment (e.g. from bathyal turbidites to deltaic sands)? Do the restorations explain the occurrence of hydrocarbons in some traps and their absence in others? The asking of, and resolution of, such questions is fundamental to generating the best possible interpretation given the data available.

More specifically, sequential restoration has a number of benefits, both practical and academic. Although most of these benefits have not been addressed in detail in this paper, they are worth enumerating.

(1) Determining the structural evolution (e.g. the initiation and growth of salt diapirs, the linking and balance of extension and contraction above an allochthonous salt sheet and the timing of structural trap formation).

(2) Measuring the rates of various processes (e.g. sedimentary accumulation, subsidence, salt flow, extension, and contraction; see Rowan 1993; Rowan *et al.* 1993).

(3) Estimating changes in paleobathymetry and sea-floor paleotopography through time (Rowan *et al.* 1995*a*).

(4) Providing a foundation for studying and understanding facies development and the sediment transport/deposition system (Rowan *et al.* 1995*b*).

(5) Producing a geometric framework for hydrocarbon generation and migration studies (e.g. impact of salt sheet emplacement, focusing of hydrocarbon flow, timing of salt weld formation; see McBride *et al.* 1995).

Despite its many benefits, however, two-dimensional section restoration has its limitations. As already stated, one of the principal uses of restoration has been to balance and validate sections by evaluating the restored geometries. Errors such as those shown in Fig. 6 have traditionally been used to infer either that the restoration algorithms are inadequate or that the interpretation is incorrect and needs to be modified. In this case, since errors are produced using both vertical simple shear and bed-length restoration, the temptation is to change the interpretation to eliminate the problems.

Making this kind of interpretational change would be a mistake in this case. It has been argued that the 'errors' are not due to a poor interpretation, but to the inappropriateness of one of the assumptions fundamental to section restoration, namely that of plane-strain deformation. The complex three-dimensional geometry of allochthonous salt systems in this portion of the Gulf of Mexico does not impact salt flow alone. It also dictates that faults develop with a myriad of orientations, and that supra-salt sequences consequently move in multiple directions (although not necessarily perpendicular to the fault traces). The result is that there is generally no proper orientation for profiles, other than very local ones, that can be restored correctly while honoring the conservation-of-area rule of section balancing.

Cross section restoration may be used to evaluate the large-scale implications of an interpretation. But, at least in this type of structural province, it generally cannot be used to validate an interpretation. The dangers inherent in testing the details of an interpretation (for example, reservoir segmentation or cross-fault correlations) in any type of terrane have been stated before (e.g. Rowan 1993; Hauge & Gray this volume). The 'errors' shown in Fig. 6 are of a larger scale: the amount of overlap (Fig. 6*a*) and the amount of shear on the faults (Fig. 6*b*) both measure about 3 km. These are no longer in the realm of interpretational 'details', and thus may be construed as real errors that need to be rectified. However, such a conclusion would be the real error.

Although the limitations inherent in using restoration as a method of section validation are especially clear and even expected for these structural geometries, they may occur in more subtle forms any time sections are restored. Even in cases in which section validation through restoration is believed to be justifiable, care must be taken. Small errors will always be suspect because of the inability of restoration algorithms to correctly model the actual rock deformation. But even large errors may be due to limitations in the method rather than to interpretational errors. For example, restoration of a properly-oriented profile that crosses a relay ramp between two planar normal faults will yield geometries that are unacceptable and that could be construed as indications of an incorrect interpretation. Evaluation of both the data and the applicability of the assumptions is required in every case in order to judge whether modification of an interpretation is warranted.

It might appear that two-dimensional restoration is not suitable for this type of area. Such a conclusion, however, is unwarranted and deprives geoscientists of the many clear benefits of section restoration. The applicability of the technique depends on the intended goals. If these are to validate and modify the interpretation, then two-dimensional restoration is not appropriate. If, on the other hand, the goals are to determine and illustrate the most likely geometric evolution, bearing in mind the three-dimensional aspects, then profiles should be restored. If analysis of the restorations and the mapped geometries suggests significant out-of-the-plane movement, the plane-strain constraint should be abandoned and material added or subtracted as needed. This should not be done lightly, nor as a way to 'fudge' the restoration; if it is done, it must be clearly identified and justified.

Ultimately, the results of this study highlight the need for three-dimensional restoration of complex structural geometries, and it is heartening that several companies (CogniSeis Development and Midland Valley) are currently developing software for this purpose. In order to apply such software to real problems, however, more research is needed to characterize and understand the three-dimensional deformation of different tectonic regimes. The focus of ongoing study at the University of Colorado is to determine the three-dimensional kinematics and evolution of both salt flow and deformation of the overburden in the Gulf Coast (McBride *et al.* 1995), and restoration will play a fundamental role in this research.

The author wishes to thank Paul Weimer and Peter Flemings for initiating this project and carrying out the stratigraphic analysis of the area. Barry McBride, Bob Ratliff, Jake Hossack and an anonymous reviewer provided many helpful comments (although they do not necessarily agree with all of my conclusions). M. R. thanks Halliburton, TGS, and Exxon for seismic data, Micro-Strat, PaleoData, Pennzoil, and Marathon for well data, and CogniSeis for GEOSEC and DLPS. The project was funded by the DOE and carried out under the auspices of the Global Basins Research Network (GBRN) program centred around Eugene Island Block 330. Additional funding was provided by an industrial consortium consisting of Amoco, Anadarko, BP, Chevron, Conoco, CNG, Exxon, Marathon, Mobil, Pennzoil, Petrobras, Phillips, Shell, Texaco, Total, Union Pacific and Unocal.

References

ALEXANDER, L. A. & FLEMINGS, P. B. 1995. Geologic evolution of a Plio-Pleistocene salt withdrawal mini-basin: Eugene Island, Block 330, offshore Lousisana. *American Association of Petroleum Geologists Bulletin* , in press.

BALLY, A. W., GORDY, P. L. & STEWART, G. A. 1966. Structure, seismic data and orogenic evolution of southern Canadian Rocky Mountains. *Bulletin of Canadian Petroleum Geology*, **14**, 337–381.

BEACH, A. & ROWAN, M. G. 1992. Techniques for the geometrical restoration of sections: an example from the Bjørnøya Basin, Barents Sea shelf. *In*: LARSEN, R. M., BREKKE, H., LARSEN, B. T. & TALLERAAS, E. (eds) *Structural and*

Tectonic Modelling and its Application to Petroleum Geology. Norwegian Petroleum Society, Special Publication **1.** Elsevier, Amsterdam, 269–276.

BISHOP, D. J., BUCHANAN, P. G. & BISHOP, C. J. 1995. Gravity-driven thin-skinned extension above Zechstein Group evaporites in the western central North Sea: an application of computer-aided section restoration techniques. *Marine and Petroleum Geology*, **12**, 115–135.

DAHLSTROM, C. D. A. 1969. Balanced cross sections. *Canadian Journal of Earth Science*, **6**, 743–757.

DIEGEL, F. A. & COOK, R. W. 1990. Palinspastic reconstructions of salt-withdrawal growth-fault systems, northern Gulf of Mexico [abs]. *Geological Society of America Abstracts with Programs* **22**, A48.

——, KARLO, J. F., SCHUSTER, D. C., SHOUP, R. C. & TAUVERS, P. R. 1995. Cenozoic structural evolution and tectono-stratigraphic framework of the northern Gulf Coast continental margin. *In*: JACKSON, M. P. A., ROBERTS, D. G. & SNELSON, S. (eds) *Salt Tectonics: a Global Perspectives for Exploration.* American Association of Petroleum Geologists, Memoirs, in press.

ELLIOTT, D. 1983. The construction of balanced cross sections. *Journal of Structural Geology*, **5**, 101.

FLETCHER, R. C., HUDEC, M. R. & WATSON, I. A. 1995. Salt glacier and composite sediment/glacier models for the emplacement and early burial of allochthonous salt sheets. *In*: JACKSON, M. P. A., ROBERTS, D. G. & SNELSON, S. (eds) *Salt Tectonics: a Global Perspectives for Exploration.* American Association of Petroleum Geologists, Memoirs, in press.

GEISER, P. A. 1988. The role of kinematics in the construction and analysis of geological cross sections in deformed terranes. *In*: MITRA, G. & WOJTAL, S. (eds) *Geometries and Mechanisms of Thrusting, with special reference to the Appalachians.* Geological Society of America, Special Papers, **222**, 47–76.

GIBBS, A. D. 1983. Balanced cross section construction from seismic sections in areas of extensional tectonics. *Journal of Structural Geology*, **5**, 153–160.

HAUGE, T. A. & GRAY, G. G. 1996. A critique of techniques for modelling normal-fault and rollover geometries. *This volume.*

HOSSACK, J. R. 1995. Geometrical rules of section balancing salt structures for *In*: JACKSON, M. P. A., ROBERTS, D. G. & SNELSON, S. (eds) *Salt Tectonics: a Global Perspectives for Exploration.* American Association of Petroleum Geologists, Memoirs, in press.

—— & MCGUINNESS, D. B. 1990. Balanced sections and the development of fault and salt structures in the Gulf of Mexico (GOM) [abs]. *Geological Society of America Abstracts with Programs*, **22**, A48.

JACKSON, M. P. A. & CRAMEZ, C. 1989. Seismic recognition of salt welds in salt tectonic regimes. *In*: *Gulf of Mexico Salt Tectonics, Associated Processes and Exploration Potential.* Tenth Annual Research Conference, Gulf Coast Section, Society of Economic Paleontologists and Mineralogists Foundation, Program with Papers, 66–71.

KOYI, H. 1994. Estimation of salt thickness and restoration of cross-sections with diapiric structures: a few critical comments on two powerful methods. *Journal of Structural Geology*, **16**, 1121–1128.

MCBRIDE, B. C., ROWAN, M. G., WEIMER, P., & RATLIFF, R. A. 1995. Restoration of allochthonous salt structures in three dimensions and its impact on understanding reservoir trends and hydrocarbon migration: preliminary results from Green Canyon and Ewing Bank, northern Gulf of Mexico. *In*: *Salt, Sediment and Hydrocarbons*. Sixteenth Annual Research Conference, Gulf Coast Section, Society of Economic Paleontologists and Mineralogists Foundation, Program with Papers, in press.

MCGUINNESS, D. B. & HOSSACK, J. R. 1993. The development of allochthonous salt sheets as controlled by the rates of extension, sedimentation, and salt supply. *In*: ARMENTROUT, J. M., BLOCH, R., OLSON, H. C. & PERKINS, B. F. (eds) *Rates of Geological Processes.* Fourteenth Annual Research Conference, Gulf Coast Section, Society of Economic Paleontologists and Mineralogists Foundation, Program with Papers, 127–139.

MORETTI, I., WU, S. & BALLY, A. W. 1990. Computerized balanced cross-section LOCACE to reconstruct an allochthonous salt sheet, offshore Louisiana. *Marine and Petroleum Geology*, **7**, 371–377.

NUNNS, A. G. 1991. Structural restoration of seismic and geologic section in extensional regimes. *American Association of Petroleum Geologists Bulletin,* **75**, 278–297.

PEEL, F. J., TRAVIS, C. J., & HOSSACK, J. R. 1995. Genetic structural provinces and salt tectonics of the Cenozoic offshore US Gulf of Mexico: a preliminary analysis. *In*: JACKSON, M. P. A., ROBERTS, D. G. & SNELSON, S. (eds) *Salt Tectonics: a Global Perspectives for Exploration.* American Association of Petroleum Geologists, Memoirs, in press.

ROWAN, M. G. 1993. A systematic technique for the sequential restoration of salt structures. *Tectonophysics*, **228**, 331–348.

—— 1995. Structural styles and evolution of allochthonous salt, central Louisiana outer shelf and upper slope. *In*: JACKSON, M. P. A., ROBERTS, D. G. & SNELSON, S. (eds) *Salt Tectonics: a Global Perspectives for Exploration.* American Association of Petroleum Geologists, Memoirs, in press.

—— & KLIGFIELD, R. 1989. Cross section restoration and balancing as an aid to seismic interpretation in extensional terranes. *American Association of Petroleum Geologists Bulletin,* **73**, 955–966.

——, MCBRIDE, B., KLIGFIELD, R., WEIMER, P. & WU, S. 1993. Sequential restoration of salt structures: a tool for measuring rates of sedimentation, subsidence, extension, and salt movement [abs]. *In*: ARMENTROUT, J. M., BLOCH, R., OLSON, H. C. & PERKINS, B. F. (eds) *Rates of Geological Processes.* Fourteenth Annual Research Conference, Gulf Coast Section, Society of Economic Paleontologists and Mineralogists Foundation, Program with Papers, 219.

——, —— & WEIMER, P. 1994*a*. Salt geometry and Plio-Pleistocene evolution of Ewing Bank and northern Green Canyon, offshore Louisiana [abs]. *American Association of Petroleum Geologists 1994 Annual Convention Program*, 247.

——, WEIMER, P., BUDHIJANTO, F. & FLEMINGS, P. B. 1994*b*. *Integrated regional sequence stratigraphic and structural framework and geologic evolution of the Eugene Island Block 330 area, offshore Louisiana*. Department of Energy Report (Dynamic Enhanced Recoveries Technologies).

——, —— & FLEMINGS, P. B. 1994*c*. Three-dimensional geometry and evolution of a composite, multi-level salt system, western Eugene Island, offshore Louisiana. *Gulf Coast Association of Geological Societies, Transactions*, **44**, 641–648.

——, ——, MCBRIDE, B.C. *ET AL.* 1995*b*. Interactions between salt deformation and sedimentation, central Louisiana upper slope and outer shelf: preliminary results. *In*: *Salt, Sediment and Hydrocarbons*. Sixteenth Annual Research Conference, Gulf Coast Section, Society of Economic Paleontologists and Mineralogists Foundation, Program with Papers, in press.

——, VILLAMIl, T., WEIMER, P., & FLEMINGS, P. B. 1995*a*. Paleobathymetry and paleotopography in the Gulf of Mexico: comparison of results from cross-section restoration and biostratigraphic analysis [abs]. *American Association of Petroleum Geologists 1994 Annual Convention Program*, 84a.

ROYSE, F., JR., WARNER, M. A. & REESE, D. L. 1975. Thrust belt structural geometry and related stratigraphic problems, Wyoming–Idaho–Northern Utah. *In*: *Symposium on Deep Drilling Frontiers in Central Rocky Mountains*. Rocky Mountain Association of Geologists, 4–54.

SCHULTZ-ELA, D. D. 1992. Restoration of cross sections to constrain deformation processes of extensional terranes. *Marine and Petroleum Geology*, **9**, 372–388.

SCHUSTER, D. C. 1995. Deformation of allochthonous salt and evolution of related salt/structural systems, eastern Louisiana Gulf Coast. *In*: JACKSON, M. P. A., ROBERTS, D. G. & SNELSON, S. (eds) *Salt Tectonics: a Global Perspectives for Exploration*. American Association of Petroleum Geologists, Memoirs, in press.

SENI, S. J. 1992. Evolution of salt structures during burial of salt sheets on the slope, northern Gulf of Mexico. *Marine and Petroleum Geology*, **9**, 452–468.

SUMNER, H. S., ROBISON, B. A., DIRKS, W. K. & HOLLIDAY, J. C. 1991. Structural style of salt/mini-basin systems: lower shelf and upper slope, central offshore Louisiana [abs]. *Gulf Coast Association of Geological Societies, Transactions*, **41**, 582.

VENDEVILLE, B .C. & JACKSON, M. P. A. 1992. The rise of diapirs during thin-skinned extension. *Marine and Petroleum Geology*, **9**, 331–353.

VERRALL, P. 1981. Structural interpretation with applications to North Sea problems. *Joint Association of Petroleum Exploration Course, Notes 3*.

WHITE, N. J., JACKSON, J. A., & MCKENZIE, D. P. 1986. The relationship between the geometry of normal faults and that of the sedimentary layers in their hanging walls. *Journal of Structural Geology*, **8**, 897–909.

WOODWARD, N. B., BOYER, S. E. & SUPPE, J. 1985. *An Outline of Balanced Cross-Sections* (2nd edn). University of Tennessee, Department of Geological Sciences, Studies in Geology **11**.

WORRALL, D. M. & Snelson, S. 1989. Evolution of the northern Gulf of Mexico, with emphasis on Cenozoic growth faulting and the role of salt. *In*: BALLy, A. W. & PALMER, A. R. (eds) *The Geology of North America: an overview*. Geological Society of America, Decade of North American Geology, **A**, 97–138.

WU, S., BALLY, A. W. & CRAMEZ, C. 1990. Allochthonous salt, structure and stratigraphy of the north-eastern Gulf of Mexico, Part II: structure. *Marine and Petroleum Geology*, **7**, 334–370.

The growth of normal faults by segment linkage

JOSEPH A. CARTWRIGHT,[1] CHRIS MANSFIELD[1] & BRUCE TRUDGILL[2]

[1] *Department of Geology, Royal School of Mines, Imperial College of Science, Technology, and Medicine, Prince Consort Road, London SW7 2BP, UK*

[2] *Department of Geological Sciences, University of Colorado at Boulder, Campus Box 250, Boulder, CO 80309-0250 USA*

Abstract: Fault growth is widely described using a scaling law between maximum displacement (D) and length (L), of the form $D = cL^n$. This expression defines a model of fault growth by radial propagation from a single seed fracture or fault. This paper presents geometrical and kinematic evidence from a set of exceptionally well exposed normal faults in Utah for an alternative model of fault growth. This model is referred to as growth by segment linkage, and involves the propagation and linkage of independent fault segments on ascending length scales. The evidence presented focuses on the geometry and displacement variation in the region of relay structures, and on local scaling relationships between D and L. The $D-L$ data from 97 faults in the study area range over three orders of magnitude, and show a general trend to increasing D for increasing L. There is a large scatter in the data, similar to that recognized in previous $D-L$ compilations. It is argued that the scatter cannot be attributed either to measurement errors or to variation in mechanical properties. Instead, we argue that the model of growth by segment linkage provides a simple explanation of this scatter, and propose that the process of segment linkage may explain scatter in other datasets.

Considerable progress has been made in the past decade towards a better understanding of the processes governing the growth of faults (Watterson 1986; Walsh & Watterson 1988; Scholz 1990; Peacock & Sanderson 1991; Cowie & Scholz 1992a, b, c; Scholz *et al.* 1993; Dawers *et al.* 1993; Anders & Schlische 1994). One of the main advances has been made by compiling quantitative data on specific fault parameters such as trace length and fault displacement. These quantitative data have been used to derive numerical growth laws for faults that can in turn be employed in predictive analysis (Watterson 1986; Walsh & Watterson 1988; Marrett & Allmendinger 1991; Cowie & Scholz 1992*c*; Gillespie *et al.* 1992). One important application of fault displacement analysis is in fault correlation, where instead of relying solely on geometrical rules for correlation, the interpreter is able to incorporate fault displacement characteristics into the correlation exercise (Freeman *et al.* 1990; Beach & Trayner 1991; Chapman & Meneilly 1991). Another recent application of fault growth modelling is in fault population analysis for strain calculation (Scholz & Cowie 1990; Walsh *et al.* 1991). Population analysis can also be used to predict numbers of faults or fractures in a certain region whose size range is smaller than the resolution limit of whatever detection system is in use (Belfield 1992; Yielding *et al.* 1992; Gauthier & Lake 1993).

In moving towards these types of predictive applications, it is obviously crucial that the numerical basis for the predictive technique is appropriate for the particular process and area of application. For fault analysis, it has been suggested that fault displacement scales with fault length according to a general equation

$$D = cL^n \qquad (1)$$

where D is maximum displacement on a fault of maximum trace length, L, and where c is a constant related to rock material properties (Watterson 1986; Walsh & Watterson 1988; Cowie & Scholz 1992*a*). There is no general agreement on the value of the exponent, n, and published figures range from 1.0 (Cowie & Scholz 1992*a*) to 2.0 (Watterson 1986; Walsh & Watterson 1988). A growth law of the general form represented by equation (1) is only strictly applicable to faults growing by a process of radial propagation, in which the rupture dimensions for each slip event or period of stable sliding involves the entire fault surface (Watterson 1986; Cowie & Scholz 1992*b*).

This paper presents evidence from a field geological study of a system of exposed normal faults in the Canyonlands Grabens of south-eastern Utah, to show that these particular faults do not grow by radial propagation but by a combined process of radial propagation and linkage of precursor segments. A model is presented that demonstrates how growth by segment linkage produces a fault population with fundamentally

From Buchanan, P. G. & Nieuwland, D. A. (eds), 1996, *Modern Developments in Structural Interpretation, Validation and Modelling*, Geological Society Special Publication No. 99, pp. 163–177.

different characteristics to one produced by models of growth by radial propagation. This model shows that fault growth cannot be expressed by a scaling law between *D* and *L*. The end result of growth by segment linkage is a fault with a specific type of segmented geometry, and a central aim of this paper is to show some of the key diagnostic features associated with fault segmentation, in order to aid recognition of this phenomenon on sub-surface datasets.

The Canyonlands Grabens

Geological setting

The Canyonlands Grabens (McGill & Stromquist 1979) refers to a complex array of extensional faults developed over an area of approximately 2000 km^2 in the centre of the Colorado Plateau in S.E. Utah (Fig. 1). The area lies at the southwestern margin of the Paradox Basin, a major Late Palaeozoic to Early Mesozoic half-graben whose fill is dominated by evaporitic cycles of the Paradox Formation (Hintze, 1988). Superimposed on this half-graben structure is the Monument Uplift, a major N–S-trending anticline that formed during the Early Cenozoic Laramide compressional phase (Fig. 1) (Lewis & Campbell 1965). The Canyonlands Grabens are located on the gently northwestward plunging limb of this anticline. The generalized fault pattern of the Canyonlands Grabens is illustrated in Fig. 2. More detailed fault maps can be found in Lewis & Campbell (1965), Trudgill & Cartwright (1994) and Cartwright *et al.* (1995).

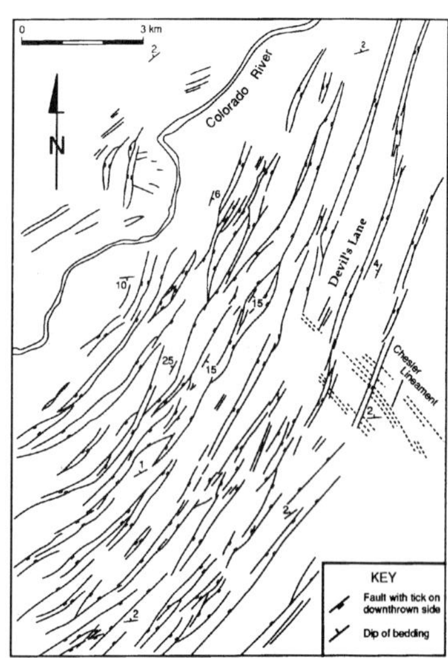

Fig. 2. A simplified structural map of the northern part of the Canyonlands Graben system, based on aerial photographs, field data and published map data.

The stratigraphy of the area is shown in a schematic section across a graben exposed in one of the transverse canyons (Fig. 3). Throughout most of the graben area, the surface rocks are cross-bedded sandstones. The deeper graben and cross canyons expose limestones and sandstones. The oldest rocks are exposed only at the base of tributary canyons to the Colorado Canyon and along the banks of the Colorado itself. These consist of deformed beds of gypsum and limestones (Paradox Member). Nearby boreholes indicate that the interbedded gypsum and limestones are underlain by a considerable thickness (> 300 m) of Paradox evaporites (Joesting *et al.* 1966).

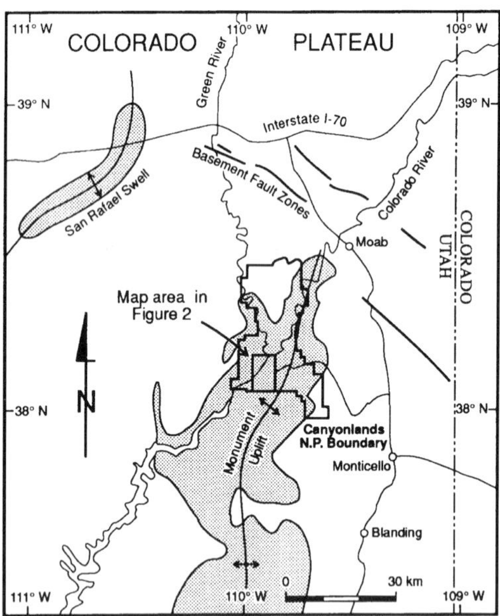

Fig. 1. Location map for the Canyonlands Graben system, which lies in the central part of the Colorado Plateau in the USA. Shaded areas define Palaeozoic outcrops, the result of compressional deformation in the Cenozoic. The boxed study area is shown in detail in Fig. 2.

General structure

The Canyonlands Grabens are located preferentially along the SE side of the Colorado River gorge (Fig. 2). The general strike of graben-bounding faults changes from NNE in the northern part of the area, to NE in the southern part. This change of strike may have been the result of minor reactivation of the pre-existing Chesler Lineament (Fig. 2),

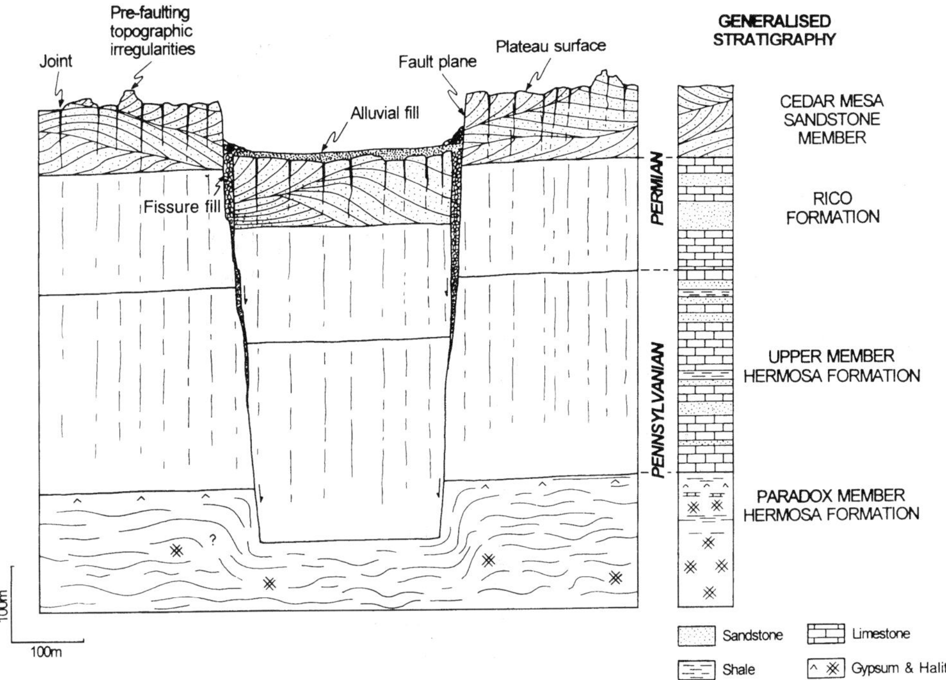

Fig. 3. A synoptic structural/stratigraphic section of a typical Canyonlands graben based on compilations of sections exposed at different structural levels in one of the transverse tributary canyons feeding into the Colorado River.

a possible sub-Paradox basement fault (Potter & McGill 1978*a*; Trudgill & Cartwright 1994).

The mapped area contains over 100 extensional faults that can be defined as discrete structures with isolated tips at both ends of the fault. The majority of the faults are arranged as pairs of graben-bounding faults, bounded on both margins by faults with approximately equivalent displacements i.e. they exhibit a full graben form (Fig. 3). Inter-graben spacing is fairly regular, with an average distance of about 1 km, centre to centre, while graben widths range from 100 to 600 m (Fig. 2). The maximum fault throw recorded on individual faults ranges from a few metres to in excess of 150 m. The preservation of fault planes at the surface is exceptionally good: only a few of the outcropping fault planes have been eroded to any degree to form fault scarps. The quality of the exposed fault planes in combination with excellent exposure of hangingwall and footwall stratigraphy allows accurate determinations of fault throw to be made along strike.

The full cross-sectional geometry of the graben-bounding faults is well constrained down to a level at the top of the evaporites sequence in a series of tributary canyons to the Colorado River (Fig. 3). The majority of graben-bounding faults are vertical at surface, and remain vertical to sub-vertical for the 400–500 m distance to the top of the evaporite sequence. Displacement is a maximum at the surface and is not seen to vary with depth in any of the transverse sections. The vertical to sub-vertical trajectory of the faults and the constant displacement-with-depth relationship observed in the tributary canyons together suggest that the faults propagated downwards from the surface (Cartwright *et al.* 1995), and terminate in salt flowage or salt dissolution structures (Fig. 3).

Mechanics of graben formation

Most previous workers in the Canyonlands consider the grabens to have been formed by a mechanism of downslope gravitational collapse into the Cataract Canyon of the Colorado River (Baker 1933; Lewis & Campbell 1965; McGill & Stromquist 1975, 1979; Huntoon 1982; Trudgill & Cartwright 1994). The incision of the Colorado River through Cataract Canyon in the late Pleistocene created a large unbounded surface along the southeastern margin of the canyon (Fig. 4). The incision of Cataract Canyon resulted in the unloading of the Paradox evaporites, and this led to the development of a salt-cored anticline whose

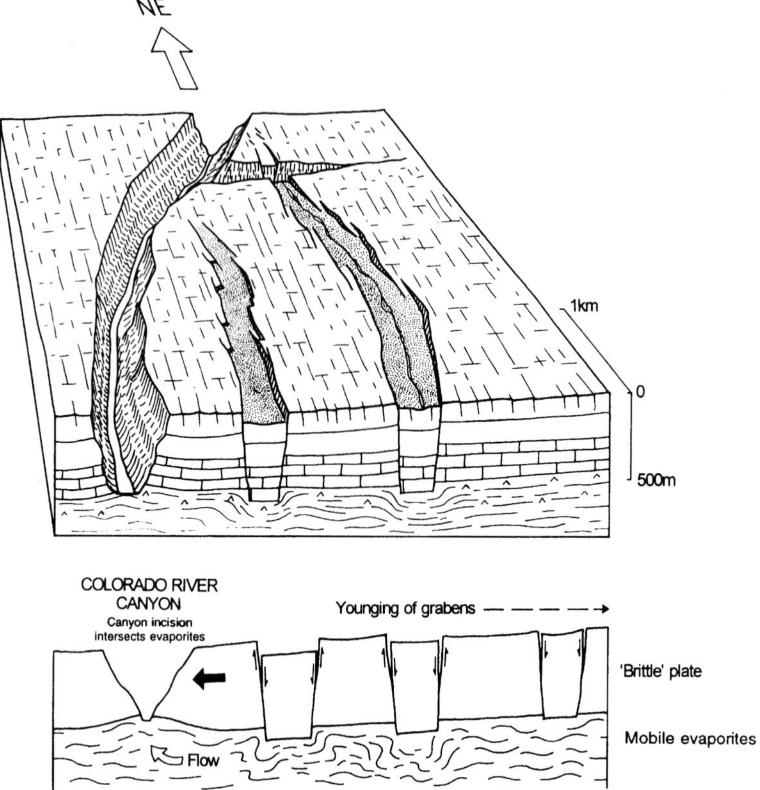

Fig. 4. Schematic block diagram showing the formation of the Canyonlands Grabens in response to late Pliocene–Pleistocene incision of the Colorado River Gorge, and the subsequent triggering of flow in the Paradox evaporites. Flow of the evaporites imposed tensile stresses on the overlying brittle plate of sandstones and limestones, and produced a set of collapse grabens, with a gently arcuate strike pattern concave towards the river.

axis follows the course of the river (Potter and McGill 1978*b*). The sub-surface flow of evaporites towards the axis of river incision induced tensile stresses in the more brittle overlying limestones and sandstones, and in turn led to the break-up of the brittle plate in the form of extensional collapse and graben formation (Fig. 4) (McGill & Stromquist 1979; Huntoon 1982; Trudgill & Cartwright 1994).

Fault segmentation

Segmentation and fault geometry

All the faults mapped in the Canyonlands are segmented in plan view i.e. they are composed of a number of component fault strands or segments. This aspect of the fault trace morphology can be seen on an aerial photograph of the central region of the fault system (Fig. 5). The grabens are recognizable in the photograph as linear areas of darker contrast, when compared to the generally lighter contrast, speckled appearance of the elevated plateau surface. Graben-bounding faults on the SE margins of the NE-striking grabens are most clearly picked out as irregular traces in shadow. The segments are recognizable on the photograph from abrupt changes in strike of the fault trace, or from small lateral offsets in the fault trace (e.g. at points X and Y, respectively on Fig. 5). By combining the photogeological interpretation of segments with field structural mapping, a number of criteria have been established for the definition of segments and segment boundaries:

(1) The presence of an abrupt change in strike or offset of adjacent segments;
(2) Recognition of a change in fault throw between adjacent segments;
(3) Recognition of a change in bedding attitude in the footwall exposed at the fault plane between adjacent segments;
(4) The development of a relay structure at the segment boundary.

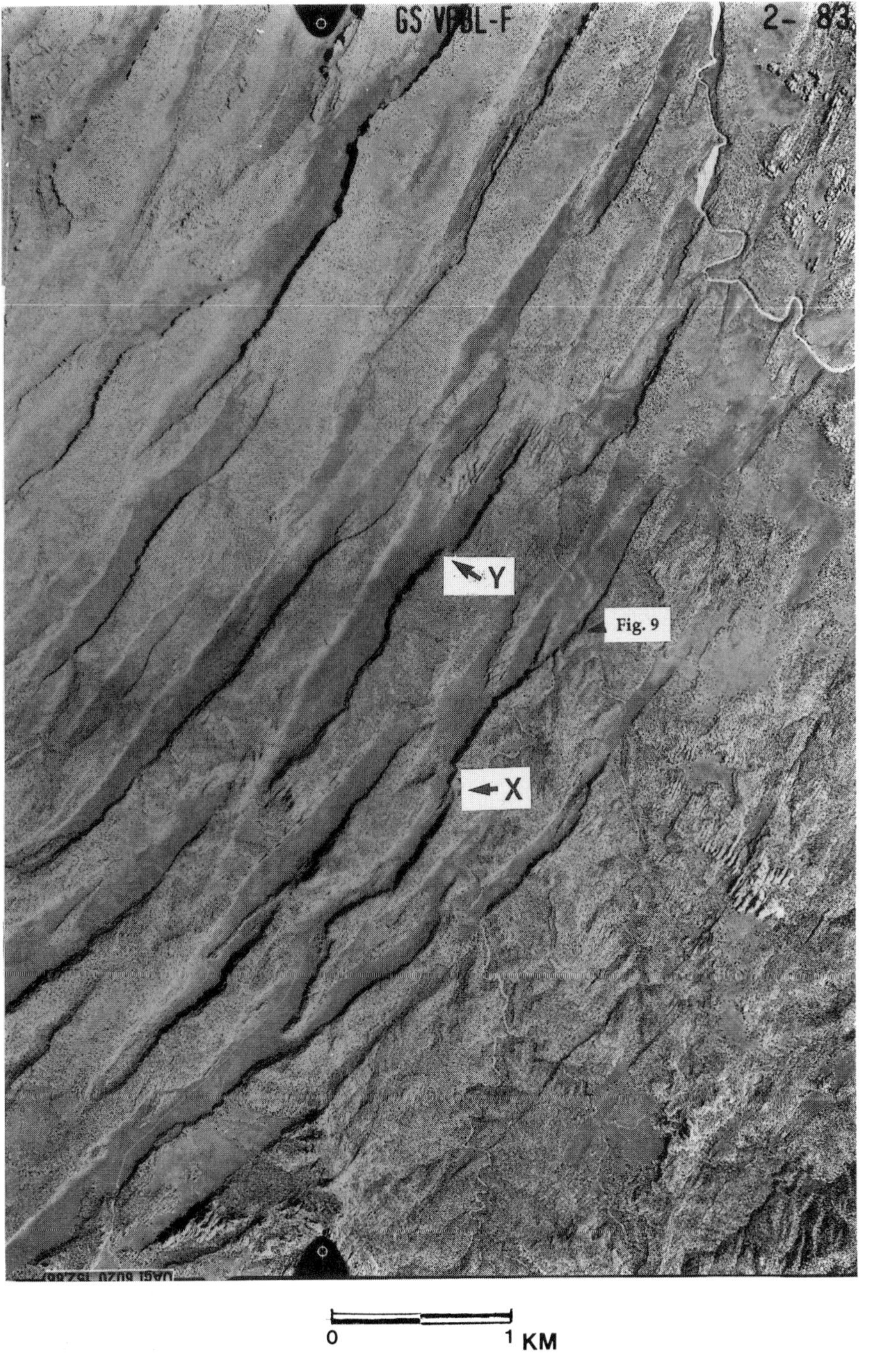

Fig. 5. A USGS 1:33 000 scale aerial photograph of the central region of the Canyonlands Grabens. Illumination is from the east (north at top of photo), with the grabens characterized by the smooth, grey tone representing the thin Quaternary alluvial and aeolian infill, and horst regions characterized by the brighter speckled pattern. The influence of the fault-controlled topography on the drainage is well illustrated by streams flowing northwest across the image. Note the sharply defined lateral tips of many of the graben-bounding faults, and the segmented morphology of all the fault traces. Points labelled X and Y are examples of segment boundaries as defined by abrupt offsets or changes in strike of adjacent fault segments. Bobby Joe Fault (Fig. 9) is located in the lower centre portion of the image.

Of these, by far the most reliable indicator of segmentation along fault strike is the recognition of relay structures.

Two basic types of relay structure are recognized in the Canyonlands (Fig. 6), although there is a large spectrum of variation in the detailed structural morphology (Trudgill & Cartwright 1994). The first type consists of two overlapping fault segments, with a relay ramp developed between the fault segments in the region of overlap (Fig. 6a). The second type (breached relay) is a more evolved stage where the relay ramp has broken down by internal faulting, and a through-going fault surface connects the overlapping segments (Fig. 6b). The relay ramps are zones of bedding continuity from the hangingwall to the footwall (Peacock & Sanderson 1991). The dips of strata in the ramps mapped in the Canyonlands range from a few degrees to over 50°. The sense of bedding rotation in the ramps ranges from longitudinal (parallel to graben axis) to transverse, with many ramps exhibiting an oblique sense of rotation.

Relay structures develop when two fault segments propagate into an overlapping configuration (Fig. 6a) (Peacock & Sanderson 1991). At this stage, further lateral propagation is inhibited by the interaction of stress fields around the overlapping fault tips (Segall & Pollard 1980). The tips are effectively locked, and displacement builds up on the overlapping segments without further radial propagation of the fault into unbroken country rock. The progressive increase in displacement gradient towards the locked fault tips results in the rotation of the ramp between the overlapped segments (Fig. 6a). In the Canyonlands, this rotation involves bending of brittle sandstones and limestones, and bedding-parallel extension above the fold hinges is accommodated by minor normal faulting.

Once the ramp folding reaches a critical limit, the distribution of minor faults developed within the ramp is such that a through-going fault can propagate and connect the two adjacent fault segments (Fig. 6b). The through-going fault connection tends to develop either at the hangingwall end of the ramp or at the footwall end (Trudgill & Cartwright 1994), which leads to a simple classification of relays as hanging-wall breached and footwall breached, respectively (Fig. 7). Once the two fault segments are linked by the interconnecting fault and displacement increases on the newly-linked segmented fault, the breached ramp will either be locked into further movement with the footwall (Fig. 7a) or the hangingwall (Fig. 7b).

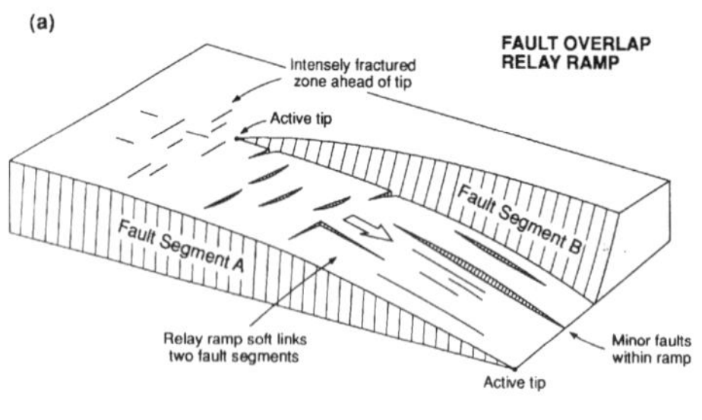

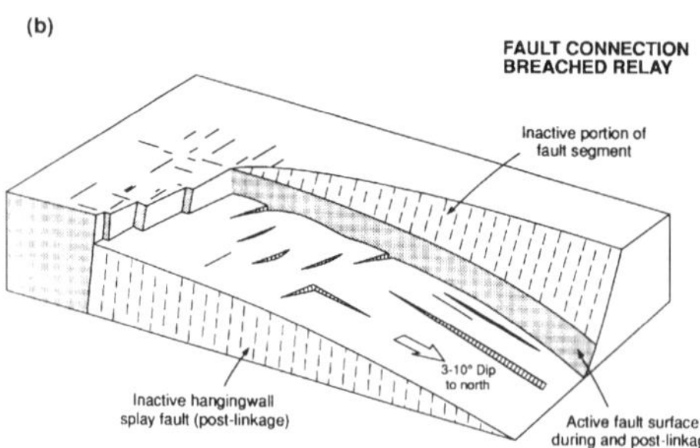

Fig. 6. Two end-member types of relay structure recognized in the Canyonlands, (**a**) overlapping fault segments with a relay ramp developed between the tip regions, and (**b**) breached relay structures with an interconnecting fault linking the two fault segments, leaving the ramp isolated from the linked fault.

Sub-surface interpretation of segmented faults

It is easy to distinguish footwall breached from hanging-wall breached relay structures in the case of exposed fault systems such as the Canyonlands. However, it is far more difficult to identify these structures correctly in the sub-surface, particularly when breaching has taken place relatively early in the development of the graben. This problem is highlighted in Fig. 7, which shows the cross-sectional geometry of hanging-wall and footwall breached relay structures at early and late stages of their structural evolution. The sections across the hanging-wall breached relay show how the ramp is left attached to the footwall during the evolution from early to late stages of development of the linked structure. The section X–X′ shows the active fault and the inactive splay, separated by what appears in section to be a small fault terrace block. The apparent dip of this terrace is towards the inactive splay, but from structures mapped in the Canyonlands, the terrace could equally exhibit dips in the opposite sense. If a structure similar to that shown in Fig. 7a was intersected by only a single seismic section, the relay structure could easily be misinterpreted as a simple terrace structure, and the important diagnostic evidence of fault segmentation would be overlooked.

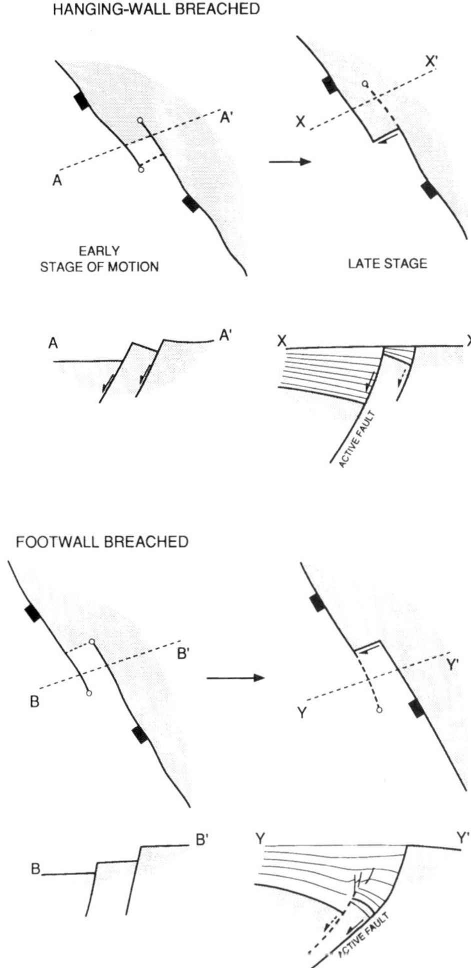

Fig. 7. Cross-sectional features typical of two types of breached relay structure. Hanging-wall breached relays (**a**) are characterized by the inactive splay fault and ramp attached to the footwall, whereas footwall breached relays (**b**) are characterized by the ramp and splay fault being coupled to the motion of the hanging wall.

In the case of the footwall breached relay structure, the inactive splay fault and the ramp are coupled to the hanging-wall, and the section Y–Y′ shows the type of cross-sectional geometry that might develop after continued motion of the hanging wall. Structural styles similar to the section Y–Y′ are quite common in rift settings (Harding 1984), and could be interpreted on a single-section basis as terrace structures, or even as oblique inversion structures (Eisenstadt & Withjack 1995). The correct identification as a breached relay structure might not be made if the section spacing was not sufficient to document the subtle variations in geometry in the region of the relay structure. In both cases, therefore, sub-surface recognition of relay structures will be critically dependant on line spacing, orientation, and precision of imaging of stratal reflections and faults in the vicinity of the ramp.

Segmentation and fault displacement

One of the key diagnostic indicators of fault segmentation involving the development of relay structures is anomalous variation in fault displacement along strike in the region of the segment boundary (Peacock & Sanderson 1991). The block diagram of a relay structure in the overlap stage (Fig. 6a) shows a pattern of displacement variation that is typical during the early stages of segment linkage. The displacement decrease on one fault segment is balanced by a sympathetic increase on the other segment, such that a sum of the displacement on the two segments measured in a strike-normal projection should be approximately equal to the displacement measured on either segment immediately beyond the relay structure (Peacock & Sanderson 1991; Walsh & Watterson 1991). In the case of breached relay structures, anomalous variations in displacement corresponding to the pre-breaching stage may be harder to recognize, particularly where a footwall breached relay is buried by a substantial graben-fill succession (Fig. 7b). Fortunately, in the Canyonlands, even when breaching is early, the limited thickness of graben fill means that breached relay structures can always be recognized as such.

Previous studies of normal faults have documented the main characteristics of displacement variation in relation to fault segmentation (Walsh & Watterson 1990; Peacock & Sanderson 1991), so this section briefly describes only two representative examples from the Canyonlands to illustrate some of these characteristics.

Devil's Lane

The structure of Devil's Lane is a major en echelon offset of a graben. Two major relay ramps define the terminations of the graben segments in the region of the offset (Fig. 8a). The graben segments overlap by approximately 600 m. The northern segment of the graben is 200 m wide, and the width of the two ramps is also about 200 m.

Displacement values were obtained at intervals of about 50–100 m along the faults by surveying the elevation difference between correlative sedimentary units in the hangingwall and footwall

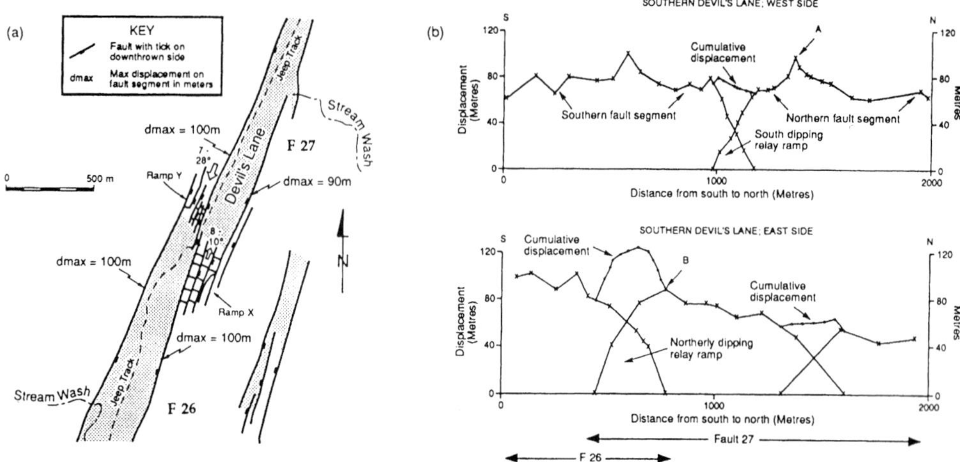

Fig. 8. Displacement variation along graben-bounding faults surveyed in the region of a major offset of a graben in Devil's Lane (located on Fig. 2). The fault map (**a**) and displacement-distance plots (**b**) for the two bounding fault systems, show the segmented fault geometry and the anomalous variation in D in the vicinity of the ramps. Note the contrast in the cumulative displacement plots for the western and eastern margin ramps.

(Cartwright *et al.* 1995). Since the fault planes at surface are vertical or sub-vertical, the stratigraphic throw is a very close approximation to the fault displacement.

Measurements for the displacement on the bounding faults in the region of the double relay structure are shown in Fig. 8b in the form of a strike projection (Barnett *et al.* 1987). Displacement on the four fault segments involved in the en echelon offset of the graben varies gradually from the positions of maximum displacement to the north for the two northern segments, and to the south for the two southern segments. The displacement gradients in the region of the two ramps are much steeper for all four fault segments (cf. Peacock & Sanderson 1991), and the ramps thus can clearly be distinguished in the displacement plots as zones of anomalous displacement gradients.

The cumulative displacement plots made by summing the displacement on each pair of segments for the two ramps shows a contrasting pattern for the two ramps developed at the graben offset. The cumulative plot for the western margin ramp matches the general regional trend in the displacement values, whereas the ramp on the eastern margin has a cumulative displacement anomaly above the general trend of values on the displacement plot. This 'excess' displacement pattern is contrary to the predicted pattern of cumulative displacement across a relay structure (Walsh & Watterson 1991). Displacement minima in the summed profiles are, however, more frequently described, and are usually explained as a proportion of displacement expressed as ductile or continuous deformation (e.g. fault drag) (Walsh & Watterson 1990; Peacock & Sanderson 1991).

Bobby Joe Fault

This fault has a highly irregular morphology consisting of eight approximately linear segments of varying length and orientation (Fig. 9). The segment boundaries are all composed of footwall breached relay structures. In several cases, the development of the interconnecting fault took place relatively late in the evolution of the structure, and the breached ramps are left elevated and exposed in the hangingwall, and are bounded by splay faults. Where breaching was early, the ramps are partly concealed beneath the thin sedimentary cover of the graben floor. These different stages of breaching can conveniently be described with reference to a breaching index, B_z, which is the ratio of maximum displacement on the inactive splay fault to the total displacement on the linked fault measured at the point of breaching (Fig. 10). Values of B_z are expressed as a percentage and shown at each of the breached relay structures along Bobby Joe Fault in Fig. 9b. It is apparent that there is no systematic age-progressive pattern in the distribution of the breached relay structures along this fault. Early breached relays are interspersed with late-stage breached relays.

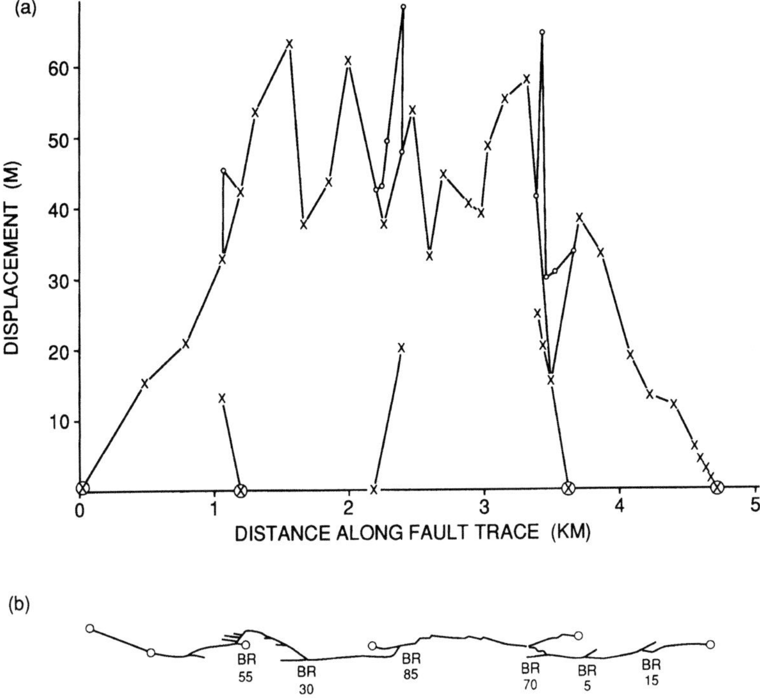

Fig. 9. (**a**) Displacement variation on a SW(left)–NE(right) traverse along Bobby Joe Fault (located on Fig. 5). The overall form of displacement along the fault is comparable to that expected for a single fault (Walsh & Watterson 1988), but in detail it is highly irregular. (**b**) Simplified outline of fault trace showing segmented structure linked at a series of breached relay structures. Numbers refer to the values of the breaching index, B_z (see Fig. 10). Anomalous variations in the displacement plot all correspond with segment boundaries as mapped along the fault trace.

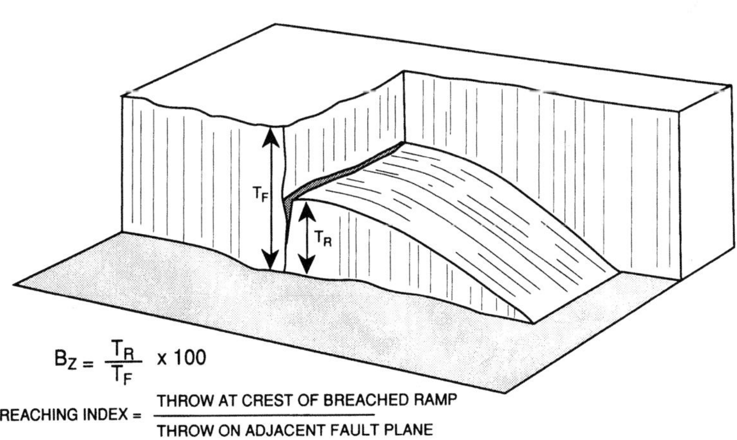

Fig. 10. Measurement of the Breaching Index, B_z, for footwall breached relay structures. This is used to convey the relative stages in the evolution of a segmented fault at which different segments linked to form a through-going fault connection. When applied to a single fault, these indices can assist in identifying the operation of a systematic or non-systematic linkage sequence (Fig. 9).

The displacement–distance plot (Fig. 9a) has an overall form that could be regarded as typical of a single fault (Walsh & Watterson 1987; Peacock & Sanderson 1991). The region of maximum displacement occupies the central region of the fault, and this decreases with a fairly constant displacement gradient to zero at both lateral tips. However, in detail, the plot is highly irregular, and this irregularity persists even when displacement is summed at all the breached relay structures. The displacement anomalies may in part be due to failure to include the ductile strains in the region of the breached ramps in the displacement sum (Walsh & Watterson 1991). Comparison of the displacement profile with the fault trace map (Fig. 9b), shows that the abrupt variations in displacement all coincide with breached relay structures.

In summary, the two examples described above provide additional evidence to that summarized previously by Walsh & Watterson (1990) and Peacock & Sanderson (1991), that relay structures are sites of anomalous displacement variation along a fault trace. The specific details of these anomalies vary, but a common facet is the steeper displacement gradients in the regions of the relay structure. In principle, displacement mapping has the potential to reveal the positions of relay structures that may not be recognizable from geometric criteria alone, even when considerable displacement is added after breaching.

D versus *L* data and the model of fault growth by segment linkage

Trace length (*L*) and maximum displacement (*D*) for 97 faults in the study area are presented in Fig. 11. Detailed description of the surveying methodology and errors are presented in Cartwright *et al.* (1995). The *L* and *D* values range over nearly three orders of magnitude (Fig. 11). Aside from the general trend towards increasing *L* for increasing *D*, the most interesting feature of the data shown in Fig. 11 is the large scatter in *D* on *L* and vice versa.

Cartwright *et al.* (1995) have explained this scatter in the $D-L$ data using a model of fault growth by segment linkage. In this model (Fig. 12), it is envisaged that individual segments grow by radial propagation according to a local, idealized scaling law (Fig. 12a) until overlap of two propagating segments occurs. As the overlap stage develops, the scaling law no longer applies because the stress conditions in the region of the interfering tips are perturbed, and further radial propagation is inhibited (Segall & Pollard 1980). Relay structures develop in the region of fault segment overlap and these are eventually breached to form a linked fault once the strain in the relay structure exceeds local thresholds for ductile accommodation (Trudgill & Cartwright 1994).

During linkage, proportionately greater lengthening occurs relative to the increase in maxi-

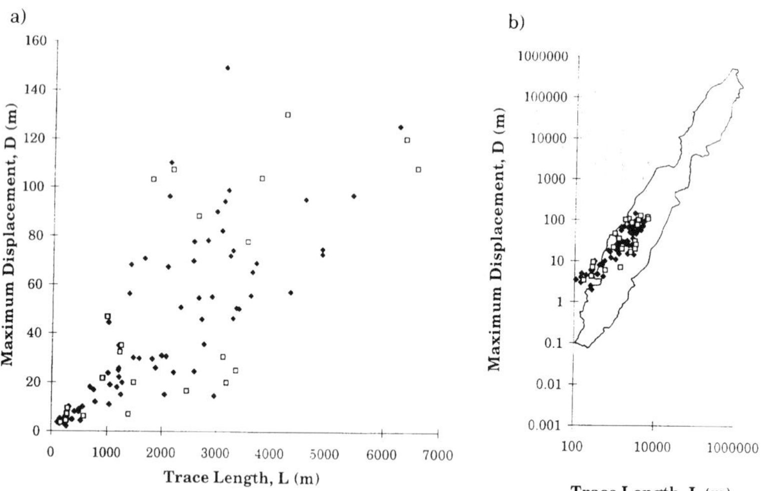

Fig. 11. Maximum displacement (*D*)–trace length (*L*) data for 97 faults from the Canyonlands, on (**a**) normal scales and (**b**) log–log scales. The data are grouped into faults from the area northeast (diamonds) and southwest (squares) of the Chesler Lineament (Fig. 2). These two areas have contrasting joint sets that are influence the mechanical properties of the two areas in different ways, but the approximately five-fold scatter in *D* and *L* in each set is comparable. An outline of the field of *D–L* data compiled for a range of fault types from different tectonic/stratigraphic settings by Gillespie *et al.* (1992) is shown on (**b**) for comparison. Note the consistent bandwidth of about an order of magnitude between *D* and *L* of the Gillespie *et al.* data.

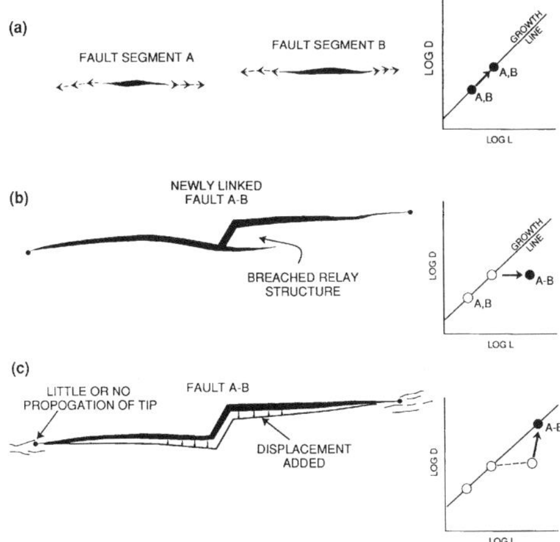

Fig. 12. Model of fault growth by segment linkage, represented by several stages in map view evolution of a segmented normal fault matched against a growth path on a log D — log L plot (see text for explanation). This model is based on observations of segmentation of the normal faults in the Canyonlands, plus conceptual framework from Segall & Pollard (1980), Pollard & Aydin (1984), Gudmundsson (1987), Martel *et al.* (1988) and Peacock & Sanderson (1991).

mum displacement, particularly if more than two segments are involved, and this results in a departure from the idealized fault growth line (Fig. 12b). At this stage the newly linked fault can be described as being 'underdisplaced'. Further propagation of the linked fault may be inhibited until a critical shear strain is re-established (Cowie & Scholz 1992*b*). To achieve this, the linked fault must accumulate displacement without increasing length to any significant degree, until the threshold for continued lateral propagation is met. The addition of displacement for limited increase in length marks the return of the fault towards the idealized growth line (Fig. 12c). The linked fault will then continue along the growth line until such time as a new overlap occurs, at which point the segment linkage cycle is repeated, and a further step-like departure from the growth line results.

No scale is given to the sequence of events shown in Fig. 12, since linkage of segments is a self-similar process that can be envisaged to influence the relationship between D and L from the scale of microfractures up to that of crustal faults (Segall & Pollard 1980). It is important to emphasize that although this model is highly schematic in the form shown in Fig. 12, and takes no account of any linkages or propagation in the vertical dimension, the stages represented by steps a–c are evident in the progressive geometrical development of relay structures found in the Canyonlands (Figs 6 & 7).

The cycle of growth shown in Fig. 12 would be repeated whenever propagation results in overlap and linkage with neighbouring fault segments (Fig. 13). Each new linkage of two or more segments would result in a segmented fault being established which is underdisplaced for its length, and a new step-like departure from the growth curve would result. In any evolving fault system, there would be a large range of faults at any one moment that were at different stages of their own particular step-wise growth path (Fig. 13). Some would be on the ideal curve, others would be 'under-displaced' to different degrees. This alone would produce a scatter in the D–L data whose range in L would depend to a large extent on the number of segments that were sufficiently co-linear to form a linked structure on a given trajectory of propagation. If linkage of multi-segmented faults occurred at a faster rate than the displacement deficit could be recovered, then significant scatter between D and L (half to one order of magnitude) could easily result. This model is therefore capable of explaining the extent of scatter observed in the $D - L$ data for the Canyonlands.

Discussion

The model of growth by segment linkage illustrated in Fig. 12 offers a simple explanation for the observed scatter of $D - L$ data from the

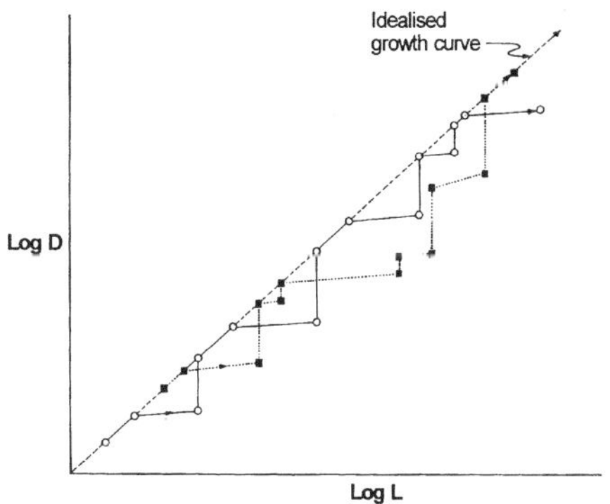

Fig. 13. Scatter in $D - L$ data resulting from segment linkage. Two faults are shown growing via a series of step-like cycles of linkage on a log D – log L plot. At any instant in this growth history, individual faults forming part of a fault array would be at different positions on these step-like paths: they may be either overlapping, or linking, or radially propagating. This will produce a scatter in the $D - L$ data, with an upper bounding limit dictated by the idealised growth curve, and a lower limit set by the deformational boundary conditions.

Canyonlands, but is it more generally applicable to other normal fault systems? This scatter in Fig. 11 is less than that noted from previous single and multi-population fault studies (Gillespie *et al.* 1992; Cowie & Scholz 1992*c*), but in these other datasets measurement errors may make up a significant component of the total scatter, which is clearly not the case for the Canyonlands data (Cartwright *et al.* 1995). The degree of scatter in the compiled field of fault data is fairly uniform over a large range of dimensions for faults from a variety of tectonic and rheological settings (Fig. 11). It is suggested, therefore, that a significant component of the scatter recorded in these data could be explained if a process of fault growth by segment linkage is common to all fault types across a broad range of length scales.

The wider applicability of this model is thus appealing as a general explanation for a significant component of the scatter in $D - L$ data, but is there any evidence that this process is an integral component of fault growth? In the Canyonlands Grabens, the evidence supporting the model of growth by segment linkage is the segmented geometry of the faults themselves, and in particular the temporal progression in the specific geometry of the relay structures (Fig. 7). If the segmentation of the Canyonlands faults can be satisfactorily attributed to this model, then it seems reasonable that any fault segmentation involving relay structures must be a consequence of this type of growth model.

Fault segmentation has been widely documented from field geological and sub-surface studies of fault systems ranging from metre-scale structures (Segall & Pollard 1983; Granier 1985; Gudmundson 1987; Martel *et al.* 1988; Peacock & Sanderson 1991, 1994), through kilometre-scale structures (McGill & Stromquist 1979; Walsh & Watterson 1990) to crustal-scale structures (Chapman *et al.* 1978; Chenet & Letouzey 1983; Pollard & Aydin 1984; Cartwright 1987; Milani & Davison 1988; Morley *et al.* 1990). These descriptions of fault segmentation deal with a spectrum of structural geometries, of which relay structures form only a small sub-set. The first specific description and mechanical analysis of relay structures is, however, fairly recent (Larsen 1988), and subsequently there have been numerous descriptions of relay structures from a variety of tectonic settings and scales (Morley *et al.* 1990; Roberts & Jackson 1991; Peacock & Sanderson 1991; 1994, Dawers *et al.* 1993; Schlische 1993; Anders & Schlische 1994; Gawthorpe & Hurst 1993; Jackson & Leeder 1994; Trudgill & Cartwright 1994). If it is accepted that the presence of a relay structure must indicate that a particular segmented fault has grown by the process of segment linkage rather than by radial propagation from a single seed structure (Watterson 1986), then the wide range of scales and settings of these published examples of relay structures suggests that segment linkage is a general process in the growth of normal faults.

Implications for sub-surface structural interpretation

Most published descriptions of relay structures are of outcrop examples, simply because seismic imaging of relay structures is so complex, and alternative interpretations are often equally plausible, particularly on widely spaced 2D grids. Many published sub-surface structure maps contain faults with curvilinear fault traces where fault strike changes abruptly at sharp bends in the fault trace. These faults are often mapped as single, continuous structures, but a tighter grid spacing and a re-interpretation might show many such structures to be composed of fault segments connected by relay structures. An excellent example of how 3D data can lead to such a re-interpretation is presented by Needham *et al.* (this volume).

As the number of published examples of relay structures increases, seismic interpreters will have an additional set of structural models or templates to consider when deciding on a particular fault correlation solution. In particular, the wider availability of 3D seismic data will probably reveal a greater degree of discontinuity in fault trace geometry than has hitherto been mappable with 2D grids (Needham *et al.* this volume), and relay structures should become easier to define correctly.

One of the most important implications of fault growth by segment linkage for sub-surface structural interpretation is the field of predictive fault population analysis (Walsh & Watterson 1992; Yielding *et al.* 1992). One objective of this type of analysis with possible applications in the mining and petroleum industries, is to predict fault size or vein size distributions over a range of scales beneath those covered by the resolution of the exploration methods in use. The foundations of this approach are built on the assumption that fracture and fault growth is by radial propagation, and that this can be expressed in the form of a scaling law (equation 1). We have shown, however, that fault growth by segment linkage necessarily results in departures from an idealized scaling law for D and L, and that the growth path is step-like, and intrinsically unpredictable. An essential part of growth by segment linkage, is that across a range of scales, smaller segments coalesce to form larger segments. A fundamental result of this process is that the size distribution of a fault population growing by segment linkage should be constantly evolving, with smaller segments being consumed to form larger units (Filbrandt *et al.* 1994). Unless an equivalent number of small segments are formed at each stage of this process, the result will be a progressive shift in the slope of any cumulative frequency/size curve. Predictive techniques should not, therefore, be based on such an inherently unstable process.

In conclusion, the assumption that fault growth scales according to a simple law expressed between *D* and *L* should be re-evaluated, particularly in view of the importance of this scaling law for all forms of predictive fault analysis. Part of this re-evaluation must involve more intensive sub-surface interpretations of the three-dimensional geometry of relay structures and segmented faults, with specific attention given to high-resolution mapping of displacement anomalies. Only with additional case-studies will it be possible to determine whether growth by segment linkage is a fundamental element in all fault growth. If this proves to be the case then a major challenge for future research in this field will be to examine how growth by segment linkage influences dimensional attributes of fault populations such as spacing of faults and length distributions.

We gratefully acknowledge the financial support of Fina Exploration Ltd (J.A.C. and B.D.T.), Shell Research B.V., and the Nuffield Foundation (J.A.C.). C.M. is the recipient of a post-graduate scholarship from Shell UK Ltd. Faisal Jaffri and Simon Harper are thanked for assistance in the field. The assistance of the staff at the Needles District Ranger Station and the Canyonlands National Park Office in Moab are gratefully acknowledged, particularly Bill Veteto for his work on our clutch! Jon Clemson, Tim Wynn, Robin Leinster and Paul Smith are thanked for draughting the figures. Early versions of this paper were greatly improved by informal reviews by John Walsh, Patience Cowie, Raymond Franssen and John Cosgrove. Patience Cowie and Bertrand Gauthier are thanked for their helpful review of the manuscript.

References

ANDERS, M. H. & SCHLISCHE, R. W. 1994. Overlapping Faults, Intrabasin Highs, and the Growth of Normal Faults. *Journal of Geology*, **102,** 165–180.

BAKER, A. A. 1933. Geology and oil possibilities of the Moab district, Grand and San Juan counties, Utah. *US Geological Survey Bulletin*, **841**, 95.

BARNETT, J. A. M., MORTIMER, J., RIPPON, J. H., WALSH, J. J. & WATTERSON, J. 1987. Displacement geometry in the volume containing a single normal fault. *Bulletin of the American Association Petroleum Geologists*, **71**, 925–937.

BEACH, A. & TRAYNER, P. 1991. The geometry of normal faults in a sector of the offshore Nile Delta, Egypt. *In*: ROBERTS, A. M., YIELDING, G. & FREEMAN, B. (eds) *The Geometry of Normal Faults.* Geological Society, London, Special Publications, **56,** 173–182.

BELFIELD, W. C. 1992. Simulation of sub-seismic faults using fractal and multifractal geometry. *Society of Petroleum Engineers,* paper 24751, Proceedings of the SPE meeting, Washington DC, October 4–7th 1992.

CARTWRIGHT, J. A. 1987. Transverse structural zones in continental rifts – an example from the Danish sector of the North Sea. *In*: BROOKS, J. & GLENNIE, K. W. (eds) *Petroleum Geology of North-West Europe.* Graham and Trotman, London, 441–452.

——, TRUDGILL, B. D. & MANSFIELD, C. M. 1995. Fault growth by segment linkage: an explanation for scatter in maximum displacement and trace length data from the Canyonlands Grabens of S.E. Utah. *Journal of Structural Geology*, **17**, 1319–1326.

CHAPMAN, T. J. & MENEILLY, A. W. 1991. The displacement patterns associated with a reverse-reactivated, normal growth fault. *In*: ROBERTS, A. M., YIELDING, G. & FREEMAN, B. (eds) *The Geometry of Normal Faults.* Geological Society, London, Special Publications, **56,** 183–191.

CHAPMAN, S. R., LIPPARD, S. J. & MARTYN, J. E. 1978. The stratigraphy and structure of the Kamasia Range, Kenya rift valley. *Journal of the Geological Society, London,* **135**, 265–281.

CHENET, P. Y. & LETOUZEY, J. 1983. Tectonique de la zone Comprise entre Abu Durba et Gebel Mezzazat dans le contexte de l'evolution rift de Suez. *Exploration Production Elf Aquitaine*, **7**, 201–215.

COWIE, P. A & SCHOLZ, C. H. 1992*a*. Physical explanation for the displacement–length relationship of faults using a post-yield fracture mechanics model. *Journal of Structural Geology*, **14**, 1133–1148.

—— & ——1992*b*. Growth of Faults by Accumulation of Seismic Slip. *Journal of Geophysical Research,* **97**, B7, 11085–11095.

—— & ——1992*c*. Displacement-length scaling relationship for faults: data synthesis and discussion. *Journal of Structural Geology,* **14,** 1149–1156.

DAWERS, N. H., ANDERS, M. H. & SCHOLZ, C. H. 1993. Growth of normal faults: Displacement–length scaling. *Geology*, **21**, 1107–1110.

EISENSTADT, G. & WITHJACK, M. 1995. Clay analogue experiments of basin inversion. *In*: BUCHANAN, P. G (eds) *Inversion Tectonics.* Geological Society, London, Special Publications, **88**, 119–136.

FILBRANDT, J. B., RICHARD, P. D. & FRANSSEN, R. C. M. W. 1994. Growth and coalescence of faults: numerical simulations and sandbox experiments. *Extended Abstracts Volume, Fault Populations.* Tectonic Studies Group, Royal Society of Edinburgh, 57–60.

FREEMAN, B., YIELDING, G. & BADLEY, M. 1990. Fault displacement mapping as an aid in fault correlation. *First Break*, **8**, 87–95.

GAUTHIER, B. D. M. & LAKE, S. D. 1993. Probabilistic modelling of faults below the limit of seismic resolution in the Pelican Field, North Sea. *American*

Association of Petroleum Geologists Bulletin, **77**, 761–777.

GAWTHORPE, R. L & HURST, J. M. 1993. Transfer zones in extensional basins: their structural style and influence on drainage development and stratigraphy. *Journal of the Geological Society*, London, **150,** 1137–1152.

GILLESPIE, P. A., WALSH, J. J. & WATTERSON, J. 1992. Limitations of dimension and displacement data from single faults and the consequences for data analysis and interpretation. *Journal of Structural Geology*, **14**, 1157–1172.

GRANIER, T. 1985. Origin, damping and pattern of development of faults in granite. *Tectonics*, **4**, 721–737.

GUDMUNDSSON, A. 1987. Geometry, formation and development of tectonic fractures on the Reykjanes Peninsular, southwest Iceland. *Tectonophysics*, **139**, 295–308.

HARDING, T. P. 1984. Graben hydrocarbon occurrences and structural style. *American Association of Petroleum Geologists Bulletin*, **68,** 333–362.

HINTZE, L. F. 1988. Geologic history of Utah. *Brigham Young University Geology Studies, Special Publication*, **7**.

HUNTOON, P. W. 1982. The Meander anticline, Canyonlands, Utah: An unloading structure resulting from horizontal gliding on salt. *Bulletin of the Geological Society of America*, **93**, 941–950.

JACKSON, J. & LEEDER, M. 1994. Drainage systems and the development of normal faults: an example from Pleasant Valley, Nevada. *Journal of Structural Geology*, **16,** 1041–1059.

JOESTING, H. R., CASE, J. E. & PLOEFF, D. 1966. Regional geophysical investigations of the Moab-Needles area, Utah. *United States Geological Survey Professional Paper*, **516-C**.

LARSEN, P. H. 1988. Relay structures in a Lower Permian basement-involved extension system, East Greenland. *Journal of Structural Geology*, **10**, 3–8.

LEWIS, R. Q., SR. & CAMPBELL, R. H. 1965. Geology and uranium deposits of Elk Ridge and vicinity, San Juan County, Utah. *United States Geological Survey Professional Paper*, **351**.

MCGILL, G. E. & STROMQUIST, A. W. 1975. Origin of graben in the Needles District, Canyonlands National Park, Utah. *Four Corners Geological Society Guidebook, 8th Field Conference Canyonlands*, 235–243.

—— & —— 1979. The grabens of Canyonlands National Park, Utah: Geometry, Mechanics and Kinematics. *Journal of Geophysical Research*, **84, B9**, 4547–4563.

MARRETT, R. & ALLMENDINGER, R. W. 1991. Estimates of strain due to brittle faulting: sampling of fault populations. *Journal of Structural Geology*, **13**, 735–737.

MARTEL, S. J., POLLARD, D. D. & SEGALL, P. 1988. Development of simple strike-slip fault zones, Mount Abott Quadrangle, Sierra Nevada, California. *Bulletin of the Geological Society of America*, **100,** 1451–1465.

MILANI, E. J. & DAVISON, I. 1988. Basement control and transfer tectonics in Reconcavo-Tucano-Jatoba rift, Northeast Brazil. *Tectonophysics*, **154,** 40–70.

MORLEY, C. K., NELSON, R. A., PATTON, T. L. & MUNN, S. G. 1990. Transfer zones in the East African Rift system and their relevance to hydrocarbon exploration in rifts. *American Association of Petroleum Geologists Bulletin*, **74,** 1234–1253.

NEEDHAM, R. A., YIELDING, G. & FREEMAN, B. 1996. Analysis of fault geometry and displacement patterns. *This volume.*

PEACOCK, D. C. P. & SANDERSON, D. J. 1991. Displacements, segment linkage and relay ramps in normal fault zones. *Journal of Structural Geology*, **13,** 721–733.

—— & ——1994. Geometry and development of relay ramps in normal fault systems. *American Association of Petroleum Geologists Bulletin*, **78,** 147–165.

POLLARD, D. D. & AYDIN, A. 1984. Propagation and linkage of oceanic ridge segments. *Journal of Geophysical Research*, **89**, B12, 10017–10028.

POTTER, D. B. & MCGILL, G. E. 1978*a*. Field analysis of a pronounced topographic lineament, Canyonlands National Park, Utah. *Proceedings of the 3rd International Conference on Basement Tectonics*, 169–176.

—— & ——1978*b*. Valley anticlines of the Needles District, Canyonlands National Park, Utah. *Bulletin of the Geological Society of America*, **89,** 952–960.

ROBERTS, S. & JACKSON, J. A. 1991. Active normal faulting in Greece: an overview. *In*: ROBERTS, A.M., YIELDING, G. & FREEMAN, B. (eds) *The Geometry of Normal Faults*. Geological Society, London, Special Publications, **56,** 125–142.

SCHLISCHE, R. W. 1993. Anatomy and evolution of the Triassic-Jurassic continental rift system, eastern North America. *Tectonics*, **12**, 1026–1042.

SCHOLZ, C. H. 1990. *Mechanics of Faulting and Earthquakes*. Cambridge University Press, Cambridge.

—— & COWIE, P.A. 1990. Determination of total strain from faulting using slip measurements. *Nature*, **346,** 837–839.

——, DAWERS, N. H., YU, J.-Z., ANDERS, M. H. & COWIE, P. A. 1993. Fault growth and fault scaling laws: Preliminary results. *Journal of Geophysical Research*, **98**, B12, 21, 951–21, 961.

SEGALL, P. & POLLARD, D. D. 1980. Mechanics of discontinuous faults. *Journal ofGeophysical Research*, **85,** B8, 4337–4350.

—— & —— 1983. Nucleation and growth of strike-slip faults in granite. *Journal of Geophysical Research*, **88,** 555–568.

TRUDGILL, B. D. & CARTWRIGHT, J. A. 1994. Relay ramp forms and normal fault linkages – Canyonlands National Park, Utah. *Bulletin of the Geological Society of America*, **106,** 1143–1157.

WALSH, J. J. & WATTERSON, J. 1987. Distributions of cumulative displacement and seismic slip on a single normal fault surface. *Journal of Structural Geology*, **9,** 1039–1046.

—— & —— 1988. Analysis of the relationship between

displacements and dimensions of faults. *Journal of Structural Geology*, **10**, 239–247.

—— & —— 1990. New methods of fault projection for coalmine planning. *Proceedings of the Yorkshire Geological Society*, **48**, 209–219.

—— & —— 1991. Geometric and kinematic coherence and scale effects in normal fault systems. *In*: ROBERTS, A. M., YIELDING, G., & FREEMAN, B. (eds) *The Geometry of Normal Faults*. Geological Society, London, Special Publications, **56,** 193–203.

WALSH, J. J., WATTERSON, J. & YIELDING, G. 1991. The importance of small-scale faulting in regional extension. *Nature*, **351**, 391–394.

WATTERSON, J. 1986. Fault dimensions, displacements and growth. *Pure & Applied Geophysics*, **124,** 365–373.

YIELDING, G., WALSH, J. J. & WATTERSON, J. 1992. The prediction of small-scale faulting in reservoirs. *First Break*, **10**, 449–460.

Kinematic modelling of normal fault geometries using inverse theory

HUGH G. KERR & NICKY WHITE

Bullard Laboratories, University of Cambridge, Madingley Road, Cambridge CB3 OEZ, UK

Abstract: The principal features of a general kinematic method which uses the shapes of deformed strata to calculate normal fault geometries at depth in two and three dimensions are summarized here. This method assumes that hanging-wall deformation can be represented by arbitrarily inclined bulk simple shear and it should be applicable to faults at any scale. In two-dimensional modelling, deformation is assumed to occur within the plane of section. Deformation in three dimensions is parameterized by the three Euler angles which describe the horizontal component of the slip vector, the rake (i.e. pitch) angle and the inclination of shear planes. Differential compaction is included in both cases. Since we use inverse theory to determine fault geometries, the quality of our solutions is easily investigated. Both two- and three-dimensional algorithms have been rigorously tested on synthetic data, on sand-box experiments, and on depth-converted and interpreted seismic reflection profiles. Modifications of this approach can be used to calculate fault geometries which have changed shape during or after fault displacement. A similar approach is applicable to reverse faulting.

There is a requirement within the oil industry for a general kinematic model which will allow fault geometries to be calculated at depth from the deformation observed within hanging and foot-walls. To date, the most widely used models are two-dimensional (2D) (Verrall 1981; see also Roberts *et al.* 1991 for a recent summary). Many published techniques have several important limitations: (a) deformation may occur out of the plane of section and so solutions could be invalid; (b) fault geometry may change shape as deformation proceeds as a result of salt or shale diapirism, flexure, or rigid-body rotation; (c) the strong emphasis upon forward modelling has meant that important issues such as the existence and uniqueness of solutions together with the resolution and trade-off of model parameters are often ignored.

In order to tackle these limitations, Kerr *et al.* (1993) have developed a general three-dimensional (3D) inverse model which allows fault geometry to be determined from deformed hanging and foot-wall beds without requiring *a priori* knowledge of the horizontal component of the slip vector (i.e. direction of extension). A crucial feature of this scheme is that it is based upon inverse theory and so the validity of different assumptions can be systematically tested. For example, it has proved possible to diagnose automatically circumstances under which fault shape has changed either during or after motion along the fault. The quality of solutions can be assessed by examining how random and systematic noise affect them (Kerr & White 1994). Trade-off and resolution of different calculated parameters can also be investigated.

The purpose of this brief review is to outline the 2D and 3D problems and to show how inverse theory can be exploited to find solutions. Throughout, a simple kinematic description of deformation is used in order to determine deformation velocity fields associated with faulting. The most important constraint is conservation of mass and we emphasize that any model which fails to obey this constraint is invalid. No attempt is made to address the more difficult dynamic problem.

The two-dimensional problem

It is now generally agreed that deformation which occurs within the hanging wall of a normal fault can be described by simple shear which is arbitrarily inclined at some angle α (Roberts *et al.* 1991). This model does not attempt to match exactly the observed finite deformation within a hanging wall since this problem cannot be easily posed in inverse form. Rather, it determines the bulk or average velocity field (Waltham 1990). The effects of differential compaction can be allowed for by using an empirical relationship between initial porosity, ϕ_0, and compaction length, λ (White *et al.* 1986). Compaction is assumed to occur in the direction of shearing. Kerr & White (1994) have shown that more sophisticated compaction models are not justified by existing data.

If the geometry of only one bed is used to calculate the fault geometry, then unique solutions cannot be obtained as long as that α, ϕ_0, and λ, are unknown. Kerr & White (1992) have subsequently

From Buchanan, P. G. & Nieuwland, D. A. (eds), 1996, *Modern Developments in Structural Interpretation, Validation and Modelling,* Geological Society Social Publication No. 99, pp. 179–188.

shown that if a single set of these unknown parameters accounts for the deformation of two or more beds, then the geometries of two or more beds can be used to obtain the correct fault geometry together with all three parameters. In other words, the forward model, which calculates bed geometries for a given fault shape, can be posed as an inverse problem where indirect measurements (i.e. hanging-wall deformation) are used to determine some property of the Earth (i.e. fault geometry and deformation parameters). This inverse problem can be solved by designing a misfit function which measures the difference between all calculated faults (Fig. 1). For an arbitrarily chosen set of values of α, ϕ_0, and λ, a different fault can be calculated from each bed. The misfit between these calculated faults is defined by

$$\mathcal{M} = \frac{1}{P}\left\{\sum_{k=1}^{P}\frac{2}{N(N-1)}\left\{\sum_{i=1}^{N-1}\sum_{j=1+i}^{N}\left|z_{ik}-z_{jk}\right|\right\}\right\} \quad (1)$$

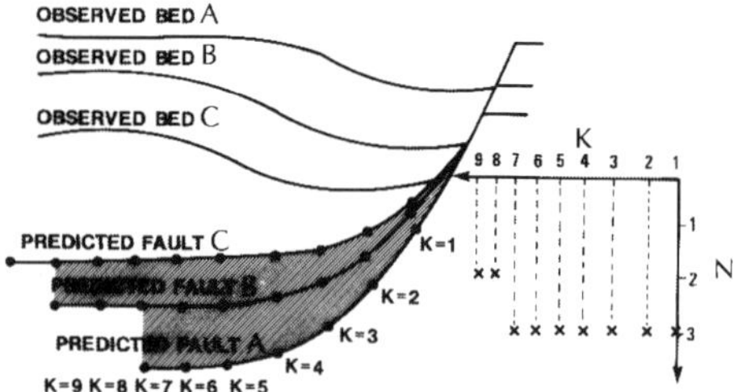

Fig. 1. Inverse problem: in general, for given values of the model parameters, faults *A*, *B*, and *C* can be calculated using beds *A*, *B*, and *C*. To find the correct fault geometry and correct values of model parameters, the misfit, $\mathcal{M}$, between the different calculated faults must be minimised by systematically varying the model parameters. *P* is the total number of points at which misfit is calculated. *N* is number of faults each point *k* is located upon. Misfit function is calculated using Equation (1) by evaluating, and then summing, the vertical difference between points for all pairs of calculated faults. Function is normalized by number of calculations at each point, $N(N-1)/2$, and by total number of points, *P*.

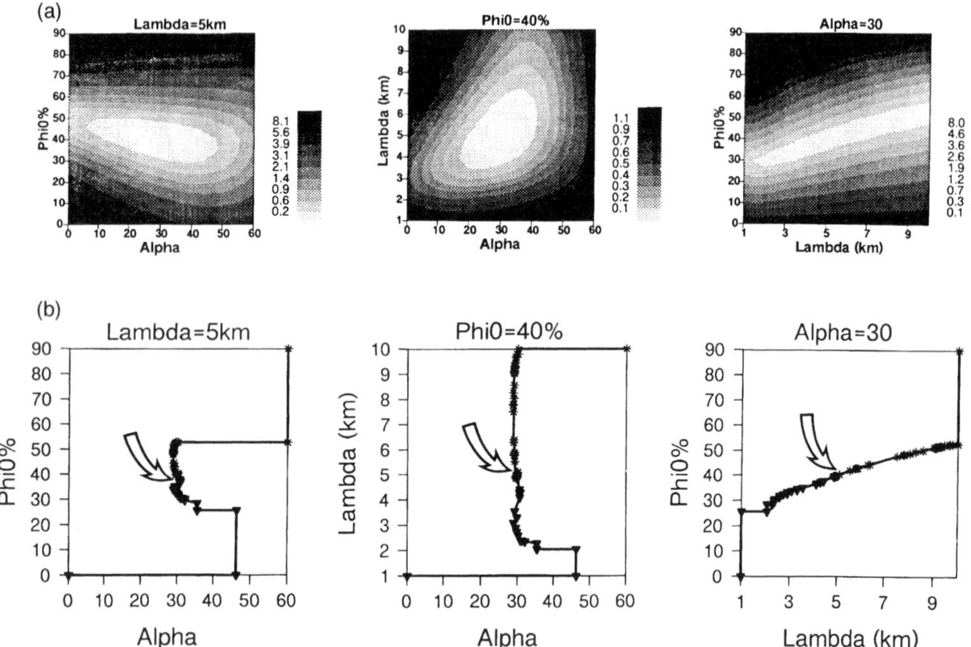

Fig. 2. Results from forward and inverse modelling of synthetic data (see Kerr & White 1992 for further details). (**a**) Three orthogonal plots of function $\mathcal{M}(\alpha, \phi_0, \lambda)$ located at automatically determined solution. All three contour plots mutually intersect at location of values used in forward model. In each case, correct solution (i.e. minimum value of $\mathcal{M}$) lies within most lightly shaded region. Contour interval indicated at right-hand-side of each plot. (**b**) Same orthogonal plots as in (**a**) but now showing path taken by two inversion runs using Powell's algorithm (Press *et al.* 1986). Solid triangles map the projected path from starting point at $\alpha = \theta°$, $\phi_0 = 90\%$, and $\lambda = 1$ km. Stars map the projected path from starting point at $\alpha = 60°$, $\phi_0 = 90\%$, and $\lambda = 10$ km. Arrows indicate that method converges to correct solution from both starting points.

Thus the misfit function, $\mathcal{M}(\alpha, \phi_0, \lambda)$, is evaluated at a series of grid points, at each of which the vertical distance between every possible pair of fault surfaces is calculated and summed. The total misfit at each grid point is normalized by the number of misfit calculations performed at that point. Each grid point misfit is then summed to yield a total misfit, which is normalized by the total number of grid points, P.

A solution to the inverse problem is found by searching 3D solution space for a minimum misfit value where α, ϕ_0, and λ represent the axes of this 3D space. The search procedure is clearly contingent upon both the existence and uniqueness of such a solution (Parker 1977). For this particular problem, it has been demonstrated, but not proven, that unique solutions exist.

In Fig. 2, the results from a synthetic modelling example are shown. The solution is located automatically, starting at some position given by an arbitrarily chosen combination of values for α, ϕ_0, and λ. Three mutually intersecting contoured slices through the misfit function are shown at the obtained solution. In this example, the misfit function varies smoothly as a function of α, ϕ_0, and λ and a global solution exists. The shape of the minimum gives a clear indication of the trade-off between the different parameters.

There are fundamental differences between forward and inverse models. In forward modelling, solutions are obtained by trial and error, starting with an 'educated guess'. Although good solutions can be obtained in this way, the process is generally inefficient and gives little idea of either the uniqueness or the quality of the solution obtained. In contrast, inversion allows solutions to be determined rapidly from, if required, a whole range of different starting points. Once a solution has been obtained, formal error analysis can provide a useful indication of the quality of a solution and of the trade-off between calculated parameters (Parker 1977; Press *et al.* 1986).

Tests using sand-box models

Published sand-box experiments of normal faulting are a convenient means for testing our 2D inversion

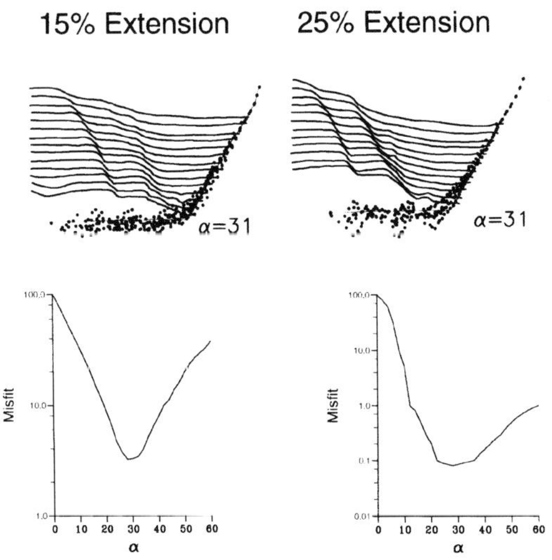

15% Extension 25% Extension

α=31 α=31

Fig. 3. Results from inverse modelling of dry sand-box experiments (McClay & Ellis 1987; see Kerr & White 1992 for further details). (**a**) Result of inverting 14 beds from model which underwent 15% extension. Solid lines, digitized bed geometries; thin line, digitized fault geometry; dots, calculated fault geometries. (**b**) Misfit plotted as function of α. Note that in this figure and in Fig. 4, the misfit function is normalized to 100 and plotted on a logarithmic scale. (**c**) Similar to (**a**) but has undergone extension of 25%. Twelve beds used. (**d**) Similar to (**b**).

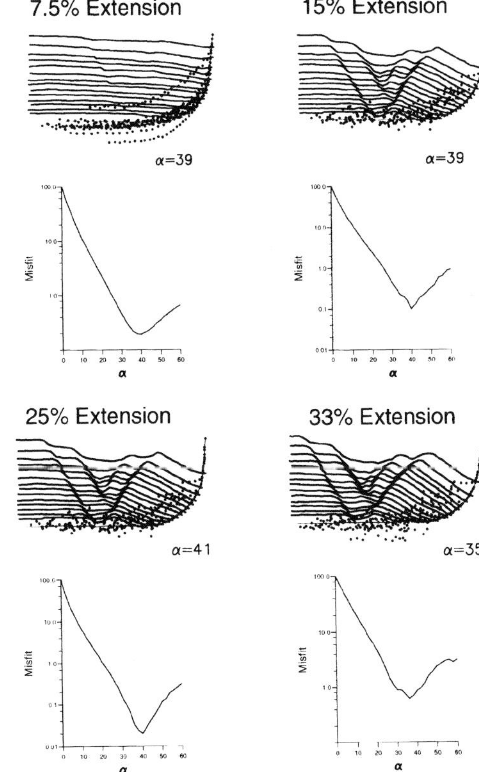

Fig. 4. Similar to Fig. 3 using a different fault geometry and greater range of extension values. Note that α, determined by inversion, does not change significantly as extension increases. See Kerr & White (1992) for further details.

algorithm (e.g. McClay & Ellis 1987). These experiments are probably not useful dynamic analogues of the faulting process, but they have several advantages for our purposes. First, the horizontal component of the slip vector is within the plane of the section and so 2D modelling can be applied with confidence. Secondly, the fault geometry and 'stratigraphy' are accurately known. Thirdly, deformation is complicated and so the experiments represent an appropriate test for simple algorithms. Kerr & White (1992) have tested the 2D inversion algorithm on many sand-box experiments. Here, we summarise the most important results and attempt to draw general conclusions.

Despite the fact that deformation is complicated, the arbitrarily inclined simple shear model is obviously a good approximation of the true velocity field (Figs 3 & 4). Unique solutions are automatically obtained even when rigid-body rotations about horizontal axes occur. More significantly, the value of α does not change with increasing amounts of extension, implying that the average velocity field remains constant as a function of time (Fig. 4). If this latter observation is generally correct, it means that the more difficult problem of predicting the history of the detailed, finite deformation of hanging walls may be tractable.

Using results obtained by inverting 25 experiments, Kerr & White (1992) suggested that there is a linear relationship between the shear angle, α, and the average dip of the fault, θ (Fig. 5). The basis for this relationship is now clear. In all of the experiments, a plastic sheet was pulled along the surface of a fixed fault (see McClay & Ellis (1987) for further details). The existence of this sheet requires the extension, ε, to be conserved along the fault

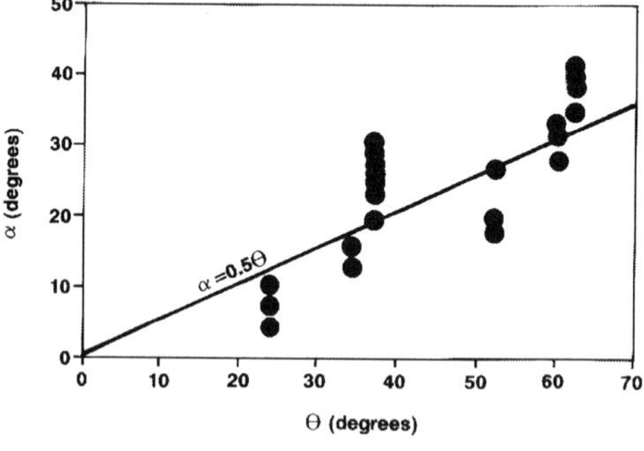

Fig. 5. Summary of results obtained by inverting 25 sand-box experiments. α = shear angle measured away from vertical; θ = average dip of fault from surface down to level of decollement. Solid line represents $\alpha = 0.5\theta$ but has not been fitted to data. See Kerr & White (1992) for further details.

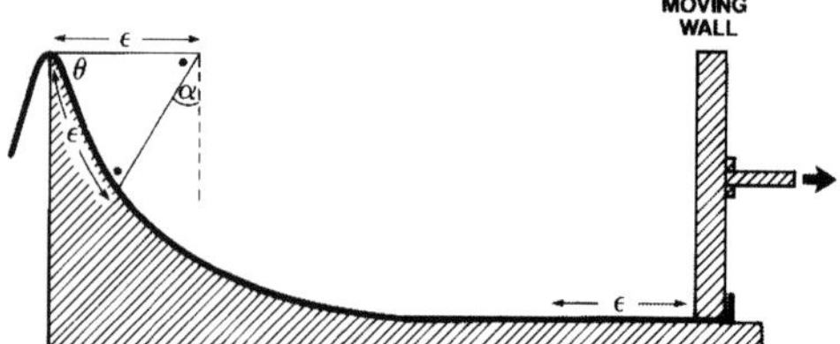

Fig. 6. Schematic drawing of deformation rig used in sand-box experiments. α = shear angle used in inverse modelling; θ = average dip of fault. The existence of plastic sheet along fault plane forces extension, ε, to be conserved along fault plane. Thus $\alpha = 0.5\theta$.

itself. By inspection of the triangle in Fig. 6, $\theta + 2(90° - \alpha) = 180°$ and so $\alpha = 0.5\theta$. The results from 25 inversions independently confirm this extremely simple prediction (Fig. 5).

At present, it is unclear whether or not this simple relationship holds for natural faults. If it does hold, it would imply that slip is conserved along fault surfaces. We emphasize that we are not concerned with arguments concerning the general relevance of sand-box experiments: our interest here is solely the kinematic description of complicated deformation fields.

The three-dimensional problem

Little progress can be made in applying 2D inverse models to natural examples so long as the horizontal component of the slip vector remains unknown. For this reason, Kerr *et al.* (1993) have developed a general 3D inverse model. In this algorithm, deformation is represented by the three Euler angles: (a) α_1 is the horizontal component of the slip vector or direction of extension within the horizontal plane; (b) α_2 is the rake angle measured on shear planes away from the horizontal; and (c) α_3 is the inclination of shear planes measured away from the vertical (Symon 1971). Note that α_3 is equivalent to α used in the 2D model. As before, a misfit function, $\mathcal{M}$, can be formulated which is necessarily a function of five parameters: α_1, α_2, α_3, ϕ_0, and λ. An important aspect of this approach is that no assumption need be made concerning α_1 which is obtained automatically by inversion.

The results of four synthetic 3D models are shown in Figs 7 & 8. Compaction effects have been ignored for clarity and so $\mathcal{M}$ is a function of only α_1, α_2 and α_3. In general, this new misfit function is more non-linear (i.e. contours are irregular and local minima exist) than that used for 2D modelling (compare Fig. 2 with Figs 7 & 8). Nevertheless, solutions can be obtained automatically from a variety of starting values of α_1, α_2 and α_3.

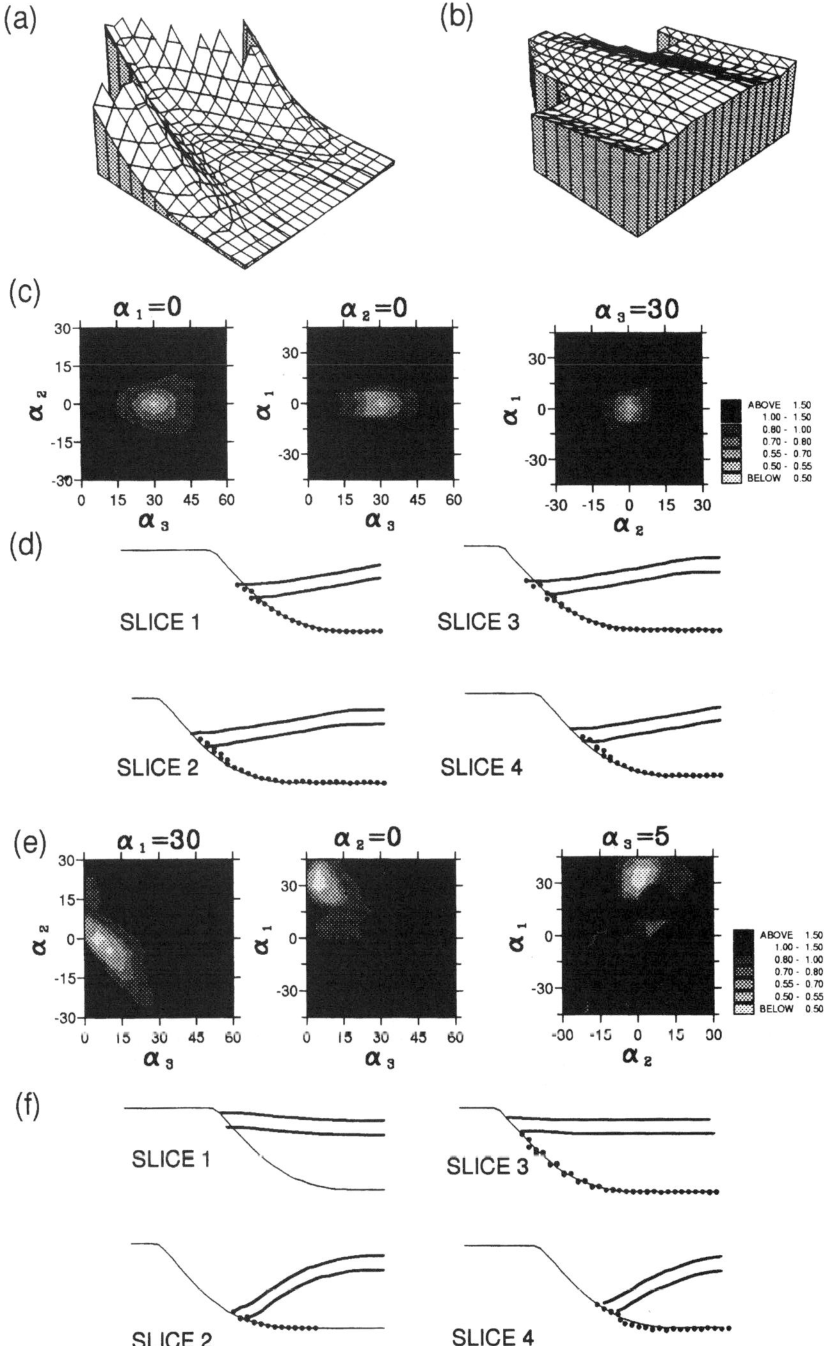

Fig. 7. Synthetic model of listric fault with considerable along-strike variation. (**a**) numerically generated fault surface used in forward model to calculate geometries of two deformed hanging wall beds. (**b**) Top surface of uppermost bed generated by forward model with $\alpha_1 = 0°$, $\alpha_2 = 0°$, and $\alpha_3 = 30°$. (**c**) Three orthogonal plots of the function $\mathcal{M}$, which mutually intersect at the location of values used in forward model. Correct solution (i.e. minimum value of $\mathcal{M}$) lies within lightest shaded region. (**d**) Four vertical 2D cross-sections cut through volume of 3D synthetic data at equal intervals. Dark solid lines are beds generated by forward modelling, thin solid line is numerically generated fault surface used in forward modelling, and dotted lines are faults calculated from the deformed beds by inverse modelling at the correct solution. Slight fluctuations in surfaces are caused by numerical interpolation. (**e**) and (**f**) Misfit function and inverse model for synthetic data using a forward model with $\alpha_1 = 30°$, $\alpha_2 = 0°$, and $\alpha_3 = 5°$. Figures 8–10 organized in similar fashion.

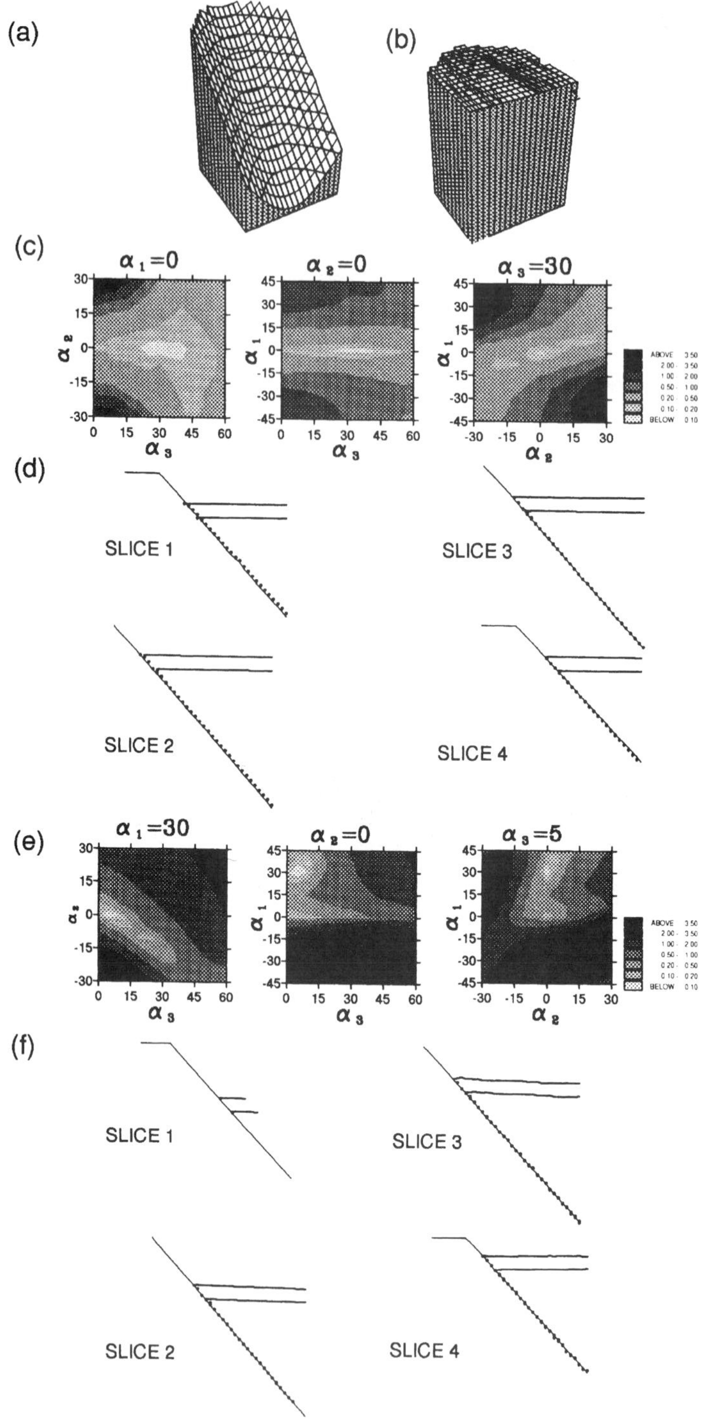

Fig. 8. Synthetic model of planar normal fault with modest along-strike variation. Misfit function and inverse model shown as in Fig. 7.

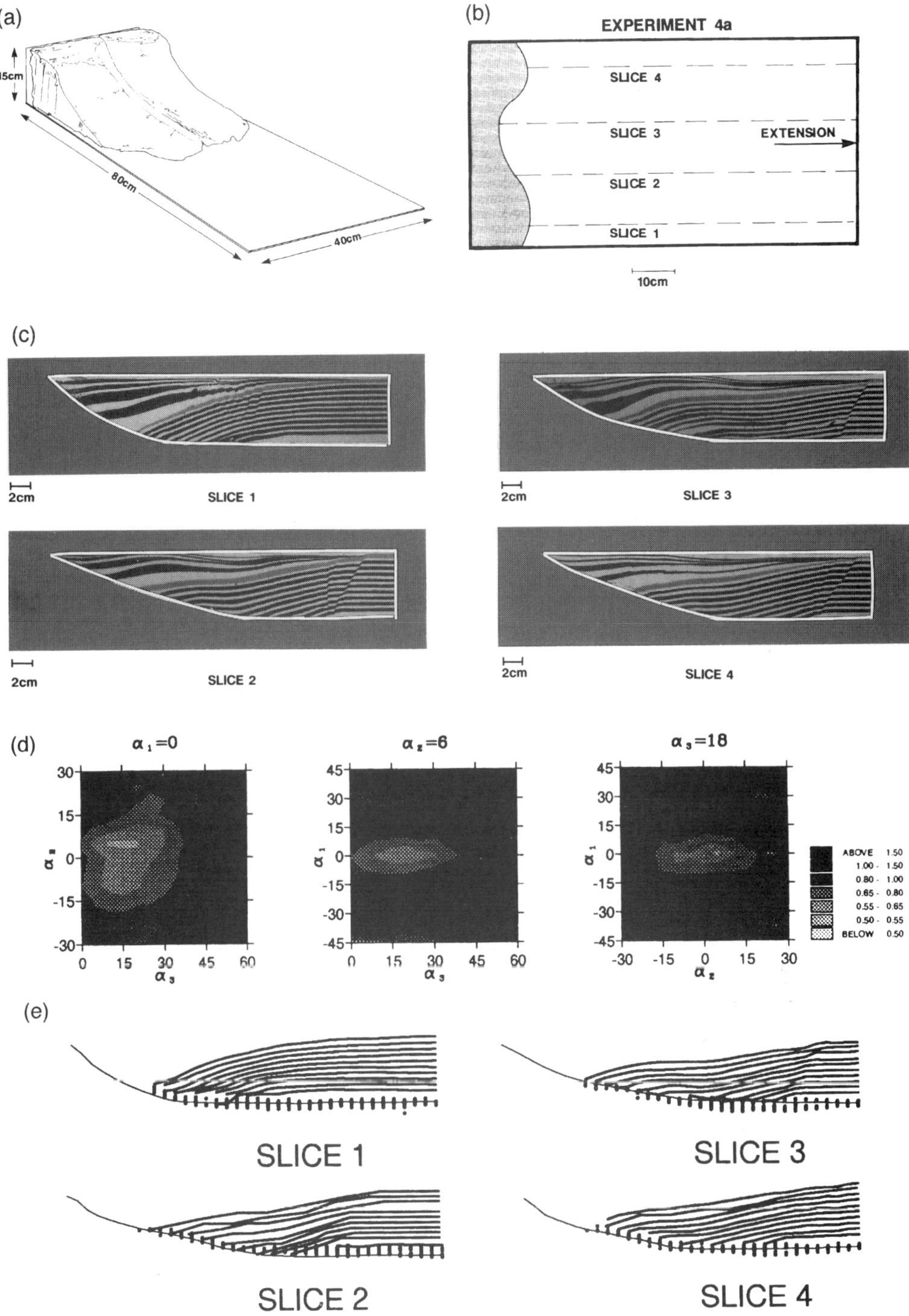

Fig. 9. Sand-box experiment using listric normal fault with considerable along-strike variation (i.e. similar to synthetic example in Fig. 7). Extension occurred parallel to vertical sections. (**a**) Main fault surface used in this experiment and in experiment shown in Fig. 10. (**b**) Plan view of model shows position of four parallel vertical sections in (c) and direction of extension. Nine beds at top of model were deposited as deformation proceeded. (**c**) Four slices through 3D volume. (**d**) Three mutually orthogonal contour plots of the misfit function, $\mathcal{M}$, which intersect at minimum located by Powell's method (Press *et al.* 1986): $\alpha_1 = 0°$, $\alpha_2 = 6°$, and $\alpha_3 = 18°$. (**e**) Four sections in similar positions to four sections in (b). Thin solid line is digitized fault surface; dark solid lines are digitized hanging wall beds; dotted lines are fault surfaces calculated at solution which was located by inverse modelling.

(a)

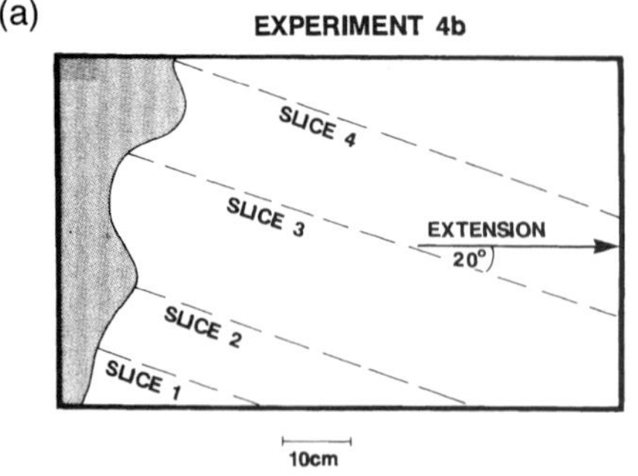

Fig. 10. Sand-box experiment which used same fault geometry as in Fig. 9 but now sections are oriented 20° clockwise to direction of extension. (**a**)–(**d**) Sand-box model and inversion results shown as before. Located minimum is at $\alpha_1 = 14°$, $\alpha_2 = 14°$, and $\alpha_3 = 15°$. Note thrust fault in SLICE 2, a result of oblique slip over an irregular fault surface.

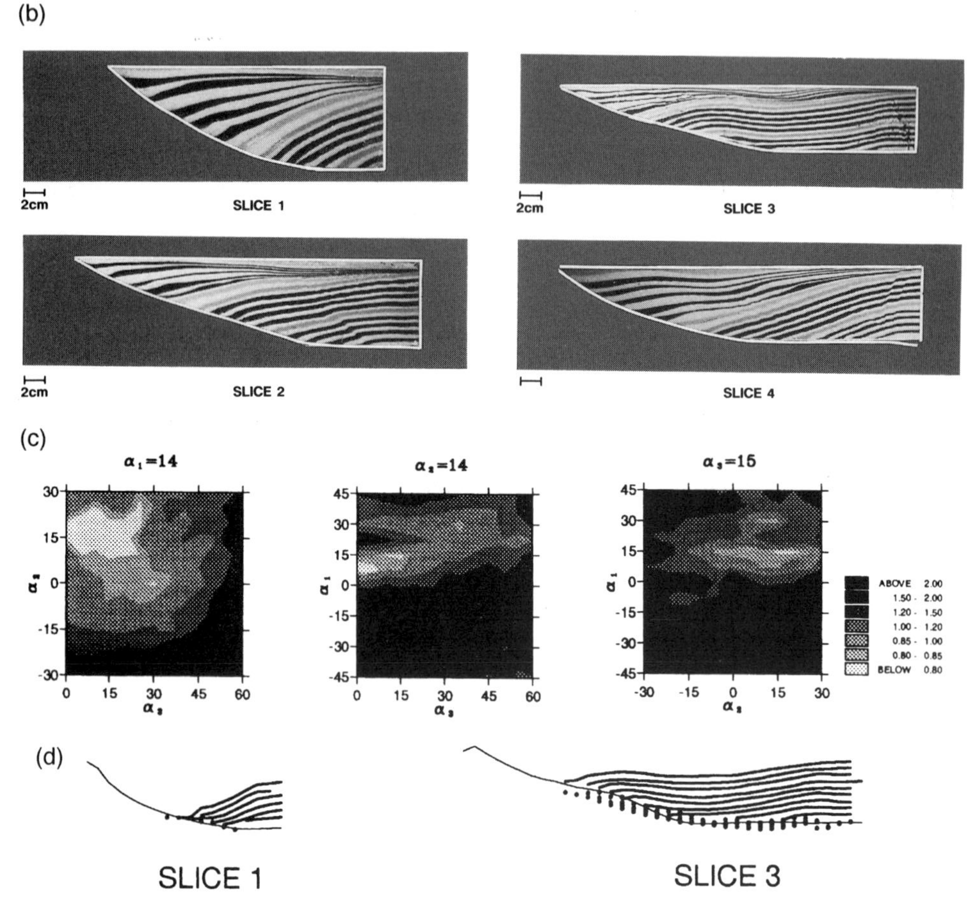

Therefore, the problem cannot be very non-linear (Scales 1985). Accurate fault shapes are obtainable but the ability to retrieve the correct value of the extension direction, α_1, is dependent upon the along-strike and down-dip variation in fault geometry. In the example shown in Fig. 7, the fault varies considerably along strike and so α_1 is well resolved on inspection of the relevant set of three orthogonal contoured slices which intersect at the global minimum of the misfit function. Planar faults with modest along-strike variation yield poorly resolved parameters although the fault shape can still be accurately determined (Fig. 8).

As before, sand-box experiments provide a useful test for this type of predictive algorithm. Most published sand-box data are only 2D but Kerr *et al.* (1993) have published a series of fully 3D experiments where fault geometries vary along strike and where horizontal-slip vectors vary. Some of their results are shown in Figs 9 & 10. It is clear

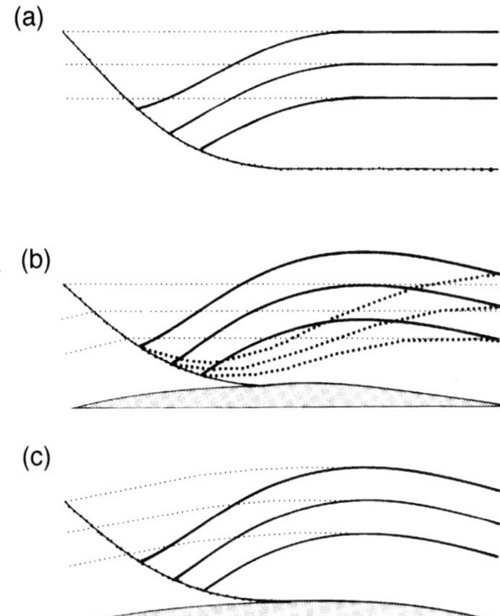

Fig. 11. Synthetic models designed to illustrate relationship between geometry of regional levels of deformed beds, and changes in geometry of main normal fault. (**a**) Forward and inverse model for 3 beds. Solid lines, deformed beds; thin line, forward-modelled fault geometry; fine dotted lines, regional levels of deformed beds prior to deformation; coarse dotted lines, calculated faults shapes for correct values of model parameters. (**b**) Fault and beds have been deformed by inhomogeneous vertical shear which simulates diapirism and/or flexural rebound. Note unacceptably large misfit for solution found by inverse modelling. (**c**) Regional levels have been perturbed until they match the deformation which has occurred. Now the solution obtained by inversion closely matches expected solution.

that unique solutions can be obtained and that the horizontal component of the slip vector is often retrievable.

Finally, Kerr & White (1995) have demonstrated that this general algorithm can be successfully applied to 3D seismic reflection data.

Future work

There are likely to be circumstances when faults change shape as deformation proceeds. This process is probably most common in regions affected by salt or shale diapirism. On the scale of the brittle upper crust, the shapes of basement faults can also be changed by flexural rebound, by rotation, or by the mutual interaction of faults. Kerr & White (1994) have made a preliminary attempt to tackle this problem by inverting synthetic data where known amounts of fault deformation have occurred. They found that when fault geometry changes shape significantly during deformation, the predicted faults for a minimum misfit were not coincident and the misfit function itself was uniformly large. Such a result is diagnostic of a breakdown in one of the underlying assumptions and is thus a useful feature of our inversion scheme: when no adequate solution exists then inferences may be drawn concerning the underlying assumptions (e.g. faults do or do not change shape).

Here, we suggest that this problem can be tackled in a more systematic fashion when both hanging-wall bed geometries and fault geometry are known. Conservation of mass is a fundamental constraint, and so any mechanism that alters the shape of a fault must also affect both the hanging-wall bed geometries and the regional levels. Diapir formation beneath a fault or flexural rebound within the footwall immediately adjacent to a fault are essentially laterally varying, vertical simple shear. Thus the correct solution to the inverse problem can be obtained if appropriately deformed regional levels are used (Fig. 11). Thus, when the fault geometry is already known, deformation within the hanging wall and that which affects the fault can both be solved for by varying the shapes of the regional levels (e.g. Fig. 11c).

Our general approach can also be used to tackle the problem of sequential palinspastic restoration. The algorithms described above yield a velocity field for hanging-wall deformation which is only an approximation. This approximate solution can be perturbed in order to retrieve a more accurate velocity field which matches precisely the observed finite deformation. Provided the resultant field is a function of time, it may be possible to carry out sequential restorations.

This work was supported by a Petroleum Science and Technology Institute (PSTI) studentship held by HGK. Computational facilities were generously provided by PSTI. We are very grateful to T. Abramovitz, G. Gray, R. Groshong, B. Hall, T. Hauge, M. Naylor, M. Verschuren and D. Waltham for discussion and comments. Sharon Capon drafted the figures. Department of Earth Sciences contribution no. 4075.

References

Kerr, H. G. & White, N. 1992. Laboratory testing of an automatic method for determining normal fault geometry at depth. *Journal of Structural Geology* **14/7**, 873–885.

—— & —— 1994. Application of an automatic method for determining normal fault geometries. *Journal of Structural Geology*, **16**, 1691–1709.

—— & —— 1995. Application of an inverse method for calculating three-dimensional fault geometries and extension directions: the Nun River field, Nigeria.

Bulletin of American Association of Petroleum Geologists, in press.

——, —— & BRUN, J. P. 1993. An automatic method for determining 3-D normal fault geometries. *Journal of Geophysical Research*, **98**, 17837–17857.

MCCLAY, K. R. & ELLIS, P. G. 1987. Analogue models of extensional fault geometries. *In*: COWARD, M.P ., DEWEY, J. F. & HANCOCK, P. L. (eds) *Continental Extensional Tectonics*. Geological Society, London, Special Publications, **28**, 109–125.

PARKER, R. L. 1977. Understanding inverse theory. *Earth & Planetary Science Letters*, **5**, 35–64.

PRESS, W. H., FLANNERY, B. P., TEUKOLSKY, S. A. & VETTERLING, W. T. 1986. *Numerical Recipes: The Art of Scientific Computing*, Cambridge University Press, New York. 294–301.

ROBERTS, A. G., YIELDING, G. & FREEMAN, B. (eds) 1991. *The Geometry of Normal Faults*. Geological Society, London, Special Publications, **56**.

SCALES, L. E. 1985. *Introduction to non-linear optimization*, Macmillan Publishers Ltd, London and Basingstoke.

SYMON, K. R. 1971. *Mechanics*, Addison Wesley, Reading, Mass.

VERRALL, P. 1981. Structural interpretation with applications to North Sea problems. *Joint Association for Petroleum Exploration Courses*, **3**, 1–156.

WALTHAM, D. 1990. Finite difference modelling of sandbox analogues, compaction and detachment-free deformation, *Journal of Structural Geology*, **12**, 375–381.

WHITE, N., JACKSON, J. A. & MCKENZIE, D. P. 1986. The relationship between the geometry of normal faults and that of the sedimentary layers in their hanging walls. *Journal of Structural Geology*, **8**, 879–909.

Analysis of fault geometry and displacement patterns

D. T. NEEDHAM, G. YIELDING & B. FREEMAN
Badley Earth Sciences Ltd, North Beck House, North Beck Lane, Hundleby, Spilsby, Lincolnshire PE23 5NB, UK

Abstract: Interpretation of the subsurface is still often performed on suites of vertical seismic sections with structures being correlated from line to line to build up a map view at a particular horizon. The advent of 3D seismic surveys and routine use of interpretation software has made it common to produce 'horizon-based' interpretations, using auto-tracking techniques. Neither of these approaches guarantees a three-dimensionally consistent interpretation because of ambiguities in lateral and vertical correlation of structures. More reliable descriptions of a faulted surface can be obtained if the three-dimensional properties of a fault network are observed directly. Fault traces picked on vertical sections can be used to model the fault surfaces, which are then viewed in perspective to give an immediate assessment of the plausibility of the interpreted fault geometry. By associating horizon terminations with their respective fault segments this cut-off information can be shown on the perspective view of the fault surface, allowing the interpreter to examine in detail how reservoir layers are disrupted and juxtaposed by the fault network. The horizon cut-off information can also be used to calculate displacement-related values (throw, heave, separation) which may be gridded over the entire picked surface in order to produce displacement maps, also in the perspective view. Isolated single faults have simple displacement patterns with a maximum near the fault centre. Fault arrays show partitioning of displacement between the various splays, with relatively abrupt changes in the displacement at branchlines. Long 'single faults' are frequently shown to be segmented into en-echelon arrays rather than being simple continuous structures. Antithetic faults exhibit a downward decrease in displacement towards a tip line near their master fault. In areas of cross-cutting fault trends, displacement patterns can discriminate between different possible kinematic histories. Fault displacement can also be used to calculate other attributes such as fault seal potential, which can also be posted onto the perspective view of the fault surface. These analyses of fault geometry and displacement allow a more objective assessment of subsurface interpretation. They do not rely on particular assumptions about deformation mechanisms, only on the observation that displacement variation on individual fault surfaces is systematic.

Interpretation of seismic data aims to generate a full three-dimensional portrayal of the structure of a horizon or series of horizons. This structure may be complicated by arrays of faults offsetting the horizons. Assessing the three-dimensional form of these faults is crucial as they control the geometry of potentially hydrocarbon-bearing closures and themselves may act as conduits or seals during migration. The three-dimensional shape of a fault, its linkage with other faults and the distribution of slip on its surface are important in that a fault is often critical in defining the geometry of the intervening beds. Displacement variations can also reveal whether a fault has developed as an isolated structure or is kinematically linked with other faults (Walsh & Watterson 1991). The degree to which apparently single faults are segmented into en-echelon arrays can also be examined by displacement analysis. Assessing the linkage/segmentation of apparently single faults is vital when determining the integrity of a prospect or investigating the degree of fault compartmentalisation within a field. Fault-seal characteristics are dependent on the magnitude of displacement at any point on a fault surface, the lithologies which have slipped past that point and the reservoir pressure regime.

Most traditional methods of determining the subsurface structure involve the interpretation of horizons and faults on vertical seismic sections. Faults are correlated from line to line to build up a map view of the horizon. This correlation may be arbitrary, involving the interpreter's judgement of the 'structural style' of a particular area. For any cross section to be valid, the three dimensional continuity and connectivity of the fault system should be assessed. Methods of section validation relying only on geometrical analysis of a single section cannot test the impact of that interpretation on adjacent sections. Fault correlation poses a particularly difficult problem in areas of two-dimensional seismic data where the fault density is high but the lines are relatively widely spaced. In this case there is a spatial aliasing problem and line-to-line correlations become difficult to make. An independent means of testing correlations is

From Buchanan, P. G. & Nieuwland, D. A. (eds), 1996, *Modern Developments in Structural Interpretation,Validation and Modelling,* Geological Society Special Publication No. 99, pp. 189–199.

necessary. The correlation problem has been mostly overcome with the routine use of three-dimensional seismic datasets with line spacings of less than 50 m. If the 3D data quality is good, the correlation of faults from line to line becomes less of a problem, although by no means non-existent. Decisions on fault linkage remain largely subjective. In areas where the data quality is sufficiently good horizons can be auto-tracked. Auto-tracking is a tool that allows the interpretation of a horizon automatically by following a pre-set character on a seismic reflection (e.g. a maximum amplitude). It aims to produce a continuously interpreted horizon surface with the faults represented by zones of higher dip. This horizon-based style of interpretation allows the recognition of small scale faults by the use of attribute maps displaying dip and azimuth variations which can reveal subtle detail (Dalley *et al.* 1989). Despite advances, three-dimensional auto-tracked interpretations produce a new set of barriers to structural interpretation. The main problem of fault correlation becomes one of vertical association between horizons rather than lateral association between lines. This is because faults are not necessarily picked on vertical sections to link the structural interpretation between horizons. Hence, when using an horizon-based approach there is the danger that the fault pattern drawn for one horizon may not correspond to that produced for adjacent surfaces. This is increasingly likely if the fault polygons on each horizon map are hand digitised during the mapping process. An additional problem with hand-interpreted polygons is that they often show heaves that are wider than the true heave on the fault that they represent. For the three-dimensional geometry of a fault to be accurately defined we need to use the information from each of the horizons cut by a fault and have some independent method to be able to assess whether the chosen correlation is viable. The best approach to this is to view the displacement pattern on the fault surface and to analyse the likely causes of variations in this pattern.

The pattern of displacement on a single fault surface is now well documented (Watterson 1986; Barnett *et al.* 1987). An ideal, blind fault in an isotropic material has a tip-line showing an elliptical form (Fig. 1). Displacement is greatest in the centre of the fault and decreases to zero at the tip. In nature, many faults show close approximations to this form when allowances are made for the likely causes of deviation from the ideal such as lithological contrasts, mechanical differences between layers and seismic velocity variations. Other, more complex patterns exist where faults interact, i.e. they are kinematically linked. Overall, fault displacement patterns vary in a systematic manner and variations should be attributable to real geological causes such as fault linkage or intersection. If non-systematic variations exist which cannot be accounted for by the geology, they should be investigated to determine whether they result from interpretation errors.

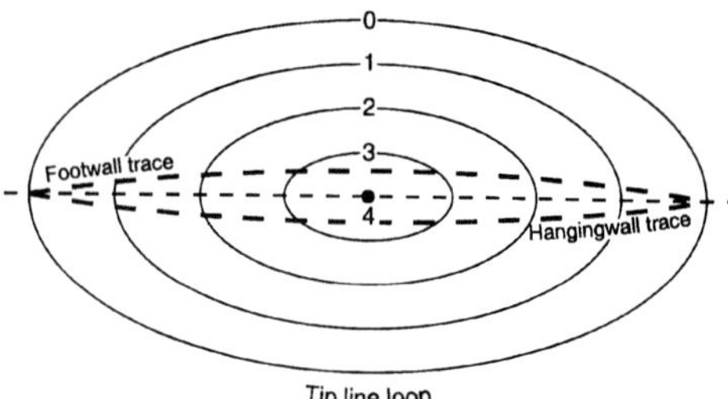

Fig. 1. Displacement contours on an idealized fault plane viewed perpendicular to the fault surface (after Barnett *et al.* 1987). The tip line has an elliptical form and displacement increases towards the centre of the fault. The horizon separation also increases towards the centre.

In this contribution we intend to show how better use can be made of existing subsurface data to obtain fault patterns that are reliable and constrain the geometry and properties of fault surfaces between the mapped horizons. We consider a number of situations of increasing complexity, building up from the analysis of a single fault to a consideration of the effects of fault interaction and intersection on displacement patterns. The examples described are taken from 2D and 3D seismic interpretations of oilfields in extensional basins.

Prior to the analysis, the interpretation itself needs to be optimized. Both the section and horizon based interpretations contain potential errors which do not allow an accurate representation of the fault surface. Figure 2 shows two seismic interpretations, one is a section-based manually picked interpretation, the other auto-tracked. The digitised horizons on the section-based interpretation stop short of the picked fault traces. This is a common feature of many interpretations and results, in part, from the inability of some seismic interpretation software easily to handle overlaps of the same horizon. The interpreter therefore stops short of the fault to minimise the chances of overlap. Making an horizon map by using the width of the uninterpreted zone around the fault to define the fault polygon would be incorrect. The fault polygons would be too wide. This leads to an over-estimation of the fault heave and consequently the amount of extension in an area. Fault polygons that are too wide also result in an incorrect calculation of

(a)

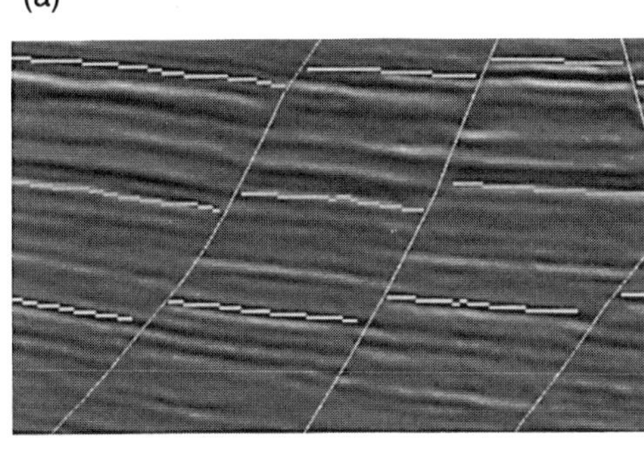

(b)

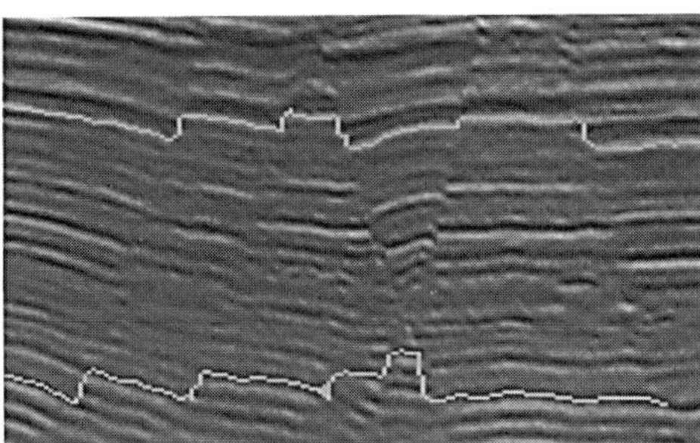

Fig. 2. Seismic sections showing (**a**) manual section-based interpretation with faults picked. Note that the horizon picks stop short of the fault traces. (**b**) Auto-tracked horizon-based interpretation with faults represented as steeply dipping segments. Note the irregularity of the auto-tracked horizon along the faults compared to the clarity with which they are imaged.

the fault dip. In turn, this can lead to an incorrect estimation of the volume of a reservoir interval. The auto-tracked interpretation is little better. Auto-tracking gives a good interpretation of the horizons between the faults, if the data quality is good. However, the auto tracking gives a very poor representation of the faults and these could be picked much more accurately by hand on the vertical section. The poorly defined fault plane geometry increases the difficulty of horizon to horizon correlation. This problem is likely to be worse in areas of poor data quality.

We suggest that interpretation of faults should be performed on vertical sections, even in a situation where the interpretation is horizon based, as this results in a more accurate definition of the structure. The horizon-to-horizon correlation problems are therefore avoided. Once the interpreted faults are in place a polygon of the correct width can be generated by projecting the horizons so that their cut-offs lie on the interpreted fault surface. The horizons are then said to be 'snapped' onto the fault. This results in the structure being much more accurately defined. Snapping can be carried out automatically by extrapolating the trace of the horizon using the last few interpreted data points as a control. The effect of horizons snapped onto the fault is illustrated in Fig. 3. The upper map (Fig. 3a) shows a subsampled 3D seismic grid with horizon breaks and interpreted faults. The fault polygons are as wide as the break in the interpreted horizon for each fault. Figure 3b shows the same map with the horizons snapped on to the interpreted faults. The resultant polygons are much narrower than those resulting from the unsnapped section-based interpretation or even hand-digitised onto an auto-tracked surface. Once the position of the horizon cut-offs have been accurately defined, analysis of fault displacement patterns can begin.

Single faults

Analysis of displacement patterns on faults provides a powerful method of testing the viability of lateral correlations (Barnett *et al.* 1987; Freeman *et al.* 1990; Walsh & Watterson 1991). Displacement data can be extracted from each horizon on the individual fault traces. A series of these fault traces form the correlated fault. The

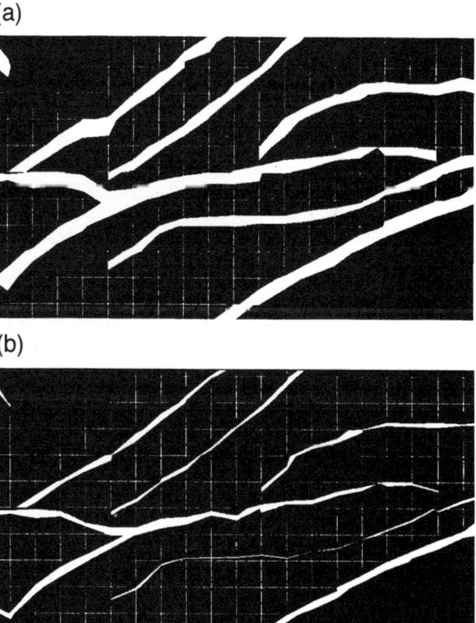

Fig. 3. Comparison of fault polygons for horizons (**a**) not snapped on to the faults and (**b**) snapped on to the fault surfaces. The polygons in (b) are much more accurately defined and therefore much narrower.

displacement data from each of these traces can then be gridded and contoured to produce a continuous map of displacement variation on the fault surface. Unless the displacement vector is well constrained, which is rarely the case in subsurface interpretation, it is usually best to use a component of the displacement such as apparent throw or apparent heave. This also overcomes the problem of working in mixed time-distance units, which would be the case when trying to analyse slip on a dipping fault plane in the time domain. Throw values can be corrected for the displacement direction if a slip vector is known or assumed. Snapping the horizons onto the fault surface not only ensures their accurate placement on the modelled fault plane but also optimises the measurement of throw for each horizon.

It is also useful to analyse the fault surface topography. Few faults are exactly planar. Figure 4 shows a series of fault traces, extracted from interpreted seismic lines, projected on to a perspective view of the fault surface. The fault is shaded using the horizontal gradient of the surface topography (shaded relief) which gives the appearance of it being illuminated from the left side. A fault grid, constructed from the raw data points on the fault traces, is also shown. The fault surface topography varies independently of the line traces and therefore is a real feature rather than an artefact of the interpretation. In general, abrupt topographic variations on the fault surface should be checked to see if they result from mis-picks. Other variations in topography, such as that shown in Fig. 4, may relate to and help constrain the slip-vector on the fault surface. In this example there is a series of consistent, steeply pitching 'corrugations' on the fault surface which we believe indicates fault slip in the direction of dip. Slip oblique to the fault surface undulations would be more difficult since it requires larger strains in the wall-rocks.

Fault displacement can be assessed by posting the hangingwall and footwall cut-offs of all the faulted horizons onto the fault surface (Fig. 5a). Using the horizons projected on to the fault surface (i.e. those that are snapped) gives the most accurate representation of the horizon traces. The resulting diagram shows that separation increases smoothly towards the centre of the fault on all horizons. The fault surface grid and the shaded topography indicate that this fault is a relatively smooth feature

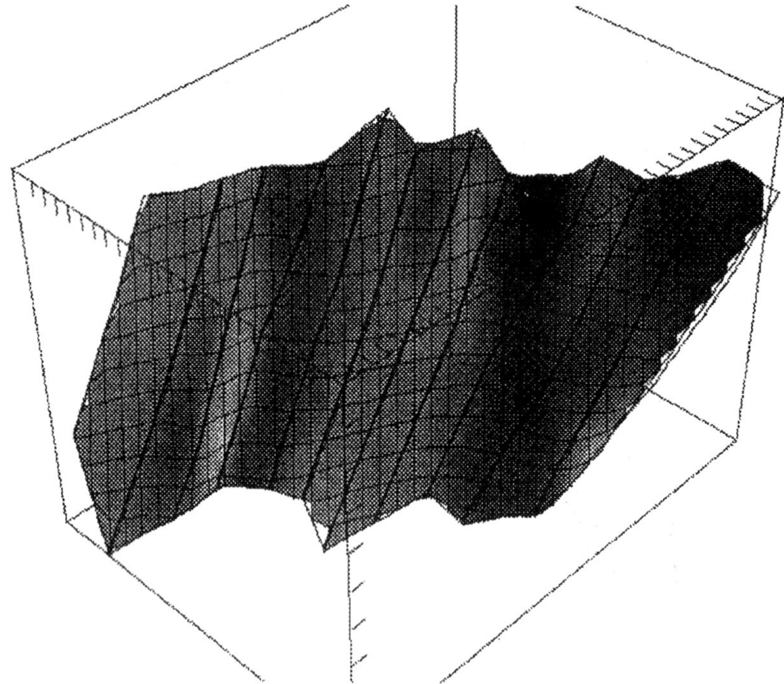

Fig. 4. Fault surface topography shown by shaded contours of the horizontal derivative of topography (giving the appearance of illumination from the left). The thick lines are the interpreted fault traces on individual seismic lines. The gridded representation of the fault surface is also shown. The shading picks out a number of corrugations that pitch steeply on the fault surface. These topographic variations may represent the displacement vector on the fault.

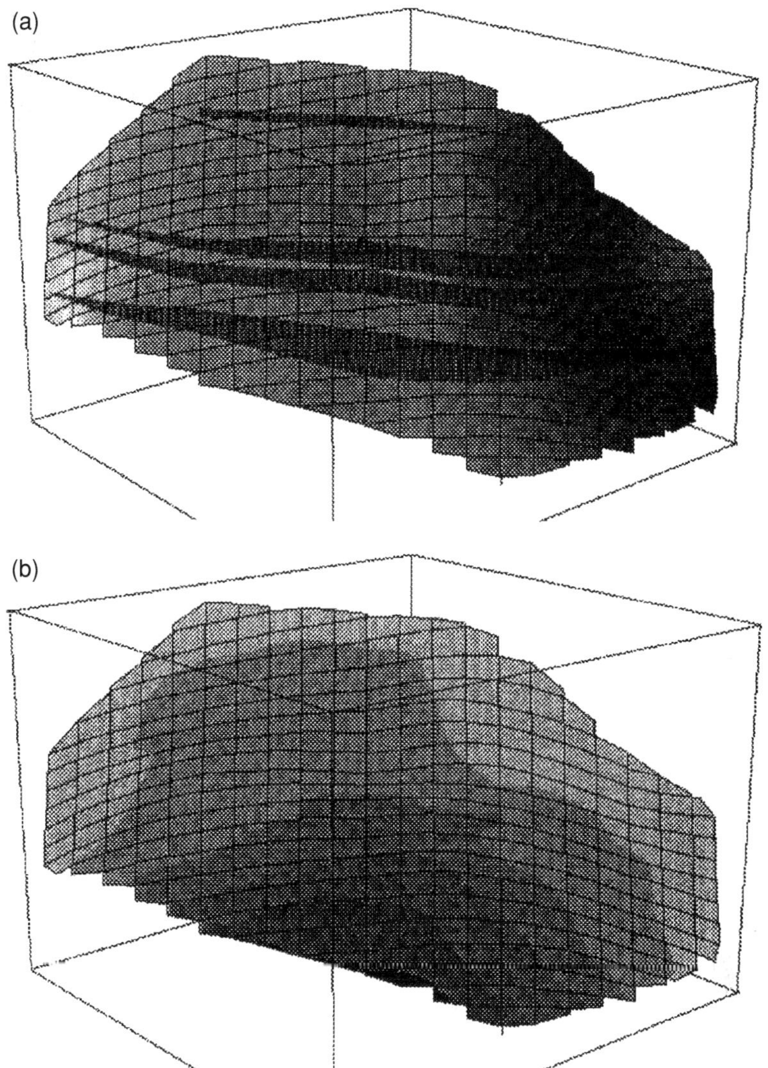

Fig. 5. Perspective view of an isolated fault surface. (**a**) The fault is shaded to show the surface topography and the horizon separations are shown as dark polygons on the fault surface. The separations increase smoothly towards the centre of the fault. (**b**) Contoured throw on the fault surface. The throw increases systematically from low values (light) to high values (dark). The fault continues below the deepest interpreted seismic horizon.

with few sharp surface undulations. The raw data provided by the horizon separations can then be gridded to produce a map of throw on the fault surface. The pattern of throw on this isolated fault shows a regular, broadly elliptical pattern (Fig. 5b). The shading (dark-high to light-low) indicates the variation in throw. Throw reaches a maximum in the centre of the fault and decreases smoothly towards the tips. Only the upper part of the roughly elliptical fault surface is visible as the lower part extends below the lowest interpreted horizon. This is a common feature and should be remembered during fault analysis. Using the lateral and vertical displacement gradients it is possible to estimate the extent of the fault. In most areas of interest, the structure is more complex and faults tend to interact.

The previous examples showed single, isolated faults. In some cases, however, even apparently single faults may show more complex displacement patterns. A fault correlation may have been made on the basis that a series of fault cuts lie along the same trend (Fig. 6a). This is a perfectly reasonable starting assumption and it is one that is readily tested by displacement analysis. The displacement pattern from this fault shows two throw maxima (Fig. 6b) which strongly suggests that the correlated structure in fact consists of two faults. It is impossible to say that the faults have not grown to a point where they have coalesced into a single structure. Even careful examination of the crucial lines on the 3D survey may not be able to resolve this question. The important point is, however, that displacement analysis has shown that continuity cannot be reliably assumed for the central part of the fault in the initial interpretation. The possibility that a migration/spill pathway exists in a relay zone between the two segments cannot be ruled out and should be considered in risk estimates. Cross-sections in the central area could include two small displacement faults rather than a single structure. The adjusted fault map with the fault split into two segments is shown in Fig. 6c.

An example using an interpreted 2D survey from the Western Desert of Egypt where spatial aliasing is a problem is shown in Fig. 7. Spatial aliasing is a

(a)

(b)

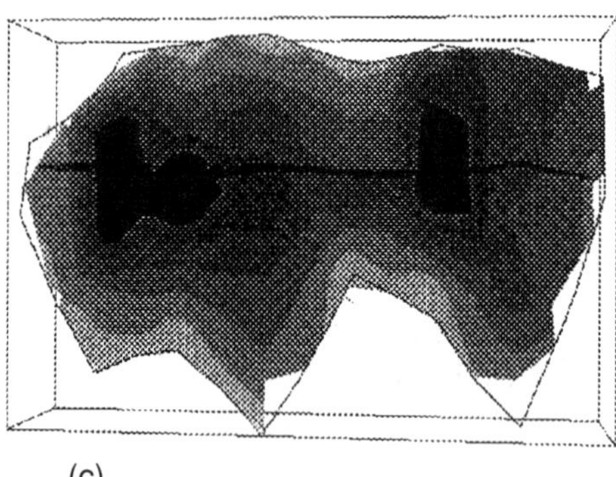

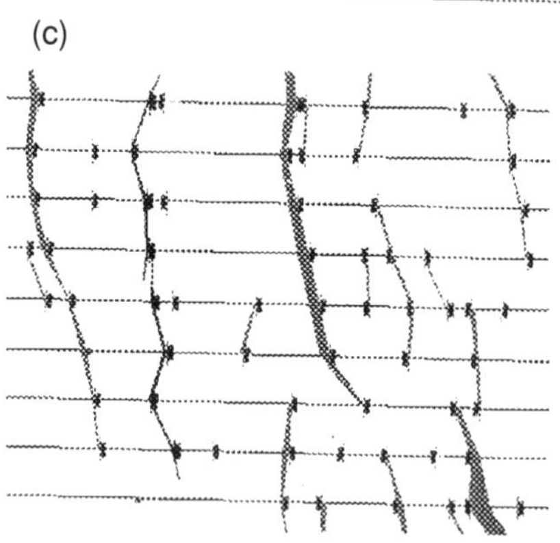

Fig. 6. (**a**) Map of initial fault interpretation showing fault cuts correlated as a single fault (highlighted in black). (**b**) The pattern of throw on the fault distinguished by banded contour pattern. Two maxima are developed suggesting that it is really two separate faults. (**c**) Fault recorrelated to show two separate structures.

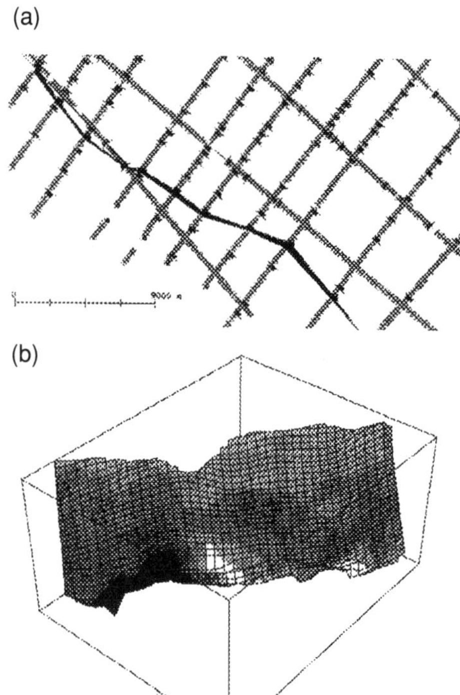

Fig. 7. (**a**) Map of fault cuts on a series of 2D-seismic lines. A set of fault cuts have been correlated to form a single, continuous structure (highlighted in black). (**b**) Perspective view of the fault surface with the pattern of throw on this fault (dark–high to light–low) displayed. The pattern is irregular with more than one area of high throw. The fault should be split into more than one segment.

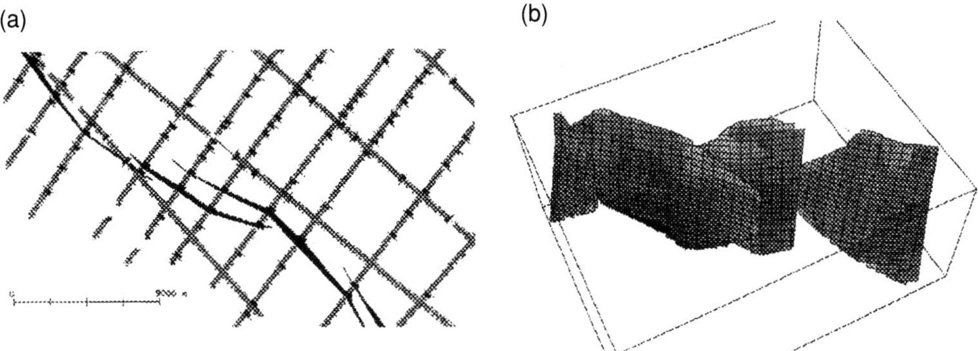

Fig. 8. (**a**) The single fault shown in Fig. 7 recorrelated as a series of en-echelon faults. (**b**) The throw pattern now shows that each fault has its own displacement maximum.

problem because of the large line spacing. A series of fault cuts which lie along the same trend are initially interpreted to form part of the same structure. This is tested by displacement analysis which suggests that the interpretation should be reassessed. The contoured throw pattern shows a series of high displacement areas separated by marked lows. This pattern suggests than more than one fault forms part of the correlated structure. Re-examination of the data and a different

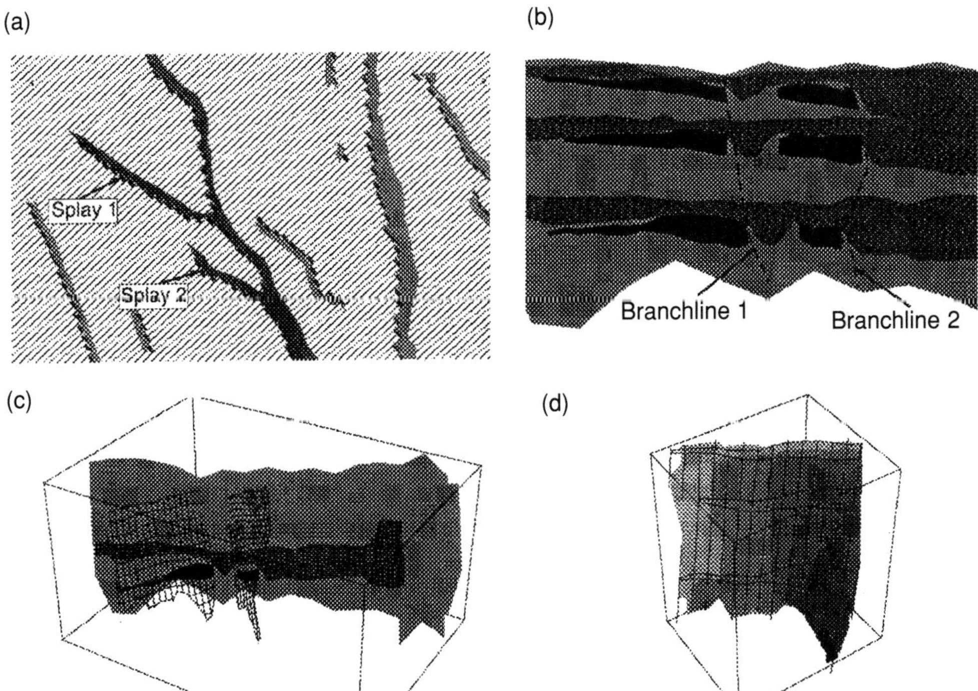

Fig. 9. Displacement patterns around splaying faults. (**a**) Map of a fault with two splays (darker than other faults) from a 3D-seismic dataset. (**b**) Strike-projection of horizon separations for three horizons on the main fault (dark grey) and the splays (black). There is a systematic decrease in separation at the branchlines with the splays. (**c**) Perspective view of the master fault (solid fill) and the splays (wireframe grids) showing the variation in separation on the lower horizon. (**d**) Throw map of the northernmost splay (splay 1) showing a systematic increase towards the branchline with the master fault (light to dark shading).

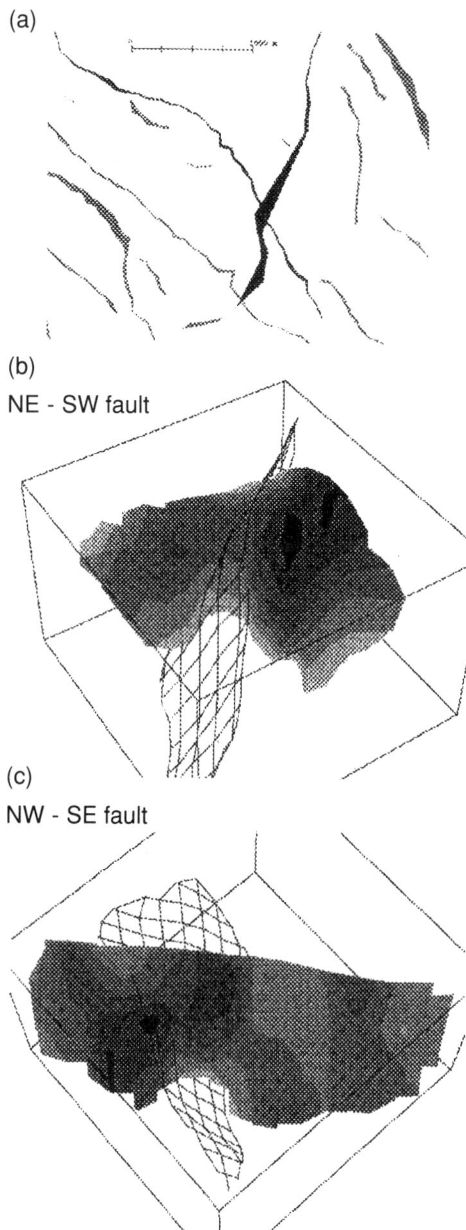

Fig. 10. (**a**) Map showing two cross-cutting faults (black). (**b**) Pattern of throw (dark–high, light–low) on the NE–SW fault. Throw decreases abruptly at the intersection with the NW–SE fault (shown as a wire-frame). There are separate displacement maxima on either side of the cross-cutting fault. (**c**) Pattern of throw on the NW–SE fault. Again, the displacement abruptly changes at the fault intersection. This indicates that both faults have been active during the same period and there is not just a simple cross-cutting relationship.

approach to the correlation gives an en-echelon pattern of four overlapping faults, each with its own displacement maximum (Fig. 8). The displacement now varies in a more systematic manner on each of the faults. The wireframe grid which displays the fault surface topography shows that the re-interpretation gives a series of smoother, more planar, fault surfaces in contrast with the more irregular fault generated by the initial correlation. This analysis does not, in itself, prove that the fault zone is en-echelon: the individual faults may have linked together to produce a now-continuous fault. However, examining displacement and geometry in this way can alert the geologist to alternative interpretations, and hence allows a more objective assessment of the risk associated with the mapping of structures. Fault topography is also valuable in assessing linkage where the displacement pattern indicates that there may be more than one segment in a particular correlated fault. Topographic variation may show that the fault consists of multiple en-echelon segments rather than being a single structure.

Fault systems

In most areas faults link or bifurcate into a series of master faults and splays. Faults which physically join are said to be 'hard-linked' but many of the observations of displacement variation made for hard linked faults are also applicable to 'soft-linked' faults. 'Soft-linked' faults are faults which are kinematically but not physically linked. Faults which intersect at high angles may be sometimes erroneously taken as an indication of polyphase faulting with one trend cross-cutting another.

The effect of splays on fault displacement is shown in Fig. 9. In the example shown, two splays link with a master fault (Fig. 9a). A strike-projection of the horizon separations on the master fault (Fig. 9b & c) shows that they decrease abruptly at the branchlines with the splay faults. This is a good criterion for recognizing the presence of linking faults. The displacements on the splays also vary in a systematic manner (Fig. 9c & d). Throw increases smoothly from zero at the splay tips and reaches a maximum at the branch-line. When viewed together the horizon separations from both main fault and splays show a systematic variation (Fig. 9b). The summed throw on the main fault and splay is equal to the throw expected if only a single, main fault had existed. Displacement is partitioned between the main fault and splays. This relationship is clearly demonstrated for all three mapped horizons cut by these faults. There appears to be a slight short-fall in the amount of summed throw on the upper horizon. Such discrepancies can be attributed to a number of causes. An incorrect seismic pick on this horizon is a possibility which has to be checked by recourse to the original data. A mis-pick can be ruled out in this case as the data quality is good. A geological cause is more likely. Some of the displacement could be accommodated by ductile deformation around these faults and so the observed slip on the faults is less than the expected total. This ductile deformation can take a number of forms. Ductility is scale dependent (Lister & Williams 1983) and deformation which appears ductile, i.e. with no discontinuities, at the seismic scale may well be accommodated entirely by sub-seismic faulting. The master fault may be surrounded by an array of small faults all below the scale of seismic resolution. There is evidence from field studies that small faults cluster around larger faults (Jamison & Stearns 1982, Gillespie *et al.* 1993) and so this is a plausible explanation. In rocks such as shales the deformation may be ductile even at the outcrop scale of observation. The main fault and the splays are a good example of kinematically linked faults where the displacement is partitioned.

Whilst splays may be recognised as part of the same kinematically linked system, faults which intersect at higher angles are often ascribed to a different phase of faulting or are labelled as 'transfer faults'. Figure 10 shows two faults intersecting at a high angle, one oriented NW–SE, the other NE–SW. Superficially, these may be interpreted as the result of two phases of faulting. The displacement patterns on these two faults reveal, however, that they formed coevally. The NW–SE-trending fault shows an abrupt variation in displacement at the intersection with the NE–SW fault. Similarly the NE–SW fault also shows a

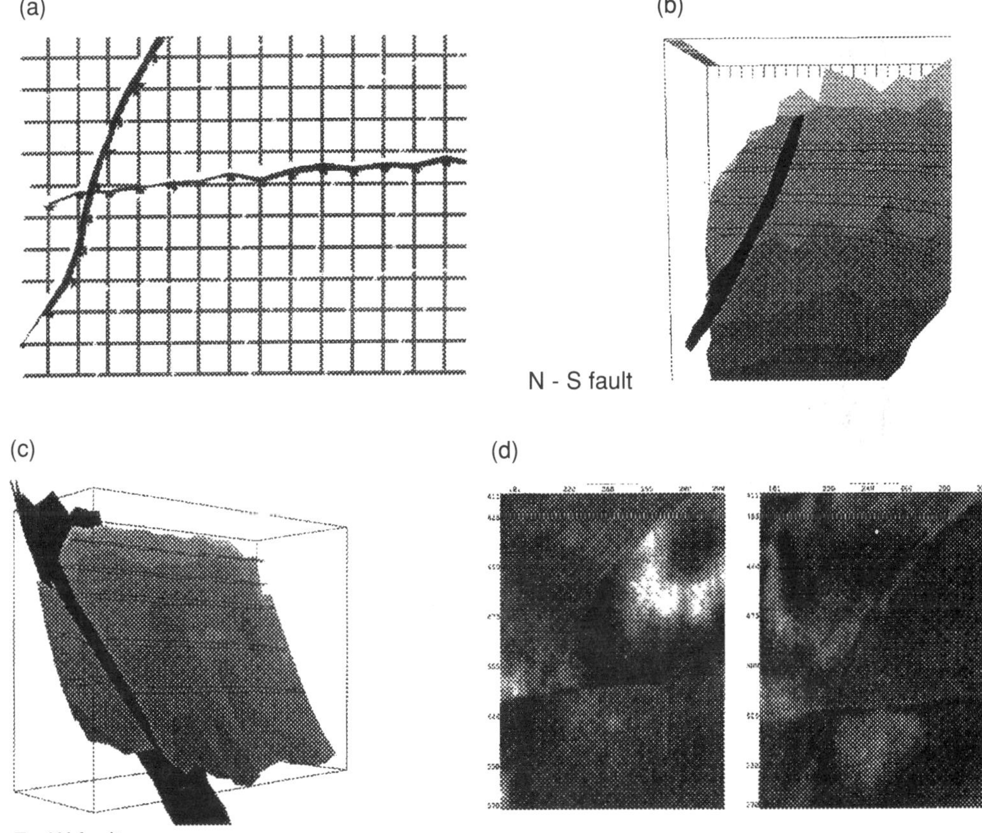

Fig. 11. (**a**) Map of cross-cutting faults. (**b**) The N–S fault shows a systematic variation in fault throw apparently unaffected by the E–W fault. (**c**) The E–W fault shows a systematic variation in throw. The throw pattern is not affected by the N–S fault but horizon traces are offset. (**d**) Two timeslices through the 3D-seismic data volume showing the cross-cutting relationship between the faults.

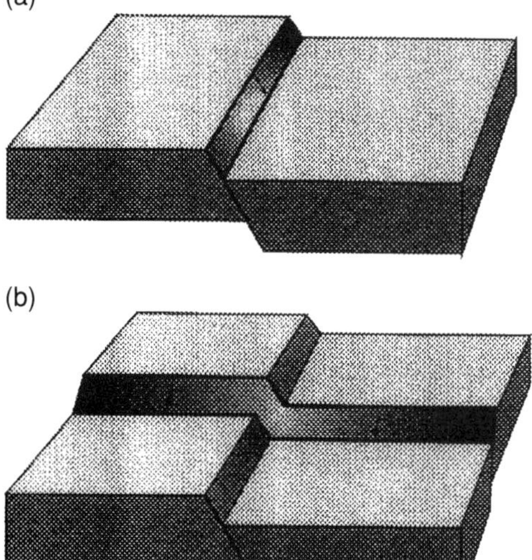

Fig. 12. Block diagram showing the effects of a cross-cutting faults. An initial fault (**a**) is cross-cut by a later fault (**b**). The displacement pattern on the early fault varies in a systematic manner but is offset by the later fault. Throw varies systematically on the younger fault but the horizon separations are offset due to the presence of the earlier fault.

variation in displacement at the branchline. This suggests that all four fault segments were interacting during the same phase of faulting with slip-events switching between faults in a non-specific manner so that a clear cross-cutting relationship cannot be defined. Both faults grew to approximately the same dimensions during this period.

The interaction shown by these two faults can be contrasted with a case in which one fault can clearly seen to post-date another. This situation is shown in Fig. 11, taken from a Gulf Coast 3D dataset. A fault trending N–S is cross-cut by an E–W-trending fault (Fig. 11a). This cross-cutting relationship can be identified on both vertical 3D seismic lines and horizontal time-slices through the data volume (Fig. 11d). There is a small apparent lateral offset of the N–S fault across the E–W fault but the displacement is believed to be dip-slip since sedimentary channel features, recognised on higher time-slices through the data volume, show no deflection across the E–W fault. The deflection is purely a consequence of faulting a dipping surface, another fault in this case. Polyphase faulting is shown in block diagram form in Fig. 12. The displacement patterns on the faults contrast with the coeval faults shown above. The later E–W fault offsets a smoothly varying displacement profile on the N–S fault. There is however, no abrupt change around the fault intersection line. There is a similar smooth pattern on the E–W fault itself. The contrasting displacement patterns between these two examples allows coeval and polyphase faults to be distinguished.

A common feature of many fault arrays is the presence of minor antithetic faults. Many interpretations show these antithetic faults linking to the master fault. This geometry is difficult to reconcile

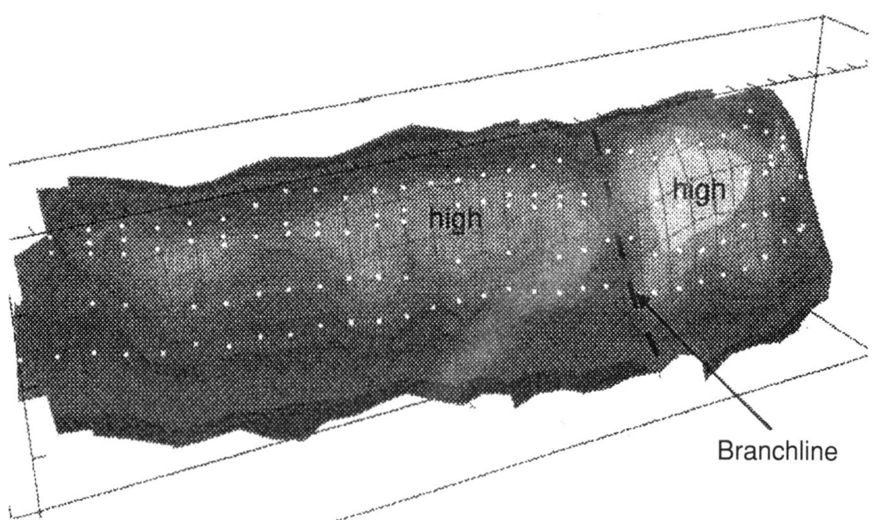

Fig. 13. Throw pattern on an antithetic fault shown by banded contour pattern. Displacement decreases downwards to a lower tip without the fault linking to the major fault. Raw throw data points are shown as white dots. The fault grid is also shown. The displacement low towards the right hand side is due to the presence of a splay.

as it gives displacement compatibility problems (i.e. the slip vector of the antithetic is typically almost perpendicular to the plane of the master fault). Our studies of antithetic faults suggest that the antithetic faults decrease in displacement towards the master fault and die out at a tip without linking to that master fault (Fig. 13). Such a geometry allows the faults to be kinematically linked without the problems of physical 'hard-linkage' of the two faults. Apparent antithetic faults that do intersect with the master fault probably form part of a conjugate fault in the area of overlapping faults with opposed dips (Nicol *et al.* 1995). In such cases the 'antithetic' fault is cross-cut by the master fault and should be present in the footwall as well as the hanging-wall.

The three-dimensional representation of the entire fault surface allows various attributes to be mapped on to that surface. The examples we have used primarily show throw or horizon separation. The fault surface topography or deviation of that topography from a best fit plane can also be displayed. The usefulness of this is in identifying areas of abrupt shape change which may be a result of mis-picks of the fault. Alternatively these shape changes may be real and represent areas in which initially isolated faults have coalesced. Topographic variations may also help constrain the displacement vector. Such interpretations can be made when the topography is viewed in conjunction with fault throw data. Other fault attributes are dependent upon displacement, particularly the ability of the fault to act as a seal.

The techniques that we have applied to analyse the validity of the interpretations are not model-driven. No particular mode of deformation or fault geometry is assumed. The broadly elliptical fault surface geometry is well established from numerous investigations including high-resolution subsurface coal mine plans and seismic data (Barnett *et al.* 1987) and the kinematic coherence of fault arrays is supported by direct observation (Walsh & Watterson 1991).

Conclusions

The use of section-based or horizon-based interpretations does not in itself guarantee three-dimensional consistency, with lateral or vertical fault correlation a particular problem.

'Snapping' of horizon terminations onto fault surfaces dramatically improves the accuracy of mapped fault polygons.

Analysis of displacement patterns on single or multiple fault surfaces is a powerful way of testing lateral fault correlations in 2D datasets and vertical correlations in 3D datasets.

Long 'single' faults are often segmented into en-echelon arrays and displacement is also partitioned between splays.

Other displacement related attributes can be calculated on parts of the fault surface between mapped horizons e.g. fault seal potential.

These techniques do not assume any particular model of deformation but make better use of the existing subsurface interpretation.

The interpretations shown are based on seismic data kindly provided by Enterprise Oil, GECO-Prakla and Statoil. J. Cartwright and C.F. Kluth are thanked for their constructive comments on the initial version of this paper.

References

BARNETT, J. A. M., MORTIMER, J., RIPPON, J. H., WALSH, J. J. & WATTERSON, J. 1987. Displacement geometry in the volume containing a single normal fault. *American Association of Petroleum Geologists Bulletin*, **71**, 925–937.

DALLEY, R. M., GEVERS, E. C. A., STAMPFLI, G. M., DAVIES, D. J., GASTALDI, C. N., RUIJTENBERG, P. A. & VERMEER, G. J. O. 1989. Dip and azimuth displays for 3D seismic interpretation. *First Break*, **7**, 86–95.

FREEMAN, B., YIELDING, G. & BADLEY, M. E. 1990. Fault correlation during seismic interpretation. *First Break*, **8**, 87–95.

GILLESPIE, P. A., HOWARD, C. B., WALSH, J. J. & WATTERSON, J. 1993. Measurement and characterisation of spatial distributions of fractures. *Tectonophysics*, **226**, 113–141.

JAMISON, W. R. & STEARNS, D. W. 1982. Tectonic deformation in the Wingate sandstone, Colorado National Monument. *American Association of Petroleum Geologists Bulletin*, **66**, 2584–2608.

LISTER, G. & WILLIAMS, P. F. 1983. The partitioning of deformation in flowing rock masses. *Tectonophysics*, **92**, 1–33.

NICOL, A., WALSH, J. J., WATTERSON, J. & BRETAN, P. G. 1995. 3-D geometry and growth of conjugate normal faults. *Journal of Structural Geology*, **17**, 847–862.

WALSH, J. J. & WATTERSON, J. 1991. Geometric and kinematic coherence and scale effects in normal fault systems. *In*: ROBERTS, A. M., YIELDING, G. & FREEMAN, B. (eds) *The Geometry of Normal Faults*, Geological Society, London, Special Publications, **56**, 193–203.

WATTERSON, J. 1986. Fault dimensions, displacement and growth. *Pure and Applied Geophysics*, **124**, 365–372.

Recent advances in analogue modelling: uses in section interpretation and validation

K. R. McCLAY

Fault Dynamics Project, Geology Department, Royal Holloway University of London, Egham, Surrey, TW20 0EX, UK

Abstract: Scaled analogue modelling is a powerful tool for the understanding of the geometric and kinematic development of extensional, strike-slip and inverted fault systems, as well as for some thrust systems. In this paper the results of 2D and 3D scaled analogue model experiments designed to simulate listric fault systems, rift fault systems, and inverted listric faults are reviewed. Simple 2D rift models are characterized by asymmetric rift graben structures with planar rift border faults and domino-style intra-rift fault systems. 2D listric fault models produce roll-over anticlines with associated crestal collapse grabens whereas the ramp-flat listric fault models produce roll-over anticlines associated with each concave-up segment of the main fault system and also a syncline above the convex-up ramp section of the main detachment. The 2D analogue models are directly comparable to natural examples of extensional fault systems. The results of simple 3D analogue models of an orthogonal and a 60° oblique rift are presented. Orthogonal rift models are characterized by both rift-border and intra-rift faults that are nearly orthogonal to the extension direction. Like-dipping intra-rift faults overlap to form relay ramps that transfer displacement between the faults. In contrast oblique rift models are characterized by en-echelon arrays of segmented extensional faults. Rift border fault systems are parallel to the underlying zone of basement stretching whereas the intra-rift faults are segmented and at a high angle to the extension direction. Oblique accommodation zones formed by interlocking conjugate fault arrays separate individual depocentres within the models. No strike-slip or oblique-slip transfer faults are found in these models. The 3D rift models produce fault patterns that are comparable with natural examples of rift systems. Analogue modelling of inverted listric faults is described and the results compared with natural examples of inverted fault systems. The results of the scaled analogue models described in this paper provide important constraints on section interpretation and restoration in that they permit an understanding of the geometric and kinematic evolution of fault (and fold) structures as well as indicating the sequences of faulting and activity on faults. They provide templates for the interpretation of seismic sections in regions of poor data quality and for projecting surface data into the sub-surface.

Ever since the pioneering model studies of Cadell (1889) who conducted sandbox experiments designed to investigate how the thrust and fold systems of the Scottish Highlands formed, scaled analogue modelling has been commonly (though somewhat intermittently) used to study the geometric, kinematic and dynamic evolution of most geological structures, from extensional fault systems (e.g. Cloos 1928; Cloos 1968; Horsfield 1977, 1980; Faugère & Brun, 1984; Withjack & Jamieson 1986; McClay & Ellis 1987; Vendeville *et al.* 1987; Ellis & McClay 1988; Mandl 1988; Vendeville & Cobbold 1988; Serra & Nelson 1989; McClay 1990*a*, *b*; McClay *et al.* 1991; McClay & Scott 1991; Tron & Brun 1991; Vendeville 1991; Brun & Tron 1993; Lemon & Mahmood 1994; McClay & White 1995; Withlack *et al.* 1995), thrust faulting (e.g. Davis *et al.* 1983; Malavielle 1984; Mulugeta & Koyi 1987; Mulugeta 1988*a*, *b*; Colletta *et al.* 1991; Dixon & Liu 1992; Lallemand *et al.* 1992; Liu 1992; Liu *et al.* 1992; Marshak & Wilkerson 1992; Marshak et al. 1992; Calassou *et al.* 1993; Malavielle *et al.* 1993; Koyi 1995), strike-slip tectonics (e.g. Emmons 1969; Tchalenko 1970; Wilcox *et al.* 1973; Bartlett *et al.* 1981; Odonne & Vialon 1983; Naylor *et al.*, 1986; Reche 1987; Mann *et al.* 1983; Richard & Cobbold, 1990; Richard 1991; Richard *et al.* 1991; Schreurs 1994; Richard *et al.* 1995), inversion (e.g. Koopman *et al.* 1987; McClay 1989; Buchanan & McClay 1991, 1992; McClay & Buchanan 1992, McClay 1995), and salt tectonics (e.g. Talbot 1992; Vendeville & Jackson, 1992*a*, *b*; Weijermars *et al.* 1993; Jackson & Vendeville 1994).

Scaled physical models provide graphic and visual images of how complex geological structures may form in time and space. Such forward models enable the geologist to visualize the geometric and kinematic pathways by which a final complex structural configuration may be reached. In addition and in contrast to most numerical models, scaled analogue models of brittle deformation in

From Buchanan, P. G. & Nieuwland, D. A. (eds), 1996, *Modern Developments in Structural Interpretation, Validation and Modelling*, Geological Society Special Publication No. 99, pp. 201–225.

the upper crust allow new faults to form. Analogue models allow the sequences of faulting to be determined and produce templates for the interpretation of complex structures. This understanding is essential if realistic and retrodeformable balanced geological cross-sections are to be constructed through deformed terranes. In addition, detailed analysis of the deformation in analogue models gives an indication of the kinematics and mechanisms of deformation. Such data are essential for the determination of algorithms that can be applied in both forward models of fault systems and particularly in section balancing and restoration software.

The analogue modelling of strike-slip faulting has been recently reviewed in detail by Richard *et al.* (1995) and the reader is referred to that paper for further information. The analogue modelling of inversion structures has been reviewed by McClay (1995) and only a brief summary of the main aspects will be presented here.

Analogue model studies have proved to be extremely valuable in developing an understanding of the geometries and progressive evolution of extensional fault systems (e.g. McClay & Ellis, 1987*a*, *b*; Ellis & McClay 1988; McClay 1990*a*; Tron & Brun 1991; Lemon & Mahmood 1994; McClay & White 1995; Withjack *et al.* 1995; and others). In the present paper 2D analogue models of simple rift systems, listric faults and ramp-flat listric faults and 3D analogue models of rift systems are discussed. The 2D listric and ramp-flat listric fault models typically simulate upper-plate deformation in detached terranes such as those found in progradational deltas or above salt and shale decollements whereas the simple rift models replicate the structural architectures in intracontinental rift systems. 3D analogue modelling of rift systems has focused upon the development of orthogonal and oblique rifts in order to understand the development of fault patterns, the transfer of extensional displacements within the rift systems and in particular the along-strike variations in fault architectures and fault linkages.

Recent advances in scaled analogue models of fault systems have included the use of X-ray tomography (e.g. Richard *et al.* 1989; Colletta *et al.* 1991; Schreurs 1994; Richard *et al.* 1995), 3D reconstruction using advanced computer software (see below), detailed analysis of hangingwall deformation patterns (Lemon & Mahmood, 1994; McClay 1995; Dula 1991; Withjack *et al*, 1995), and improved modelling materials, particularly the use of silicone polymers to simulate plastic deformation in the lower crust (e.g. Weijermars 1986; Davy & Cobbold 1988, 1991; Tron & Brun 1991; White 1993; Schreurs 1994; and others). In particular the modelling of structures in 3D has provided valuable insights into the 3D geometries and evolution of complex fault systems (e.g. Calassou *et al.* 1993; McClay & White 1995; Richard *et al.* 1995) that can be applied to section construction and restoration.

This paper reviews the recent advances in analogue model studies and illustrates how they have been used to determine the geometric and kinematic evolution of complex fault systems in extensional and inverted terranes.

Experimental methods

The 2D analogue modelling experiments were carried out in a glass-sided deformation rig (Fig. 1a) with dimensions of 150 cm long, 20 cm wide and up to 20 cm deep. Initial model dimensions were typically $30 \times 20 \times 10$ cm. The sand models were constructed between the two end walls, one of which was kept fixed and the other was moved by a motor driven worm screw at a constant displacement rate of 4.16×10^{-3} cm sec^{-1}. Models were constructed either on a plastic detachment above a footwall block with fixed geometry (Fig. 1a) or above a rubber sheet that was stretched. For listric fault models two fundamental footwall blocks were used: a simple listric fault and a ramp-flat listric fault, with all models having an upper fault cut-off angle of 60°. The models are constrained by the plastic sheet which simulates a low friction decollement and the hangingwall is translated over the footwall at a constant displacement rate. The models were constructed by carefully sieving alternating layers of coloured sand (190 μm grain size) 2–5 mm thick, or where appropriate, alternating layers of coloured sand 2–5 mm thick with intercalated layers of vermiculite mica flakes (layers approximately 0.5–1 mm thick) into the deformation rig. During extension of the models, sand layers (or sand and mica layers) were added incrementally into the accommodation space created by the extensional deformation in order to maintain a constant, horizontal upper free surface and to simulate syn-extensional sedimentation. The glass side walls of the models were coated with a polymer that reduced frictional drag and minimized smearing of the model. The experiments were recorded by time-lapse 35 mm photography and 16 mm cinematography through the glass side walls of the apparatus. Serial sections of the completed and resin impregnated models were used to determine the detailed final geometry of the deformation.

The 3D analogue modelling experiments were carried out in a glass-sided deformation rig (Fig. 1b) with dimensions of 200 cm long, 60 cm wide and up to 10 cm deep. Initial model dimensions were typically $120 \times 60 \times 10$ cm. As in the 2D experiments a layered sandpack was constructed

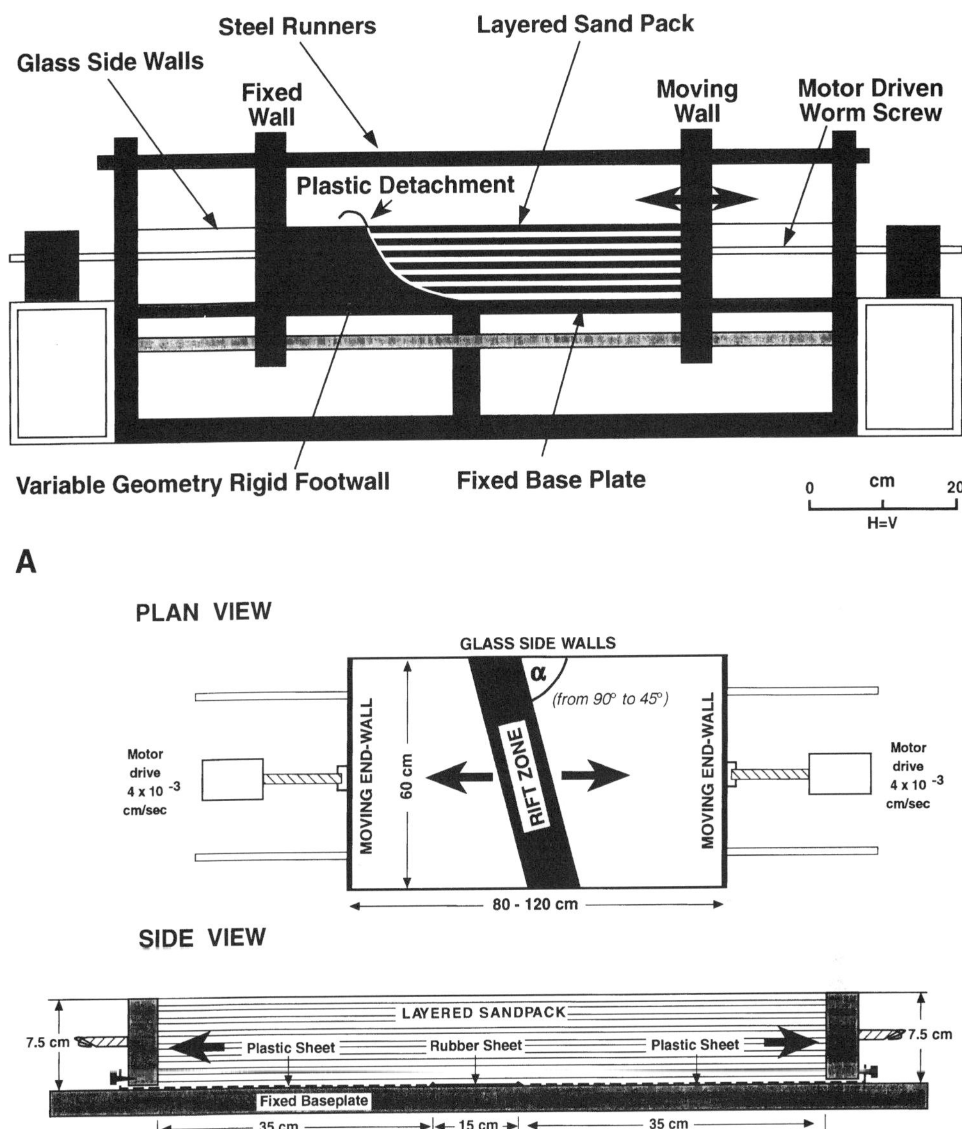

Fig. 1. Experimental apparatus. (**a**) Schematic diagram of the 2D experimental deformation rig. (**b**) Schematic diagram of the 3D experimental apparatus, plan view and side view. Extension can be achieved by moving either one end wall (asymmetric extension) or by moving both endwalls simultaneously (symmetric extension).

between two moveable end walls, one or both of which were moved by a motor driven worm screw at a constant displacement rate of 4.16×10^{-3} cm s^{-1}. The models were constructed of alternating coloured and white sand layers above a basement plastic sheet that contained a central stretching zone formed by a rubber sheet oriented at angles that varied from 90° to 45° with respect to the extension direction (Fig. 1b). During the experiments, syn-extensional sand layers were added incrementally in order to maintain a constant, horizontal upper free surface and to simulate syn-rift sedimentation that kept pace with extension. Serial sections of the completed and resin impregnated models were used to determine the detailed final geometry of the deformation and they were also digitized for 3D reconstruction on a Sun workstation. All of the experiments summarized in this paper have been repeated at least twice with similar results produced each time.

Extensional deformation within sedimentary basins developed within the upper crust is characterized by Navier-Coulomb brittle behaviour. In order to model this, extremely weak, near cohesionless model materials with a similar rheology must be used (Horsfield 1977; Ellis & McClay 1988; McClay 1990*b*). Dry quartz sand is such a material and has been widely used in analogue models studies (Horsfield 1977; Naylor *et al.* 1986; Vendeville *et al.* 1987; McClay & Ellis 1987*a*, *b*; Ellis & McClay 1988; Ellis 1988; McClay 1990*b*; McClay & White 1995). Dry quartz sand is isotropic, and in order to introduce anisotropy and competency contrasts into experiments, dry vermiculite mica and dry china clay may be added respectively (Ellis & McClay 1988). Ellis (1988) and McClay (1990*b*) have demonstrated that these materials have essentially linear Navier-Coulomb rheologies – dry quartz sand (190–275 μm) having an angle of friction of 31°, dry china clay 35° and dry vermiculite mica 30°. In the experiments described in this review homogeneous sand models or sand models with intercalated mica layers were used. If models to prototype ratios of 10^{-5} are used (McClay 1990*a*), 1 cm in the models represents approximately 1 km in nature.

Extensional models: results

The general geometric styles of the analogue model experiments have been described in detail in previous papers (see McClay & Ellis 1987*a*, *b*; Ellis & McClay 1988; McClay 1990*a*; McClay & Scott 1991; McClay & White 1995) and only a brief review is given here.

Simple rift model: 2D

Extension above a horizontal rubber sheet may simulate the development of a simple rift system in 2D. Experiment E-39 (Fig. 2D) shows a typical asymmetric rift model produced by 50% extension

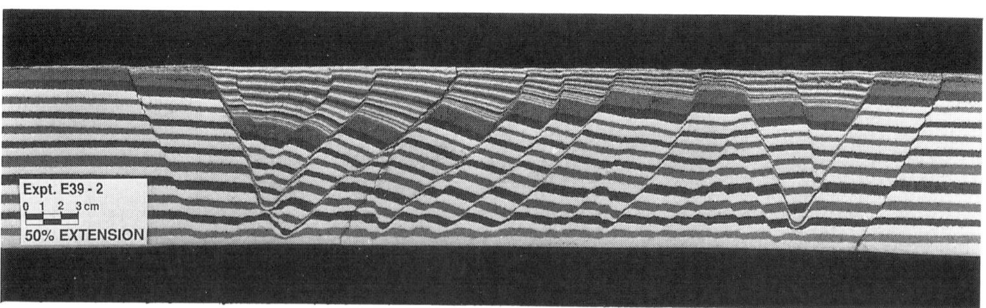

E-39 SIMPLE RIFT

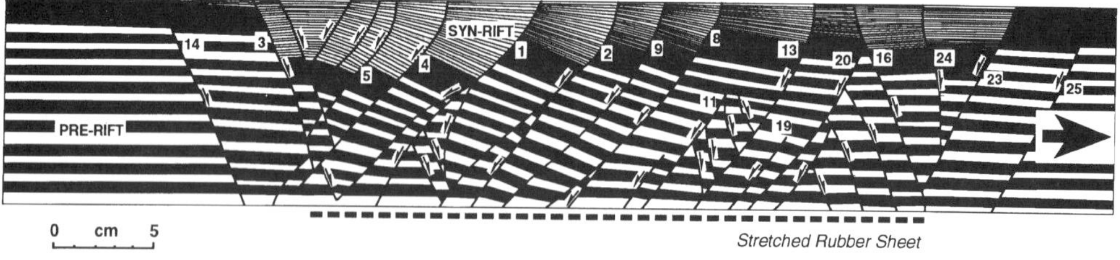

B

Fig. 2. 2D rift model. (**a**) Experiment E39 after 10 cm of extension (50% with respect to the initial length of the stretched rubber sheet at the base of the model). (**b**) Line diagram of Experiment E39 showing the fault patterns and sequence of faulting.

of the basal rubber sheet (right hand end wall moved away to the right). This produces a characteristic half graben bounded by planar rift border faults (Fig. 2a) with domino style intra-rift faults. In the pre-rift strata the intra-rift faults are typically planar but in the syn-rift sequence they are listric growth faults (Fig. 2a). At the left-hand side of the model a small keystone graben (cf. Jackson 1987) has developed (Fig. 2a). The displacement on all of the main extensional faults in this model dies out at the horizontal basal detachment surface that underlies the zone of rifting. The fault sequence diagram (Fig. 2b) shows initial nucleation of faults in the central section of the model with the left-hand rift boundary fault forming early in the deformation history followed by the progressive development of the intra-rift domino fault system and the right-hand rift border fault system (Fig. 2b). Note that there is not a simple sequence of faulting but that the intra-rift domino faults broadly develop from left to right in the direction of extension as increased stretching is progressively taken up upon the basal rubber sheet. This is characteristic of these asymmetric stretching models. The syn-rift growth strata show that many of the intra-rift domino faults were operating synchronously (Fig. 2a). The simple rift model illustrated here is limited in that it does not incorporate footwall uplift associated with individual fault blocks nor as yet can the analogue models incorporate the thermal effects of rift processes.

Simple listric faults: 2D

Extension above a simple listric fault with a horizontal basal detachment is shown for two models in Fig. 3. Hanging-wall deformation above a simple listric fault produces a characteristic rollover anticline with associated crestal collapse graben. In experiment E-80 where the extension is only moderate (10 cm approximately 50% of the initial length of the model) two simple crestal collapse graben are formed (Fig. 3a) each consisting of a keystone graben bounded by planar antithetic faults (e.g. Nos 2, 6 & 11) and sigmoidal synthetic faults (e.g. Nos 1 & 10). Syn-extensional strata have wedge shaped geometries against the main fault surface (Fig. 3a). The fault sequence diagram (Fig. 3b) shows nucleation of new faults into the crestal collapse graben and the formation of a second crestal collapse graben that is superposed upon the first. Figure 3c shows a model with identical footwall geometry to that shown in Fig. 3a but with 30 cm of extension (approximately 100% of the initial length of the model). In this model the geometry is more complex, with the development of many superposed crestal collapse grabens that migrate towards the main breakaway with increased extension (Fig. 3c). A fan of listric faults synthetic to the main detachment surface form in the syn-rift strata with a panel of highly rotated strata adjacent to the main fault surface (Fig. 3a). The fault sequence diagram for this model (Fig. 3d) shows the progressive superposition of the crestal collapse grabens and their migration towards the main detachment. It should be noted that the simple listric fault models described above have limitations in that the footwall shape is fixed during extension, compaction is not modelled, and that the displacement on the basal detachment surface is constant throughout the deformation history. Despite these limitations many natural examples of listric faults in detached terranes such as progradational delta systems show remarkably similar geometries to these simple analogue models (see below).

Ramp-flat listric faults: 2D

Extension above a ramp-flat fault is shown in Fig. 4. In this style of model thin mica interbeds are placed between the sand layers in order to promote bed-parallel slip and folding as the hangingwall moves down and over the complex ramp-flat surface. A typical model with an intermediate sloping detachment in the middle of the listric footwall block is shown in Fig. 4a. This consists of an upper listric fault segment, an intermediate 20 cm long gently sloping segment which is separated from the lower listric part of the main fault surface by a convex-upwards ramp (Fig. 4a). The upper listric part of the detachment produces a roll-over anticline with crestal collapse graben (Fig. 4a) that superpose in a manner similar to the simple listric faults described above (e.g. Fig. 3c). The lower listric part of the detachment, where it links with the horizontal basal section, has an associated roll-over anticline and crestal collapse graben (Fig. 4a). Where the hanging-wall is translated over the ramp (convex upwards) segment of the detachment a zone of folding and reverse faulting occurs. As a result a marked roll-over anticline with a strong ramp syncline is produced (Fig. 4a). Small reverse faults commonly initiate from the tips of buried crestal collapse graben faults. The fault sequence diagram (Fig. 4b) illustrates the superposition of crestal collapse grabens above the upper listric section of the detachment, and the complexities of fault nucleation and growth over the ramp section of the fault where new reverse faults (e.g. fault numbered 34 & 43) nucleate at the tips of early extensional faults as they pass down the ramp systems. As in the simple listric fault models there are limitations of fixed footwall geometries and constant displacement along the detachment

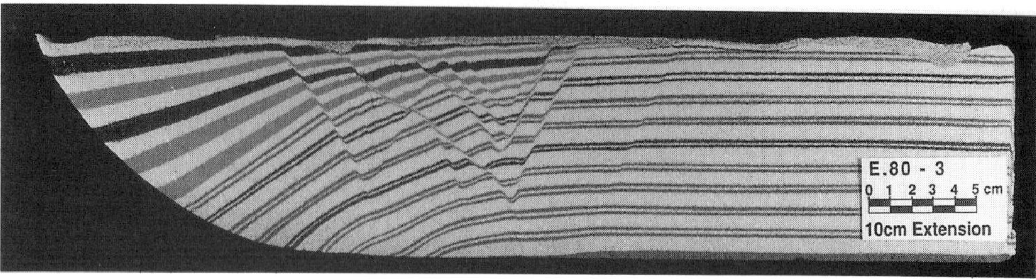

A

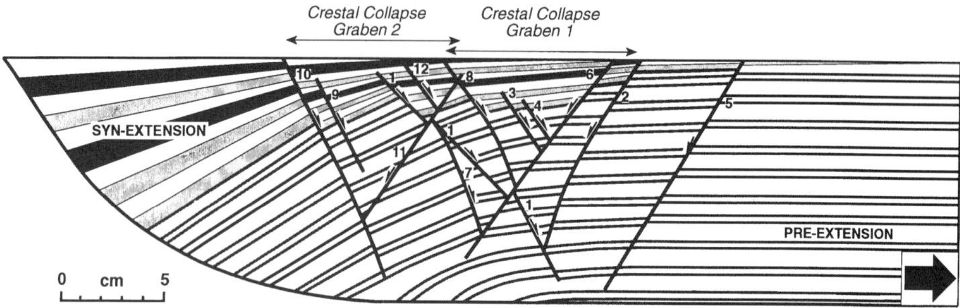

B

C

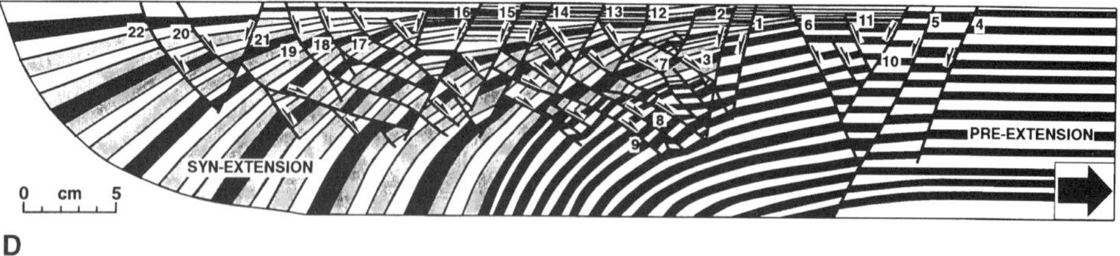

D

Fig. 3. 2D Listric fault models. (**a**) Experiment E 80 – simple listric fault – 10 cm of extension (approximately 50% with respect to the initial length of model). (**b**) Line diagram of Experiment E 80 showing the fault patterns and sequence of faulting. (**c**) Experiment E 30 – simple listric fault – 30 cm of extension (approximately 100% with respect to the initial length of model). (**d**) Line diagram of Experiment E 30 showing the fault patterns and sequence of faulting.

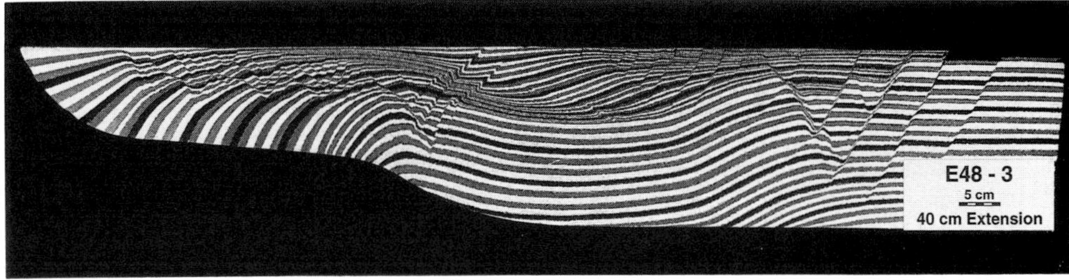

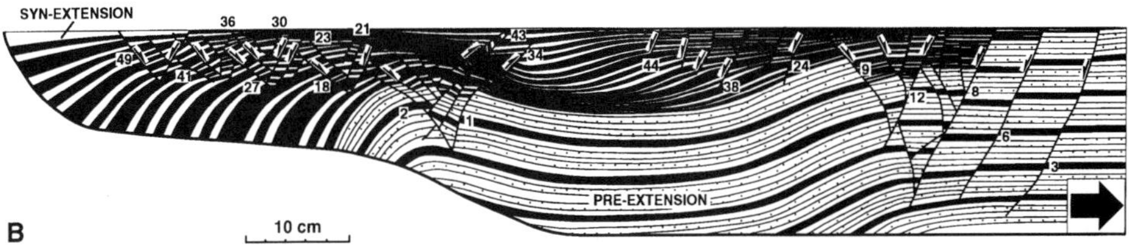

Fig. 4. 2D ramp-flat listric fault models. (**a**) Experiment E 48 after 40 cm of extension (approximately 50% with respect to the initial length of model).(**b**) Line diagram of Experiment E 48 showing the fault patterns and sequence of faulting.

surface for the ramp-flat fault models. These limitations are discussed below.

Particle displacement paths

Hangingwall deformation in 2D extensional fault systems may be analysed by tracking the displacement of individual marker points in the model during deformation (McClay 1995). Triangular marker points formed by coloured sand (with the same material properties as that in the rest of the model) were placed at 1–2 cm intervals in the sand models. Particle paths were monitored during the experiments and analysed. Particle displacement paths for a typical listric extensional fault are shown in Fig. 5. Figure 5a shows the particle paths measured with respect to a reference point in the footwall. These show trajectories parallel to the main detachment surface. However when the particle displacements are calculated with respect to a point in the hanging-wall (Fig. 5b) the relative trajectories of the hangingwall collapse are revealed. For listric fault surfaces these are characteristically curved from moderate dips in the upper part of the model to steep dips near the detachment surface (Fig. 5b). These curved shear trajectories for hanging-wall collapse are markedly different from the vertical or inclined planar shear trajectories commonly assumed for hanging-wall shear during extensional deformation (e.g. Gibbs 1983; 1984; Dula 1991; Withjack & Petersen 1993; Withjack *et al.* 1995)

Synoptic models for 2D extensional fault systems

One of the powerful results from analogue modelling is an understanding of the progressive evolution of extensional fault systems in 2D. This permits the development of forward synoptic models which explain the progressive geometric and kinematic evolution of the faults and fault blocks as well as the sequences of faulting. Such models can then be applied to the interpretation and analysis of deformed terranes such that realistic and valid cross-sections can be constructed. Figure 6a shows the synoptic model for the progressive evolution of a simple rift system where at low extensions the rift border faults and intra-rift faults are dominantly planar. With increased extension new intra rift domino faults develop and deform synchronously. In the syn-rift sequence these intra-rift faults become listric as a result of synchronous rotation of both the hangingwall and the footwall during syn-rift sedimentation (cf. Vendeville & Cobbold 1988). Intra-rift faults may become sigmoidal within the pre-rift sequence as a result of rotation and internal deformation as extension increases.

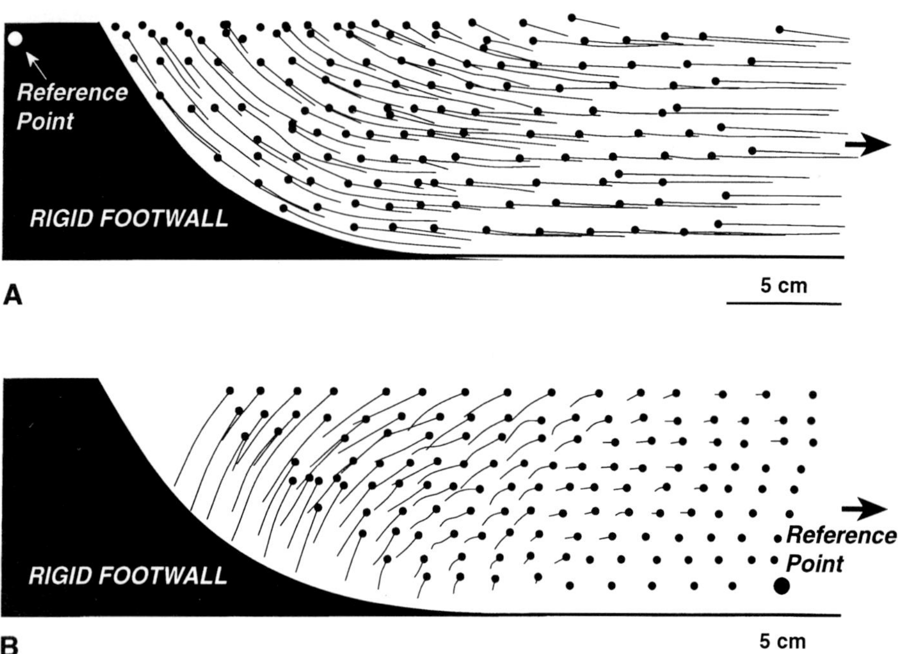

Fig. 5. Hanging-wall deformation in a 2D listric fault model E 80. (**a**) Particle displacement paths measured relative to a fixed point in the footwall block. (**b**) Particle displacement paths measured relative to a fixed point in the hanging-wall block.

Figure 6b shows a synoptic model for a simple listric fault. At low values of extension a simple roll-over and crestal collapse graben develops (Fig. 6b) whereas at high values of extension successive crestal collapse grabens are formed migrating towards the fault breakaway as the hangingwall extends. The superposition of crestal collapse graben produces complicated fault patterns that are difficult to interpret in detail and difficult to restore. The synoptic model for the ramp-flat fault (Fig. 6c) shows the characteristic development of an upper roll-over and crestal collapse graben system together with the ramp syncline and lower roll-over and crestal collapse graben system. With increased extension the ramp syncline may become unfolded (cf. Fig. 4) and reverse faults may nucleate from the tips of the upper crestal collapse graben extensional faults as they pass over the ramp section of the detachment (Fig. 6c). The results of the analogue models and the synoptic diagrams discussed above may be used as templates for the interpretation of seismic sections in areas where data quality is poor.

Examples of extensional fault systems in 2D

Cross-sections through many continental rift systems show many of the features found in the simple 2D analogue models of rift systems (e.g. Fig. 2; see also McClay 1990*a*). In spite of the limitations of sandbox modelling (see discussion below) the 2D sections (e.g. Fig. 2a) show many similarities to cross-sections through the Mozambique rift (McClay 1990*a*), to sections across the East Africa rift lakes (cf. Rosendahl *et al.* 1986; Rosendahl 1987), to sections across the Gulf of Suez (Bosworth 1995) and to sections across the Viking Graben, North Sea (Fig. 7a) (also Scott & Rosendahl 1989; Kusznir *et al.* 1995). In particular in the Northern Viking Graben (as in other intra-continental rifts) the rift is characterised by planar rift border faults and domino style intra-rift fault blocks typically 8–15 km wide (Fig. 7a). The Viking Graben cross-section shows the effects of isostasy and flexure (Kusznir *et al.* 1995) as well as post-rift thermal sag – features which as yet cannot be incorporated into the analogue models (see discussion below).

A typical listric extensional fault with a roll-over and crestal collapse graben is shown in Fig. 7b. This fault system is characterized by wedge geometries of syn-extensional strata in the roll-over together with keystone crestal collapse grabens and well developed planar antithetic faults (Fig. 7b). The architecture is very similar to that developed in the analogue models (Fig. 3) and this style of

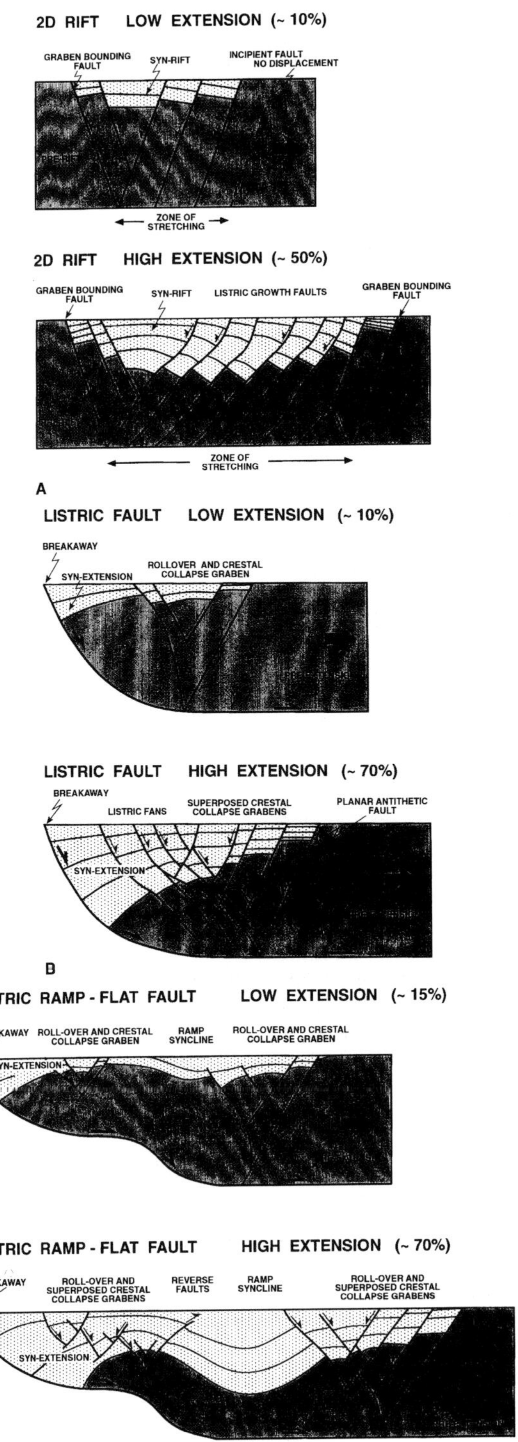

Fig. 6. Synoptic models for 2D extensional fault systems. (**a**) Simple rift model. (**b**) Simple listric fault model. (**c**) Ramp-flat listric fault model.

extensional faulting is characteristic of regional growth faults in shale and salt terranes such as progradational delta systems (e.g. Mississippi Delta: Bruce 1973; Bally 1983; Lopez 1990; Nelson 1991), Nile Delta (Beach & Trayner 1991), Niger Delta (Whiteman 1982) or the Barram Delta, Brunei (Ellenor & James 1984). Excellent seismic examples of this style of listric growth faulting are given in Bally (1983), Ellenor & James (1984), Lopez (1990) and are also discussed in McClay (1990*a*).

Natural examples of ramp-flat extensional faults are not well illustrated in the literature but examples have been interpreted in the Gulf of Mexico (McClay 1990*a*; Withjack *et al.* 1995), and from the Jeanne d'Arc basin, offshore Newfoundland (Tankard *et al.* 1989). Figure 7c shows a section across the ramp-flat Mermaid fault from the North West Shelf of Australia (after Lemon & Mahmood 1994). In this case complex fault systems similar to those in the analogue models (Fig. 4) are developed in the pre-extensional and in the syn-extensional Permian strata (Fig. 7c).

3D rift models: orthogonal rifting

In Experiment M54 (Fig. 8) the zone of rifting was oriented at 90° to the direction of extension and the model was extended to the right achieving 50% extension of the basal rubber sheet (Fig 8a). The surface fault pattern is characterized by long, relatively straight rift border faults (Fig. 8a) and shorter intra-rift faults. Both sets of faults form at, and remain at high angles *c.* 90° to the direction of extension. Along strike the intra-rift faults die out and are overlapped by other faults and form distinct relay ramp structures (Fig. 8a). At either end of the model the tips of the faults are curved due to frictional drag against the sidewalls. Figure 8b shows an interpretation of the surface fault pattern where the rift border faults at each side extend across the whole of the model. The left hand border fault system is the most strongly developed and is terraced (Fig. 8b). Three depocentres separated by indistinct accommodation zones are developed in the model (Fig. 8b). The main sub-basin is bounded by the left hand border fault system and two separate sub-basins are formed towards the end of the right hand border fault system. Within the central part of the rift zone the intra-rift faults remain perpendicular to the extension direction and develop long overlap zones forming relay ramps (Fig. 8b).

3D rift models: oblique rifting

In Experiment M55 the zone of rifting was oriented at 60° to the direction of extension. A distinct left-stepping en-echelon pattern (Fig. 9) of shorter, segmented rift border faults develop (in contrast to the orthogonal rift system in Fig. 8). The rift border faults, developed adjacent to the right hand moving margin of the rift system, are segmented and parallel to the underlying zone of stretching. In contrast the intra-rift faults are arranged in an en-echelon array and the individual faults are at high angles to the extension vector (Fig. 9a). These intra-rift faults are slightly curved, concave towards the extension direction. They nucleate at high angles to the extension direction, rotate slightly during deformation and at the end of the experiment have strikes that are approximately 80° to the extension vector. Both the rift border faults and the intra-rift faults appear to be dominantly dip-slip extensional faults with only small components of oblique slip (it is extremely difficult to determine exact displacement vectors on individual faults in sand). It is probable that the component of lateral motion required to accommodate the oblique rifting is taken up by both the slightly oblique extensional faults and by internal sub-fault scale deformation within the models. The intra-rift faults propagate along strike to form strongly developed overlapping antithetic faults to the main border fault system. Three distinct extensional sub-basins are developed and these are separated by broad accommodation zones oriented at an angle of approximately 45° to the direction of extension (Fig. 9b). In these zones, a fault system with one dip polarity merges into another fault system with opposite dip polarity. As in the orthogonal rift experiments no strike-slip transfer faults were observed. Similar evolutionary patterns of rift border fault systems and intra-rift fault systems were observed in all of the oblique rift experiments (i.e. $\alpha < 90°$ to $\alpha = 45°$). As in the simple 2D rift models described above the 3D rift models are limited in that they do not incorporate footwall uplift nor do they incorporate the thermal effects of rift processes.

Examples of extensional fault systems in 3D

The 3D analogue models of extensional fault systems provide templates for the interpretation of map patterns of extensional faulting. Orthogonal rifting (i.e. the extension direction is perpendicular to the underlying zone of basement stretching produces intra-rift and rift border faults that are 90° to the stretching direction and with large and long zones of overlap. Natural rift systems such as Gregory's Rift, Kenya (Baker 1986; Bosworth *et al.* 1986) where the extension direction is dominantly E–W, show similar fault patterns (Fig. 10a) to the models with long overlapping extensional faults and no strike-slip or oblique slip

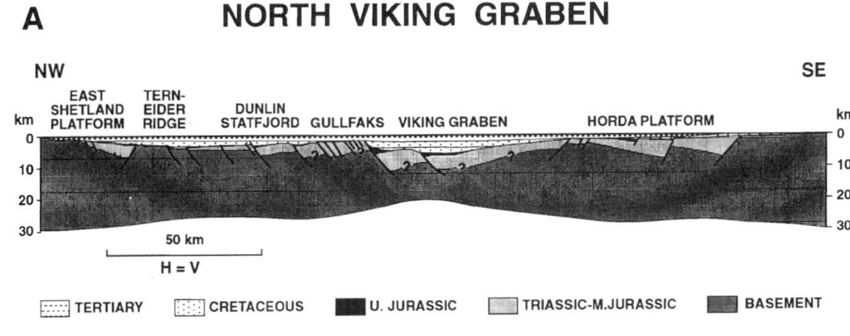

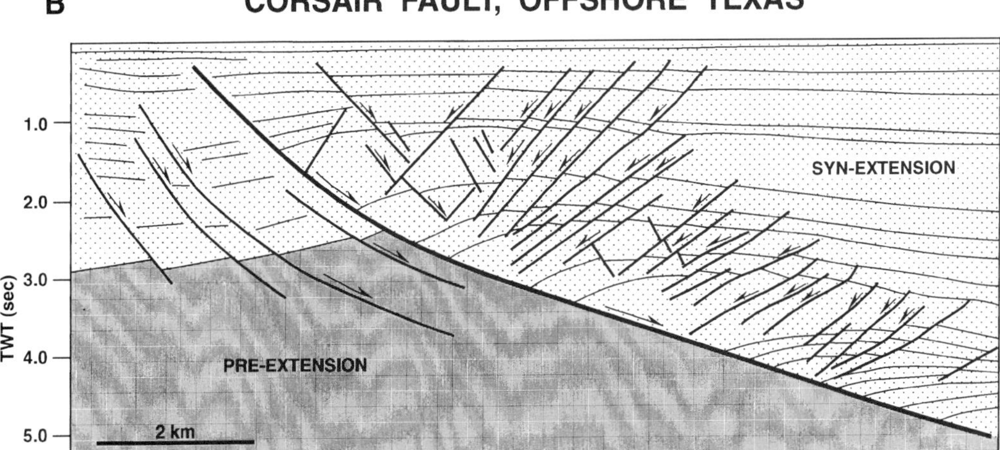

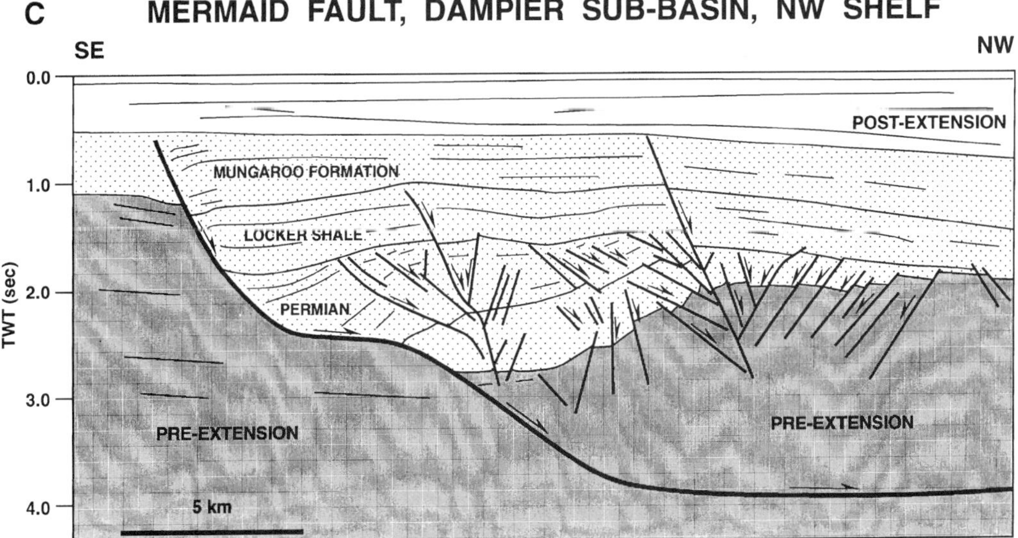

Fig. 7. Examples of extensional fault systems. (**a**) Cross-section through the northern Viking Graben, North Sea (after Kusznir et al. 1995). (**b**) Listric extensional fault, Gulf of Mexico (line drawing of a seismic section in Christiansen 1983). (**c**) Ramp-flat listric extensional fault, Dampier sub-basin, North West Shelf, Australia (line drawing and interpretation of a seismic section in Lemon & Mahmood 1994).

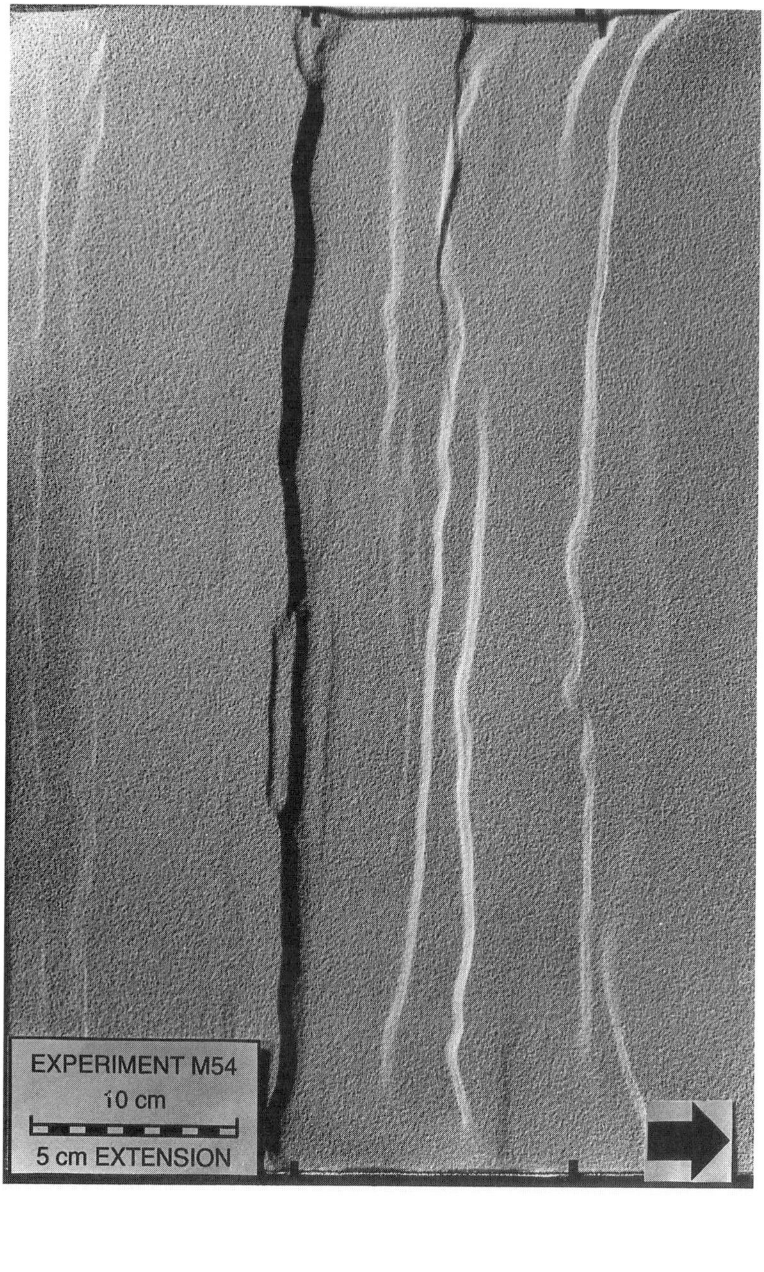

transfer faults. Similarly in regions where the rift axis is oblique to the extension direction such as in parts of Afar (Kronberg 1991) and the East African Rift system (e.g. Morley *et al.* 1992) the rift border faults are segmented and parallel to the rift axis whereas the intra-rift faults are oriented at a high angle to the extension direction (Fig. 10b). In the above cases and in other rift systems (cf. Gulf of Suez) no trans-rift strike-slip or oblique-slip transfer fault systems are found, rather zones of complex faulting generally known as accommodation zones (Bosworth *et al.*, 1986; Morley *et al.* 1990; McClay & White 1995; also see discussion below). It is probable that in oblique rift systems

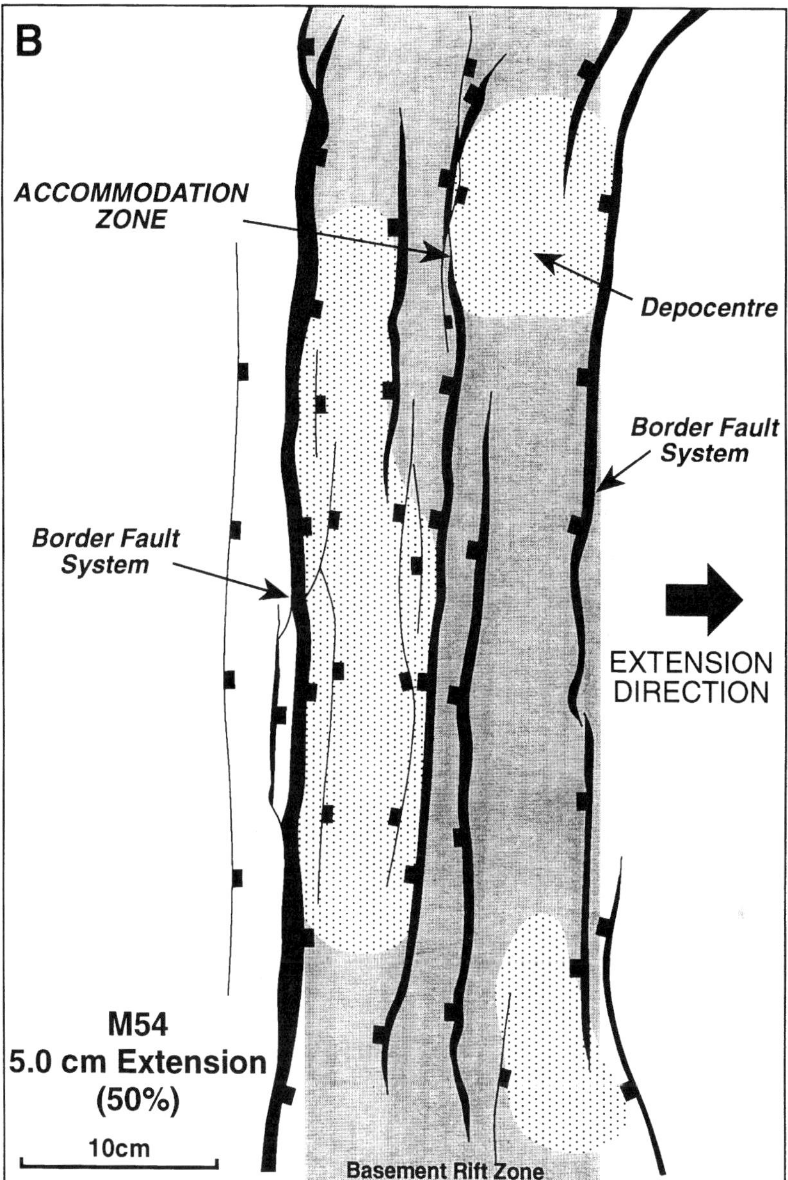

Fig. 8. 90° Orthogonal rift model M54. (**a**) View of the upper surface after 5 cm of extension (50% with respect to the initial length of the stretched rubber sheet at the base of the model). Lighting direction is from the left hand side of the model. (**b**) Map pattern of faults after 5 cm of extension. (from McClay & White 1995, and published with permission of *Marine and Petroleum Geology*)

many of the extensional faults would have components of oblique slip but determination of exact displacement vectors on these fault surfaces is usually extremely difficult. The 3D development of extensional faults and the architectures of accommodation zones are discussed in detail below.

Inversion models – results

Analogue modelling of inverted fault systems involves first extending the models (as described for extensional experiments above) and then horizontally recompressing them such that the extensional faults are reactivated (or partially

A

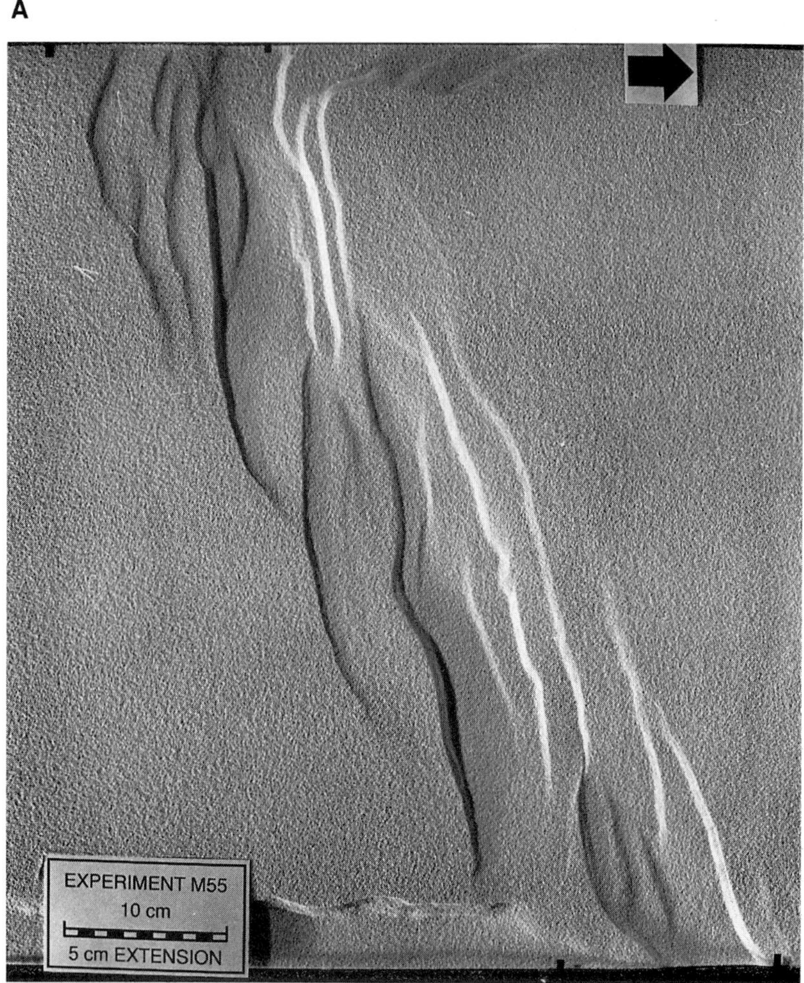

reactivated) as contractional faults (Buchanan & McClay 1991, 1992; McClay & Buchanan 1992; also see McClay 1995 for details). Syn-inversion sedimentation is usually added incrementally during the inversion deformation (cf. McClay, 1995).

Inversion of listric faults

Extension above a listric detachment fault produces a characteristic roll-over structure with an associated crestal collapse graben (cf. descriptions above and Fig. 3a). Inversion of the listric extensional fault model was achieved by horizontal recompression such that the hanging wall is pushed upwards and rotated by reactivation of the main detachment surface to produce a fault-bounded wedge above the main extensional fault breakaway (Fig. 11a). The main detachment fault propagated into the overlying post-rift and syn-inversion strata with the same geometry as the listric extensional segment (Fig. 11a). Small displacement short-cut faults form in the post-rift and syn-rift section in the footwall of the main fault. These short-cut thrust faults flatten upwards in the post-rift and syn-inversion strata. During inversion the crestal collapse graben becomes tighter (i.e. shortened) and small thrust faults nucleate from the tips of the pre-existing crestal-collapse graben extensional faults. This produces a characteristic 'pop-up' structure above the inverted crestal-collapse graben

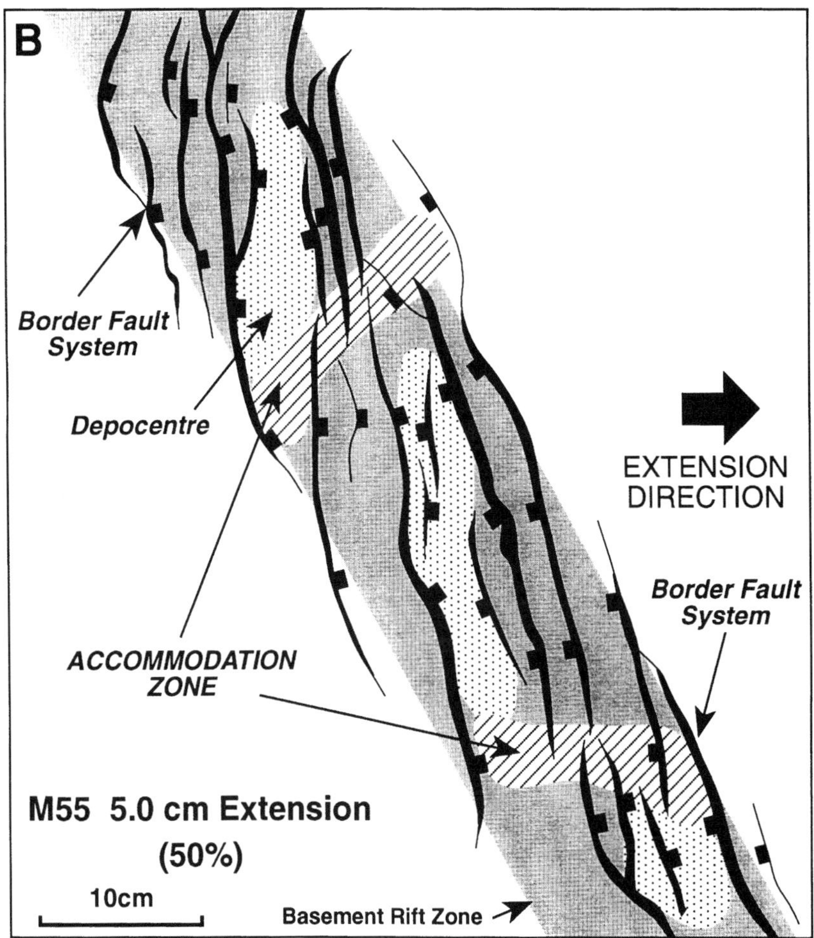

Fig. 9. 60° Oblique rift model M55. (**a**) View of the upper surface after 5 cm of extension (50% with respect to the initial length of the stretched rubber sheet at the base of the model). Lighting direction is from the left hand side of the model. (**b**) Map pattern of faults after 5 cm of extension. (from McClay & White 1995, and published with permission of *Marine and Petroleum Geology*)

(Fig. 11a). Syn-inversion strata thin over the crest of the main fault-bounded wedge and thicken away to either side (Fig. 11a). The fault sequence diagram shows the generation of the extensional fault architecture followed by the inversion architecture. An important aspect to note for section balancing is the changes in displacement history along the faults and in particular the nucleation of thrust faults from the tips of extensional faults (Fig. 11b). An important limitation of this style of inversion model (see also Buchanan & McClay, 1991) is that the rigid footwall block does not permit the detachment to change shape during both extension and inversion and in particular it does not permit footwall short-cut faults to develop in the pre- and syn-extensional sequences during inversion. In natural inverted fault systems footwall shortcut faults are observed (e.g. in southwest Wales, Powell 1989).

Particle displacement paths for inversion

Particle displacement paths for extension above a simple listric fault are shown in Fig. 5. During inversion the particle displacement paths relative to a fixed point in the footwall are parallel the main detachment surface (Fig. 12a) whereas particle displacement paths calculated relative to a fixed point in the hangingwall give trajectories for

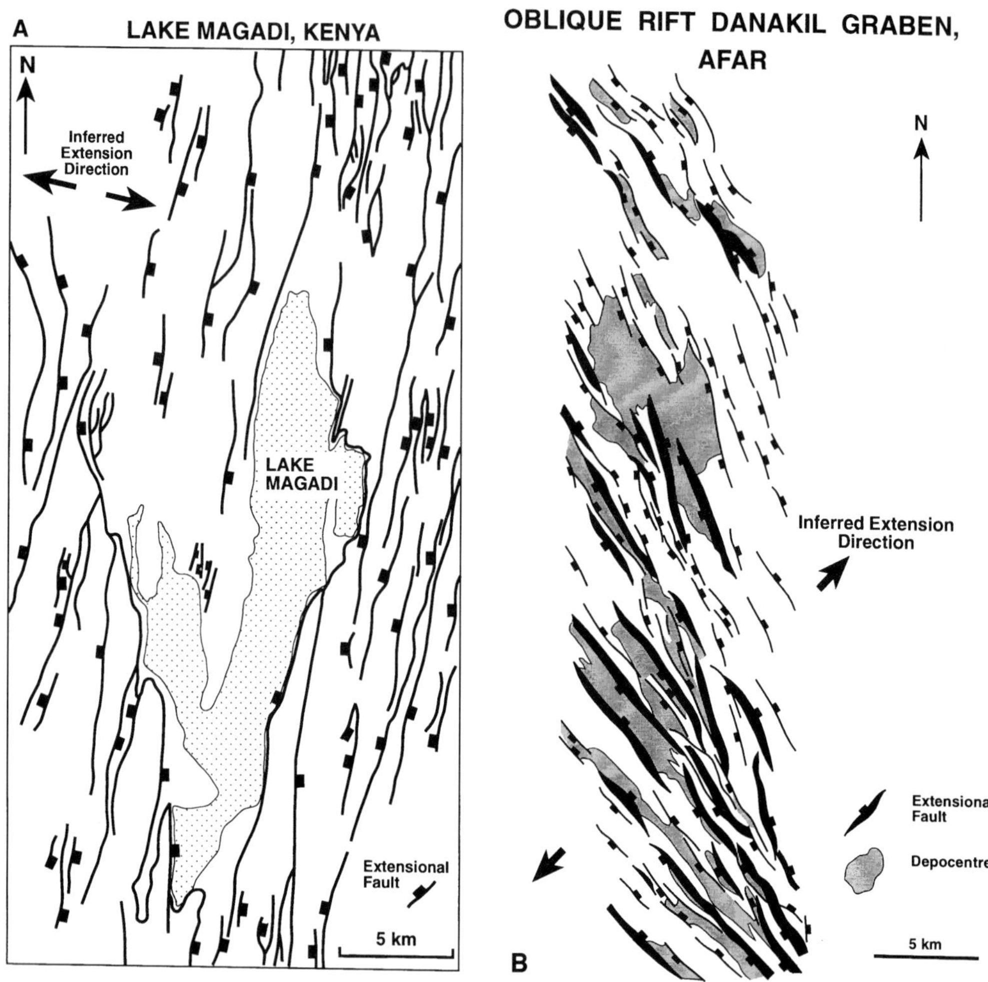

Fig. 10. Examples of rift fault patterns (**a**) Fault pattern in orthogonal rift, Lake Magadi, Gregory Rift Kenya (adapted after Baker 1986). The inferred extension direction is slightly north of E–W. (**b**) Map pattern of faults in an oblique rift, Afar (adapted after Kronberg 1991). The inferred extension direction is NE–SW and the faults are strongly segmented and at a high angle to the extension direction.

inversion that are initially shallow and then steepen upwards with increased contraction (Fig. 12b). The initial flat or shallow apparent shear trajectories reflect an important component of dilation (volume contraction) at the early stages of inversion. It is also important to note that the apparent shear trajectories during inversion are less steep than those during extension thus indicating a significant change to the angle of shear failure between extension and inversion. This has important implications for numerical methods of fault surface predictions and for algorithms used in balancing and restoring inverted terranes (see discussion below).

Synoptic models and examples of inverted listric faults

A synoptic model for an inverted listric fault is presented in Fig. 13a. Note the characteristic arrowhead or harpoon geometry of the inverted syn-extensional sedimentary wedge and the development of a growth anticline above the reactivated main detachment. Natural examples of inverted listric faults are discussed in Buchanan & McClay (1991), McClay & Buchanan (1992) and in McClay (1995). An example of an inverted listric fault from the Natuna Sea is shown in Fig. 13b. This shows many features similar to those developed in the

A

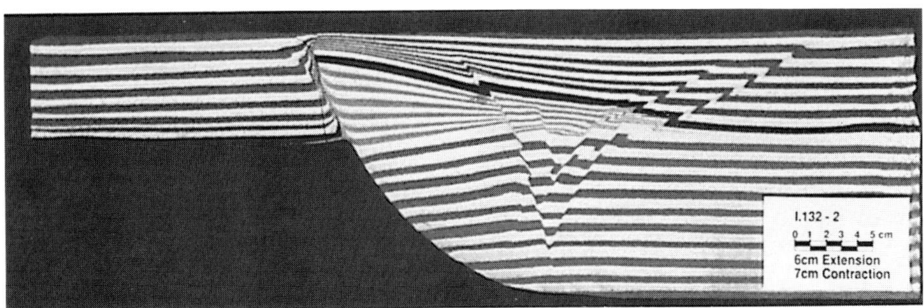

I-132 INVERTED LISTRIC FAULT

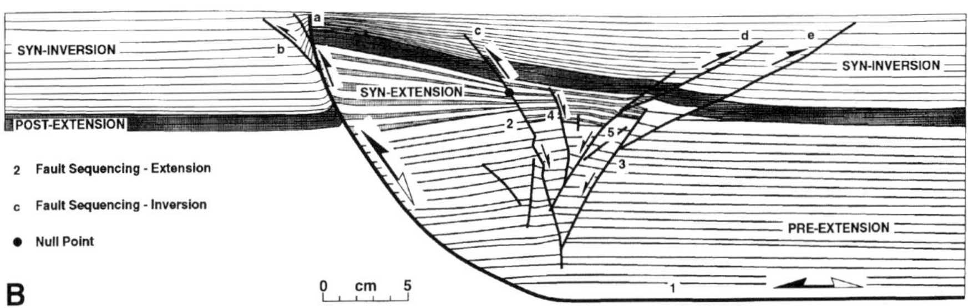

Fig. 11. 2D inversion of a simple listric fault. (**a**) Analogue model I132 after 6 cm of extension (approximately 25% with respect to the initial length of model), followed by 7 cm of inversion. (**b**) Fault sequence diagram for model I132.

analogue models. In particular note the erosion on top of the anticline and growth and onlap of syn-inversion strata onto the anticline (Fig. 13b). The complex structures in the footwall of the inverted listric fault make seismic interpretation difficult and the analogue models are valuable templates for understanding the structural styles in this area.

Discussion

The power of the analogue models that have been briefly reviewed above is shown by their ability to replicate the geometries of many natural examples of extensional and inverted fault systems (e.g. McClay 1990*a,* 1995; Buchanan & McClay 1991, McClay & Buchanan 1992). In particular they show the progressive evolution of structures such that an understanding of both the geometric and kinematic evolution can be attained (McClay 1990*a,* 1995). Similarly the analogue models of strike-slip fault systems (e.g. Naylor *et al.* 1986; Richard *et al.* 1995) and of thrust fault systems/Coulomb wedges (e.g. Malavielle 1984; Mulugeta & Koyi 1987; Colletta *et al.*1991; Dixon & Liu 1992; Liu *et al.* 1992; Calassou *et al.* 1993; Koyi 1995) have provided great insight as to the geometric and kinematic evolution of these fault systems.

Detailed analysis of the hanging-wall deformation in the analogue models offers the potential for understanding hanging-wall deformation mechanisms in both extensional and inverted fault systems. Most fault reconstruction techniques assume that during extension the hanging-wall deforms either by vertical simple shear (Verrall 1981; Gibbs 1983, 1984) or by inclined simple shear (Dula 1991; Nunns 1991; Kerr & White 1992; de Matos 1993; Withjack & Petersen 1993; Withjack *et al.* 1995). The results of the analysis of hanging-wall deformation in the analogue models, however, shows that above curved extensional faults the shear trajectories are curved. In addition the shear trajectories change from extension to compression becoming less steeply dipping (cf. Fig. 12 and also McClay 1995). These variations in hanging-wall shear trajectories need to be taken into account not only in fault trajectory predictions but also in the algorithms used for section construction and restoration (cf. Gibbs 1983; Rowan & Kligfield 1989).

Advances in modelling techniques, modelling materials, in the use of X-ray tomography, together

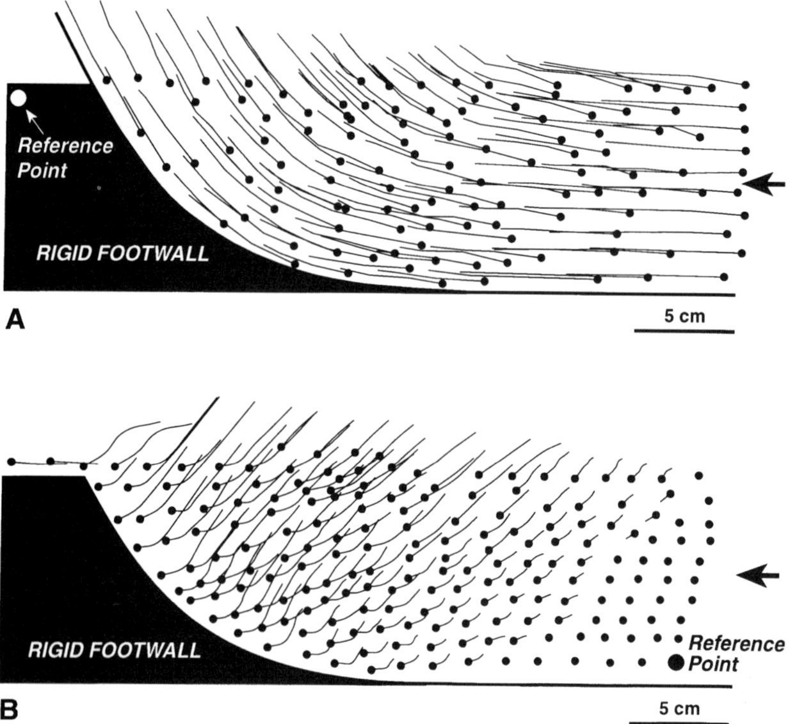

Fig. 12. Hanging-wall deformation in a 2D inverted listric fault. (**a**) Particle displacement paths measured relative to a fixed point in the footwall block. (**b**) Particle displacement paths measured relative to a fixed point in the hangingwall block.

with the analysis of hanging-wall deformation and 3D reconstructions have greatly improved the understanding and applicability of scaled analogue modelling to natural 3D structures. In particular 3D reconstruction using digitized serial sections has greatly enhanced the 3D visualization of complex fault geometries. Figure 14 shows two examples of rift models that have been reconstructed from serial sections cut parallel to the direction of extension. In the reconstruction of the top of the pre-rift surface in 60° oblique rift model M30 (Fig. 14a) the en-echelon arrangement of the rift border faults and the overlap of faults in relay ramp structures is clearly visible. In addition this display shows the elliptical displacement pattern along individual faults and the nature of along-strike fault linkages. These features are characteristic of the analogue models of orthogonal and oblique rift systems (McClay & White 1995). In the orthogonal, oblique and offset rift models no strike-slip fault systems were found – instead displacement transfer occurred by overlapping relay ramp structures (e.g. Fig. 14a and discussed in Larsen 1988; Peacock & Sanderson 1991; McClay & White 1995). The fault geometries in the 3D analogue models of rift systems show close similarities to those found in natural rifts (e.g. Tron & Brun 1991; Brun & Tron 1993; McClay & White 1995). In particular for oblique rifts both the rift border and intra-rift faults are segmented and overlap in relay ramp systems. Further detailed research and analysis required in order to understand the strain partitioning on both the border and intra-rift fault systems within the analogue models and to investigate the effects of ductile substrates on rift fault development (cf. Tron & Brun 1991; Brun & Tron 1993). The analogue models show how the faults grow along strike and how they link and interact and thus place important constraints upon section construction and validation in extensional terranes.

Figure 14b shows the 3D reconstruction of surfaces within an offset orthogonal rift model but with the addition of ray paths for a synthetic seismic line across this 3D structure (cf. Wang & Waltham 1995). 3D analogue modelling and reconstruction of complex fault systems, although still in its infancy (cf. Tron & Brun 1991; McClay

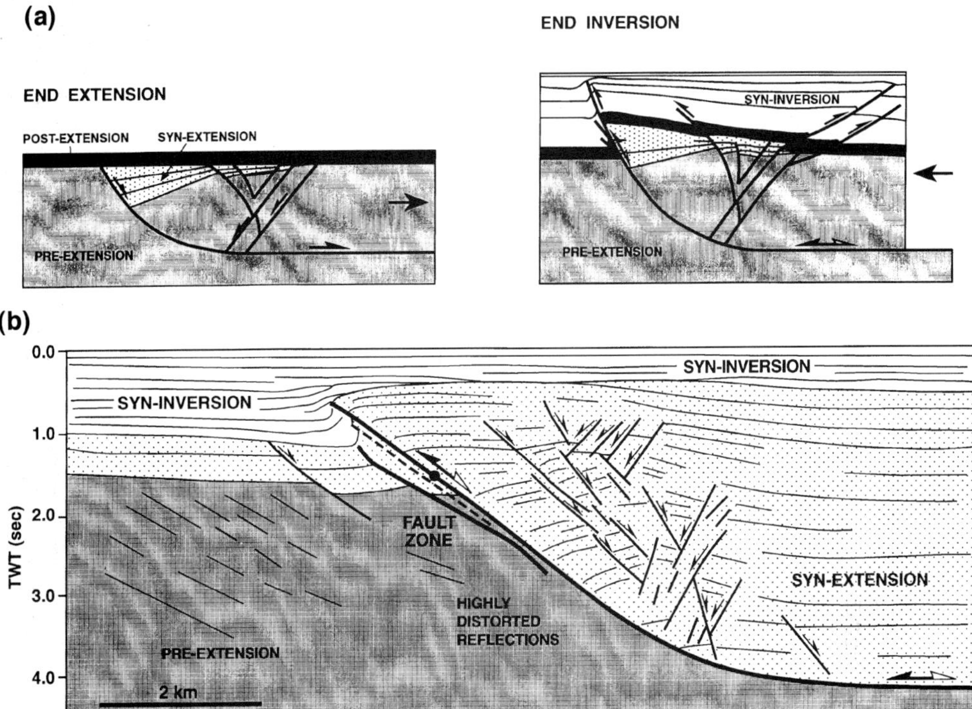

Fig. 13. Listric Fault inversion structures. (**a**) Synoptic model of extension followed by inversion of a simple listric fault. (**b**) Example of an inverted listric fault showing Miocene syn-inversion growth strata in the Natuna Sea, Southeast Asia (interpretation of a seismic section in Bunks *et al.* 1993)

& White 1995), are likely to provide important constraints on fault geometries, kinematics and fault linkages both in plan and in section. In particular the 3D models will provide data on fault displacements that can be tested against natural examples of fault systems and provide important constraints for fault displacement and fault population analyses (e.g. Walsh & Watterson, 1988).

In addition to providing interpretation templates for complex fault systems the analogue models may be used as the starting geometry for seismic modelling of complex structures (e.g. Fig. 14b). An important aspect of the Fault Dynamics research at Royal Holloway, University of London, is seismic modelling of complex structures in order to understand how they image and to determine the limitations of data processing and interpretation in complexly faulted areas (e.g. McClay *et al.* 1991; Waltham *et al.* this volume). Forward analogue modelling combined with synthetic seismic simulations of complex structural geometries will be a powerful tool to aid the exploration geologist in constructing and restoring cross-sections in areas such as inverted and contractional terranes where seismic imaging is commonly difficult.

Strengths and limitations of the analogue models

Scaled physical models are powerful visual tools for the forward modelling of brittle deformation in the upper crust. In particular they can model a wide range of tectonic settings from extension, inversion, strike-slip and contractional terranes as well as mobile salt and shale systems. Analogue models have distinct advantages over numerical models in that they are visual and 3D and are relatively easy to construct. In particular sandbox models allow new faults to form and whole new fault arrays develop and undergo progressive deformation – a feature that is difficult to replicate in most numerical models. Preservation of the models and serial sectioning permit the full 3D geometries of the deformation to be determined. Analysis of 3D structural geometries permits

(a)

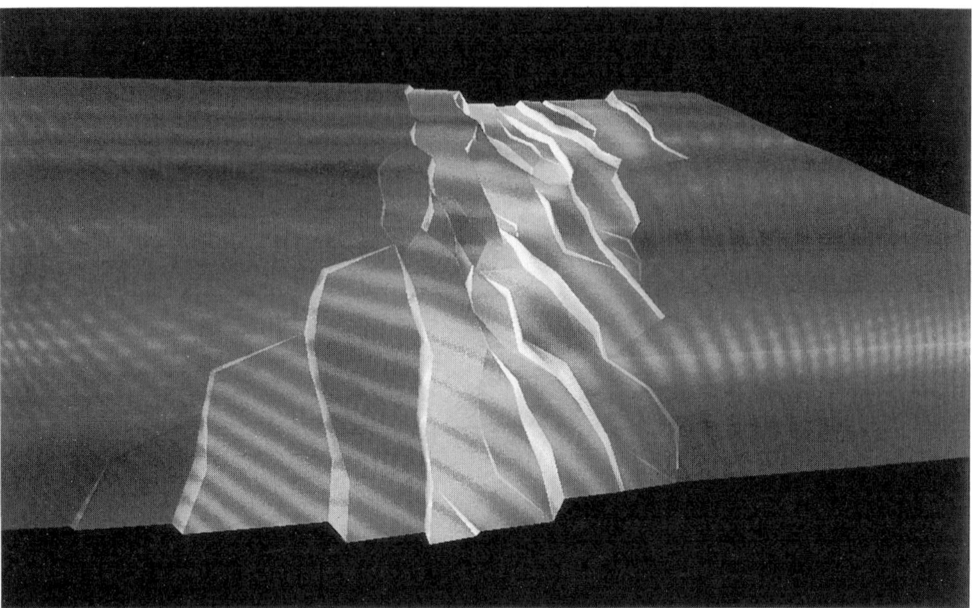

(b)

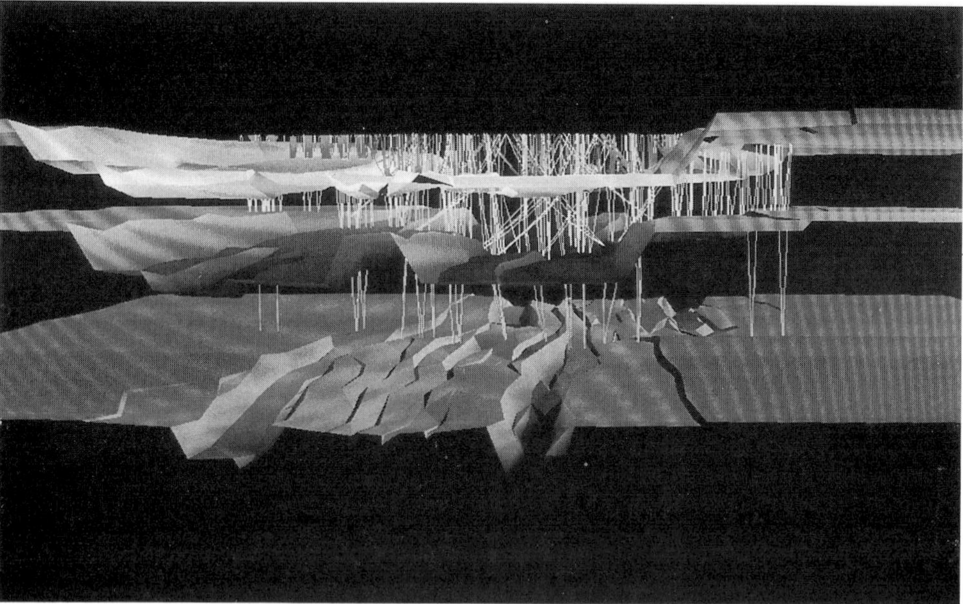

Fig. 14. 3D reconstructions of analogue models of rift structures. (**a**) Top of the pre-rift surface in analogue model M30 (60° oblique extension). The direction of extension is parallel to the front edge of the surface and the zone of rifting is approximately 25 cm wide. (**b**) Three surfaces in analogue model M26 (90° orthogonal rift with a central offset). The top surface is the pre-rift/syn-rift boundary and the lower two surfaces are within the pre-rift section. The direction of extension is parallel to the front edge of the surface and the zone of rifting is approximately 25 cm wide. The reflected rays from each surface generated by a synthetic 3D seismic survey line (calculated using the 3D Gaussian Beam method (Wang & Waltham 1995)) across this model are also shown.

assessment of fault linkages, fault displacements and complex 3D fault geometries in a manner that the exploration geologist can visualize in 3D in structurally complex areas. X-ray tomography of models whilst they are being deformed allows the internal deformations of the models to be monitored but the small size of the deformation rigs that can be used in the X-ray scanner place serious limitations on the boundary conditions and sizes of models that can be analysed in this way. In addition the resolution of X-ray tomography falls short of that which can be achieved by serially sectioning the models and reconstructing them on a workstation.

There are however important limitations to scaled physical models and these must always be born in mind when interpreting and applying the results of analogue models to natural deformation. Deformation in sandboxes or other experimental apparatus will always be strongly controlled by the boundary conditions at the base and sides of the model. Great care is needed in the construction of physical models to ensure that the boundary conditions in the model replicate those in nature. This feature is commonly a major limitation in the application of scaled physical models to natural deformation. Other important limitations to scaled analogue models include difficulties in correctly scaling the grain sizes and physical properties of natural prototypes, i.e. introducing realistic competency contrasts, elasticities and slip mechanisms. In particular the faults in the analogue models are formed by granular shear processes without grain breakage – a process that does not replicate fault mechanics in nature (cf. Scholtz 1990). Analogue models cannot as yet adequately model flexural and isostatic, and thermal effects of deformation. Compaction of sediments, fluid flow and the development of overpressures are also limitations of the modelling methods that need to be addressed.

Conclusions

From the examples described above it is evident that scaled analogue modelling is a powerful tool for the understanding of the geometric and kinematic development of extensional, and inverted fault systems, as well as for some thrust systems (e.g. Dixon & Liu 1992; Liu *et al.* 1992) and for strike-slip fault systems (e.g. Richard *et al.* 1995). Recent advances in modelling techniques, in modelling materials, in analysis of deformation within the models, and utilization of X-ray tomography and 3D reconstruction have enabled scaled physical models to be more widely applied to complex tectonic regimes and to provide constraints on structural styles, geometries and kinematics.

In particular the results of scaled analogue models can provide significant constraints on the construction, validation and restoration of geological cross sections in extensional and inverted terranes in that they:

(1) Provide an understanding of the geometric and kinematic evolution of fault (and fold) structures during the brittle deformation of the upper crust;
(2) Permit an understanding of the sequences of faulting and activity on faults (i.e. relative slip rates and also synchronous slip on two or more faults);
(3) Provide templates for the interpretation of seismic sections in regions of poor data quality and for projecting surface data into the subsurface;
(4) Can potentially provide an understanding of the mechanisms of hangingwall deformation and the relationships between hanging-wall shear regimes, the shape of the basal detachment surface and the displacement conditions along the fault surface;
(5) Provide templates for seismic modelling of complex structural regimes and templates for numerical modelling of tectonics and sedimentation;
(6) Show that the hangingwall shear orientations change from extension to inversion – an important constraint for section construction in mixed mode tectonic terranes.

The ability of the models to replicate real geological structures is shown by the close geometrical similarities between the models and the natural examples described above and in the literature (e.g. examples in Bally 1983; Lowell 1985; and also discussed in McClay 1990*a*, 1995).

For extensional fault systems the models may simulate structures on a number of scales, from entire rift basins to hanging-wall deformation above a single fault or detachment. The extensional fault models are expected to be particularly powerful for the prediction of structural styles associated with individual detachments, i.e. at the scale of an individual field.

Although still in its infancy, 3D modelling is likely to prove to be extremely valuable in the study of the nucleation, propagation and interaction of extensional fault systems along their strike and at depth and thus provide important constraints on section construction in 3D.

The models reviewed in this paper, nevertheless, have limitations which must be borne in mind when interpreting the results. In the models, the accommodation space in the basins formed by extension is always infilled with syn-rift sediment, a condition which does not always occur in nature. In addition, the listric fault models have constant displacement along the whole length of the detachment whereas in nature displacement rate gradients may be expected to occur on some faults. The models cannot, as yet, incorporate the isostatic, compaction or thermal effects expected for deformation in natural fault systems.

Research summarized in this paper was supported by the Fault Dynamics Research Project sponsored by Arco British Limited, BP Exploration, BRASOIL (UK.) Ltd, Conoco (UK) Ltd, Mobil North Sea Limited and Sun Oil Britain. Kevin D'Souza and Keith Denyer kindly printed the photographs. M. White, N. Deeks, M. Simmons, P. Buchanan and P. Hollings assisted with the analogue modelling and Chris Willacy assisted with the AVS reconstructions. T. K. Wang and X. Wang carried out the synthetic seismic modelling. D. Waltham and P. Richard are thanked for constructive criticism of the manuscript. Fault Dynamics Publication No. 52.

References

BAKER, B. H., 1986. Tectonics and volcanism of the southern Kenya Rift Valley and its influence on rift sedimentation. *In*: FROSTICK, L. E., REBAUT, R. W., REID, I. & TIERCELIN, J.-J. (eds) *Sedimentation in the African Rifts*. Geological Society, London, Special Publications, **25**, 45–58.

BALLY, A. W. 1983. *Seismic expression of structural styles*. American Association of Petroleum Geologists, Studies in Geology, **15**.

BARTLET, W. L., FRIEDMAN, M. & LOGAN, J. M. 1981. Experimental folding and faulting of rocks under confining pressure: Part IX. Wrench faults in limestone layers. *Tectonophysics*, **79**, 255–277.

BEACH, A. & TRAYNER, P. 1991. The geometry of normal faults in a sector of the offshore Nile Delta, Egypt. *In*: ROBERTS, A. M., YIELDING, G. & FREEMAN, B. (eds) *The Geometry of Normal Faults*. Geological Society, London, Special Publications, 56, 173–182.

BOSWORTH, W., 1986. Comment and Reply on Detachment faulting and the evolution of passive continental margins. *Geology*, **14**, 890–892.

——, 1995. The southern Gulf of Suez: a high strain rift. *In*: LAMBIASE, J. J. (ed.) *Hydrocarbon Habitat in Rift Basins*. Geological Society, London, Special Publications, **88**, 91–103.

——, LAMBIASE, J. & KEISLER, R., 1986. A new look at Gregory's Rift: The structural style of continental rifting. *EOS*, **67**, 577–583.

BRUCE, C. H., 1973. Pressured shale and related sediment deformation: Mechanism for development of regional contemporaneous faults. *Bulletin American Association of Petroleum Geologists*, **57**, 878–886.

BRUN, J-P. & TRON, V. 1993. Development of the North Viking Graben: inferences from laboratory modelling. *Sedimentary Geology*, **86**, 31–51.

BUCHANAN, P. G. 1991. *Geometries and kinematic analysis of inversion tectonics from analogue model studies*. PhD thesis, University of London.

—— & MCCLAY, K. R., 1991. Sandbox experiments of inverted listric and planar faults systems. *In*: Cobbold, P. E. (ed.) Experimental and Numerical Modelling of Continental Deformation. *Tectonophysics*, **188**, 97–115.

—— & —— 1992. Experiments on basin inversion above reactivated domino faults. *Marine and Petroleum Geology*, **9**, 486–500.

BUNKS, C., PICHON, P. L., BEN BRAHOM, L. & SAKTI, S. 1993. 2D prestack depth migration: Application to inverted grabens. *Proceedings Indonesian Petroleum Association*, **93**, 1, 22–44.

CADELL, H. M. 1889. Experimental researches in mountain building. *Transactions of the Royal Society of Edinburgh*, **35**, 337–357.

CALASSOU, S., LARROQUE, C. & MALAVIEILLE, J. 1993. Transfer zones of deformation in thrust wedges: an experimental study: *Tectonophysics*, **221**, 325–344.

CHRISTIANSEN, A. F. 1983. An example of a major syndepositional listric fault. *In*: BALLY, A. W. (ed.) *Seismic expression of structural styles*. American Association of Petroleum Geologists, Studies in Geology, **15**.

CLOOS, E. 1968. Experimental analysis of Gulf Coast fracture patterns. *Bulletin American Association of Petroleum Geologists*, **52**, 420–444.

CLOOS, H. 1928. Experimente zur inneren tektonik. *Cetralblatt fur Mineralogie*, Abt. B, 609–621.

COLLETTA, B., LETOUZEY, J., PINEDO, R., BALLARD, J. F. & BALÉ, P. 1991. Computerized X-ray tomography analysis of sandbox models: examples of thin skinned thrust systems. *Geology*, **19**, 1063–1067.

DAVIS, D. M., SUPPE, J. & DAHLEN, F. A., 1983. Mechanics of fold-and-thrust belts and accretionary wedges: *Journal of Geophysical Research*, **88**, 1153–1172.

DAVY, P & COBBOLD, P. R. 1988. Indentation tectonics in nature and experiment, 1. *Bulletin Geological Institution University of Uppsala, N.S*, **14**, 129–147.

—— & —— 1991. Experiments on shortening of a 4-layer model of the continental lithosphere. *Tectonophysics*, **188**, 1–25.

DIXON, J. M. & LIU, S., 1992. Centrifuge modelling of the propagation of thrust faults: *In*: MCCLAY, K. R. (ed.) *Thrust Tectonics*. Chapman & Hall, London, 53–69.

DULA, W. F. 1991. Geometric models of listric normal faults and rollover folds. *American Association of Petroleum Geologists Bulletin*, **75**, 1609–1625.

ELLENOR, D. W. & JAMES, D. M. D. 1984. The oil and gas resources of Brunei. *In*: JAMES, D. M. D. (ed.) *The geology and hydrocarbon resources of Negra Brunei Darussalam*. Muzium Brunei, Kota Batu., 130–139.

ELLIS, P. G. 1988. *Analogue model studies of the kinematics of extensional faulting*. PhD thesis, University of London.

—— & MCCLAY, K. R. 1988. Listric extensional fault systems – results of analogue model experiments. *Basin Research*, **1**, 55–70.

EMMONS, R. C. 1969. Strike-slip rupture patterns in sand models. *Tectonophysics*, **7**, 71–87.

FAUGERE, E. & BRUN, J. P. 1984. Modelisation experimentale de la distention continentale. *Compte Rendu Acadamie des Sciences, Paris*, Series **11**, 299, 365–370.

GIBBS, A. D. 1983. Balanced section constructions from seismic sections in areas of extensional tectonics. *Journal of Structural Geology*, **5**, 2, 153–160.

— 1984. Structural evolution of extensional basin margins. *Journal of the Geological Society, London*, **141**, 609–620.

HORSFIELD, W. T. 1977. An experimental approach to basement controlled faulting. *Geologie en Mijnbouw*, **56**, 363–370.

—— 1980. Contemporaneous movement along crossing conjugate normal faults. *Journal of Structural Geology*, **2**, 305–310.

JACKSON, J. A. 1987. Active normal faulting and crustal extension. *In*: COWARD, M. P., DEWEY, J. F. & HANCOCK, P. L. (eds) *Continental Extension Tectonics.* Geological Society, London, Special Publications, **28**, 3–17.

JACKSON, M. P. A. & VENDEVILLE, B. C., 1994. Regional extension as a trigger for diapirism. *Geology Society of America Bulletin*, **106**, 57–73.

KERR, H. G. & WHITE, N. 1992. Laboratory testing of an automatic method for determining normal fault geometry at depth. *Journal of Structural Geology*, **14**, 873–885.

KOOPMAN, A., SPEKSNIJDER, A. & HORSFIELD, W. T. 1987. Sandbox model studies of inversion tectonics. *Tectonophysics*, **137**, 379–388.

KOYI, H. 1995. Mode of deformation in sand wedges. *Journal of Structural Geology*, **17**, 293–300.

KRONBERG, P. 1991. Geometries of extensional fault systems, observed and mapped on aerial and satellite photographs of Central Afar (Ethiopia/ Djibouti). *Geologie en Mijnbouw*, **70**, 145–161.

KUSZNIR, N. J., ROBERTS, A. M. & MORLEY, C. K. 1995. Forward and reverse modelling of rift basin formation. *In*: LAMBIASE, J. J. (ed.) *Hydrocarbon Habitat in Rift Basins.* Geological Society, London, Special Publications, **88**, 33–56.

LALLEMAND, S. E., MALAVIEILLE, J. & CALASSOU, S., 1992. Effects of oceanic ridge subduction on accretionary wedges: Experimental modeling and marine observations: *Tectonics*. **11**, 1301–1313.

LARSEN, P 1988. Relay structures in a lower Permian basement-involved extension system, East Greenland. *Journal of Structural Geology*, **10**, 3–8.

LEMON, N. M. & MAHMOOD, T. 1994. Two and three dimensional analogue modelling of extensional fault systems with applicayion to the North West Shelf, Western Australia. *Australian Petroleum Exploration Association Journal*, **34**, 555–565.

LIU, H. 1992. *Analogue modelling of thrust wedges.* PhD thesis, University of London.

——., MCCLAY, K. R. & POWELL, D. 1992. Physical models of thrust wedges: *In*: McClay. K. (ed.) *Thrust Tectonics.* Chapman & Hall, London, 71–81.

LOPEZ, J. A. 1990. Structural styles of growth faults in the US Gulf Coast Basin. *In*: BROOKS, J. (ed.) *Classic Petroleum Provinces.* Geological Society, London, Special Publications, **50**, 203–220.

LOWELL, J. D., 1985. *Structural Styles in Petroleum Exploration.* Oil and Gas Consultants International Inc., Tulsa.

MCCLAY, K. R. 1989. Analogue models of inversion tectonics. *In*: COOPER, M. A. & WILLIAMS, G. D., (eds) *Inversion Tectonics.* Geological Society, London, Special Publications, **44**, 41–59.

—— 1990*a*. Extensional fault systems in sedimentary basins. A review of analogue model studies. *Marine and Petroleum Geology*, **7**, 206–233.

—— 1990*b*. Deformation mechanics in analogue models of extensional fault systems. *In*: RUTTER, E. H. & KNIPE, R. J. (eds) *Deformation Mechanisms, Rheology and Tectonics.* Geological Society of London Special Publication, **54**,. 445–454.

——, 1995. The geometries and kinematics of inverted fault systems: a review of analogue model studies. *In*: BUCHANAN, J., G. & BUCHANAN, P. G. (eds) Basin Inversion. Geological Society, London, Special Publications, **88**, 97–118.

—— & BUCHANAN, P. G. 1992. Thrust faults in inverted extensional basins. *In*: MCCLAY, K. R. (ed.) *Thrust Tectonics.* Chapman & Hall, London, 419–434.

—— & ELLIS, P. G. 1987*a*. Analogue models of extensional fault geometries. *In*: COWARD, M. P., DEWEY, J. F. & HANCOCK, P. L. (eds) *Continental extension tectonics.* Geological Society, London, Special Publications, **28**, 109–125.

—— & —— 1987*b*. Geometries of extensional fault systems developed in model experiments. *Geology*, **15**, 341–344.

—— & SCOTT, A. D. 1991. Hangingwall deformation in ramp-flat listric extensional fault systems. *Tectonophysics*, **188**, 85–96.

—— & WHITE, M. 1995. Analogue models of orthogonal and oblique rifting. *Marine and Petroleum Geology*, **12**, 137–151.

——, WALTHAM, D., SCOTT, A. D. & ABOUSETTA, A., 1991. The geometries of listric extensional fault systems. *In*: ROBERTS, A. M., YIELDING, G. & FREEMAN, B. (eds) *The Geometry of Normal Faults.* Geological Society, London, Special Publications, **56**, 231–239.

MALAVIEILLE, J. 1984. Modélisation expérimentale des chevauchements imbriqués: Application aux chaînes de montagnes: *Geological Society of France Bulletin*, **7**, 129–138.

——, LARROQUE, C. & CALASSOU, S. 1993. Modelisation experimentale des relations tectonique/sedimentation entre bassin avant-arc et prisme d'accretion: *Compte Rendu Acadamie Sciences, Paris*, **316**, II, 1131–1137.

MANDL, G. 1988. *Mechanics of tectonic faulting. Models and basic concepts.* Elsevier, Amsterdam.

MANN, P., HEMPTON, P. R., BRADLEY, D. C. & Burke, K., 1983. Development of pull-apart basins. *Journal of Geology*, **91**, 529–554.

MARSHAK, S. & WILKERSON, M. S. 1992. Effect of overburden thickness on thrust belt geometry and development: *Tectonics*, **11**, 560–566.

——, —— & HSUI, A.T. 1992. Generation of curved fold-thrust belts: Insight from simple physical and analytical models: *In*: MCCLAY, K. R. (ed.) *Thrust Tectonics.* Chapman & Hall, London, 83–92.

DE MATOS, R. M. D. 1993. Geometry of the hangingwall above a system of listric normal faults. *Bulletin American Association of Petroleum Geologists*, **77**, 1839–1859.

MORLEY, C. K., CUNNINGHAM, S. M., HARPER, R. M. & WESCOTT, W. A. 1992. Geology and geophysics of the Rukwa Rift, East Africa. *Tectonics*, **11**, 69–81.

——, NELSON, R. A., PATTON, T. L. & MUNN, S. G. 1990. Transfer zones in the East African Rift system and their relevance to hydrocarbon exploration in rifts. *American Association of Petroleum Geologists Bulletin*, **74**, 1234–1253.

MULUGETA, G. 1988*a*. Squeeze box in a centrifuge: *Tectonophysics*, **148**, 323–335.

—— 1988*b*. Modelling the geometry of Coulomb thrust wedges: *Journal of Structural Geology*, **10**, 847–859.

—— & KOYI, H. 1987. Three-dimensional geometry and kinematics of experimental piggyback thrusting: *Geology*, **15**, 1052–1056.

NELSON, T. H. 1991. Salt tectonics and listric normal faulting. *In*: SALVADOR, A. (ed.) *The Gulf of Mexico Basin.* Geological Society of America, The Geology of North America, **J**, 73–89.

NAYLOR, M. A., MANDL, G. & SIJPESTEIN, C. H. K. 1986. Fault geometries in basement induced wrench faulting under different initial stress states. *Journal of Structural Geology*, **8**, 737–752.

NUNNS, A. G. 1991. Structural restoration of seismic and geologic sections in extensional regimes. *Bulletin American Association of Petroleum Geologists*, **75**, 278–297.

ODONNE, F. & VIALON, P. 1983. Analogue models of folds above a wrench fault. *Tectonophysics*, **99**, 31–46.

PEACOCK, D. C. P. & SANDERSON, D. J. 1991. Displacements, segment linkage and relay ramps in normal fault zones. *Journal of Structural Geology*, **13**, 721–733.

POWELL, C. 1989. Structural controls on Plaeozoic basin evolution and inversion in Southwest Wales. *Journal of the Geological Society, London*, **146**, 439–446.

RECHES, Z. 1987. Mechanical aspect of pull-apart basins and push-up swells with applications of the Dead Sea transform. *Tectonophysics*, **141**, 75–88.

RICHARD, P. D., 1991. Experiments on faulting in a two layer cover sequence overlying a reactivated basement fault with oblique (normal-wrench or reverse-wrench) slip. *Journal of Structural Geology*, **13**, 459–469.

—— & COBBOLD, P. R. 1990. Experimental insights into partitioning of fault motions in continental convergent wrench zones. *Annales Tectonicae*, **4**, 35–44.

——, BALLARD, J. F., COLLETTA, B. & COBBOLD, P. R. 1989. Fault initiation and development above a basement strike-slip fault: analogue modelling and tomography. *Compte Rendu Acadamie des Sciences, Paris*, **309** (II), 2111–2118.

——, MOQUET, B. & COBBOLD, P. R. 1991. Experiments on simultaneous faulting and folding above a basement wrench fault. *Tectonophysics*, **188**, 133–141.

——, NAYLOR, M. A. & KOOPMAN, A. 1995. Experimental models of strike-slip tectonics. *Petroleum Geoscience*, **1**, 71–80.

ROSENDAHL, B. R. 1987. Architecture of continental rifts with special reference to east Africa. *Annual Review of Earth and Planetary Sciences*, **15**, 445–503.

——, REYNOLDS, D., LORBER, P., BURGESS, C., MCGILL, J., SCOTT, D., LAMBIASE, J. & DERKSEN, S. 1986. Structural expressions of rifting: lessons from Lake Tanganyika. *In*: FROSTICK, L. E., REBAUT, R. W., REID, I. & TIERCELIN, J.-J. (eds) *Sedimentation in the African Rifts.* Geological Society, London, Special Publications, **25**, 29–43.

ROWAN, M. G. & KLIGFIELD, R. 1989. Cross section restoration and balancing as aid to seismic interpretation in extensional terranes. *Bulletin American Association of Petroleum Geologists*, **73**, 955–966.

SCHOLZ, C. H. 1990. *The mechanics of earthquakes and faulting.* Cambridge University Press, New York.

SCHREURS, G. 1994. Experiments on strike-slip faulting and block rotation. *Geology*, **22**, 567–570.

SCHULTZ-ELA, D. D. 1992. Restoration of cross-sections to constrain deformation processes of extensional terranes. *Marine and Petroleum Geology*, **9**, 372–388.

SCOTT, D. L. & ROSENDAHL, B. R. 1989. North Viking graben: An East African perspective. *Bulletin American Association of Petroleum Geologists*, **73**, 155–165.

SERRA, S. & NELSON, R. A. 1989. Clay modelling of rift asymmetry and associated structures. *Tectonophysics*, **153**, 307–312.

TALBOT, C. J. 1992. Centrifuge models of the Gulf of Mexico profile. *Marine and Petroleum Geology*, **9**, 412–432.

—— 1993. Spreading of salt structures, upward-moving gravity structures. *Tectonophysics*, **228**, 151–165.

TANKARD, A. J., WELSINK, H. J. & JENKINS, W. A. M. 1989. Structural styles and stratigraphy of the Jeanne d'Arc basin, Grand Banks of Newfoundland. *In*: BALKWILL & TANKARD (eds) *Extensional Tectonics and Stratigraphy of the North Atlantic Margins.* American Association of Petroleum Geologists, Memoir, **46**, 95–110.

TCHALENKO, J. S. 1970. Similarities between shear zones of different magnitudes. *Geological Society of America Bulletin*, **81**, 1625–1640.

TRON, V. & BRUN, J-P. 1991. Experiments on oblique rifting in brittle-ductile systems. *Tectonophysics*, **188**, 71–84.

VENDEVILLE, B. 1991. Mechanisms generating normal fault curvature: a review illustrated by physical models. *In*: ROBERTS, A. M., YIELDING, G. & FREEMAN, B. (eds) *The Geometry of Normal Faults.* Geological Society, London, Special Publications, **56**, 241–250.

—— & COBBOLD, P. R. 1988. How normal faulting and sedimentation interact to produce listric fault profiles and stratigraphic wedges. *Journal of Structural Geology*, **10**, 7, 649–659.

—— & JACKSON, M. P. A. 1992*a*. The rise of diapirs during thin-skinned extension. *Marine and Petroleum Geology*, **9**, 331–353.

—— & —— 1992*b*. The fall of diapirs during thin-skinned extension. *Marine and Petroleum Geology*, **9**, 354–371.

——, COBBOLD, P. R., DAVY, P., BRUN, J-P. & CHOUKROUNE, P. 1987. Physical models of extensional tectonics at various scales. *In*: COWARD, M. P., DEWEY, J. F. & HANCOCK, P. L. (eds) *Continental Extensional Tectonics*. Geological Society, London, Special Publications, **28**, 95–107.

VERRALL, P. 1981. *Structural Interpretation with Applications to North Sea Problems*. Course Notes No. 3, Joint Association for Petroleum Exploration Courses (UK).

WALSH, J. J. & WATTERSON, J. 1988. Analysis of the relationship between displacements and dimensions of faults. *Journal of Structural Geology*, **10**, 239–248.

WANG, T. K. & WALTHAM, D. A. 1995. 3D Gaussian beam seismic modelling. *Journal of Geophysical Prospecting*, in press.

WANG, X. & WALTHAM, D. A. 1995. The Stable Beam Ray Tracing Method. *Journal of Geophysical Prospecting*, in press.

WEIJERMARS, R. 1986. Flow behaviour and physical chemistry of bouncing putties and related polymers in view of tectonic laboratory applications. *Tectonophysics*, **124**, 325–358.

——, JACKSON, M. P .A. & VENDEVILLE, B. C. 1993. Rheological and tectonic modelling of salt provinces. *Tectonophysics*, **217**, 143–174.

WHITE, M. J. 1993. *Physical modelling and analysis of the geometries and kinematics of orthogonal and oblique rift systems*. PhD thesis, University of London.

WHITEMAN, A. 1982. *Nigeria: Its petroleum geology, resources and potential* (Vol. 2). Graham & Trotman, London.

WILCOX, R. E., HARDING, T. P. & SEELY, D. R., 1973. Basic wrench tectonics. *American Association of Petroleum Geologists Bulletin*, **57**, 74–96.

WITHJACK, M. O. & JAMISON, W. R., 1986. Deformation produced by oblique rifting. *Tectonophysics*, **126**, 99–124.

—— & PETERSEN, E. T. 1993. Prediction of normal fault geometries – A sensitivity analysis. *Bulletin American Association of Petroleum Geologists*, **77**, 1860–1873.

——, ISLAM, Q. T. & LA POINTE, P. R. 1995. Normal faults and their hangingwall deformation: An experimental study. *Bulletin American Association of Petroleum Geologists*, **79**, 1–18.

Multiple detachment levels in thrust tectonics: Sandbox experiments and palinspastic reconstruction

MARK VERSCHUREN[1], DICK NIEUWLAND[1] & JIM GAST[2]

[1] *Shell Research, Postbus 60, 2280 AB Rijswijk, The Netherlands*

[2] *Shell Expro, Shell Mex House, Strand, London WC2 0DX, United Kingdom*

Abstract: Sandbox models of thrust tectonics with multiple detachment levels allow alternative interpretations of fault patterns and evolution, even though they are well controlled experiments. Palinspastic reconstruction resolves these ambiguities and leads to refined interpretations. It reveals that if there are thick viscous interlayers, all non-fault related contraction can be accommodated by folding, without tectonic compaction. It also confirms the complete decoupling of faulting above and below a thick weak viscous layer, and the initial straightness and regular spacing and dips of thrusts above a decollement of low basal friction. A weak elasto-plastic interlayer or decollement merely has a lubricating effect, without complete decoupling.

Many natural thrust systems are underlain by a ductile substratum and have more than one decollement level. These examples almost always exhibit a complex thrust belt geometry and are apparently formed by different thrusting events that can be difficult to interpret and to time relative to one another.

Weak decollement materials and interlayers have a strong effect on the structural style of thrust belts at the moment of detachment. However, other material properties, in particular the viscosity, play a dominant role during the structural development following detachment (the overthrust phase).

The objective of this study was to improve structural interpretation of data from areas of thin-skinned thrust tectonics with weak ductile detachment horizons (i.e. excluding large scale basement involvement). The approach involved:

- the use of sandbox experiments to provide controlled structural models with different decollement rheologies, and
- the refinement and validation of selected cross-section interpretations with palinspastic reconstruction.

The result is a set of better constraints on structural interpretation of cross-sections in thrust tectonics. In natural cases, the most common weak decollement lithologies are over-pressured shales and evaporites, in particular rock salt. In our experiments we have simulated rock salt by using silicone putty ('SP'), a viscous material that can flow slowly as a Newtonian fluid. We modelled overpressured and/or very soft clay with an oil–water emulsion ('OWE'), which is a ductile, weak elasto-plastic, non-viscous material in which localized shear zones can form, as in any other elasto-plastic solid.

Experimental configuration

A large 3D sandbox (approximately 1.5 m × 1.5 m) was used to study the effects on the structural style and internal geometry of a thrust belt. The experiments were built up in four layers from bottom to top: OWE or SP (1 cm)/sand (1 cm)/OWE or SP (0.5 cm)/1 cm sand. Compression was applied with a vertical backstop which, in the case of SP, moved at a carefully controlled speed of 2 cm/hr. For experiments with OWE, the speed was approximately 15 cm/hr to deform the decollement in the strain rate insensitive elasto-plastic regime. The experimental strain-rates were chosen to realistically scale the material properties (strength and viscous behaviour of the ductile interlayers (Hubbert 1937, Horsfield 1977, Mandl .*et al.* 1977, Davy & Cobbold 1991). The models initially developed a critical taper immediately next to the backstop. Once the taper reached its critical shape, thrusts propagated further forward into the foreland.

The internal structures of the models were examined by vertically slicing the wetted sand pack at the end of an experiment. Before wetting the model, a protective additional sand-layer (of a contrasting colour) was added to preserve the surface topography of the model.

From Buchanan, P. G. & Nieuwland, D. A. (eds), 1996, *Modern Developments in Structural Interpretation,Validation and Modelling,* Geological Society Special Publication No. 99, pp. 227–234.

Material properties of the ductile interlayers

As an analogue for rock salt, silicone putty (SP) has been used by several workers (Davy & Cobbold 1991; Vendeville *et al.* 1987). The experiments can be scaled because rock salt and silicone putty are both Newtonian viscous fluids (and therefore very weak at low strain rates). Overpressured (or very soft) clay shows a fundamentally different behaviour: although it is weak it is not Newtonian viscous and at shallow crustal levels its strength is not strongly dependent on strain rate. Silicone putty is therefore not a good analogue for clay. Instead we explored the use of an oil-water emulsion (OWE). The OWE proved to be a more appropriate analogue for clays and shales at shallow crustal levels, where the strength of these rocks is commonly not strongly strain-rate dependent. More accurate scaling is presently not possible due to a lack of data on the mechanical properties of decollement shales.

The strength of the two ductile materials is comparable at the strain rates of our experiments (Fig. 1). At faster rates SP is stronger than OWE, while this relationship is inverted towards lower rates. The most important difference in behaviour lies in the fact that the viscous SP can flow and does therefore not develop a localised shear zone, whereas the OWE does not flow and forms a localized shear zone, just as in any other elasto-plastic solid.

Sandbox experiments with multiple detachments

In many cases, detachment may occur at more than one level in the stratigraphy. Figure 2 (Kulander & Dean 1986) is an example from the Appalachians with one minor and two major levels of detachment. This cross section can be compared with the sandbox experiments (Fig. 3 top view and Fig. 4 cross-sections) with a weak basal layer and another one embedded in the sediment. Although the field example is not an exact analogue of the experiments that we are presenting, it shows a similar style of deformation, a sequence with two decollements with the thrusted lower units folding the shallower ones.

In the experiment with SP (Fig. 4d), the complete decoupling of thrusts, as well as the greater number of thrusts above the second viscous weak layer, is very similar to the Appalachian example. The long

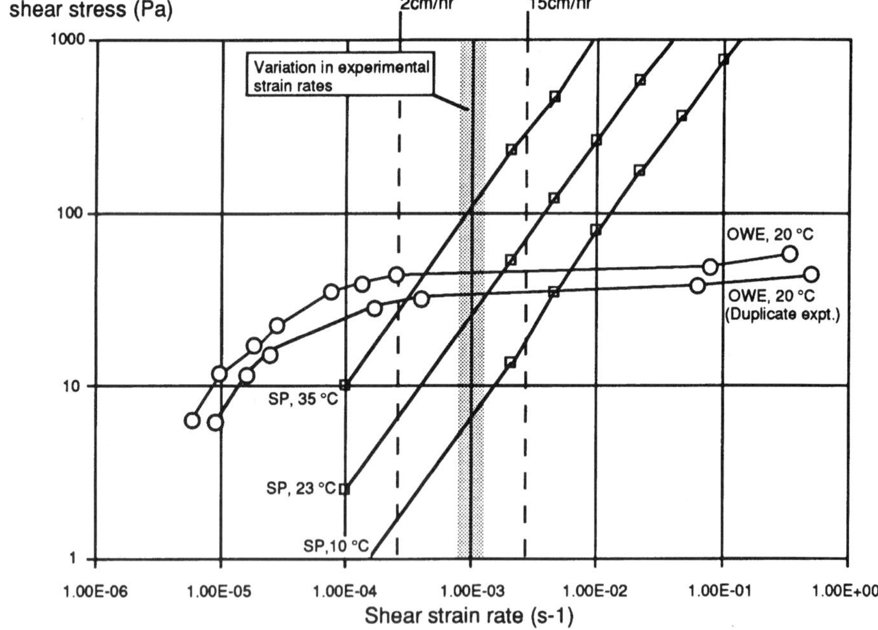

Fig. 1. Rheology of the viscous silicone putty (SP) and the non-viscous oil–water–emulsion (OWE). The SP is strain-rate sensitive, whereas, at the strain-rates used in the experiments (1.00E-03), the OWE is not. At these strain rates OWE and SP have very similar strengths (30–40 Pa).

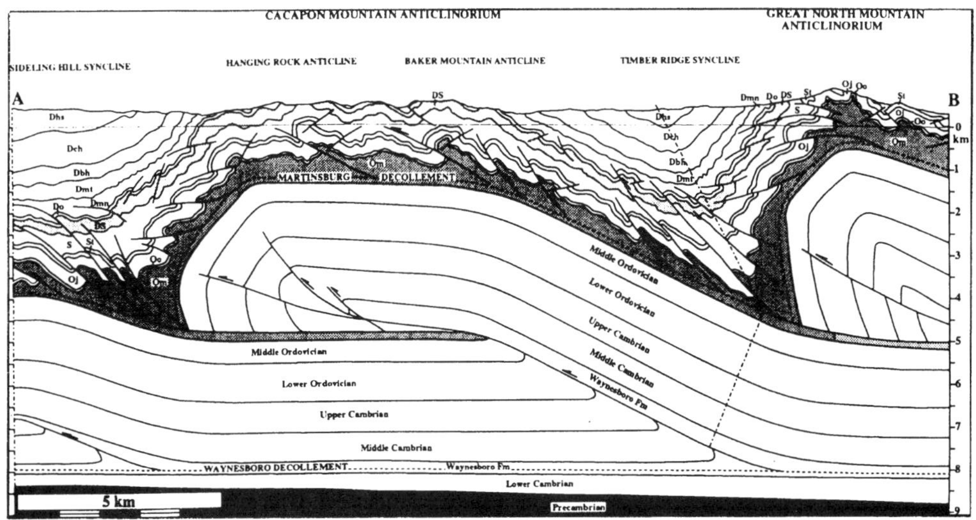

Fig. 2. Cacapon Mountain Anticlinorium, Appalachians (from Kulander & Dean 1986). Cross section based on outcrop and well data, showing multiple detachment levels and increasing thrust length with depth.

deeper thrusts follow a flat-ramp-flat trajectory, climbing up into the second SP layer, thereby deforming the overlying shorter thrusts. This is caused by the differences in thrust sheet lengths and is not the result of two deformation phases. In contrast, decoupling across the intermediate elasto-plastic OWE (Fig. 4a) plays an insignificant role and is hardly noticeable at first sight.

For a detailed interpretation and palinspastic reconstruction, the cross sections in Fig. 4 were selected on the basis of overhead views (Fig. 3) to ascertain that they would be exactly parallel to the direction of shortening and unaffected by out-of-plane strain. GeoSec 3.0 (from CogniSeis Dev. Inc.) was used for interpretation and palinspastic reconstruction.

Palinspastic reconstruction: model with weak plastic detachments

Although the plastic intermediate OWE couples the upper and lower thrust sheets quite well, a careful interpretation (Fig. 4a) revealed small faults confined to the layers of sand either above or below the upper decollement (white in Fig. 4b), as well as disharmonic folding. The two sheets of sand were therefore reconstructed separately with the flexural slip mechanism, which conserves bed length (Fig. 4c). The reconstruction proceeded from a pin line at the right end of the section towards the left, with upper and lower fault blocks restored in parallel to common throughgoing thrusts. Area balancing was used to derive the initial thickness of the two weak layers and of the upper sand layer that was partially re-sedimented at the emergent fault tips. The total bed lengths of the five reconstructed sheets are equal, which lends confidence to the interpretation and its restoration. However, the amount of contraction derived from this reconstruction (28% in Fig. 4b) is different from the experimentally applied shortening (34%). While most of the contraction was accommodated by thrusting and folding, one-sixth was due to tectonically induced, bed parallel compaction of the sand. The flexural slip mechanism worked well for most of this section, except for one fault bend fold in the centre that could only be restored acceptably assuming vertical shear. In addition, the most proximal back thrust (BT in Fig. 4c) could not be restored to a straight initial fault, in contrast with all other through going thrusts in this section. The reason for these anomalies probably lies in heterogeneous tectonic compaction.

There remain some problems with the interpretation of the experiments. Figure 5 shows a detail that could be interpreted in two different ways: either with a thrust 'T' that flattened out towards its branch point on thrust 'B' (Fig. 5a), or with an almost straight thrust with a wedge of sand above the branch, presumably sliced off the footwall of thrust B (Fig. 5b). Palinspastic reconstruction rules out the first interpretation, because it implies (Fig. 5c) that the lower part of T was originally listric and that B had an abnormally large dip to start with.

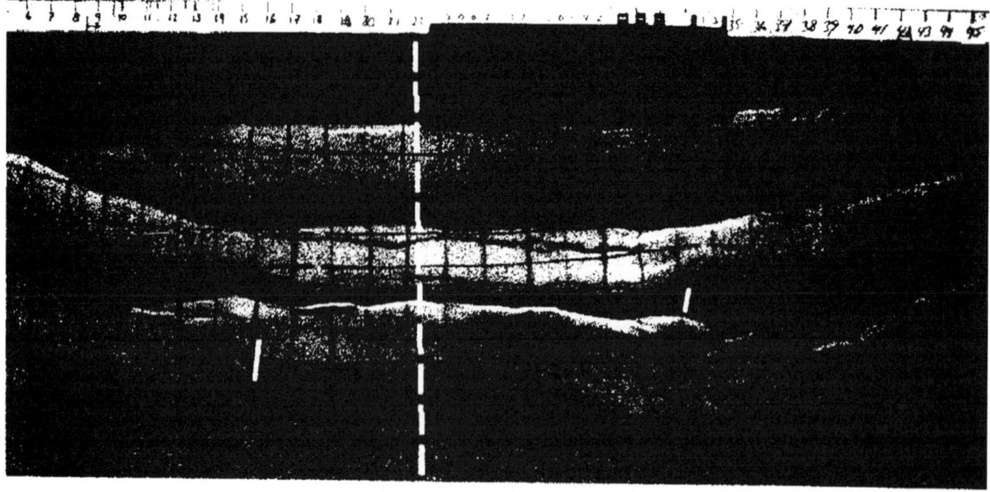

(a)

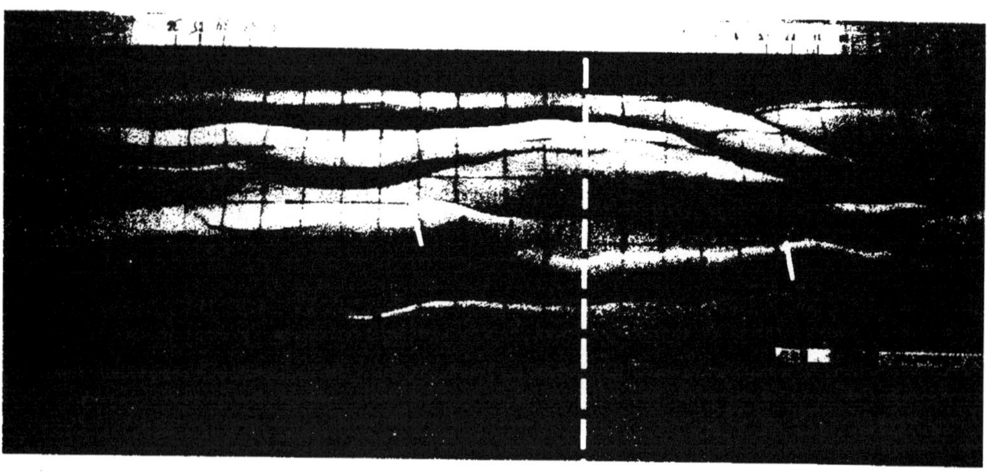

(b)

Fig. 3. Top views of sandbox experiments with two detachment layers of (**a**) an oil-water emulsion and (**b**) silicone putty. The sections for Fig. 4a and d respectively (dashed lines) were chosen to be away from the lateral thrust branches, which are affected by out-of-plane rotation (white). Grid spacing is 3 cm.

The second interpretation is arguably better: the lower parts of T and B appear as straight conjugate faults with symmetric dips (Fig. 5d). The thrusts to the left of B deformed the sand sheet above the upper detachment into disharmonic folds, also folding the upper parts of both T and B. This illustrates how palinspastic reconstruction can rule out unlikely interpretations and unravel the relative timing of some structures. However, in the absence of syn-tectonic sedimentation the sequential propagation of these non-intersecting faults cannot be resolved (Fig. 4).

From observations of the development of the surface topography of the experiments, the main thrust forms shortly prior to the related backthrust, which acts as an accommodation structure for the thrust sheet moving up the ramp. Finite element simulations (Mäkel & Walters 1993) have shown

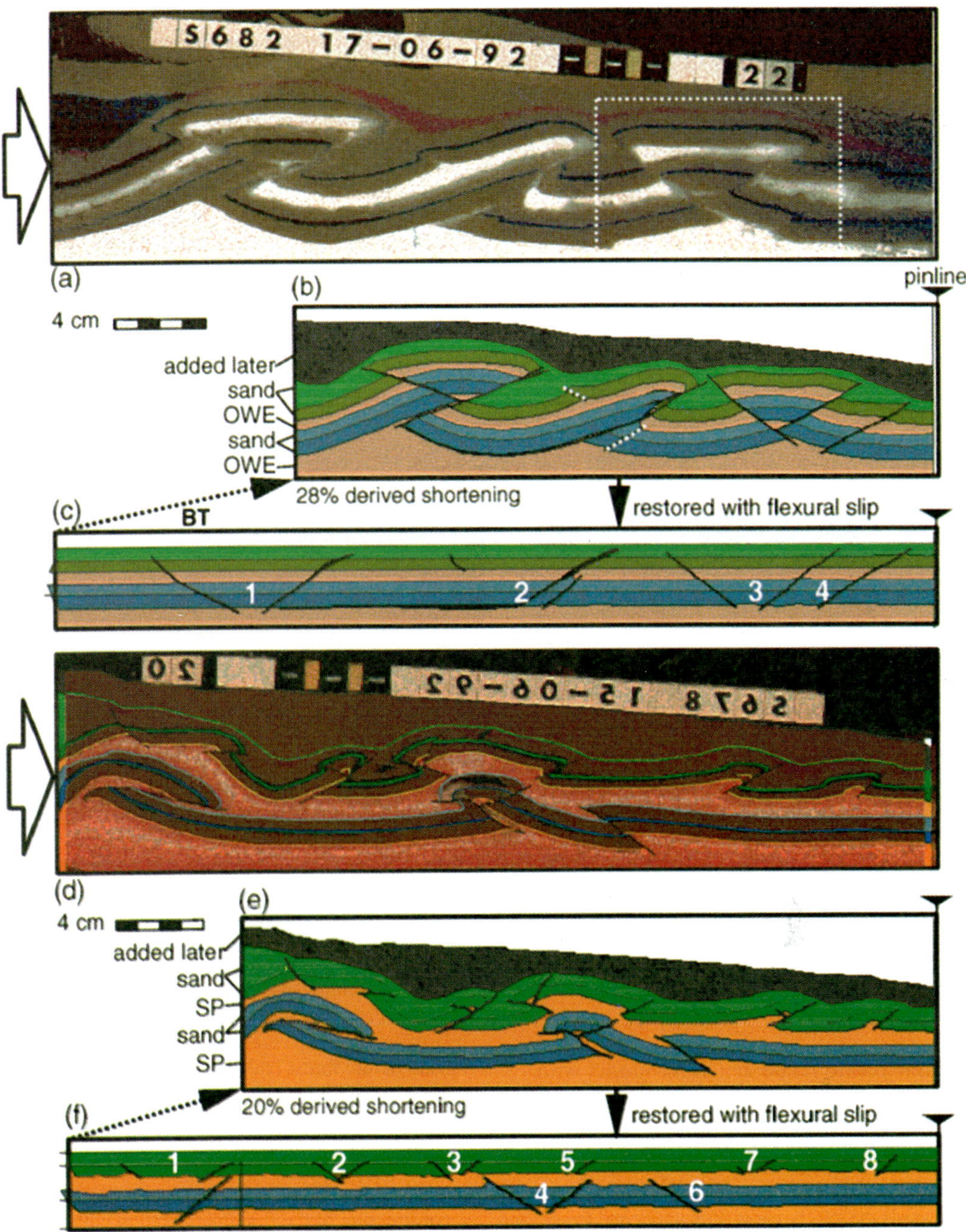

Fig. 4. Sandbox experiments of thrusting with double detachment in weak plastic material (**a**) and viscous material (**d**), together with interpretations (**b,e**) that were validated by palinspastic reconstruction (**c,f**) in GEOSEC (from CogniSeis Dev. Inc.). Coupling across the intermediate plastic layer (oil–water emulsion, OWE) is fairly strong, with most faults initiated through the complete sediment section. In contrast, a relatively thick viscous detachment (silicone putty, SP) uncouples fault initiation across it. Numbers (**c,f**) indicate sequential propagation of the thrust front, as derived from photographs taken during the experiment.

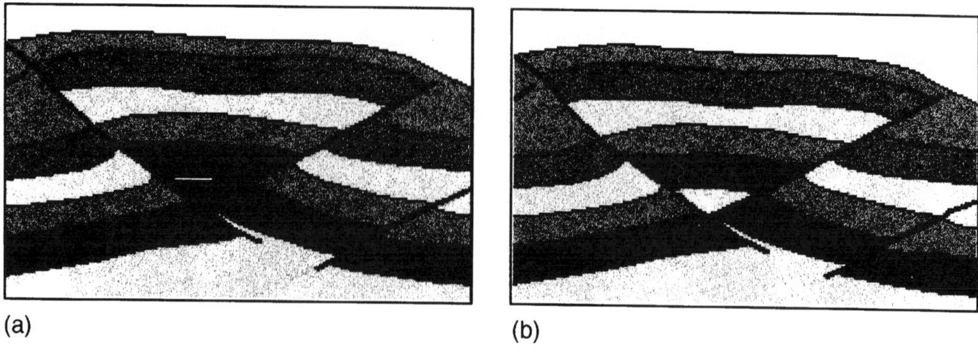

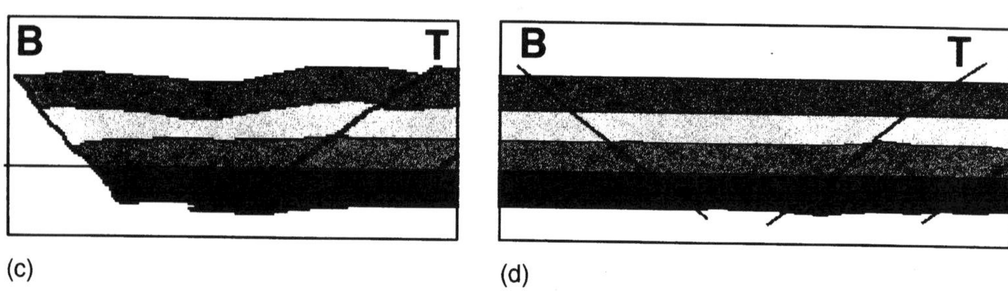

Fig. 5. Ambiguity in interpretation of sandbox experiment in Fig. 4a (white box) and a solution by palinspastic restoration. Interpretation (**a**) is questionable because of the asymmetric dip of fault *T* versus *B* and the inferred curved part of T at initiation (**c**). The preferred interpretation of this fault branch (**b**) leads to symmetric fault dips and straight initial thrusts (**d**).

that under undrained conditions in a tapered wedge the backthrust forms later than the main thrust, whereas under drained conditions, the backthrust forms just prior to the main thrust.

Palinspastic reconstruction: model with weak viscous detachments

The reconstruction of the section with a double viscous detachment layer (Fig. 4d) again proceeded from a pin line at the right end of the section towards the left, but now with upper and lower sand blocks restored completely separately with the flexural slip mechanism (Fig. 4f). All layers ended up with similar restored bed lengths. Small deviations from parallel bedding planes were left in the reconstructed section, in order to signal remaining imperfections in the interpretation, its restoration and applicability of the algorithm.

The amount of contraction derived from this reconstruction (20% in Fig. 4e) is identical to the applied shortening, which implies no measurable tectonic compaction. The contraction was completely accommodated by thrusting and folding. Indeed in contrast to the previous experiment, the relatively thin top sand layer was observed to buckle before thrust faults broke through it.

Another ambiguity in the interpretation is illustrated in Fig. 6. It is unclear how to make the distinction between a double angular fold (as defined by Suppe (1985)) and a broad shear zone that could be simplified as a fault. Palinspastic reconstruction of the first interpretation with a fold (thin white dashed line in Fig. 6a) results in a dip reversal of the adjacent fault (white dotted line in Fig. 6b), making it a normal fault, which is highly unlikely given the known deformation history and its conjugate relation to the next thrust fault to the left. This highlights the inability of GeoSec's flexural-slip algorithm to deal with slight thickness variations. A line length balance learns that this

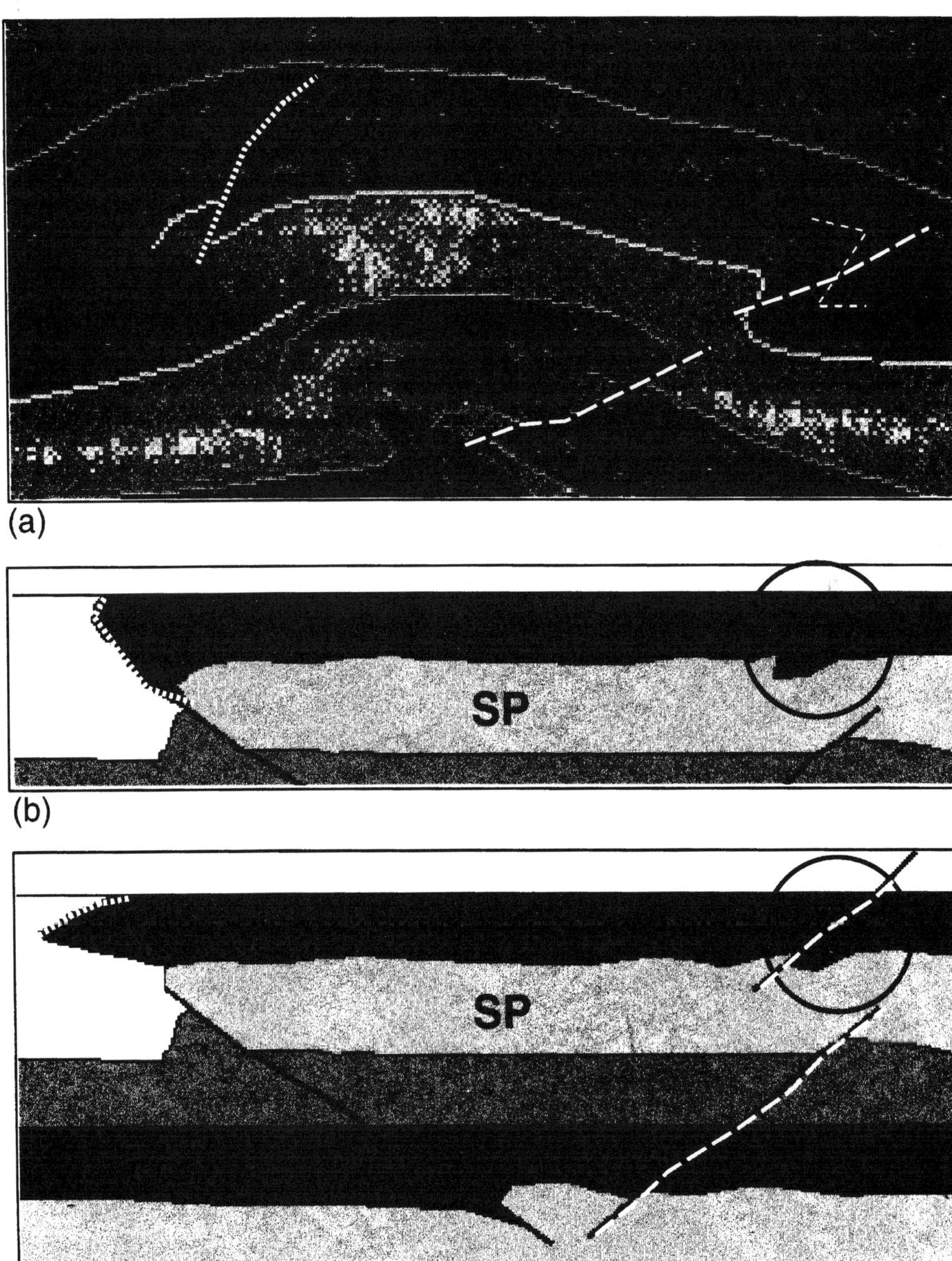

Fig. 6. Ambiguity in interpretation of sandbox experiment in Fig. 4d. A palinspastic restoration of the angular fold (white dashes in **a**) with the flexural slip mechanism leads to a dubious dip reversal of the adjacent fault (white dots in **b**). An interpretation of the same structure as a shear zone (fault in circle) does not produce this dip reversal (white dots in **c**), and highlights the complete detachment of fault initiation above and below the thick viscous detachment (dashed white faults above and below the SP).

interpretation does not lead to a dip reversal. However, the second interpretation of Figure 6a, with a reverse fault (thick dashed line) and some drag on the horizon cut-offs, restores much better with no dip reversal of the bounding fault to the left (Fig. 6c). This example serves to make a second point. One might be tempted to interpret a through-going fault across the weak silicone putty layer. However, the restoration highlights the complete decoupling of fault initiation across the thick viscous detachment.

Conclusions

Sandbox experiments provide models of thrust tectonics with multiple detachment levels of various weak lithologies. A relatively thick viscous decollement layer allows the formation of buckle folds which precede faulting, and thereby minimizes tectonic compaction. Deep seated thrust sheets tend to be longer than shallower ones.

An elasto-plastic decollement and interlayer hardly decouples layers on either side, permits much less folding and can therefore be accompanied by significant tectonic compaction.

Sandbox experiments allow alternative interpretations even though they are well constrained. Palinspastic reconstruction resolves these ambiguities in the presented cases and leads to a refined interpretation and the detection of differing tectonic compaction. In areas with poor quality seismic data, the power of palinspastic reconstruction is a fortiori greater still.

Reconstruction also confirmed the total decoupling of fault initiation above a thick viscous weak layer, and the initial straightness and regular spacing and dips of thrusts with low basal friction.

Palinspastic reconstruction can unravel the relative timing of some structures, but it fails to resolve the sequential propagation of non-intersecting thrusts in the absence of syn-tectonic sedimentation.

References

DAVY, Ph., COBBOLD, P. R. 1991. Experiments on shortening of a 4-layer model of the continental lithosphere. *Tectonophysics*, **188**, 1–25.

FERMOR, P. R. & MOFFAT, J. W. 1992. Tectonics and structure of the Western Canada foreland basin. *In*: American Association of Petroleum Geologists Memoir **55**, 81–105.

HORSFIELD, B. 1977. An experimental approach to basement controlled faulting. *Geologie en Mijnbouw*, **56/4**, 363–370.

HUBBERT, M. K. 1937. Theory of scale models as applied to the study of geological structures. *GSA Bulletin*, **48**.

KULANDER, B. R. & DEAN, S. L. 1986. Structure and tectonics of Central and Southern Appalachian Valley and Ridge and Plateau Provinces, West Virginia and Virginia. *Bulletin of the American Association of Petroleum Geologists*, 70/11, 1674–1684.

MÄKEL, G. & WALTERS, J. V. 1993. Finite-element analyses of thrust tectonics: computer simulation of detachment phase and development of thrust faults. *Tectonophysics*, **226**, 167–185.

MANDL, G., DE JONG, L. N. J. & MALTHA, A. 1977. Shear zones in granular material. *Rock Mechanics*, **9**, 95`–144.

SUPPE, J. 1985. *Principles of Structural Geology*, Prentice-Hall, New Jersey.

VENDEVILLE, B., COBBOLD, P. R., DAVY, P., CHOUKROUNE, P. & BRUN, J. P. 1987. Physical models of extensional tectonics at various scales. *In*: COWARD, P. P., DEWEY, J. F. & HANCOCK, P. L. (eds) *Continental Extensional Tectonics*. Geological Society, London, Special Publications, **28**, 95–107.

Finite element modelling of the competition between shear bands in the early stages of thrusting: Strain localization analysis and constitutive law influence

J. D. BARNICHON & R. CHARLIER

Département MSM, Université de Liège, 6 Quai Banning, 4000 Liège Belgium

Abstract: Finite element simulation of a sandbox model of thrusting is performed using large strain analysis and two different non-associated elastoplastic constitutive laws (namely Drücker–Prager and Van Eekelen criteria). The analysis of strain localization using the Rice bifurcation criterion coupled with a kinematic indicator shows that, in the early stages of imbricated thrusting, the development of major shear bands can be influenced by some competition with second order bands. The influence of the vertical/horizontal stress ratio is checked, as well as the influence of Lode angle in the constitutive law. Significant differences are found between a classical plasticity criterion (Drücker–Prager) and a more realistic one (Van Eekelen) regarding the resistance of the model and the stress paths. These differences may result in erroneous fault type prediction.

Sandbox experiments have been used extensively for many years to assist structural interpretation and validation of tectonic models. Main interests of this method come from the very large range of initial geometries which can be investigated and from the large amount of strain which can be applied. Physical parameters obtained are mainly displacement fields at different stages, especially since non destructive methods are used, e.g. X-ray tomography (Colletta *et al.* 1991).

The finite element method is also an attractive method for geological modelling as it enables computation in several parameters, examples of which are displacement, strain and stress fields, and also more recently strain localization analysis. Moreover it allows the explicit choice of mechanical characteristics, constitutive laws and boundary conditions. Abilities of the finite element method for the investigation of strain localisation along shear bands and particularly thrusting mechanism have already been shown, see, for example, Mäkel & Walters (1993).

The present paper aims to show the relative influence of the constitutive law in the results and also to analyse information obtained from localization analysis.

In the first part, some theoretical considerations are presented: basics of the large strain formulation used and principle of the finite element method are given; the choice of a representative constitutive law for sand between Mohr–Coulomb, Drücker–Prager and Van Eekelen models is discussed; a short literature overview of the strain localization analysis performed in finite element method is presented; eventually the localisation analysis used in the present paper is detailed.

The second part is mainly concerned with the simulation of a sandbox model of thrusting carried out with the LAGAMINE finite element code. Early stages of thrust development are analysed in a first simulation in the light of the Rice bifurcation criterion and of a kinematic indicator of localization. A non-associated Drücker–Prager yield criterion is used. A remeshing procedure is also performed in this first simulation to increase the applied loading value. The influence of the vertical/ horizontal stress ratio is checked in a second simulation. The last simulation investigates the influence of the third stress invariant in the plasticity surface definition using a more sophisticated yield criterion, namely the Van Eekelen.

In the last part, finite element simulation results obtained are compared with some experiments performed on sandbox models.

Theory

Simulation of geological deformation often requires taking into account the occurrence of large strains, i.e. strains larger than 1–5%. Highlights of the large strain finite element formulation are presented. These concepts have been applied for one decade to metal forming processes. They have been presented in a set of papers, see for instance Cescotto (1989) and Cescotto *et al.* (1989). The choice of a constitutive law for sand and the strain localization analysis are also discussed.

Large strains

Modelling of a solid undergoing large strains can be described in different configurations: the initial

From Buchanan, P. G. & Nieuwland, D. A. (eds), 1996, *Modern Developments in Structural Interpretation,Validation and Modelling,* Geological Society Special Publication No. 99, pp. 235–250.

unstrained configuration, the current deformed configuration, or any other configuration between the initial and the present state. In the present work, equilibrium is expressed using the updated Lagrangian formulation in the current configuration. The Cauchy stress tensor is adopted. The constitutive relation can be written in a general form

$$\overset{\nabla}{\underline{\sigma}} = fct(\underline{D}, \underline{\sigma}) \quad (1)$$

It is related to the symmetric part $\underline{D}$ of the velocity gradient $\underline{L}$

$$\underline{D} = \frac{1}{2}\ (\underline{L} + \underline{L}^T) = \frac{1}{2}\left(\frac{\partial \underline{v}}{\partial \underline{x}} + \frac{\partial \underline{v}^T}{\partial \underline{x}}\right) \quad (2)$$

where x are the current co-ordinates and $\underline{v}$ are the current velocities.

The Jaumann objective stress rate is adopted

$$\overset{\nabla}{\underline{\sigma}} = \underline{\dot{\sigma}} - \underline{\omega}\underline{\sigma} - \underline{\sigma}\underline{\omega}^T \quad (3)$$

with the rotation rate ω which is equal to the anti-symmetric part of the velocity gradient $\underline{L}$

$$\underline{\omega} = \frac{1}{2}\ (\underline{L} - \underline{L}^T) = \frac{1}{2}\left(\frac{\partial \underline{v}}{\partial \underline{x}} - \frac{\partial \underline{v}^T}{\partial \underline{x}}\right) \quad (4)$$

It must be noticed that equation (2) cannot generally be integrated, therefore strains associated with the Cauchy stress tensor do not exist. When the strains produced by the whole loading processes analysed later in this paper, the natural strain tensor will be used

$$\underline{G} = \ln \underline{U} \quad (5)$$

where the stretching tensor $\underline{U}$ is obtained by the polar decomposition of the transformation Jacobian $\underline{F}$ from the initial to the current configuration, $\underline{R}$ being the rigid body rotation tensor

$$\underline{F} = \underline{R}\underline{U} \quad (6)$$

This strain measure $\underline{G}$ coincides with the integral of (2) in principal axis if they remain unchanged.

Finite element principle

The finite element method is based on the virtual work principle (Zienckiewicz & Taylor 1988). It assumes that the local volume and surface equilibrium conditions are verified if the virtual work equation

$$\int_v \underline{\sigma}{:}\underline{\delta\varepsilon}\, dv = \int_v \rho \underline{g}_{\mathrm{r}}^{\,T}\, \underline{\delta u}\, dv + \int_a \underline{t}^{-T}\, \underline{\delta u}\, da \quad (7)$$

where $\underline{\delta\varepsilon}$: virtual strain
ρ: volumic mass
$\underline{g}_{\mathrm{r}}$: gravity acceleration
$\underline{t}$: surface forces

is satisfied for any virtual displacement $\underline{\delta u}$. The large transformation effect is here taken into account because the volume integral is computed in the current configuration. The virtual strains are

$$\underline{\delta\varepsilon} = \frac{1}{2}\left(\frac{\partial \underline{\delta u}}{\partial \underline{x}} + \frac{\partial \underline{\delta u}^T}{\partial \underline{x}}\right) \quad (8)$$

The virtual work equation is computed in the current configuration at the end of each load step.

In the finite element method, equation (7) is only verified for a limited number of virtual displacement fields associated with the mesh degrees of freedom. Hereafter isoparametric finite elements with eight nodes (parabolic displacement field) and four Gauss integration points (under-integration of the virtual work) are used.

The virtual work equation is non-linear since it is using an integration on a configuration and on stresses which are defined at the end of the step and unknown at the beginning of the step. This non linearity is solved using the Newton Raphson iterative method.

Constitutive laws for geomaterials

Regarding the choice of a representative constitutive law for sand, the question is open as the actual behaviour of the sand under small confining pressure is not very well known. However if we consider mainly internal frictional characteristics of sand, several constitutive models can be put forward. In order to avoid confusion, the choice of stress sign follows geomechanics convention, i.e. compressive ones are counted as positive.

Mohr–Coulomb criterion

The well known Mohr–Coulomb failure criterion expresses a linear relationship between the shear stress τ and the normal stress σ_n acting on a failure plane

$$\tau = C + \sigma_n \tan\phi \quad (9)$$

Parameters ϕ (friction angle) and C (cohesion) are commonly identified from conventional triaxial compression tests (i.e. $\sigma_1 > \sigma_2 = \sigma_3$) on cylindrical specimens by construction of the Mohr envelope at failure. Therefore the friction angle ϕ found corresponds to a compressive stress path, it can be called 'friction angle in compression' and noted ϕ_C. Mohr–Coulomb theory assumes that this friction angle does not depend on the stress path, for instance friction angles in compression ϕ_C and in tension ($\sigma_1 < \sigma_2 = \sigma_3$) ϕ_E are equal (Desai & Siriwaradane 1984).

However this model is not convenient to use in numerical analysis mainly because its representation in the principal stress space (σ_1, σ_2, σ_3) is an irregular hexagonal pyramid leading to singularities (the normal of the surface at intersection lines is not unique). This feature can be seen on Fig. 1 which represents the trace of this plasticity surface in the deviatoric plane.

Drücker–Prager criterion

An alternative solution to overcome this difficulty has been proposed by Drücker & Prager (1952) who defined the yield function F using a linear relationship between the first stress tensor invariant and the second deviatoric stress tensor invariant

$$F = \sqrt{II_{\hat{\sigma}}} - mI_\sigma - k = 0 \qquad (10)$$

with

$$I_\sigma = \sigma_{kk} \qquad (11)$$

$$II_{\hat{\sigma}} = \frac{1}{2}\,\hat{\sigma}_{ij}\hat{\sigma}_{ij} \quad \text{where } \hat{} \text{ represents the deviatoric part of the tensor} \qquad (12)$$

In the principal stress space, the plasticity surface becomes a cone which is much easier to use in numerical algorithms. The trace of this plasticity surface on the Π plane (i.e. the deviatoric plane) is then a circle (see Fig. 1).

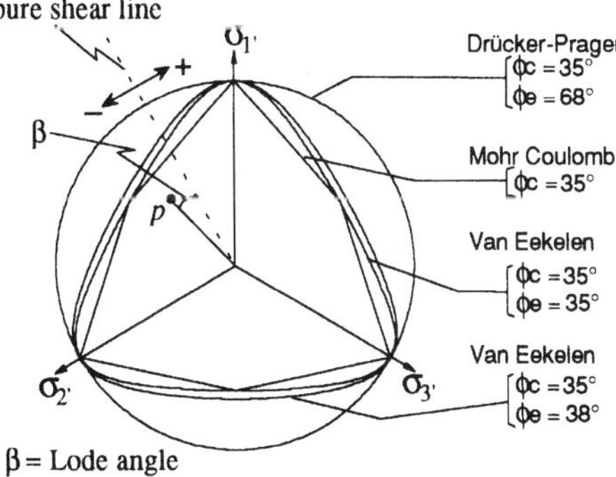

Fig. 1. Limit surface for Mohr–Coulomb, Drücker–Prager and Van Eekelen models in the deviatoric plane. Note that Drücker–Prager criterion can either be chosen circumscribed or inscribed to the Mohr–Coulomb. Lode angle $\beta = 30°$ with respect to the pure shear line corresponds to triaxial compression and $\beta = -30°$ to triaxial extension. Van Eekelen model is a smoothing of Mohr–Coulomb criterion which can respect both compression and extension friction angles, but which can also take different values for these two angles.

However this quite simple criterion has one main disadvantage very often neglected and which must be outlined: it does not incorporate a dependence on the third stress invariant and thus on the Lode angle β

$$\beta = -\frac{1}{3}\sin^{-1}\left[\frac{3\sqrt{3}}{2}\,\frac{III_{\hat{\sigma}}}{II_{\hat{\sigma}}^{3/2}}\right] \quad \text{with } III_{\hat{\sigma}} = \frac{1}{3}\,\hat{\sigma}_{ij}\hat{\sigma}_{jk}\hat{\sigma}_{ki} \qquad (13)$$

which is the angular position of the projection p of a stress state on the Π plane with respect to the pure shear line (see Fig. 1). From Fig. 1, β factor is in the range –30° to 30° triaxial compressive stress paths correspond to $\beta = 30°$, whereas triaxial extensive ones correspond to $\beta = -30°$. Consequently mobilized friction angles are either over or underestimated depending on the stress path, leading to discrepancy in the results obtained (Schweiger 1994): a compression cone (i.e. a circumscribed cone) induces high friction angle for extension paths, whereas the internal cone leads to low equivalent friction angles.

This can be illustrated considering the case where the compression cone is chosen:

– identification of parameters m and k on the Mohr envelope as a function of internal friction angle in compression ϕ_C and cohesion C then gives (Desai & Siriwaradane 1984)

$$m = \frac{2\sin\phi_C}{\sqrt{3}(3-\sin\phi_C)}, \quad k = \frac{6C\cos\phi_C}{\sqrt{3}(3-\sin\phi_C)} \qquad (14)$$

– with the definition of reduced radius r

$$r = \frac{II_{\hat{\sigma}}}{I_\sigma}, \qquad (15)$$

expressions of the reduced radius in compression (r_C) and in extension (r_E) for triaxial tests can be deduced from Mohr circle and intrinsic curve, leading to

$$r_C = \frac{1}{\sqrt{3}}\left(\frac{2\sin\phi_C}{3-\sin\phi_C}\right), \quad r_E = \frac{1}{\sqrt{3}}\left(\frac{2\sin\phi_E}{3+\sin\phi_E}\right) \qquad (16)$$

which, putting $r_E = r_C$ as the radius is constant, gives a simple relation between ϕ_C and ϕ_E. For instance if a value $\phi_C = 35°$ is chosen, then ϕ_E takes the value $\phi_E = 68°$ which is much higher than ϕ_C and far from being realistic.

This can also be seen graphically on Fig. 1: the Drücker–Prager cone is a only a poor representation of the Mohr–Coulomb hexagon as one can either chose a circumscribed cone (Fig. 1), an inscribed cone, or any intermediate choice between the last two. None of these choices is actually satisfactory for any stress path.

Classically, the flow rule is associated if $G = F$, where G represents the flow potential, which implies that the plastic flow is normal to the yield surface and leads to an excess of dilatancy. A non-associated flow rule G can be defined in an identical fashion as the yield function F (see relations (10) and (14)) simply substituting ϕ by ψ in equation (14), ψ being the dilatancy angle.

Van Eekelen criterion

More sophisticated models can be built from the Drücker–Prager cone by introduction of a dependence on the Lode angle β in order to match more closely Mohr–Coulomb criterion. Several models of this type have been described, including the Matsuoka–Nakaï and Van Eekelen. The Van Eekelen plasticity criterion (Van Eekelen 1980)

$$F = aI_{\sigma}(1 - \mathrm{b} \sin 3\beta)^{n} - \sqrt{II_{\hat{\sigma}}} = 0 \qquad (17)$$

allows an independent choice for ϕ_C and ϕ_E via coefficients a and b

$$b = \frac{\left(\dfrac{r_C}{r_E}\right)^{1/n} - 1}{\left(\dfrac{r_C}{r_E}\right)^{1/n} + 1}, \quad a = \frac{r_C}{(1 + b)^{n}} \qquad (18)$$

where r_C and r_E are given by equation (16). The exponent n actually controls the convexity of the yield surface, condition generally required in classical plasticity framework. A parametric study showed that the convexity is always verified with $n = -0.229$ provided that $\phi_E \geq \phi_C$ which is a realistic assumption for sand.

The trace of this plasticity surface in the Π plane is shown on Fig. 1. Such a model is actually a smoothing of the Mohr Coulomb hexagon, but it fits much better the Mohr–Coulomb criterion and experimental data than the Drücker–Prager failure criterion (Matsuoka & Nakai 1982). Moreover this criterion allows for an independent choice of friction angles in compression and extension (see Fig. 1 with $\phi_C = 35°$, $\phi_E = 38°$).

As for Drücker–Prager model, the flow rule G can either be defined associated ($G = F$) or non-associated ($G \neq F$).

A noteworthy feature of this more sophisticated criterion is that no additional material parameters are required compared to a Drücker–Prager criterion. Simply the friction angle in extension ϕ_E is explicit here whereas it was implicit in the Drücker–Prager criterion.

Contact friction law

For rigid wall/sand contacts, the Coulomb friction law (Charlier & Cescotto 1988) is generally considered

$$F = \sqrt{\tau_S^2 + \tau_T^2} - C - \mu P = 0 \qquad (19)$$

where μ is the friction coefficient, C the cohesion, P the contact pressure, τ_S and τ_T the shear stresses. The corresponding boundary condition is implemented using interface finite elements (Charlier & Cescotto 1988; Charlier & Habraken 1990; Cescotto & Charlier 1993).

Bifurcation and shear band localization of deformation

Bifurcation phenomenon can be viewed as alternative stress and strain paths under a given loading condition, which traduces mathematically by the loss of uniqueness of the solution. Strain localization along shear bands, which represents one potential bifurcation mode, is known to occur frequently in frictional materials.

The study of problems involving shear band localization by the finite element method faces several difficulties.

(1) The constitutive relation considered in the finite element model must allow for bifurcation. For elastoplastic laws it has been demonstrated that localization requires at least either a hardening/softening law or a non-associated one (Rice 1976). Bifurcation can occur at the peak load for strain softening models, but also in the hardening range if a non-associated law is considered (Rice 1976). This last feature has actually been observed in experiments on dense sands (Desrues 1984) in which localization occurs before the stress peak.

(2) Unique solution at the bifurcation point must be determined. If strain softening is introduced in a classical continuum model, the condition for localization coincides with the loss of ellipticity of equations. As a result, the finite element solution shows pathological mesh dependence because the band thickness is undetermined (De Borst 1993). In classical finite element computations, the band thickness has usually the size of an element (Ortiz *et al.* 1987; De Borst & Sluys 1991). This difficulty is still a very active field of research, and up to now 3 main approaches have been developed to overcome this problem by introduction of an internal length scale in the problem, which causes the problem to remain well-posed at localization: higher order strain gradients (De Borst 1992) introduced in the constitutive equation; non local constitutive relations in which some variables are taking into account the spatial variation of strain and micro-polar (or Cosserat) continuum in which rotational degrees of freedom are added in the element formulation (De Borst 1993). However a meaningful physical interpretation for the internal

scale has still not been found. Therefore identification of this parameter is an open question and obviously the value chosen in numerical analyses is arbitrary.

(3) The post-bifurcation behaviour must be correctly computed, and here several approaches have been described in literature. They aim to provide an improved strain computation in shear bands (which are characterized by high strain gradients): mesh refinement around the band, higher order shape functions of finite elements (Ortiz *et al.* 1987) are some of the possible strategies.

The simulations presented in this paper are performed in the framework of classical continuum theory. Non-associated constitutive relations are considered to allow for bifurcation. A local analysis of strain localization is performed at the element level based on a criterion for shear band bifurcation (Rice 1976; Wang 1993). A kinematic indicator is also found useful to follow accurately the evolution of strain localization, especially initiation. An adaptive remeshing technique has also been used here to avoid large distortion of elements and thus enables the modelling to be carried out until larger loading values. All these features are described more explicitly in the next three sections.

Local shear band bifurcation criterion

The Rice criterion (Rice 1976) analyses the stress state and investigates the possibility of a bifurcation by formation of a shear band in the stress and strain paths. The theoretical scheme of a shear band is presented in Fig. 2. Its development has become classical. It is based on a kinematic condition, a static condition and on the constitutive equation. We note ^{0} variables outside the band and ^{1} variables inside the band.

The static condition expresses the surface equilibrium at the interface between the inner and outer band

$$n_j(\dot{\sigma}^1_{ij} - \dot{\sigma}^0_{ij}) = 0 \qquad (20)$$

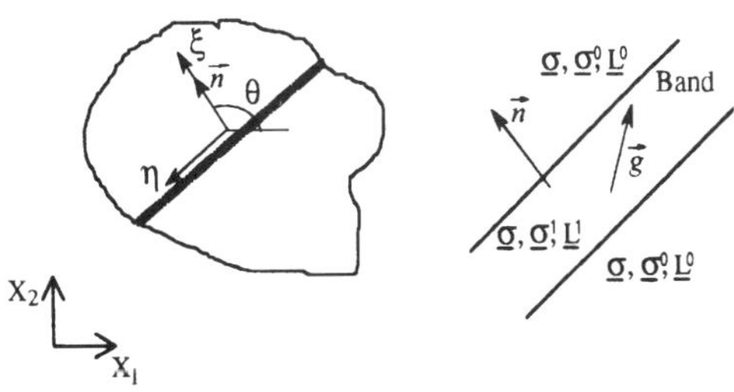

Fig. 2. Theoretical scheme of a shear band.

The kinematic condition expresses the strain jump across the band interface by a dilatant strain jump and a shear strain jump, but without any longitudinal strain jump

$$\underline{L}^1 = \underline{L}^0 + \vec{g} \otimes \vec{n} \qquad (21)$$

where $\underline{L}$ is the velocity gradient, $\underline{n}$ is the normal to the band, and $\underline{g}$ is a vector describing the band mode (from shearing to dilatation, from mode II to mode I).

The third equation introduced in the Rice criterion is the constitutive law. For an elastoplastic or an elastic law, one has

$$\dot{\underline{\sigma}} = \underline{\underline{D}}\underline{L} \qquad (22)$$

where the constitutive tangent tensor $\underline{\underline{D}}$ includes the objective stress rate correction, usually the Jaumann one.

For an elastoplastic law this equation becomes

$$\dot{\underline{\sigma}} = \underline{\underline{C}}^{ep} \frac{1}{2}(\underline{L} + \underline{L}^T) + \underline{\sigma}\frac{1}{2}(\underline{L} - \underline{L}^T)^T + \frac{1}{2}(\underline{L} - \underline{L}^T)\,\underline{\sigma} \qquad (23)$$

It is important to point out that equation [23] assumes loading or unloading stress paths. The obtained bifurcation criterion is therefore a lower bound (if the loading tensor is used) and not an exact one.

Introducing (23) into (20) and (21), one obtains a third order equation system which unknowns are the components of the $\underline{g}$ vector. The trivial solution $\underline{g} = 0$ is always possible but it means that not shear band can appear. The condition $\underline{g} \neq 0$ can be transformed in a fourth order equation in $\tan(\theta) = t$, with θ being the angle between the band normal and the x axis (Wang 1993)

$$at^4 + bt^3 + ct^2 + dt + e = 0 \qquad (24)$$

At the beginning of loading, this equation does not have any real solution. After some load steps, the first real solution is a double one. More generally one or two solutions are possible.

The Rice bifurcation criterion indicates the possibility of a shear band appearance from a local stress point of view. It actually does not show developing shear bands effectively. From a practical point of view, this criterion is computed at every iteration for all integration points.

Kinematic indicator of localization

Analysis of the strain field can actually show strain localization. The simplest analysis of a shear band apparition is based on the visualization of the cumulated equivalent strain map

$$\varepsilon_{eq} = \sqrt{\hat{G}_{ij}\hat{G}_{ij}} \qquad (25)$$

where G is given by relation (5). However, maps of this type show well achieved shear bands when localization has strongly developed, but does not give the possibility of showing early stages of localization.

In order to overcome this deficiency, a kinematic scalar indicator α can be defined, based on some propositions by Vilotte *et al.* (1990)

$$\alpha = \frac{\Delta\dot{\varepsilon}_{eq}\,\Delta t}{\varepsilon_{eq}} \tag{26}$$

where Δt is the numerical time step increment and $\Delta\dot{\varepsilon}_{eq}$ is the incremental deviatoric strain rate.

This indicator represents the incremental equivalent strain related to cumulated equivalent strain. It enables localization to show much earlier than relation (26).

Adaptive remeshing

Under the intense shearing occurring in shear bands, elements are distorted and quickly fail. Remeshing techniques are powerful methods to overcome this difficulty and to enable the simulation to carry on.

Adaptive remeshing procedures have been initially developed in fluid dynamics and in metal forming modelling (see for example Cheng (1988) and Habraken & Cescotto (1990)). The general procedure is in three steps:

(1) evaluation of the remeshing need, based on some indicator. One can use an error estimate algorithm (Zienkiewicz & Zhu 1987) or a localization indicator. In the present case, remeshing is based more arbitrarily on a convergence criterion.
(2) creation of a new mesh, based on a meshing algorithm with variable element density in order to produce a mesh adapted to the deformations to be modelled.
(3) transfer of state variables (mainly stresses) and kinematics variables (nodal velocities) from the old mesh to the new one using the following interpolation algorithm

$$a(\underline{x}^J) = \begin{cases} a(\underline{x}^i) & \text{if } r_{iJ} \le r_0 \\ \dfrac{\sum_{\text{NINT}} \dfrac{a(\underline{x}^i)}{r^p_{iJ}}}{\sum_{\text{NINT}} \dfrac{1}{r^p_{iJ}}} & \text{if } r_{iJ} > r_0 \end{cases} \tag{27}$$

where a is the quantity to transfer, ^{i} relates to the old mesh and ^{J} to the new one, $\underline{x}$ are co-ordinates, r_{iJ} is the distance between point ^{i} and ^{J}, p is the order of interpolation and the summation sign indicates a summation over all integration points of the old mesh.

It must be pointed out that after these operations, the simulation re-start is always slightly difficult because the new configuration is never perfectly equilibrated. This difficulty appears as a small discontinuity on the load/displacement curve at remeshing points (indicated with arrows on Fig. 5).

Application: simulation of sandbox thrusting model

Geometry, boundary conditions and discretization

Plane strain two dimensional simulation of thrusting in a sandbox model has been performed using the LAGAMINE finite element code. The basic model consists of a 20 cm length/1.4 cm thick horizontal sand layer laying on a horizontal rigid basement (see Fig. 3). The right hand side of the model is horizontally fixed. A horizontal displacement u is applied to the rigid wall at the left hand side of the model using incremental displacements Δu.

Contact between rigid wall/basement and sand is managed via 58 frictional interface elements. The basal contact actually represents a basal detachment plane whose position is given a priori, i.e. which represents a major rheological discontinuity.

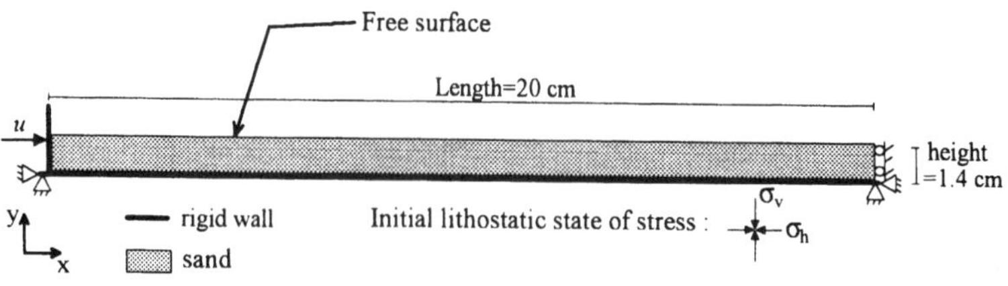

Fig. 3. Geometry and boundary conditions for the basic model.

Discretization of the sand massif is achieved using 400 elements (0.4×0.175 cm).

The convergence criterion for the Newton–Raphson iterative method is based on a displacement and a force criterion. An adaptive step size method based on convergence criteria is used in which size of loading increments is allowed to vary between 10^{-5} and 10^{-1} cm.

A balanced lithostatic stress field is introduced in the initial conditions of the model. Considering the orientation of the y axis shown in Fig. 3, vertical stresses σ_v are given by

$$\sigma_v = -\rho g_r y \tag{28}$$

and horizontal stresses σ_h are defined by the ratio K_0 with

$$K_0 = \frac{\sigma_h}{\sigma_v} \tag{29}$$

Physical parameters

Three sets of numerical simulations have been carried out. Rigid wall/sand contacts are simulated using non-associated Coulomb law; the parameters are given in Table 1. A non-associated elastoplastic law without dilatancy ($\psi = 0°$) and without hardening/softening is chosen for the sand in all three sets of experiments. A Drücker–Prager (DP) criterion is considered for experiments 1 and 2 whereas a Van Eekelen (VE) criterion is chosen for experiment 3. The major difference between those two models comes mainly from the friction angle in extension ϕ_E (see Table 1).

Results

All the maps presented hereafter represent only the left part of the model as in the right part almost nothing occurs (far field limit).

Simulation 1 (sm1): initial model

In this simulation, a remeshing procedure has been used twice in order to carry on simulation when convergence could no longer be achieved, respectively at $u = 0.5$ cm and $u = 0.8$ cm, u being the displacement applied to the rigid wall. Therefore this simulation consists of three consecutive runs from which several computed results can be analysed.

Equivalent strain. Figure 4 represents isovalue maps of the evolution of equivalent strain cumulated (see equation [25]) over each remeshing phase (i.e. between 0 and 0.5 cm, 0.5 and 0.8 cm, 0.8 and 0.9 cm respectively), i.e. equivalent strain is reset to zero at the begin of each remeshing. As this deformation measurement is cumulated over one run, it gives rather rough indication of what actually happens within the model. However it clearly shows the formation of three main shear bands.

(a) From $u = 0$ to 0.5 cm (Fig. 4a), a major synthetic reverse shear band (called band n. 1) develops in front of the rigid wall. This band is linked to the free surface but does not reach the bottom part of the model.

(b) From $u = 0.5$ to 0.8 cm (Fig. 4b), a second major synthetic reverse shear band (called band n. 2) develops in front of the band n. 1, with the same orientation than the latter. However an antithetic reverse shear band (band n. 3) has also developed to a lesser extent; these two reverse faults define a pop-up structure. Clearly band n. 1 has been inactive during this stage.

(c) From $u = 0.8$ to 0.9 cm (Fig. 4c), most of the deformation is achieved along the band n. 2 in which cumulated equivalent strain reaches locally 30%. Little deformation is achieved along band n. 3, and a new antithetic reverse band (band n. 4) is developing on the left of band n. 3. The close left boundary does not allow to argue for relevance of this latter band.

Strain localization analysis. Localization of deformation can be analysed using information obtained from the Rice criterion and from the kinematic indicator α. It is also interesting to incorporate in this analysis the force/displacement curve given in Fig. 5.

The localization of deformation along band n. 1 is quite clear, four phases can be distinguished.

(1) From $u = 0$ to 0.247 cm

Table 1. *Physical parameters for the three sets of experiments*

Simulation	Contact C (Pa)	Contact μ	Sand Law type	E (Pa)	ν	C (Pa)	ϕ_C (°)	ϕ_E (°)	ψ (°)	ρ (kg/m^3)	K_0
1	0	1	DP	$5\ 10^4$	0.2	20	35	68	0	2250	0.25
2	0	1	DP	$5\ 10^4$	0.2	20	35	68	0	2250	1.0
3	0	1	VE	$5\ 10^4$	0.2	20	35	35	0	2250	0.25

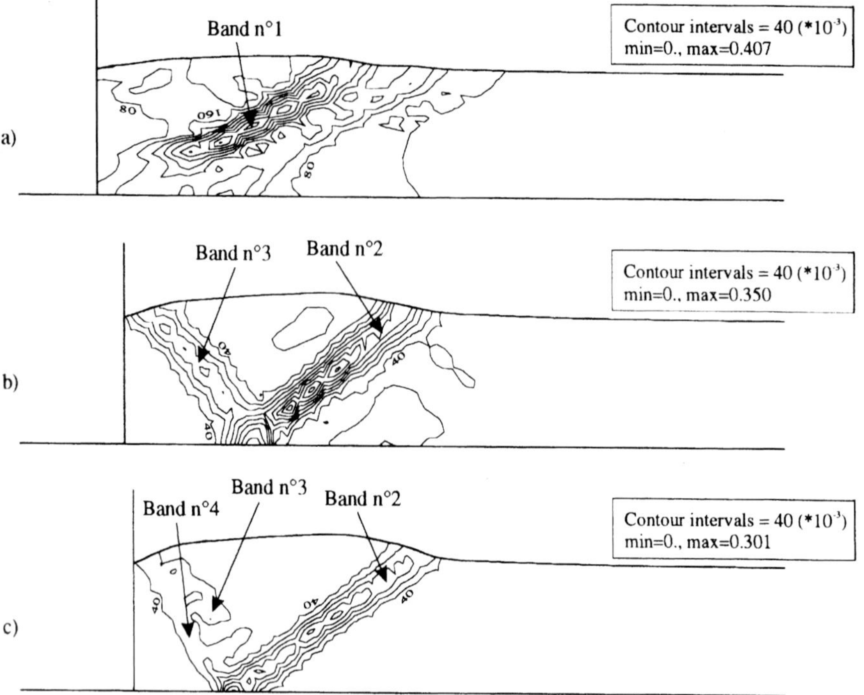

Fig. 4. Isovalues maps of cumulated equivalent strain for each respective mesh of simulation 1. (**a**) $u = 0$ to 0.5 cm: mainly the synthetic reverse shear band n°1 develops.(**b**) $u = 0.5$ to 0.8 cm: the synthetic reverse shear band n°2 and in a lesser extent the antithetic shear band n°3 have developed during this phase. (**c**) $u = 0.8$ to 0.9 cm: mainly the band n°2 is active, band n°3 is only a little active. A new antithetic reverse shear band (n°4) starts to develop.

The loading curve slope is positive (Fig. 5), the Rice criterion and the kinematic indicator give rather diffuse information. However as the imposed displacement gets closer to $u = 0.247$ cm, the slope of the loading curve decreases progressively and the kinematic indicator shows that localisation starts along band n. 1.

(2) From $u = 0.247$ to 0.257 cm

The loading curve slope is negative (Fig. 5) corresponding to an unloading phase. It is worth mentioning that such phenomenon can be referred to as structural (or geometrical) softening because no hardening/softening is included in the constitutive equation. Therefore in the model, a global 'softening' response does not even require introduction of softening in the constitutive law.

The Rice criterion (Fig. 6a) is represented at each potentially bifurcated point using arrows which indicate the potential two directions for bifurcation (computed from θ given by equation [24]). During this phase, the two localization indicators give almost the same result, i.e. all the deformation which occurs in the model is localized along band n. 1 (see Figs 6a and b). This phase corresponds to an active phase of localization. One must point out that during this phase of active localization, numerical convergence could only be achieved with very small loading increments ($\approx 10^{-5}$ cm).

(3) From $u = 0.257$ to 0.29 cm

The loading curve slope becomes positive again (Fig. 5). The Rice indicator gives again a rather diffuse information, whereas the kinematic indicator shows that two 'second order' shear bands appear in the basal area and compete with the main band n. 1 ('second order' refers to the point that these bands develops from a major band where almost the whole strain is achieved). One of these second order shear is shown at $u = 0.27$ cm on Fig. 6c. These second order bands indicate a clockwise rotation of the base of the main shear band n. 1.

(4) For $u > 0.29$ cm

As the applied displacement u increases further, band n. 2 and 3 develop. Formation of these 2 faults does not appear in a very active localization phase as band n. 1: for example see Fig. 5 where the negative slope phase at $u = 0.325$ cm is quite reduced compared with the one at $u = 0.25$ cm.

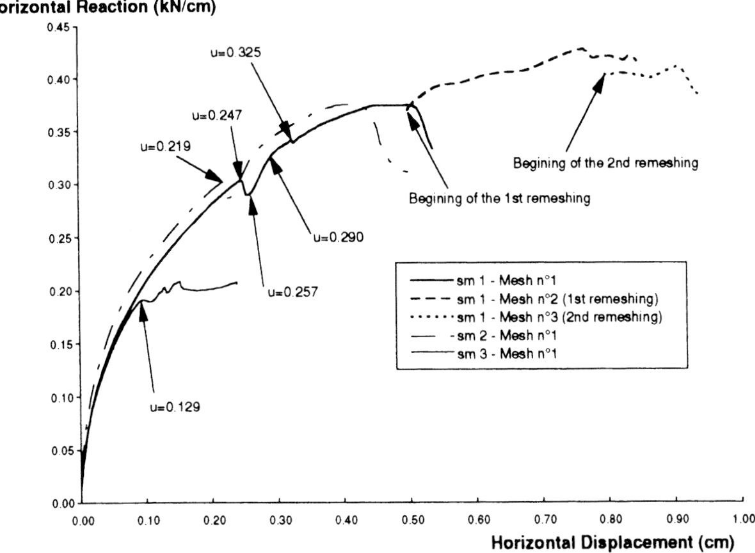

Fig. 5. Plot of the horizontal displacement versus horizontal reaction for the three simulations. For sm1 and sm2, bifurcation appears as an unloading phase at $u = 0.247$ and 0.219 cm respectively. For sm3, bifurcation occurs earlier, i.e. at $u = 0.129$ cm. Note also that prior localization, the reaction is lower for sm3 compared with sm1 and sm2.

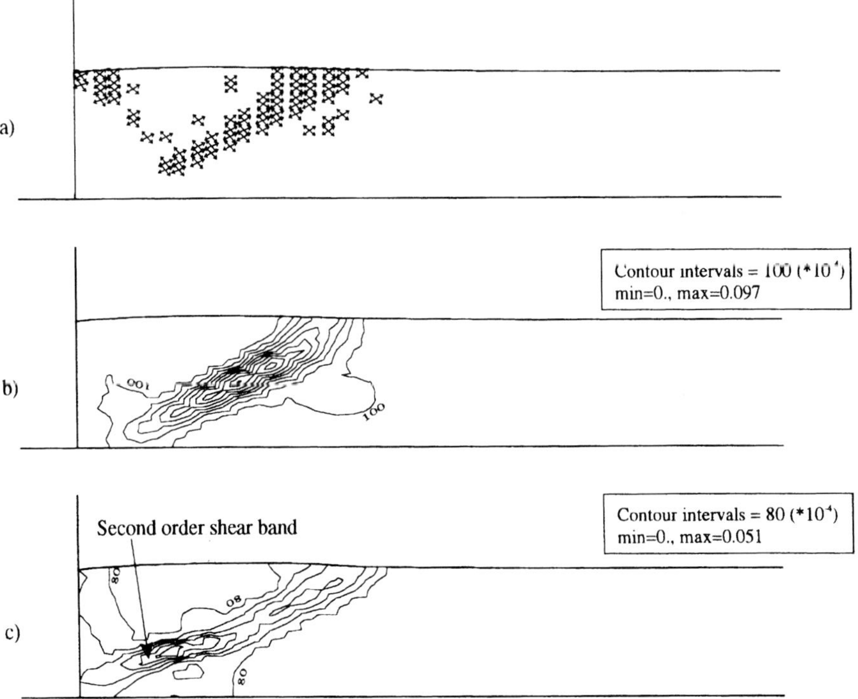

Fig. 6. Strain localization analysis for simulation 1. (**a**) Rice bifurcation at $u = 0.25$ cm: the two potential directions for bifurcation show localization along band n°1. (**b**) Isovalues map of the kinematic indicator α at $u = 0.25$ cm: localization occurs clearly along band n°1. Note that information is comparable to (a). (**c**) Isovalues map of the kinematic indicator α at $u = 0.27$ cm: a 'second order band' is developing at the bottom part of the model and indicates a clockwise rotation of band n°1.

The 'fault activity' (which is represented by the kinematic indicator α) shows non negligible variations as bands n. 2 and 3 are initiated. It actually illustrates transient faulting phases (i.e. which do not accommodate important deformation) and shear band competition. It also shows locking of some major bands (see for example band n. 1 for $u > 0.5$ cm). A sketch of the overall fault activity between $u = 0$ and $u = 0.9$ cm is given in Fig. 7. In this diagram, locations referred as 'low fault activity' are sketched to record that localization has occurred in these areas at some time during the loading process. No localisation is actually occurring at the respective loading stage they are sketched.

This competition between bands is likely to be partly induced by the elastoplastic parameters chosen, especially regarding softening: as no hardening/softening is taken into account in the constitutive relation, the material remains 'virgin' or unweakened even when it experiences deformation. However such phenomenon of competition between shear bands has also been observed in experiments performed on sand samples (Desrues 1984), i.e. on strain hardening/softening materials.

The localization analysis enables study at a small loading scale where localization actually occurs and allows visualization of competition phenomenon between bands as well as second order faults, phenomenon which were not seen with classical equivalent strain maps.

Shear band orientation. Principal stress tensors are plotted on Fig. 8a. Maximum principal stress (σ_1) directions deduced from the stress field at $u = 0.2$ cm are represented in dotted lines on Fig. 8b. Note the clockwise rotation of σ_1 direction close to the basal interface, which is induced by shear stresses on the sand/rigid basement interface.

This principal stress rotation has two main effects on orientation of shear bands, whose positions are illustrated on Fig. 8b:

- bands n. 1 and 2 exhibit concave upward shapes, the actual dip (angle with respect to horizontal) for band 1 and 2 varies vertically between 25 and 38°,
- dip for band n. 3 is approximately 46°, which is a much higher angle than for the other 2 bands.

Identical effects of basal shear stresses on shear band orientation have already been suggested by some previous authors (Mandl & Shippam 1981). This has also been verified in some finite element analysis (Makel & Walters 1993).

If angles of shear bands with respect to σ_1 direction are now considered, it is found that all the bands develop with an approximate angle $\Theta = 36°$ with respect to σ_1. This angle value is compatible with the range theoretically predicted for granular media

$$\frac{\pi}{4} - \frac{\phi}{2} \leq \Theta_{theo} \leq \frac{\pi}{4} - \frac{\phi + \psi}{4} \quad (30)$$

where the lower bound is obtained from classical Coulomb prediction and the upper bound is given by Arthur *et al.* (1977). Here with $\phi = \phi_C = 35°$ and $\psi = 0°$, equation (30) leads to

$$25.7 \leq \Theta_{theo} \leq 36.25°$$

The value obtained $\Theta = 36°$ is compatible with Θ_{theo}. If the upper bound for friction angle is

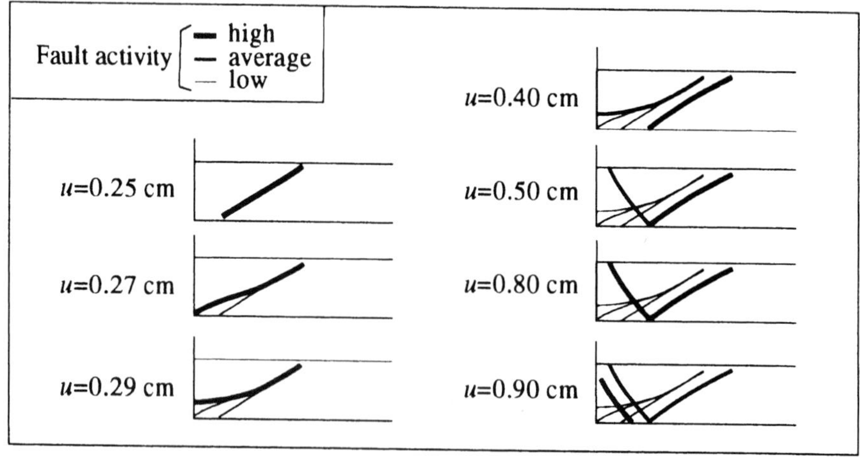

Fig. 7. Schematic relative fault activity (based on kinematic indicator α) for simulation 1.

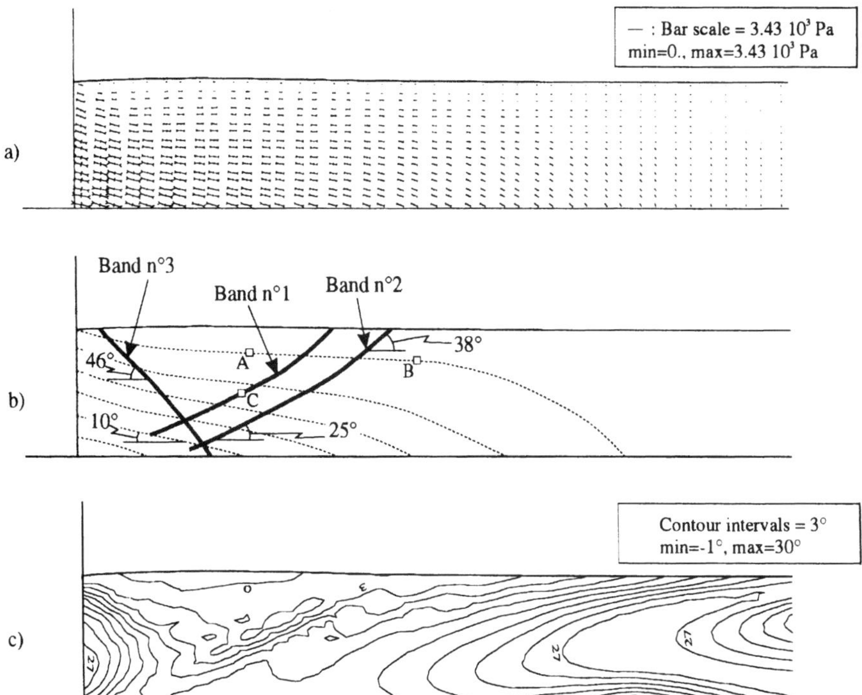

Fig. 8. Simulation 1. (**a**) Stress tensor representation at $u = 0.20$ cm, i.e. prior localization. (**b**) Dotted lines: maximum principal stress (σ_1) direction built from the stress field given in (a) at $u = 0.20$ cm; in the left part of the model, stresses are clockwise rotated due to shear stresses which develop at the model basement. Bold lines: schematic location of the future shear bands with approximate angles values at initiation (location of points A, B and C refer to Fig. 12). (**c**) Lode angle (β) isovalues map at $u = 0.25$ cm: in most of the left part, β falls in the range $-1 < \beta < 9°$, which indicates a stress path with a shearing component.

considered (i.e. $\Theta_E = 68°$), the predicted friction angle given by equation (30) is lower than the values obtained.

As expected, the thickness of the band is found to be approximately the size of one element. It must also be pointed out that no evident relation has been found to determine the position of the band. Clearly there is some competition between shear stresses on the basal interface and internal shear resistance of the material. It must be pointed out that in the early stages of localization (see Figs 6a–b), the shear band n. 1 does not initiate from the base of the model (see Figs 6a–b), whereas this is the case when no frictional interface is given in initial conditions (Makel & Walters 1993).

Lode angle. Lode angle value β (see equation (13)) indicates whether the current state of stress follows compressive ($\beta = 30°$) or extensive ($\beta = -30°$) stress paths. From the isovalue map of β presented in Fig. 8c, it is clear that Lode angle values fall in the range $-1 < \beta < 30°$ if the left part of the model is considered. As we get closer to the bands, β actually ranges between -1 and 9° which indicates a stress path with shear component, which is far from being purely compressive. As a result, mobilized friction angle ϕ_m deduced from the DP criterion in this part of the model is much higher than the value $\phi_C = 35°$ (see description of Drücker–Prager criterion) and ranges between 35 and 55°. Effects of this friction angle over-estimation will be investigated in simulation 3 using Van Eekelen criterion.

Simulation 2 (sm2): influence of the K_0 ratio

The aim of this second simulation is only to check the influence of K_0 ratio on localisation. Therefore the difference between this simulation and sm1 comes from the K_0 value which here is equal to one, that is the initial state of stress is hydrostatic. Computations are only performed for displacement values up to $u = 0.5$ cm without remeshing.

Regarding the loading/displacement curve (see Fig. 5), one can observe that localization occurs here for an identical loading value than for sm1, i.e. approximately 0.30 kN cm^{-1}. For identical displacement values, the horizontal reaction is slightly

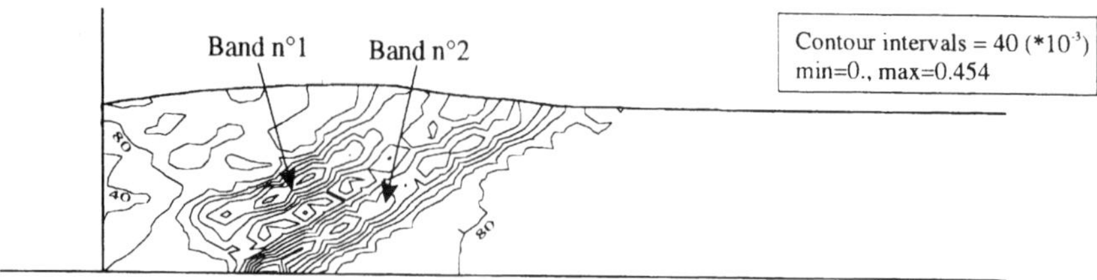

Fig. 9. Isovalues maps of cumulated equivalent strain for simulation 2 for $u = 0$ to 0.5 cm. Bands n°1 and 2 have developed as in sm1. However as localization occurred earlier in this simulation than in sm1, these 2 bands are better developed, especially band n°2 (compare with Fig. 4a at identical loading).

higher here than for sm1. This results directly from the increase of confining pressure induced by K_0 ratio. The two curves have almost identical shapes apart that localization occurs earlier in sm2 than in sm1 ($u = 0.219$ cm and $u = 0.247$ cm respectively).

Regarding displacement field and strain localization computed at the end of the simulation, direct comparison of equivalent strain at $u = 0.5$ cm between sm2 (Fig. 9) and sm1 (Fig. 4a) can be made. Few differences are found with sm1: the main difference comes actually from the band n. 2 which is markedly more developed here. This results from the earlier localization which occurs in this simulation.

Simulation 3 (sm3): influence of the Lode angle

The difference between this simulation and sm1 comes from the plasticity criterion which here is a Van Eekelen one whereas a Drücker–Prager was used in sm1. Computations are performed for displacement values up to $u = 0.22$ cm without remeshing.

Strain localization. Localization occurs earlier in this model ($u = 0.129$ cm), i.e. for lower loading values than in sm1 and sm2 (see Fig. 5). Map of the kinematic indicator a presented on Fig. 10a shows that two imbricated synthetic shear bands have started to develop in the left part of the model at $u = 0.2$ cm, which is qualitatively similar to sm1 and sm2. However the analysis of the equivalent strain map (Fig. 10b) shows that the main part of deformation (locally up to 80%) has occurred close to the basement model along a basal slip. This directly results from the mobilised friction angle ϕ_m which is lower here ($35 < \phi_m < 40°$) than in sm1 and sm2 ($35 < \phi_m < 55°$). Thus, as the interface

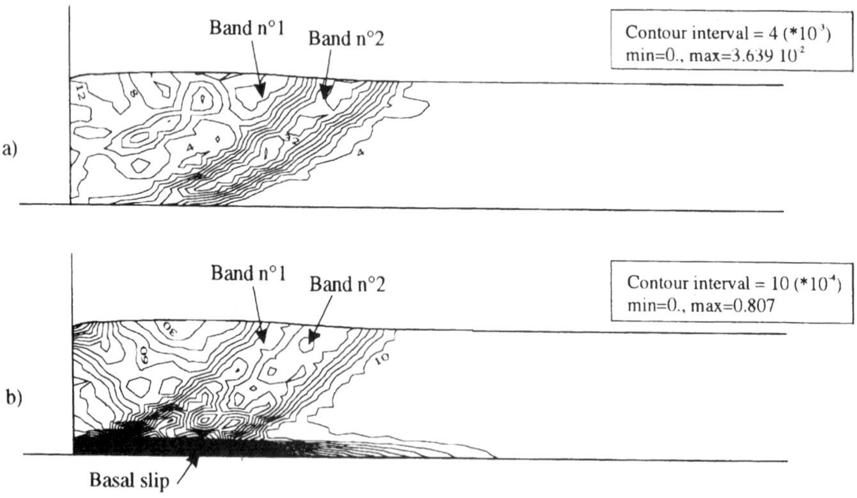

Fig. 10. Isovalues maps for simulation 3 at $u = 0.2$ cm. (**a**) Kinematic indicator α. (**b**) Equivalent strain: higher strain values are located along the basal slip.

friction angle corresponds to 45°, plastic failure preferentially occurs in the sand massif instead of occurring along the frictional interface.

Additional differences with sm1. Another difference can be expected between sm3 and sm1: as mobilized friction angle is lower here than in sm1, sm3 model should b e less resistant than sm1. This is in fairly good agreement with finite element results obtained at $u < 0.129$ cm: the horizontal reaction is found to be slightly lower for sm3 than for sm1 (see Fig. 5).

However the Lode angle influence has even induced more subtle differences between sm1 and sm2, i.e. some important changes in the stress path followed. Figures 11a–b present, in the deviatoric plane, the stress paths followed by the 3 points A, B and C (previously located on Fig. 8) between $u = 0$ and $u = 0.2$ cm (i.e. prior localization), for simulation 1 and simulation 3 respectively.

In the first stages, in both cases the stress paths are close to the hydrostatic axis and follow triaxial compression ones: this is quite normal as at the beginning of computation, σ_1 is vertical and $\sigma_2 = \sigma_3$ are horizontally oriented.

As the loading increases, the stress paths for sm1 and sm3 are no longer similar as shown in the deviatoric plane (see Fig. 10a–b): there is an important stress rotation in sm1 whereas almost none occurs in sm3, and the σ_1 and σ_2 components are greater in sm1 than in sm3. These differences might be explained as follow considering the left part of the model: for the two simulations, $\sigma_1 \approx \sigma_x$ is almost horizontal (see Fig. 8a–b), $\sigma_3 \approx \sigma_v$ is vertical and constant due to the body forces (see equation (28)); as the loading increases, σ_1 and σ_2 will increase by moving onto the yield surface; however the different shapes of the two yield surfaces with σ_3 constant will lead to different stress paths and stress states, which will result in a higher increase of σ_1 and σ_2 in sm1 than in sm3 and thus which will traduce in a larger rotation of stress path in deviatoric plane for sm1. This important change in stress distribution can be seen clearly on Figs 11c–d showing the evolution of the stress shape factor R_S

$$R_S = \frac{\sigma_2 - \sigma_1}{\sigma_3 - \sigma_1} \quad (\text{with } 0 < R_S < 1) \qquad (31)$$

and for which the R_S values are lower in sm1 (Fig. 11c) than in sm3 (Fig. 11d).

As a result, the predicted type of faulting at a given loading will be different in the two cases. For instance at $u = 0.2$ cm and according to the classification given in Sassi & Faure (1995), in sm1 the faulting would consist of pure reverse faulting, whereas in sm3 it would rather consist of a

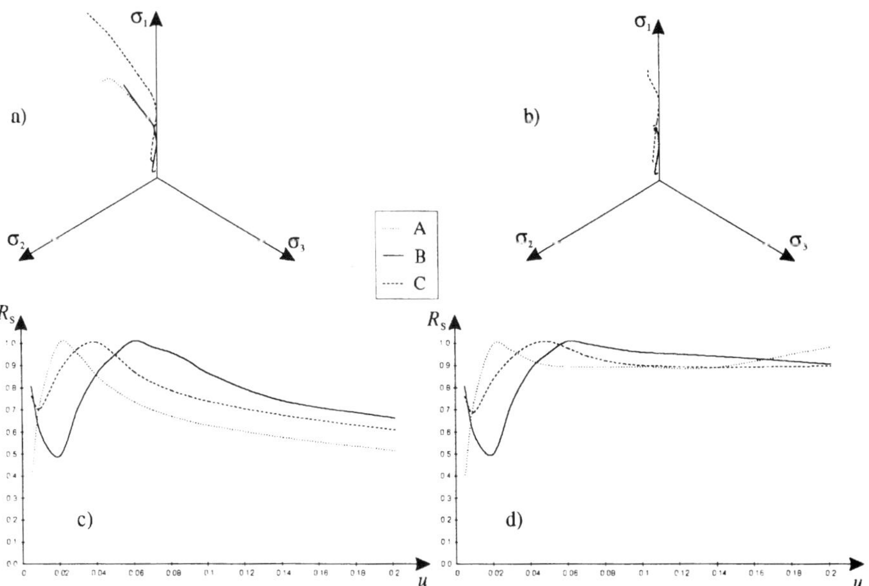

Fig. 11. Stress path and shape ratio for points A, B and C, for $0 < u < 0.2$ cm (for exact location, refer to Fig. 8). (**a, b**) Stress paths in the deviatoric plane for simulation 1 and simulation 3, respectively. (**c, d**) Stress shape ratio, R_S, versus horizontal displacement, u, for simulation 1 and simulation 3, respectively.

combination of reverse strike-slip and pure reverse faulting.

Thus the influence of the Lode angle in the constitutive relation is non-negligible on the stress path and on the stress shape ratio R_S, which might be of some importance when these are used to interpret faulting regimes (Sassi *et al.* 1993; Sassi & Faure 1995).

Comparison with analogue experiments

Although a very large number of physical experiments have been performed using sandbox models, it is only recently that the computer assisted tomography has improved significantly the amount of information which can be obtained from these experiments. One of the published experiments (Colletta *et al.* 1991) is of great interest here as the initial geometry used is almost identical to the one described in Fig. 3. However, the loading displacement values at which experimental results are given ($u = 3.6$ cm) are much higher than in the numerical simulations performed here ($u = 0.9$ cm for sm1).

The structural interpretation of the physical experiment and the present numerical simulation 1 is presented on Figs 12a–b: fault locations are built in the first case from computed tomography pictures (Colletta *et al.* 1991) and in the second case from the cumulated equivalent strain for $0 < u < 0.9$ cm (summation of values given in Figs 4a, b, c).

Strong similarities exist between the physically observed and numerically predicted structures: in both cases some imbricated forward thrusts develop in front of the left boundary whereas one backthrust develop close to the left boundary. The foreland migration of thrusting observed by Colletta et al. (1991) is also observed here as band n. 1 develops prior band n. 2. The pop-up structure observed is in front of the thickened wedge is not present here due to the smallest amount of applied displacement.

Conclusion

The three numerical simulations of a sandbox compression experiment performed in this study lead to several comments.

Two simulations performed using a 'classical' Drücker–Prager yield criterion lead to the formation of several shear bands, which can be interpreted as reverse faults (forward thrusts and backthrusts). Especially the phenomenon of localization could be followed in a very attentive way using Rice criterion and a kinematic indicator, which enable the second order bands associated with shear band competition to be visualized. The orientation of those shear bands is compatible with theoretical predictions; the backthrust dip is found greater than the forward one, which is explained by the clockwise rotation of principal stresses induced by the basal shear. A comparison with a similar sandbox experiment shows that numerical displacement fields and shear band localization are in fairly good agreement with what is observed in experiments.

An interesting feature is that a structural (or geometrical) softening in the force/displacement curve could be observed even if no softening was introduced in the constitutive relation.

The effect of initial stress state has been checked using two different values for the horizontal/vertical stress ratio K_0. Results show that, for the hydrostatic ($K_0 = 1$) and the deviatoric ($K_0 = 0.25$) state of stress, the resulting displacement fields and shear band localization are almost identical; the main difference consists only in the slightly lower displacement required to obtain bifurcation in the hydrostatic case compared with the deviatoric one.

Regarding the friction angle effects, some significant differences between Drücker–Prager and Van Eckelen criterion are observed: (a) Drücker–Prager model in which extension friction angle is excessively high (e.g. $\phi_E = 68°$ if $\phi_C = 35°$) leads to an overestimation of the sand resistance, which results in an overestimation of forces, (b)

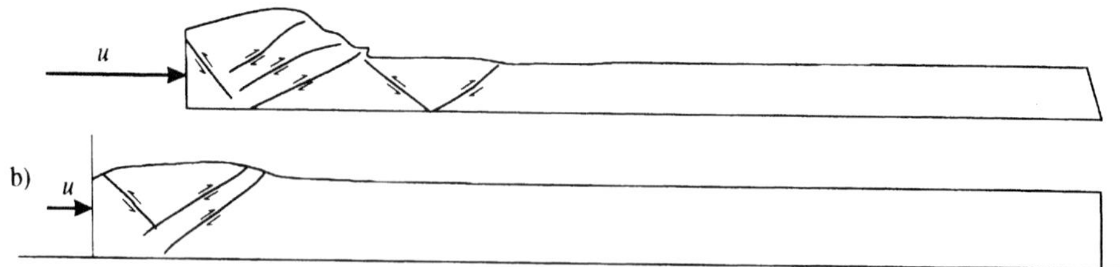

Fig. 12. (**a**) Structural interpretation of thrust propagation experiments performed on sandbox models at $u = 3.6$ cm (adapted after Colletta *et al.* 1991): three imbricated forward thrusts have developed close to the vertical boundary, with only one associated backthrust. A new pop-up structure starts to develop in the foreland at the right part of the thickened wedge. (**b**) Structural interpretation from cumulated equivalent strain for numerical simulation 1 at $u = 0.9$ cm: two imbricated forward thrusts and one backthrust have developed.

strain localization occurs in both cases, although an additional basal slip plane develops in the VE model due to the relative friction angle between the sand and interface, (c) the 2 models lead to very different stress paths and thus may lead to a quite strong modification the stress shape ratio, which has a non-negligible influence when faulting analysis based on the stress shape factor is performed.

Clearly the Van Eekelen criterion seems a quite promising criterion for simulation of frictional material as sand. It is also worth pointing out that this criterion actually does not necessarily require laboratory identification of any additional material parameters as ϕ_E becomes an explicit parameter, whereas ϕ_E is a hidden one in the Drücker–Prager criterion.

Further investigations should show the effect of other parameters (basal friction, cohesion, elastic module) on the Lode angle, the position and orientation of faults. In that sense, a kind of inverse analysis of the significant material parameters could be done. Some improvements of the finite elements with regards to localization should be incorporated in such analysis in the future (see e.g. Wang 1993). Three dimensional aspects and hydromechanical coupling and other localization considerations could also be analysed in future work.

The authors thank Jean Chéry, Dick Nieuwland and an anonymous reviewer for their critical comments on the first manuscript. Dominique Fourmaintraux who initiated this project and the research group Geofrac (Elf Aquitaine, Institut Français du Pétrole and Total, which gives permission for publication are also gratefully acknowledged).

References

ARTHUR, J. R. F., DUNSTAN, T., AL-ANI, Q. A. J. L. & ASSADI, A. 1977. Plastic deformation and failure in granular media. *Géotechnique*, **27**, 53–74.

CESCOTTO, S. 1989. General strategy for non-linear finite element formulation. *In*: GRUBER, R. *et al.* (eds) *Proceedings of the 5th International Symposium on Numerical Methods in Engineering*. Springer Verlag, **1**, 155–164.

—— & CHARLIER, R. 1993. Frictional contact finite elements based on mixed variational principles. *International Journal for Numerical Methods in Engineering*, **36**, 1681–1701.

——, HABRAKEN, A. M., RADU, J. P. & CHARLIER, R. 1989. Some recent developments in computer simulation of metal forming processes. *Proceedings of the 9th International Conference on Computer Methods in Mechanics*, **4**, 19–52, Krakow-Rytrov, Poland.

CHARLIER, R. & CESCOTTO, S. 1988. Modélisation du phénomène de contact unilatéral avec frottement dans un contexte de grandes déformations. *Journal de Mécanique Théorique et Appliquée*, special issue, Suppl. **1**, 7.

CHARLIER, R. & HABRAKEN, A. M. 1990. Numerical modelisation of contact with friction phenomena by the finite element method. *Computers and Geotechnics*, **9**, 1 & 2.

CHENG, C. H. 1988. Automatic adaptive remeshing for finite element simulation of forming processes. *International Journal for Numerical Methods in Engineering*, **26**, 1–18.

COLLETTA, B., LETOUZEY, J., PINEDO, R., BALLARD, J. F. & BALE, P. 1991. Computerized X-ray tomography analysis of sandbox models: examples of thin-skinned thrust systems. *Geology*, **19**, 1063–1067.

DE BORST, R. 1992. Gradient-dependent plasticity: formulation and algorithmic aspects. *International Journal for Numerical Methods in Engineering*, **35**, 521–539.

—— 1993. A generalisation of J_2-flow theory for polar continua. *Computer Methods in Applied Mechanics and Engineering*, **103**, 347–362.

—— & SLUYS, L. J. 1991. Localisation in a Cosserat continuum under static and dynamic conditions. *Computer Methods in Applied Mechanics and Engineering*, **90**, 805–827.

DESAI, C. S. & SIRIWARADANE, H. J. 1984. *Constitutive laws for engineering materials with emphasis on geologic materials*. Prenctice-Hall.

DESRUES, J. 1984. *La localisation de la déformation dans les matériaux granulaires*. PhD Thesis, USMG & INPG Grenoble.

DRÜCKER, D. C. & PRAGER, W. 1952. Soil mechanics and plasticity analysis or limit design. *Quarterly Applied Mathematics*, **10** (2), 157–165.

HABRAKEN, A. M. & CESCOTTO, S. 1990. An automatic remeshing technique for finite element simulation of forming processes. *International Journal for Numerical Methods in Engineering*, **30**, 1503–1525.

MÄKEL, G. & WALTERS, J. 1993. Finite-element analyses of thrust tectonics: computer simulation of detachment phase and development of thrust faults. *Tectonophysics*, **226**, 167–185.

MANDL, G. & SHIPPAM, G. K. 1981. Mechanical model of thrust sheet gliding and imbrication. *In*: *Thrust and Nappe Tectonics*, Geological Society, London, Special Publication, **9**, 79–97.

MATSUOKA, H. & NAKAI, T. 1982. A new failure condition for soils in three-dimensional stresses. *Proceedings IUTAM Conference Deformation and Failure of Granular Materials*, Delft, 253–263.

ORTIZ, M., LEROY, Y. & NEEDLEMAN, A. 1987. A finite element method for localized failure analysis. *Computer Methods in Applied Mechanics and Engineering*, **61**, 189–214.

RICE, J. R. 1976. The localisation of plastic deformation. *In*: KOITER, W. (ed.) *Theoretical and Applied Mechanics*. North-Holland.

SASSI, W. & FAURE, J. L. 1995. Role of faults and layer interfaces in the spatial variation of stress regime in basins: inference from numerical modelling, *Tectonophysics*, in press.

——, COLLETTA, B., BALÉ, P. & PAQUREAU, T. 1993. Modelling of structural complexity in sedimentary basins: the role of pre-existing faults in thrust tectonics. *Tectonophysics*, **226**, 97–112.

SCHWEIGER, H. F. 1994. On the use of Drücker–Prager failure criteria for earth pressure problems. *Computers and Geotechnics*, **16**, 223–246.

VAN EEKELEN, H. A. M. 1980. Isotropic yield surfaces in three dimensions for use in soil mechanics. *International Journal for Numerical and Analytical Methods in Geomechanics*, **4**, 98–101.

VILOTTE, J. P., POZZI, J. P., PHILIPPE, C., PASTOR, M., ZIENCKIEWICZ, O. C., DAUDRE, B. & BARDES, P. 1990. Modes localisés de la déformation dans les matériaux crustaux. *Rapport scientifique, GRECO Géomatériaux.*

WANG, X. 1993. *Modélisation numérique des problèmes avec localisation de la déformation en bandes de cisaillement.* PhD thesis, Université de Liège, Faculté des Sciences Appliquées.

ZIENKIEWICZ, O. C. & ZHU, J. Z. 1987. A simple error estimator and adaptive procedure for practical engineering analysis. *International Journal for Numerical Methods in Engineering*, **24**, 337–357.

ZIENCKIEWICZ, O. C. & TAYLOR, R. L. 1988. *The finite element method* (4th edn), MacGraw-Hill Book Company.

Crustal fault reactivation facilitating lithospheric folding/ buckling in the central Indian Ocean

F. BEEKMAN,[1] J. M. BULL,[2] S. CLOETINGH[1] & R. A. SCRUTTON[3]

[1] *Institute of Earth Sciences, Vrije Universiteit, De Boelelaan 1085, 1081 HV Amsterdam, The Netherlands*

Present address: Institut Français du Pétrole, BP 311, 92506 Rueil-Malmaison Cedex, France

[2] *Department of Geology, University of Southampton, Highfield, Southampton SO9 5NH, UK*

[3] *Department of Geology and Geophysics, Edinburgh University, West Mains Road, Edinburgh EH9 3JW, UK*

Abstract: High-quality, normal-incidence seismic reflection data confirm that tectonic deformation in the central Indian Ocean occurs at two spatial scales: whole lithosphere folding with wavelengths varying between 100 and 300 km, and compressional reactivation of crustal faults with a characteristic spacing of *c.* 5 km. Faults penetrate through the crust and probably into the upper mantle. Both types of deformation are driven by regional large intraplate stresses originating from the Indo-Eurasian collision. Numerical modelling of the spatial and temporal relationships between these two modes of deformations shows that, in agreement with geophysical observations, crustal faults are reactivated first with stick-slip behaviour. Subsequent lithospheric folding does not start until horizontal loading has significantly reduced the mechanical strength of the lithosphere, as predicted by elasto-plastic buckling theory. Modelling suggests that lithospheric folding does not develop in the absence of fault reactivation. Crustal fault reactivation, therefore, appears to be a key facilitating mechanism for oceanic lithospheric buckling in the central Indian Ocean.

Recent single-channel and multi-channel seismic reflection studies (e.g. Bull & Scrutton 1990, 1992; Pilipenko & Sivukha 1991; Chamot-Rooke *et al.* 1993) have demonstrated the occurrence of two spatial scales of active tectonic deformation in the central Indian Ocean (Fig. 1), confirming earlier findings from single-channel seismic reflection data and compilations of earthquake, heatflow, geoid and gravity data (Weissel *et al.* 1980; Geller *et al.* 1983). The tectonic deformation is characterized by long wavelength undulations of oceanic basement and overlying sediments, and superimposed small scale faulting and folding (Fig. 2). The active intraplate deformation in the Indian Ocean is thought to be driven by large regional compressional stresses (e.g. Curray & Moore 1971; Patriat & Achache 1984). The large stresses may originate from the intense plate boundary processes (Indo-Eurasion collision; Indonesian subduction) surrounding this intraplate section of the Indo-Australian plate (Cloetingh & Wortel 1985, 1986), or be related with a possible (nascent) diffuse plate boundary separating the Australian plate from the Indo-Arabian plate (Petroy & Wiens 1989; DeMets *et al.* 1990; Gordon *et al.* 1990; Royer & Chang 1991).

Several tectonic models have been proposed to explain the undulations of the oceanic basement, from the interpretations of geoid and gravity anomalies, seismic reflection and refraction studies, and numerical and analogue modelling. Most support is given to flexural folding/buckling of the mechanically strong upper part of the lithosphere (Weissel *et al.* 1980; McAdoo & Sandwell 1985; Karner & Weissel 1990; Bull *et al.* 1992; Martinod & Davy 1992; Molnar *et al.* 1993). Other proposed models are inverse boudinage of the oceanic crust caused by hydrodynamic viscous flow (Zuber 1987; Leger & Louden 1990), whole-crustal block faulting (Neprochnov *et al.* 1988), and a flexurally folded oceanic crust decoupled from the subcrustal lithosphere (Verzhbitsky & Lobkovsky 1993). Theoretical studies (e.g. Timoshenko & Woinowsky-Krieger 1959; Martinod & Davy 1992) show that some kind of perturbation is required to destabilize the homogeneous shortening of a plate under compression in order to initiate other modes of deformation. In the previously mentioned

From Buchanan, P. G. & Nieuwland, D. A. (eds), 1996, *Modern Developments in Structural Interpretation, Validation and Modelling,* Geological Society Special Publication No. 99, pp. 251–263.

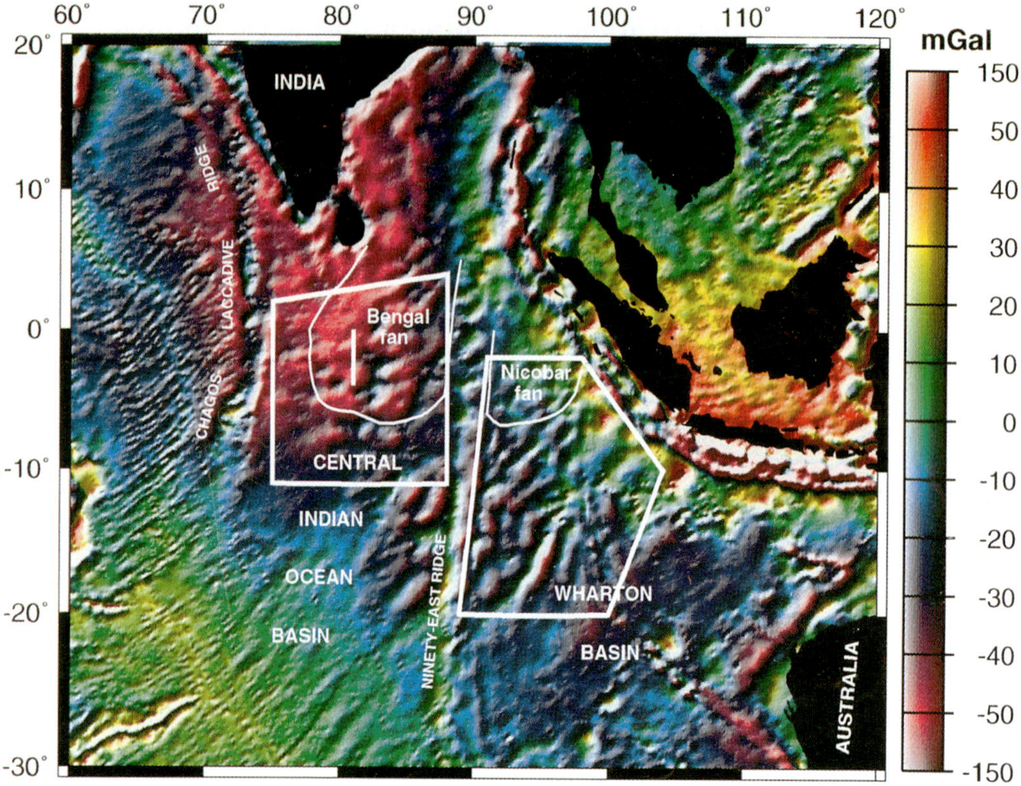

Fig. 1. Shaded relief image of ERS/GEOSAT/SEASAT derived free-air gravity anomalies (Sandwell & Smith 1992) within the central Indian Ocean, and location map of main features. Boxes denote areas with consistently oriented linear trends in the gravity anomalies coinciding with axes of undulations of oceanic basement. Short line in left box denotes position of long seismic reflection line shown in Fig. 2.

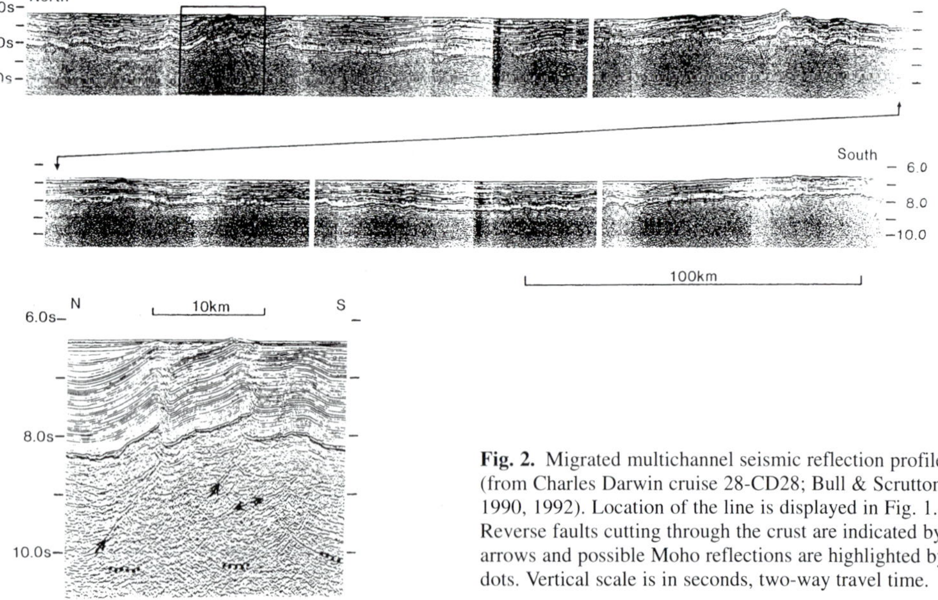

Fig. 2. Migrated multichannel seismic reflection profile (from Charles Darwin cruise 28-CD28; Bull & Scrutton 1990, 1992). Location of the line is displayed in Fig. 1. Reverse faults cutting through the crust are indicated by arrows and possible Moho reflections are highlighted by dots. Vertical scale is in seconds, two-way travel time.

modelling studies of intraplate deformation, various perturbation mechanisms have been assumed, such as line loading by the aseismic Ninety-East Ridge (McAdoo & Sandwell 1985), the growth of mantle instabilities (Zuber 1987), point loading by seamounts (Karner & Weissel 1990), or surface sediment loading (Molnar *et al.* 1993).

However, regardless of the tectonic model and perturbation mechanism, all studies concentrate on reproducing only the long wavelength undulations of oceanic basement, neglecting the crustal faulting. This paper presents results of a numerical analysis using a nonlinear, elasto-plastic finite element method to analyse the temporal and spatial relationship between the two modes of tectonic deformation (lithospheric buckling and crustal fault reactivation) observed in the central Indian Ocean. The main objective is to investigate the role of crustal fault reactivation as a potential perturbation mechanism.

Intraplate deformation in the central Indian Ocean

In the central Indian Ocean Basin, undulations of oceanic basement and, if present, sedimentary cover (Figs 2 and 3) trend E–W to NE–SW, with an amplitude of 1–2 km and a wavelength that increases from less than 100 km in the south to more than 300 km towards the north, averaging around 200 km (Weissel *et al.* 1980; Zuber 1987; Neprochnov *et al.* 1988; Bull & Scrutton 1992). The increase in wavelength is possibly related to the northward increase in age from 40 to 70 Ma of the oceanic lithosphere, and the associated increase in flexural rigidity (McAdoo & Sandwell 1985). Within the Wharton Basin the undulations have similar wavelengths and amplitudes, but strike more consistently NE–SW (Stein *et al.* 1989; Pilipenko & Sivukha 1991). In both basins, the axes of undulations are roughly perpendicular to directions of maximal horizontal compressional stress, as inferred from earthquake focal mechanism studies (Bergman & Solomon 1985; Levchenko 1989; Petroy & Wiens 1989) and numerical computations of the Indo-Australian plate stress field (Cloetingh & Wortel 1985, 1986). The basement undulations are not symmetrically shaped, but have broad lows and pronounced crests. In the southern parts of the area of intraplate deformation there is negligible sediment thickness. To the north, there is a progressive infill of the undulation lows by the sediments of the Bengal and Nicobar fans (Fig. 1).

Geoid and free-air gravity anomalies contour maps also exhibit undulating character (Fig. 1), correlating with oceanic basement undulations. Folding axes appear to be discontinuous and even seem to undergo rotation across old fracture zones (Bull 1990; Bull & Scrutton 1992). The offset of undulations across the fracture zones is consistent with N–S compression. Some rotation is to be expected as the 3-dimensional intraplate stress field is unlikely to be uniformly oriented over this wide area. This is compatible with stress orientations derived from earthquake focal mechanism studies (Bergman & Solomon 1984, 1985; Petroy & Wiens 1989). Numerical computations of the Indo-Australian plate stress field (Cloetingh & Wortel 1985, 1986) show how the unique geometry of the plate boundaries and the distribution of dynamic plate boundary processes (collision, subduction, spreading) led to rotated directions of principal stresses with variable magnitudes in the interior of the Indo-Australian plate.

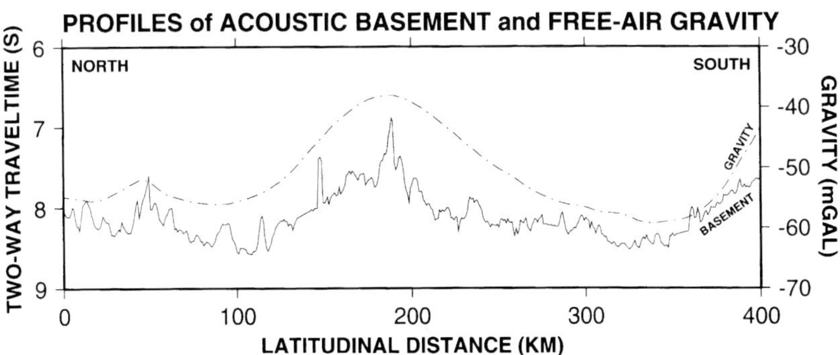

Fig. 3. North–south running profiles (for location see Fig. 1) of digitized acoustic basement (solid line) and free-air gravity anomaly (dashed line), demonstrating both spatial modes of tectonic deformation: long wavelength folding and small scale faulting.

Superimposed on the long wavelength undulations of oceanic basement is intense short wavelength folding and faulting in the crust and overlying sediments. Recent multi-channel seismic profiles (Bull & Scrutton 1990, 1992; Chamot-Rooke *et al.* 1993) provide new evidence that the faulted blocks are bounded by high-angle reverse faults penetrating through the crust and possibly into the upper mantle (Fig. 2). Fault activity is recognized on the seismic images by displacements of oceanic basement and sedimentary horizons. Some unreactivated pre-existing basement faults have also been identified, as well as some intra-crustal low-angle faults and sub-horizontal reflectors. The new seismic data also reveals the widespread existence of open anticlinal folds of 5–10 km wavelength within the sedimentary cover, most of them occurring with limited faulting in the deeper parts or with faulting extending from seafloor to basement.

The fault pattern is very complex with variations in fault type, fault dip, dip direction, penetration depth, fault spacing and length, and fault activity. This is, for example, illustrated by the north–south running profile of acoustic basement (Fig. 3), where the small scale effects of varying fault throws are seen to be superimposed on the long wavelength undulations of the oceanic basement. The vast majority of high-angle faults display reverse displacements (Bull & Scrutton 1990; Pilipenko & Sivukha 1991). Basement faults can be divided into those that dip > 40° to the south, and those that dip 35–40° to the north. If penetrating into the sedimentary cover, faults dramatically steepen (40–90°), presumably due to the change in rheology (Melosh 1990). Some of the N-dipping faults can be resolved to Moho depth (Bull & Scrutton 1990). The S-dipping faults are less well resolved, as they dip more steeply. Both sets strike 90–100 °E, roughly perpendicular to the strike of fracture zones in the area. The fault length averages ≈ 10 km, with lengths up to 40 km. Average fault spacing is 5–6 km. The fault amplitude, in terms of the vertical throw of the basement/sedimentary cover interface, averages several hundreds of meters, ranging from very small (detection limit) to over more than 1000 meters (Chamot-Rooke *et al.* 1993). Some of the basement faults on the seismic profiles seem to penetrate the entire crust and possibly into the subcrustal lithosphere. Bull and Scrutton (1992) suggest that these faults may nucleate in the upper mantle at the brittle–ductile transition and propagate upwards, reactivating the ridge-parallel fault fabric in the oceanic crust.

These geometrical characteristics, together with other factors, have led to the interpretation that the faults result from the reactivation of the oceanic structural fabric that originated at the midocean spreading centres (Bull 1990). Steep dipping syn- and antithetic faults are formed in the transition of the rift valley into the rift mountains, creating fault blocks with widths ranging from 1–5 km (Macdonald 1982; Kaz'min & Borisova 1992). A similar complex fault fabric (in pristine state), although, in general, with lower dips, has also been observed on seismic images of oceanic crust in the Atlantic (White *et al.* 1990).

Excellent resolution of faults is attributed to the presence of hydrothermal alteration fronts along the fault planes (Shipboard Scientific Party 1990; Bull & Scrutton 1992). The hydrothermal activity within the crustal fault blocks is thought to be responsible for the locally anomalous high surface heat flow measured within the intraplate deformation area (Geller *et al.* 1983; Stein & Weissel 1990). Regionally, heat flow is normal (as predicted by cooling plate models), indicating that there is no deep thermal origin for the driving mechanism of tectonic deformation. Furthermore, the presence of fluids in the fault planes may have a weakening effect on the frictional resistance against sliding (Brace & Kohlstedt 1980) and facilitate fault reactivation.

Seismic stratigraphy and DSDP and ODP drilling have demonstrated the presence of two major unconformities within the sediments of the Bengal and Nicobar distal fans over much of the central and northeastern part of the Indian Ocean (Cochran 1989; Curray & Munasinghe 1989). An early Eocene (*c.* 55 Ma) unconformity is correlatable with the first stage of collision of the Indo-Australian plate with the Eurasian plate. A latest Miocene (7 Ma) unconformity has been correlated with the collision of the Indian continent with the Eurasian plate (Peirce 1978; Patriat & Achache 1984; Klootwijk *et al.* 1985; Cochran 1989; Molnar *et al.* 1993). The Miocene unconformity separates pre- and syn-deformational sediments, dating the onset of the intraplate deformation at 7 Ma. There is little evidence for deformation before this time. Sedimentary onlap patterns suggest that at 7 Ma there was a major shortening event followed by several smaller pulses of intense fault block rotation and fault motion (Bull & Scrutton 1992). The relative timing of crustal faulting versus long wavelength folding is unclear, and requires more detailed stratigraphic analysis. Analysis of offset sedimentary horizons shows that on several faults displacement decreases steadily upwards: these faults grow by upwards propagation. Other faults have a roughly constant offset through time, and either formed rapidly at one of the deformation pulses, or are faults on which propagation is complete.

Statistical analysis shows that there is no relationship between magnitude of fault throw and position relative to the long wavelength basement undulations (Bull 1990). Chamot-Rooke *et al.* (1993) report a gradual increase in fault throw from south to north over the area of intraplate deformation in the Central Indian Ocean Basin, likely related with the northward increase in sediment loading by the Bengal fan. These authors find for the same area a very long wavelength (600–700 km) trend in fault downthrow direction. The southernmost area is affected primarily by north-dipping faults, the central area by south-dipping faults, whereas the more extensively deformed northernmost area has a mixture of both polarities, although with a majority of north-dipping faults. There is, however, no general relationship between fault downthrow direction and basement undulations (wavelength of *c.* 200 km), suggesting a flexural origin for the undulations. In some cases faults tend to downthrow downslope, accentuating the crests of the undulations.

Shortening of this part of the Indo-Australian plate, subject to large compressional stresses, is accommodated by both long wavelength folding and short wavelength faulting and folding. Digitization of oceanic basement from seismic profiles shows that the long wavelength contribution is equivalent to less than 0.1% shortening. Bull and Scrutton (1992) estimate from seismic sections the horizontal shortening due to the reverse faulting to be 1.2 (±0.4)%, corresponding to 18 (±6) km of shortening over 1500 km north–south extent of deformation. Chamot-Rooke *et al.* (1993) also assess a possible contribution from very small-scale faulting. Their estimates of the shortening range from 2.5% to maximal 4.3% when small-scale faulting is taken into account. When distributed over the 900 km of deformed basement observed along their seismic profile, these estimates correspond with absolute shortening values of 22 km and 62 km, respectively. At the same longitude, plate tectonic reconstructions (Royer & Chang 1991) predict a finite shortening of 23 (±45) km. Assuming that the shortening has taken place in a steady way from the onset of deformation (7 Ma) until present, corresponding rates of shortening are 2.5 (±0.9) mm a^{-1} (Bull & Scrutton 1992), 3 mm a^{-1} (Royer & Chang 1991), and 6 (±3) mm a^{-1} (Chamot-Rooke *et al.* 1993). This last is consistent with the 6 to 7 mm a^{-1} rate inferred from plate kinematic models (Gordon *et al.* 1990). The pulselike character of shortening through time, as suggested by the deformation patterns in the sedimentary cover on the seismic profiles, may have led to higher shortening rates at certain times.

Active intraplate deformation is still taking place in the central Indian Ocean, as evidenced by the present-day occurrences of large intraplate earthquakes (Bergman & Solomon 1980, 1985; Levchenko 1989; Petroy & Wiens 1989) and reported substantial micro-seismicity in this area (Levchenko & Ostrovsky 1993).

A finite element model

In order to investigate the temporal and spatial relationship between the two modes of tectonic deformation, numerical analyses are carried out using the finite element technique. The numerical analysis assumes a two-dimensional plane-strain approach, which is validated by the E–W continuity of the axes of the undulations of oceanic basement (Weissel *et al.* 1980; Bull & Scrutton 1992). A finite element model (Fig. 4a) is constructed comprising a 2000 km long, N–S section through the central Indian Ocean part of the Indo-Australian plate. To correctly compute displacements, finite elements in the faulted upper part of the plate must be small enough to allow deformation within the faulted blocks, and are therefore sized 1 × 1 km triangular in the uppermost 10 km, increasing in size to 1 × 5 km rectangular in the lower parts (Fig. 4b). For the full mesh with dimension 2000 × 40 km this becomes, including spring elements, 60 000 finite elements, connected by more than 42 000 nodal points, and with a total of more than 80 000 degrees of freedom. Correct behaviour of the elements is

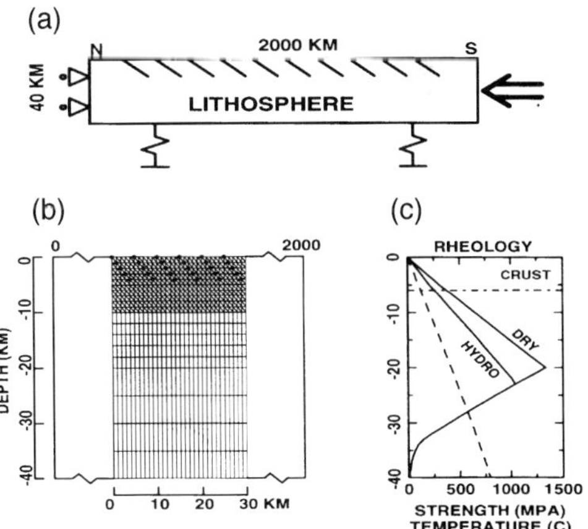

Fig. 4. (**a**) Geometry and boundary conditions of the mechanical model. (**b**) Enlarged part of the finite element mesh showing geometry of the pre-existing crustal faults.(**c**) Geotherm (dashed) and yield-strength envelopes (continuous) for a 50 Ma oceanic lithosphere. Assuming hydrostatic pore pressure reduces the strength of the lithosphere.

ensured by using higher-order elements, in this case linear triangulars and quadrilaterals which have been expanded quadratically using 'incompatible modes of deformation' (e.g. Beekman 1994).

The complex fault geometry in the oceanic crust is approximated by including a simplified set of faults in the upper part of the plate model, which dip 45° and have a penetration depth and fault spacing of 5 km (average values). For the computations the finite element code TECTON is used (Melosh & Raefsky 1981), which contains the slippery node technique to model fault behaviour (Melosh & Williams 1989), and which is modified to also account for friction along fault planes and for elasto-plastic deformation (Beekman 1994).

The model boundaries are allowed to move freely in both the horizontal and vertical direction, except for the vertical outer edges which are restricted to move horizontally. The vertical buoyancy of the asthenosphere, which can be modelled as a fluid on geological timescales and which counteracts vertical motions of the overlying lithosphere, is included by attaching an array of linear elastic Hookean springs to the base of the plate. The springs will exert forces on the base of the plate, proportional to, but directed opposite to, any vertical deflection of the base of the plate. Horizontal boundary conditions at the N and S edge imitate the effects of the India–Himalayan collision and resistance associated with subduction along the Banda and Sunda arcs, and mid-ocean ridge push, respectively (Cloetingh & Wortel 1985, 1986). The plate is shortened with a velocity of 6.3 mm a^{-1}, according to the currently most accurate available shortening estimate for the area of deformation (Chamot-Rooke *et al.* 1993). For a 2000 km long plate, this velocity is equivalent to a strain-rate of $\dot{\varepsilon} = 10^{-16}\ s^{-1}$, which is a value characteristic for the slow deformation of oceanic lithosphere (Carter & Tsenn 1987). Furthermore, the model is loaded with gravitational body forces.

The Indian oceanic lithosphere comprises a 6 km thick basaltic crust, overlying an olivine upper mantle. The age of the lithosphere in the areas of intraplate deformation in the Indian Ocean varies from 40 Ma in the south to 70 Ma in the north. An average age of 50 Ma is assumed for the entire plate model. The temperatures applied to the model are computed using this model geometry for which an initial steady-state temperature solution can be derived from the heat-conduction equation. Subsequently, the structure is allowed to cool through conduction of heat. Transient temperature fields are computed by solving the heat-conduction equation using a finite difference scheme (third order Runge-Kutta). Thermal properties are listed in Table 1. The assumed thermal boundary conditions are a constant surface temperature of 0 °C, and an initial surface heatflux of 80 mW m^{-2}, which is a characteristic value for young oceanic lithosphere (Sclater *et al.* 1980). The geotherm after 50 Ma of conductive cooling within the multi-layered oceanic lithosphere is computed for the adopted thermal initial and boundary conditions and plotted in Fig. 4c (dashed line).

The rheological behaviour of the rocks which constitute the lithosphere is derived from laboratory experiments, and subsequently extrapolated to geologically relevant timescales (assuming that the empirically derived rheological laws remain valid). At low confining pressures and low temperatures, brittle failure is predominant. The corresponding frictional yield criterion is given by Byerlee's law (Byerlee 1978), which can be rewritten in terms of effective principal stress difference, lithostatic overburden pressure, and pore fluid pressure (Ranalli, 1987):

$$\sigma_{\text{brittle}} = \sigma_1 - \sigma_3 = \alpha\rho g z(1-\lambda) \text{ with}$$

$$\alpha = \begin{cases} \dfrac{R-1}{R} & \text{for normal faulting} \\ R-1 & \text{for thrust faulting} \\ \dfrac{R-1}{1+\beta(R-1)} & \text{for strike-slip faulting} \end{cases} \quad (1)$$

and where σ_1 and σ_3 are the maximum and minimum principal stress, respectively, ρ is density, g is gravitational acceleration, z is depth, and where

$\lambda = \dfrac{p_{fl}}{\rho g z} = \dfrac{\rho_w}{\rho} \approx 0.35$ is the pore fluid factor, here for hydrostatic pore pressure

$0 < \beta < 1$ denotes magnitude of intermediate stress: $\sigma_2 = \sigma_3 + \beta(\sigma_1 - \sigma_3)$

$R = [(1+\mu^2)^{1/2} - \mu]^{-2}$ where μ is the (static) friction coefficient

Steady-state creep of a wide variety of rocks is empirically described by a ductile flow law which relates the critical principal stress difference necessary to maintain a given steady-state strain-rate of deformation to a power of the strain-rate, and which is, therefore, called power-law creep (Kirby 1983):

$$\sigma_{creep} = \sigma_1 - \sigma_3 = \left(\frac{\dot{\varepsilon}}{A_P}\right)^{1/N} \exp\left[\frac{E_P}{NRT}\right] \quad (2)$$

where A_P, N, and E_P are empirically determined material 'constants', assumed not to vary with stress and *P–T*-conditions (Ranall, 1987). Microphysically this type of creep mainly involves dislocation climb. The ductile flow law (equation 2) shows that the variation of the creep stress with

Table 1. *Physical parameters and material properties*

Definition	Symbol	Units	Value	
Gravitational acceleration	g	m s^{-2}	9.81	
Universal gas constant	R	J mol^{-1} K^{-1}	8.314	
Seawater density	ρ_w	kg m^{-3}	1030	
Sediment density	ρ_s	kg m^{-3}	2200	
Moho depth	z_m	km	6	
Mantle melt temperature	T_m	°C	1300	
Mantle basal heatflux	q_b	mW m^{-2}	23	
Static friction coefficient	f_s	—	0.6	
Bulk strain-rate	$\dot{\varepsilon}$	s^{-1}	10^{-15}	
			Upper Crust	Mantle
Petrology	—	---	granite (wet)*	olivine (dry)
Density	ρ	kg m^{-3}	2700	3300
Youngís modulus	E	GPa	50	70
Poissonís ratio	ν	—	0.25	0.25
Thermal conductivity	k	W m^{-1} K^{-1}	3.1	3.5
Specific heat	c_P	J kg^{-1} K^{-1}	1050	1050
Heat production	A	μW m^{-3}	2.5	—
Exponential heat decay rate	h	km	10	—
Power law exponent	N	—	1.9	3.0
Power law activation energy	E_P	kJ mol^{-1}	140	510
Power law strain rate	A_P	Pa $^{-N}$ s^{-1}	7.94 10^{-16}	7.0 10^{-14}
Dorn law activation energy	E_D	kJ mol^{-1}	—	535
Dorn law strain rate	A_D	s^{-1}	—	5.7 1011
Dorn law stress	σ_D	GPa	—	8.5
Failure/creep functions			**Sources**	
Brittle failure	$\sigma_{brittle} = \alpha\rho(1-\lambda)$		Goetze & Evans 1979	
Power law creep	$\sigma_{creep} = (\dot{\varepsilon}/A_P)^{1/N} \exp[E_P/NRT]$		Carter & Tsenn 1987	
Dorn law creep	$\sigma_{creep} = \sigma_D(1-[1-(RT/E_D)\ln(\dot{\varepsilon}/A_D)]^{1/2})$		Tsenn & Carter 1987	
			Carmichael 1989	

* 'wet' refers to rock samples that contain variable amounts of structural water

depth is strongly controlled by rock-type and temperature. A steady-state bulk strain-rate of 10^{-16} s^{-1} is assumed, induced from the applied velocity boundary condition.

Above a critical stress level (≈ 200 MPa), power law creep breaks down into high-stress, low-temperature plasticity (dislocation glide), which is phenomenologically described by a Dorn law (Goetze & Evans 1979; Tsenn & Carter 1987):

$$\sigma_{creep} = \sigma_1 - \sigma_3 = \sigma_D\left(1 - \left[-\frac{RT}{E_D}\ln\left(\frac{\dot{\varepsilon}}{A_D}\right)\right]^{1/2}\right) \quad (3)$$

with A_D, E_D, and σ_D material properties, assumed, as before, to be constant.

A depth-varying rheology has been incorporated in the finite element model with elasto-brittle deformation in the upper parts of the oceanic lithosphere and ductile deformation in the lower parts. The associated yield envelope has been constructed using the rheological equation (1) for brittle failure, and equations (2) and (3) for creep flow (at a constant strain-rate of 10^{-16} s^{-1}) of rocks, respectively, and assuming a basaltic crust and a olivine subcrustal lithosphere. Adopted material properties and physical parameters are listed in Table 1. The empirical rheological equations and laboratory derived material properties are used under the assumption they are also valid when extrapolated to geological strain-rates and macroscopic scales. The brittle-ductile transition lies at a depth of 24 km. The depth at which the yield envelope has reduced to an arbitrary small stress level in the order of 10 MPa (Ranalli 1987) indicates the base of the mechanically strong upper part of the lithosphere, and is used in the modelling to control the depth of the lower boundary of the numerical plate model, here selected at 40 km depth.

Based on the presence of hydrothermal fluid convection in the upper part of the oceanic lithosphere (Shipboard Scientific Party 1990) a hydrostatic pore fluid pressure is assumed, which significantly weakens the brittle strength of the lithosphere (equation (1), parameter λ). Hydrothermal fluid alteration fronts seem to coincide with pre-existing fault planes in the crust, and are thought to lower the frictional resistance against renewed sliding along the fault planes (Brace & Kohlstedt 1980).

Brittle deformation in the depth range over which the pre-existing faults extend is assumed to be carried out explicitly by these predefined faults. Therefore, no (brittle) yield stress is assigned to the finite elements in this upper part of the plate. Over the remaining depth range down to the brittle–ductile transition there is no information on whether brittle deformation occurs as fracturing of rocks or as fault reactivation. What is known, as evidenced by the existence of deep earthquakes, is that it does occur. Below the resolution depth of the seismic profiles, which is approximately Moho depth, there is no geometrical information of faults (if existing). Therefore, no faults are incorporated at depths exceeding Moho depth. The elements in this depth range are nevertheless characterized by a brittle yield stress assigned on the basis of the empirical brittle deformation equation (1), although any associated brittle failure is modelled numerically as plastic flow (Beekman 1994).

Results

At time zero the plate undergoes an instantaneous displacement downwards in response to the gravitational loading of the model (see, for example, Fig. 6a). This is followed by a slow linear uplift of the plate surface in response to continuing horizontal shortening. After several million years of shortening (0.3% of shortening occurs in a period of 1 Ma), the models start to develop a folding phase, as illustrated by the evolution of the surface deflection of the plate (Fig. 5a). Fourier spectral analysis of the calculated surface folding demonstrates a dominant wavelength between 200 and 350 km (Fig. 5b).

The folding evolves nonlinearly through time: after several Ma, surface deflections (Fig. 6a) begin to deviate from the homogeneous uplift trend and accelerate upward or downward depending on their position inside a fold (crest or trough). The onset of the deviations corresponds to the onset of lithospheric buckling.

Comparison of Figs 6a and b shows that fault activity starts before the onset of the large scale folding. The pre-defined faults in the models exhibit typical stick-slip behaviour through time (Fig. 6b). Jumps in fault slip (at the surface of the plate) are in the order of several meters, accumulating to throws of more than 100 m (Fig. 6b). After long wavelength folding has started to develop, fault activity becomes more complex with some faults exhibiting accelerated reversed faulting,

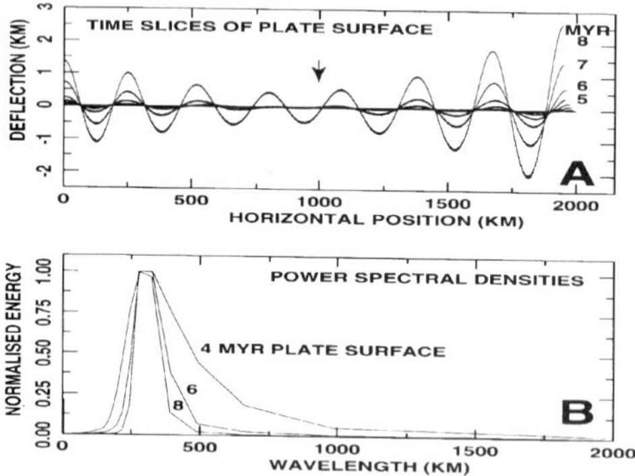

Fig. 5. (**a**) Vertical deflection of the plate model surface through time (1 Ma ≈ 0.3% shortening). Wavelength of developing folding is in agreement with observations. (**b**) Normalized Fourier power spectra of the plate surface at several times of loading.

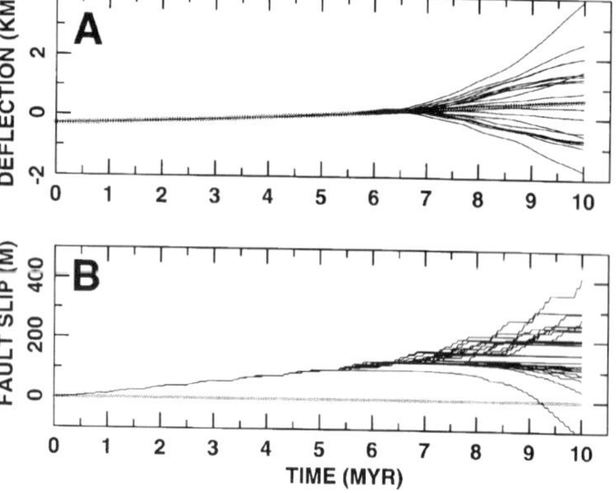

Fig. 6. (**a**) Vertical deflection through time of distinct points at the surface of a horizontally compressed oceanic lithospheric plate which has pre-existing faults in its crust which get reactivated through time in a stick-slip way. Curves are for surface points separated by a horizontal distance of 50 km. The nonlinear time response is typical for a buckling mode of plate deformation. Thick, grey line illustrates behaviour of a plate with locked faults: no buckling.(**b**) Amount of net slip through time at the surface appearance of a fault. Curves are for every tenth fault (horizontal separation of 50 km). Reverse slip is positive, normal slip is negative. Note the typical stick-slip response. Comparison of fig. (6a) and (6b) indicates that crustal fault reactivation occurs before lithospheric buckling starts.

some faults stay locked, and some faults even show normal faulting.

The spatial relationship between long wavelength folding and intensity of faulting is demonstrated by plotting the surface deflection (black line) of the plate in combination with the course of the amount of surface fault slip (grey line) over the plate, both after 8 Ma of shortening (Fig. 7a). Clearly, maxima in throw coincide with fold, whereas minima in fault surface throw coincide with fold highs. Furthermore, the surface slip of all faults over the distance from 900 km to 1100 km, which roughly comprises one fold, has been plotted versus time (Fig. 7a). The black curves show slip for faults located between 900 km and 1000 km, roughly coinciding with a fold low. Grey curves are for faults between 1000 km and 1100 km, coinciding with a fold high. All 'low positioned' faults exhibit accelerating reverse faulting, thus producing the maxima in fault slip, whereas the 'high positioned' faults show fault locking or even an inversion to normal faulting, both responsible for the fault slip minima.

This link between fault activity and fold geometry suggests that the amount of slip of a fault at its surface appearance may also vary with its position within a fold. This is confirmed by Fig. 7b, which displays the surface deflection of the plate after 8 Ma of shortening (black line) as well as the course of the amount of surface slip over the plate

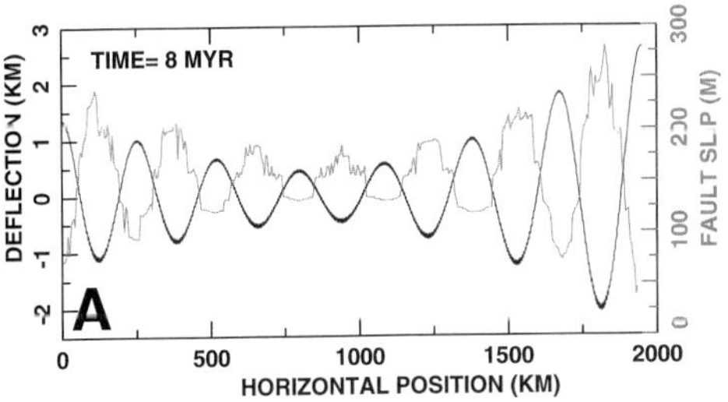

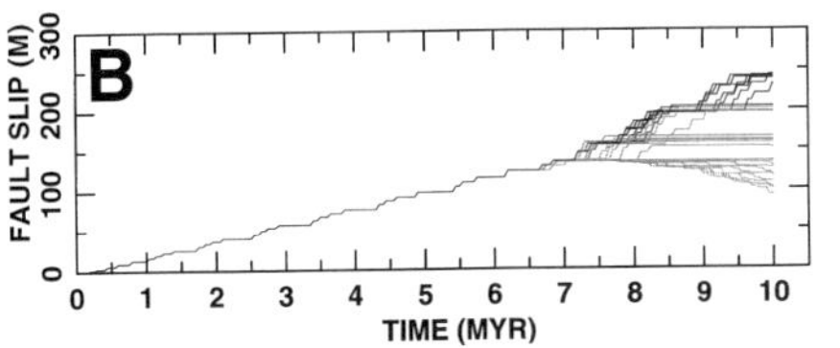

Fig. 7. (**a**) Deflection of the surface of the plate model (black line) and the amount of fault slip (at the plate surface) as distributed over the plate, both after 8 Ma of shortening (2.4%). (**b**) Surface slip versus time for all faults located in the long wavelength fold low between 900 km and 1000 km (black lines), and for all faults located in the long wavelength fold high between 1000 km and 1100 km (grey lines).

(grey line). As expected, maxima in throw coincide with fold lows where accelerating reverse faulting takes place. Minima in fault surface throw coincide with fold highs where faults get locked or throw decreases through normal faulting.

Figure 8a depicts vertical effective stress* profiles inside the plate (at the position denoted by the arrow in Fig. 5a) at several stages of compression. The adopted yield stress envelope limits the effective stresses in the upper brittle and lower ductile part of the lithosphere. In the faulted upper part of the plate, stress profiles develop that have a steeper slope than the theoretical brittle yield envelope. Furthermore, high effective stresses appear at a depth of 5 km (Fig. 8a), which is the depth at which the predefined faults in the model end. The steeper slope, equivalent to a higher 'fault reactivation' stress, arises from pre-setting the dip angle of the predefined fault planes. This results, in general, in an angle between fault planes and maximal principal stress that is not the optimal angle for faulting (according to a Mohr diagram) (Jaeger & Cook 1976). More effective stress is required for reactivation than in the case of an optimal angle, on which the yield stress value is based.

Immediately below the layer which contains the faults effective stresses reach the yield stress limit over an increasingly larger depth range. At depths where this occurs, rocks will deform in a brittle way. Stress also increasingly reaches the yield limit in the lowermost part of the plate model, where rock deformation will occur by ductile flow.

Also illustrated by Fig. 8a is the decrease in thickness of the strong elastic core of the oceanic lithosphere and the associated saturation of the yield envelope, until after c. 8 Ma the elastic core has almost vanished and is not able to mechanically sustain the stresses anymore. The lithosphere starts to fail completely, which will result in extensive deformation, as is numerically indicated by the acceleration of the surface deformation of the plate model and by the increase in magnitude of fault slip (Figs 6a and b).

The stick-slip behaviour of faults is also reflected in the temporal fluctuations of effective stress within crustal finite elements (Fig. 8c). Curves are for the first 5 upper-triangular elements in the column below the location denoted by the arrow in Fig. 5a. The 'lowest stress' curve belongs to shallowest element (depth range 0–1 km), whereas the 'highest stress' curve belongs to the deepest considered element (depth range 4–5 km).

* Effective stress is a useful scalar representation of the deviatoric state in a solid or fluid, and is the square root of the second invariant of the deviatoric stress tensor: $\sigma_E = (J_2)^{1/2} = (s_{ij} : s_{ij})^{1/2}$.

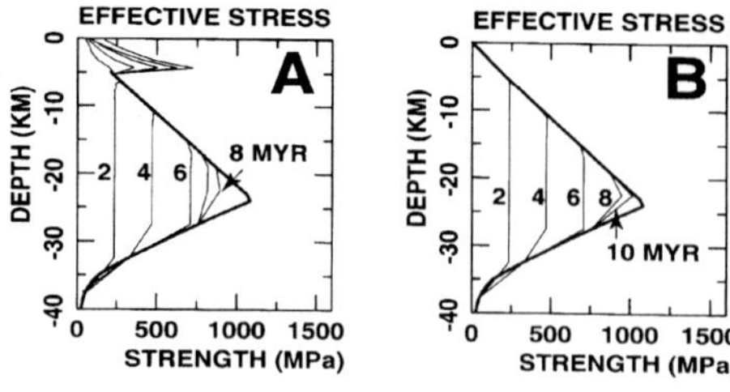

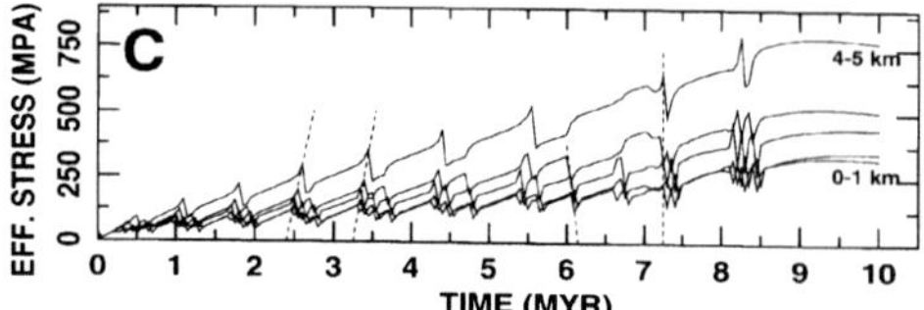

Fig. 8. (**a**) Vertical stress profiles at several stages of compression for the position marked by arrow in Fig. 5a. Solid line is the limiting yield envelope. (**b**) Same as (a), but now for a plate with locked faults. (**c**) Effective stress through time in the uppermost five 'crustal' upper-triangular finite elements in the column at the location denoted by the arrow in Fig. 5a. The 'lowest stress' curve belongs to shallowest element (depth range 0–1 km), whereas the 'highest stress' curve belongs to the deepest considered element (depth range 4–5 km). Fault response is typical stick-slip with cycles of stress buildup followed by stress relaxation as the fault becomes active. Lining up stress drops belonging to the same event, shows that activation fronts propagate downwards as well as upwards along a fault plane. Both the stick–slip behaviour and up- and downward propagating reactivation of crustal faults are confirmed by seismic stratigraphic observations.

Repeated periods of stress buildup are followed by unlocking of the fault and displacement of the fault blocks resulting in stress relaxation. Stress drops associated with the stick-slip fault activity are in the order of a few tens of mega pascals. Comparison of stress drops from the same event yields information on propagation direction of fault reactivation. Although not always clearly distinguishable it is possible to recognise downwards propagating stress drop fronts (for example, between 3 and 3.5 Ma) as well as upwards propagating stress drop fronts (e.g. just after 6 Ma), while the event after 7.3 Ma seems to reactivate the entire fault almost instantenously. The upwards propagating event, occurring slighly after 6 Ma, starts in the middle of the crust (not originating from a deeper level).

An additional numerical experiment has been undertaken to investigate whether it is indeed the reactivation of the faults, and associated local changes in stress and strain, that ultimately cause the large scale buckling of the plate. To ensure that the faults do not reactivate they have been assigned a very high friction coefficient. Furthermore, the elements in the 'faulted' upper part of the plate have now a yield stress, according to the regular yield envelope (Fig. 4c), to avoid this layer behaving as an elastic layer. Thus, failure is still included in this layer with a brittle yield stress derived from the brittle failure equation (1), but the associated deformation is now simulated numerically by plastic flow. The results show that in this case even after 10 Ma of convergence (equivalent to a horizontal shortening of 3%), the initially flat plate is still flat and has not developed a folding mode of deformation. Deflections and power spectra of the plate surface at several times are included in Figs 5a and b, but no harmonic deviations are evident. This is also demonstrated by Figs 6a and b, where the temporal course of surface point deflection and surface fault slip are shown by the grey curves. The small, straight slope of the vertical deflection curves in Fig. 5a indicates the homogeneous thickening of the plate in response to the horizontal shortening. Folding/buckling is absent, even when the elastic core of the oceanic lithosphere has almost vanished. This demonstrates that it is the crustal fault reactivation and associated changes in stress and strain, that in these finite element models acts as the perturbing mechanism that facilitates the large scale mode of tectonic deformation.

Discussion

Although quite advanced in the treatment of frictional faulting and the incorporation of a non-linear depth-varying rheology, the numerical models that have been used are relatively simple when considering the complex features and processes within the oceanic lithosphere of the central Indian Ocean. Nevertheless, the models successfully predict stress and strain patterns associated with the long wavelength folding and small scale faulting occurring in this intraplate part of the Indo-Australian plate, which is subjected to considerable compressional stress loading and associated tectonic shortening.

The dominant wavelength of large scale folding that is developed in the models lies well within the range of observed wavelengths in the central Indian Ocean (Zuber 1987). It is slightly larger than the average observed wavelength, but this can be attributed to the simplified character of the model, not accounting for weakening effects introduced by intra-lithospheric inhomogeneities. The yield envelope, which exerts a major control on the flexural rigidity of the lithosphere, is also a theoretical idealization, probably resulting in a mechanically too strong lithosphere. The sharp brittle–ductile transition, for example, will probably be transitional, as argued by Carter & Tsenn (1987). Furthermore, recent rock mechanic experiments (Ord & Hobbs 1989) indicate that the linear brittle branch of the yield envelope may even break down, and maximum sustainable stresses may be as low as 300 MPa, substantially reducing the maximum strength of the lithosphere.

The nonlinear behaviour through time is consistent with the 'buckling' response of a horizontally compressed plate, characterized by rapidly accelerating vertical displacements prior to full lithospheric failure (Cloetingh *et al.* 1989; Stephenson *et al.* 1990; Stephenson & Cloetingh 1991). Analysis of vertical effective stress profiles shows that buckling starts when the remaining elastic core of the oceanic lithosphere has almost vanished, as predicted by elasto-plastic plate deformation theory (McAdoo & Sandwell 1985).

Fault activity starts before the onset of the large scale folding. The increasingly complex activity of the pre-defined faults after long wavelength deformation has started to develop, seems to be caused by an interaction of developing bending stresses with the tectonic compressional stress. Support for this is given by the relationship of fault type and folding magnitude with fault position within a long wavelength fold. Bending would produce additional compressive stress in the downflexed areas and tensile stress in the upflexed areas. These compressive or tensile bending will, respectively, assist or counteract the tectonic stresses. This leads to a decrease in effective stress in upflexed regions, accompanied by permanent locking of faults, or even leads to net tensile effective stresses and normal fault reactivation, as indeed demonstrated by the model results. Statistical analysis of bathymetric and seismic data is required to see if the effects of bending are present.

The pre-defined faults in the models exhibit typical stick–slip behaviour through time. Jumps in fault slip at the plate surface are in the order of several meters, cumulating to throws at the plate's surface of more than 100 m. These values fit with measured fault slips on seismic profiles (Bull & Scrutton 1992; Chamot-Rooke *et al.* 1993). The stick–slip fault activity is in agreement with observations of substantial microseismicity in the crust of the central Indian Ocean (Levchenko & Ostrovsky 1993). Deep earthquakes, which have been observed in this intraplate area (Bergman & Solomon 1980, 1985), are also correctly predicted by the models with stress within the plate reaching the brittle failure limit over considerable depth ranges down to the brittle-ductile transition (Govers *et al.* 1992). Reactivation fronts along a single fault can propagate downwards as well as upwards, or reactivate the entire fault instantaneously. Upwards propagating faults and possibly rapidly formed faults have been identified on the seismic stratigraphic record (Bull & Scrutton 1992). More analysis is required to check if there are relationships of type of fault reactivation with the position of the fault concerned relative to the long wavelength folds.

The numerical models develop high effective stresses at the tips of the faults. These stress concentrations can be interpreted physically as being responsible for propagation of faults. Such stress concentrations at fault tips are predicted theoretically by the Griffith Crack Theory (Scholz 1990), and are consistent with rock-mechanics experiments showing that stress concentrations do occur at fault tips (Jaeger & Cook 1976).

Modelling also shows that no long wavelength mode of tectonic deformation develops when there is no fault activity. This demonstrates that in this study the adopted 'simplified' model geometry and boundary conditions the crustal fault reactivation and associated changes in stress and strain act as the essential perturbing mechanism that facilitates the large scale mode of tectonic deformation.

The complex fault fabric in the Indian Ocean crust has been approximated by a simplified set of faults with a geometry based on average values. Additional numerical analyses have to be carried out to investigate if variations in one or more of the fault parameters (such as dip, dip direction, penetration depth) have a major influence on the long wavelength deformation of the Indian Ocean lithosphere. The successful predictions of the numerical model with the 'average' fault set do suggest, however, that it is merely the reactivation of faults that is of essential importance rather than the geometrical features of the faults.

Another model simplification which may affect the model results is the assumption of a constant lithospheric thermal age (and thus lithospheric thickness) for the entire model, whereas the the thermal age of the area of interest ranges from approximately 40 Ma in the south to 70 Ma in the north. The increase in age is probably to gradual to initiate the intraplate deformation. It may, however, affect the wavelength of the large scale folding, as with age also the flexural rigidity of the lithosphere increases, which will lead to a larger wavelength of deformation (McAdoo & Sandwell 1985). Zuber (1987) indeed observed a gradual northward increase in the wavelength of lithospheric folding in the central Indian Ocean.

Conclusions

Substantial compressional stresses in the central Indian Ocean reactivate pre-existing crustal faults, and, in a later phase, initiate folding and buckling of the entire oceanic lithosphere. Stress and strain patterns, spatial as well as temporal, are successfully predicted by the numerical models, and are in agreement with geophysical observations.

The modelling also demonstrates that in the absence of fault reactivation no lithospheric folding develops. Crustal fault reactivation, therefore, appears to be essential to facilitate long-wavelength oceanic lithospheric buckling in the central Indian Ocean.

We thank G. Ranalli and R. A. Stephenson for their constructive reviews of an early draft of this paper. Comments by F. Nieuwland and an anonymous reviewer are also gratefully acknowledged.

References

BEEKMAN, F. 1994. *Tectonic modelling of thick-skinned compressional intraplate deformation*. PhD thesis. Vrije Universiteit, Amsterdam.

BERGMAN, E. A. & SOLOMON, S. C. 1980. Oceanic intraplate earthquakes: Implications for local and regional intraplate stress. *Journal of Geophysical Research*, **85**, 5389–5410.

—— & —— 1984. Source mechanisms of earthquakes near mid-ocean ridges from body waveform inversion: Implications for the early evolution of oceanic lithosphere. *Journal of Geophysical Research*, **89**, 11415–11441.

—— & —— 1985. Earthquake source mechanisms from body waveform inversion and intraplate tectonics in the northern Indian Ocean. *Physics of the Earth and Planetary Interiors*, **40**, 1–23.

BRACE, W. F. & KOHLSTEDT, D. L. 1980. Limits on lithospheric stress imposed by laboratory experiments. *Journal of Geophysical Research*, **85**(B11), 6248–6252.

BULL, J. M. 1990. Structural style of intra-plate deformation, Central Indian Ocean Basin: evidence for the role of fracture zones. *Tectonophysics*, **184**, 213–228.

—— & SCRUTTON, R. A. 1990. Fault reactivation in the central Indian Ocean and the rheology of oceanic lithosphere. *Nature*, **344**, 855–858.

—— & —— 1992. Seismic reflection images of intraplate deformation, central Indian Ocean, and their tectonic significance. *Journal of the Geological Society, London*, **149**, 955–966.

——, MARTINOD, J. & DAVY, PH. 1992. Buckling of the oceanic lithosphere from geophysical data and experiments. *Tectonics*, **11**, 537–548.

BYERLEE, J. D. 1978. Friction of rocks. *Pure and Applied Geophysics*, **116**, 615–626.

CARMICHAEL, R. S. 1989. *Practical handbook of physical properties of rocks and minerals*, CRC Press, Boca Raton, Florida.

CARTER, N. L. & TSENN, M. C. 1987. Flow properties of continental lithosphere. *Tectonophysics*, **136**, 27–63.

CHAMOT-ROOKE, N., JESTIN, F., DE VOOGD, B. & PHEDRE WORKING GROUP 1993. Intraplate shortening in the central Indian Ocean determined from a 2100-km-long north–south deep seismic reflection profile. *Geology*, **21**, 1043–1046.

CLOETINGH, S. & WORTEL, R. 1985. Regional stress field of the Indian Plate. *Geophysical Research Letters*, **12**, 77–80.

—— & —— 1986. Stress in the Indo-Australian plate. *Tectonophysics*, **132**, 49–67.

——, KOOI, H. & GROENEWOUD, W. 1989. Intraplate stress and sedimentary basin evolution *In*: PRICE, R. A. (ed.) *Origin and evolution of Sedimentary Basins and their energy and mineral resources*. AGU Geophysical Monograph, **48**, 1–16.

COCHRAN, J. R. 1989. Himalayan uplift, sealevel and the record of the Bengal Fan sedimentation at the ODP Leg 116 Sites. *In*: COCHRAN, J. R., STOW, D. A. V., *et al.* (eds) *Proceedings of the Ocean Drilling Program, Scientific Reports*, College Station, Texas, 397–414.

CURRAY, J. R. & MOORE, D. G. 1971. Growth of the Bengal Deep-Sea Fan and Denudation in the Himalayas. *GSA Bulletin*, **82**, 563–572.

—— & MUNASINGHE, T. 1989. Timing of intraplate deformation, northeastern Indian Ocean. *Earth and Planetary Science Letters*, **94**, 71–77.

DEMETS, C., GORDON, R. G., ARGUS, D. F. & STEIN, S. 1990. Current plate motions. *Geophysical Journal International*, **101**, 425–478.

GELLER, C. A., WEISSEL, J. K. & ANDERSON, R. N. 1983. Heat transfer and intraplate deformation in the central Indian Ocean. *Journal of Geophysical Research*, **88**, 1018–1032.

GOETZE, C. & EVANS, B. 1979. Stress and temperature in the bending lithosphere as constrained by experimental rock mechanics. *Geophysical Monograph*, **59**, 463–478.

GORDON, R. G., DEMETS, C. & ARGUS, D. F. 1990. Kinematic constraints on distributed lithospheric deformation in the equatorial Indian Ocean from present motion between the Australian and Indian plates. *Tectonics*, **9**(3), 409–422.

GOVERS, R., WORTEL, M. J. R., CLOETINGH, S. A. P. L. & STEIN, C. A. 1992. Stress magnitude estimates from earthquakes in oceanic plate interiors. *Journal of Geophysical Research*, **97**(B8), 11749–11760.

JAEGER, J. C. & COOK, N. G. W. 1976. *Fundamentals of Rock Mechanics*, Chapman & Hall, London.

KARNER, G. D. & WEISSEL, J. K. 1990. Factors controlling the location of compressional deformation of oceanic lithosphere in the central Indian Ocean. *Journal of Geophysical Research*, **95**(B12), 19795–19810.

KAZ'MIN, V. G. & BORISOVA, I. A. 1992. Mechanisms of rift valley formation for slowly spreading ridges. *Geotectonics*, **26**(4), 344–352.

KIRBY, S. H. 1983. Rheology of the lithosphere. *Reviews of Geophysics and Space Physics*, **21**, 1458–1487.

KLOOTWIJK, C. T., CONAGHAN, P. J. & Powell, C. MCA. 1985. The Himalayan Arc: large-scale continental subduction, oroclinal bending and back-arc spreading. *Earth and Planetary Science Letters*, **75**, 167–183.

LEGER, G. T. & LOUDEN, K. E. 1990. Seismic refraction measurements in the Central Indian Basin: Evidence for crustal thickening related to intraplate deformation. *In*: COCHRAN, J. R. & STOW, D. A. V. (eds) *Proceedings of the Ocean Drilling Program, Scientific Reports*, **116**, College Station, Texas.

LEVCHENKO, O. V. 1989. Tectonic aspects of intraplate seismicity in the northeastern Indian Ocean. *Tectonophysics*, **170**, 125–139.

LEVCHENKO, O. V. & OSTROVSKY, A. O. 1993. Seismic seafloor observations: A study of 'anomalous' intraplate seismicity in the northeastern Indian Ocean. *Physics of the Earth and Planetary Interiors*, **74**, 173–182.

MACDONALD, K. C. 1982. MID-OCEAN RIDGES: Fine scale tectonics, volcanic and hydrothermal processes within the plate boundary zone. *Annual Review of Earth and Planetary Science*, **10**, 155–190.

MARTINOD, J. & DAVY, P. 1992. Periodic instabilities during compression or extension of the lithosphere, 1. Deformation modes from an analytical perturbation mode. *Journal of Geophysical Research*, **97**(B2), 1999–2014.

MCADOO, D. C. & SANDWELL, D. T. 1985. Folding of oceanic lithosphere, *Journal of Geophysical Research*, **90**, 8563–8569.

MELOSH, H. J. 1990. Mechanical basis for low-angle normal faulting in the Basin and Range Province. *Nature*, **343**, 331–335.

—— & RAEFSKY, A. 1981. A simple and efficient method for introducing faults into finite element computations. *Bulletin of the Seismological Society of America*, **71**, 1391–1400.

—— & WILLIAMS, JR., C. A. 1989. Mechanics of graben formation in crustal rocks: A finite element analysis. *Journal of Geophysical Research*, **94**(B10), 13961–13973.

MOLNAR, P., ENGLAND, PH. & MARTINOD, J. 1993. Mantle dynamics, uplift of the Tibetan Plateau, and the Indian monsoon. *Reviews of Geophysics*, **31**(4), 357–396.

NEPROCHNOV, Y. P., LEVCHENKO, O. V., MERKLIN, L. R. & SEDOV, V. V. 1988. The structure and tectonics of the intraplate deformation area in the Indian Ocean. *Tectonophysics*, **156**, 89–106.

ORD, A. & HOBBS, B. E. 1989. The strength of the continental crust, detachment zones and the development of plastic instabilities. *Tectonophysics*, **158**, 269–289.

PATRIAT, P & ACHACHE, J. 1984. India–Eurasia collision chronology has implications for crustal shortening and driving mechanism of plates. *Nature*, **311**, 615–621.

PEIRCE, J. W. 1978. The northward motion of India since the Late Cretaceous. *Geophysical Journal of the Royal Astronomical Society*, **52**, 277–311.

PETROY, D. E. & WIENS, D. A. 1989. Historical seismicity and Implications for Diffuse Plate Convergence in the Northeast Indian Ocean. *Journal of Geophysical Research*, **94**(B9), 12301–12319.

PILIPENKO, A.I. & SIVUKHA, N. M. 1991. Geological structure and geodynamics of the West Australian Basin (Wharton Basin). *Geotectonics*, **25**(1), 84–93.

RANALLI, G. 1987. *Rheology of the Earth. Deformation and flow processes in geophysics and geodynamics*, Allen & Unwin, Boston.

ROYER, J.-Y. & CHANG, T. 1991. Evidence for relative motions between the Indian and Australian Plates during the last 20 m.y. from plate tectonic reconstructions: implications for the deformation of the Indo-Australian Plate. *Journal of Geophysical Research*, **96**(B11), 11779–11802.

SANDWELL, D. T. & SMITH, W. H. F. 1992. Global marine gravity from ERS-1, Geosat, and Seasat reveals new tectonic fabric. *EOS*, **73**(43), Suppl., 1992 Fall Meeting.

SCHOLZ, C. 1990. *The Mechanics of Earthquakes and Faulting*, Cambridge University Press, Cambridge.

SCLATER, J. G., JAUPART, C. & GALSON, D. 1980. The heat flow through oceanic and continental crust and the heat loss of the Earth. *Reviews of Geophysics and Space Physics*, **18**, 269-311.

SHIPBOARD SCIENTIFIC PARTY 1990. ODP Leg 116 Survey. *In*: COCHRAN, J. R. & STOW, D. A. V. (eds) *Proceedings of the Ocean Drilling Program, Scientific Reports*, **116**, College Station, Texas.

STEIN, C. A. & WEISSEL, J. K. 1990. Constraints on the Central Indian Basin thermal structure from heat flow, seismicity and bathymetry. *Tectonophysics*, **176**, 315–332.

——, CLOETINGH, S. & WORTEL, R. 1989. SEASAT-derived gravity constraints on stress and deformation in the northeastern Indian Ocean. *Geophysical Research Letters*, **16**(8), 823–826.

STEPHENSON, R. A. & CLOETINGH, S. 1991. Some examples and mechanical aspects of continental lithosphere folding. *Tectonophysics*, **188**, 27–37.

——, RICKETTS, B. D. CLOETINGH, S. A. P. L. & BEEKMAN, F. 1990. Lithosphere folds in the Eurekan Orogen, Arctic Canada? *Geology*, **18**, 103–106.

TIMOSHENKO, S. P. & WOINOWSKY-KRIEGER, S. 1959. *Theory of Plates and Shells* (2nd edn), McGraw-Hill, Singapore.

TSENN, M. C. & CARTER, N. L. 1987. Upper limits of power law creep of rocks. *Tectonophysics*, **136**, 1–26.

VERZHBITSKY, E. V. & LOBKOVSKY, L. I. 1993. On the mechanism of heating up the Indo-Australian plate. *Journal of Geodynamics*, **17**(1/2), 27–38.

WEISSEL, J. K., ANDERSON, R. N. & GELLER, C. A. 1980. Deformation of the Indo-Australian plate. *Nature*, **287**, 284–291.

WHITE, R. S., DETRICK, R. S., MUTTER, J. C., BUHL, P., MINSHULL, T. A. & MORRIS, E. 1990. New seismic images of oceanic crustal structure. *Geology*, **18**, 462–465.

ZUBER, M. T. 1987. Compression of the oceanic lithosphere: An analysis of intraplate deformation in the Central Indian Basin. *Journal of Geophysical Research*, **92**(B6), 4817–4825.

Mathematical modelling of growth strata associated with fault-related fold structures

STUART HARDY, JOSEP POBLET, KEN McCLAY & DAVE WALTHAM,
Fault Dynamics Project, Department of Geology, Royal Holloway University of London, Egham, Surrey TW20 OEX, UK

Abstract: Previous geometric models of fault-related folding have been limited in their ability to model sedimentation contemporaneous with contractional deformation. Waltham's 1992 general forward modelling equation allows more realistic models of sedimentation to be combined with mathematical models of fault-bend folding, fault-propagation folding and limb-rotation associated with detachment folding. Using average rates of fault slip and sedimentation derived from the literature, the mathematical models are used to predict stratal geometries associated with each of the major styles of fault-related folding in both sub-marine and sub-aerial settings. Both background sedimentation and local erosion, transport and deposition as a result of fold growth are modelled. Comparison is made with natural examples and the utility of growth strata architectures in the assessment of fault-fold kinematics assessed.

Continental compressive deformation results in a wide variety of structures in the brittle upper crust, in particular thrusts and folds (e.g. Rich 1934; Dahlstrom 1970; Suppe *et al.* 1991; Avouac *et al.* 1993). During deformation thrusting and folding are often intimately linked resulting in a variety of different styles of fault-fold structure. These structures are extremely important in economic terms, being the sites of many important oil and gas fields (e.g. Suppe 1983; Mitra 1990; Medwedeff 1989, 1992). They are also very useful in understanding the spatial and temporal evolution of a basin, particularly in the external parts of foreland fold and thrust belts where they can directly affect the sedimentary fill of the basin (e.g. Deramond *et al.* 1993; Doglioni 1993; Zoetemeijer *et al.* 1993). As a result, fault-related folds have been the focus of much research and there is a voluminous literature on their description and particularly their use in section balancing and modelling (e.g. Suppe 1983; Mitra 1990; Suppe *et al.* 1991; Zoetemeijer *et al.* 1992; Shaw *et al.* 1994).

There has been much quantitative investigation of these structures using geometric (e.g. Suppe 1983), mechanical (e.g. Johnson & Berger 1989; Braun & Sambridge 1994) and analogue modelling techniques (e.g. Calassou *et al.* 1993). In particular, there has been a wealth of geometric modelling, particularly of fault-bend folds and, to a lesser extent, fault-propagation folds (e.g. Suppe 1983; Endignoux & Mugnier 1990; Mount *et al.* 1990; Suppe & Medwedeff 1990; Suppe *et al.* 1991; Zoetemeijer *et al.* 1992; Al Saffar 1993; Shaw & Suppe 1994; Shaw *et al.* 1994). Using constraints of conservation of area, bed-length and bed-thickness, these geometric models have quantified the relationship between slip, fault-shape and fold-shape. These geometric models have, in the main, been used to constrain structural interpretations in areas where sub-surface or seismic data is poor, but have also been used as forward models to investigate the structural and stratigraphic evolution of fault-fold structures. These forward modelling studies have, in general, used basic models of syn-tectonic sedimentation (e.g. Suppe *et al.* 1991).

This paper uses the general tectono-sedimentary forward modelling equation of Waltham (1992) to study the interaction of tectonics and sedimentation in three types of fault-related folds often seen in compressive settings: fault-bend folds (Suppe 1983), fault-propagation folds (Suppe & Medwedeff 1990) and detachment folds (Jamison 1987). In each case previously published velocity models of deformation are linked with models of syn-tectonic erosion, transport and deposition. The mathematical models are used to examine syn-tectonic stratigraphic architectures associated with each style of fault-related folding in both sub-aerial and sub-marine environments using typical rates of deformation and sedimentation from compressive settings. Modelled growth strata architectures are then compared with natural examples. Differences in growth strata architectures are identified and their utility as indicators of fault-fold kinematics discussed.

Fault-related folding in compressive settings

This paper will focus on fault-fold structures which form in the brittle upper crust during compressive

From Buchanan, P. G. & Nieuwland, D. A. (eds), 1996, *Modern Developments in Structural Interpretation, Validation and Modelling*, Geological Society Special Publication No. 99, pp. 265–282.

deformation. This deformation results in a wide variety of structures which may directly affect the sedimentary fill of a basin, particularly in the external parts of foreland fold and thrust belts. Structural interpretations in such belts are often aided by section-balancing techniques, which require the reconstruction of the fold and thrust belt to its original configuration (Dahlstrom 1969).

Fold and thrust belts are often characterized by complex structures, with surface and sub-surface data often being of poor quality preventing unique structural interpretations. However, while any individual structure may be more complex, three types of fault-fold interaction have been successfully used to explain many structural geometries seen in fold and thrust belts: fault-bend folds, fault-propagation folds and detachment folds (e.g. Suppe 1985; Jamison 1987; Mitra 1990, 1992). The characteristics of these structures are shown in Figure 1. The relationships between slip, fault-shape and fold-shape for each of these structures have been quantified in a number of geometric models (Suppe 1983; Jamison 1987; Suppe & Medwedeff 1990) and these relationships used to constrain structural interpretations in many fold and thrust belts. More complex structures such as duplexes, imbricate thrust systems or detachment fold trains are often considered to be systems of these structures (e.g. Mitra 1992; Zoetemeijer 1993).

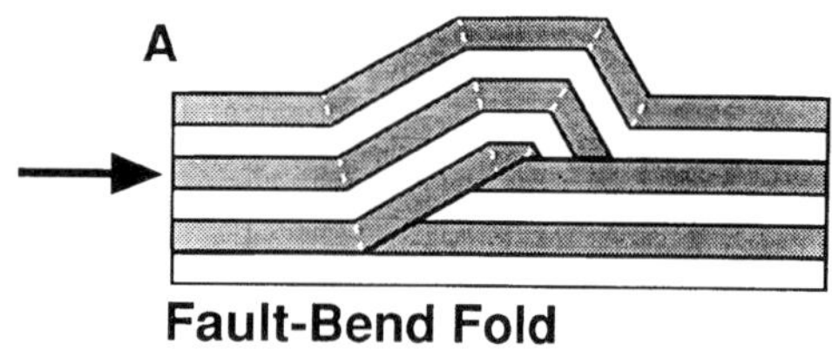

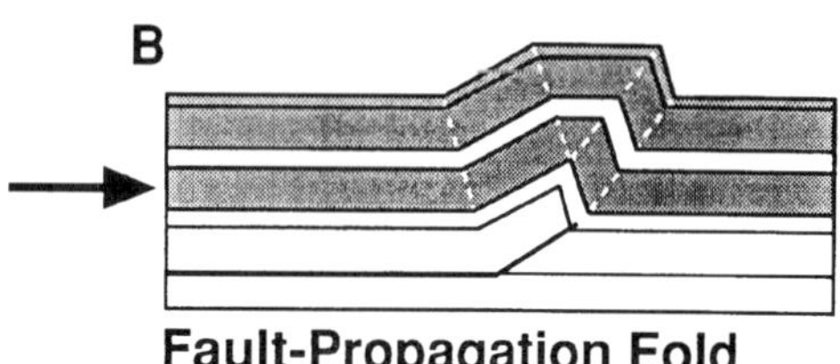

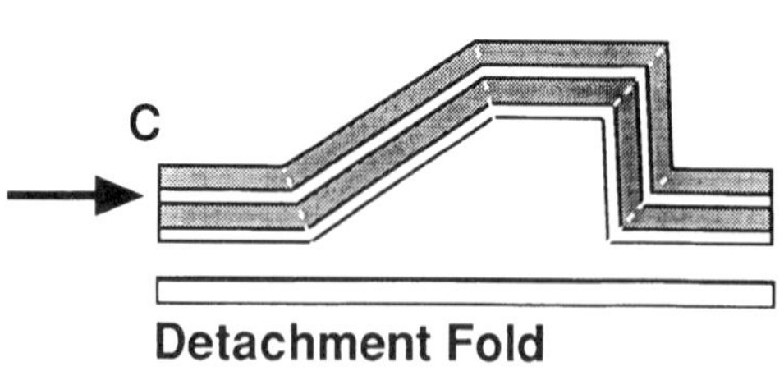

Fig. 1. The three types of fault-fold interaction commonly observed in fold and thrust belts, A fault-bend folds, B fault-propagation folds and C detachment folds.

Fault-bend folds characteristically occur where a detachment steps up along a ramp from one incompetent horizon to another (Rich 1934). Movement of the thrust sheet over this ramp causes folding of the rocks in the hangingwall (Fig. 1a). Folds of this type have been widely recognized and interpreted in a variety of areas (Dahlstrom 1970; Suppe 1983; Medwedeff 1989; Mount *et al.* 1990). Suppe (1983) first proposed a quantitative, area balanced, geometric model for the development of such structures based on kink-band migration which has been widely applied since its introduction.

Fault-propagation folds are similar to fault-bend folds in that they occur where a ramp steps up from a lower decollement, but differ in that the thrust fault does not exist as a through-going structure from the start of deformation but gradually propagates with continued slip, deformation up-section of the fault-tip being taken up in folding (Dahlstrom 1969; Faill 1973) (Fig. 1(b)). Such folds have been recognized and interpreted in a wide variety of fold and thrust belts (e.g. Suppe & Medwedeff 1990; Mitra 1990; Deramond *et al.* 1993; Jordan *et al.* 1993). Suppe & Medwedeff (1990) proposed a quantitative geometric model of fault-propagation folding for the analysis of such structures, analogous to that developed for fault-bend folding, which has been widely used and adapted since its introduction (Mitra 1990; Mosar & Suppe 1991; Al Saffar 1993).

Detachment folds differ from the other two styles in that they are not associated with a ramp in an underlying thrust fault, but generally form as a result of buckling of a competent layer overlying a less competent layer such as shales or evaporites (Fig. 1c). Detachment folds may have a parallel geometry in their outer arc and disharmonic and non-parallel geometries in the core of the anticline (disharmonic detachment folds) or may form by the folding of strata above a basal detachment which is isoclinally folded in the core of the anticline (lift-off folds). Detachment folds have been widely recognized in the Alps and the Pyrenees (e.g. Buxtorf 1916; Laubscher 1962; Burbank & Verges, 1994), the Ventura Basin, California (Yeats 1983; Rockwell *et al.* 1988) and Taiwan (Mitra & Namson 1989). Various geometric models for the analysis of both disharmonic and lift-off detachment folds have been proposed in the literature (De Sitter 1956; Jamison 1987; Mitra & Namson 1989; Dahlstrom 1990; Mitra 1992; Groshong & Epard 1994; Hardy & Poblet 1994).

While these types of fault-fold interaction are idealized end-members, and many complex natural structures may be formed by the simultaneous or sequential occurrence of two or more of these mechanisms of folding (e.g. Medwedeff 1989; Mitra 1992), they have proved extremely useful in analysing a great many structures in fold and thrust belts (e.g. Suppe *et al.* 1991; Mitra 1992; Zoetemeijer *et al.* 1992). In recent years the interaction of such structures and sedimentation has been the focus of much research (e.g. Suppe *et al.* 1991; Zoetemeijer *et al.* 1993) which has highlighted the importance of syn-tectonic stratigraphic architectures in understanding the evolution of both individual structures and deformation within a basin. However, the models of sedimentation used in these studies have been rudimentary, preventing a comprehensive analysis of tectonics and sedimentation in this setting. This paper will present mathematical models for each of these types of fault-fold structures which allow more realistic modelling of sedimentation.

Mathematical modelling of tectonics and sedimentation

The mathematical models of tectonics and sedimentation discussed in this paper have been developed using the general tectono-sedimentary forward modelling equation (GFME) of Waltham (1992), which has been successfully used to model tectonics and sedimentation in a wide variety of environments: carbonate platforms (Bosence *et al.* 1994), deltas (Hardy & Waltham 1992; Hardy *et al.* 1994), domino fault blocks (Waltham *et al.* 1993; Hardy 1993), fault-bend and fault-propagation folds (Hardy & Poblet 1995) and detachment folds (Hardy & Poblet 1994). This section will review the application of this modelling methodology to contractional structures.

The modelling approach is based upon the concept that the height of a geological surface can be modified in four ways: (i) material can be added to, or removed from, the surface; (ii) material can be moved from one part of the surface to another; (iii) the surface can be moved; (iv) the surface can be deformed. The first two of these mechanisms are sedimentary whilst the second two are tectonic. Using the equation of continuity of an open system as a starting point, Waltham (1992) derived the following partial differential equation (PDE) for use in modelling tectonics and sedimentation in two dimensions:

$$\underbrace{\partial h/\partial t = [p - \partial F/\partial x]}_{\text{Sedimentary processes}} + \underbrace{[v - u.\partial h/\partial x]}_{\text{Tectonic process}} \quad (1)$$

where:

h is the height of a geological surface at a fixed horizontal location
t is time
p is a source term
F is the sediment flux
x is a horizontal co-ordinate
v is the vertical velocity (uplift positive, subsidence negative)
u is the horizontal velocity (left to right positive, right to left negative)

This equation combines sedimentary and tectonic processes in an Eulerian co-ordinate system, that is a fixed co-ordinate system which does not move as a result of flow or tectonic deformation. The strength of this approach is that it allows tectonics and sedimentation to be combined in a single mathematical formulation ensuring that they are modelled as simultaneous rather than sequential processes (Waltham 1992; Waltham & Hardy 1995). An alternative method is the Langrangian, or material, description in which the displacement of particles is described in terms of a moving co-ordinate system (for a geological example see Contreras & Suter 1990). The main drawback of the Lagrangian description is that several processes cannot be modelled as occurring simultaneously, they must be modelled as occurring after one another. The relationship of the Eulerian description to the Langrangian description is discussed fully in Waltham & Hardy (1995).

As with any application of the GFME to the modelling of tectonics and sedimentation, two distinct steps must be undertaken: (1) the specification of the horizontal and vertical velocities which result from a given deformation mechanism, and (2) the specification of the models of sedimentation applicable to the geological setting under consideration.

The three types of fault-related folding discussed above will be considered in this paper. Until recently, these structures have been treated in a purely geometric manner (e.g. Suppe 1983; Jamison 1987; Suppe & Medwedeff 1990) and therefore one of the principal tasks in applying the GFME in this setting is the derivation of equivalent velocity descriptions of deformation for the kinematics of such fault-fold structures (Hardy & Poblet 1994). These velocity descriptions of deformation provide the horizontal and vertical velocities which are used in Equation (1) to model the structural evolution of a fault-related fold. The specific velocity descriptions of deformation for fault-bend folding (Suppe 1983), fault-propagation folding (Suppe & Medwedeff 1990) and limb rotation associated with detachment folding are given in two previously published papers (Hardy &

Poblet 1994, 1995) to which the interested reader is referred for the detailed derivations. More complex deformation than these simple structures, such as imbrications and duplexes, can also be dealt with by the modelling scheme, providing the deformation mechanisms are well-specified (Waltham & Hardy 1995).

As highlighted by Suppe *et al.* (1991), strata deposited during the development of a contractional structure (growth strata) can provide a record of the evolution of the structure and help elucidate deeper fault geometry where surface exposure or seismic data is poor. As such it is important that erosion, transport and sedimentation in the region of a growing fault-fold structure are quantitatively modelled. Previous geometric models of fault-fold structures have been limited in their ability to investigate the interaction of tectonics and sedimentation and have used basic models of sedimentation. In these studies, simple 'fill-to-the-top' sedimentation models have often been used often preventing any structural relief or bathymetry from developing in the region of a growing structure. The work of Zoetemeijer (1993) was more sophisticated in that it allowed erosion of all material above a specified level and allowed topography to develop. However, the values of fault slip reported in the literature for thrust faults (Table 1) are associated with uplift rates of c. 1–10 m ka^{-1} (Yeats 1983; Rockwell *et al.* 1988; Leeder 1991; Avouac *et al.* 1993), whereas average rates of sedimentation in both sub-aerial and submarine compressive settings rarely exceed 1 m ka^{-1} (Table 2) and can be several orders of magnitude less (Johnson *et al.* 1986; Kukal 1990; Suppe *et al.* 1991). We would therefore expect topographic or bathymetric expression of fault-fold structures to be the norm. The modelling methodology must be able to model such situations and deal with the local sedimentary effects of topographic or bathymetric relief.

To model the evolving sediment geometries in this system a base level must first be defined, this is the datum below which sediments may be deposited. Base level in many previous studies has risen as a proportion of ramp uplift, and has thus not been independent of deformation, although the work of Zoetemeijer (1993) is an exception. Here base level is independent of deformation and may rise or fall during a model run in order to create or destroy accommodation space.

Having defined a base level, models of erosion, transport and sedimentation may be included in the modelling scheme. First, a background sedimentation rate is introduced, which is considered to represent non-locally derived material such as fine siliciclastics, a hemi-pelagic drape or alluvial sedimentation. This additional sediment input can be regarded, on this scale, as approximately constant across the fold and may thus be simulated by introducing a constant value for p in equation (1). This sediment will be deposited everywhere across the model at a given rate, except where the fold is uplifted above the specified base level. Although constant across the model, this background sedimentation rate may vary with time during the model run.

In addition to any background sedimentation,

Table 1. *Selected slip rates from compressional settings*

Slip rate (m ka^{-1})	Structure	Reference
0.30	Beartooth fault, Wyoming and Montana	DeCelles *et al.* (1991)
0.82	Loth Hills Thrust, San Joaquin Valley, California	Medwedeff (1989)
1.25	Tugulu-Dushanzi Thrust, Northern Tien Shan.	Avouac *et al.* (1993)
1.3	Channel Islands Thrust, Santa Barbara Channel, California	Shaw & Suppe (1994)
1.3	Pitas Point Thrust, Santa Barbara Channel, California	Shaw & Suppe (1994)
0.0–2.0	Precordillera Thrust Belt, West-central Argentina	Jordan *et al.* (1993)
1.0–2.0	Santa Barbara Channel and Los Angeles Basin, California	Suppe *et al.* (1991)
2.1	Sierras Marginales Thrust, south-central Pyrenees	Burbank *et al.* (1992)
1.4–2.3	Perdido Fold Belt, Gulf of Mexico	Mount *et al.* (1990)
4.3	Wheeler Ridge Thrust, California	Medwedeff (1992)
1.9–5.2	Pliocene Thrusts, Los Angeles Area	Davis *et al.* (1989)
max. 2.5	N. Apennines Foreland	Zoetemeijer *et al.* (1993)

Table 2. *Selected continental and marine accumulation rates from compressional settings*

Accummulation rate (m ka^{-1})	Setting	Reference
0.06–0.13	Scala Dei Group, Ebro Foreland Basin	Colombo & Verges (1992)
0.07–0.20	Mediano anticline, south Pyrenean foreland basin	Holl & Anastasio (1993)
0.14–0.23	Perdido fold belt, Gulf of Mexico.	Mount *et al.* (1990)
0.06–0.25	Syn-orogenic alluvial sediments, South Central Unit, Pyrenees	Burbank & Verges (1994)
0.1–0.5	Siwalik Molasse, Pakistan	Johnson *et al.* (1986)
0.01–0.5	North Appenine foreland basins, Italy	Ricci Lucci (1986)
0.5–1.3	North Appenine foreland basins, Italy	Zoetemeijer *et al.* (1993)
0.25–0.5	Escanilla Formation, Ainsa Basin, Spain.	Bentham *et al.* (1993)
0.1–0.9	Continental sediments, Precordileran foreland basin, Argentina.	Jordan *et al.* (1993)
1.8	Wheeler Ridge, San Joaquin Valley, California.	Medwedeff (1992)

the evolving fold will generate a structural high (a sub-aerial ridge or sub-marine high) which may lead to local erosion and sedimentation. To account for this, a simple model of sediment transport which can be built into Equation (1) is the diffusion equation approach in which the sediment flux, F, is assumed to be proportional to slope and directed down the slope, i.e.

$$F = -\alpha\ \partial h/\partial x \quad (2)$$

where α is the diffusion coefficient. This leads to the diffusion equation

$$\partial F/\partial x = -\alpha\ \partial^2 h/\partial x^2 \quad (3)$$

provided a is assumed constant. A full discussion of the use of the diffusion model is given in Waltham (1992) and Waltham *et al.* (1993). The value of the diffusion coefficient is very difficult to assess since published estimates vary from 9×10^{-4} m^2 a^{-1} for arid fault scarps (Colman & Watson 1983) to 5.6×10^5 m^2 a^{-1} for a prograding delta (Kenyon & Turcotte 1985). It appears to depend on a variety of factors such as climate, lithology and scale (Kooi & Beaumont 1994). Assuming that fault-related folds lie somewhere between these extremes suggests an intermediate value of approximately 1 m^2 a^{-1} might be appropriate. The values used in the examples presented in this paper are close to unity and have been chosen because they illustrate well the distinctive features caused by the interaction of tectonics and sedimentation in this setting.

Once the tectonic and sedimentary functions in equation (1) are specified the equation is solved using explicit finite difference techniques. There are however some numerical problems which relate specifically to this geological setting which need to be addressed. The first of these problems arises specifically because of the sharp, angular boundaries often encountered in geometric fault-fold models (cf. Fig. 1). Such boundaries can cause numerical dispersion, and ultimately an unstable solution of equation (1), if the advective term in the equation is treated using centred finite differencing (Fletcher 1991). In order to ensure stable solutions of the equation, the technique of upwind differencing is often used when tackling such problems (Press *et al.* 1988; Fletcher 1991; Waltham & Hardy 1995) and is used in the mathematical models presented here. In addition, the Courant condition and stability restrictions arising from the use of the diffusion equation must also be adhered to (Fletcher 1991).

Typical growth strata geometries associated with fault-related folds

In this section the evolution of simple fault-bend, fault-propagation and a rotating limb of a detachment fold under a variety of conditions will be examined using the mathematical approach described above. For the fault-bend and fault-propagation folds the starting configuration is 10 km wide with a simple step in decollement. This ramp dips at 29° to the right and steps up from a decollement located 2 km beneath the top of the model. Slip is constant, and from right to left, throughout all the model runs at 1.5 m ka^{-1}, over a total run time of 1 Ma. The initial configuration for the rotating limb of the detachment fold is 10 km wide with a limb length of 2 km. In the detachment model runs a slip rate of 0.75 m ka^{-1} (equivalent to 1.5 m ka^{-1} for a symmetric detachment fold) was applied for a total of 1 Ma. These values are chosen as they correspond approximately to values reported in the literature for the evolution of fault-related folds by a variety of authors (Table 1).

Fill to the top sedimentation

The mathematical models are first used to assess growth strata geometries formed when all available accommodation space is filled. Thus, these models are comparable with the 'fill to the top' approach used in many previous geometric models of growth folding (e.g. Suppe *et al.* 1991). Base level was defined to start at the level of the top pre-growth horizon and to rise at a constant rate of 2.0 m ka^{-1} throughout the model runs. The background sedimentation rate was also set to 2.0 m ka^{-1} in order to fill any accommodation space created in this manner. These values are thought to represent upper limits of sedimentation in active collisional orogenies (cf. Table 2). The results are illustrated for the fault-bend, fault-propagation and detachment folds in Figs 2, 3 and 4 respectively, with growth strata recorded at intervals of 200 ka.

Several features of note are seen in the fault-bend fold model (Fig. 2). Firstly, as base level rise is always greater than the creation of structural relief, growth strata are continuous, but thin, across the structure as a result of the decrease in accommodation space across the growing fault-bend fold. Kink bands appear as narrowing upwards triangles within the growth strata. The end of deformation is marked by the apexes of these triangles. The triangles record the locus of particles which have passed through the active axial surfaces, while the flat-lying growth strata on the crest of the fold represent material which has passed through neither of the active axial surfaces. This model compares very well with those produced by Suppe *et al.* (1991) and confirms the equivalence of the geometric and velocity models of fault-bend folding (Hardy & Poblet 1995).

Similar features are seen in the fault-propagation fold model (Fig. 3). Growth strata are continuous, but thin, across the structure as a result of the decrease in accommodation space across the fold. The kink bands delimit the hanging-wall material which has passed through the active axial surfaces, while the horizontal growth strata on the crest and backlimb of the fold represent material which has passed through neither of the active axial surfaces. This model compares very well with those produced by Suppe *et al.* (1991) and confirms the equivalence of the geometric and velocity fault-propagation fold models (Hardy & Poblet 1995).

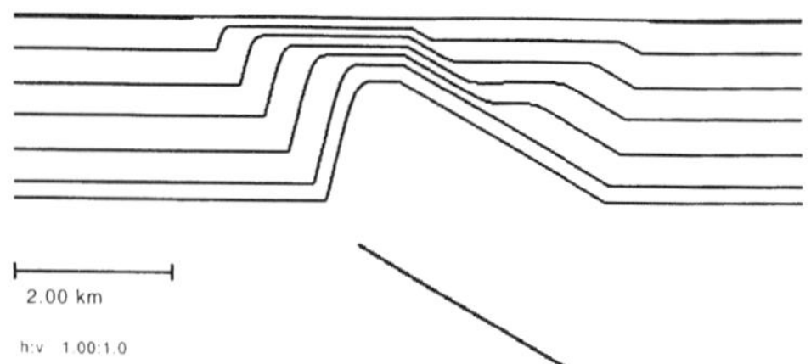

Fig. 3. Fault-propagation fold model with the inclusion of growth strata recorded at 200 ka intervals. Slip is constant from right to left during model run at a constant rate of 1.5 m ka^{-1}. Base level rise during the model run is 2.0 m ka^{-1}, background sedimentation rate is 2.0 m ka^{-1}.

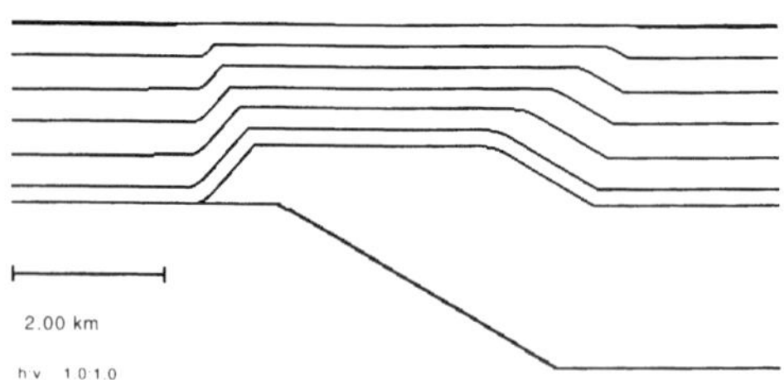

Fig. 2. Fault-bend fold model with the inclusion of growth strata recorded at 200 ka intervals. Slip is constant from right to left during the model run at 1.5 m ka^{-1}, total run time is 1 Ma. Base level rise during model run is 2.0 m ka^{-1}, and background sedimentation rate is 2.0 m ka^{-1}.

A distinct fanning of growth strata is seen in the detachment model with beds thickening away from the rotating fold limb (Fig. 4). The fold wavelength increases up-section, and a well-defined hinge line separates constant-thickness growth strata from beds that thin towards the rotating limb. As a result of the decreasing uplift rate with continued slip inherent in limb rotation (Hardy & Poblet 1994), and the constant sedimentation rate, the growth strata firstly onlap the fold limb and eventually overlap the crest. This records the very rapid uplift and emergence which occurs at the start of limb rotation. The amount of rotation between successive growth strata is not constant but decreases up-section as a result of the decrease in rotation rate with continued slip (Hardy & Poblet 1994).

Base level rise greater than sedimentation rate

A condition not considered in the majority of previous studies (with the exception of Zoetemeijer

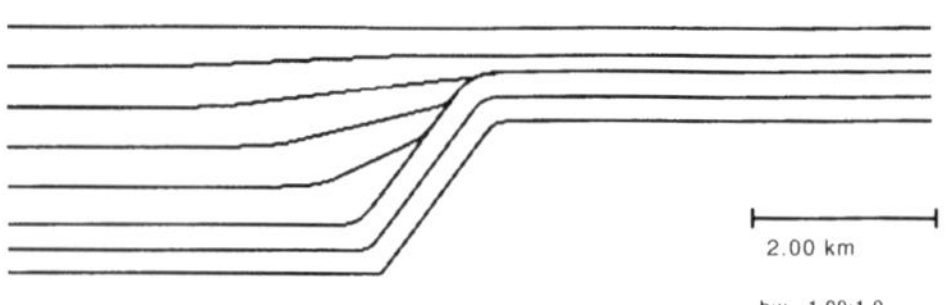

Fig. 4. Model of progressive limb rotation with the inclusion of growth strata recorded at 200 ka intervals. Limb length is 2 km. A constant slip rate of 0.75 m ka^{-1} was applied for 1 Ma. Background sedimentation rate and base level rise were both 2.0 m ka^{-1}.

1993) is that of base level rise being greater than any background sedimentation rate, thus preventing growth strata from filling all of the available accommodation space. This will result in sediments draping a structure, and it is likely that the steep dips on the limbs of a structure will cause slumping and transport of this material downslope (Hardy & Poblet 1995). Indeed, complex seismic facies ('denudation complexes') have been recognized in front of the steep front limbs of fault-fold structures in the Apenninic-Adriatic foredeep (Ori *et al.* 1986) and are thought to be related to erosion of local thrust highs. It is this geologically more realistic situation which is considered in the following models.

The models described so far only include background sedimentation, that is they assume that the growing fault-bend fold does not have a direct effect upon the amount of sediment. The inclusion of the diffusion model of erosion, transport and sedimentation allows the effects of the growth of the fault-related folds on local sedimentation to be evaluated. Examples are given in Figs 5, 6 and 7. All modelling parameters are identical to the previous cases except that the background sedimentation rate has been reduced to 1.3 m ka^{-1} and a diffusion model has been added with a diffusion coefficient of 3.0 $m^2 a^{-1}$.

In the case of fault-bend folding (Fig. 5) it can be seen that this results in a growth fold which is much broader than the previous case. Growth strata can be seen to thicken towards the structure, but to thin over it. This is a result of the diffusion mechanism moving sediment away from the steep limbs of the fold. The growth axial surfaces, although less sharply defined, are similar in orientation to those seen in Fig. 2. Thus, although the geometries seen in the fault-bend fold are somewhat different, the orientations of the growth axial surfaces could still yield useful information concerning sub-surface structure.

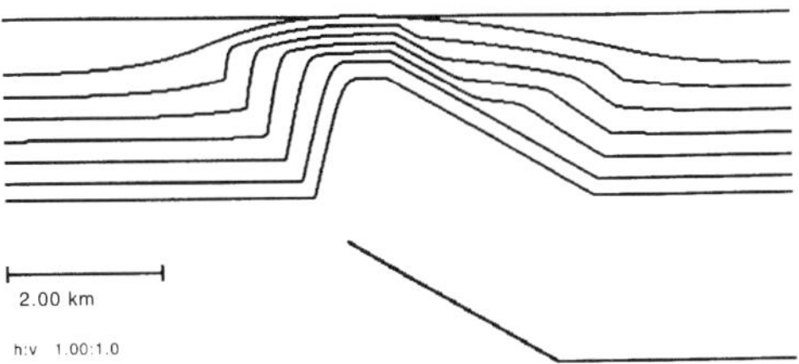

Fig. 6. Fault-propagation fold model which includes a diffusion model of erosion, transport and deposition. Slip is constant from right to left during model run at a constant rate of 1.5 m ka^{-1}. Base level rise during the model run is 2.0 m ka^{-1}, background sedimentation rate is 1.3 m ka^{-1}. The diffusion model uses a diffusion coefficient of 3.0 $m^2 a^{-1}$.

In the case of fault-propagation folding (Fig. 6) it can be seen that the growth fault-propagation fold is also much broader than the previous case. Growth strata can be seen to thicken towards the structure, but to thin over it . In particular, over the uplifting crest of the structure strata are noticeably thinned. The growth axial surfaces, although less sharply defined, are similar in orientation to those seen in Fig. 3. Thus, although the geometries seen in the fault-propagation fold are somewhat different, the orientations of the growth axial surfaces could still yield useful information concerning sub-surface structure and slip.

In the case of detachment folding (Fig. 7) the inclusion of the diffusion model results in the erosion of the limb and crest of the structure and deposition of this eroded material downdip as growth strata. Note that the oldest growth strata onlap the un-eroded fold limb, whereas upsection the onlap surface is also an erosion surface. This is a result of the progressive development of the structure which produces steeper slopes, and therefore more erosion, with time.

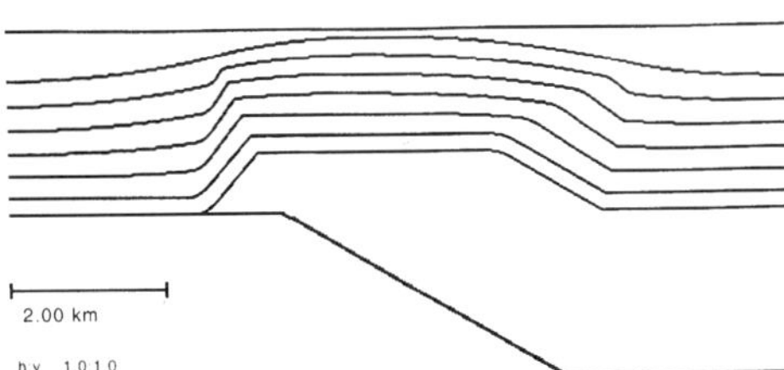

Fig. 5. Fault-bend fold model which includes a diffusion model of erosion, transport and sedimentation. Slip is constant from right to left during the model run at 1.5 m ka^{-1}, total run time is 1 Ma. Base level rise during model run is 2.0 m ka^{-1} and background sedimentation rate is 1.3 m ka^{-1}. The diffusion model uses a diffusion coefficient of 3.0 $m^2 a^{-1}$. Growth strata are recorded at 200 ka intervals.

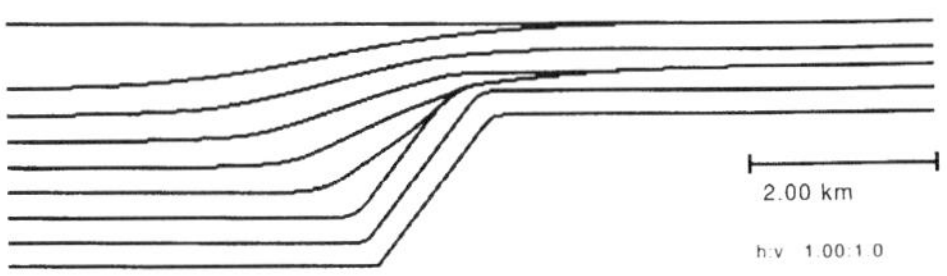

Fig. 7. Model of progressive limb rotation which includes a diffusion model of erosion, transport and sedimentation. Limb length is 2 km. A constant slip rate of 0.75 m ka^{-1} was applied for 1 Ma. Background sedimentation rate and base level rise were, 1.3 and 2.0 m ka^{-1} respectively. A diffusion coefficient of 3.0 $m^2 a^{-1}$ was used. Growth strata are displayed at 200 ka intervals.

Low background sedimentation rate

The effect of fold growth on sedimentation is more clearly seen if the background sedimentation rate is reduced. Examples of this are given in Figs 8–10. Here all modelling parameters are as before except that the background sedimentation rate is reduced to 0.5 m ka^{-1}, a value more typical of sedimentation rates in compressional settings (cf. Table 2). Thus in these examples rates of fold growth are high compared to rates of sedimentation.

This has some dramatic effects in the case of fault-bend folding (Fig. 8). Sedimentation across the uplifting fold is much reduced leading to onlap of strata on the forelimb, and a thinned and condensed sequence on the crest of the fold. The onlap surface on the forelimb can be seen to climb up through the stratigraphy and to end up as the topmost surface on the backlimb of the fold. Similar unconformities have been predicted in previous geometric models and have been observed in the Lost Hills anticline from the Temblor fold belt in California (Medwedeff 1989).

Distinctly different relationships occur in the case of fault-propagation folding (Fig. 9). Sedimentation across the growing fold is much reduced leading to onlap of growth strata on the backlimb, and thinning and truncation of growth strata on the forelimb. There is a no growth sequence on the crest of the fold and pre-growth strata are quite deeply eroded. The onlap surface on the backlimb can be seen to climb up through the growth strata and to end up as a truncation surface on the forelimb. Thus, time-equivalent growth strata which are above the unconformity on the backlimb are below the unconformity on the forelimb. Similar unusual unconformities have been predicted in previous geometric models and have been observed in outcrop and seismic examples (e.g. Suppe *et al.* 1991).

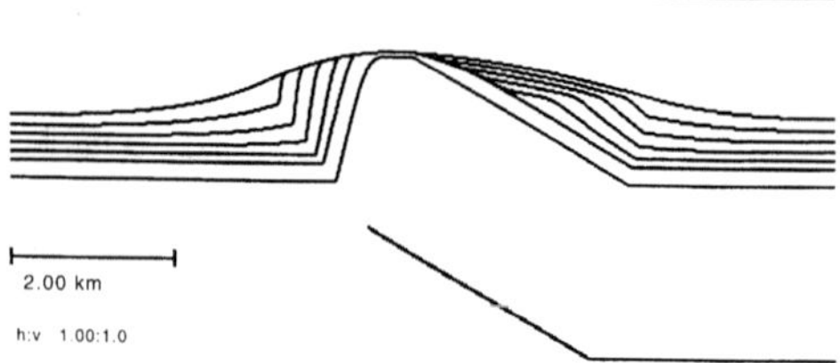

Fig. 9. Fault-propagation fold model which includes a diffusion model of erosion, transport and deposition. Slip is constant from right to left during model run at a constant rate of 1.5 m ka^{-1}. Base level rise during the model run is 2.0 m ka^{-1}, background sedimentation rate is 0.5 m ka^{-1}. The diffusion model uses a diffusion coefficient of 3.0 m^2 a^{-1}.

In the case of limb rotation (Fig. 10) the decrease in background sedimentation rate produces dramatic thinning and erosion on the limb and crest of the structure and redeposition of material downdip. Maximum thinning and stratigraphic condensation occur at the highest point on the rotating limb. The growth strata still display a characteristic fanning away from the rotating limb.

Fold growth in a sub-aerial setting

Many growth anticlines do not develop in marine conditions (e.g. Medwedeff 1989; Jordan *et al.* 1993). Thus it is important to investigate the types of growth strata architectures that may develop in response to fold growth in sub-aerial settings. To model this situation, the background sedimentation rate is set to zero and only a diffusion model is used. The diffusion model uses a diffusion coefficient of 3.0 m^2 a^{-1}. Thus only local erosion, transport and deposition due to fold growth are being modelled. The modelling parameters are otherwise identical to those used in the previous examples.

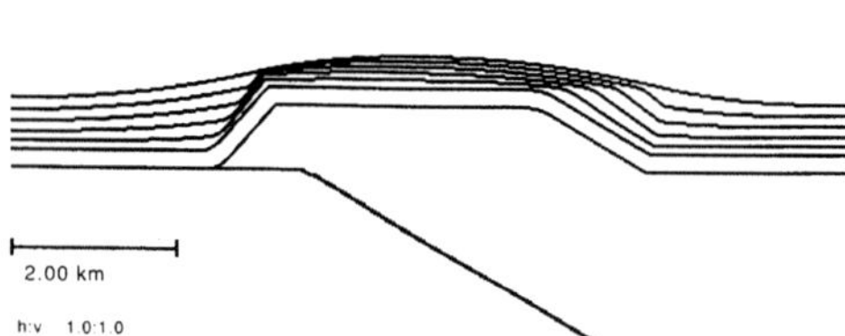

Fig. 8. Fault-bend fold model which includes a diffusion model of erosion, transport and sedimentation. Slip is constant from right to left during the model run at 1.5 m ka^{-1}, total run time is 1 Ma. Base level rise during model run is 2.0 m ka^{-1} and background sedimentation rate is 0.5 m ka^{-1}. The diffusion model uses a diffusion coefficient of 3.0 m^2 a^{-1}. Growth strata are recorded at 200 ka intervals.

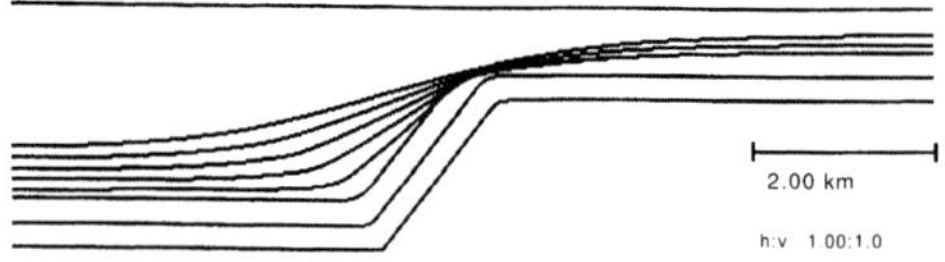

Fig. 10. Model of progressive limb rotation which includes a diffusion model of erosion, transport and sedimentation. Limb length is 2 km. A constant slip rate of 0.75 m ka^{-1} was applied for 1 Ma. Background sedimentation rate and base level rise were 0.5 and 2.0 m ka^{-1} respectively. A diffusion coefficient of 3.0 m^2 a^{-1} was used. Growth strata are displayed at 200 ka intervals.

The resultant models and details of the growth strata are shown in Figs 11–13. From these figures it can be seen that the growth strata and the structures are quite different to the marine examples.

In the case of fault-bend folding, the pre-growth strata are deeply eroded, especially on the forelimb of the structure (Fig. 11a). As the forelimb always has a steeper dip than the backlimb in fault-bend fold theory, and as the diffusion model is based primarily on slope, this asymmetric erosion is a natural consequence of the model. In addition, two wedges of growth strata are seen on the limbs of the structure. It is apparent that although these wedges of sediment were both derived from the growing and uplifting fault-bend fold, their stratal architectures are quite different. The growth strata on the forelimb consist of series of foreland dipping and thinning beds which onlap the structure (Fig. 11b). In contrast, the growth strata on the backlimb consist of a series of hinterland dipping and thinning beds which offlap the structure and are themselves eroded and truncated up-dip. This is a consequence of the growth strata on the backlimb being transported and uplifted as part of the developing fault-bend fold. This should be contrasted with the growth strata on the forelimb which are passively translated during the growth of the structure. These important differences are a natural consequence of the fault-bend fold model.

In the case of fault-propagation folding the pre-growth strata are also deeply eroded, especially on the crest of the structure (Fig. 12a). In addition, two wedges of growth strata are seen on the limbs of the structure. It is apparent that although these wedges of sediment were both derived from the growing and uplifting fault-propagation fold, their stratal architectures are different. The growth strata on the forelimb consist of a series of foreland dipping and thinning beds which have been deformed into a tight syncline adjacent to the forelimb of the fold (Fig. 12b). The growth strata have been rotated to a near vertical orientation adjacent to the forelimb. This is a result of the propagation of the fault tip and the progressive incorporation of undeformed growth strata into the fold. As a result, growth strata are uplifted, deformed and eroded during fold growth. In contrast, the growth strata on the backlimb consist of a series of hinterland dipping and thinning beds which have been deformed into a more open syncline. The growth strata on the backlimb offlap the structure and are eroded and truncated up-dip. This is a consequence of them being transported and uplifted as part of the developing fault-propagation fold.

In the case of detachment folding there is deep erosion on the limb and crest of the structure. The growth strata occur as a wedge of sediment which firstly thickens and then thins away from the rotating limb (Fig. 13a). It can be seen that the growth strata derived from the uplifting and rotating fold limb are themselves uplifted, rotated and eroded during fold growth and offlap the structure (Fig. 13b).

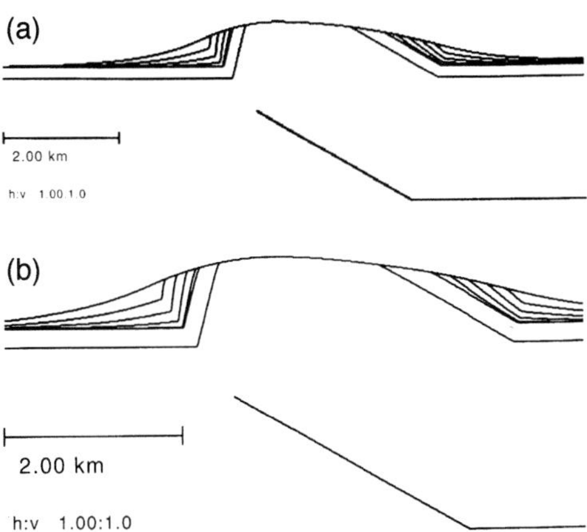

Fig. 12. (**a**) Fault-propagation fold developed in a sub-aerial setting. Slip is constant from right to left during model run at a constant rate of 1.5 m ka^{-1}. A diffusion model of erosion, transport and sedimentation is included, there is no background sedimentation or base level rise. The diffusion model uses a diffusion coefficient of 3.0 m^2 a^{-1}. (**b**) Detail of growth strata.

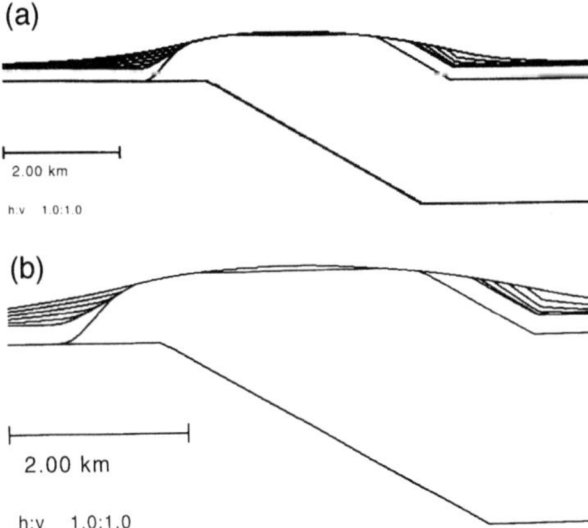

Fig. 11. (**a**) Fault-bend fold model developed in a sub-aerial setting. Slip is constant from right to left during the model run at 1.5 m ka^{-1}, total run time is 1 Ma. A diffusion model of erosion, transport and sedimentation is included, and there is no background sedimentation or base level rise. The diffusion model uses a diffusion coefficient of 3.0 m^2 a^{-1}. Growth strata are recorded at 200 ka intervals, (**b**) Detail of growth strata.

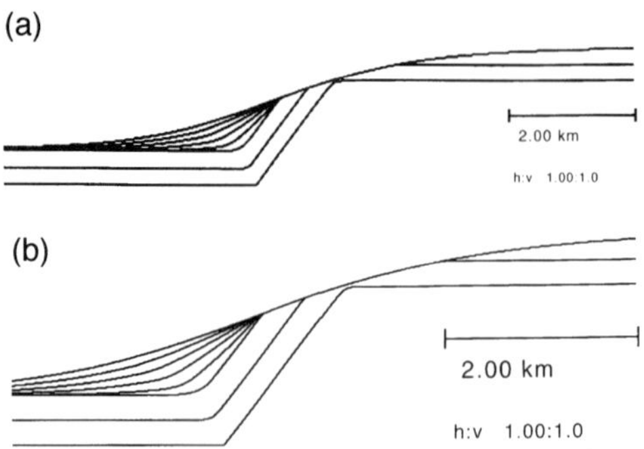

Fig. 13. (**a**) Model of progressive limb rotation in a sub-aerial setting, which includes a diffusion model of erosion, transport and sedimentation. Limb length is 2 km. A constant slip rate of 0.75 m ka^{-1} was applied for 1 Ma. There was no background sedimentation rate or base level rise. A diffusion coefficient of 3.0 m^2 a^{-1} was used. Growth strata are displayed at 200 ka intervals, (**b**) Detail of growth strata.

Comparison with natural examples

The model results described above represent a variety of different possibilities of the interaction of fault-related folding and sedimentation. It is now useful to compare the predicted growth strata relationships to those seen in a number of natural examples.

A well-studied growth fault-bend fold, the Lost Hills anticline from the Temblor fold belt in California (Medwedeff 1989), is shown in Fig. 14. Medwedeff (1989) presented a detailed analysis of this structure based upon a combination of seismic data, well data and fault-bend fold geometric models. The structure is thought to have started growing during the late Miocene and continued growing to the late Pleistocene or Holocene. Growth strata of the Etchegoin and San Joaquin formations are shallow marine to brackish water deposits, while the youngest unit, the Tulare Formation, consists of fluvial and lacustrine siltstone, sandstone and conglomerate. This structure contains many of the key relationships between pre-growth and growth strata predicted in the fault-bend fold model runs. Growth strata onlap the forelimb of the structure at the same time as equivalent growth strata on the backlimb of the structure are uplifted and eroded. This is particularly noticeable within the Etchegoin and San Joaquin Formations. A distinctive feature of this structure is the un conformity within the growth strata which can be traced up the northeast limb of the fold to the crest and which then becomes the unconformity at the base of the Tulare Formation above the southwest limb of the fold (Medwedeff 1989). Another

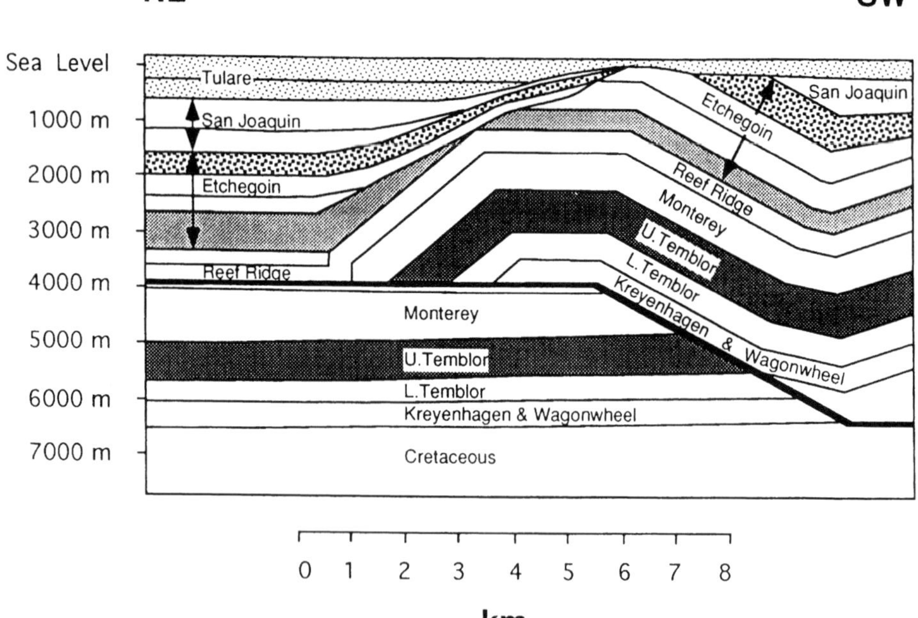

Fig. 14. The Lost Hills structure, redrawn from Medwedeff (1989).

distinctive feature is the deep erosion of the lowermost Etchegoin Formation on the northeastern limb and crest of the structure. Comparison with the modelled relationships suggest that this structure grew rapidly compared with rates of local base level rise (cf. Figs 5, 8 and 11) and that the crest of the structure may have been emergent during growth.

An interpretation of a growth fold structure imaged on seismic data (Prieri 1989) from the Po Basin is shown in Fig. 15. The sediments of the Po Basin consist of Tertiary and Quaternary clastics which were deformed during the late Miocene to Recent (Vai 1987). The deformation is a response to the southwestward subduction of the northern part of the Apulo-Adriatic lithospheric plate underneath the Apennines (Royden 1988). The Late Miocene and Pliocene sediments involved in the deformation are marine (Ricci Lucci 1986) and therefore the structures imaged in this basin may be good analogues to the sub-marine modelled geometries. The fault-fold structure shown in Fig. 15 exhibits a steep front limb and a more gentle back limb which shows prominent onlap of growth strata within the middle and late Pliocene sequences. Displacement of strata decreases up-section, with upper growth strata being folded into a gentle anticline. Thus this structure exhibits many of the features predicted by the fault-propagation fold model (cf. Fig. 6). By comparison with Figs 6, 9 and 12, the onlap of growth strata on the backlimb of the structure suggests that the fold grew rapidly compared to local rates of sedimentation. This interpretation may be contrasted with that of Zoetemeijer *et al.* (1992) where this structure was modelled using kinematics based on fault-bend folding, using a ramp which cut through the sedimentary cover to the surface during the Pliocene. As a result, the models presented by Zoetemeijer *et al.* (1992) predicted offlap on the back limbs of the structure. The results presented here suggest that the structure shown in Fig. 15 is better interpreted and modelled as a fault-propagation fold in which the local sedimentation rate is less than the rate of fold growth. If this is done the gross structural and growth strata relationships predicted by the model are consistent with those observed in Fig. 15.

The Wheeler Ridge, California may offer a good modern-day analogue to the modelled sub-aerial fault-propagation fold geometries (Medwedeff 1992; Burbank & Verges 1994). The Wheeler Ridge occurs in the southern San Joaquin Valley, California and represents the most external fold in an active fold and thrust belt (Fig. 16). It is an eastward plunging anticline which is buried beneath the modern depositional plain at its eastern end. The crest of the Wheeler Ridge is uplifting at a rate of 3.2 m ka^{-1} which is very high compared to the surrounding alluvial plain which is aggrading at a rate of 1.8 m ka^{-1} (Medwedeff 1992). As a result the structure is emergent and deflects drainage systems around its surface expression. On the northern flank of the structure, short-radius (< 1 km) locally sourced alluvial fans are being

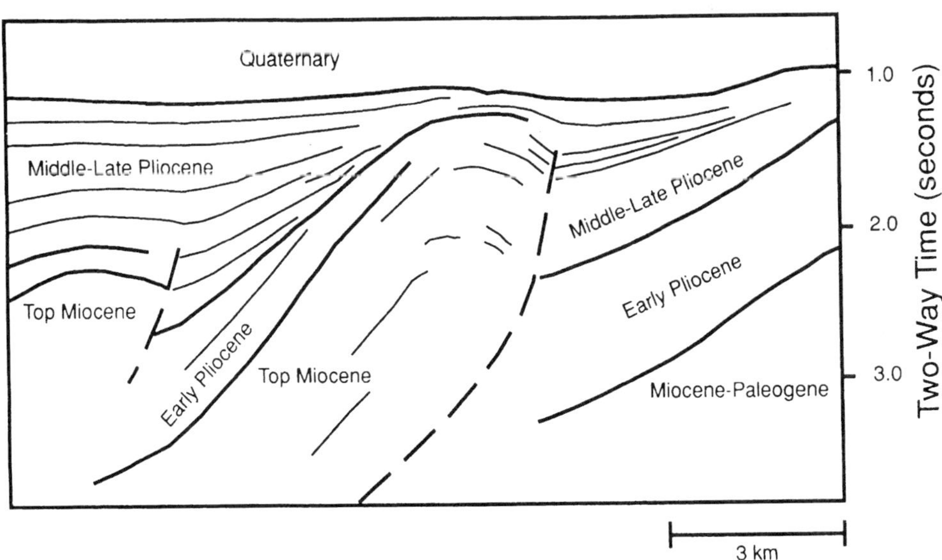

Fig. 15 . Line drawing of part of seismic section from the Po Basin, Italy (after Prieri 1989) showing major reflectors in the the immediate vicinity of a thrust-related fold, stratigraphic data are taken from Prieri (1989).

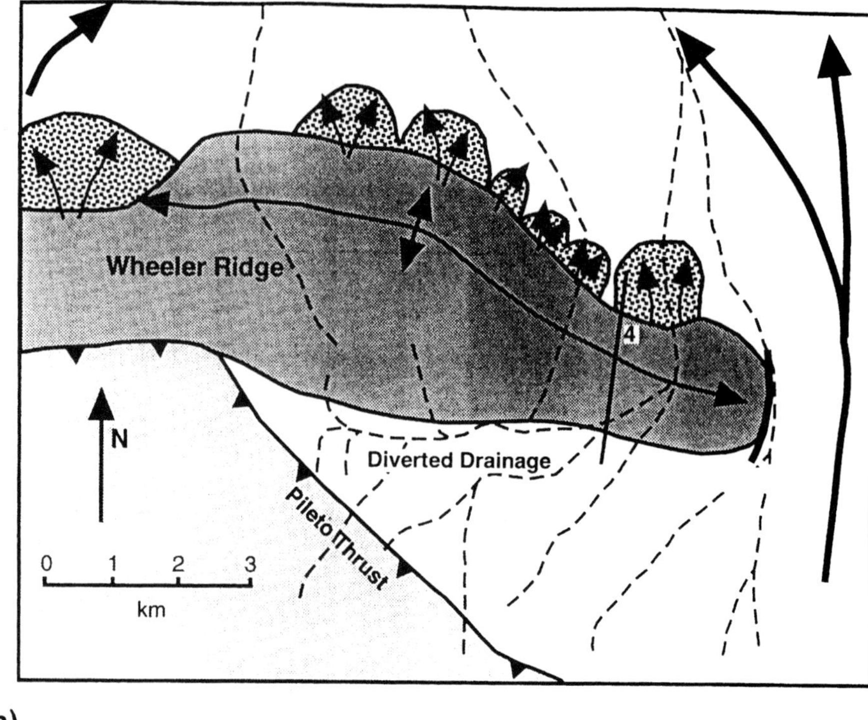

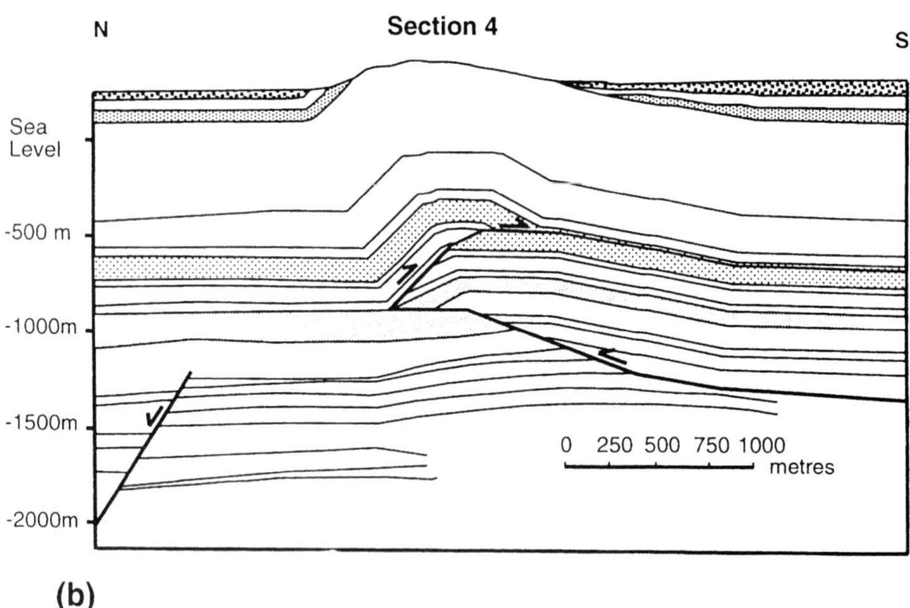

Fig. 16. The Wheeler Ridge anticline, California. (**a**) Simplified geologic map redrawn from Burbank & Verges (1994). (**b**) Cross-section 4 of Medwedeff (1992) showing inferred structural and stratigraphic relationships using a wedge-thrust solution.

actively deformed in their proximal parts by the growth of the fold (Burbank & Verges 1994). The Wheeler Ridge appears to be structurally complex and while Medwedeff (1992) interpreted this structure as a wedge-thrust, he acknowledged that the density of structural and stratigraphic data available were insufficient to uniquely identify the nature of the structure. A fault-propagation fold solution for the structure is equally reasonable. Indeed Zoetemeijer (1993) has shown that wedge-thrusts produce similar growth strata architectures to fault-propagation folds. It is reasonable therefore to compare the modelled geometries with those seen in the Wheeler Ridge.

On the northern limb of the fold, growth strata have been uplifted and deformed as they are progressively incorporated into the fold limb (Fig. 16b), these growth strata consist of both locally-derived alluvial fans and the alluvial plain itself. To the south of the fold crest, the alluvial plain between the Pieto thrust and the fold is being incorporated into the uplifting backlimb of the anticline. The forelimb growth strata geometries shown by Medwedeff (1992) in his Section 4 are very similar to those predicted by the fault-propagation models developed in sub-aerial settings (cf. Fig. 12). However, backlimb geometries are different because in addition to locally derived material, the alluvial plain is burying the gently dipping backlimb of the Wheeler Ridge as it grows. Thus this alluvial aggradation is equivalent to a local base level rise which is insufficient to completely envelop the structure. Therefore while growth strata derived solely from the erosion of the fold crest offlap the backlimb of a fault-propagation fold (Fig. 12b), when alluvial aggradation rates are high growth strata may onlap the backlimb of the structure.

Similar relations to those seen in the detachment fold models can be seen in the Mediano anticline in the Spanish Pyrenees. The Mediano anticline is a transverse fold with a wavelength of *c*. 8 km and a strike length of *c*. 20 km cored by Triassic shales and evaporites. Fold growth is recorded by angular and progressive unconformities with syn-tectonic Eocene sedimentary deposits (Holl & Anastasio 1993). A cross section of the western limb of this fold is shown in Fig. 17. The growth strata display many of the characteristics seen in the modelled folds and suggest that progressive limb rotation, punctuated by periods of rapid offlap, was the dominant mode of fold growth in this anticline. This agrees well with the conceptual model and palaeomagnetic data presented by Holl & Anastasio (1993). Indeed, the tilting rates derived by Holl & Anastasio (1993) suggest that limb rotation was greatest in the early stages of the development of this fold and decreased with time, in agreement with the model predictions (Hardy & Poblet, 1994).

Discussion

Near surface fold shapes may be traced to depth using a variety of geometric constructions based upon fault-bend, fault-propagation and detachment fold geometric models. However, unless the data base is very good uniquely determining the

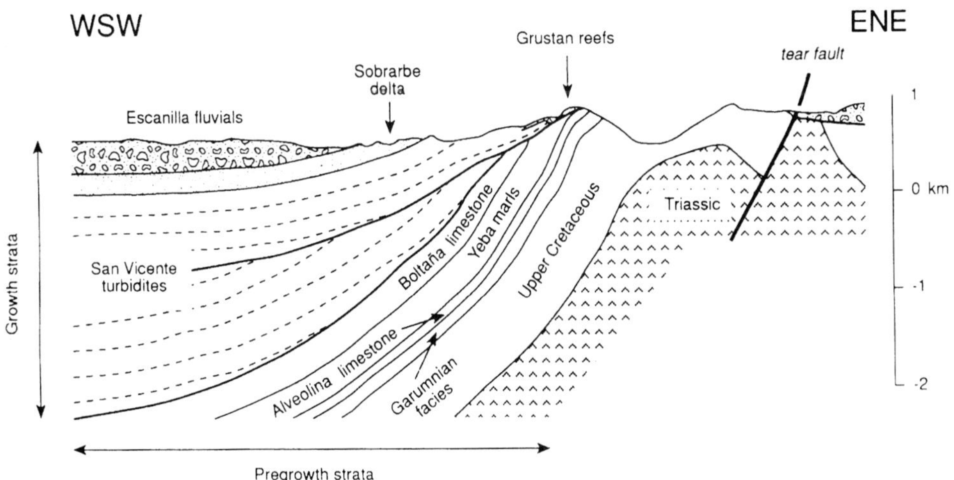

Fig. 17. Cross section of the western limb of Mediano anticline, southern Pyrenees, Spain, derived from field mapping by Josep Poblet. No vertical exaggeration.

specific type of fault-related fold which caused the near surface structure is often very difficult (Mitra 1990).

The stratigraphic architectures observed in syn-tectonic (growth) strata can help to constrain the most appropriate structural model. The deformation that growth sediments undergo depends on both where they are deposited on a given structure and on the nature of the structure. For this reason the stratigraphic relationships seen within the growth sediments can help constrain the kinematics of deformation and deeper fault geometries. As the various types of fault-related folds have different kinematics, the modelling has shown that many fault-related folds can be distinguished on the basis of growth strata architectures alone. Importantly growth sediments are typically shallow and are often the best-imaged part of a seismic line. Conversely, however, they are the first material to be lost during erosion in an uplifted fold and thrust belt.

The mathematical models of fault-bend, fault-propagation and limb rotation and detachment folding discussed in this paper have shown the range of growth strata architectures that may result from the development of fault-related folds. In particular, the models have shown the types of relationships that may be expected when fold growth rates are greater than local rates of sedimentation in both sub-aerial and sub-marine settings. Reported rates of fault-slip and sedimentation from compressive settings indicate that such situations are more typical than previous 'fill-to-the-top' approaches where sedimentation keeps ahead of deformation and there is no topographic or bathymetric expression of a structure. The growth strata architectures developed in such typical situations may be compared in order to distinguish the stratigraphic responses to the different types of fault-fold interaction (Fig. 18).

In fault-bend folding, backlimb growth strata are uplifted and deformed during fold growth, while those on the forelimb are passively translated above the thrust. When the fold is uplifted above base level or there is erosion and transport of material away from the fold crest, growth strata onlap the forelimb with contemporaneous uplift, thinning or truncation of growth strata on the backlimb (cf. Figs 5 and 8). In sub-aerial settings, locally-derived growth strata onlap the forelimb of the structure while those on the backlimb offlap the structure (cf. Fig. 11).

In fault-propagation folding, growth strata on both the backlimb and forelimb of the structure are

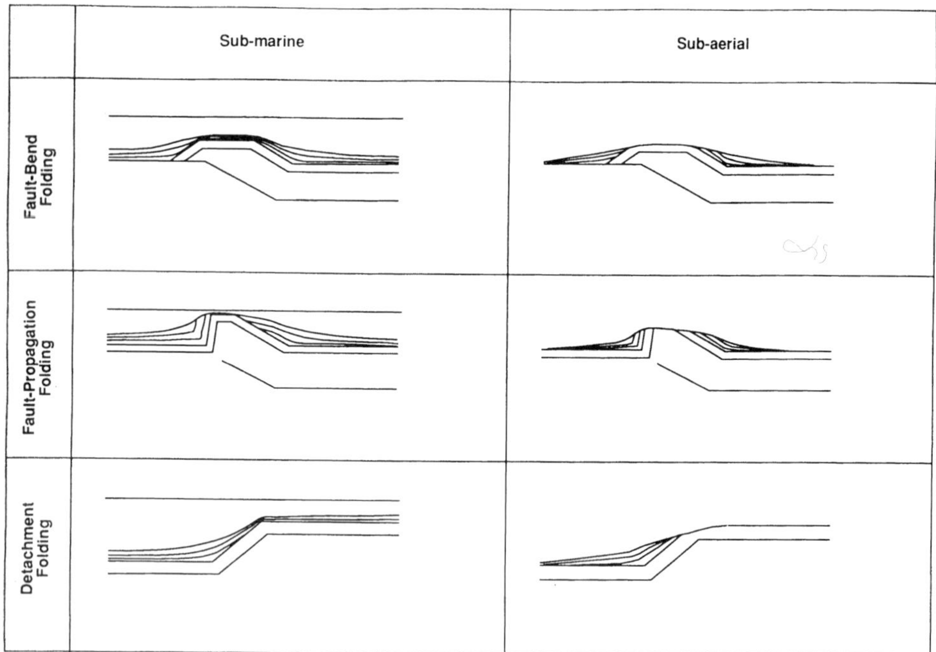

Fig. 18. Synoptic diagram illustrating typical stratal architectures associated with fault-bend, fault-propagation and detachment folds

uplifted and deformed during fold growth. When the fold is uplifted above base level or there is erosion and transport of material away from the fold crest, growth strata onlap the backlimb with contemporaneous uplift, thinning and truncation of growth strata on the forelimb (cf. Figs 6 and 9). In sub-aerial settings, locally-derived growth strata offlap both the backlimb and the forelimb of the structure (cf. Fig. 12).

In disharmonic detachment folding growth strata on the limbs of the structure are rotated and deformed during fold growth. This leads to a characteristic fanning of growth strata away from the rotating limb (a 'progressive unconformity') (Fig. 7). Due to the rapid uplift at the start of fold growth, detachment folds often become emergent with growth strata onlapping and then overlapping the structure as the uplift rate decreases. This decrease in uplift rate is a natural consequence of the kinematics of limb rotation. Such 'kick-start' unconformities may be characteristic of detachment folds. When the fold is uplifted above base level or there is erosion and transport of material away from the fold crest, growth strata thin towards the fold crest. In sub-aerial settings, locally-derived growth strata offlap the rotating limb of the structure (Fig. 13).

The problems of multiple interpretations of surface folds based solely on the configuration of pre-growth strata are discussed by Mitra (1990) but not satisfactorily solved. This study indicates that, in the absence of good seismic or outcrop data, the presence of the features described above within growth strata can be used to distinguish between these three end-member styles of fault-related folding.

The three types of fault-fold interaction discussed above are not mutually exclusive, either through time in a particular structure or spatially within a basin. Thus if a particular structure has evolved from being a detachment fold, through a fault-propagation stage to a fault-bend fold, we would expect characteristic variations in growth strata geometries. In addition, while the three end-member types of structure may represent different responses to compressive deformation that may be occurring simultaneously within a basin, they have very different kinematic behaviours. These differences in kinematics lead to very different rates of uplift for a given slip rate. This difference is illustrated in Fig. 19, where uplift rates over 1 Ma are shown for a fault-bend and fault-propagation fold both stepping up from the same decollement and for a detachment fold with a limb length of 2.0 km. The slip rate used in these examples was 1.5 m ka^{-1}. These values correspond to those used in the examples given in this paper. From Fig. 19 we see that, for a given slip rate, fault-bend folds produce much less uplift than fault-propagation folds, while detachment folds uplift extremely rapidly initially and then decrease with time. This is confirmed by the uplift rates of 4–20 m ka^{-1} reported by Yeats (1983) and Rockwell *et al.* (1988) for detachment folds from the Ventura Basin in California. These results indicate that these differences in kinematics can have a profound effect on local topography or bathymetry.

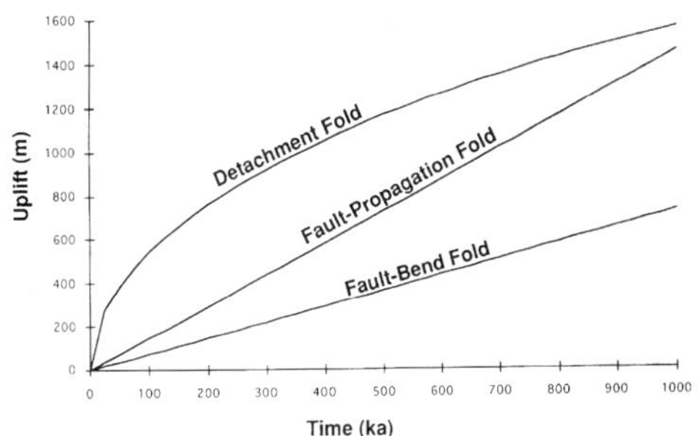

Fig. 19. Difference in rates of uplift for detachment, fault propagation and fault-bend folding.

The mathematical models described in this paper are a first-order approach to modelling stratal architectures and fault-related folding. While they improve upon previous approaches, they could be enhanced by the inclusion of compaction. This is, in principle, a straightforward matter as it involves the inclusion of an additional vertical velocity in Equation (1). More realistic models of sedimentation, perhaps using a variable diffusion coefficient (e.g. Rivenaes 1992), would also be a valuable addition to the model.

Conclusions

This paper has shown the applicability of the general forward modelling equation to contractional structures formed as a result of compressive deformation. In particular, the modelling methodology has been applied to fault-related folds.

Three major types of fault-fold structures encountered in many fold and thrust belts, fault-bend, fault-propagation and disharmonic detachment folds, have been examined in this paper. In each case, as a prerequisite to the modelling of syn-tectonic strata, velocity models of deformation have been derived from geometrical constructions. Models of background sedimentation and local erosion, transport and deposition and then easily included in the modelling scheme. The models have allowed a wider range of relationships between fold growth and sedimentation to be investigated than previous geometric models, in particular those situations in which the folds have a topographic or bathymetric expression. Reported fault-slip and sedimentation rates from compressional settings indicate that such situations are the norm.

The mathematical models show that distinctive stratal architectures occur in both sub-marine and sub-aerial settings as a result of the different kinematics of each of the types of fault-fold interaction. These stratigraphic relationships may be used to distinguish between the different modes of fault-related folding when pre-growth geometric relationships are very similar or are unclear.

The different styles of fault-related folding also lead to very different rates and patterns of uplift for equal amounts of displacement. Thus, fault-bend folds produce small amounts of uplift over the whole of a ramp, while a fault-propagation fold produces much more localized rapid uplift on an equivalent ramp. Detachment folds, by their nature, have the capacity to produce very rapid rates of uplift and rotation and thus topographic expression. These differences may be expected to be found in different parts of a basin simultaneously.

The mathematical models presented here allow the relationships of the different styles of fault-related folding, base level rise and rates of erosion, transport and sedimentation to be assessed. It is only through the consideration of the manner in which these different factors interact that the growth strata architectures associated with a fault-fold structure can be better understood.

We are grateful to R. Zoetemeijer for a very thorough review of this paper. Thanks to Gary Nichols, Andrea Argnani, Fabrizio Storti, Josep Anton Muñoz, Pete Talling and many other colleagues for discussions on growth folding. This work was done with the help of the Fault Dynamics Project (Sponsored by ARCO British Limited, BRASOIL UK Ltd, BP Exploration, Conoco (UK) Limited, Mobil North Sea Limited and Sun Oil Britain) whose support is gratefully acknowledged. Josep Poblet acknowledges an EEC fellowship from the Human Capital and Mobility Program.

References

AL SAFFAR, M. 1993. Geometry of fault-propagation folds: method and application. *Tectonophysics*, **223**, 363–380.

AVOUAC,J-P., TAPPONNIER, P., BAI, M., YOU, H. & WANG, G. (1993) Active thrusting and folding along the Northern Tien Shan and Late Cenozoic rotation of the Tarim relative to Dzungaria and Kazakhstan. *Journal of Geophysical Research*, **98**(B4), 6755–6804.

BENTHAM, P. A., TALLING, P. J. & BURBANK, D. W. 1993. Braided stream and flood-plain deosition in a rapidly aggrading basin: the Escanilla formation, Spanish Pyrenees. *In*: BEST, J. L. & BRISTOW, C. S. (eds) *Braided Rivers*. Geological Society, London, Special Publication, 75, 177–194.

BOSENCE, D. W. J., POMAR, L., WALTHAM, D & LANKESTER, T. 1994. Computer modelling a Miocene carbonate platform, Mallorca, Spain. *American Association of Petroleum Geologists Bulletin*, **78**(2), 247–266.

BRAUN, J. & SAMBRIDGE, M. 1994. Dynamic Lagrangian Remeshing (DLR): A new algorithm for solving large strain deformation problems and its application to fault-propagation folding. *Earth and Planetary Science Letters*, **124**, 211–220.

BURBANK, D. W. & VERGES, J. 1994. Reconstruction of topography and related depositional systems during active thrusting. *Journal of Geophysical Research*, **99**, 20 281–20 297.

——, ——, MUNOZ, J. A. & BENTHAM, P. (1992) Coeval hindward- and forward-imbricating thrusting in the south-central Pyrenees, Spain: Timing and rates of shortening and deposition. *Geological Society of America Bulletin*, **104**, 3–17.

BUXTORF, A. (1916) Prognosen und befunde beim Hauenstembasis-und Grenchenberg-tunnel und die Bedeutung der letzeren fur die geologie des Juragebirges. *Verhandlungen der Naturforschenden Gesellschaft in Basel*, **27**, 184–205.

CALASSOU, S., LARROQUE, C. & MALAVIELLE, J. 1993. Transfer zones of deformation in thrust wedges: an experimental study. *Tectonophysics*, **221**, 325–344.

CHESTER, J. S. & CHESTER, F. M. 1990. Fault-propagation folds above thrusts with constant dip. *Journal of Structural Geology*, **12**(7), 903–910.

COLMAN, S. M. & WATSON, K. 1983. Ages estimated from a diffusion equation model for scarp degradation. *Science*, **221**, 263–265.

COLOMBO, F. & VERGES, J. 1992. Geometria del margen S.E. de la Cuenca del Ebro: discordancias progresivas en el Grupo Scala Dei. Serra de La Llena. (Tarragona). *Acta Geologica Hispanica*, **27** (1–2), 33–53.

CONTRERAS, J. & SUTER, M. 1990. Kinematic modelling of cross-sectional deformation sequences by computer simulation. *Journal of Geophysical Research*, **95**, 21913–21929.

DE SITTER, L. V. 1956. *Structural geology*, McGraw Hill, New York.

DAHLSTROM, C. D. A. 1969. The upper detachment in concentric folding. *Bulletin of Canadian Petroleum Geology*, **17**, 326–346.

—— 1970. Structural geology in the eastern margin of the Canadian Rocky Mountains. *Bulletin of Canadian Petroleum Geology*, **18**, 332–406.

—— 1990. Geometric constraints derived from the law of conservation of volume and applied to evolutionary models of detachment folding. *American Association of Petroleum Bulletin*, **74**, 336–344.

DAVIS, T. L., NAMSON, J. & YERKES, R. F. 1989. A cross

section of the Los Angeles area: seismically active fold and thrust belt, the 1987 Whittier Narrows earthquake and earthquake hazard. *Journal of Geophysical Research*, **94** (B7), 9644–9664.

DECELLES, P. G., GRAY, M. B., RIDGWAY, K. D., COLE, R. B., SRIVASTAVA, P., PEQUERA, N. & PIVNIK, D. A. 1991. Kinematic history of a foreland uplift from Paleocene synorogenic conglomerate, Beartooth Range, Wyoming and Montana. *Geological Society of America Bulletin*, 103, 1458–1475.

DERAMOND, J., SOUQUET, P., FONDECAVE-WALLEZ, M-J. & SPECHT, M. 1993. Relationships between thrust tectonics and sequence stratigraphy surfaces in foredeeps: model and examples from the Pyrenees (Cretaceous–Eocene, France, Spain). *In*: WILLIAMS, G. D. & DOBB, A. (eds) *Tectonics and Seismic Sequence Stratigraphy*. Geological Society, London, Special Publications, **71**, 193–291.

DOGLIONI, C. 1993. Some remarks on the origins of foredeeps. *Tectonophysics*, **228**, 1–20.

ENDIGNOUX, L. & MUGNIER, J-L. 1990. The use of a forward kinematic model in the construction of balanced cross sections. *Tectonics*, 9, 1249–1262.

FAILL, R.T. 1973. Kink-band folding, Valley and Ridge Province, Pennsylvania. *Geological Society of America Bulletin*, **84**, 1289–1341.

FLETCHER, C. A. J. 1991. *Computational Techniques for Fluid Dynamics (Volume 1), Fundamentals and General Techniques* (2nd edn), Springer Verlag.

GROSHONG, R. H. & EPARD, J-L. (1994) The role of strain in area-constant detachment folding. *Journal of Structural Geology*, **16**(5), 613–618.

HARDY, S. 1993. Numerical modelling of the response to variable stretching rate of a domino fault block system. *Marine and Petroleum Geology*, **10**(2), 145–152

—— & POBLET, J. 1994. Geometric and Numerical Model of Progressive Limb Rotation in Detachment Folds. *Geology*, **22**, 371–374.

—— & —— 1995. The Velocity Description of Deformation. Paper 2. Sediment Geometrics Associated with Fault-Bend and Fault-Propagation Folding. *Marine and Petroleum Geology*, **12**(2), 165–176.

—— & WALTHAM, D. 1992. Computer modelling of tectonics, eustacy, and sedimentation using the Macintosh. *Geobyte*, **7**(6), 42–52.

——, DART, C. & WALTHAM, D. 1994. Computer modelling of the influence of tectonics upon sequence architecture of coarse-grained fan-deltas. *Marine and Petroleum Geology*, **11**, 561–574.

Holl, J. E. & Anastasio, D. J. 1993. Paleomagnetically derived folding rates, southern Pyrenees, Spain. *Geology*, **21**, 271–274.

Jamison, J. W. 1987. Geometrical analysis of fold development in overthrust terranes. *Journal of Structural Geology*, **9**(2), 207–219.

JOHNSON, A. M. & BERGER, P. 1989. Kinematics of fault-bend folding. *Engineering Geology*, **27**, 181–200.

JOHNSON, N. M., JORDAN, T. E., JOHNSSON, P .A., & NAESER, C. W. 1986. Magnetic polarity stratigraphy, age and tectonic setting of fluvial sediments in an eastern Andean foreland basin, San Juan Province, Argentina. *In*: ALLEN, P. & HOMEWOOD, P. (eds) *Foreland Basins*. International Association of Sedimentologists, Special Publications, **8**, 63–75.

JORDAN, T. E., ALLMENDINGER, R. W., DAMANTI, J. F. & DRAKE, R.E. 1993. Chronology of Motion in a Complete Thrust Belt: The Precordillera, 30–31°S, Andes Mountains. *Journal of Ecology*, 101, 135–156.

KENYON, P. M. & TURCOTTE, D. L. 1985. Morphology of a delta prograding by bulk sediment transport. *Geological Society of America Bulletin*, **96**, 1457–1465.

KOOI, H. & BEAUMONT, C. 1994. Escarpment evolution on high-elevation rifted margins: Insigts from a surface process model that combines diffusion, advection, and reaction. *Journal of Geophysical Research*, **99**(B6), 12191–12209.

KUKAL, Z. 1990 The rate of geological processes. *Earth Science Reviews*, **28**, 8–284.

LAUBSCHER, H. P. 1962. Die Zwiephasenhypothese der Jurafaltung. *Eclogae Geologicae Helvetiae*, **55**, 1–22.

MCCLAY, K. R. 1991. *Thrust Tectonics*, Chapman & Hall, London.

MEDWEDEFF, D .A. 1989. Growth Fault-Bend Folding at Southeast Lost Hills, San Joaquin Valley, California. *American Association Geologists Bulletin*, **73**(1), 54–67.

—— 1992. Geometry and Kinematics of an Active, Laterally Propagating Wedge Thrust, Wheeler Ridge, California. *In*: SHANKAR MITRA, & FISHER, GEORGE W. (eds) *Structural Geology of Fold and Thrust Belts*. Baltimore, Johns Hopkins University Press, 3–28.

MITRA, S. 1990. Fault-propagation folds: Geometry, kinematic evolution, and hydrocarbon traps. *American Association of Petroleum Geologists Bulletin*, **74**, 921–945.

—— 1992. Balanced Structural Interpretations in Fold and Thrust Belts. *In*: SHANKAR MITRA & FISHER, GEORGE W. (eds) *Structural Geology of Fold and Thrust Belts*. Baltimore, Johns Hopkins University Press, 53–77.

—— 1993. Geometry and Kinematic Evolution of Inversion Structures. *American Association of Petroleum Geologists Bulletin*, **77**(7), 1159–1191.

—— & NAMSON, J. 1989. Equal-area balancing. *American Journal of Science*, **298**(5), 563–599.

MOSAR, J. & SUPPE, J. 1991. Role of shear in fault-propagation folding. *In*: MCCLAY, K. R. (eds) *Thrust Tectonics*. Chapman & Hall, London, 123–132.

MOUNT, V. S., SUPPE, J. & HOOK, S. 1990. A Forward Modelling Strategy for Balancing Cross Sections. *American Association of Petroleum Geologists Bulletin*, **74**(5), 521–531.

ORI, G. G., ROVERI, M. & VANNONI, F. 1986. Plio-Pleistocene sedimentation in the Apenninic–Adriatic foredeep (Central Adriatic Sea, Italy). *In*: ALLEN, P. & HOMEWOOD, P. (eds) *Foreland Basins*. International Association of Sedimentologists Special Publications, **8**, 183–198.

PRESS, W. H., FLANNERY, B. P., TEUKOLSKY, S. A. & VETTERLING, W. T. 1988. *Numerical Recipes in C*. Cambridge University Press, Cambridge.

PRIERI, M. 1989. Three seismic profiles through the

PO Plain. *In*: BALLY, A. W. (ed.) *Atlas of Seismic Stratigraphy* (3). American Association of Petroleum Geologists Studies, **27**, 90–110.

RAMSAY, J. G. 1991. Some geometric problems of ramp-flat thrust models. *In*: MCCLAY, K. R. (ed.) *Thrust Tectonics*. Chapman & Hall, London, 191–200.

RICCI LUCCI, F. 1986. The Oligocene to Recent foreland basins of the northern Apennines. *In*: ALLEN, P. & HOMEWOOD, P. (eds) *Foreland Basins*. International Association of Sedimentologists, Special Publications, **8**, 105–139.

RICH, J. L. 1934. Mechanics of low-angle overthrust faulting as illustrated by the Cumberland Thrust Block, Virginia, Kentucky, and Tennessee. *American Association of Petroleum Geologists Bulletin*, **18**, 1584–1596.

RIVENAES, J. C. 1992. Application of a dual-lithology, depth-dependent diffusion equation in stratigraphic simulation. *Basin Research*, **4**, 133–146.

ROCKWELl, T. K., KELLER, E. A. & DEMBROFF, G. R. 1988. Quaternary rate of folding of the Ventura Avenue Anticline, western Transverse Ranges, southern California. *Geological Society of America Bulletin.*, **100**, 850–858.

ROYDEN, L. 1988. Flexural behaviour of the continental lithosphere in Italy: constraints imposed by gravity and deflection data. *Journal of Geophysical Research*, **93**, 7747–7766.

SHAW, J. H. & SUPPE, J. 1991. Slip Rates on Active Blind Thrust Faults in the Santa Barbara Channel, Western Transverse Ranges Fold-and-Thrust Belt. *Eos, Transactions of the American Geophysical Union*, **72**(44), 443.

—— & —— 1994. Active faulting and growth folding in the eastern Santa Barbara Channel, California. *Geological Society of America*, 106, 607–626.

——, HOOK, S. C. & SUPPE, J. 1994. Structural trend analysis by axial surface mapping. *American Association of Petroleum Geologists Bulletin*, **78**(5), 700–721.

SUPPE, J. 1983. Geometry and kinematics of fault-bend folding. *American Journal of Science*, **283**, 684–721.

—— 1985. *Principles of Structural Geology*. Prentice Hall, Englewood Cliffs, New Jersey.

—— & MEDWEDEFF, D. A. 1984. Fault-propagation folding. GSA Annual Meeting With Abstracts, **16**, 670.

—— & —— 1990. Geometry and kinematics of fault-propagation folding. *Eclogae Geological. Helvetiae*, **83**/3, 409–454.

——, CHOU, G. T. & HOOK, S. C. 1991. Rates of folding and faulting determined from growth strata. *In*: MCCLAY, K. R. (ed.) *Thrust Tectonics*. Chapman & Hall, London, 105–121.

VAI, G. B. 1987. Migratione complessa del sistema fronte deformativo-avanfossa-cercine periferico, il caso dell'Appennino settentrionale. *Memorie della Società Geologica Italiana*, **38**, 95–105.

WALTHAM, D. 1992. Mathematical modelling of sedimentary basin processes. *Marine and Petroleum Geology*, **9**, 265–273

—— & HARDY, S. 1994. The Velocity Description of Deformation. Paper 1: Theory. *Marine and Petroleum Geology*, **12**(2), 153–163.

——, —— & ABOUSETTA, A. 1993. Sediment geometries and domino faulting. *In*: WILLIAMS, G. D. & DOBB, A. (eds) Tectonics and Seismic Sequence Stratigraphy. Geological Society, London, Special Publications, 71. 67–85.

YEATS, R. S. 1983. Large-scale Quaternary detachments in Ventura, southern California. *Journal of Geophysical Research*, **88**(B1), 569–583.

ZOETEMEIJER, R. 1993. *Tectonic Modelling of Foreland Basins, thin skinned thrusting, syntectonic sedimentation and lithospheric flexure*. PhD Thesis, Vrije Universiteit, Amsterdam.

—— & SASSI, W. 1991. 2-D reconstruction of thrust evolution using the fault-bend fold method. *In*: MCCLAY, K.R. (ed.) *Thrust Tectonics*. Chapman & Hall, London, 133–140.

——, CLOETINGH, S., SASSI, W. & ROURE, F. 1993. Modeling of piggyback basin stratigraphy: record of tectonic evolution. *Tectonophysics*, **226**, 253–269.

——, SASSI, W., ROURE, F. & CLOETINGH, S. 1992. Stratigraphic and kinematic modeling of thrust evolution, northern Apennines, Italy. *Geology*, **20**, 1035–1038.

Numerical modelling of extension in faulted crust: effects of localized and regional deformation on basin stratigraphy

M. TER VOORDE & S. CLOETINGH

Institute of Earth Sciences, Vrije Universiteit, De Boelelaan 1085, 1081 HV Amsterdam, The Netherlands

Abstract: A numerical model for localized and regional deformation during crustal extension is presented. In this model, the behaviour of the faulted upper crust is coupled with ductile deformation in the lower crust and upper mantle. The incorporation of finite extension rates makes the model an appropriate tool for analysing the effects of synrift processes on the basin structure and stratigraphy.

Modelling results indicate that the shape of a basin depends not only on the fault geometry and the deformation mechanism, but also on the behaviour of the lower lithosphere. The authors demonstrate that onlap and offlap patterns can be caused by successive activation of a series of faults.

As shown by the modelling, the amount of footwall uplift of the fault blocks is determined by the integrated effects of deep-lithospheric and near-surface processes. Of particular importance are the spacing of the faults, the amount of extension, the depth of necking, the depth of the Moho, and the lithospheric rigidity. The rate of extension turns out to be a very important but undervalued factor for the postrift to synrift sediment thickness ratio. It is shown that this ratio increases with the extension velocity.

Following the introduction of the simple uniform stretching model for the development and evolution of rifted sedimentary basins (McKenzie 1978), substantial progress has been made in the refinement and expansion of this concept. Regional instead of local isostatic compensation has been included (Beaumont *et.al.* 1982), thermal calculations have been sophisticated (Stephenson 1989; Burrus & Audebert 1990), and models including faults have been introduced (Kusznir *et al.*, 1987; Waltham 1989). The application of these models has mainly focused on predictions for paleobathymetry, heatflow and subsidence (e.g. Issler & Beaumont 1985; Kooi *et al.* 1992), and regional scale stratigraphy modelling (e.g. Janssen *et al.* 1993; Reemst *et al.* 1994).

The fundamental importance of major basement faulting is shown by many deep seismic reflection profiles. Half grabens, linked together in various ways, form the building blocks of the architecture of rifts. This is illustrated in Fig. 1, giving an example of a horst-like feature observed in Lake Tanganyika in the east African rift zone (Fig. 1a, after Scott & Rosendahl 1989) and an example of a number of half grabens with similar polarity, forming the Potiguar Basin in the Northeast Brazilian rift system (Fig. 1b, after de Matos 1992). Like most rifted basins, these structures are developed by 'pulsed' extension. Movement occurs along one fault system for some period of time, and then shifts to another, often adjacent, fault system. Though there is often an overlap in the periods of activation of different faults, basins rarely, if ever, develop by simultaneous subsidence of all the fault blocks. However, to our knowledge, no quantitative models exist predicting basin evolution by alternating activation of different faults. In this paper, a numerical model for crustal extension is presented, including deformation along faults, and allowing for different periods of activation for every separate fault.

As shown by various authors, active extensional faulting and sedimentation are intimately linked (White *et al.* 1986; Barr 1987; Underhill 1991; Prosser 1993; Steel 1993). Underhill (1991) demonstrated that depositional sequences recognized in the Inner Morray Firth basin are controlled by regional subsidence associated with local extensional fault activity. Prosser (1993) showed that in settings tectonically active during sediment deposition, the synrift and postrift stratigraphy depends on the rate of faulting and basin formation. Not only the accommodation space for sediments is controlled by fault activity, but also the sediment supply: an uplifted footwall in a region with tilted fault blocks mostly erodes, and its erosion products accumulate in the most adjacent basins.

Features of synrift stratigraphy that might be explained by tectonic mechanisms, are for example, (i) the existence of unconformities, (ii) the different

From Buchanan, P. G. & Nieuwland, D. A. (eds), 1996, *Modern Developments in Structural Interpretation, Validation and Modelling,* Geological Society Special Publication No. 99, pp. 283–296.

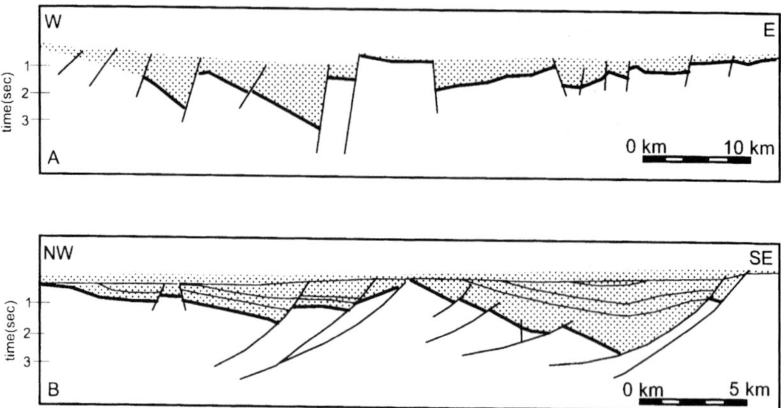

Fig. 1. (**a**) Schematic profile across Lake Tanganyika in the East African rift system. (After Scott & Rosendahl, 1989.) (**b**) Interpreted cross section across the onshore Portguar basin in the northeast Brazilian rift system. (After de Matos 1992.)

amounts of footwall uplift in tilted fault block regions, and (iii) the variation in the ratio between synrift and postrift subsidence.

(i) Onlap and offlap patterns observed in stratigraphic sequences have been explained by various mechanisms, including glacio-eustatic sealevel changes (Vail *et al.* 1977) and temporal fluctuations in the intraplate stress field, changing the flexural shape of the basin (Cloetingh 1986). By incorporating faults and a finite duration of the synrift phase in numerical models for lithospheric deformation, some additive tectonic mechanisms can be studied, especially for local, small scale (third order) patterns.

(ii) The amount of footwall uplift of a fault block depends on the size of the block (i.e. the spacing of the faults) and the amount of rifting (Yielding 1990). If we start from the notion that footwall uplift is caused by the integrated effect of faulting and flexural compensation (e.g. Kusznir *et al.* 1991), the amount of footwall uplift should be determined not only by the characteristics of faulting, but also by the properties of the lithosphere. These are for example the lithospheric rigidity, the depth of the Moho and the depth around which the lithospheric thinning occurs.

(iii) The pure shear model of McKenzie (1978) predicts that the ratio of the synrift to postrift sediment thickness is constant for given crustal and lithosphere thickness. Kusznir & Ziegler (1992) developed a more sophisticated model including faults, and concluded that the synrift to postrift sediment thickness ratio is controlled by the location (and thus spacing) of the faults and their individual amounts of displacement. In both these studies an instantaneous stretching phase was assumed. In this paper the importance of the velocity of extension as a third and critical parameter controlling the synrift to postrift ratio will be demonstrated.

To study the topics mentioned above a new numerical model was developed, combining brittle tectonics at shallow levels with ductile deep lithosphere behaviour, and incorporating a finite duration of the synrift phase. This offers the opportunity to quantify the influence of time on various processes acting during and after extensional basin formation. Furthermore, the model can be utilized to predict the interaction between faulting at the near surface and pure shear deformation in the deeper lithosphere.

Lithosphere rheology

The model assumes that the lithosphere deforms by localized deformation in the upper crust, and by diffuse deformation in the lower lithosphere (Fig. 2). This is in accordance with observations from deep seismic sections (e.g. Marsden *et al.* 1990; Bois 1993) and can be explained by the rheological profile of the continental lithosphere (Fig. 3a), showing a relatively strong, mostly brittle upper crust, underlain by a ductile, mostly weak lower crust. This rheological stratification has implications not only for the deformation mechanism (normal faulting in the upper crust versus diffuse deformation in the lower lithosphere), but also for the flexural response of the lithosphere to loading. This response, as will be shown later, is largely dependent on the so-called 'depth of necking', defined as the level which, in the absence of isostatic forces, would remain horizontal during extension (Braun & Beaumont 1989; Weissel & Karner 1989; Kooi *et al.* 1992).

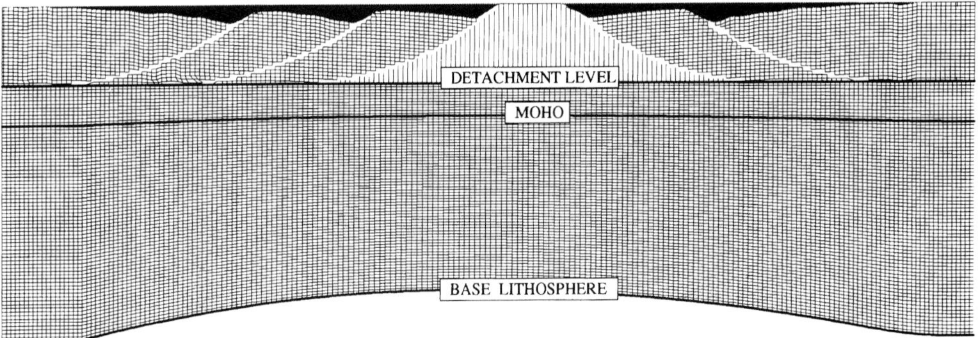

Fig. 2. Model configuration. The upper part extends by deformation along faults, the lower part by a diffuse, sinusoidal thinning of which the location and wavelength can be varied.

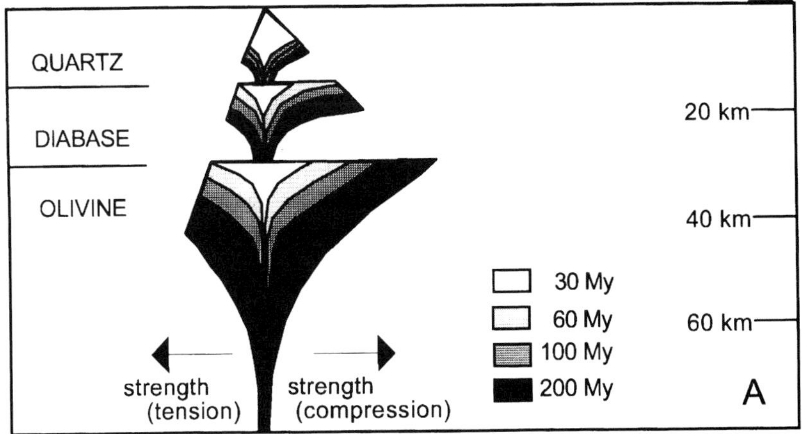

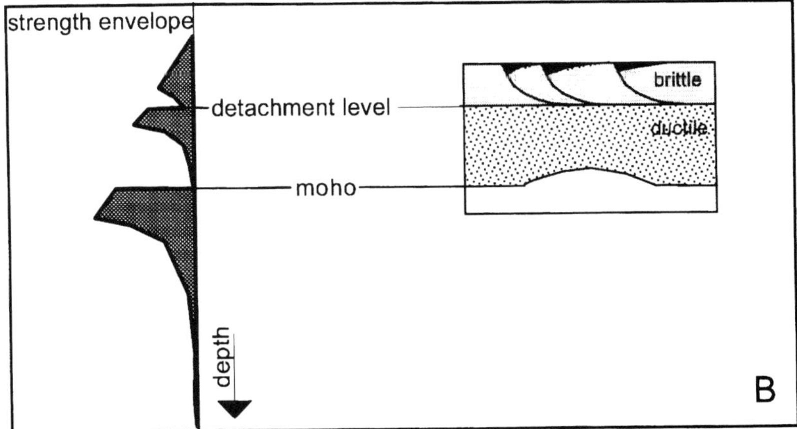

Fig. 3. (**a**) Strength profiles for a number of thermal ages and adapting a ductile strain-rate of 10^{-15} s^{-1} for a stratified continental lithosphere (after Beekman 1994). (**b**) In the model, the necking level and the detachment zone where the faults flatten out, are assumed to coincide, and are located in the lower, weak part of the upper crust.

Finite-element modelling of extension in the lithosphere (e.g. Bassi *et al.* 1993), shows that the lithosphere necks around the level of maximum strength. However, the necking depth as estimated from forward basin modelling shows large variations for different case studies (see Cloetingh *et al.* in press). The values vary from 25 km in the Tyrrhenian Sea (Spadini *et al.* 1995) to 7 km in the Pannonian Basin (Van Balen & Cloetingh 1994) and often do not coincide with the position of the strong layers in the lithosphere. This can be explained in different ways. If more than one strong layer is present in the lithopshere, the 'effective' level of necking will be positioned between these layers. A multi-layer, temperature-dependent rheology, with strong layers in the upper crust and upper mantle, can thus be cast in terms of a level of necking at lower crustal levels – which are the levels in which major faults sole out (Van Balen & Cloetingh 1994; Van der Beek *et al.* 1994). Alternatively, as a result of the crust being detached from the strong upper mantle, intracrustal necking can be the preferred mechanism (Van der Beek *et al.* 1994).

In this model, we assume that the lithosphere thins around the detachment zone in the lower weak part of the upper crust, where the normal faults sole out (Fig. 3b).

Description of the model

The 2D forward model is time dependent and calculates the deformation of the crust and lithosphere during extension. The model geometry consists of a rectangular area of which the upper part extends by localized deformation, i.e. deformation along faults, and the lower part by diffuse deformation (Fig. 2). With this principle, adopted from Kusznir *et al.* (1987), a coupling between the brittle near surface and the ductile lower lithosphere is accomplished.

The total deformation time is divided into small timesteps. In each timestep, the velocity field is calculated on a rectangular grid. A second, moving grid, representing the extending basin, is deformed according to this velocity field. The resulting basin is filled with sediments up to sealevel by defining a new horizontal line in the moving grid after each timestep. The vertical position of this line is equal to the position of the sealevel. Subsequently, at the end of each timestep, the temperature field and the flexural deformation of the basin are calculated.

The deformation in the upper part of the model is calculated according to a method described by Waltham (1989). In this 'brittle' region a fault is defined (Fig. 4). On one side of the model a constant horizontal velocity, the velocity of extension, is imposed. The direction field of the velocity

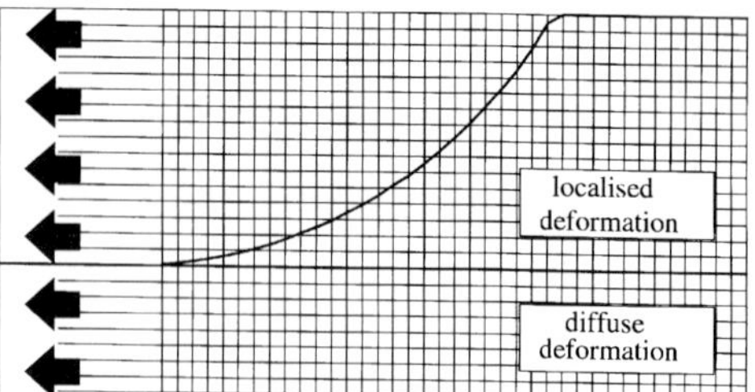

Fig. 4. Principle of the model. In the rectangular area a fault is defined. On one side of the area a constant horizontal velocity is imposed. The velocity in each gridpoint is then calculated from volume conservation.

is prescribed by defining 'isogons', lines along which the movement-direction is constant, and equal to the dip of the fault at the point of intersection (Fig. 5). The magnitude of this velocity field is calculated from the condition of volume conservation

$$\nabla \cdot \vec{v} = 0 \quad (1)$$

where v is the velocity. This equation is solved using an implicit finite difference technique (Waltham 1989).

The deformation in the ductile lower layer is defined as a sinusoidal varying pure shear thinning, also satisfying the condition of volume conservation. The horizontal location of maximum thinning, l_{max}, can be varied, eventually leading to an asymetric pattern. In this case, the wavelength of thinning is different on both sides of lmax, while the amplitude is the same. Also the width of the region over which the thinning occurs can be varied. As a result, the model is capable of quantifying the effects of a large range of different thinning patterns in the ductile layer.

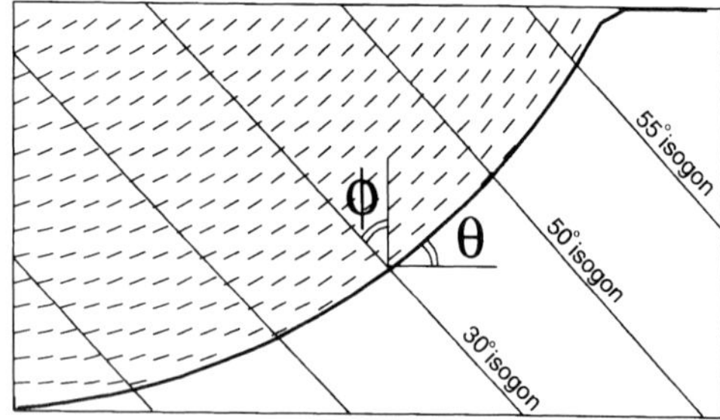

Fig. 5. Isogons are the lines along which the direction of the velocity, Θ, is constant. In this model isogons are straight lines intersecting the fault and dipping ϕ degrees anticlockwise from vertical. Θ is equal to the dip of the fault at the point of intersection. The short lines indicate the velocity directions.

To simulate multiple faults in an area, we simply add the velocity fields resulting from each separate fault. In this way, the interaction between several faults that are activated in specified periods can be modelled. These periods may either coincide, leading to simultaneous activity of the faults, be seperated, yielding successive fault activation, or partly overlap.

The temperature during and after the deformation is calculated from the heat equation

$$\frac{\partial T}{\partial t} = \nabla(\kappa \nabla T) - \vec{v}\nabla T + \frac{F}{\rho C} \quad (2)$$

where

T = temperature (°C)
t = time(s)
κ = thermal diffusivity ($m^2\ s^{-1}$)
F = function of heat production ($W\ m^3$)
ρ = density ($kg\ m^{-3}$)
C = specific heat ($J\ kg^{-1}\ °C^{-1}$)

as described by Ter Voorde & Bertotti (1994). The flexure is calculated by means of the thin plate approximation (Turcotte 1979) given by equation (3):

$$\frac{d^2}{dx^2}\left[D(x)\frac{d^2w}{dx^2}\right] + F\frac{d^2w}{dx^2} + (\rho_a - \rho_s)\, g\, w(x) = q(x) \quad (3)$$

where

w = deflection (m)
D = rigidity (Nm)
F = horizontal force ($N\ m^{-1}$)
ρ_a = density asthenosphere ($kg\ m^{-3}$)
ρ_s = density sediments ($kg\ m^{-3}$)
g = gravitational acceleration ($m\ s^{-2}$)
q = vertical load ($N\ m^{-2}$)

The load q is calculated by vertical integration of the density contrast $\Delta\rho$, caused by deformation of the crust, temperature changes and deposited sediments, relative to the undeformed crust:

$$q(x) = \int \Delta(\rho(x, z)\,(1 - \alpha T(x, z))\, g\, dz \quad (4)$$

where

α = thermal expansion coefficient (K^{-1})

Equation (3) is solved using a finite difference method (e.g. Bodine *et al.* 1981).

We assume that after each timestep regional isostatic equilibrium is reached.

During its development, the modelled basin is assumed to be continuously filled with sediments up to sealevel. As a result, if the sealevel is assumed to remain constant, the sedimentation rate is linearly dependent on the rate of extension. As sedimentation patterns are in general dependent on several other parameters, for example sediment supply, topography, the transport medium and the sediment type, this assumption is a strong simplification. However, this modelling approach is capable of isolating the effects of fault tectonics from the effects of other mechanisms.

Effects of normal faulting on basin configuration

Extensional faulting controls the synrift stratigraphy in various ways. Fault configuration and deformation mechanism of the hanging wall have a very straightforward effect on basin shape, whereas the extension rates and time intervals of activation determine the sedimentation patterns in the basin.

The velocity description for hangingwall deformation, as adopted from Waltham (1989), allows any fault shape to be modelled, as long as it flattens out at depth. Figure 6 shows examples for a straight fault, a listric fault and a ramp-flat-ramp fault. As shown, the effect of the different fault geometries on the basin shape is quite pronounced. The requirement that the faults have to flatten out at a certain level is often, but not always, met in natural structures, and thus leads to restrictions on the application of the model. Examples of faults

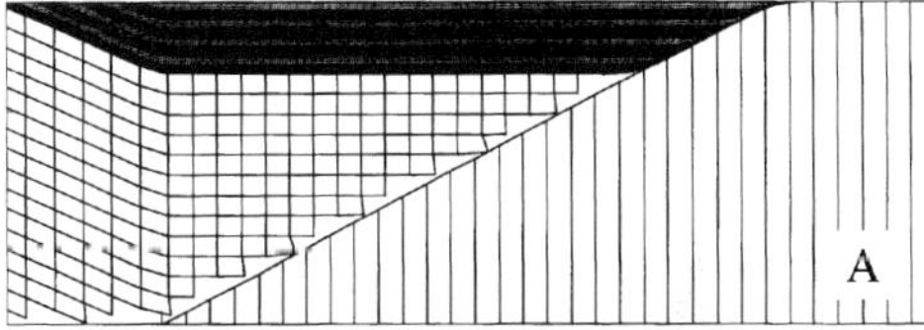

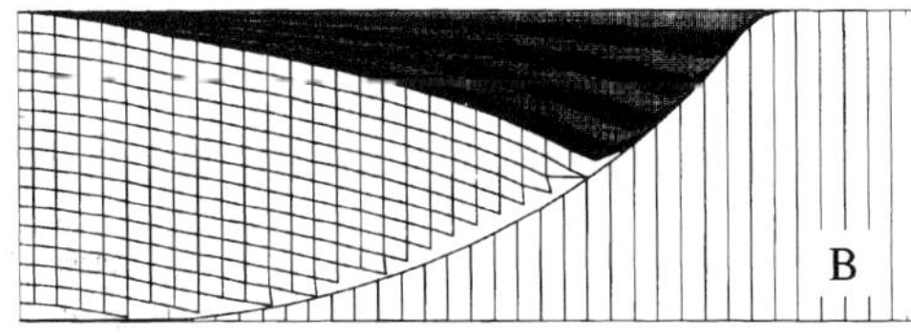

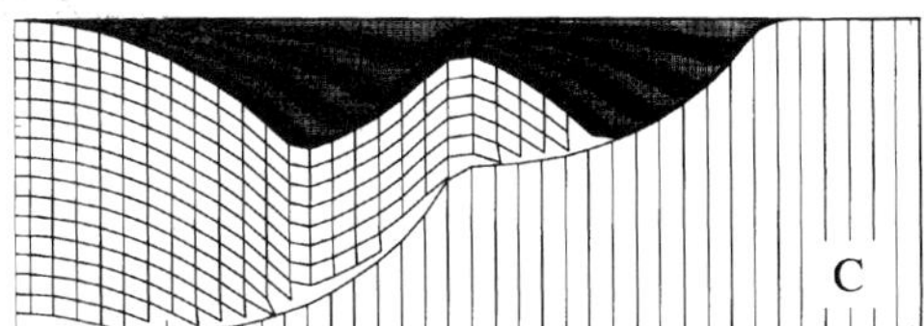

Fig. 6. Crustal deformation resulting from 20% of extension, for different fault geometries. (**a**) Straight fault, (**b**) listric fault, (**c**) ramp-flat-ramp fault. Isogons are assumed to be vertical.

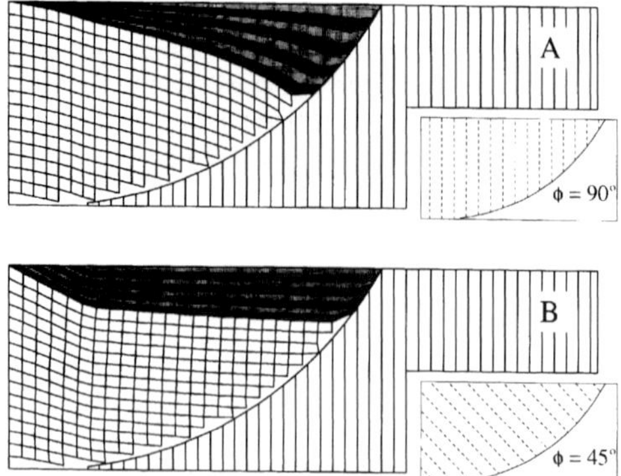

Fig. 7. Crustal deformation due to 20% of extension along a listric fault, for different isogon dips. The isogons are shown on the inset figures. (**a**) $\phi = 90°$, (**b**) $\phi = 45°$.

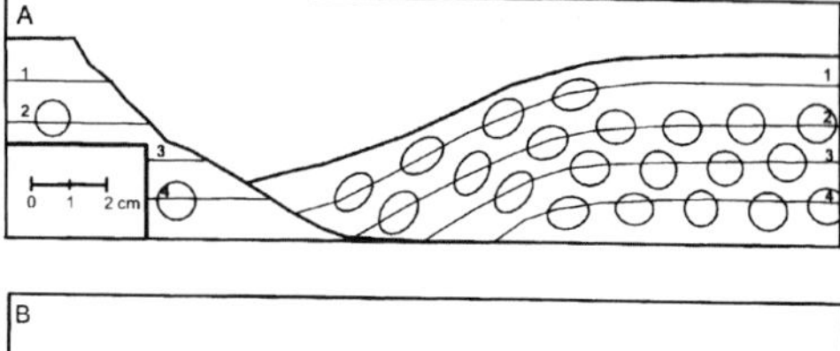

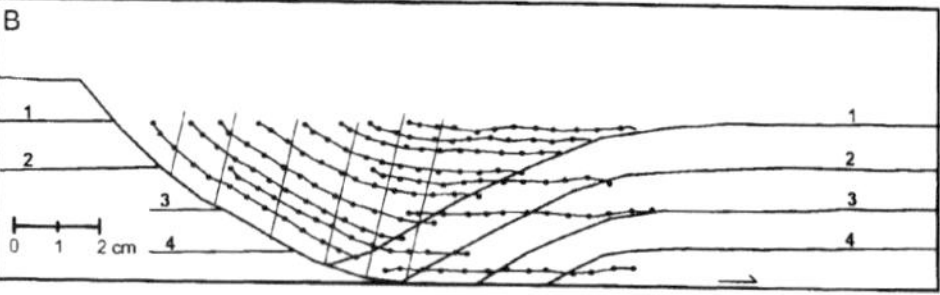

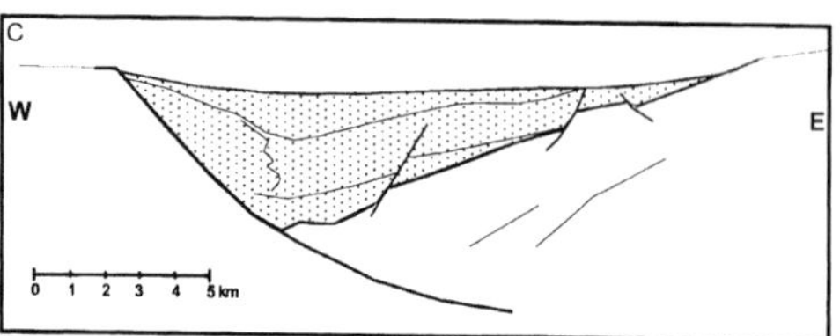

Fig. 8. (**a**) Result of a clay-model experiment in which the clay hanging wall has been extended 6 cm from the clay-clad aluminium footwall. (After Dula 1991.). (**b**) Displacement paths of material points, determined from the experiment shown in Fig. 8a. Positions of selected marker points were plotted at 0.5 cm increments of extension. (After Dula 1991.). The result can be explained by inclined simple shear, with a shear angle of 20°. (**c**) Interpretation of SOHIO Spring Valley seismic line, showing the Schell Creek fault. (After Gans *et al.* 1985). No vertical exaggeration.

flattening out with depth are the Murre Fault, offshore Newfoundland (Tankard & Welsink 1987) and the Corsair fault, offshore Texas (Christiansen 1983).

The deformation characteristics of the hanging wall can be changed in the model by varying the shape and dip of the isogons. This paper is restricted to parallel, non-curved isogons, resulting in hanging wall deformation according to the inclined simple shear mechanism. The influence of the dip of the isogons (thus the shear angle) is shown in Fig. 7, where isogon dips are 90° and 45° respectively. As can be observed in Fig. 7, the deepest point of the basin moves towards the fault with steepening isogons.

Information about deformation mechanisms of faulting can be obtained by dynamical numerical modelling (e.g. Van Wees 1994) and sandbox modelling (e.g. McClay 1989; Dula 1991; Withjack *et al.* 1995). Figures 8a and b show a result from sandbox modelling by Dula (1991). The inclined shear construction (with a shear angle of 20°) produces the best approximation to the observed hanging wall deformation. A natural example is provided by the Schell Creek Range – Spring Valley near the northern Snake Range, east-central Nevada (Fig. 8c, after Gans *et al.* 1985). The Schell Creek fault flattens out at a depth of 10–15 km. A wedge of west-dipping refectors has been observed, indicating synrift sedimentation. Also this basin can be explained by an inclined, almost vertical, shear mechanism.

If not stated otherwise, the isogons will be vertical in all the following modelling results. The other parameters used in the model are summarized in Table 1.

Table 1. *Standard parameters used in the model*

Crustal density	2700 kg m^{-3}
Mantle density	3300 kg m^{-3}
Surface temperature	0°C
Asthenosphere temperature	1330°C
Base lithosphere	125 km
Thermal diffusivity	10^{-6} m^2 s^{-1}
Flexural rigidity	7×10^{20} Nm
Young's modulus	7×10^{10} N m^{-2}
Poisson's ratio	0.25
Thermal expansion coefficient	3.4×10^{-5} °C^{-1}
Moho depth	35 km
Necking (detachment) depth	30 km

Normal faulting and synrift stratigraphy

Patterns of onlap and offlap as observed in stratigraphic sequences can be explained by eustatic sealevel changes (Vail *et al.* 1977), large scale changes in the flexural shape of the basin due to intraplate stresses (Cloetingh 1986), and on a more local scale by fault-tectonics.

Hardy (1993) examined the effects of variations

in stretching rates during rifting on the synrift stratigraphy in a domino fault block system, using a diffusion model for erosion and sedimentation, combined with a constant background sedimentation rate. He predicted complex synrift sequences, showing offlap at the fault block crest at increasing stretching rates, and onlap at decreasing stretching rates.

An important advantage of the model presented here above the domino fault block model (Hardy 1993) is that the fault blocks can move sequentially and in different orders, while the domino fault block model assumes that all the dominoes move simultaneously.

Figure 9 demonstrates the effects of extension along three listric faults for two different sequences of fault activation. In case (a) the faults are activated from the left to the right, leading to small scale offlap patterns (indicated by the boxes). In case (b) the faults are activated from the right to the left, resulting in onlap patterns. The condition for the arising of the onlap/offlap patterns is that the spacing between the faults is narrow, so that the half-grabens bounded by the different faults overlap.

An example of case (a) is observed in the Shetland region, north of Scotland (Earle *et al.* 1989). The upper crust of this area consists of half-graben basins formed by extension and subsidence during the Devonian to Carboniferous and Permian to Jurassic periods. The half-graben basins developed above listric faults which dip southeast and sole out into a detachment surface which is a reactivated Caledonian thrust fault.

Figure 10a shows an interpreted seismic section beneath the Hebrides shelf. In each of these basins, sedimentary fill thickens westward, indicating that the faults were active during deposition (Earle *et al.* 1989). The fault activation in this area must have started in the southeast and moved to the northwest, causing the offlap of Permian–Triassic sediments on the Devonian-Carboniferous sediments. In this case, however, some overlap in the time intervals and later rotation and erosion must have occurred. Figure 10b shows the result of a modelling simulation carried out for this scenario (fault activation from the SE to the NW, followed by uplift and erosion).

Apparently, the successive, non-simultaneous activation of a number of faults can provide a new explanation for observed synrift onlap and offlap patterns, provided that the spacing between the faults is narrow, so that the affected areas overlap. To the authors' knowledge, no attention has been given to this mechanism sofar. Especially on seismic sections with poorly imaged faults, this mechanism for the origin of onlap and offlap patterns could easily be overlooked.

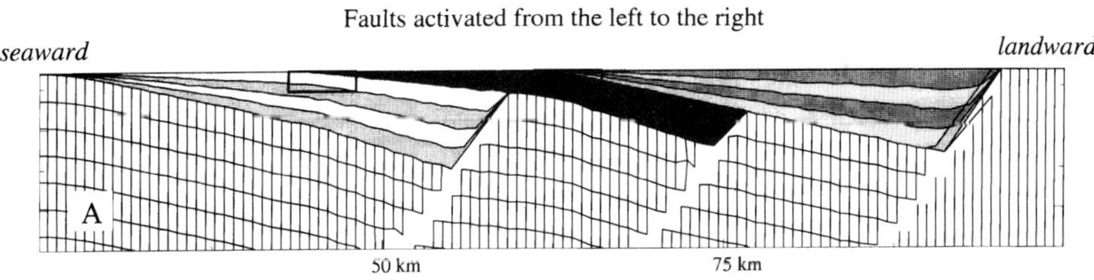

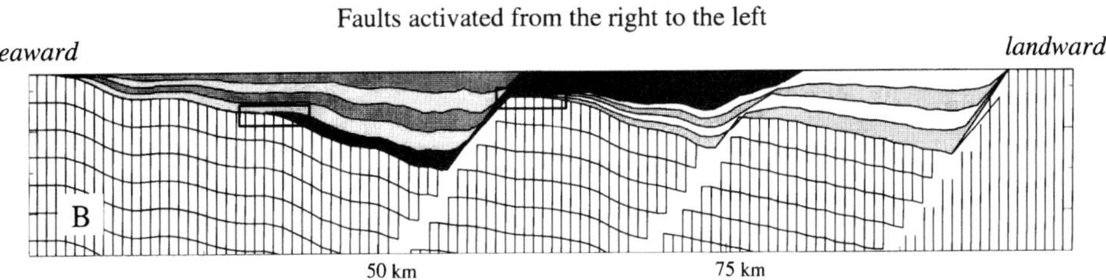

Fig. 9. Crustal deformation due to 12 km of extension along three listric faults, activated successively. (**a**) Faults are activated from the left to the right. Boxes mark offlap patterns. (**b**) Faults are activated from the right to the left. Boxes mark onlap patterns. Small scale undulations are produced numerically, as a result of the deformation of the faults themselves. Vertical exaggeration = 2.8

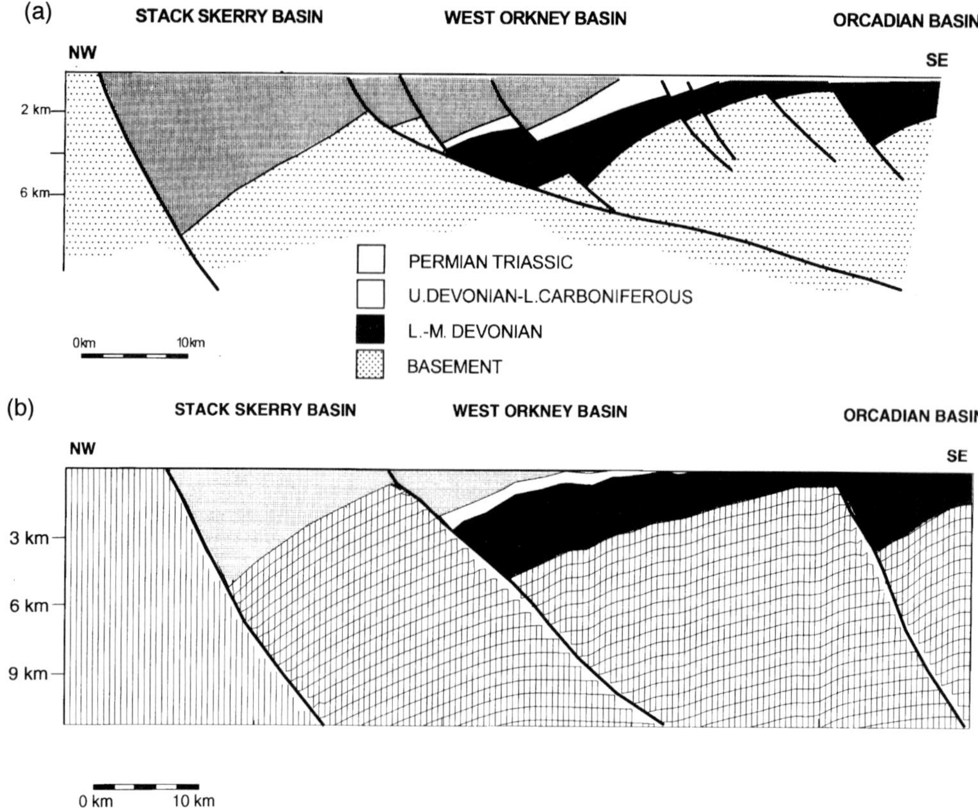

Fig. 10. (**a**) Geological section across the half grabens beneath the Hebrides Shelf, north of Scotland (after Earle *et al.* 1989). The faults must have been active succesively, from the SE to the NW. (**b**) Modelling result, obtained by activating the faults succesively, but with some overlap in the time intervals, from the SE to the NW, followed by flexural uplift and erosion.

Footwall uplift and the properties of the lithosphere

The redistribution of loads due to the deformation along faults results in a flexural bending of the lithosphere. As a consequence, the shape of the basin changes. The final shape of a basin is thus an integrated effect of the small scale faulting and the large scale flexural response. The features of this response depend on various lithosphere properties.

The wavelength of the thinning of the lower lithosphere influences the basin shape: thinning over a large area results in a relatively wide and shallow basin, while thinning over a smaller area results in a basin that is deep and narrow. The deformation of the lower lithosphere is thus reflected in the shape of the basin at the crustal surface (see also Kusznir *et al.* 1987).

In case of upward flexural bending, an uplift of the footwall can be generated (Kusznir & Egan 1989). The total amount of footwall uplift is then a summation of the footwall subsidence due to the fault activity, and the uplift caused by the flexural response. This implies that the amount of footwall uplift is not only a function of fault block size and extension (Yielding 1990), but also of lithosphere properties and the position of the fault block.

The overall state of flexure depends on the density of the infilling sediments, the depth of the Moho and the depth of necking during lithospheric extension. As explained before, the depth of necking is the level of zero vertical motion in the absence of isostatic forces, and is assumed here to coincide with the level at which the faults sole out. This level determines the ratio between thinning of the upper crust, where crustal material is replaced by sediments with, in general, low densities, and thinning of the lower lithosphere, where crustal material is replaced by dense mantle material. A deep level of necking thus results in an upward

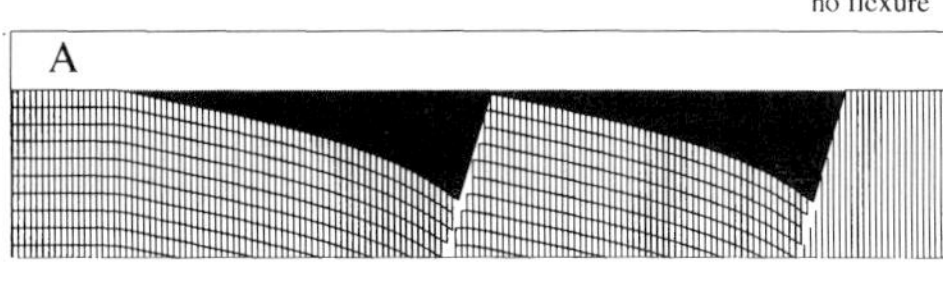

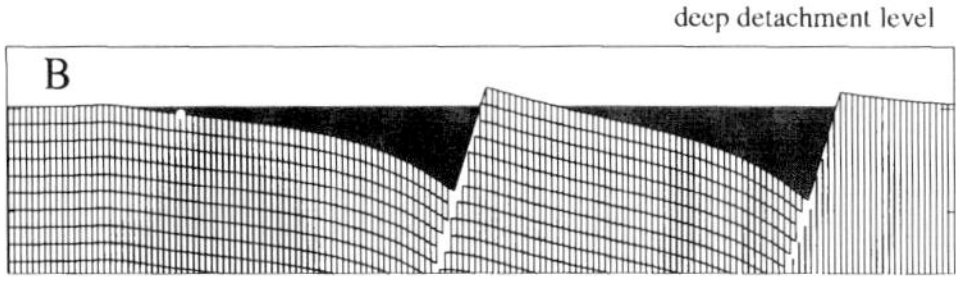

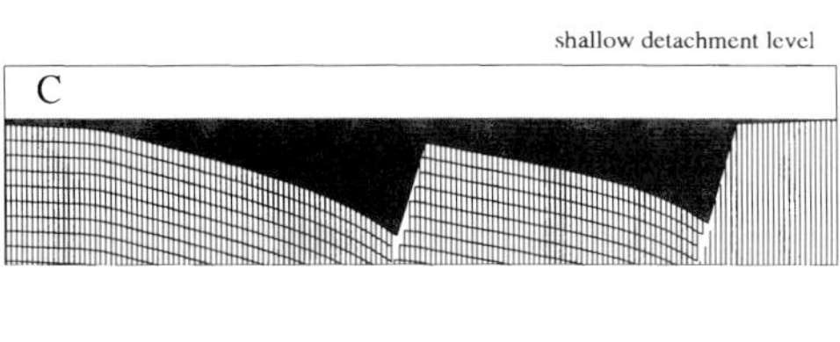

Fig. 11. Lithosphere deformation after 10 km of extension with different values for depth of necking and sediment density. Flexural rigidity is 6×10^{20} Nm, other parameters as in Table 1. (**a**) Zero flexure. (**b**) Flexure included. Detachment depth is 30 km, sediment density is 1.8 g cm^3. (c) Flexure included. Detachment depth is 15 km, sediment density is 2.2 g cm^3. Vertical exaggeration = 3.

state of flexure accompanied by footwall uplift, while a shallow level of necking results in a downward state of flexure (Kooi *et al.* 1992). The different states of flexure are illustrated in Fig. 11, which shows lithospheric deformation after 10 km of extension, for a case without flexure (a), with a necking depth of 30 km and a sediment density of 1.8 g cm^{-3} (b) and with a necking depth of 15 km and a sediment density of 2.2 g cm^{-3} (c). The results displayed in these figures reflect the situation after complete thermal relaxation (i.e. at the end of the postrift phase).

The relation between footwall uplift and necking depth after 10 km of extension along a listric fault is depicted in Fig. 12b. We have adopted the fault configuration of Fig. 12a, with model parameters as given in Table 1. Again, complete thermal relaxation occurred.

As explained above, the amount of footwall uplift increases with the depth of necking. The relationship turns out to be non-linear, the increase in footwall uplift reduces at larger necking depths.

Figure 12c shows the predicted footwall uplift as a function of extension. As can be observed, the amount of footwall uplift increases linearly with the amount of extension. This can easily be understood by considering that equation (3) is a linear equation, and the total load $q(x)$ depends linearly on the amount of extension. This implies that for constant extension rates the footwall uplift is linear with time, if thermal effects are neglected.

The curves describing the relation between footwall uplift and Moho depth (Fig. 12d) are piecewise linear and can be divided into three declining parts. The part with the highest gradient is the part where the Moho depth is less than 20 km. This is the area where the fault-movements occur – an increase in the Moho depth thus means that more crustal instead of mantle material is replaced by the sediments. In the interval between 20 km and 30 km (which is the depth of necking), the position of the Moho is only of limited influence, due to the minor movements here. For a Moho deeper than the necking depth, an increase in Moho depth means that extra crustal material is replaced by mantle material during the thinning, and the gradient increases again.

Figure 12e illustrates the relationship between the amount of footwall uplift and the effective elastic thickness (EET) of the lithosphere. The EET is a measure for the flexural rigidity of the lithosphere, the relation is given in equation (5).

$$D = \frac{E(EET)^3}{12(1-\nu)^2} \qquad (5)$$

where

E = Young's modulus (N m^{-2})
ν = Poisson's ratio

Though the amplitude of $w(x)$ decreases with the flexural rigidity, we observe an increase followed by a decrease in the curve describing the footwall uplift. This can be explained by the notion that not only the amplitude changes with the effective elastic thickness, but also the flexural wavelength and the distance at which the flexural response diminishes. Both increase with higher values for rigidity (e.g. Turcotte & Schubert 1982), resulting in a relatively higher footwall uplift.

To conclude, the properties of the lithosphere, such as necking depth, Moho depth and flexural rigidity, exert a strong influence on the amount of footwall uplift.

The synrift to postrift ratio

The subdivision of the subsidence history of an extensional basin into an active rifting phase and a thermal cooling phase is a distinction that cannot be made very strictly in many cases. It was already

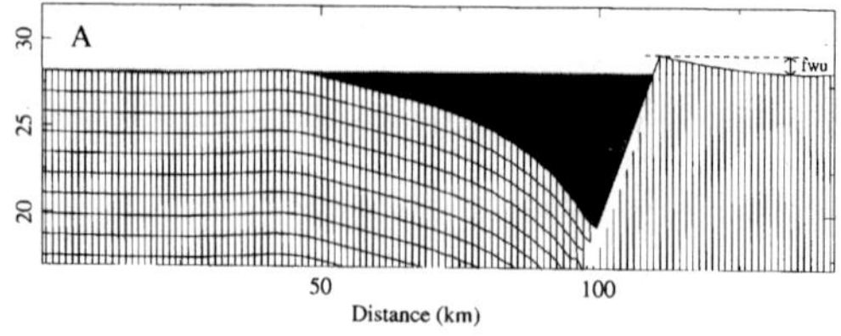

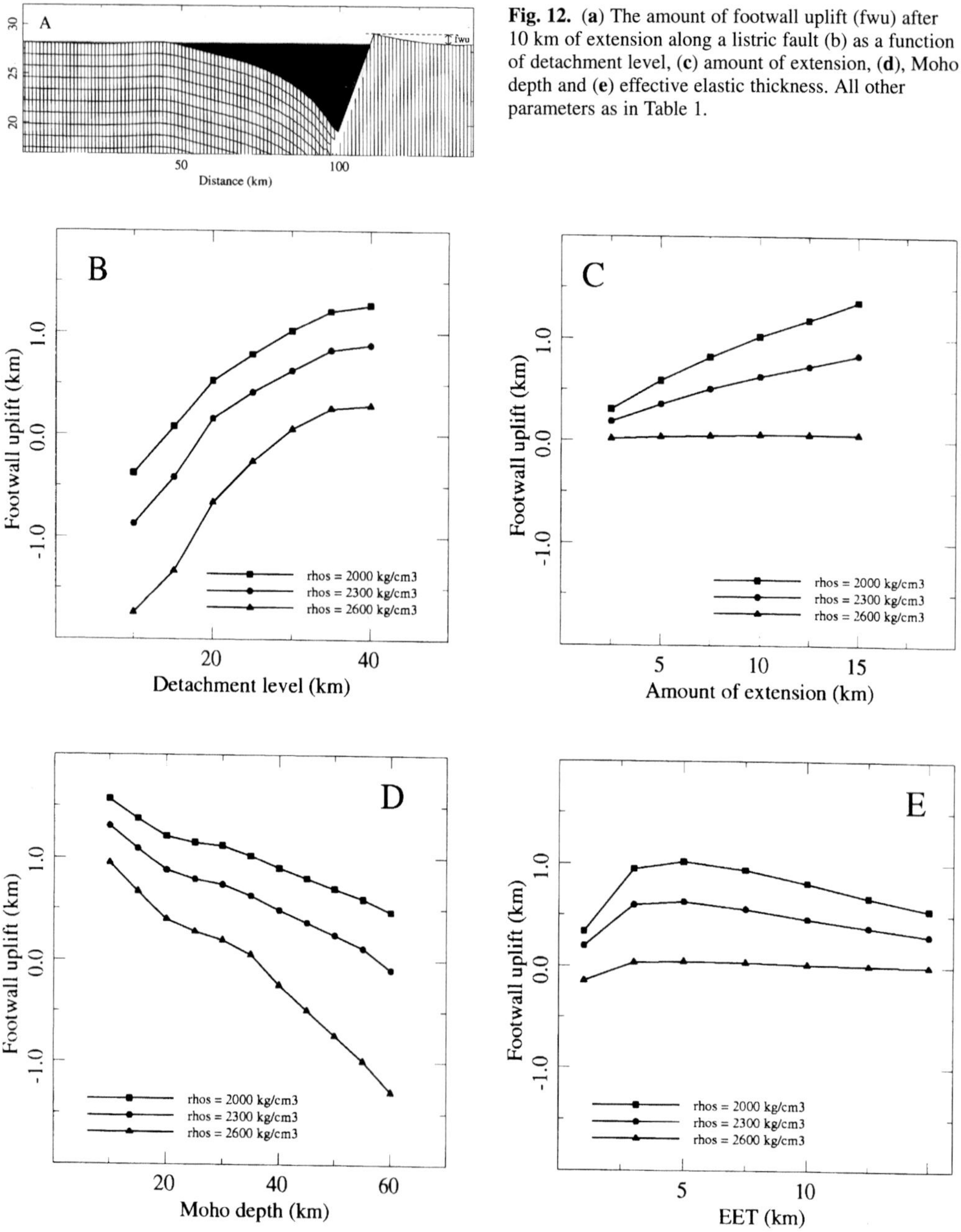

Fig. 12. (**a**) The amount of footwall uplift (fwu) after 10 km of extension along a listric fault (b) as a function of detachment level, (**c**) amount of extension, (**d**), Moho depth and (**e**) effective elastic thickness. All other parameters as in Table 1.

pointed out by Jarvis & McKenzie (1980) that the finite duration of rifting should be taken into account in subsidence-history calculations if the rifting time is larger than $60/\beta^2$ Ma if $\beta \leq 2$, or $60(1 - 1/\beta)^2$ Ma if $\beta \geq 2$ (where β is the extension factor). Their statement was based on analytical calculations. Ter Voorde & Bertotti (1994) and Bertotti & ter Voorde (1994) found, from a numerical modelling study, that at low extension rates the largest part of the temperature re-equilibration occurs already during rifting instead of during the postrift phase. Therefore, we propose that the ratio of synrift to postrift subsidence is mainly dependent on the rate of stretching.

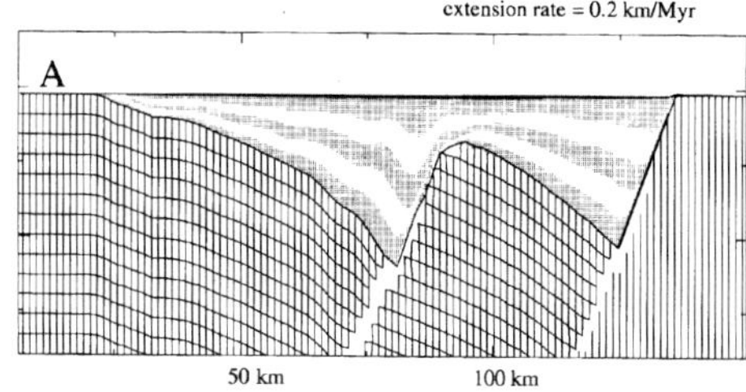

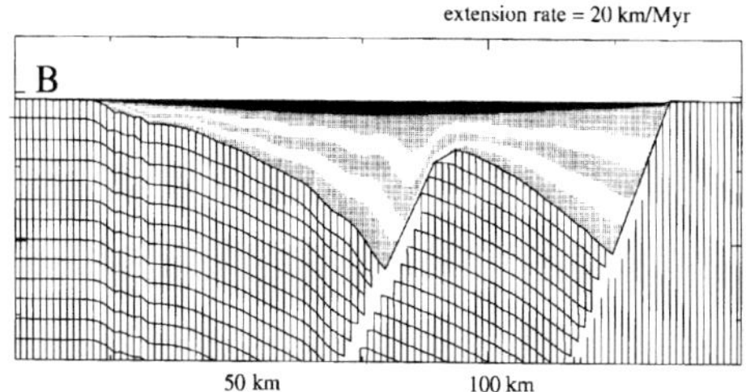

Fig. 13. Lithosphere deformation after 20 km of extension with different extension rates. In light-grey: synrift sediments; in black: postrift sediments. (**a**) Extension rate is 0.2 km Ma^{-1}, (**b**) extension rate is 20 km Ma^{-1}. Sediment density is 2600 kg m^{-3}, other model parameters as in Table 1. Small scale undulations are produced numerically, as a result of the deformation of the faults themselves. Vertical exaggeration = 2.8.

Figures 13a and b show modelling results for 20 km of extension along two listric faults, occurring over a period of 10 Ma and 100 Ma respectively. The results indicate that the synrift sediment thickness becomes larger with decreasing extension rates at the expense of the postrift thickness, which is a consequence of the thermal relaxation during extension. In case (a), where the extension rate is 20 km Ma^{-1}, the maximum postrift thickness is 0.9 km, while in case (b), where the extension rate is 0.2 km Ma^{-1}, the maximum postrift sediment thickness is only 0.23 km. Note that the total (synrift + postrift) sediment thickness is equal for both cases. We applied a sediment density of 2600 kg m^{-3}, other parameters as in Table 1.

Figure 14a shows the calculated maximum amplitude with time, for different extension rates. Fault configuration as in Fig. 13. The corresponding β-value is 1.13. The following patterns can be observed:

(i) The amount of flexural uplift during the synrift phase is not linear with time, but shows a decreasing gradient because of the cooling that already takes place during rifting. The total amount of flexural uplift at the end of the synrift phase is lower for cases with lower extension rates.

(ii) If the synrift phase is terminated, only the thermal subsidence remains, decreasing exponentially with time.

(iii) Lithospheric cooling occurs slower at low deformation rates than at high deformation rates, because of the lower temperature gradients that arise in the synrift phase. After 200 Ma, the modelled cases with extension rates higher than 0.40 km Ma^{-1} hardly subside anymore, while the slowest case (0.125 km Ma^{-1}) will still subside some 60 m more.

(iv) The maximum postrift thickness is given by the difference between the amplitude immediately after the synrift phase and the amplitude at the end

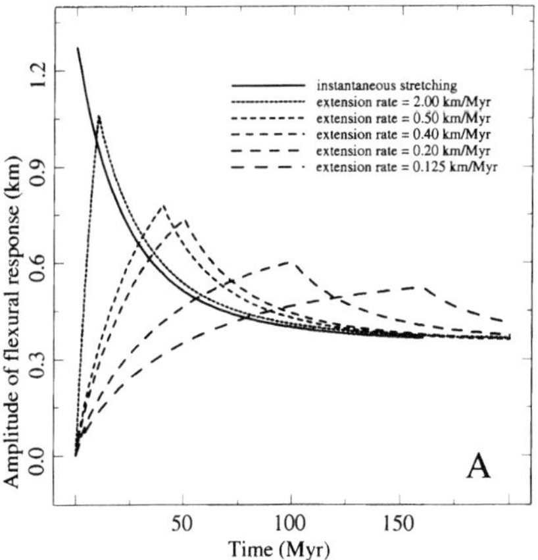

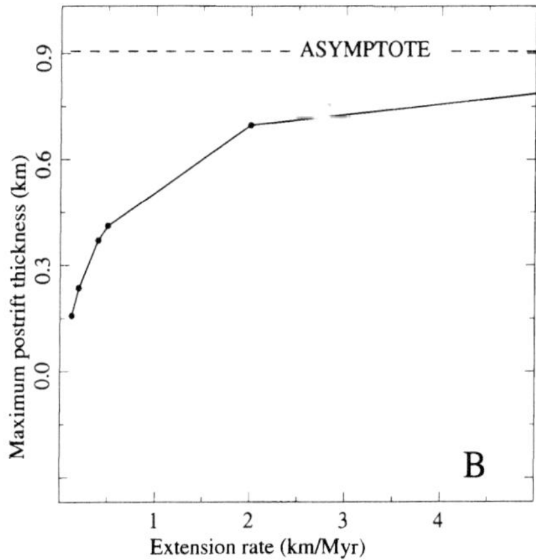

Fig. 14. (**a**) The maximum flexural uplift of the lithosphere as a function of time for various extension rates. Fault configuration as in Fig. 13, sediment density 2600 kg m^{-3}, other parameters as in Table 1. (**b**) The maximum postrift thickness of sediments as a function of extension rate. Fault configuration as in Fig. 13. The asymptote gives the maximumpostrift thickness for the case of instantaneous stretching (i.e. infinite extension rate).

of the thermal subsidence. The amplitude at the end of thermal subsidence is not dependent on the extension rate, whereas the amplitude immediately after the synrift phase increases with the extension rate towards an asymptotic value. As a consequence, as shown in Fig. 14b, the maximum postrift thickness also increases with the extension rate.

The conclusion, therefore, is that the extension rate has a substantial influence on the synrift/postrift ratio of sediment thicknesses: for the same amount of total extension, high stretching rates cause thicker postrift sediment piles. This effect becomes even stronger if the sedimentation rate is constant instead of dependent on the extension rate, like in this model. In that case, upon fast extension, subsidence will outpace sedimentation, causing a relatively thinner synrift and thicker postrift sedimentary package.

Conclusions

From the modelling results the following conclusions can be drawn.

(i) A strong coupling exists between small scale tectonics and the synrift stratigraphy in extensional basins. The fault configuration and deformation mechanism have a direct effect on basin shape, and the order of fault activation is a controlling factor for onlap/offlap patterns in stratigraphic sequences, which are commonly interpreted in terms of eustatic sealevel changes.

(ii) To evaluate the structural evolution of an extensional basin, it is necessary to couple the small scale deformation along faults at the near surface with the large scale diffuse deformation in the deeper lithosphere. The wavelength of thinning of the ductile deep lithosphere is reflected directly in the basin shape, while the amount of flexural uplift or subsidence is affected by the depth of necking, the depth of the Moho and the flexural rigidity. These are, apart from the spacing of the faults and the amount of extension, the key factors that determine the amount of footwall uplift. The amount of footwall uplift increases with necking depth, decreases with Moho depth, and shows an increase followed by a decrease with the effective elastic thickness.

(iii) The rate of extension turns out to be a very important factor for the evolution of a rifted basin. The ratio of synrift to postrift sedimentary thickness is largely dependent on stretching rates. For the same amount of extension, the postrift sediment thickness increases exponentially with the stretching rate, whereas the total sediment thickness remains constant.

The authors wish to thank Giovanni Bertotti for many useful suggestions and discussions. Comments of Ronald van Balen and reviews of Lidia Lonergan and an anonymous reviewer are appreciated, and improved the paper substantially. Dick Nieuwland is thanked for comments, and for editorial effort. This research was supported by the IBS (Integrated Basin Studies) project, part of the Joule II research programme funded by the Commission of European Communities (contract No. JOU2-CT 92-0110). Publication No. 950301 of the Netherlands School of Sedimentary Geology.

References

BARR, D. 1987. Structural/stratigraphic models for extensional basins of half-graben type. *Journal of Structural Geology*, **9**, 491–500.

BASSI, G., KEEN, C. E. & POTTER, P. 1993. Contrasting styles of rifting: Models and examples from the eastern Canadian margin. *Tectonics*, **12**, 639–655.

BEAUMONT, C., KEEN, C. E. & BOUTILLIER, R. 1982. On the evolution of rifted continental margins: Comparison of models and observations for the Nova Scotia margin, *Geophysical Journal of the Royal Astronomical Society*, **70**, 667–715.

BEEKMAN, F. 1994. *Tectonic modelling of thick-skinned compressional intraplate deformation.* PhD thesis, Vrije Universiteit, Amsterdam.

BERTOTTI, G. & TER VOORDE, M. 1994. Thermal effects of normal faulting during rifted basin formation. 2: The Lugano Normal Fault and the role of pre-existing thermal anomalies. *Tectonophysics*, **240**, 145–157.

BODINE, J. H., STECKLER, M. S. & WATTS, A. B. 1981. Observations of flexure and the rheology of the oceanic lithosphere, *Journal of Geophysical Research*, **86**, 3695–3707.

BOIS, C. 1993. Initiation and evolution of the Oligo-Miocene rift basins of southwestern Europe: contibution of deep seismic reflection profiling. *Tectonophysics*, **226**, 217–226.

BRAUN, J. & BEAUMONT, C. 1989. A physical explanation of the relation between flank uplifts and the breakup unconformity at rifted continental margins. *Geology*, **17**, 760–764.

BURRUS, J. & AUDEBERT, F. 1990. Thermal and compaction processes in a young rifted basin containing evaporites: Gulf of Lions, France. *American Association of Petroleum Geologists Bulletin*, **74**, 1420–1440.

CHRISTIANSEN, A. F. 1983. An example of a major syndepositional listric fault. *In*: BALLY, A. W. (ed.) *Seismic Expression of Structural Styles*: American Association of Petroleum Geologists Studies in Geology, **15**, 2.3.1-36-40.

CLOETINGH, S. 1986. Intraplate stress: a new tectonic mechanism for fluctuations in relative sealevel. *Geology*, **14**, 617–620.

——, VAN WEES, J. D., VAN DER BEEK, P.A. & SPADINI, G. Role of pre-rift rheology in kinematics of extensional basin formation: constraints from thermomechanical models of Mediterranean and intracratonic basins. *Marine and Petroleum Geology*, Special IBS Volume, in press.

DULA, W. F. 1991. Geometric models of listric normal faults and rollover folds. *American Association of Petroleum Geologists Bulletin*, **75**, **10**, 1609–1625.

EARLE, M. M., JANKOWSKI, E. J. & VANN, I. R., 1989. Structural and stratigraphic evolution of the Faeroe-Shetland Channel and Northern Rockall Trough. *In*: TANKARD, A. J. & BALKWILL, H. R. (eds) *Extensional Tectonics of the North Atlantic Margins*. American Association of Petroleum Geologists Memoirs, **46**, 461–469.

GANS, P. B., MILLER, E. L., MCCARTHY, J. & OULDCOTT, M .L. 1985. Tertiary extensional faulting and evolving ductile-brittle transition zones in the northern Snake Range and vicinity: New insights from seismic data. *Geology*, **13**, 189–193.

HARDY, S. 1993. Numerical modelling of the response to variable stretching rate of a domino fault block system. *Marine and Petroleum Geology*, **10**, 145–152.

ISSLER, D. R. & BEAUMONT, C. 1985. A finite element model of the subsidence and thermal evolution of extensional basins: Application to the Labrador Continental Margin. *In*: NAESER, N. D. & MCCULLOH, T. H. (eds) *Thermal history of sedimentary basins, methods and case histories*. Springer-Verlag, New York, 239–267.

JANSSEN, M.E., TORNÉ, M., CLOETINGH, S. & BANDA, E. 1993. Pliocene uplift of the eastern Iberian margin: Inferences from quantitative modelling of the Valencia Trough. *Earth and Planetary Science Letters*, **119**, 585-597.

JARVIS, G. T. & MCKENZIE, D. P. 1980. Sedimentary basin formation with finite extension rates. *Earth and Planetary Science Letters*, **48**, 42–52.

KOOI, H., CLOETINGH, S. & BURRUS, J. 1992. Lithospheric necking and regional isostasy at extensional basins. 1. Subsidence and gravity modelling with an application to the Gulf of Lions Margin (SE France). *Journal of Geophysical Research*, **97** **(B12)**, 17553–17571.

KUSZNIR, N. J. & EGAN, S. S. 1989. Simple-shear and Pure-shear models of extensional sedimentary basin formation: application to the Jeanne d'Arc Basin, Grand Banks of Newfoundland. *In*: TANKARD, A. J. & BALKWILL, H. R. (eds) *Extensional Tectonics of the North Atlantic Margins*. American Association of Petroleum Geologists Memoir **46**, 305–322.

—— & ZIEGLER, P. 1992. The mechanics of continental extension and sedimentary basin formation: A simple shear/pure-shear flexural cantilever model. *Tectonophysics*, **215**, 117–131.

——, KARNER, G. D. & EGAN, S. 1987. Geometric, thermal and isostatic consequences of detachments in continental lithosphere extension and basin formation. *In:* BEAUMONT, C. & TANKARD, A. J. (eds) *Sedimentary Basins and Basin-Forming Mechanisms*, Canadian Society of Petroleum Geologists Memoir, **12**, 185–203.

——, MARSDEN, G. & EGAN, S. S. 1991. A flexural-cantilever simple-shear/pure-shear model of continental lithosphere extension: applications to the Jeanne d'Arc Basin, Grand Banks and Viking Graben, North sea. *In*: ROBERTS, A. M., YIELDING, G. & Freeman, B. (eds) *The Geometry of Normal Faults*. Geological Society, London, Special Publications, **56**, 41–60.

MARSDEN, G., YIELDING, G., ROBERTS, A. M. & KUSZNIR, N. J. 1990. Application of the flexural cantilever simple-shear/pure-shear model of continental lithosphere extension to the formation of the North Sea basin. *In*: BLUNDELL, D. J. & GIBBS, A. D. (eds) *Tectonic Evolution of the North Sea Rifts*. Oxford University Press, 236–257.

MATOS, R. M. D. DE, 1992. The Northeast Brazilian rift system. *Tectonics*, **11**, 766–791.

MCCLAY, K. R. 1989. Physical models of structural styles during extension. *In*: Tankard, A. J. & Balkwill, H. R. (eds) *Extensional Tectonics of the North Atlantic Margins*. American Association of Petroleum Geologists Memoir, **46**, 95–110.

MCKENZIE, D. P. 1978. Some remarks on the development of sedimentary basins. *Earth and Planetary Science Letters*, **40**, 25–31.

PROSSER, S. 1993. Rift-related linked depositional systems and their seismic expression. *In:* WIILIAMS, G. & DOBB, A. (eds) *Tectonics and Seismic Sequence Stratigraphy*. Geological Society, London, Special Publications, **71**, 35–66.

REEMST, P., CLOETINGH, C. & FANAVOLL, S. 1994. Tectono-stratigraphic modelling of Cenozoic uplift and erosion in the SW Barents Sea. *Marine and Petroleum Geology*, **11**, 478–490.

SCOTT, D. L. & ROSENDAHL, B. R. 1989. North Viking Graben: An East African perspective. *The American Association of Petroleum Geologists Bulletin*, **73**, 155–165

SPADINI, G., CLOETINGH, S. & BERTOTTI, G. 1995. Thermo-mechanical modeling of the Tyrrhenian Sea: lithospheric necking and kinematics of rifting, **14**, 629–644.

STEEL, R. J. 1993. Triassic–Jurassic megasequence stratigraphy in the Northern North Sea: rift to post-rift evolution. *In*: PARKER, J. R. (ed.) *Petroleum Geology of Northwest Europe: Proceedings of the 4th Conference*. Geological Society, London, 299–315.

STEPHENSON, R. 1989. Beyond first-order thermal subsidence models for sedimentary basins? *In*: CROSS, T. A. (ed.) *Quantitative Dynamic Stratigraphy*. Prentice Hall, 113–125.

TANKARD, A. J. & WELSINK, H. J. 1987. Extensional tectonics and stratigraphy of Hibernia oil field, Grand Banks, Newfoundland. *American Association of Petroleum Geologists Bulletin*, **71**, 1210–1232.

TER VOORDE, M. & Bertotti, G. 1994. Thermal effects of normal faulting during rifted basin formation. 1: A finite difference model. *Tectonophysics*, **240**, 133–144.

TURCOTTE, D. L. 1979. Flexure. *In*: SALTZMAN, B. (ed.) *Advances in Geophysics,* Academic Press, Inc., New York, **21**, 51–86.

—— & SCHUBERT, G. 1982. *Geodynamics, applications of continuum physics to geological problems.* L.Wiley and Sons, USA.

UNDERHILL, J. R. 1991. Controls on Late Jurassic seismic sequences, Inner Moray Firth, UK North Sea: a critical test of a key segment of Exxon's original global cycle chart. *Basin Research*, **3**, 79–98.

VAIL, P., MITCHUM, JR., R. M. & THOMPSON, S. 1977. *Global cycles of relative changes of sea level.* The American Association of Petroleum Geologists Memoir, **26**, 83–97.

VAN BALEN, R. & Cloetingh, S. 1994. Tectonic control of the sedimentary record and stress induced fluid flow: constraints from basin modelling. *In*: PARNELL, J. (ed.) *Geofluids: Origin, Migration and Evolution of Fluids in Sedimentary Basins.* Geological Society, London, Special Publications, **78**, 9–26.

VAN DER BEEK, P., CLOETINGH, S. & ANDRIESSEN, P. 1994. Mechanisms of extensional basin formation and vertical motions at rift flanks: Constraints from tectonic modelling and fission-track thermochronology. *Earth and Planetary Science Letters*, **121**, 417–433.

VAN WEES, J. D. 1994. *Tectonic modelling of basin deformation and inversion dynamics: the role of pre-existing faults and continental lithosphere rheology in basin evolution.* PhD Thesis, Vrije Universiteit, Amsterdam.

WALTHAM, D. 1989. Finite difference modelling of hanging wall deformation. *Journal of Structural Geology*, **11**, 433–437.

WEISSEL, J. K. & KARNER, G. D. 1989. Flexural uplift of rift flanks due to mechanical unloading of the lithosphere during extension. *Journal of Geophysical Research*, **94 (B10)**, 13919–13950.

WHITE, N. J., JACKSON, J. A. & MCKENZIE, D. P. 1986. The relationship between the geometry of normal faults and that of the sedimentary layers in their hanging walls. *Journal of Structural Geology*, **.8**, 897–909.

WITHJACK, M. O., ISLAM, Q. T. & LA POINTE, P. R. 1995. Normal faults and their hanging-wall deformation: An experimental study. *American Association of Petroleum Geologists Bulletin*, **79**, 155–165.

YIELDING, G. 1990. Footwall uplift associated with Late Jurassic normal faulting in the northern North sea. *Journal of the Geological Society, London*, **147**, 219–222.

The role of pre-existing faults in basin evolution: constraints from 2D finite element and 3D flexure models

J. D. VAN WEES[1], S. CLOETINGH[1] & G. DE VICENTE[2]

[1] *Vrije Universiteit, Instituut van Aardwetenschappen, De Boelelaan 1085, 1081 HV Amsterdam, The Netherlands*

[2] *Universidad Complutense de Madrid, Facultad de Ciencias Geologicas, Departamento de Geodinamica, 28040 Madrid, Spain*

Abstract: The lithospheric driving forces which cause intraplate basin deformation are relatively constant over large areas. Consequently, lateral variations in deformation and stress and strain concentrations seem to be primarily caused by (pre-existing) heterogenities in the rheological signature of the continental lithosphere underlying the sedimentary basins. In this paper, we explore the weak character of upper crustal faults and their control on basin shape for a number of case studies on intraplate Phanerozoic basin settings, using a 2D finite element and a 3D flexure model. Of key importance is the integration of seismic data and field observations with the tectonic modelling, allowing the investigation of deformation processes and their expressions on different scales, operating on different levels of the lithosphere. Finite element models for sub-basin scales, incorporating weak upper crustal faults, predict strong control of these weak zones on local stress distributions and subsequent deformation, in agreement with observed deformation patterns. Slip along upper crustal faults control stress distribution and subsequent faulting in overlying sedimentary rocks. The effect of weak upper crustal fault movements on basin-wide (regional) upper crustal flexure is looked at in two case studies on: (1) extensional tectonics in the Lake Tanganyika Rift Zone (East Africa); and (2) compressional tectonics in the Central System and Tajo Basin (Central Spain). Both settings indicate that basement warpings are controlled by large amounts of slip along so-called weak crustal-scale border faults, which are mostly planar. Adopting border fault displacements in the 3D flexure model, the results indicate low effective elastic thickness (EET) values in a range of 3–7 km for induced basement deflection patterns in accordance with observations. The low EET values most likely reflect a (partly) decoupling of upper crustal and subcrustal deformation, facilitated by the weak lower crust, and in agreement with standard rheological assumptions for Phanerozoic lithosphere. In contrast, the inferred weakness of upper crustal faults relative to surrounding rock is not evident from uniform rheological assumptions. However, observations of reactivations of faults which are not preferably aligned with the stress field, and reactivations of basin deformation on long time-scales are in support of this feature.

The lithospheric driving forces which cause intraplate basin deformation are relatively constant over large areas (Zoback *et al.* 1989; Müller *et al.* 1992; Zoback 1992). Therefore, lateral variations in deformation and stress and strain concentrations seem to be primarily caused by heterogenities in the rheological signature of the continental lithosphere underlying the sedimentary basins (Zoback *et al.* 1993). Following widely accepted concepts on lithospheric rheology (e.g. Brace & Kohlstedt 1980), numerical modelling studies on rift dynamics confirm that rheology strongly controls the distribution of rift activity and vice versa (e.g. Bassi *et al.* 1993). The actual trigger of rift localization can be small perturbations in an initially homogeneous material configuration and homogeneous thermal state of the lithosphere (e.g. Braun & Beaumont 1989). Then, further localization and the associated lithospheric weakening is a function of (a) rift induced lateral variations in the thermal state of the lithosphere and (b) rift induced lateral variations in material distribution (England 1983; Kusznir & Park 1987). The latter found that rift induced heating weakens the lithosphere, whereas crustal thinning has an opposite effect. The relative contributions heavily depend on the rift velocity, the distribution of crustal and subcrustal deformation, and the initial thermal and material state of the lithosphere (Buck 1991; Bassi *et al.* 1993). Thermo-mechanical models for a single extension phase, show that predicted styles of extension are in good agreement with observations on lithospheric scales (Buck 1991; Bassi *et al.* 1993).

For older basins, characterized by multiple phases of extension and an extremely low mean rift

From Buchanan, P. G. & Nieuwland, D. A. (eds), 1996, *Modern Developments in Structural Interpretation, Validation and Modelling,* Geological Society Special Publication No. 99, pp. 297–320.

velocity, these types of models, tend to predict locking or widening of rift activity for each new phase (Braun 1992). This is in contradiction with observations for a large number of repeatedly reactivated basins, indicating that lithospheric deformation remains localized in a restricted domain over time-spans of more than 300 Ma (e.g. Western Europe, Ziegler 1990; Kooi 1991). It has been noted by many authors that deformation on these time-scales involves reactivation of tectonic structures at various scales, ranging from lithospheric scale suture zones to basement faults (e.g. Wilson 1966; Hayward & Graham 1989; Ziegler 1990; Yielding *et al.* 1991).

Some authors have stressed the importance of pre-existing heterogeneities in rheology and weak zones in the distribution of deformation on lithospheric scales (Braun & Beaumont 1987; Dunbar & Sawyer 1988). Recent studies on a lithospheric scale (Van Wees & Stephenson 1995) show that repeated basin reactivation on long time-scales can be related to the existence of weak zones, deviating from standard rheological assumptions (Van Wees & Stephenson 1995). On smaller scales, stress and strain localizations typically observed in basin-scale faulting, have been studied by analogue and numerical models, incorporating movement along pre-existing structural elements like pre-existing faults. These models have been used very successfully to explain observed deformation distributions (e.g. Mandl 1988; McClay 1989; Kusznir & Ziegler 1992; Sassi *et al.* 1993). The modelling results also indicate that the pre-existing structures exhibit a certain weakness, deviating from standard rheological assumptions.

The characteristics of weak zones in perspective of standard assumptions on the rheological framework of the lithosphere have been the focus of very few studies (e.g. Ranalli & Yin 1990; Le Pichon & Chamot-Rooke 1991; Van Wees & Stephenson 1995). Furthermore, modelling studies incorporating the effect of weak zones, either focus on large crustal and lithospheric processes or small-scale basement and fill processes.

In this paper, we examine the effect which (pre-existing) faults, have on the structural evolution of sedimentary basins on various scales. First, we will review continental lithosphere rheology and show that upper crustal faults are most likely marked by a pronounced weakness relative to surrounding rock, and in the framework of a stratified lithosphere rheology are bound to have a strong influence on basin shape. Subsequently, the controlling role of these weak zones, affecting the structural expression of basins, will be explored with a 2D finite element model and a recently developed 3D flexure model (Van Wees & Cloetingh 1994), with applications in a number of case studies on Phanerozoic basin structures, offshore and onshore, integrating seismic data and field studies.

Lithospheric-scale rheological properties of the continental lithosphere

Understanding of the rheological behaviour of rocks that constitute continental lithosphere stems from laboratory experiments in the brittle regime (i.e. strain-rate independent deformation) and ductile regime (i.e. strain-rate dependent) (e.g. Ranalli 1987). At geologically relevant time-scales these regimes are characterized by a Coulomb-Navier criterion and various creep laws, respectively.

At low confining pressures and temperatures, fracturing is predominant. Fracturing is described by a Coulomb-Navier criterion, given the compressive stresses predominant in the lithosphere (Jaeger & Cook 1976; Ranalli 1987). Commonly, it is assumed that cohesion is negligible and Andersonian stresses are adopted, which means that one of the principal stresses is vertical. In this case, the critical stress difference for sliding along new faults, with friction angle ø and which form at an angle of $\alpha = \frac{1}{4}\pi - \frac{1}{2}\phi$ with respect to the largest principal stress direction, is given by (Sibson 1974; Ranalli & Yin1990):

$$(\sigma_1 - \sigma_3) = (R-1)\,\rho g z(1-\lambda) \qquad \text{(1a, thrusting; } \sigma_1 \text{ horizontal, } \sigma_3 \text{ vertical)}$$

$$(\sigma_1 - \sigma_3) = \frac{(R-1)}{R}\,\rho g z(1-\lambda) \qquad \text{(1b, normal faulting; } \sigma_1 \text{ vertical, } \sigma_3 \text{ horizontal)}$$

$$(\sigma_1 - \sigma_3) = \frac{R-1}{1+\delta(R-1)}\,\rho g z(1-\lambda) \qquad \text{(1c, strike slip faulting, } \sigma_1, \sigma_3 \text{ horizontal)}$$

where ρ is density, g acceleration of gravity, z depth, λ is the ratio of pore fluid pressure to overburden pressure, R is principal effective stress ratio $(\tilde{\sigma}_1/\tilde{\sigma}_3) = (\sigma_1 - \lambda\rho g z)/(\sigma_3 - \lambda\rho g z)$, and factor δ comes from substituting the intermediate principal stress $\sigma_2 = \sigma_3 + \partial(\sigma_1 - \sigma_3)$, $0 < \partial < 1$. The R is related to the coefficient of friction $\mu (= \tan(\phi))$ as:

$$R = [(\mu^2 + 1)^{1/2} + \mu]^2 \qquad (2)$$

Friction parameters hardly depend on rock type. Following Ranalli (1987), we adopt a value of $\phi = 37°$ ($\mu = 0.75$) and we assume hydrostatic pore fluid pressure ($\lambda \approx 0.4$).

At high temperatures creep deformation mechanisms dominate in the lithosphere. Numerous laboratory studies show that the critical principal differential stress necessary to maintain a given steady-state strain-rate, is a function of a power of the strain-rate and varies strongly with rock type and temperature (e.g. Goetze & Evans 1979; Carter & Tsenn 1987; Tsenn & Carter 1987). Here we adopt power-law creep (e.g. Carter & Tsenn 1987):

$$(\sigma_1 - \sigma_3) = A^{-1/n}\,\dot{\varepsilon}^{1/n}\,e^{(E/nRT)} \qquad (3)$$

where $\dot{\varepsilon}$ is strain-rate, A, n and E flow parameters, T is absolute temperature and R the gas constant. Compilations of flow parameters for various rock-forming minerals show that power-law creep flow parameters are strongly controlled by silica content (e.g. Carter & Tsenn 1987). Felsic rocks (e.g. granite) show low critical differential stress values compared to mafic rocks (e.g. olivine), under similar conditions of strain-rate and temperature.

Strength profiles

For a given tectonic environment (thrusting, normal faulting or strike-slip faulting), depth, flow properties, temperature and strain-rate, the lowest critical stress difference $\sigma_1 - \sigma_3$ comparing equations (1) and (3), gives a rheological strength. For principal stress difference below this strength value, the imposed strain-rate will not occur. As a function of depth, the strength constitutes a yield envelope, or a so-called rheological profile, which provides much information on the spatial distribution of lithospheric deformation (e.g. Ranalli 1987). Vertically integrated lithospheric strength, gives an estimate of intraplate force required to overcome lithospheric deformation (e.g. Kusznir & Park 1987; Ranalli & Murphy 1987).

The construction of rheological profiles requires knowledge of the thermal regime, strain-rate, lithospheric material distribution and its flow properties. To construct rheological profiles typical for a large number of phanerozoic basins, we adopt steady-state geotherms with surface heat flow ranging from Q_s = 60–80 mW m^{-2} and for a crustal thickness of 32 km (cf. Pollack & Chapman 1977; Cermak & Bodri 1991). In rheological models of the crust, adopted flow properties commonly correspond to 2 or 3 layers of different rock types (e.g. Ord & Hobbs 1989; Braun & Beaumont 1989). Here we adopt 2 layers, a wet quartzite rheology for the upper part of the crust and a wet diorite rheology for the lower crust (Carter & Tsenn 1987). For the subcrustal lithosphere we incorporate a wet dunite rheology (Carter & Tsenn 1987). For strain-rate a value of 10^{-15} s^{-1} is used, corresponding to intermediate values for intraplate deformation (e.g. Cloetingh & Banda 1992). The resulting rheological profiles for extension and compression (Fig. 1), are marked by upper crustal brittle strength, lower crustal ductile strength and brittle and/or ductile strength for the subcrustal lithosphere (Fig. 1). The relatively low strength of the lower crust predicted by these models, is a

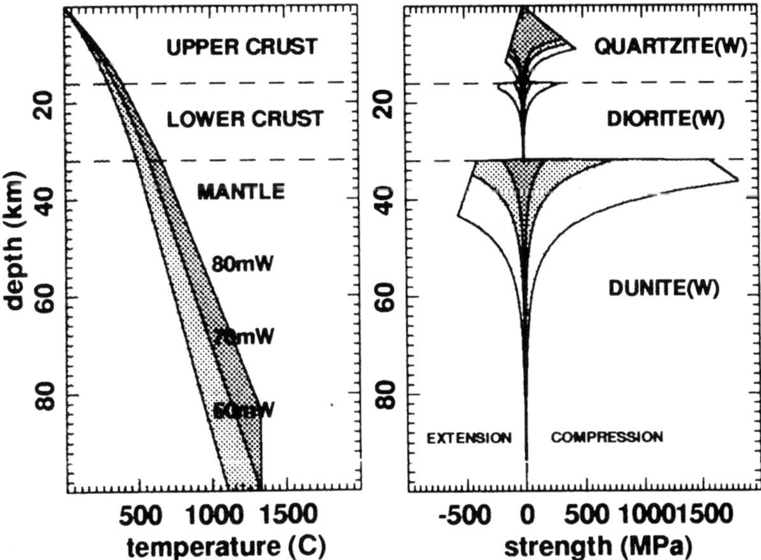

Fig. 1. Steady-state geotherms and rheological profiles for various surface heat flow values (60, 70, 80 mW m^{-2}) and 32 km crustal thickness, representative of Phanerozoic continental lithosphere. Adopted strain-rate is 10^{-15} s^{-1}.

consequence of the adopted wet diorite rheology, which is characterized by relatively low estimates of ductile strength compared to other rock types. The weak ductile rheology in the lower crust generally agrees well with shallow earthquake depth distributions and high values of seismic attenuation in the lower crust (Meissner 1986).

Rheological stratification

The rheological profiles in Fig. 1, show a pronounced layering of relatively strong (mostly brittle) upper crust, a weak (mostly ductile) lower crust, and a strong (brittle and ductile) subcrustal lithosphere. This layering predicted from extrapolation of rock mechanics data agrees well with interpretation of geophysical and geological data.

Deep seismic images of rifted continental lithosphere for different areas (e.g. Meissner 1986; Brun *et al.* 1991; Bois 1993) support the existence of a rheological stratification. They show that, acoustically, the upper part of the continental crust has a rather transparent character with localized bands of dipping reflectors, interpreted as fault zones. On the other hand the lower crust is characterized by high reflectivity of discontinuous horizontal or gently dipping reflectors, inferred to be layered lower crust. The subcrustal lithosphere shows few reflective events; however, in some cases clear dipping reflectors offsetting the layered lower crust are observed (Flack *et al.* 1990). Kinematic deformation models for lithospheric extension derived from seismic profiles (e.g. Pinet & Coletta 1990; Reston 1990; Brun *et al.* 1991) are characterized by faulting along relatively steep faults in the upper crust, relatively low angle shear in the lower crust, and sometimes include activity of subcrustal lithospheric faults or shear zones (Fig. 2). Geological field observations also indicate activity of localized deformation in upper mantle rocks (Vissers *et al.* 1991).

The transition in the crust from steep faults to low angle shear is abrupt and can be explained as a transition from brittle to ductile deformation mechanisms (Jackson 1987; Pinet & Coletta 1990). This is supported by shear stress balance considerations (Mandl 1988) at the interface of low angle sheared lower crust and the overlying high angle sheared upper crust, indicating an abrupt transition from relatively strong upper crust to relatively weak lower crust. Both finite element models (Melosh 1990) and analogue models (Vendeville *et al.* 1987; McClay & Ellis 1987) support this feature.

The subcrustal deformation characteristics are poorly constrained from continental lithosphere data. However, oceanic and continental lithosphere flexure studies assessing flexural parameters from rheological assumptions of mantle material (e.g. Goetze & Evans 1979; Kooi *et al.* 1992), clearly support the validity of extrapolation from rock mechanics data, predicting a relatively strong subcrustal lithosphere.

Weakness of crustal scale faults and shear zones

The rheological stratification of continental lithosphere corresponds well to its main deformation characteristics. However, upper crustal faults and subcrustal faults and shear zones are marked by rheological properties which probably correspond to considerably lower strength values than those predicted from bulk properties of the lithosphere.

According to equation (1), faults in the upper crust form at an angle of $\alpha = \frac{1}{4}\pi - \frac{1}{2}\phi$ with respect to the largest principal stress direction σ_1. For pre-existing faults, which have a different orientation, the critical principal stress difference for frictional sliding is higher than for new faults (Mandl 1988; Ranalli & Yin 1990). Seismic data indicate that most upper crustal faults are planar during their

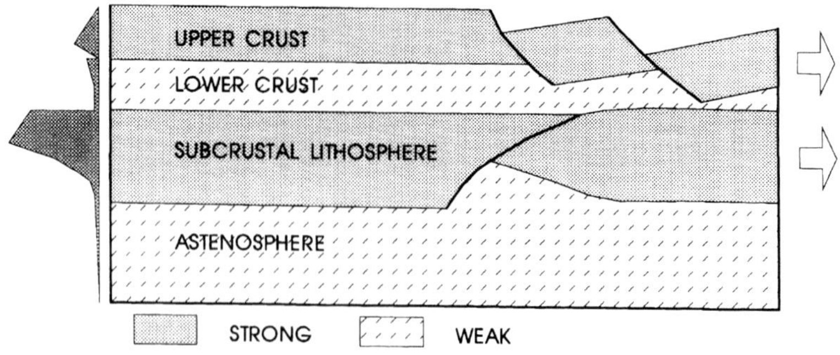

Fig. 2. Reconstruction of continental lithospheric extension, based on deep seismic reflection profiles (after Reston 1990).

formation (McGeary & Warner 1986; Jackson 1987; Kusznir *et al.* 1991) and can progressively rotate during extension according to the domino fault model (Ransome *et al.* 1910; Le Pichon & Sibuet 1981; Wernicke & Burchfiel 1982; Barr 1987). Finite element and analogue models for extension support formation of steeply dipping upper crustal normal faults, which are remarkably planar in origin (Vendeville *et al.* 1987; McClay & Ellis 1987; Cundall 1990). Both observations and model results show that listric basin faults can be generated due to progressive rotation of the crustal planar faults and growth faulting, interacting with buoyancy forces (Vendeville & Cobbold 1988; Buck 1988, Vendeville 1991). However, in some cases, inferred listric crustal faults do not agree with the domino fault model and require formation of initially listric faults (Wernicke & Burchfiel 1982; Gibbs 1984). In a similar fashion to the domino model, initially listric faults probably rotate during extension (Le Pichon & Chamot-Rooke 1991).

The rotation of crustal faults and continuous strain localization during extension, require either a change of frictional properties once these faults are formed, or a change in orientation or magnitude of principal stresses close to the rotating faults. These conditions also apply for reactivation of normal faults during inversion (e.g. Hayward & Graham 1989), and transpressional and transtensional reactivation of strike slip faults (e.g. Ziegler 1990). Sibson (1985) explained the continuous activation of extensional faults in terms of reduction of cohesion. However, for low initial cohesion values, like the ones adopted in equation (1), this is not a plausible mechanism. Alternatively, reduction of friction angle ϕ has been considered (Ranalli & Yin 1990; Yin & Ranalli 1992). Sand box experiments on reactivation of extensional faults during inversion, support decrease of friction angle, in the models induced by alignment of mica flakes along extensional faults (McClay & Ellis 1987). In addition, numerical mechanical models for extension demonstrate a decrease of magnitude of the principal stresses close to previously formed faults (Cundall 1990). These indicate that stress deviations also may play an important role in localization. Evidently, the decrease in principal stresses close to such faults can be described in terms of an additional reduction of friction angle in equation (1).

The weakness of subcrustal faults or shear zones is indicated by localized deformation. If behaving brittle, subcrustal faults can be characterized by weakening similar to the upper crustal faults. In the ductile regime, various strain softening mechanisms support weakening (e.g. Drury *et al.* 1991). Furthermore, recent studies on the evolution of integrated strength of the lithosphere (Van Wees & Stephenson 1995) indicate that reduction of subcrustal strength is neccessary in order to explain localization of basin deformation on long time-scales of the order of 300 Ma.

Sub-basin-scale (local) stress and strain concentrations: 2D finite element models

As outlined in the previous section, upper crustal faults, the lower crust, subcrustal faults and shear zones can be defined as (pre-existing) zones of weakness, in the sense that they indicate low strength relative to surrounding rocks. Linkage and discontinuities of these weak zones and their actual shape, are bound to have a large impact on styles of lithospheric deformation. Assuming a weak coupling of subcrustal and upper crustal deformation, which is suggested for a large number of Phanerozoic basins (Fig. 2) (Van Wees & Cloetingh 1994; Burov & Diament 1995), the structural expression of basins seems therefore to be primarily controlled by the movements along weak upper crustal faults, interacting with the rheological behaviour of the upper and lower crust and basin infill. In this and the following section, we will explore respectively the effects of these interactions on local stress and strain concentrations with a 2D finite element technique and basement warping on a more regional scale with a recently developed 3D flexure model (Van Wees & Cloetingh 1994).

Stress and strain concentrations

After a stage of tectonic quiescence and sedimentation, weak upper crustal faults, grading into unfaulted sediments and soling in the lower crust, will induce large stress concentrations occurring at the tip of the weak faults when reactivated (Fig. 3). This will particularly occur during multiple phases of rifting or inversion. In addition, large bending stresses can occur in the case of listric faults (Fig. 3). The stress concentrations are likely to have a strong control on the locus and signature of new faults during progressive basin deformation for both extension and compression (inversion).

Here, we examine the effects of stress and strain concentrations induced by movements along pre-existing faults, using a 2D elasto-plastic plane-strain finite element model, and compare them with observations and alternative modelling results.

The finite element model we use is an extension of the TECTON code originally developed by Melosh & Raefsky (1981). A typical model configuration and pertinent model parameters are given in Fig. 4. Pre-existing weak zones are represented by frictionless slippery nodes (Melosh & Raefsky 1981). Failure induced by incremental movements

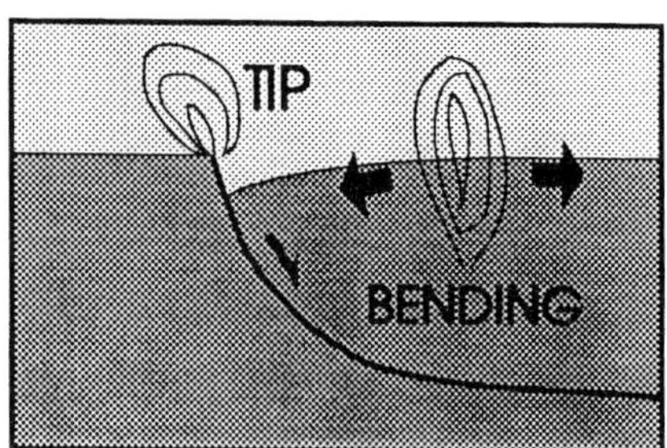

Fig. 3. Stress concentrations due to slip along weak upper crustal planar and listric faults.

along the slippery nodes obeys the Coulomb-Navier criterion, which is incorporated by adopting a non-associated flow-law with hardening of the elastic moduli in the element stiffness matrix (Zienkewicz & Taylor 1991). For the non-associated flow-law, friction angle $\phi = 35°$ and dilatation angle $\psi = 5°$. The adoption of a non-associated flow-law with relatively low dilatation angle agrees with other finite element models on rock-mechanics (e.g. Cundall 1990; Nieuwland & Walters 1993), preventing an unrealistically large plastic volume increase. Predicted stresses which according to the yield criterion are outside the yield surface are corrected to the yield surface adopting a return algorithm (e.g. Zienkiewicz & Taylor 1991).

Finite element model results

To study the effects of extension along a pre-existing weak planar basement fault, we adopt a fault dip of *c.* 30°, grading into 1000 m of unfaulted sediments, and apply 54 m of hangingwall displacement in 700 increments (first 200 increments account for 4 m of displacement). The results depicted in Fig. 5a indicate a symmetric pattern of basin floor warping, bounded by basin faults (plastic shear bands in Fig. 6a) rooting in the pre-existing fault. These results agree both with observations (Yielding *et al.* 1991; Fig. 5b), and predictions from analogue and distinct element models (Buchanan & McClay 1991; Saltzer & Pollard 1992). It is noted that the orientation of the shear bands are sensitive to the mesh orientation.

For extension along a listric fault, we adopt a curved geometry for the basement fault, dipping from *c.* 70° to subhorizontal and grading into 500 m of unfaulted sediments. We apply a horizontal hanging-wall displacement of 69 m in 1500 increments (first 200 increments account for 4 m). The results in Fig. 6a show a more asymmetric

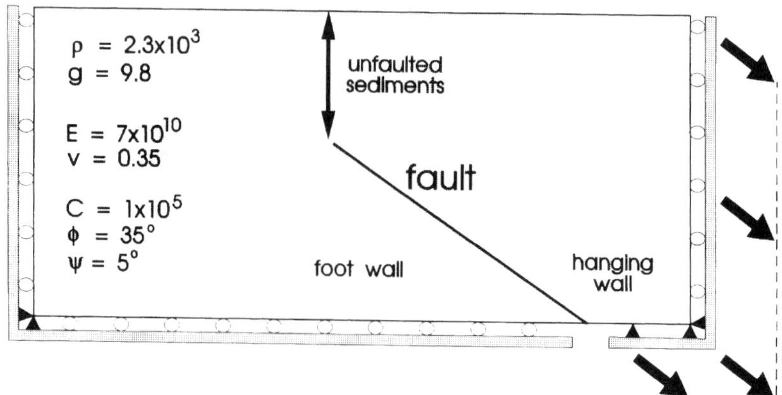

Fig. 4. Cross-section showing finite element model parameters and boundary conditions. g = gravity, ρ = density of sediments and basement. Elastic parameters: E = Young's Modulus, v = Poison's Ratio. Plastic parameters: C = Cohesion, ϕ = friction angle, ψ = dilatation angle.

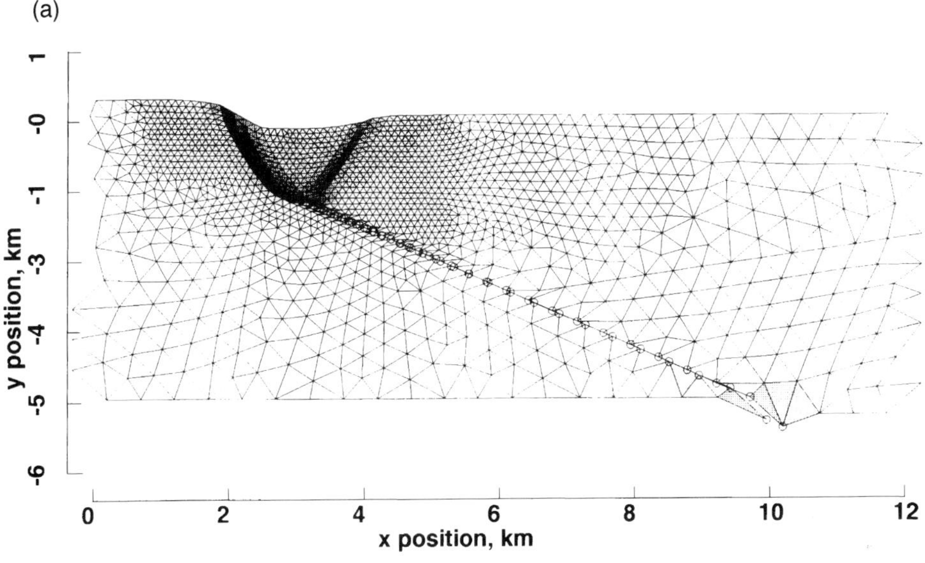

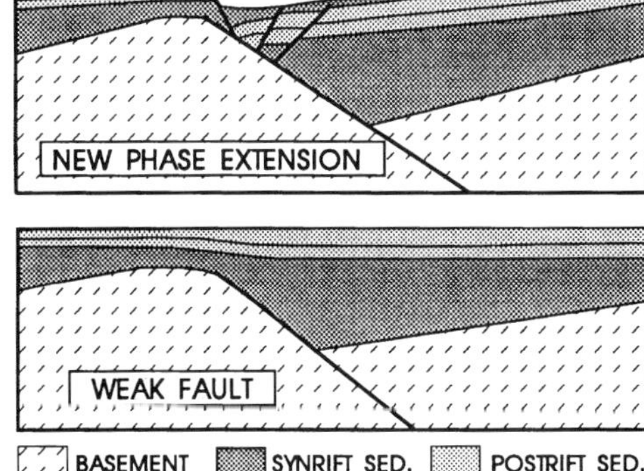

Fig. 5. (**a**) Finite element model for extensional plastic deformation at the tip of a frictionless planar fault. Intensity of subsequent plastic flow is indicated by gray shading. Deformation is exaggerated 10 ×. Model and material parameters as in Fig. 4. (**b**) Typical deformation features observed in rifted basins (after Yielding *et al.* 1991); compare with finite element results in (a).

pattern of basement warping, compared to planar faulting and the development of hanging-wall collapse structures. These features correspond to observed fault patterns and basement bathymetry evolving above listric upper crustal faults (Wernicke & Burchfiel 1982; Gibbs 1984) and agree with results from sand box models (McClay 1989). To examine the effects of inversion on listric faults, we use the same fault geometry as for extension but now apply opposite hanging-wall displacements (170 m in 900 increments, of which first 10 m in 100 increments). The results (Fig. 6b), show buttress effects and footwall shortcuts, which agree with geological observations (Hayward & Graham 1989) and sand box models (McClay 1989; Buchanan & McClay 1991).

Basinwide (regional) basement warping: 3D flexure models

In general, rifted basins are formed at a much larger scale than individual basement fault blocks. Basement bathymetry inferred from seismic reflection profiles reveals that rifted basins are characterized by complex block structures, intersected by a large number of steeply dipping faults, with spacing in the order of 10 km (Rosendahl 1987).The basin block structure is bounded by so-called border faults (Versfelt & Rosendahl 1989), which relative to the other rift block faults show markedly greater fault displacements (Fig. 7). The displacements vary along-strike of the border faults and in map view they have a slightly to pronounced

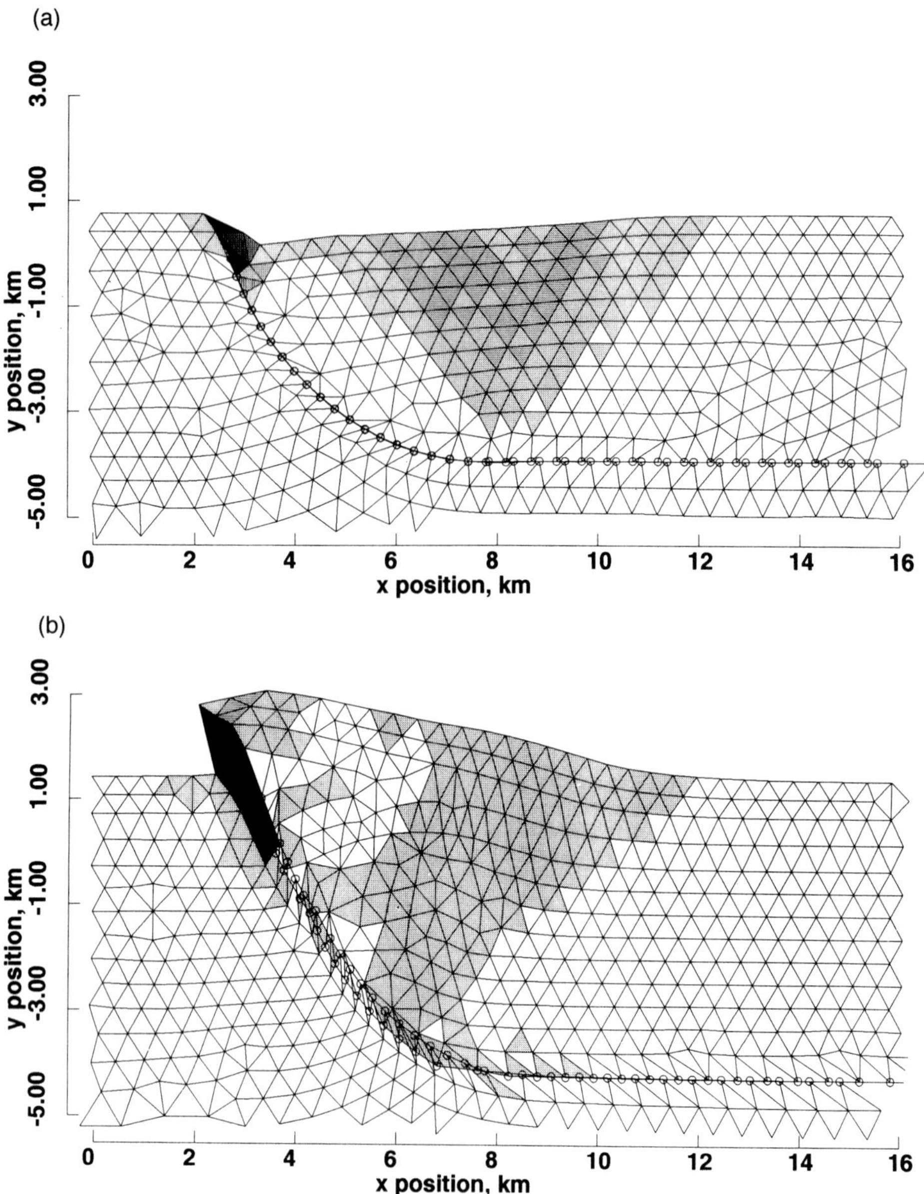

Fig. 6. Finite element models for (**a**) extensional and (**b**) compressional plastic deformation at the tip of a frictionless listric detachment. Model, material and figure conventions as in Figs 4, 5.

arcuate shape (Rosendahl 1987; Fig. 7). It has been noted that the alignment of border faults has a large influence on patterns of basement warping (e.g. Rosendahl 1987). The most elementary basin geometry is a half-graben, in which only one side of the basin is bounded by a border fault. More complex basin geometries are the result of linkage of a number of border faults and associated half-grabens. As pointed out by Versfelt & Rosendahl (1989), rift block faults, outside the border faults, play a passive role, and merely accommodate deformation.

Some authors have suggested that large basin floor geometries are controlled by crustal-scale listric faults (e.g. Kusznir *et al.* 1987). However, more recent two-dimensional basin modelling

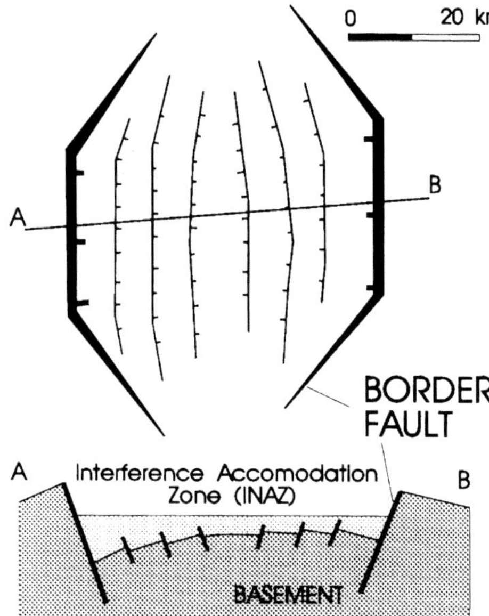

Fig. 7. Example of linked half-grabens bounded by border faults (modified from Rosendahl 1987). Opposing geometries create INterference Accommodation Zones (INAZ).

studies indicate that the basement warpings are most likely a result of planar faulting with tilting of the basement induced by buoyancy forces (Fig. 8) (Marsden *et al.* 1990; Kusznir & Ziegler 1992). In the latter studies, the upper crustal rift blocks are considered as an elastic entity, floating on the lower crust and/or subcrustal ductile substratum. 2D modelling results show that the wavelength of curvature of basement topography between border faults are closely related to the flexural characteristics of the upper crust in terms of effective elastic thickness (EET) (e.g. Kusznir *et al.* 1991).

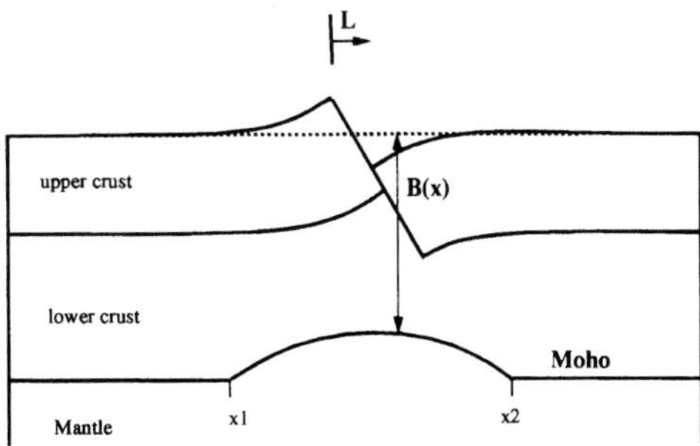

Fig. 8. Representation of lithospheric extension by faulting in the upper crust and by ductile flow in lower crust and mantle (after Kusznir *et al.* 1991).

Evident basin features in this respect are convex curvatures of basement topography, which are commonly referred to as Low Relief Accommodation Zones or Interference Accommodation Zones (Fig. 7) (Rosendahl 1987; Versfelt & Rosendahl 1989).

The relative weakness of crustal-scale faults compared to surrounding rock, most likely provides a driving mechanism to induce the observed basement flexure. Consequently, border normal faults, and indeed any type of weak crustal-scale fault would be expected to exert a strong control over upper crustal flexure during extension and compression.

Here we will analyse basin flexure patterns in both extensional and compressional settings, using a recently developed 3D basin modelling technique, incorporating flexure and planar faulting (Appendix; Van Wees & Cloetingh 1994). This model, focusing on the interaction of border faulting and flexure of the upper crust and its controls on three-dimensional basement warping, will allow us to appraise the full three-dimensional complexity of basin shape, evident from high quality datasets.

Extensional setting: Lake Tanganyika Rift Zone (east Africa)

To study the interaction of faulting and 3D flexure in an extensional regime, we focus on the northern part of the Lake Tanganyika Rift Zone (Fig. 9) which forms part of the north–south trending Western Branch of the East African Rift System. Seismic data show that the rifted area is composed of a number of mostly Cenozoic half-grabens, bounded by planar faults, linked in a complex way by accommodation zones (Fig. 11) (Rosendahl *et al.* 1992). The grabens were formed under E–W extensional stress conditions, with minor components of oblique slip (Delvaux *et al.* 1992; Ring *et al.* 1992). The N–S trending border faults indicate mainly dip-slip behaviour, whereas the NW–SE trending border faults, most likely rooting in Proterozoic dislocation zones (Rosendahl *et al.* 1992), accommodate half-graben polarity changes (Rosendahl *et al.* 1992).

For a three-dimensional flexure model of the northern part of Lake Tanganyika (Fig. 9), we adopt a 60×80 grid (spacing $\Delta x = \Delta y = 5$ km). Focusing on the wavelength of basement tilting patterns, displacements along the minor rift block faults are neglected. Minimum throw values for the border faults have been constrained from water depth and sediment thickness derived from seismic data (Morley 1988; Rosendahl *et al.* 1992; Fig. 9a). In the depth conversion a low mean seismic velocity of 2.75 km s is adopted from Morley

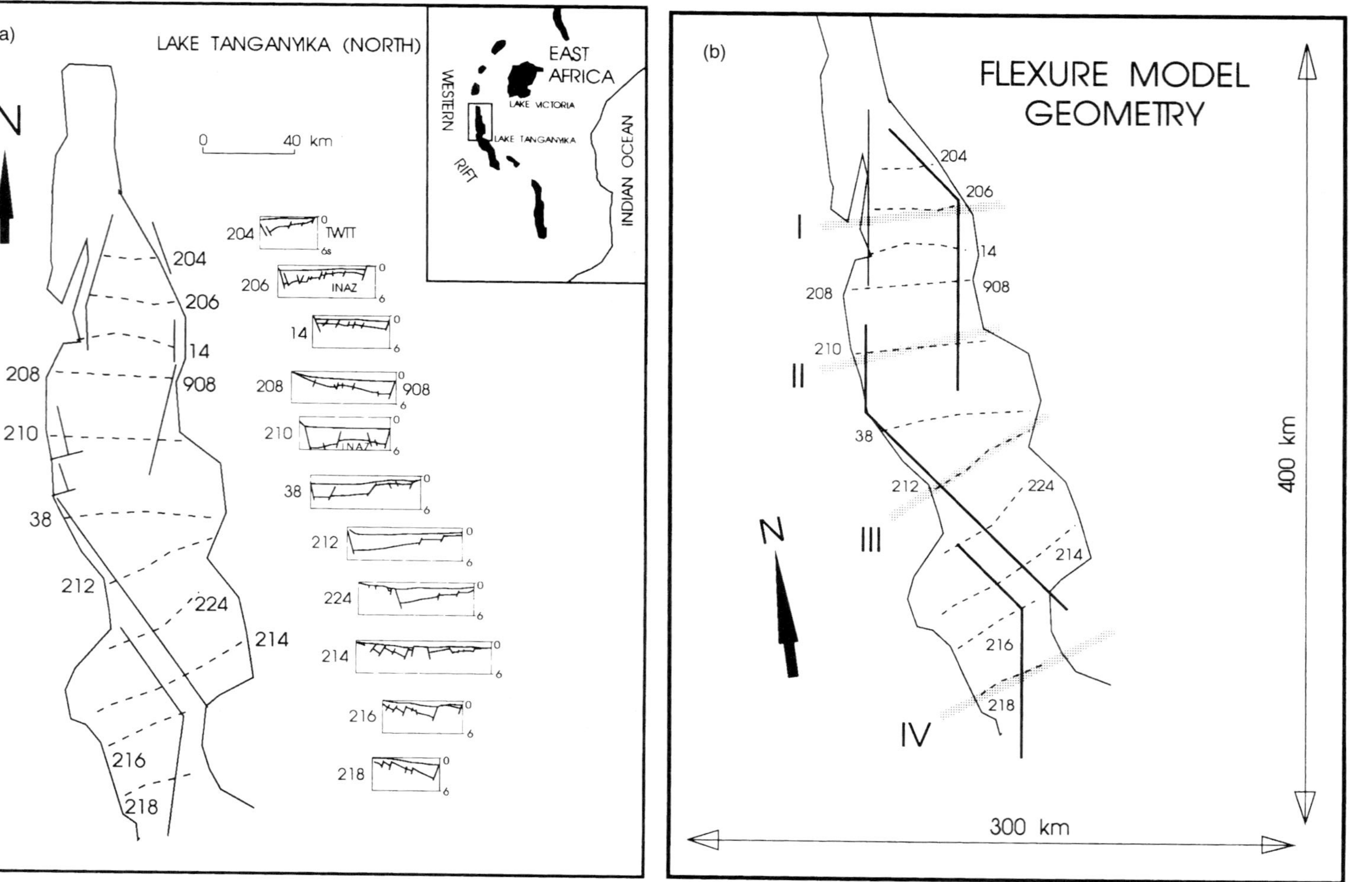

Fig. 9. (**a**) Basin configuration of the northern half of Lake Tanganyika, showing the architecture of half-graben and accommodation zones (after Morley 1988; Rosendahl *et al.* 1992). Bold solid lines mark border faults. Top-sediment and top-basement reflectors inferred from seismic sections are depicted. Interference accommodation zones are indicated by INAZ. (**b**) Three-dimensional finite difference fault model geometry (300 × 400 km, grid spacing $\Delta x = \Delta y = 5$ km) for northern half of Lake Tanganyika, corresponding to basin architecture depicted in (a). Bold solid lines denote border faults.

(1988), and effects of moderate dip of reflectors have been neglected. Additional amounts of throw along the border faults are reflected by present-day topographic elevations of footwall blocks bounding Lake Tanganyika, which range up to 1000 m or more above lake level. In the absence of data on erosion of footwall blocks, precise estimates of this component cannot be made.

To incorporate effects related to crustal thinning, fault dips of 60° are adopted, and we assume a crustal thickness of 40 km, values that agree with data published by Ebinger *et al.* (1991). For the β distribution, we assume a radius r_1 for the spherical cap function (see Fig. A1) of 50 km. It is assumed that depressions ($w < 0$) are filled with low density material ($\rho = 2.0 \times 10^3$ kg m^{-3}), accounting for water and sediment infill, and that no erosion of basement culminations ($w \geq 0$) occurs.

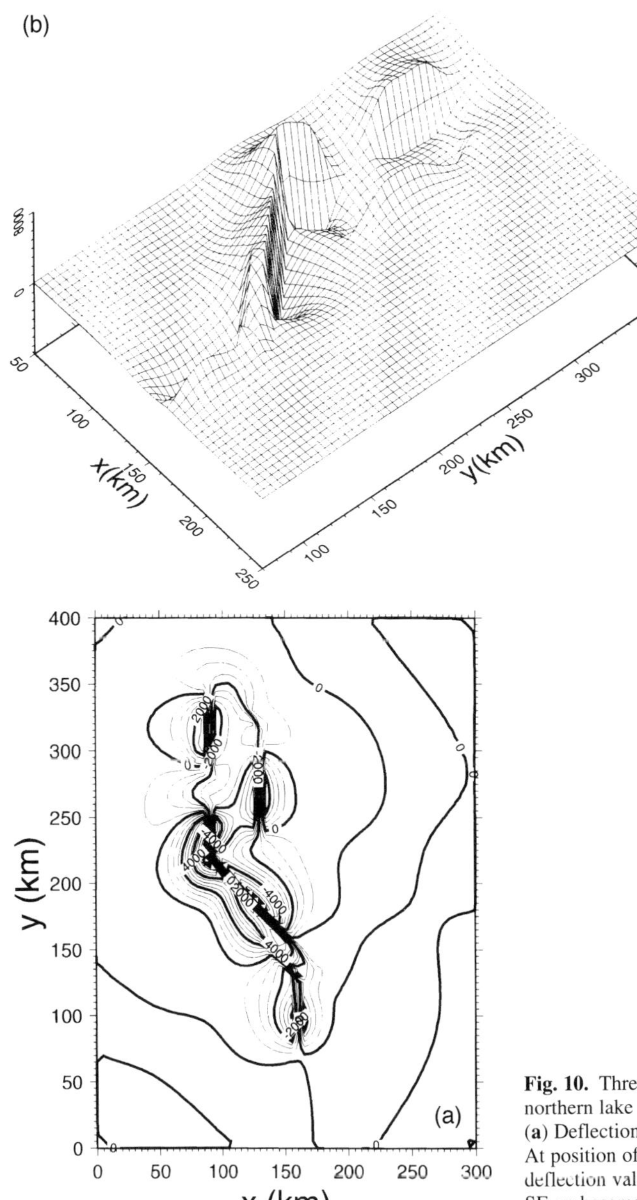

Fig. 10. Three-dimensional basement flexure for northern lake Tanganyika for an EET of 5 km. (**a**) Deflection contour map, interval is 500 m. At position of faults mean deflection is adopted for deflection value. (**b**) Three dimensional view from SE on basement deflection depicted in (a).

Starting from the minimum throw estimates derived from seismic data, throw values have been increased to fit migrated basement depths for four seismic sections (Figs 9, 10 & 11). Figure 11 shows flexural response results for different values of EET for a fixed throw distribution which agrees closely with observed basement depths. The three-dimensional flexural response for an EET of 5 km is shown in Fig. 10a and 10b.

The results show that the flexure model incorporating the effects of border faulting is well capable of reproducing observed 3D basement warping patterns. As is evident from Fig. 11 the *wavelength* of calculated basement warping is largely controlled by EET.In profile 206, the wavelength of the observed interference accommodation zone, marked by a convex shape in basement tilting, corresponds best to an EET of 3 km. For larger values of EET the flexural wavelength is too large to predict a convex shape. Also for the southern lines, the lateral variations of basement tilting best fit an EET of 3 km.

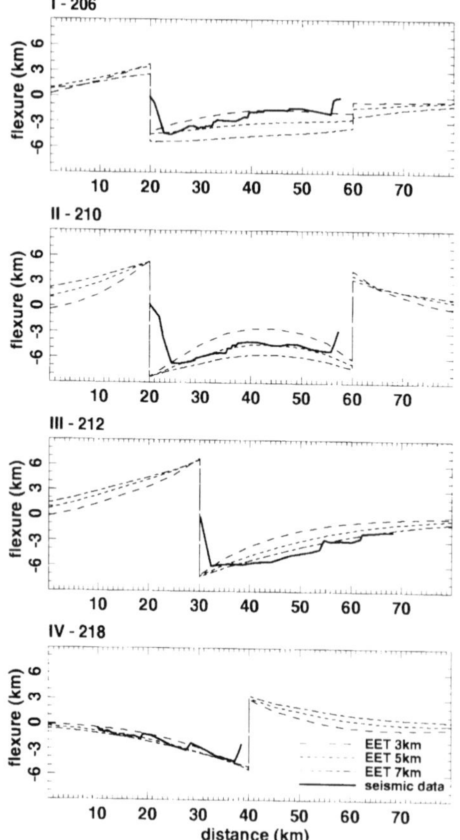

Fig. 11. Sections of three-dimensional flexure models of northern Lake Tanganyika for three values of EET (3, 5 and 7 km) and a fixed throw distribution, which agrees closely to observed basement depths in seismic sections 206, 210, 212 and 218 (after Van Wees & Cloetingh 1994). Locations of sections are indicated in Fig. 9b.

From a comparison of *amplitudes* of calculated deflection with observed ones, the best fitting EET values tend to be higher: 5 km for profile 210 and 7 km for profile 212. These values differ from the EET values derived from the flexural wavelength. This discrepancy can probably be explained by the fact that the model is not taking into account plastic deformation in between border faults, characterized by displacements along minor faults (Fig. 9a). Furthermore, the model fails to predict present-day submersion of the horst block left, right and in the centre of seismic sections 224, 216 and 224 respectively. This effect can be explained by the lack of basement erosion in the model and the relative large displacements along minor faults, adjacent to the border fault.

Summarizing, the modelling results indicate that the EET for upper crustal flexure in the northern part of Lake Tanganyika is characterized by a low value of about 3–5 km. The low EET values correspond well to estimates of EET for basement warping patterns in 2D flexure studies on extensional basins (Buck 1988; Stein *et al.* 1988; Kusznir & Ziegler 1992). Kusznir *et al.* (1991), indicated that the relatively low values of EET probably reflect a decrease relative to values estimated for the lithosphere from basin margins, as a result of effects of plastic bending. As argued by Van Wees & Cloetingh (1994), it is more likely that the relatively high EET estimates (e.g. Kooi *et al.* 1992; Karner *et al.* 1992) relate to deep crustal or subcrustal lithosphere flexure, (partly) decoupled from the upper crust, characterized by much lower EET values (Van Wees & Cloetingh 1994).

Compressional setting: the Central System and adjacent Tajo Basin (central Spain)

Tertiary uplift and exhumation of the Central System occurred simultaneously with subsidence and infill of the adjacent Tajo and Duero Basins in central Spain (Figs 12 & 13). During this process NE–SW trending Late Variscan crustal-scale faults, bounding the Central System, were reactivated in an overall convergent setting (Vegas & Banda 1982; Sopeña *et al.* 1988; Vegas *et al.* 1990). Balanced sections show a throw of several kilometres for these faults, and indicate a NW–SE shortening of at least 11%, taking into account internal thrusting and strike slip deformation in the Central System (De Vicente *et al.* 1992). In agree-

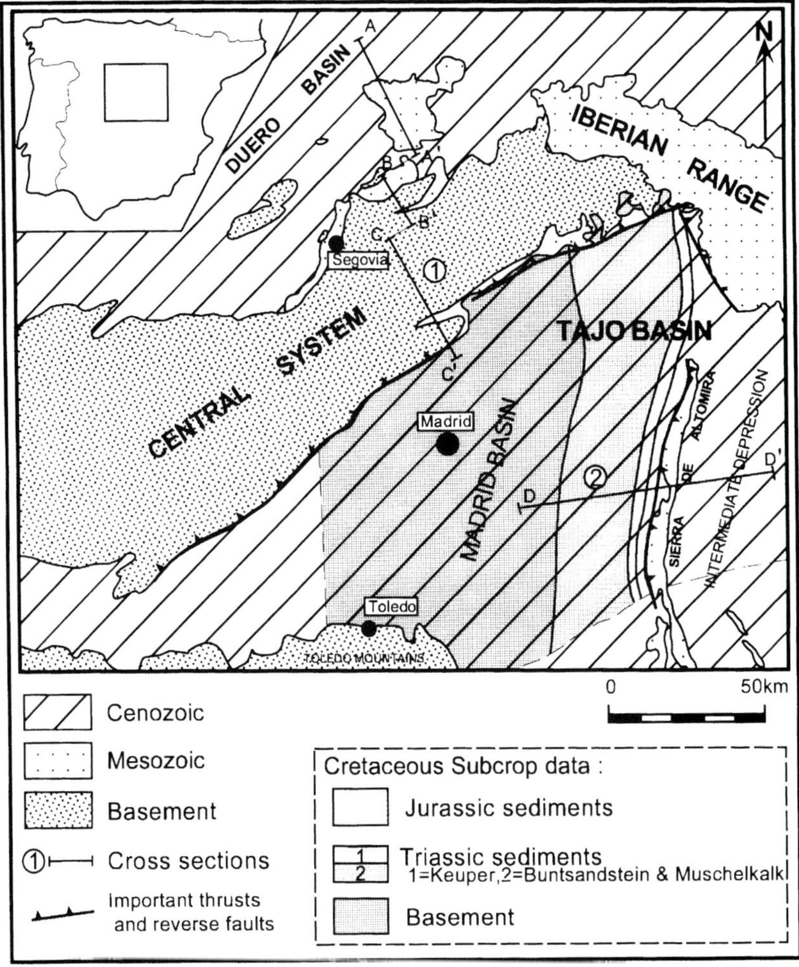

Fig. 12. Geological map of the Central System and the adjacent Duero and Tajo Basins in Central Spain. Basement consists of pre-mesozoic sediments and crystalline rocks (after De Vicente *et al.* 1992). The Tajo Basin is subdivided into the Madrid Basin and the Intermediate depression, separated by the Sierra de Altomira, a thin-skinned thrust belt where Mesozoic sediments are exposed. Cretaceous subcrop information (after Querol Muller 1989) reflects eastward deepening of the inverted Mesozoic Iberian Basin, presently exposed in the Iberian Range and Sierra de Altomira.

ment with shortening is the crustal thickening of 5 km inferred from seismic refraction below the Central System (Suriñach and Vegas 1988). Contemporary basin subsidence in the Duero Basin (e.g. Portero *et al.* 1983) and in the Tajo Basin (e.g. Calvo *et al.* 1990), provide an excellent setting for a study of 3D signature of basement deflections in response to fault-bounded uplift and crustal thickening in the Central System. Unfortunately, the geometry of the Duero Basin is not sufficiently constrained to compare 3D flexure modelling results with observations. We, therefore, focus the modelling on the Tajo Basin.

Seismic data of the Tajo Basin show a Mesozoic and Cenozoic sequence up to about 3500 m thickness (Querol Muller 1989; Figs 13 & 14b). Based on its stratigraphic development, the Tajo Basin is subdivided into the western Madrid Basin and the eastern Intermediate Depression, which are separated by the N–S trending Sierra de Altomira. The Sierra de Altomira pops up as an isolated thin-skinned thrust belt of latest Oligocene–Early Miocene age. Its formation is related to the final stages of inversion of the Mesozoic Iberian Basin, in which the sedimentary cover thickens gradually eastward from the Altomira Range (Fig. 13b) (Salas & Casas 1993; Van Wees & Stephenson 1995). The location of the Sierra de Altomira

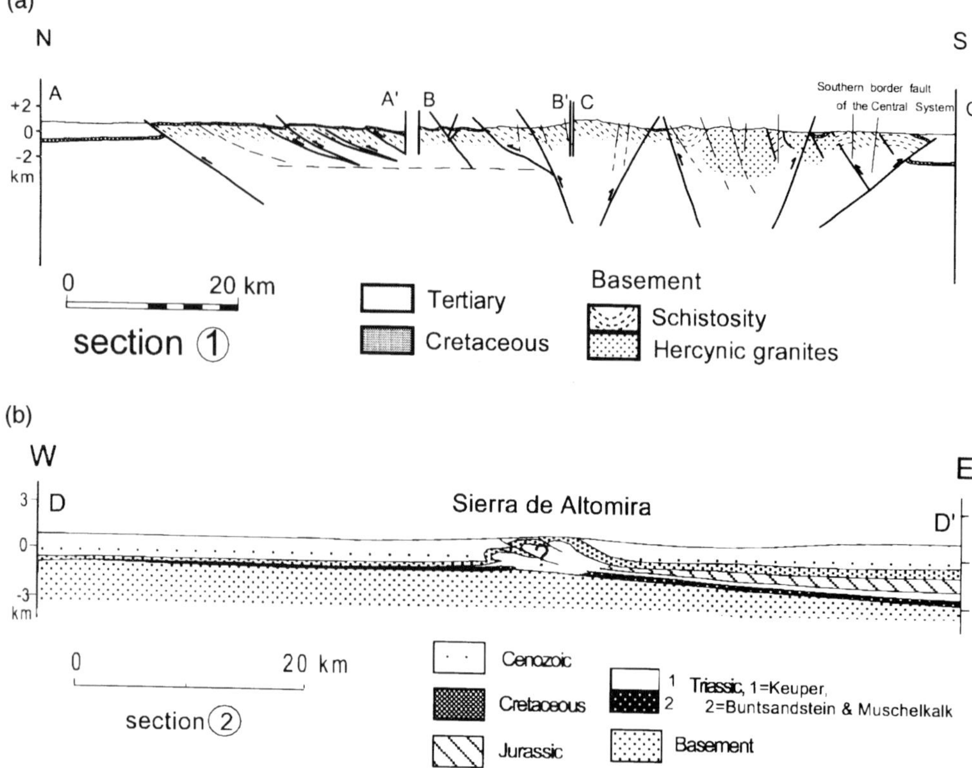

Fig. 13. Geological cross-sections 1 and 2 depicted in Fig. 12, illustrating the structural grain of the Central System and adjacent Tajo Basin. (**a**) Cross-section 1 through the Central System and adjacent basins, demonstrating border fault controlled uplift of the Central System in a compressive setting (after De Vicente *et al.* 1992). (**b**) Cross-section 2 across the Tajo Basin. The Sierra de Altomira forms a pop-up structure separating the western edge of the Iberian Basin to the east (Intermediate Depression) from the Madrid Basin in the west. As is evident from the section and Cretaceous subcrop information in Fig. 12, the location of the Sierra de Altomira is intimately linked to the westward termination of a sedimentary detachment (keuper evaporites).

coincides with the westward termination of eastward dipping Keuper evaporites (Figs 12 & 13b) (Querol Muller 1989), which form the most eminent cover detachment involved in inversion of the Iberian Basin (Viallard 1989). Evidently, the basement warping east of the Sierra de Altomira reflects mainly Mesozoic extension. On the other hand the Madrid Basin is hardly affected by Mesozoic extension, and Tertiary continental sediments range in thickness from 1500 m close to the Toledo Mountains and the Sierra de Altomira, up to about 3500 m close to the Central System (Fig. 14b) (Megias *et al.* 1983; Calvo *et al.* 1990).

In the Madrid Basin, Tertiary strata are not affected by large-scale folding and faulting (Racero Baena 1988). However, stratigraphic analysis shows a subdivision into a number of sequences, separated by (angular) unconformities. These show significant lateral facies contrasts as a result of relative uplift of the Central System, the Sierra de Altomira, and the Toledo Mountains (Megias *et al.* 1983; Calvo *et al.* 1989; 1990). The timing of uplift of the Toledo Mountains is not well constrained but probably of Paleogene age (Racero Baena 1988; Calvo *et al.* 1990). Micromammiferous and palynological data from borehole and outcrop data indicate that uplift of the Sierra de Altomira and Central System, marked by proximal alluvial sedimentation at the margins of the Madrid Basin, started in the Late Oligocene (Racero Baena 1988; Calvo *et al.* 1990). In the Sierra de Altomira, the uplift ceased in the Early Miocene, whereas the Central System shows continuous uplift and sediment supply into the Middle Miocene (Calvo *et al.* 1990).

The basement depth distribution in the Madrid Basin, showing a Tertiary downwarping in the vicinity of the Central System (Fig. 14b), indicates

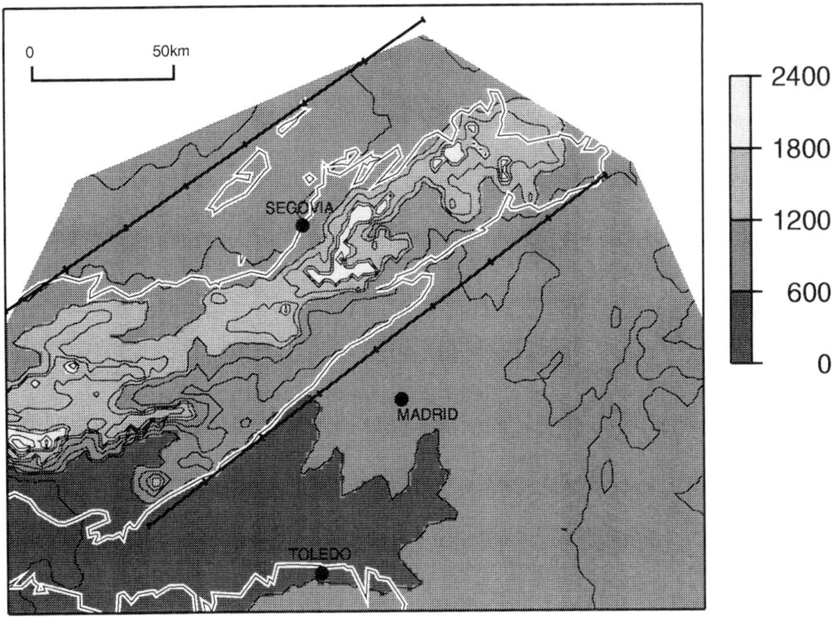

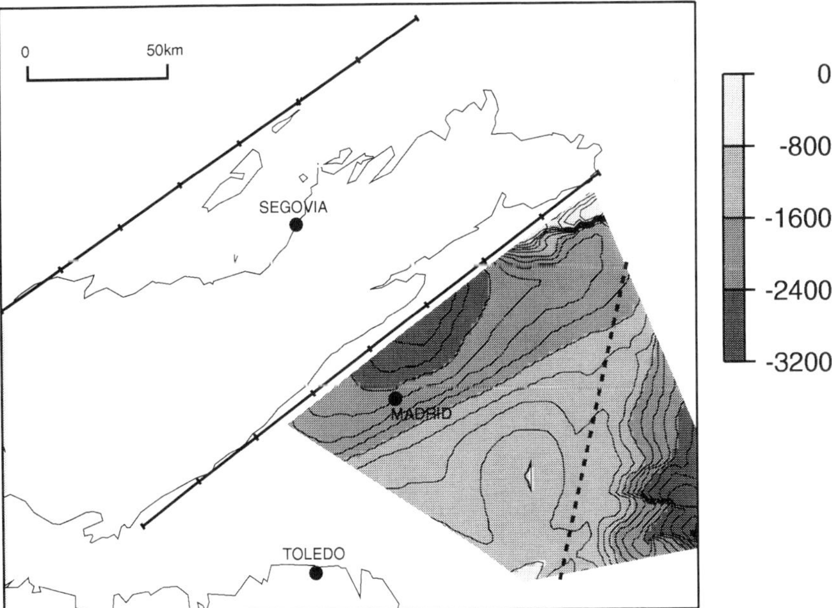

Fig. 14. Basement (pre-Mesozoic) elevation maps of the Central System and adjacent Tajo Basin. (**a**) Topographic elevations, contoured in 200 m interval. Inside areas of exposed basement (denoted by lines) basement minimum values of uplift correspond to present-day topography up to 2400 m. (**b**) Elevation of migrated top-basement reflector in the Tajo Basin, contoured in 200 m interval (after Querol Muller 1989). The dashed line marks the transition from thick Mesozoic cover (east, Sierra de Altomira and Intermediate Depression) to thin Mesozoic cover (west, Madrid Basin). In (a) and (b), the two straight lines bordering the Central System represent border faults adopted in the flexure model.

that basement subsidence is related to uplift of the Central System.

Tectonic subsidence

The basement subsidence values in the Madrid Basin constitute two components, one related to isostatic compensation of topographic loads and associated sediment infill and a tectonic component, which may reflect flexural compensation.

The contribution of topographic loading and sediment infill, can be eliminated by calculating tectonic subsidence, using the backstripping technique (cf. Steckler & Watts 1978). Therefore, we calculated the tectonic component of vertical movements of the basement for the cumulative sediment thickness data constrained by the topography and basement depth data available in the Tajo Basin (Fig. 14). It is noted that effects of topography can be incorporated in tectonic subsidence calculations by interpreting topographic elevations as a relative rise in sea-level. For the sediment infill, we incorporated an exponential porosity–depth equation, adopting parameters for a sand lithology according to Sclater & Christie (1980).

The resulting air loaded tectonic subsidence, as depicted in Fig. 15, shows large values up to 600 m at the depocentre of the Madrid Basin, and in the eastern part of the Intermediate Depression. On the other hand, the Sierra de Altomira and northwestern part of the Madrid Basin indicate tectonic uplift of 200 m or more, whereas large parts of the Madrid Basin are characterized by moderate uplift.

The subsidence values in the Intermediate Depression are in agreement with Mesozoic extension. For other parts of the Tajo Basin, the subsidence and uplift values, are either a result of Tertiary crustal deformation, or a result of uncertainties in the adopted lithology, or they can be attributed to flexural loading.

To test the sensitivity of the tectonic subsidence values for variations in constant lithological parameters, we adopted a constant density sediment infill ρ_s = 2300 kg m^{-3}. The resulting maximum tectonic subsidence show values approximately 200 m lower than in Fig. 15. However, they do not indicate large deviations in the subsidence patterns. On the other hand, taking into account observed lateral facies contrasts, incorporating a relatively large abundance of low density evaporitic sediments in the centre of the Madrid Basin, results in a slight increase of subsidence values in the basin centre.

Alternatively variations in tectonic uplift and subsidence values in the Madrid Basin could be attributed to the isostatic consequences of Tertiary crustal thickening and thinning. Extensional crustal thinning explaining Tertiary tectonic subsidence, seems not likely, since extensional structures are lacking. The role of crustal thickening, explaining the tectonic uplift, is limited to the northwestern margin of the Madrid Basin and the Sierra de

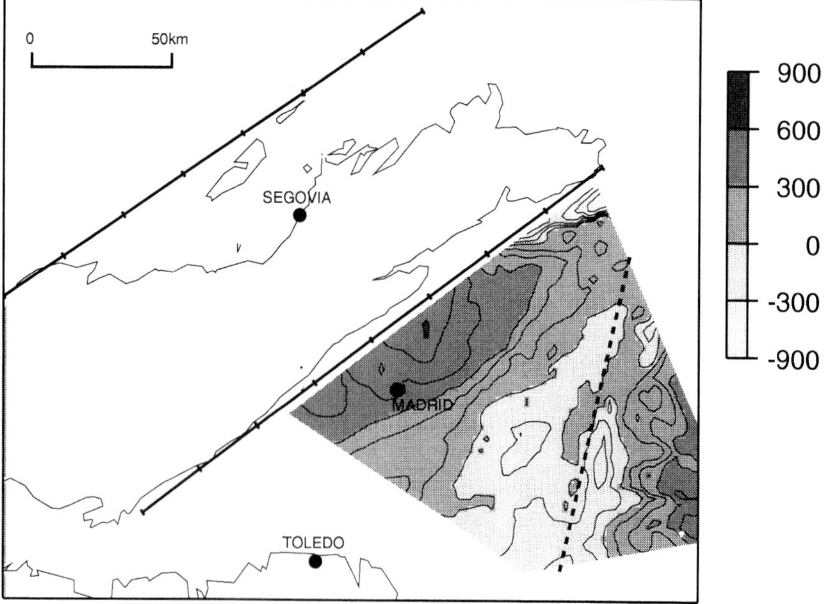

Fig. 15. Air loaded tectonic subsidence values for the Tajo Basin. Contour interval is 200 m.

Altomira region. In the Sierra de Altomira, thin-skinned shortening observed at the surface (Fig. 8), supports crustal thickening, whereas for the north-western margin, thickening can be attributed to a southward migration of the faults bordering the uplifted Central System (Querol Muller 1989). These results indicate that large tectonic subsidence values in the depocentre of the Madrid Basin cannot be related to crustal deformation. In the present study, it is assumed that they reflect a flexural loading effect.

Flexural modelling

For a three-dimensional flexure model of the Central System and adjacent basins, we adopt the same crustal and subcrustal parameters and boundary conditions as used in the Lake Tanganyika model. To incorporate border faults bounding the Central System, we assume planar faulting, adopting two linear discontinuities depicted in Fig. 14. In order to ensure that the faults coincide with grid nodes and grid connections, we use a non-equally spaced grid, mixed with rectangular and triangular nodal connections. In the vicinity of the faults, the grid spacing corresponds to approximately 3 km.

To incorporate effects of crustal thickening related to the northern and southern border faults, we assume for the β distribution,a spherical cap radius r_1 of 85 km (see Appendix). An initial crustal thickness of 32 km is adopted (Banda *et al.* 1983; Suriñach & Vegas 1988). To obtain a maximum crustal thickening in the centre of the Central System (Suriñach & Vegas 1988), the centre of the spherical cap is shifted perpendicular to the faults to positions which align along the central axis of the Central System. For the border faults, we adopt very low (thrust) fault angles of 30°, which represent a minimum value compared to estimates from balanced reconstructions (Fig. 13a) (Banks & Warburton 1991; De Vicente *et al.* 1992). Lacking well documented basement data for the Duero Basin, a constant vertical displacement of 2000 m is assumed for the northern fault. This is a relatively large value according to balanced reconstructions (Banks & Warburton 1991; De Vicente *et al.* 1992). For the southern fault adjacent to the Madrid Basin, minimum throw estimates are given by the topography and seismic data. Adopting these values for faulting, the crustal thickening shows a maximum of about 3 km, which rapidly decreases towards the boundaries of the Central System. In contrast to the spherical cap β distribution, crustal thickening values deduced from the refraction data (Surinach & Vegas 1988) show a more constant thickening below the Central System, most likely related to distributed reverse faulting and thrusting within the Central System (Fig. 13a). We therefore include an additional crustal thickening $\beta = 0.9$ in a quadrilateral area, of which the northern and southern border faults form two opposing sides.

In the flexure calculations, it is assumed that depressions below topography (Fig. 16a) are infilled by a sand lithology with equal grain density and porosity–depth parameters as adopted in calculating tectonic subsidence (Sclater & Christie 1980). Elevated areas above topography are eroded, and we adopt average crustal density (Table A1) for eroded material.

Starting from minimum throw estimates, the throw values for the southern fault have been increased to fit basement depths in the Madrid Basin and exposure of basement rocks in the Central System. For the northern border fault, throw values have been varied to fit predicted basement elevations closely to near surface exposure observed in adjacent parts of the Central System (Figs 13b & 14). Results for an EET value of 7 km, depicted in Fig. 16a, show the flexural response for a throw distribution, which closely agrees with the pattern of observed basement depths in the Madrid Basin, and which corresponds to basement uplift and erosion in the adjacent Central System. The β distribution corresponding to these throw values is depicted in Fig. 16a. The associated predictions for crustal thickening in the Central System range from about 4 to a maximum of about 8 km, which is in reasonable agreement with the 5 km thickening deduced from refraction data (Suriñach & Vegas 1988).

In Fig. 17, we compare predicted basement deflections and observed basement depth and topography for two cross sections, illustrating the predicted deflections for various values of EET (5, 7, and 10 km), using the best fit throw values for EET = 7 km. The cross-sections indicate that the large-scale wavelength of convex down warping of the basement in the Madrid Basin corresponds to relatively low EET values of about 7 km. Some smaller wavelength concave deflections occur close to the southern border fault. These probably reflect plastic drag close to the fault and most likely small-scale normal faulting basinward. The latter feature can be attributed to extensional bending stresses induced by flexural downwarping.

The magnitude of relative uplift and erosion, predicted in the flexure model, is largely controlled by the throw of the border faults. Relatively high amounts of uplift and erosion are predicted adjacent to the Tajo Basin whereas the amount of predicted basement erosion adjacent to the Duero basement is approximately zero. These predictions agree well with the geological data (see Figs 13 & 14),

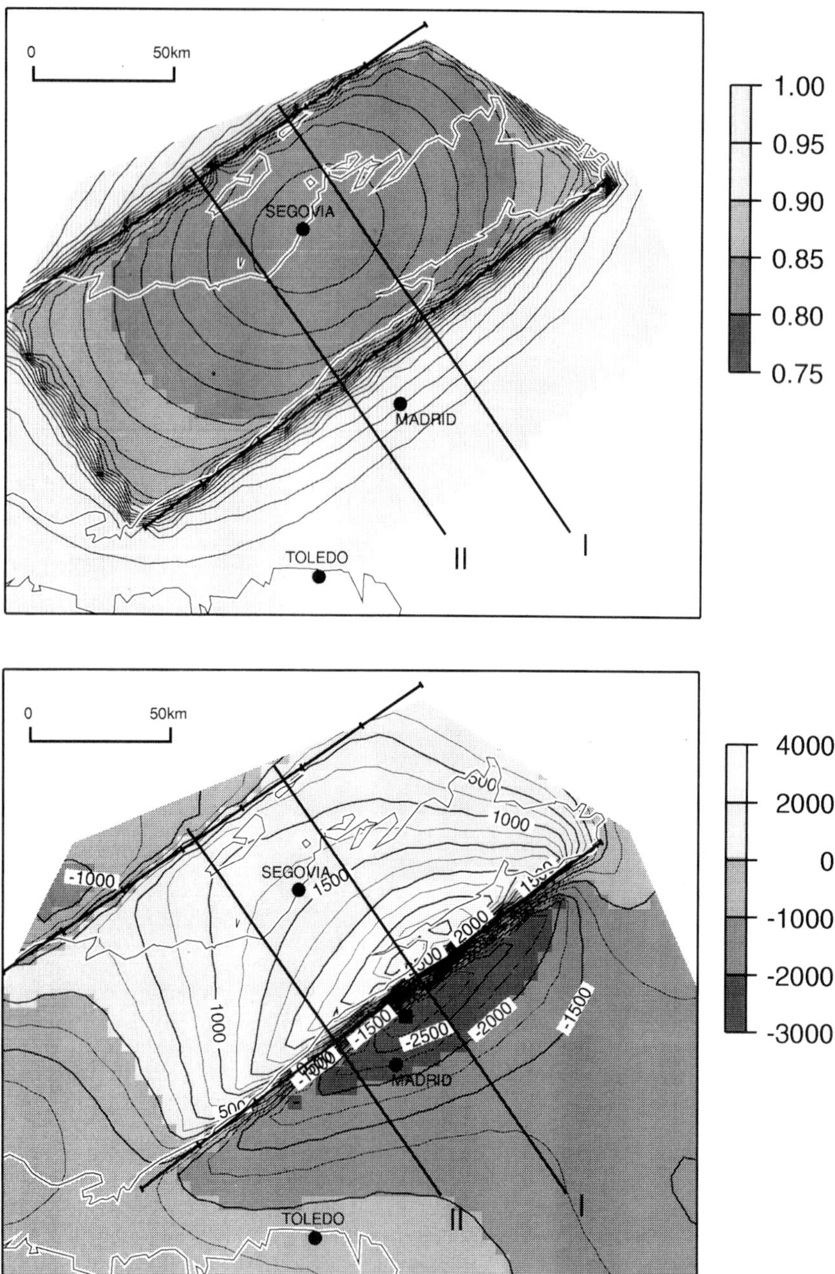

Fig. 16. Three-dimensional flexure model of the Central System and adjacent Duero and Tajo basins, for an EET of 7 km and a throw distribution for the border faults, which agrees closely with observed basement depths in the Madrid Basin. (**a**) β distribution corresponding to fault heaves, assuming a (thrust) fault dip Θ = 30° and radius r_1 = 85 km for the spherical cap function (see Appendix), which is centred along the central axis of the Central System. Fault-related β values have been augmented by β = 0.9 for a quadrilateral area bounded by the border faults. Contour interval is 0.01. (**b**) Deflection distribution, contour interval is 250 m. At position of faults mean deflection is adopted for deflection value.

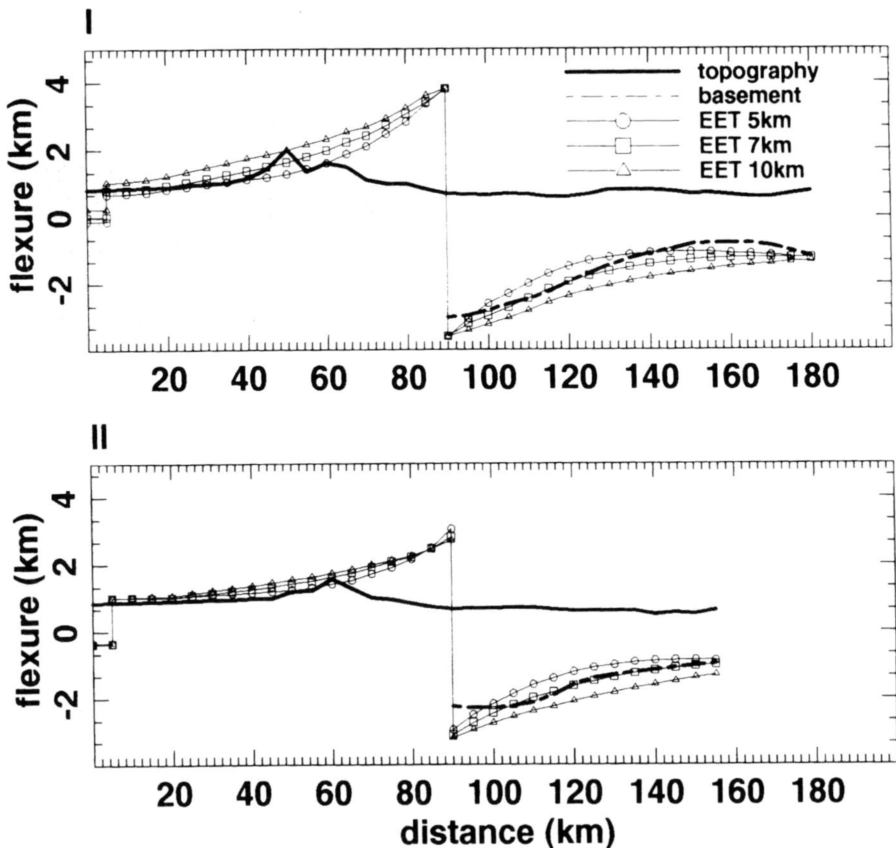

Fig. 17. Predicted and observed basement deflection for two cross-sections (I, II), indicated by solid lines on Fig. 16. Sections are for three values of EET (5, 7 and 10 km) and a fixed throw distribution, corresponding to values adopted in Fig. 16. Note that predicted uplift and associated erosion of basement above topography is in agreement with observed exposure of basement rocks.

indicating deep truncation of basement rocks close to the Tajo Basin and nearly no erosion adjacent to the Duero Basin. The latter is also evident from the partial exposure of Mesozoic cover. Furthermore, the asymmetric distibution of erosion, predicted by the flexure model, agree well with observations of asymmetric topography relative to the border faults. The drainage divide line parallel to the border faults, is located closer to the Tajo Basin than to the Duero Basin. Such a conspicious relationship between asymmetry in topography and erosion agrees very well with findings from tectonic modelling and fission track analysis of large-scale escarpment evolution at rifted margins (Van der Beek *et al.* 1995).

The flexure model results for the Central system and Tajo Basin show that observed basement warpings can be well explained adopting border fault activity interacting with an upper crust marked by a low EET of about 7 km. As for the Lake Tanganyika case study, these results indicate (partial) decoupling of upper crustal deformation and subcrustal deformation. Both case studies clearly demonstrate that border faults have a large influence on basin shape. The 3D flexure model we used provides an effective quantitative approach in demonstrating this control by fitting observed basement geometries. It also provides a useful tool for prediction of basement warping where data are not available or unreliable. This maybe particularly relevant in the appraisal of three-dimensional exploration studies (Gabrielson & Strandenes 1994).

Conclusions

The 2D and 3D modelling results clearly show that observed basin geometries at various scales can be succesfully explained by adopting a pronounced weakness for upper crustal faults. 2D finite element models, incorporating movements on weak upper crustal faults, show that predicted local stress and strain distributions and subsequent faulting in overlying rocks, agree with a large number of observations and other analogue and numerical modelling results. On a basin-wide scale, the 3D flexure model, incorporating weak planar border faults, successfully explains regional basement geometries for the Lake Tanganyika Rift Zone (extension) and the Central System of Spain uplift (compression). These model results suggest that upper crustal flexure is marked by low values of effective elastic thickness (EET), ranging from 3–7 km, and agrees with values from other 2D upper crustal flexure studies (e.g. Marsden *et al.* 1990; Kusznir *et al.* 1991).

The low EET values most likely reflect a (partial) decoupling of upper crustal and subcrustal deformation, facilitated by the weak lower crust, and in agreement with standard rheological assumptions for Phanerozoic lithosphere. In contrast, the inferred weakness of upper crustal faults is not evident from rock mechanics data. However, observations of reactivations of faults which are not preferably aligned with the stress field, and reactivations of basin deformation on long time-scales are in support of this feature.

Appendix

The three-dimensional flexure model for border faulting and basement warping (for detailed description, see Van Wees & Cloetingh 1994), uses a finite difference technique solving the bi-harmonic equation for the upper crustal deflection w:

$$D\,\frac{\partial^4 w}{\partial x^4} + D\,\frac{\partial^4 w}{\partial y^4} + 2D\,\frac{\partial^4 w}{\partial x^2\,\partial y^2} = q \qquad \text{(A1)}$$

where D is flexural rigidity and q represents buoyancy load (pertinent constants and parameters are listed in Table A1). In the finite difference scheme, upper crustal planar faults are represented by linear discontinuities, dissecting the elastic plate, being marked by an additional nodal degree of freedom for vertical offset (throw) and a constant tilt condition across the fault in accordance to planar faulting concepts (Van Wees & Cloetingh 1994). The load q, is obtained by vertical integration of the density contrasts $\Delta\rho_L$ of the deformed lithosphere and overlying water and sediments relative to the undeformed lithosphere column (e.g. Kooi1991):

$$q = -\int_{z_{topo}}^{z_a} \Delta\rho_L\, g dz \qquad \text{(A2)}$$

where g is the acceleration of gravity, z_{topo} is topographic elevation, z_a is the compensation depth. The latter often corresponds to the base of the undeformed lithosphere.

Following two-dimensional concepts (Kusznir *et al.* 1991), it is assumed that lower crustal flow locally compensates the faulting movements. Extending 2D concepts to 3D, it is assumed that for a single fault node the summation of different types of deformation in the brittle and ductile levels of the lithosphere, lead to a smooth spherical cap β distribution of Moho deformation relative to the

Table A1. *Constants and parameters for 3D flexure model*

Constant	Name	Units
E	Young's modulus	7×10^{10} Nm^2
v	Poisson's ratio	0.25
g	Gravity	9.8 m s^{-2}
ρ_c	Crustal density	2800 kg m^{-3}
ρ_m	Mantle and Asthenosphere density	3300 kg m^{-3}
Parameter	**Name**	**Units**
h	effective elastic thickness (EET)	m
D	flexural rigidity, equal to $[Eh^3/12(1-v^2)]$	Nm
w	deflection	m
q	load	Nm^{-2}
β	lithospheric stretching factor	
r_1	radius of spherical cap function	m

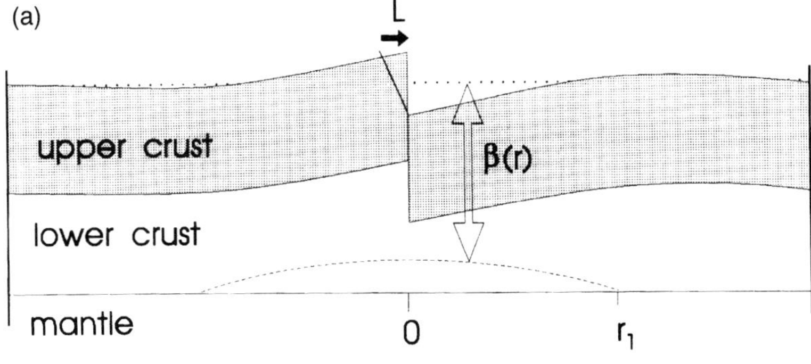

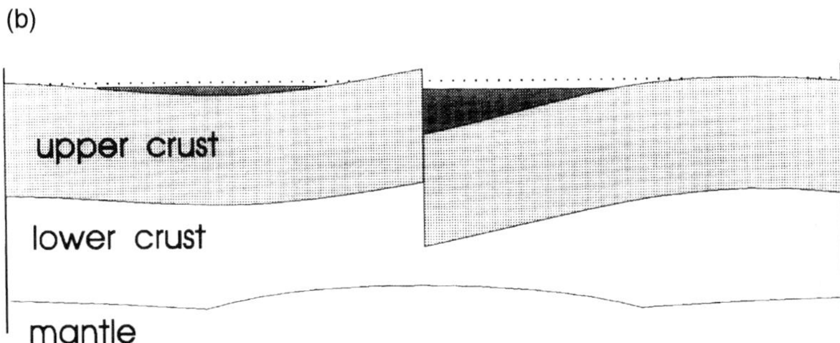

Fig. A1. Incorporation of faulting and buoyancy forces in two steps in the 3D flexure model. (**a**) First step, faulting and flexure compensated by lower crustal flow. (**b**) Next, flexural rebound in response to infill of sediments and crustal deformation.

zero surface level (see Fig. A1a):

$$L\Delta f = -\int_0^{r_1} (\beta(r) - 1)\,\mathrm{d}r \qquad \text{(A3)}$$

where L is fault extension, Δf is fault-node spacing, r is distance from the fault node and r_1 is radius of the domain of the β distribution, corresponding to a spherical cap function. Contributions of the individual fault nodes are multiplied together to obtain the β distribution for the fault.

To avoid compensation of crustal thinning or thickening and erosion or sediment infill by lower crustal flow out or into the model, fault deformation is accomplished in two steps. First (Fig. A1a) fault-induced deflections are calculated in absence of erosion or sediment infill and lithospheric thickening or thinning, adopting lower crustal flow. In the next step incremental deflections are calculated in accordance to lithospheric thinning and sediment infill and erosion, adopting asthenospheric flow (Fig. A1b). Since we focus on upper crustal structures, which are generally smaller in wavelength than those involved in lithospheric thermal perturbations, thermal effects of lithospheric deformation are neglected in calculating lithospheric buoyancy forces.

References

BANDA, E., UDIAS, A., MUELLER, ST., MEZCUA, J., BOLOIX, M., GALLART, J. & APARICIO, A. 1983. Crustal structure beneath Spain from deep seismic sounding experiments. *Physics of the Earth and Planetary Interiors*, **31**, 277–280.

BANKS, C. J. & WARBURTON, J. 1991. Mid crustal detachment in the Betic system of southeast Spain, *Tectonophysics*, **191**, 275–289.

BARR, D. 1987. Lithospheric stretching, detached normal faulting and footwall uplift. *In*: COWARD, M. P., DEWEY, J. F. & HANCOCK, P. L. (eds) *Continental extensional tectonics*. Geological Society, London, Special Publications, **28**, 75-94.

BASSI, G., KEEN, C. E. & POTTER, P. 1993. Contrasting styles of rifting: Models and examples from the eastern Canadian margin. *Tectonics*, **12**, 639–655.

BOIS, C. 1993. Initiation and evolution of the Oligo-Miocene rift basins of southwestern Europe: contribution of deep seismic reflection profiling. *Tectonophysics*, **226**, 217–226.

BRACE, W. F. & KOHLSTEDT, D. L. 1980. Limits on lithospheric stress imposed by laboratory experiments, *Journal of Geophysical Research*, **85**, 6248–6252.

BRAUN, J., 1992. Postextensional Mantle Healing and Episodic Extension in the Canning Basin, *Journal of Geophysical Research*, **97**, 8927–8936.

—— & BEAUMONT, C. 1987. Styles of continental rifting: Results from dynamic models of lithospheric extension, *In*: BEAUMONT, C. & TANKARD, A. J. (eds) *Sedimentary Basins and Basin-Forming Mechanisms.* Canadian Society of Petroleum Geologists Memoir, **12**, 241–258.

—— & —— 1989. A physical explanation of the relation between flank uplifts and the breakup unconformity at rifted continental margins. *Geology*, **17**, 760–764.

BRUN, J. P., WENZEL, F. & ECORS-DEKORP team. 1991. Crustal-scale structure of the southern Rhinegraben from ECORS-DEKORP seismic reflection data. *Geology*, **19**, 758–762.

BUCHANAN, P. G. & MCCLAY, K. R. 1991.Sandbox experiments of inverted listric and planar fault systems. *Tectonophysics*, **188**, 97–115.

BUCK, W. R. 1988. Flexural rotation of normal faults, *Tectonics*, **7**, 959–973.

—— 1991. Modes of Continental Lithospheric Extension. *Journal of Geophysical Research*, **96**, 20161–20178.

BUROV, E. B. & DIAMENT, M. 1995. The effective elastic thickness (T_e) of continental lithosphere: What doas it really mean, *Journal of Geophysical Research*, in press.

CALVO, J. P., ALONSO, A. M. & GARCIA DE CURA, M. A. 1989. Models of Miocene marginal lacustrine sedimentation in the Madrid Basin (Central Spain). *Palaeogeography, Palaeoclimatology, Palaeoecology*, **70**, 199–214.

——, HOYOS, M., MORALES, J. & ORDONEZ, S. 1990. Neogene stratigraphy, sedimentology and raw materials of the Madrid Basin. *In*: AGUSTI, J., DOMENECH, R., JULIA, R. & MARTINELL, J. (eds) *Iberian Neogene Basins.* Field Guidebook IX Congress of the Regional Comitte on Mediterranean Neogene Stratigraphy. Institut de Paleontologia Miquel Crusafont, Sabadell, 63–95.

CARTER, N. L. & TSENN, M. C. 1987. Flow properties of continental lithosphere. *Tectonophysics*, **136**, 27–63.

CERMAK, V. & BODRI, L. 1991. A heat production model of the crust and upper mantle. *Tectonophysics*, **194**, 307–323.

CLOETINGH, S. & BANDA, E. 1992. Europe's lithosphere – physical properties: mechanical structure. *In*: BLUNDELL, D., FREEMAN, R. & MUELLER, S. (eds) *A continent revealed – the European Geotraverse.* Cambridge University Press–European Science Foundation, Cambridge, 80–91.

CUNDALL, P. A. 1990. Numerical modelling of jointed and faulted rock. *In*: ROSSMANITH, A. (ed.) *Mechanics of Jointed and Faulted rocks.* Balkema, Rotterdam, 11–18.

DE VICENTE, G., GONZALEZ CASADO, J. M., BERGAMIN, J. F., TEJERO, R., BABIN, R., RIVAS, A. & ENRILE, J. L. 1992. Alpine structure of the Spanish Central System. *Actas III Congresso Geologico de España (Salamanca), Part I*, Madrid. 284–288.

DELVAUX, D., LEVI, K., KAJARA, R. & SAROTA, J. 1992. Cenozoic paleostress and kinematic evolution of the Rukwa–North Malawi rift valley (East African rift). *Bulletin Centres Recherche Exploration-Production Elf Aquitaine*, **16**, 383–406.

DRURY, M. R., VISSERS, R. L. M., VAN DER WAL, D. & HOOGERDUIJN STRATING, E. H. 1991. Shear localization in upper mantle peridotites. *Pure Applied Geophysics*, **137**, 439–460.

DUNBAR, J. A. & SAWYER, D. S. 1988. Continental rifting at pre-existing lithospheric weaknesses. *Nature*, **333**, 450–452.

EBINGER, C. J., KARNER, G. D. & WEISSEL, J. K. 1991. Mechanical strength of extended continental lithosphere: constraints from the Western Rift System, Africa. *Tectonics*, **10**, 1239–1256.

ENGLAND, P. 1983. Constraints on extension of continental lithosphere. *Journal of Geophysical Research*, **88**, 1145–1152.

FLACK, C. A., KLEMPERER, S. L., MCGEARY, S. E., SNYDER, D. B. & WARNER, M. R. 1990. Reflections from mantle fault zones around the British Isles. *Geology*, **18**, 528–532.

GABRIELSEN, R. H. & STRANDENES, S. 1994. Dynamic basin development – a complete geoscientific tool for basin analysis. P*roceedings of the 14th World Petroleum Congress*, John Wiley & Sons, pp. 13–21.

GIBBS, A. D. 1984. Structural evolution of extensional basin margins. *Journal of the Geological Society, London,* **141**, 609–620.

GOETZE, C. & EVANS, B. 1979. Stress and temperature in the bending lithosphere as constrained by experimental rock mechanics. *Geophysical Journal of the Royal Astronomical Society*, **59**, 463–478.

HAYWARD, A. B. & GRAHAM, R. H. 1989. Some geometrical characteristics of inversion. *In*: COOPER, M. A. & WILLIAMS, G. D. (eds) *Inversion Tectonics.* Geologcal Society, London, Special Publications, **44**, 17–40.

JACKSON, J.A., 1987. Active normal faulting and crustal extension. *In*: COWARD, M. P., DEWEY, J. F. & HANCOCK, P. L. (eds) *Continental extensional tectonics.* Geological Society, London, Special Publications, **28**, 3–17.

JAEGER, J. C. & COOK, N. G. W. 1976. *Fundamentals of Rock Mechanics.* Chapman & Hall, London.

KARNER, G. D., EGAN, S. E. & WEISSEL, J. K. 1992. Modeling the tectonic development of the Tucano and Sergipe-Alagoas rift basins, Brazil. *Tectonophysics*, **215**,133–160.

KOOI, H. 1991. *Tectonic modelling of extensional basins: the role of lithospheric flexure, intraplate stress and relative sea-level change.* PhD thesis, Vrije Universiteit, Amsterdam.

——, CLOETINGH, S. & BURRUS, J. 1992. Lithospheric necking and regional isostasy at extensional basins, 1, Subsidence and gravity modelling with an application to the Gulf of Lions Margin (SE

France). *Journal of Geophysical Research*, **97**, 17553–17572.

KUSZNIR, N. J. & PARK, R. G. 1987. The extensional strength of the continental lihosphere: its dependence on geothermal gradient, and crustal composition and thickness. *In*: COWARD, M. P., DEWEY, J. F., & HANCOCK, P. L. (eds) *Continental Extensional Tectonics*. Geological Society, London, Special Publications, **28**, 75–94.

—— & ZIEGLER, P. A. 1992. The mechanics of continental extension and sedimentary basin formation: a simple-shear/pure-shear flexural cantilever model. *Tectonophysics*, **215**, 117–131.

——, KARNER, G. D. & EGAN, S. S. 1987. Geometric, thermal and isostatic consequences of detachments in continental lithosphere extension and basin formation. *In*: BEAUMONT, C. & TANKARD, A. J. (eds) *Sedimentary Basins and Basin Forming Mechanisms*. Canadian Society of Petroleum Geologists Memoir, **12**, 185–203.

——, MARSDEN, G. & EGAN, S. S. 1991. A flexural-cantilever simple-shear/pure shear model of continental lithosphere extension: application to the Jeanne d'arc Basin, Grand Banks and Viking Graben, North Sea. *In*: ROBERTS, A. H., YIELDING, G. & FREEMAN, B. (eds) *The Geometry of Normal Faults*. Geological Society, London, Special Publications, **56**, 41–60.

LE PICHON, X. & CHAMOT-ROOKE, N. 1991. Extension of continental crust. *In*: MULLER, D. W., MCKENZIE, J. A. & WEISSERT, H. (eds) *Controversies in Modern Geology, Evolution of geological theories in sedimentology, earth history and tectonics*. Academic Press, London, 313–338.

—— & SIBUET, J.-C. 1981. Passive Margins: A model of formation. *Journal of Geophysical Research*, **86**, 3708–3720.

MANDL, G. 1988. *Mechanics of Tectonic Faulting. Models and Basic Concepts*. Elsevier, Amsterdam.

MARSDEN, G., YIELDING, G., ROBERTS, A. M. & KUSZNIR, N. J. 1990. Application of a flexural cantilever simple-shear/pure-shear model of continental extension to the formation of the northern North Sea basin. *In*: BLUNDELL, D. J. & GIBBS, A. D. (eds) *Tectonic Evolution of the North Sea Rifts*. Oxford Science Publishers, Oxford, 240–261.

MCCLAY, K. R. 1989. Analogue models of inversion tectonics. *In*: COOPER, M. A. & WILLIAMS, G. D. (eds) *Inversion Tectonics*. Geologcal Society, London, Special Publications, **44**, 41–62.

—— & ELLIS, P. G. 1987. Analogue models of extensional fault geometries. *In*: COWARD, M. P., DEWEY, J. F. & HANCOCK, P. L. (eds) *Continental Extensional Tectonics*. Geological Society, London, Special Publications, **28**, 75–94.

MCGEARY, S. & WARNER, M. R. 1986. Seismic profiling the continental lithosphere. *Nature*, **317**, 795–797.

MEGIAS, A. G., ORDONEZ, S. & CALVO, J. P. 1983. Nuevas aportaciones al conocimiento geologico de la Cuenca de Madrid. *Rev. Mat. Proc. Geol.*, I, Universidad Complutense, Madrid, 163–191.

MEISSNER, R. 1986. *The continental crust – A geophysical approach*. Academic Press.

MELOSH, H. J. 1990. Mechanical basis for low-angle normal faulting in the Basin and Range Province, *Nature*, **343**, 331–335.

MELOSH, H. J. & RAEFSKY, A. 1981. A simple and efficient method for introducing faults into finite element computations. *Bull. Seism. Soc. Am.*, **71**, 1391–1400.

MORLEY, C. K. 1988. Variable extension in Lake Tanganyika. *Tectonics*, **7**, 785–801.

MÜLLER, B., ZOBACK, M. L., FUCHS, K., MASTIN, L., GREGERSEN, S. *et al.* 1992. Regional patterns of tectonic stress in Europe. *Journal of Geophysical Research*, **97**, 11783–11804.

NIEUWLAND, D. A. & WALTERS, J. V. 1993. Geomechanics of the South Furious field. An integrated approach towards solving complex structural geological problems, including analogue and finite-element modelling. *Tectonophysics*, **226**, 143–166.

ORD, A. & HOBBS, B. E. 1989. The strength of the continental crust, detachment zones, and the development of plastic instabilities. *Tectonophysics*, **158**, 269–289.

PINET, B. & COLLETA, B. 1990. Probing into extensional sedimentary basins: comparison of recent data and derivation of tentative models. *Tectonophysics*, **173**, 185–197.

POLLACK, H. N. & CHAPMAN, D. S. 1977. On the regional variation of heat flow, geotherms and lithospheric thickness. *Tectonophysics*, **38**, 279–296.

PORTERO, J. M., DEL OLMO, P. & OLIVE, A. 1983. El Neogeno de la transversal Norte-Sur de la Cuenca del Duero. *Geologia de Espana, Libro Jubilar J.M. Rios, 2*. Instituto Geologico y Miñero de España, Madrid, 494–502.

QUEROL MULLER, R. 1989. *Geologia del subsuelo de la Cuenca del Tajo*. Instituto Tecnologico Geominero de Espana, Madrid.

RACERO BAENA, A. 1988. Condsideraciones acerca de la evolucion geologica del margen NW de la Cuenca del Tajo durante el Terciario a partir de los datos de subsuelo. *Actas II Congresso Geologico de España (Granada)*, Madrid, 213–221.

RANALLI, G. 1987. *Rheology of the Earth*. Allen and Unwin, Boston.

—— & MURPHY, D. C. 1987. Rheological stratification of the lithosphere. *Tectonophysics*, **132**, 281–295.

—— & YIN, Z.-M. 1990. Critical stress difference and orientation of faults in rocks with strength anisotropies: the two-dimensional case. *Journal of Structural Geology*, **12**, 1067–1071.

RANSOME, F. L., EMMONS, W. H. & GARREY, G. H. 1910. Geology and ore deposits of the Bullfrog District, Nevada. *United States Geological Survey Bulletin*, **407**, 1–130.

RESTON, T. J. 1990. Shear in the lower crust during extension: not so pure and simple. *Tectonophysics*, **173**, 175–183.

RING, U., BETZLER, C. & DELVAUX, D. 1992. Normal vs. strike-slip faulting during rift development in East Africa: The Malawi rift. *Geology*, **20**, 1015–1018.

ROSENDAHL, B. R. 1987. Architecture of continental rifts with special reference to East Africa. *Annual Reviews of Earth and Planetary Science*, **15**, 445–503.

——, KILEMBE, E. & KACZMARICK, K. 1992. Comparison of the Tanganyika, Malawi, Rukwa and Turkana Rift zones from analyses of seismic reflection profiles. *Tectonophysics*, **213**, 235–256.

SALAS, R. & CASAS, A. 1993. Mesozoic Extensional Tectonics, Stratigraphy and Crustal Evolution during the Alpine Cycle of the Eastern Iberian Basin. *Tectonophysics,* **228**, 33–56.

SALTZER, S. D. & POLLARD, D. D. 1992. Distinct element modeling of structures formed in sedimentary overburden by extensional reactivation of basement normal faults. *Tectonics*, **11**, 165–174.

SASSI, W., COLLETTA, B., BALE, P. & PAQUEREAU, T. 1993. Modelling of structural complexity in sedimentary basins: the role of pre-existing faults in thrust tectonics. *Tectonophysics*, **226**, 97–112.

SCLATER, J. G. & CHRISTIE, P. A. F. 1980. Continental Stretching: An explanation of the post-Mid-Cretaceous subsidence of the Central North Sea Basin. *Journal of Geophysical Research*, **85**, 3711–3739.

SIBSON, R. H. 1974. Frictional constraints on thrust, wrench, and normal faults. *Nature*, **249**, 542–544.

—— 1985. A note on fault reactivation. *Journal of Structural Geology*, **7**, 751–754.

SOPEÑA, A., LOPEZ, J., ARCHE, A., PEREZ-ARLUCEA, M., RAMOS, A., VIRGILI, C. & HERNANDO, S. 1988. Permian and Triassic rift basins of the Iberian Peninsula. *In*: Manspeizer, W. (ed.) *Triassic–Jurassic Rifting, Part B.* Elsevier, Amsterdam, 757–786.

STECKLER, M. S. & WATTS, A. B. 1978. Subsidence of the Atlantic-type continental margin off New York. *Earth and Planetary Science Letters*, **41**, 1–13.

STEIN, R. S., KING, G. C. P. & RUNDLE, J. B. 1988. The growth of geological structures by repeated earthquakes 2. field examples of continental dip-slip faults. *Journal of Geophysical Research*, **93**, 213319–213331.

SURIÑACH, E. & VEGAS, R. 1988. Lateral inhomogeneities of the Hercynian crust in central Spain. *Physics of Earth and Planetary Interiors*, **51**, 226–234.

TSENN, M. C. & CARTER, N. L. 1987. Upper Limits of power law creep of rocks. *Tectonophysics*, **136**, 1–26.

VAN DER BEEk, P. A., ANDRIESSEN, P. A. M. & CLOETINGH, S. 1995. Morpho-tectonic evolution of rifted continental margins: Inferences from a coupled tectonic-surface processes model and fission track thermochronology. *Tectonics*, in press.

VAN WEES, J. D. & CLOETINGH, S. 1994. A finite-difference technique to incorporate spatial variantions in rigidity and planar faults into 3-D models for lithospheric flexure. *Geophysical Journal International*, **117**, 179–195.

—— & STEPHENSON, R. A. 1995. Quatitative modelling of basin and rheological evolution of the Iberian Basin (Central Spain): implications for lithospheric dynamics of intraplate extension and inversion. *Tectonophysics,* in press.

VEGAS, R. & BANDA, 1982. Tectonic framework and Alpine evolution of the Iberian Peninsula. *Earth Evolution Sciences*, **4**, 320–343.

——, VAZQUEZ, R., SURINACH, E. & MARCOS, A. 1990. Model of distributed deformation block rotations and crustal thickening for the formation of the Spanish Central System. *Tectonophysics*, **184**, 367–378.

VENDEVILLE, B. 1991. Mechanisms generating normal fault curvature: a review illustrated by physical models. *In*: ROBERTS, A .M., YIELDING, G. & FREEMAN, B. (eds) *The Geometry of Normal Faults.* Geological Society, London, Special Publications, **56**, 241–249.

—— & COBBOLD, P. R. 1988. How normal faulting and sedimentation interact to produce listric fault profiles and stratigraphic wedges. *Journal of Structural Geology*, **10**, 649–659.

——, ——, DAVY, P., BRUN, J. P. & CHOUKROUNE, P. 1987. Physical models of extensional tectonics at various scales, *In*: COWARD, M. P., DEWEY, J. F. & HANCOCK, P. L. (eds) *Continental Extensional Tectonics.* Geological Society, London, Special Publications, **28**, 95–107.

VERSFELT, J. & ROSENDAHL, B. R. 1989. Relationships between pre-rift structure and rift architecture in Lakes Tanganyika and Malawi, East Africa. *Nature*, **337**, 354–357.

VIALLARD, P., 1989. Decollement de couverture et decollement medio-crustal dans une chaine intra-plaque: variations verticales du style tectonique des Iberides (Espagne). *Bulletin Societe Geologique Francaise*, **5**, 913–918.

VISSERS, R. L. M., DRURY, M. R., HOOGERDUIJN STRATING, E. H. & VAN DER WAL, D. 1991. Shear zones in the upper mantle: A case study in an alpine lherzolite massif. *Geology*, **19**, 990–993.

WERNICKE, B. & BURCHFIEL, B. C. 1982. Modes of extensional tectonics. *Journal of Structural Geology*, **4**, 105–115.

WILSON, J. T. 1966. Did the Atlantic close and re-open? *Nature*, **211**, 676–681.

YIELDING, G., BADLEY, M. E. & FREEMAN, B. 1991. Seismic reflections from normal faults in the northern North Sea. *In*: ROBERTS, A. M., YIELDING, G. & FREEMAN, B. (eds) *The Geometry of Normal Faults.* Geological Society, London, Special Publications, **56**, 79–89.

YIN, Z.-M. & RANALLI, G. 1992. Critical stress difference, fault orientation and slip direction in anisotropic rocks under non-Andersonian stress systems. *Journal of Structural Geology*, **14**, 237–244.

ZIEGLER, P. 1990. *Geological Atlas of Western and Central Europe* (2nd edn). Geological Society, London/Shell Internationale Petroleum Maatschappij, London.

ZIENKIEWICZ, O. C. & TAYLOR, R. L. 1991. *The Finite Element Method, Volume 2.* McGraw-Hill, New York.

ZOBACK, M. D., STEPHENSON, R. A., CLOETINGH, S., LARSEN, B. T., VAN HOORN, B., *et al.* 1993. Stresses in the lithosphere and sedimentary basin formation. *Tectonophysics*, **226**, 1–14.

ZOBACK, M. L. 1992. First- and second-order patterns of stress in the lithosphere: The World Stress Map Project. *Journal of Geophysical Research*, **97**, 11703–11728.

—— *et al.* 1989. Global patterns of tectonic stress. *Nature*, **341**, 291–298.

The use of satellite imagery in the validation and verification of structural interpretations for hydrocarbon exploration in Pakistan and Yemen

M. W. INSLEY,[1] F. X. MURPHY, D. NAYLOR & M. CRITCHLEY

ERA-Maptec Ltd, 5 South Leinster Street, Dublin 2, Ireland

[1] *Present address: NRSC Ltd, Delta House, Southwood Crescent, Southwood, Farnborough, Hants UK*

Abstract: Landsat TM and stereo-SPOT satellite imagery has an important application in the validation and verification of structural models for on-shore hydrocarbon exploration. The lithological discrimination of Landsat TM can be merged with the higher resolution panchromatic stereo-SPOT imagery to produce detailed geological maps at scales of 1:25 000 and 1:50 000 from which to 'hang' structural cross-sections. In addition, the calculation of surface dip data and fault off-sets, as well as the construction of structure contour and topographic maps is now possible using SPOT-derived digital elevation models (DEMs), thus providing the necessary information for the construction of structural cross-sections.

Examples of the use of satellite imagery are presented from the Sulaiman Ranges of Pakistan and southeast Yemen. For the Sulaiman Ranges of Pakistan, satellite imagery has been used to constrain the construction of a balanced deformed and restored cross-section from relatively poor quality seismic data. The satellite imagery provides information on structural style and along-strike structural continuity. The construction of a balanced cross-section is used to highlight the contrast between the position of structural closure at surface and subsurface.

In southeast Yemen, satellite imagery is combined with both field and seismic data to develop a viable structural interpretation. The model is complicated by the superposition of Late Tertiary extensional structures on an earlier Mesozoic rift system. Earlier structural trends have been inferred from the interpretation and analysis of the Late Tertiary fault patterns observed on the Landsat MSS and TM and stereo-SPOT imagery. Stereo-SPOT imagery is used at 1:50 000 scale to map out extensional faults, to examine their displacement gradients and identify potential closures at surface. Quantitative analysis of displacement patterns around faults is possible using structure contour maps derived from stereo-SPOT imagery and DEMs. Satellite imagery also enables the validation of structural continuity between seismic lines.

The aim of this paper is to demonstrate the role of satellite imagery in the provision of both two-dimensional and three-dimensional geological information for the development of accurate structural interpretations. Of particular interest is the use of stereo imagery, such as SPOT (Systéme Pour l'Observation de la Terre), for the three dimensional visualisation and interpretation of geological structures. In addition, the generation of digital elevation models (DEMs) and a series of derived products, now make quantitative analysis possible. Satellite image related components of the validation process are shown in Fig. 1. However, the integration of other data sets including seismic, gravity, magnetic and field measurements also forms an essential part of any satellite image interpretation and leads to a greater understanding of the relationships between the surface and subsurface structure.

Satellite imagery has a number of advantages which make it particularly valuable as an exploration tool.

(1) It can provide geometrically corrected, up-to-date images from which detailed geological maps can be presented without setting foot on the ground. Investigation can be made of adjacent areas or licence blocks not just for structure, but also for assessing the location of existing seismic lines and wells in relation to surface structure.

(2) Different types of imagery allow structural analysis from 1:250 000 to 1:1 000 000 scale regional structural appraisals using Landsat Thematic Mapper (TM) or Multi-spectral Scanner (MSS) data, down to 1:25 000 to 1:50 000 scale studies of individual prospects using SPOT or Russian data.

(3) Because stereo-satellite imagery such as SPOT, is available, a whole range of additional information can be derived including DEMs, orthoimages, topographic and structure contour

From Buchanan, P. G. & Nieuwland, D. A. (eds), 1996, *Modern Developments in Structural Interpretation, Validation and Modelling*, Geological Society Special Publication No. 99, pp. 321–343.

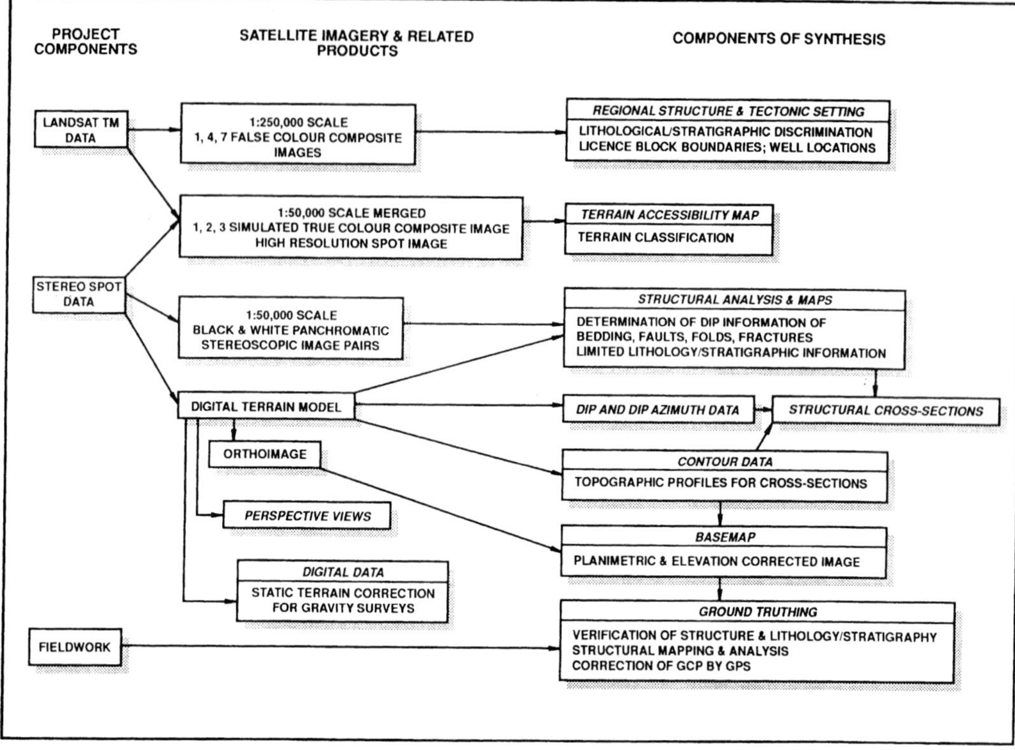

Fig. 1. Satellite imagery and derived products used in the validation process.

maps and surface dip data which increase its usefulness in the validation process.

(4) Most satellite imagery are available in digital format which makes it amenable not only to merging with other types of imagery, but more significantly, to the integration with different types of digital data such as gravity, magnetic and seismic.

(5) Digital imagery can also be processed to highlight certain characteristics; edge enhancement to emphasize structure, and contrast stretching to aid the identification of lithological/stratigraphic units.

Types of satellite imagery

The two most common types of satellite imagery used for detailed structural and lithological mapping are panchromatic SPOT and Landsat TM. The specifications of these two forms of satellite imagery and others used for geological interpretation purposes are summarized in Table 1.

Panchromatic SPOT imagery consists of one spectral band with a ground resolution of 10 m. An individual scene covers an area 60 km N–S × 60–85 km E–W (SPOT Image Corporation 1984). The SPOT satellite has a tilting mirror system enabling the satellite to look ± 27° from vertical resulting in an image up to 85 km E–W. The oblique viewing capability allows image acquisition of the same area every 2.4–4 days and also provides an option of stereo coverage (Fig. 2), for the generation of DEMs (Gugan & Dowman 1988). Multispectral SPOT (XS) imagery (Table 1), with its 20 m ground resolution and three spectral bands (SPOT Image Corporation 1984), shares some of the spectral capabilities of Landsat TM but at slightly better resolution.

A Landsat image is a vertical look image approximately 185 km E–W × 185 km N–S. Landsat TM data consist of seven spectral bands, six in the visible to near infrared range and one in the thermal range (NASA 1976; Freden & Gordon 1983). All bands have a 30 m ground resolution except the thermal band 6, which has a 120 m ground resolution (Table 1). An image can be produced from individual bands or combination of bands or band ratios. The main use of Landsat TM is lithological/stratigraphic mapping where discrimination of individual lithologies/minerals is often possible due to their different spectral signatures. Landsat MSS (NASA 1976) has lower

Table 1. *Specifications of different forms of satellite imagery used in geological interpretation*

Satellite system	No. spectral bands	Path width (km)	Spatial resolution (m)	Working scale of image	Radiometric resolution	Viewing	Repeat time (days)
NOAA AVHRR	4.5	3000	1100	1:3 000 000	256	Nadir	
Landsat TM	7	185	30 (bands 1–5, 7)	1:100 000	256	Nadir	16
			120 (band 6)	1:250 000			
Landsat MSS	4	185	79 x 56	1:200 000	64	Nadir	16
SPOT Pan	1	60–85	10	1:25 000	256	Up to ±27° off nadir	4 to 26
SPOT XS	3	60–85	20	1:50 000	256	Up to ±27° off nadir	4 to 26
KFA-3000	1	21	2 to 3	1:10 000	Photographic	Nadir	
KFA-1000	1	80	5	1:20 000	Photographic	Nadir	
MK-4	1 or 3	117–173	7.5	1:25 000	Photographic	Off nadir capability	
KATE 200	1 or 3	216 to 243	20	1:50 000	256	Nadir	
ERS-1	SAR C-band	102.5	<30 azimuth <26 range	1:100 000	256	23° mid range incidence angle	3 to 35
JERS-1	SAR L-band	75	18	1:50 000	256	38.5° mid range incidence angle	44

Optimum scales of satellite imagery for geological interpretation based on Kramber *et al.* (1990) and Light (1990).

Fig. 2. Example of stereo panchromatic SPOT image pair from the Gulf of Suez. Stereo image pairs improve the accuracy of the interpretation by enabling the geology and terrain to be visualised in three-dimensions. © CNES 1991 – Distribution Spot Image.

spatial resolution as well as poorer spectral range than TM data, but offers a relatively cheap option for carrying out regional tectonic/stratigraphic studies (Tyler 1993).

Other types of satellite imagery include the recently available high resolution KFA and MK-4 series of Russian imagery (Table 1). Although the data has the advantage of higher spatial resolution, coverage is often limited and supply in digital format is via the scanning of photographic prints. Multispectral MK-4 data also require individual ortho-rectification of the three spectral bands before they can be accurately registered and merged to form a composite image.

Active imaging systems employing radar, are frequently used in areas of adverse weather conditions and/or dense vegetation cover. This is due to the inability of passive sensors to penetrate cloud or detect structures expressed by variations in the vegetation cover. Radar imaging systems have the advantage of supplying images irrespective of illumination conditions and allow detection of geological structures mimicked by textural variations in the vegetation cover or canopy. Synthetic aperture radar (SAR) imagery is now available from the ERS-1, ERS-2 (NASDA 1990; Attema Evert 1991) and JERS-1 (Nemoto *et al.* 1991; RESTEC 1992; Nishidai 1993) satellites (Table 1) which may provide an alternative to airborne SAR and Side-looking airborne radar (SLAR) data. Although SAR imagery is associated with inherent terrain distortions related to variations in relief, this problem is usually far out weighed by the necessity to obtain details of previously unknown or uncertain geology and physiography.

Stereo-SPOT-derived products

Accurate structural interpretation of an area is also improved by combining quantitative information from a range of stereo-SPOT-related products which include DEMs, topographic and structure contours and bedding dip data (Table 1).

Digital elevation models (DEMs)

Stereo-image pairs are a pre-requisite for generating digital elevation models (DEMs) (Sasowsky *et al.* 1992). These are created by computing the x, y and z coordinates for each image pixel. By calculating the parallax through the systematic matching of individual pixels between stereo image pairs, the result is a DEM consisting of a matrix of relative height values. Ground control points, established from maps or supplied from GPS (ground positioning systems), are then used to geometrically correct and calibrate the relative height values of the DEM. Contouring the height values enables the production of topographic contour maps from which profiles can be generated to provide a topographic template for the construction of cross-sections. DEMs also provide a source of data for applying better static terrain corrections to gravity surveys.

The draping of a satellite image over the DEM allows the three dimensional visualisation of a structure and its relationship to stratigraphy to be analysed through the generation of a series of perspective views or simulated fly-throughs.

Ortho-images

Ortho-images are images derived from DEMs in which both the planimetric and elevation distortions of the stereo oblique looking images have been removed. They are commonly used in the production of merged imagery such as SPOT/Landsat TM to give an image combining the lithological discrimination of TM with the improved spatial resolution of SPOT. Ortho-images can also be used as highly accurate base maps on an image background and are useful for combining other map referenced data sets.

Structure contour maps

In a similar way, height information can be extracted from the DEM for a selected stratigraphic boundary. This information can then be contoured to produce a structure contour map. The accuracy and coverage of the contour map is dependent on there being sufficient outcrop and knowledge of the stratigraphy. The level of structural complexity is also an important control on the ability to construct structure contours.

Calculation of surface bed dips

Measurement of dip and dip azimuth (Fig. 3), using three-point elevation calculations, is a vital source of data for constraining the construction of cross-sections. Quantative measurement of surface dip data is possible by overlaying the SPOT ortho-image with the DEM on screen. The method, involves the computation of dip and dip azimuth from the x, y and z coordinates of three individual points selected on a single bedding plane identified on the orthoimage. Points are usually selected at least 500 m apart to minimise the inaccuracy of the measurements due to limitations of planimetric and vertical resolution of the DEM and irregularities in the bedding surface.

The following two case studies illustrate how satellite imagery can provide accurate data for the structural interpretation of areas of compressional

Fig. 3. Measurement of dip and dip azimuths for bedding using three-point elevation technique. Three individual points for the calculation are located on a single bedding plane identified on the satellite image. *x*, *y*, and *z* coordinates for each point are automatically determined from the DEM co-registered with the satellite image. © CNES 1991 – Distribution Spot Image.

and extensional tectonics from Pakistan and Yemen respectively. In both examples, interpretation of the surface geology from satellite imagery helps to constrain the subsurface structure identified from seismic data.

Structural validation of the Pirkoh anticline, Pakistan

In this example, satellite image derived data are used to construct a balanced cross-section for the structural validation of the Pirkoh anticline in the Sulaiman Ranges of Pakistan. Components of the analysis include: interpretation of 1:50 000 scale panchromatic stereo-SPOT for detailed structure; Landsat TM false colour composite of bands 1, 4 and 7 for maximum lithological/stratigraphic discrimination; generation of a DEM from digitised topographic contour maps and production of structure contours, perspective views and measurement of surface bed dips; regional seismic data and stratigraphic information from the Pirkoh-1 well.

Regional tectono-stratigraphic setting

The Pirkoh study area lies in the actively deforming (Quittmeyer *et al.* 1979, 1984) frontal portion of the Sulaiman fold belt in the Kirthar–Sulaiman Ranges of the western Himalayas (Fig. 4). The area is highly prospective (Tainish *et al.* 1959; Raza *et al.* 1989*a*, *b*) with significant gas fields associated with the Sui, Loti and Pirkoh fold structures (Fig. 4). The Sulaiman fold belt consists of an arcuate series of folds and imbricate thrusts which become rotated to a north–south orientation along the margins of the belt (Kazmi & Rana 1982). The Pirkoh anticline is the third fold structure north of the deformation front of the Sulaiman fold belt (Fig. 4). It is also interpreted as the foreland-most fault bend fold (Banks & Warburton 1986; Jadoon *et al.* 1992, 1994).

Existing interpretations based on regional seismic profiles and surface mapping suggest that the structure of the Sulaiman fold belt can be interpreted in terms of a passive roof duplex system (Banks & Warburton 1986; Izatt 1990; Humayon *et al.* 1991; Jadoon *et al.* 1992, 1993, 1994). The duplex consists of a foreland directed, piggy-back series of horses containing thick Jurassic limestones and older strata (Fig. 5). The floor thrust of the duplex is assumed to lie above, but near top crystalline basement, whereas the roof thrust lies within Cretaceous shales. This implies that the majority of folds and thrusts exposed at surface within the Sulaiman fold belt are developed within the hangingwall or roof complex of the duplex (Fig. 5). Interpretation of the area in terms of a passive roof duplex is supported by structural thickening of the platformal section, structurally elevated stratigraphic section, and the apparent absence of large displacement thrusts at surface.

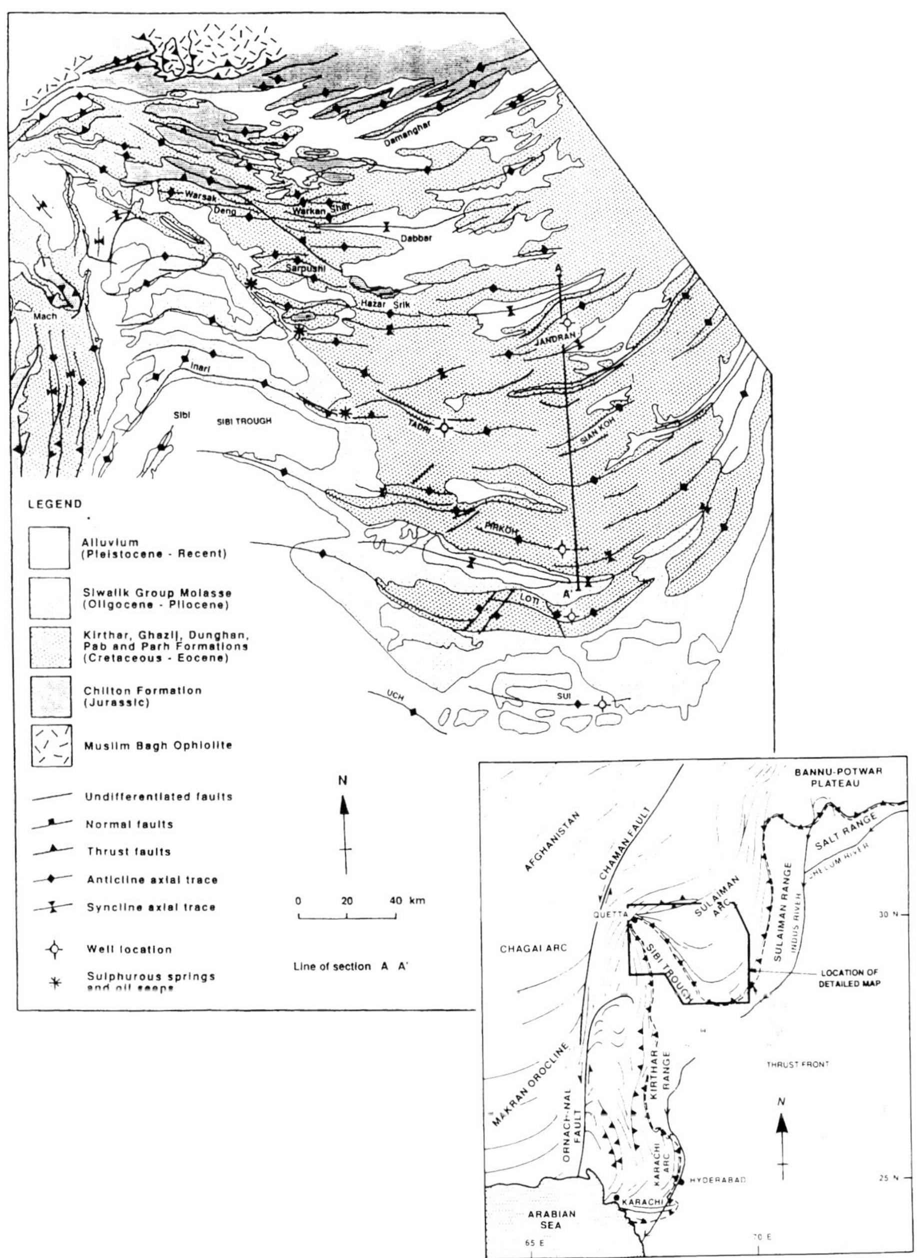

Fig. 4. Geological map of the Sulaiman fold belt showing location of Pirkoh study area (modified from Banks & Warburton 1986).

Structural style of the Pirkoh area

The Pirkoh area (Fig. 6) contains a well exposed sequence of complexly folded and thrusted Eocene age Kirthar and Ghazij Formations, overlying large anticlinal structures of Paleocene to Upper Cretaceous age Dunghan and Ranikot Formations (Shah 1977). Progressively older strata are exposed in the eroded cores of anticlinal fold hinges towards the hinterland of the belt. A significant strike swing occurs in the area with major folds trending east-west in the west and northeast-southwest in the

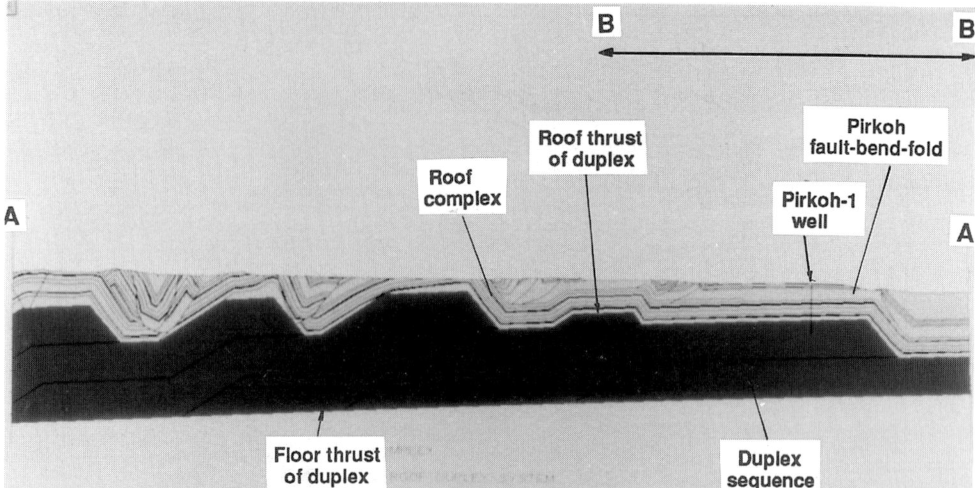

Fig. 5. Passive roof duplex model constructed from the satellite image interpretation for the Pirkoh study area using GEOSEC section balancing software. Mid grey colour corresponds to duplex sequence, overlying layered sequence represents 'passive-roof' complex.

east. This change in strike coincides with a north-northwest trending structural low defined by converging noses of plunging folds. Anticlines appear to be relatively simple structures with coherent stratigraphic section, whereas synclines are structurally complex areas consisting of tightly folded and imbricated sequences of Ghazij and Kirthar strata.

Two thin-skinned imbricate systems of tightly folded and imbricated Ghazij and Kirthar strata are exposed to the north of the Pirkoh structure (Fig. 6). Closely spaced imbricates, diverging splays and branch points, lateral ramps and hangingwall anticlines are visible on the imagery (Fig. 7). Individual thrusts appear to share a common detachment close to the base of the Ghazij Formation. It is difficult to assess the significance and lateral continuity of the detachment on the satellite imagery, where it is parallel to bedding and there is no apparent change in stratigraphy. The stratigraphically higher Kirthar sequence is preserved along strike and is deformed by moderate to asymmetric anticline–syncline pairs developed at the tip lines of the imbricate thrusts (Fig. 7).

There is a clear contrast in structural style between the moderate to tight folds and close spaced thrusts developed in the Kirthar sequence and the large wavelength, rounded folds expressed in the top Dunghan Formation (Fig. 6a). This contrast in structural style, is accommodated within the intervening incompetent Ghazij shale by complex small scale folding and thrusting above a basal Ghazij detachment.

Structure of Pirkoh Anticline

The Pirkoh Anticline is a complex periclinal structure developed within Pirkoh limestone and Eocene marls at surface (Fig. 6a). The crest of the fold is coincident with a topographic high (Fig. 8b) on which the Pirkoh-1 well is located. Both the well site and access road are clearly visible on the Landsat TM and SPOT imagery (Figs 6a and 8a).

Perspective views of the Pirkoh Anticline (Fig. 9) help to highlight the complexity of the fold surface. Viewed from the west, the fold appears to be a broad composite structure containing several smaller scale culminations (Fig. 9). In part, the fold has a pseudo box-style geometry approximately 15 km wide. The southern margin is associated with a narrow monoclinal fold hinge with a 35–70° south-dipping forelimb. The northward dip of the back limb of the fold is variable, but generally less than 30°. Towards the east the amplitude of the fold decreases and is replaced by an en echelon series of smaller scale periclines with curvilinear axial surface traces (Fig. 6).

A distinct saddle occurs along the crest of the fold towards the west and is associated with a dramatic reduction in the width of the structure as

Fig. 6. (**a**) Oblique right look panchromatic SPOT image of Pirkoh study area © CNES 1991 – Distribution Spot Image. (**b**) Geological interpretation of the Pirkoh study area based on stereo-SPOT imagery.

(a)
scale 0 10km
© CNES 1991
(b)
TADRI ANTICLINE
N
Figure 7
PIRKOH
FOLD
ANTICLINE
CREST
ANTICLINAL MONOCLINE
FOLD HINGE
Figure 8
BUGTI SYNCLINE
LEGEND
Pleistocene -Recent
Kirthar Group
Ghazij Formation
Dunghan Formation
Thrust Faults
Anticlines
Synclines
0 10 km

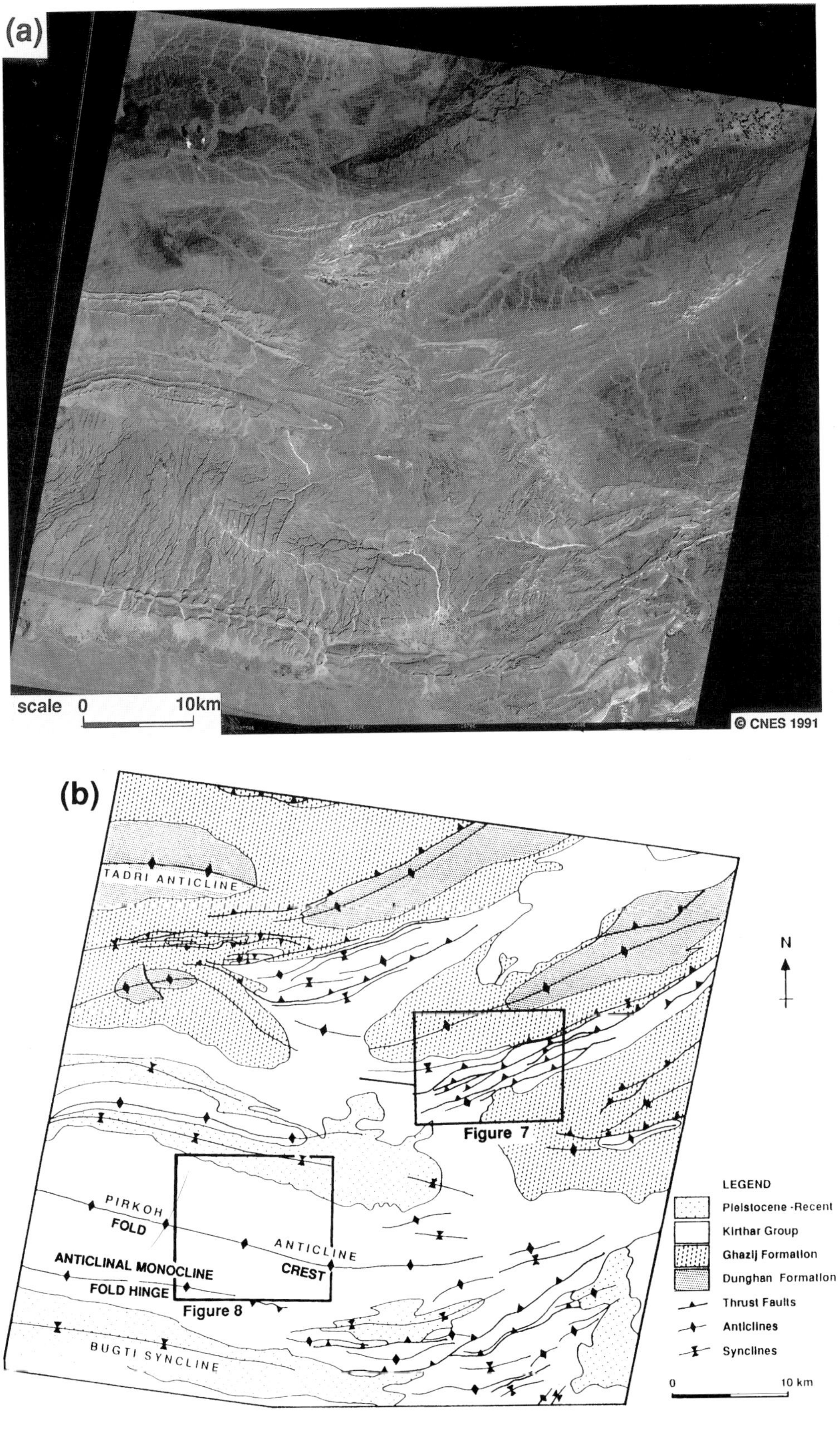

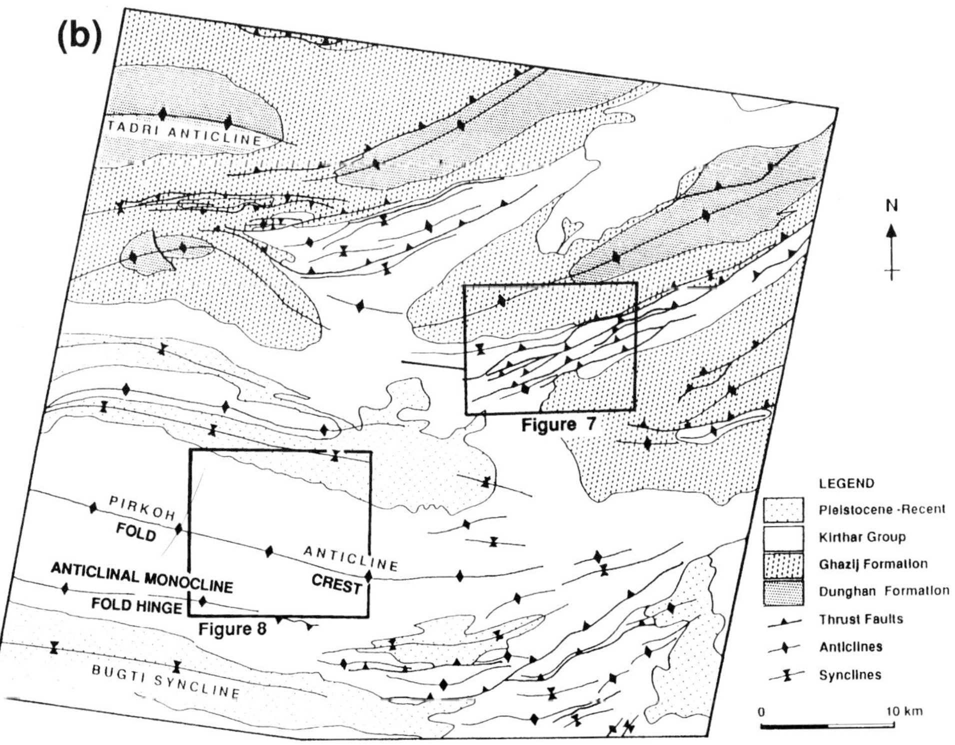

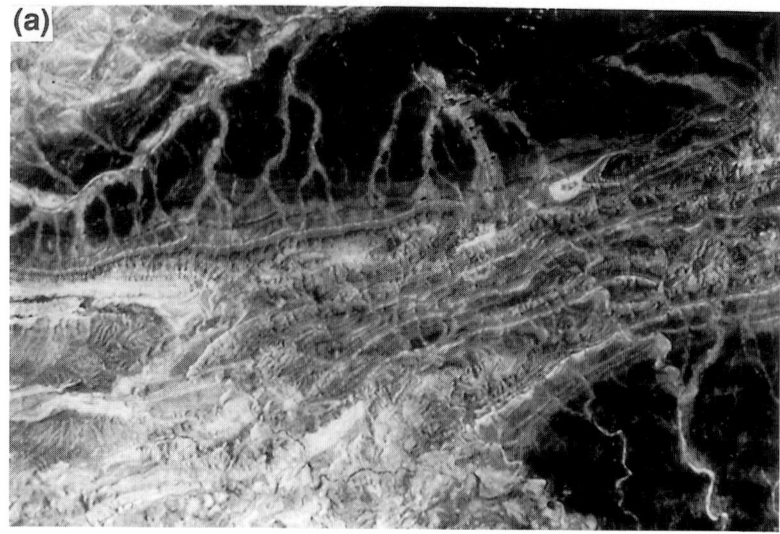

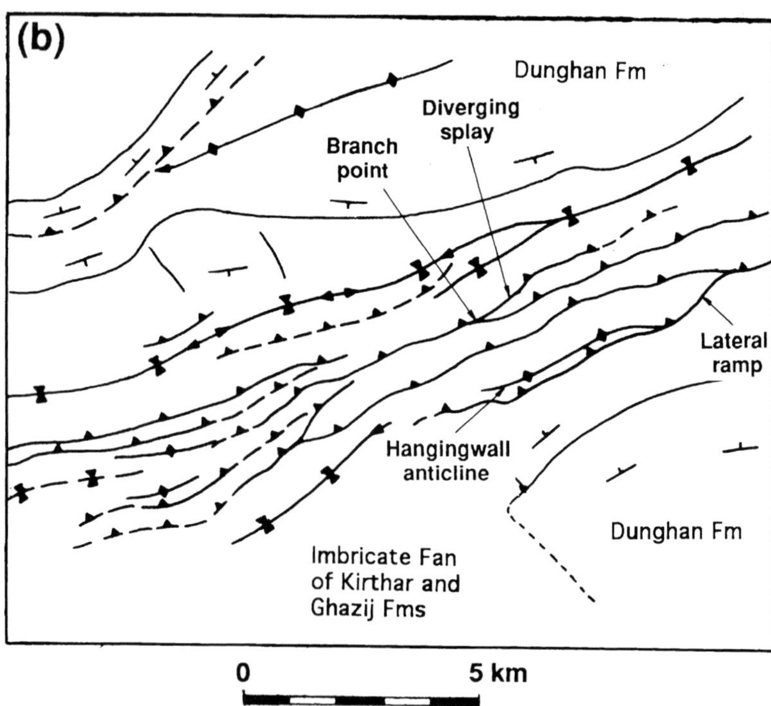

Fig. 7. (**a**) Landsat TM image of south verging imbricate fan developed in Ghazij and Kirthar strata, north of the Pirkoh anticline. (**b**) Structural interpretation of the Landsat TM image. Location of imbricate thrusts can be interpreted from the repetition of stratigraphy. Dip direction of bedding is determined from flat irons (triangular facets of exposed bedding surfaces).

it changes from a box-style to parallel-style fold geometry (Fig. 9). This large scale variation in fold geometry may be related to an along-strike change from fault-bend folding to fault-propagation folding in the subsurface. Smaller scale parasitic folds are probably related to disharmonic buckle folding of the Kirthar Formation.

The Pirkoh Anticline is transected by numerous NE- to NNE-trending, vertical to steep, west dipping normal faults (Fig. 8). Some faults show arcuate surface traces due to the intersection of the planar fault surfaces with the folded bedding surface. A complimentary set of more northerly trending minor faults show abutting relationships to the major faults. Examination of structure contours (Fig. 8b) indicate that displacements are in the order of tens to hundreds of metres.

Section validation of the Pirkoh anticline

A regional cross-section was constructed through the Pirkoh area (Fig. 5) close to the Pirkoh-1 well using GEOSEC section balancing software. The section was constructed using structural and lithological information interpreted from stereo-SPOT and Landsat TM satellite imagery respectively. The subsurface structure was constrained by published regional seismic profiles (Jadoon *et al.* 1992, 1994), and additional seismic data specific to the Pirkoh structure (Fig. 10). Thicknesses for the Kirthar–Chiltan sequence have been incorporated into the section from the Pirkoh-1 well and published data (Banks & Warburton 1986; Jadoon *et al.* 1992, 1994).

Seismic profiles through the Pirkoh and Loti Anticlines, show that there is a contrast in structural style between the two structures. The Pirkoh structure is interpreted as a fault-bend fold so that the surface structure broadly mimics the geometry of the underlying thrust sheet which repeats Jurassic and older strata. The frontal monoclinal fold along the southern margin of the Pirkoh Anticline corresponds to the position of the fault-bend fold at the leading edge of the underlying thrust sheet. Pirkoh also appears to mark the front of the buried thrust system as no stratigraphic repetition appears to occur below the Loti Anticline (Figs 10 and 11). In contrast Loti is more symmetrical and is interpreted as a buckle fold.

The Pirkoh area is interpreted in terms of a thin-skinned, 'passive-roof' duplex system developed above a 2.5° north-dipping basement. Chiltan-Kirthar stratigraphy at Pirkoh, is approximately 5000 m above its regional level in the undeformed foreland.

Structural continuity at depth is implied by the presence of 'layer-cake' stratigraphy (Fig. 11). However, a major problem is encountered when attempting to model the scale of structures developed in the Ghazij Formation. This is further complicated by internal deformation below the resolution of the imagery and is therefore assumed to approximate to ductile deformation at the 1:50 000 scale of the interpretation. The Ghazij Formation does not appear to have maintained constant bed length or layer thickness.

A fault-bend-fold mechanism was used to model horses within the duplex, and parallel concentric style folding was assumed for the top Dunghan to top Chiltan part of the roof complex (Figs 5 and 11a). This information was then combined with the surface structure as interpreted from the satellite imagery. Disharmony between the Kirthar and top Dunghan fold geometries is therefore assumed to have been accommodated by 'ductile' deformation within the Ghazij Formation. This is supported by thickened Ghazij section in the Pirkoh-1 well located near the fold hinge (Fig. 11a). This creates a problem in relating the topographic and structural closure defined at surface for the top Pirkoh limestone with closure in the top Dunghan. Internal deformation within the Ghazij shale is implied by the very irregular trajectory of the well after area balancing of the Ghazij shale back to its restored state (Fig. 11b). The Ghazij Formation was restored by area balancing after it proved impossible to achieve a valid restoration assuming constant line length alone.

Structural modelling of the Pirkoh anticline poses serious questions about the siting of the Pirkoh-1 well. The well appears to have been located on a surface closure defined by the topographically highest point on the fold surface. Therefore, although this may be an ideal placing for closure in the Habib Rahi play, it is not consistent with the position of closure in the sub-Ghazij section, such as the Pab Sandstone play (Fig. 11a).

Validation of a reactivated extensional fault model, Yemen

In this example, development of a well constrained structural interpretation of an area of extensional faulting in southeast Yemen was carried out in two stages. The first stage was a regional interpretation based on 1:250 000 and 1:500 000 scale Landsat TM and 1:2 000 000 scale MSS (USGS 1981) to define the overall structural trends and identify large scale structural features. The regional geology map of Beydoun (1964) was used for the correlation of lithological units mapped out from the satellite imagery. The second stage involved detailed structural and lithological analysis of an area of interest identified from the regional study, using a combination of stereo-SPOT and Landsat TM imagery and DEM-derived structure contours.

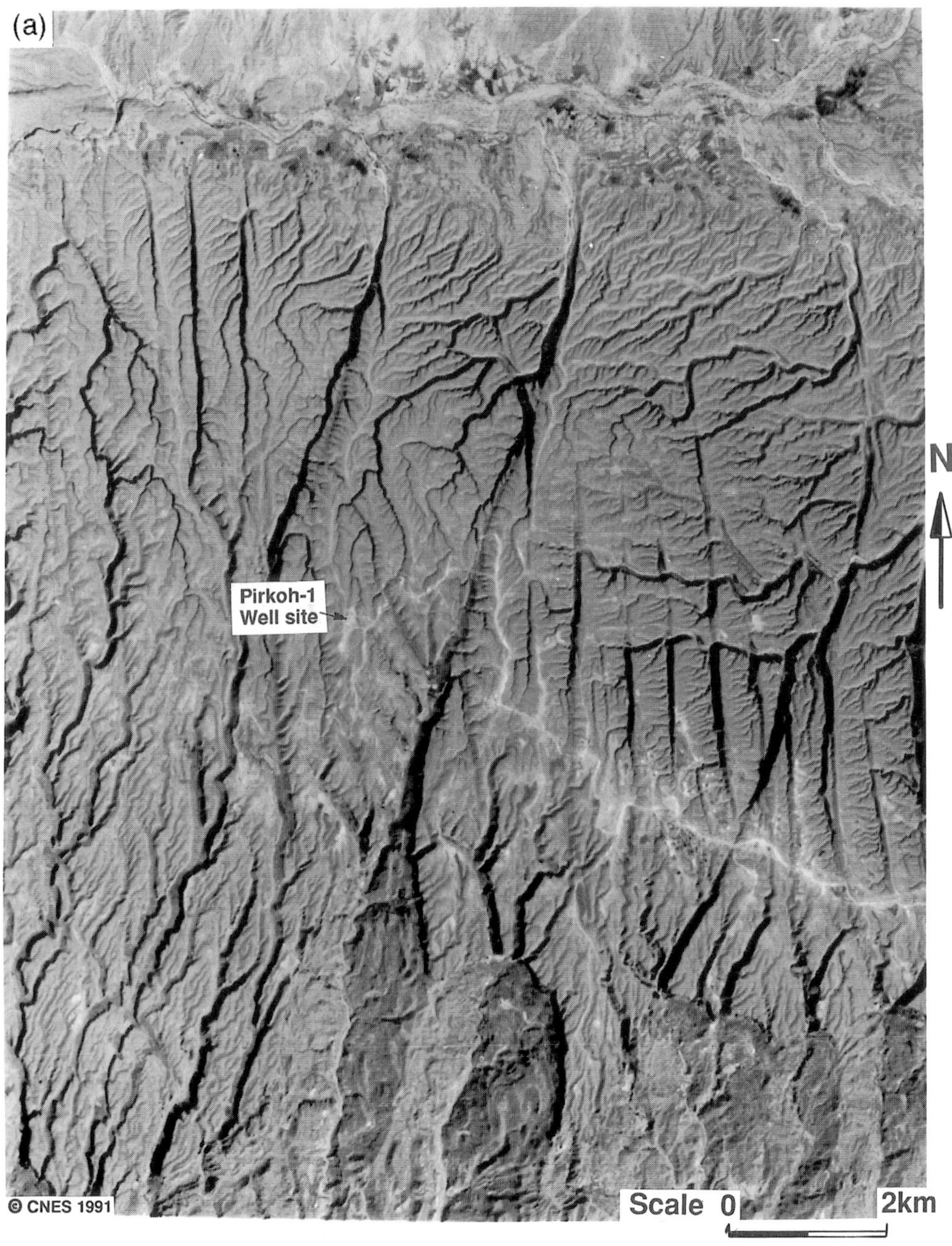

Fieldwork was carried out to verify the accuracy of the image interpretation and to collect kinematic information and surface dip data from features below the spatial resolution of the imagery. Seismic data was used to assess the relationship between the Tertiary faults at surface and the Mesozoic graben architecture buried below the surface, and to relate the surface structure to subsurface Mesozoic faults which controlled the distribution of reservoir and source rocks.

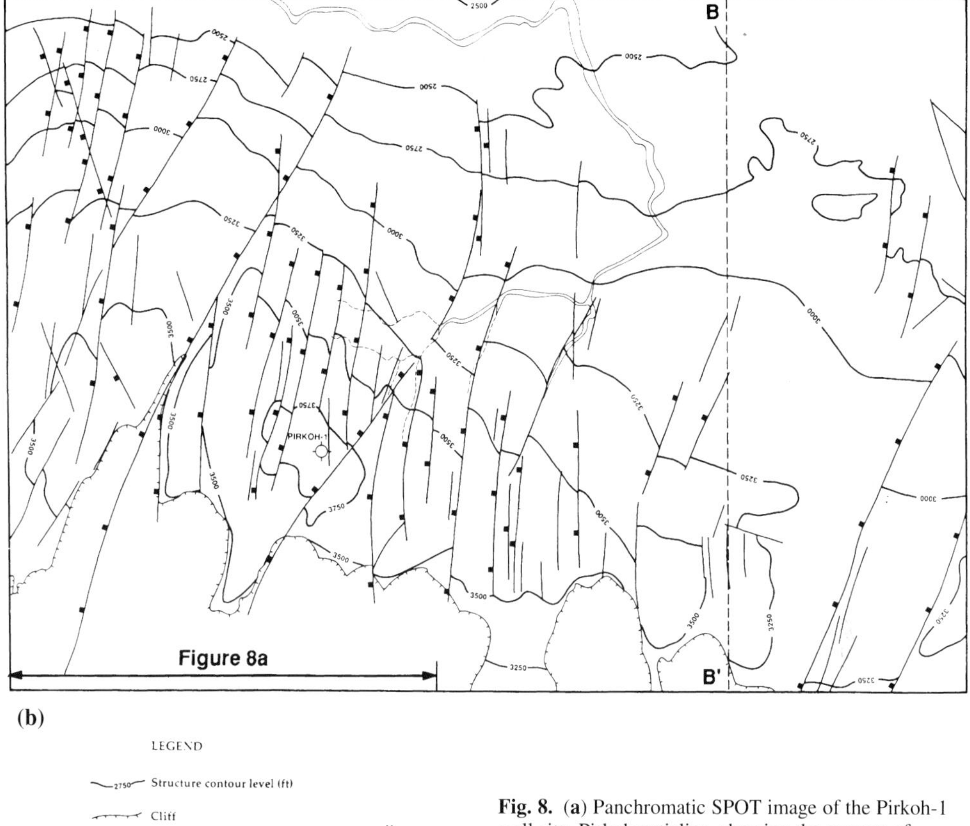

Fig. 8. (**a**) Panchromatic SPOT image of the Pirkoh-1 well site, Pirkoh anticline, showing the amount of structural detail visible in the SPOT imagery © CNES 1991 – Distribution Spot Image. (**b**) Structure contour map for the top Pirkoh limestone corresponding to SPOT image of the Pirkoh fold in (a). Note position of the Pirkoh-1 well located on surface topographic high.

Regional scale structural analysis

The regional structure (Fig. 12) of the Hadramaut region is dominated by an E–W system of Oligocene–Miocene normal faults related to the opening of the Gulf of Aden (Cochran 1981; Bohannon *et al.* 1989; Hughes *et al.* 1991; Bott *et al.* 1992). The Tertiary faults are superimposed on an earlier Mesozoic rift system which controls the petroleum prospectivity of the area (Beydoun 1989, 1991; Haitham & Nani 1990; Paul 1990; Bott *et al.* 1992; Mills 1992; Beydoun *et al.* 1993).

The faults were mapped out from the interpretation of textural discontinuities on the imagery and their relative sense of movement determined from the facing direction of footwall fault scarps, the position of shadows and presence of incised wadis along the up-thrown side. The faults are predominantly steeply inclined, planar, dip-slip structures and have linear to slightly arcuate surface traces. The northern limit of extensional faulting observed on the imagery marks the on-shore limits of late Tertiary extension i.e. the rift margin of the Gulf of Aden. The rift shoulder can also be identified just inland from the southern coast of Yemen (Fig. 12).

Two northerly trending zones occur to the east and west of Al Mukalla forming the margins of an area in which extensional faults, associated with the Canadian Occidental oil fields, down throw consistently to the south (Fig. 12). Outside this zone, faults down throw dominantly to the north. This reversal in fault polarity is probably due to the presence of regional transfer zones in the basement. A relatively narrow NNW- to NW-trending zone of en echelon normal faults also occurs to the west of the Fartaq High (Fig. 12). The orientation of the

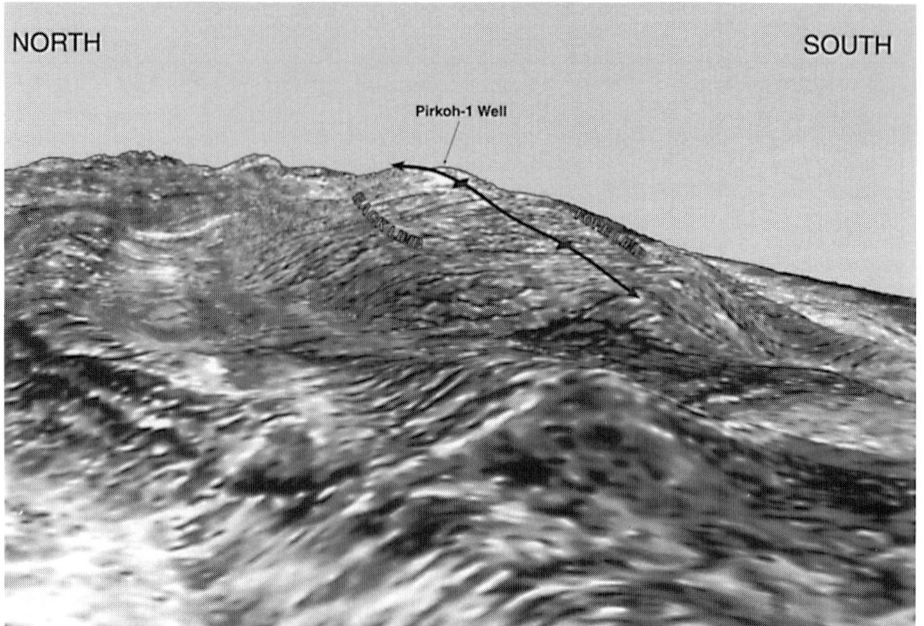

Fig. 9. Perspective view of the Pirkoh anticline viewed along the axis of the fold from the west. View highlights the broad undulating nature of the fold surface.

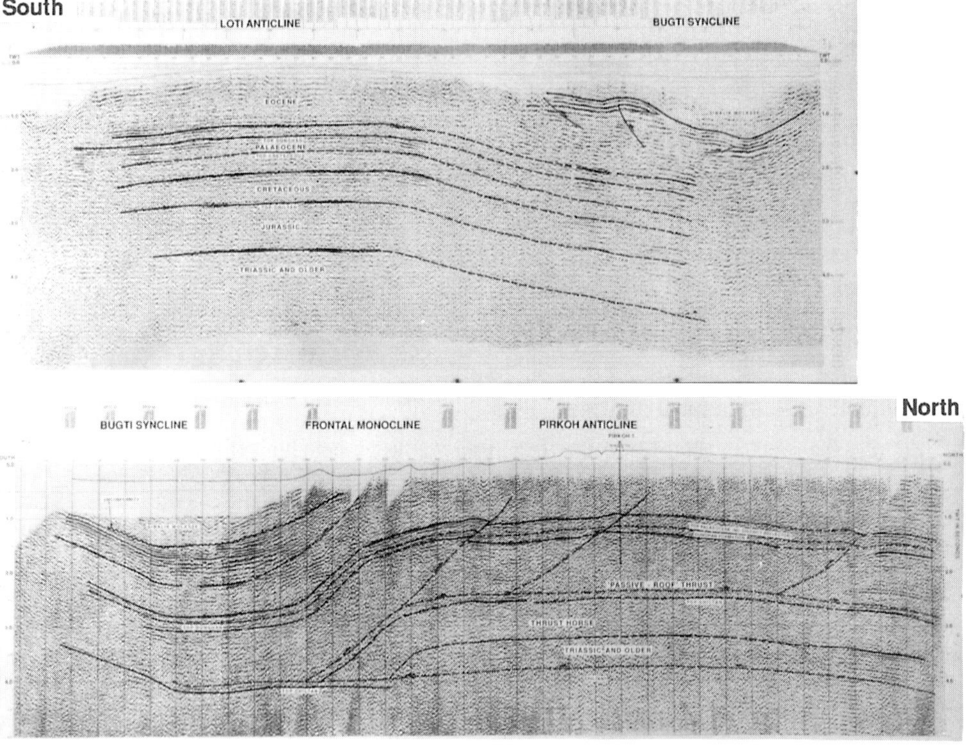

Fig. 10. Interpreted seismic profiles through the Pirkoh and Loti anticlines, showing contrast in fold style and interpretation of Pirkoh as the frontal fault-bend-fold.

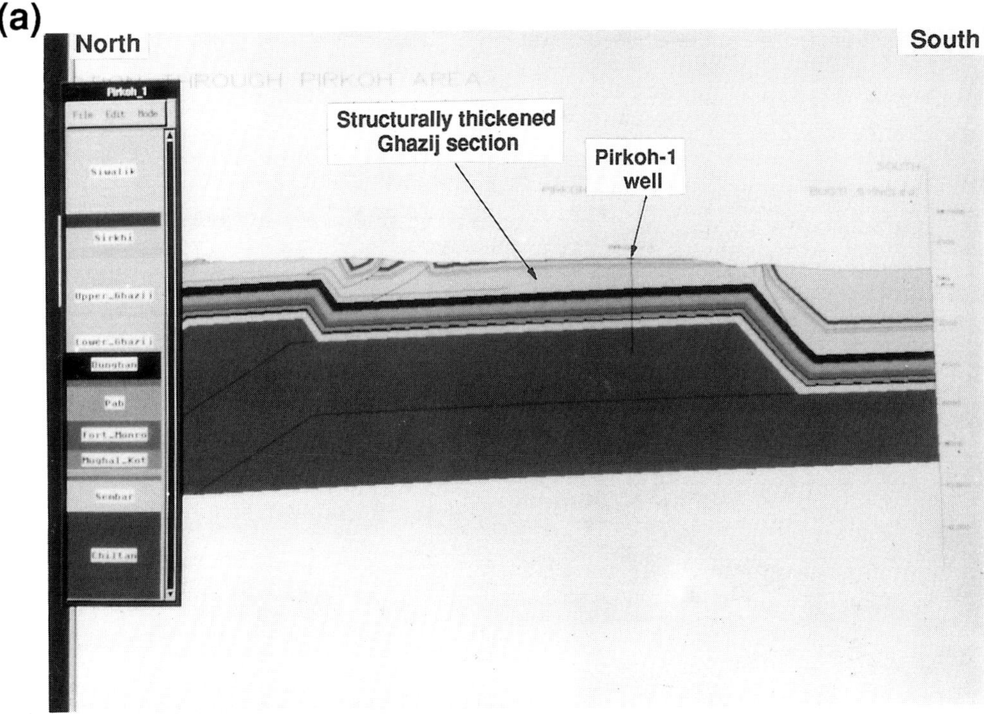

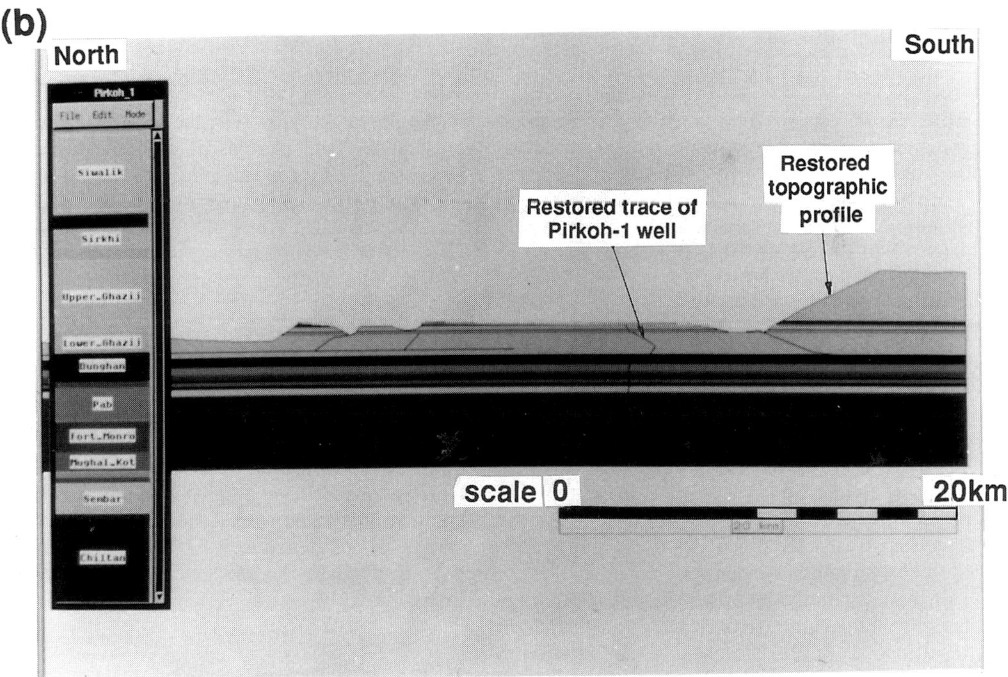

Fig. 11. Balanced deformed (**a**) and restored (**b**) cross-sections of the Pirkoh anticline constructed from satellite image interpretation using GEOSEC. Note irregular shape of the Pirkoh-1 well within the Ghazij Formation of restored section. This suggests that line length and area has not remained constant during deformation for this unit.

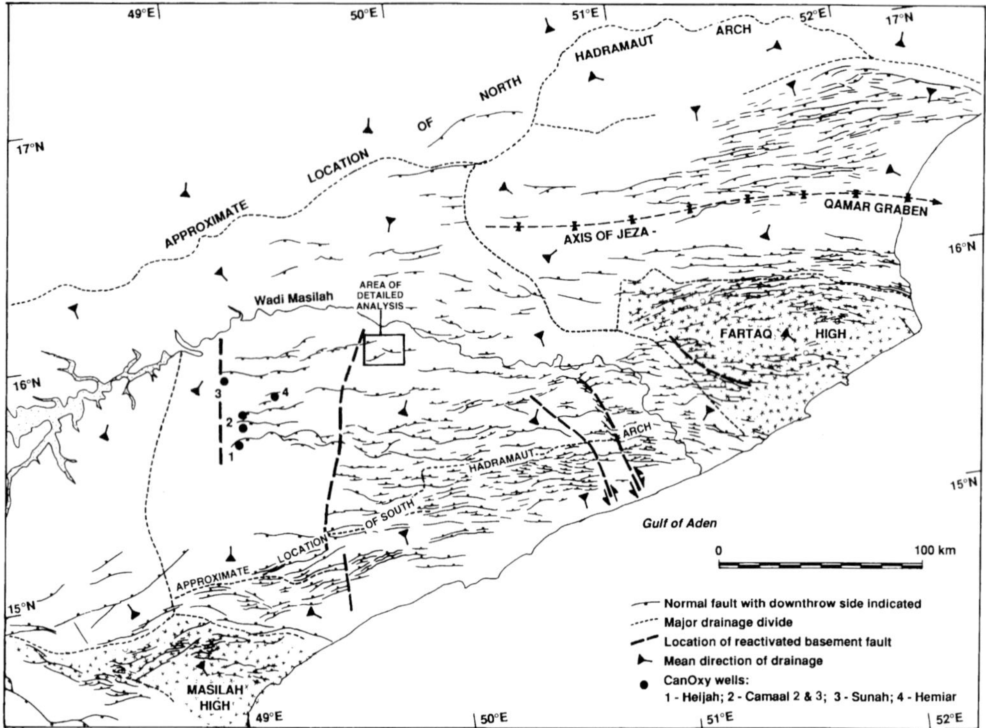

Fig. 12. Regional Landsat TM interpretation of the Hadramaut region of southern Yemen highlighting surface trace of Late Tertiary faults and major structural elements.

faults to the shear zone boundary indicates a component of relative sinistral movement across the zone. The presence of these inferred transfer zones has led to an extrapolation of the Nadj Fault System into southern Yemen (Moore 1979; Brown *et al.* 1989; Husseini & Husseini 1990; Bott *et al.* 1992; Beydoun *et al.* 1993).

A number of structural domains can be interpreted from the imagery which are characterized by relatively sparse Late Tertiary faulting and dendritic drainage patterns. These areas are coincident with the Mesozoic age Hadramaut, Mukalla and Fartaq structural highs (Bott *et al.* 1992; Beydoun *et al.* 1993) (Fig. 12). Regional drainage divides appear to define the margins of these domains, whereas the dominant drainage directions reflect the dip slopes of the larger scale Mesozoic structural elements.

In contrast, areas of relatively intense normal faulting define Late Tertiary half-graben systems superimposed on the Mesozoic basins, referred to as the Jeza Trough–Qamar Basin and East Shabwa Basin. The distribution of surface faults suggests that the Jeza Trough narrows to the west (Fig. 12) as the presence of major normal faults diminishes in that direction. However, the significance of the present day fault patterns at surface can only be determined by understanding the control of the Mesozoic extensional structures on their development. This can only be achieved through integration of additional data sets such as seismic, with the satellite image interpretation.

Prospect level structural analysis

An area was selected where the detailed interpretation of structure and stratigraphy using 1:50 000 scale stereo panchromatic SPOT (Fig. 13a) and band 2, 4, 7 false colour Landsat TM could be integrated with seismic data. Band 2 was chosen in preference to band 1 due to a problem with data quality of the latter on the particular image.

Stratigraphy

The stratigraphy exposed at surface ranges from Paleocene to Middle Eocene (Fig. 13b). As the strata have very low dips, generally less than 3°, the outcrop pattern of the stratigraphic boundaries (Fig. 13b) mimics the topographic contours (Fig. 14c). In general, all of the stratigraphic divisions identified on the basis of colour on Landsat TM, can also be readily differentiated on the SPOT imagery (Fig. 13a and b) by grey tone, style of weathering, surface texture and slope profile.

The oldest exposed unit is the Paleocene Umm er Radhuma Formation which is typically cliff-forming, and exposed along the edges of wadis and ravines and along fault scarps (Fig. 13b). It consists of an upper section of nodular carbonate and a lower section of bedded dolomitic limestone. The contact between the Umm er Radhuma Formation and the overlying Jeza Formation, is marked by a distinct break in slope, and was used for constructing the structure contour map (Fig. 13d).

The Lower Eocene Jeza Formation is approximately 100 m thick and consists of alternating units of shale and argillaceous limestone. The resistant weathering limestone intervals create a terraced profile allowing the Jeza Formation to be divided into three subdivisions (Fig. 13a and b).

The Jeza Formation is overlain by the anhydrite-dominated Rus Formation which is characterized by an uneven, dendritic weathering pattern. The anhydrite is characterized by a distinct blue spectral signature on the false colour Landsat TM. Bedding or layering is generally below the spatial resolution of both Landsat TM and SPOT so that the Rus Formation appears to be massive on the imagery.

Structure

The selected area contains a number of E–W-trending extensional fault blocks, 5–10 km in width, bounded by north downthrowing, steeply inclined normal faults with slightly arcuate surface traces (Fig. 13a and b). The major normal faults can be readily identified on the imagery by the presence of large fault scarps. These scarps become less pronounced towards the fault tips. Faults are also expressed at surface by linear erosion features, abrupt stratigraphic terminations and juxtaposition of contrasting lithologies.

In detail the block bounding faults are often composed of a number of segments, linked by means of relay zones. Second order synthetic and antithetic normal faults are also developed adjacent to the major faults. Backward rotation in the hanging walls has produced shallow southward dips towards the fault surfaces which, when combined with hanging-wall drag immediately adjacent to the fault surface, has resulted in the generation of hangingwall synclines. Large displacement faults are associated with pronounced fault scarps. Small scale monoclinal folds also occur within the Jeza Formation and are probably a surface manifestation of underlying 'blind' faults within the Umm er Radhuma. Comparison of the fault systems over a range of scales from the regional to prospect level image interpretation, as well as in the field show that normal faulting is consistent with a 'soft-domino' model described by Walsh & Watterson 1991).

Indication of the relative timing of structures and movement along faults cannot often be determined from the satellite imagery alone. Consequently additional information from the field, on the orientation, dimensions, amount of off-set, and sense of displacement of the fault surfaces was used to improve the accuracy of the interpretation. However, surface dip measurements derived from the DEM, were often found to be more accurate than bedding measurements taken in the field, particularly where the bedding surfaces were poorly exposed, rotated by slumping, and highly irregular and pitted due to careous weathering. A structure contour map (Fig. 13d) was constructed to provide a semi-quantitative analysis of the displacement patterns around the faults. The map highlights the regional dip slopes as well as the location of 'local' surface closures, footwall drag and hangingwall roll-over adjacent to the normal faults (Fig. 13d).

Integration of seismic data

The structural continuity and trend of faults interpreted from the satellite imagery helps to constrain the correlation of faults between seismic lines. Seismic profiles show that faults at surface with vertical throws of 25 m or more usually form the upward continuation of Mesozoic faults (Figs 13e and 14) reactivated as normal faults during the Late Tertiary extension. This suggests, for at least this part of the Jeza Trough, that the ESE–WNW en echelon arrays of smaller scale faults at surface are likely to be related to larger displacement, laterally more continuous Mesozoic faults in the subsurface. The mapping of surface faults on satellite imagery may therefore help to establish the distribution and polarity of the buried Mesozoic structures.

Although unconformities exist in the stratigraphic section at the top of the Jurassic syn-rift (Naifa) and within the Cretaceous post rift (base Harshiyat and Umm er Radhuma) sequences (Mills 1992), the 'layer-cake' nature of the post-Mesozoic rift sequence (Fig. 14) suggests that the geometry of the top Qishn may be at least partially determined from the surface structure. This is reflected by the general correlation between the DEM-derived structure contours for the top Umm er Radhuma and the TWT seismic map for the top Qishn (Fig. 13e).

(a)
© CNES 1991

(b)
El
Pau
(25)
(25)
Pau
El
El
Pau
El
Pau

(d)

700
650
600
700 M
650
600
BASEMENT
SEISMIC LINE 1
SEISMIC LINE 2
SEISMIC LINE 3

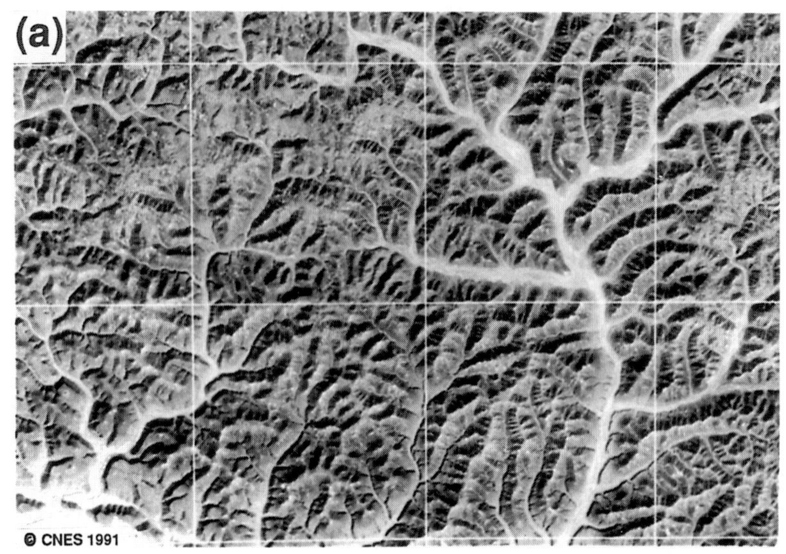

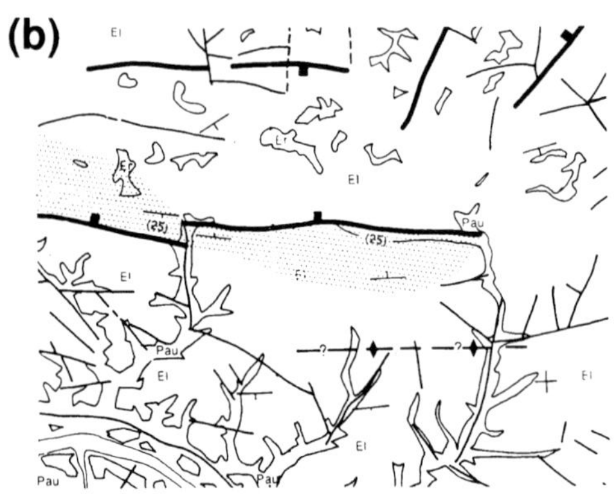

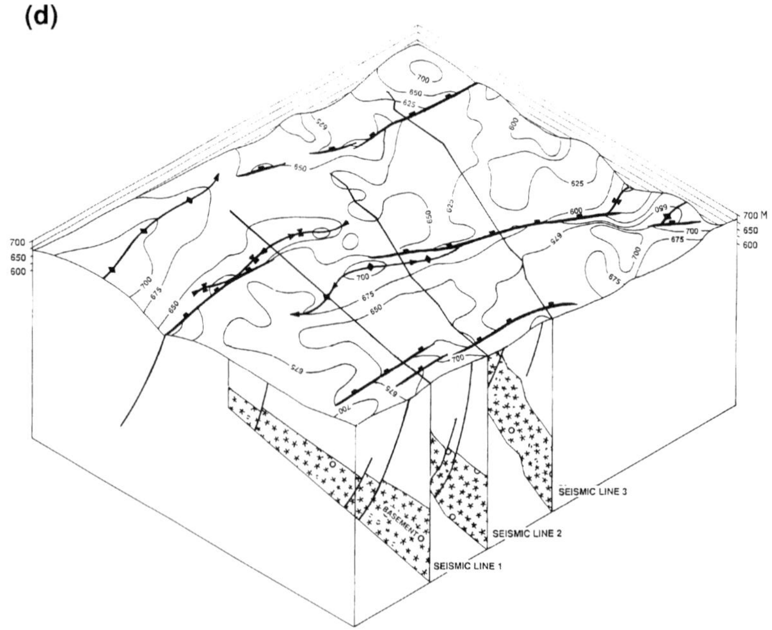

(e)

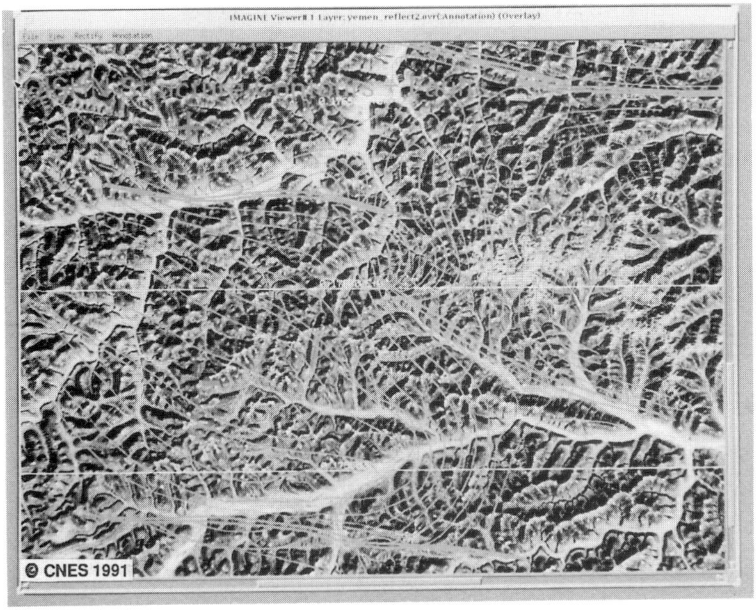

© CNES 1991

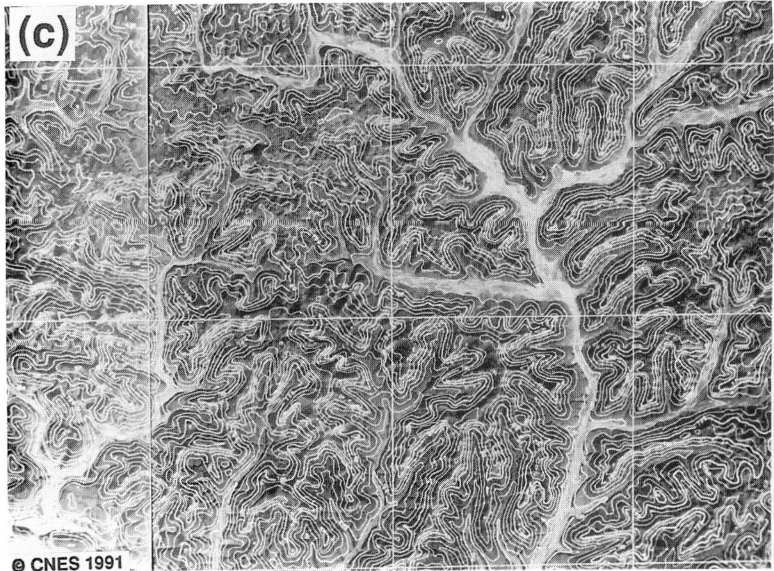

Fig. 13. Example of panchromatic SPOT imagery and derived products, together with interpretation and data integration for a detailed prospect scale analysis. (**a**) Example of ortho-rectified SPOT image. (**b**) Geological interpretation of stereo-SPOT imagery.(**c**) Topographic contours derived from stereo-SPOT generated DEM. (**d**) Isometric diagram of structure contours for top Umm er Radhuma using height information derived from DEM. (**e**) Seismic TWT map for top Qishn. (a), (c) & (e) © CNES 1991 – Distribution Spot Images.

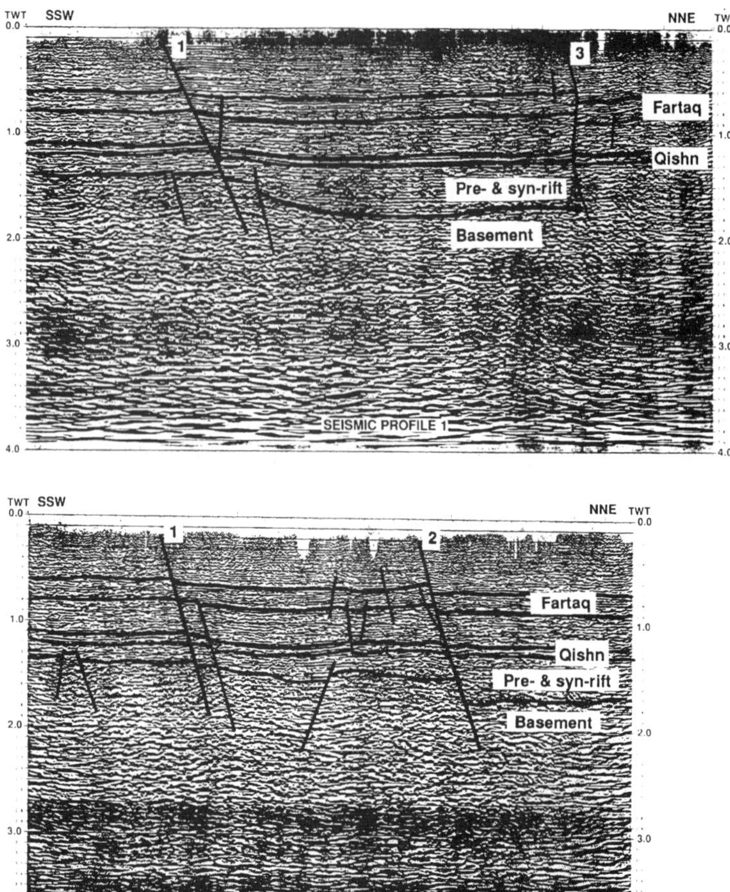

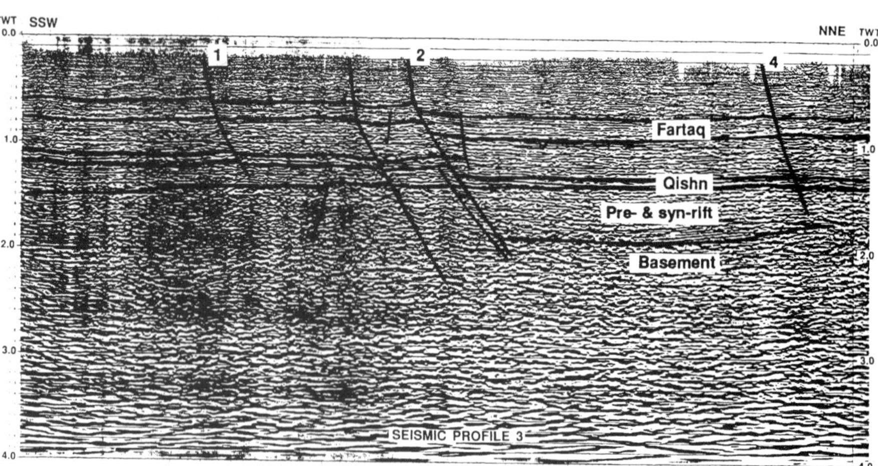

Fig. 14. Interpreted seismic sections for study area showing relationship of Late Tertiary normal faults to reactivated Mesozoic faults.

Fig. 15. Photograph showing typical outcrop pattern and style of faulting present in the study area. Top UER, top Umm er Radhuma Formation (Palaeocene); Top J1, top lower Jeza Formation (lower Eocene).

Implication for interpretation of subsurface structure

The main focus of exploration interest in Southeast Yemen is in the identification of the subsurface Jurassic basins. However, the analysis of the surface structure can provide important information on the underlying regional and local scale Mesozoic structure. The position, trend and relative sense of downthrow of the major Tertiary extensional faults are controlled by the presence of Jurassic fault blocks at depth, with Tertiary half-grabens often developed above extensionally reactivated Jurassic tilt blocks. Comparison with known hydrocarbon discoveries in the region, such as CanOxy's Sunah, Camaal and Heijah and Hemiar fields (Mills 1992; Gurney 1994), can also play an important part of the validation process. For example a stereo-SPOT interpretation of the surface structural features of these fields might provide a useful analogue for evaluating similar structures elsewhere in southern Yemen.

Summary

The examples from Pakistan and Yemen demonstrate how remotely sensed data can be applied to regional and prospect scale structural interpretations. Of particular significance is the range of products which can now be generated from readily available stereo imagery, which has undoubtedly led to the increasing use of satellite imagery for exploration. Significant developments are the ability to construct geological cross-sections and structure contour maps from DEM-derived data. DEMs can be used for the three-dimensional visualisation of structures and are also proving useful for applying accurate terrain corrections to geophysical surveys.

There is also an increasing trend toward merging geophysical data sets with the satellite imagery so that the image interpretation can also incorporate information relating to subsurface features. The satellite image interpretation should also be verified by fieldwork. Field data can provide information on the kinematics and relative timing of structures which cannot be ascertained from satellite imagery.

Selection of the correct type of satellite imagery for a particular project is important and is dependent on wide a range of factors which include: the size of the study area, scale at which the imagery is to be used; whether panchromatic or multispectral data are required; geographical location and climate; extent of vegetation cover or cultivation relative to the amount of outcrop; whether a regional or prospect level analysis is required.

The tectonic regime and intensity of deformation also influences the usefulness of imagery in the interpretation of geological structures. In general, compressional environments are associated with the best surface expression of structure as they are associated with positive changes in surface relief.

In contrast, extensional terrains are associated with negative changes in relief and are usually buried below more recent sediments, unless the area is currently undergoing active extension. Strike-slip terrains and reactivated structures have variable surface expressions on satellite imagery.

Satellite imagery also provides a cost effective method of obtaining a detailed stratigraphic and structural interpretation of an area, in comparison to a detailed field mapping programme. The acquisition and interpretation of satellite imagery at an early stage in an exploration campaign can have important implications for subsequent seismic surveys. It allows a preliminary, pre-seismic, structural interpretation to be ascertained, which can then be used to constrain the location of seismic lines and provide a base map from which seismic surveys can be planned.

It is hoped that this paper has helped to demonstrate the value of satellite imagery in the development of accurate structural interpretations. The interpretation process should involve the integration of as much information as possible in order to generate a valid structural model. Satellite imagery should be considered more than just a pretty picture and rather a valuable source of data for the structural validation of an area.

We gratefully acknowledge the permission of British Gas Exploration and Production, Coplex, Hoodoil, Idemitsu, LASMO and Tullow Oil for giving us permission to publish relevant extracts of satellite imagery and seismic data. SPOT Image is thanked for permission to publish extracts of SPOT satellite imagery. Special thanks are also due to Gary Bowes, Anita Dolan, Paul Grimes and Richie Toomey for their assistance in the draughting and digitising of figures. T. Dessauvagie, D. Nieuwland, and colleagues at NRSC and ERA-Maptec are thanked for their critical reviews of the manuscript.

References

ATTEMA EVERT, P. W. 1991. The active microwave instrument on board the ERS-1 satellite. *Proceedings of the Institute Electronical & Electronic Engineers,* **79**, 791–799.

BANKS, C. J. & WARBURTON, J. 1986. 'Passive-roof' duplex geometry in the frontal structures of the Kirthar and Sulaiman mountain belt, Pakistan. *Journal of Structural Geology,* **8**, 229–237.

BEYDOUN, Z. R. 1964. *The stratigraphy and structure of the Eastern Aden Protectorate.* Overseas Geology Mineral Resources Bulletin Supplement, **5**. HMSO, London.

—— 1989. The hydrocarbon prospects of the Red Sea – Gulf of Aden: A review. *Journal of Petroleum Geology,* **12**, 125–144.

—— 1991. *Arabian plate hydrocarbon geology and potential – a plate tectonic approach.* American Association of Petroleum Geology Study Geologists, **33**.

——, BAMAHMOUD, M. O. & NANI, A. S. O. 1993. The Qishn Formation, Yemen: lithofacies and hydrocarbon habitat. *Marine and Petroleum Geology,* **10**, 364–372.

BOHANNON, R. G., NAESER, C. W., SCHMIDT, D. T. & ZIMMERMAN, R. A. 1989. The timing of uplift, volcanism and rifting peripheral to the Red Sea: a case for passive rifting? *Journal of Geophysical Research,* **94**, 1683–1701.

BOTT, W. F., SMITH, B. A., OAKES, G., SIKANDER, A. H. & IBRAHAM, A. I. 1992. The tectonic framework and regional hydrocarbon prospectivity of the Gulf of Aden. *Journal of Petroleum Geology,* **15**, 211–243.

BROWN, G. F., SCHMIDT, D. L. & HUFFMAN, A. C. JR. 1989. Shield area of western Saudi Arabia. *In*: *Geology of the Arabian Peninsula.* US Geological Survey Professional Paper, **560-A**.

COCHRAN, J. R. 1981. The Gulf of Aden: structure and evolution of a young ocean basin and continental margin. *Journal of Geophysical Research,* **86**, B1, 263–287.

FREDEN, S. C. & GORDON, F. 1983. Landsat Satellites. *In*: COLWELL, R. N. (ed.) *Manual of Remote Sensing.* 2nd edition, American Society of Photogrammetry and Remote Sensing, Falls Church, Virginia, **1**, 517–578.

GESS, G. J., CHROWICZ, B., BECUE, B., CURNELLE, R., DEROIN, J. P., HUGER, J., PERRIN, G. & RONFOLA, D. 1986. Methodology for the use of SPOT imagery in petroleum exploration. *In*: *SPOT 1 image utilization assessment, results.* Centre National d'Etudes Spatiales, Toulouse, 811–819.

GUGAN, D. J. & DOWMAN, I. J. 1988. Topographic mapping from SPOT imagery. *Photogrammetric Engineering and Remote Sensing,* **54**, 1409–1414.

GURNEY, J. 1994. Canadian Occidental joins Hunt as Yemen oil producer. *Petroleum Review,* 72–75.

HAITHAM, F. M. S. & NANI, A. S. O. 1990. The Gulf of Aden Rift: Hydrocarbon potential of the Arabian Sector. *Journal of Petroleum Geology,* **13**, 211–220.

HUGHES, G. W., VAROL, O. & BEYDOUN, Z. R. 1991. Evidence of Middle Oligocene rifting of the Gulf of Aden and for Late Oligocene rifting of the Southern Red Sea. *Marine Petroleum Geology,* **8**, 354–358.

HUMAYON, M., LILLIE, R. J. & LAWRENCE, R. D. 1991. Structural interpretation of eastern Sulaiman foldbelt and foredeep, Pakistan. *Tectonics,* **10**, 299–324.

HUSSEINI, M. I. & HUSSEINI, S. I. 1990. Origin of the Infracambrian Salt basins of the Middle East. *In*: Brooks, J. (ed.) *Classic Petroleum Provinces.* Geological Society, London, Special Publications, **50**, 279–292.

IZATT, C. N. 1990. *Variation in thrust front geometry across the Potwar Plateau and Hazara/Karachitta hill ranges, northern Pakistan.* PhD thesis, Imperial College of Science Technology and Medicine, London.

JADOON, I. A. K., LAWRENCE, R. D. & LILLIE, R. J. 1992. Balanced deformed geological cross-section from the frontal Sulaiman Lobe, Pakistan: Duplex development in thick strata along the western margin of the Indian Plate. *In*: McClay, K. R. (ed.) *Thrust Tectonics*. Chapman and Hall, 343–356.

——, —— & —— 1993. Evolution of foreland structures, an example from the Sulaiman thrust lobe, Pakistan, SW of the Himalayas. *In*: Treloar, P. M. & Searle, M. (eds) *Himalayan Tectonics*. Geological Society, London, Special Publications, **74**, 589–603.

——, —— & —— 1994. Seismic data, geometry, evolution, and shortening in the active Sulaiman fold-and-thrust belt of Pakistan, southwest of the Himalayas. *Bulletin of American Association of Petroleum Geologists*, **78**, 758–774.

Kazmi, A. H. & Rana, R. A. 1982. *Tectonic map of Pakistan, 1:2,000,000*. Geological Survey of Pakistan.

Kramber, W. J., Morse, A., Zarriello, T. J. & Britton, B. R. 1990. Idaho's Centennial Poster: Producing a State-Wide Digital Mosaic From Landsat. *In*: *Technical Papers 1990*. American Congress on Surveying and Mapping – American Society for Photogrammetry and Remote Sensing Annual Convention, **4**, 253–259.

Light, D. L. 1990. Characteristics of remote sensors for mapping and earth science applications. *Photogrammetric Engineering and Remote Sensing*, **57**, 1613–1623.

Mills, S. J. 1992. Oil discoveries in the Hadramaut: how CanadianOxy scored in Yemen. *Oil and Gas Journal Special*, 49–52.

Moore, J. M. 1979. Tectonics of the Najd transcurrent fault system, Saudi Arabia. *Journal of the Geological Society of London*, **136**, 441–454.

NASA 1976. *Landsat data users handbook*. Goddard Space Flight Centre, Doc. **76SDS-4258**, Greenbelt, M.D.

NASDA 1990. *Earth Resources Satellite-1*. RESTEC/NASDA, Japan.

Nemoto, Y., Nishino, H., Ono, M., Mitzutamari, H, Nishikawa, K. & Tanaka, K. 1991. Japanese earth resources Satellite-1 synthetic aperture radar. *Proceedings Institute Electronical & Electronic Engineers*, **76**, 800–809.

Nishidai, T. 1993. Early results from Japan's Earth Resources Satellite (JERS-1). *International Journal of Remote Sensing*, **14**, 1825–1833.

Paul, S. K. 1990. People's Democratic Republic of Yemen: a future oil province. *In*: Brookes, J. (ed.) *Classic Petroleum Provinces*. Geological Society, London, Special Publications, **50**, 329–339.

Quittmeyer, R. C., Farah, A. & Jacob, K. H. 1979. The seismicity of Pakistan and its relation to surface faults. *In*: Farah, A. & De Jong, K. A. (eds) *Geodynamics of Pakistan*. Geological Survey of Pakistan, Quetta, 351–358.

——, —— & —— 1984. The focal mechanism and depths of earthquakes in central Pakistan: a tectonic interpretation. *Journal of Geophysical Research*, **89**, 2459–2470.

Raza, H. A., Ahmed, R., Alam, S. & Ali, S. M. 1989a. Petroleum zones of Pakistan. *Journal of Hydrocarbon Research*, **1**, 1–20.

——, ——, Ali, S. M. & Alam, S. 1989b. Petroleum prospects: Sulaiman sub-basin, Pakistan. *Pakistan Journal of Hydrocarbon Research*, **1**, 21–56.

RESTEC 1992. *User's Guide for JERS-1 SAR Data Format*. Hiki-gun, Saitama-ken, Japan. Earth Observation Centre.

Sasowsky, K. C., Peterson, G. W. & Evans, B. M. 1992. Accuracy of SPOT digital elevation model and derivatives: utility for Alaska's north slope. *Photogrammetric Engineering and Remote Sensing*, **58**, 815–824.

Shah, S. M. I. 1977. *Stratigraphy of Pakistan*. Geological Survey of Pakistan, Memoir **12**.

SPOT Image Corporation 1984. *SPOT data user's handbook*. Toulouse, France.

Tainish, H. R., Stringer, K. V. & Azad, J. 1959. Major gas fields of west Pakistan. *American Association of Petroleum Geologists Bulletin*, **43**, 2675–2700.

Tyler, W. A. 1993. The Multispectral Scanner—Landsat's other sensor. *Earth Observation Magazine*, May, 28–36.

USGS 1981. *Landsat image mosaic of the Arabian Peninsula, Scale 1:2,000,000*. Open File Report, **USGS-OF-02-11**.

Walsh, J. J. & Watterson, J. 1991. Geometric and kinematic coherence and scale effects in normal fault systems. *In*: Roberts, A. M., Yielding, G. & Freeman, B. (eds) *The Geometry of Normal Faults*. Geological Society Special Publications, **56**, 193–203.

Gravity-driven nappes and their relation to palaeobathymetry: examples from West Africa and Cardigan Bay, UK

JONATHAN P. TURNER

University of Birmingham, School of Earth Sciences, Birmingham B15 2TT, U.K.

Abstract: Gravity-tectonics serves to reduce gravitational potential compared to the undeformed state. Consequently, a common setting of gravity-driven faulting is at actively prograding passive margin basins where abrupt shelf-edge relief promotes potentials sufficiently large to induce and sustain large-scale deformation. In the West African Rio Muni basin, Albian gravity-tectonics predates the Senonian drift unconformity and owes its existence to a significant bathymetric deep that characterized the late syn-rift stage of the basin. Thin-skin nappes, detaching on a syn-rift salt and/or shaly horizon, emplace shallow water shelf carbonates in a deep, basinal setting. Seismic mapping indicates that 'upslope' extension occurred contemporaneously with 'downslope' contraction and that the nappes behaved as small coulomb wedges. It is suggested that structural restoration of these large-scale nappes, retro-deforming them to their pre-deformational position at the palaeo-shelf edge, yields a minimum estimate of palaeobathymetry. In the Rio Muni basin, variations in the inferred palaeobathymetry accord with intra-basinal differences in Bouguer gravity anomaly, here interpreted as an expression of varying crustal thickness and uplift/subsidence rate. Comparable structures in the Jurassic of the UK Cardigan Bay basin occur at the margin of a basin characterized, similarly, by a major negative Bouguer gravity anomaly and which, from evidence on seismic profiles, clearly experienced deep bathymetry during Jurassic deposition.

Gravity-driven deformation reduces gravitational potential compared to the undeformed state (Ramberg 1981) and small-scale gravity-driven faults are a common feature of many syn-rift basins. In the North Viking Graben of the North Sea, for example, swarms of closely spaced, planar normal faults accommodated the gravitational collapse of prominent, rotated footwall blocks during the late Jurassic. They stood proud of the basin floor due to rapid subsidence accompanying restricted sedimentation of the condensed Kimmeridge Clay Formation syn-rift sequence (e.g. the Cormorant structure: Speksnijder 1987; the Brent structure: Livera & Gdula 1990; the Magnus structure: Potts & Henderson 1993). From seismic data, these faults appear to terminate downward rapidly and they are interpreted as superficial structures forming at a late stage in the evolution of the syn-rift basin system.

Such faults are not the principal subject of this paper which focuses on larger scale gravity-driven faulting causing the detachment and emplacement of thick shelfal successions in a deep basinal setting. They develop in response to significant gravitational potential generated by major intra-basin relief. The origin of the relief may be due to rapid subsidence, restricted sedimentation and/or, localized, halokinetic uplift. This paper investigates links between large-scale gravity-faulting and the magnitude of the relief (i.e. palaeo-water depth, or bathymetry). In particular, a method is described which retro-deforms cross-sections through large-scale gravitational nappes as a means of estimating the magnitude of the intra-basin relief.

Rationale

Bathymetry is an important parameter to constrain if rift basin evolution is to be fully understood. It is of particular relevance to basin modellers seeking to reconstruct the history of varying rates of subsidence (and uplift) during the progression from the pre- to syn- and post-rift stages of a basin. In the absence of independent evidence to the contrary, burial history modelling often assumes that sedimentation keeps pace with subsidence throughout basin evolution. Consequently, the common method for computing the contribution of sediment load to bulk subsidence history, following the back-stripping of a well, simply replaces the sediment column with an equal column of water (Chadwick 1985). Clearly, the reliability of this method will depend largely on the accuracy of palaeobathymetric information, that is, an assessment of the degree of 'underfilling' or starvation of the basin (Allen & Allen (1990) and Bertram & Milton (1989) provide useful discussions of the sensistivity of results from backstripping to variations in water depth).

Because of the relative paucity of geologic indices by which palaeobathymetry can be reliably inferred, however, it is a notoriously difficult parameter to constrain. Bertram & Milton (1989) highlight the scope for error in using only interval

From Buchanan, P. G. & Nieuwland, D. A. (eds), 1996, *Modern Developments in Structural Interpretation, Validation and Modelling,* Geological Society Special Publication No. 99, pp. 345–362.

thickness maps to infer palaeobathymetry. Their use as palaeobathymetric indicators is restricted to those settings where sedimentation demonstrably kept pace with subsidence, that is, where negligible bathymetry was maintained over geological time periods. Palaeontological bathymetric indicators are often insufficiently narrowly constrained to be diagnostic of a specific bathymetric environment. Further, such fossils will often be reworked such that they cease to be representative of the palaeobathymetry of the host rocks in which they are found. Additionally, few sub-wave base sedimentary facies types are sufficiently diagnostic of a particular bathymetry attaining during their deposition (Allen 1967; Hallam 1967). Hence, independent evidence of palaeobathymetry, such as that from structural geology being used here, is of particular value in basin analysis, especially when used as part of an integrated study combining biostratigraphy, sedimentology, tectonics and burial history modelling.

Methodology

The method employed in inferring palaeobathymetry from gravity-driven structures is summarized in Fig. 1. It requires: (i) negligible lateral spreading of the nappes (i.e. plane strain); (ii) that the nappes moved as more or less coherent, discrete units in which deformation at the upslope 'head' was penecontemporaneous with that at the downslope 'toe' region, that is, they behaved as small coulomb wedges, analogous to deformation observed in modern thrust belts (cf. Davis *et al.* 1983), and (iii) that a near-sea level marker can be restored to its pre-deformational elevation. If these prerequisites are satisfied, then the nappes may be treated as discrete units whose movement is subject chiefly to the mass balance between the upslope head and downslope toe. Thus, measurement of the vertical difference between the absolute post- and pre-nappe elevation of the sea level marker provides a minimum estimate of the intra-basin relief generating the gravitational potential that drove the nappes' movement.

With the aid of the Geosec structural modelling software, computer restorations of large-scale gravity-driven nappes have been carried out, using the following examples from the Apto-Albian of offshore West Africa and the Liassic of the Cardigan Bay basin, offshore west Wales, UK. They yield estimates of palaeobathymetry which are both geologically reasonable, and whose local variation, in West Africa, equates with variations in basin floor depth, as interpreted qualitatively from Bouguer gravity Anomalies.

West Africa

Setting

To date, the most notable documented example of gravity-driven deformation in the West African

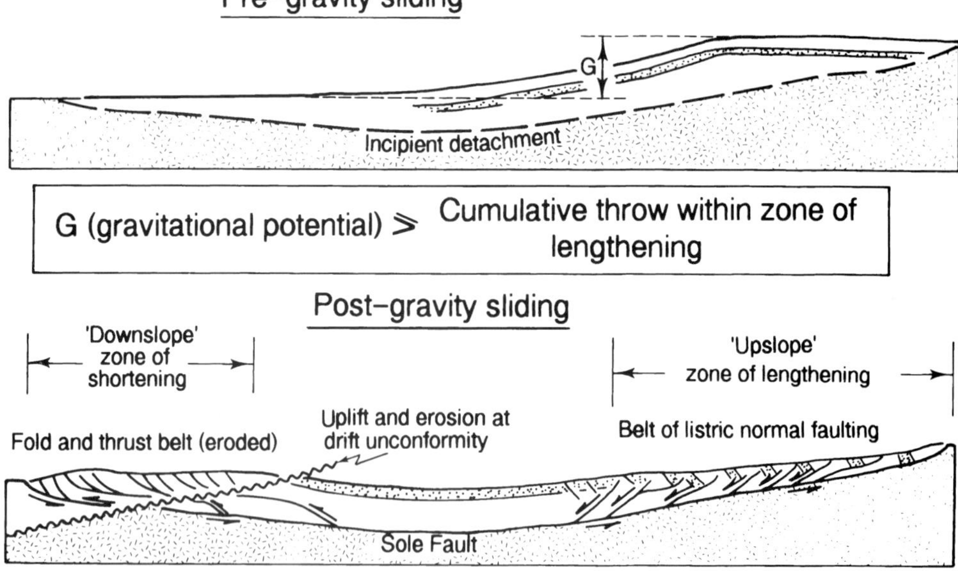

Fig. 1. Schematic illustrating the relation between basin margin relief (palaeobathymetry) and the gravity-driven movement of a large-scale submarine nappe.

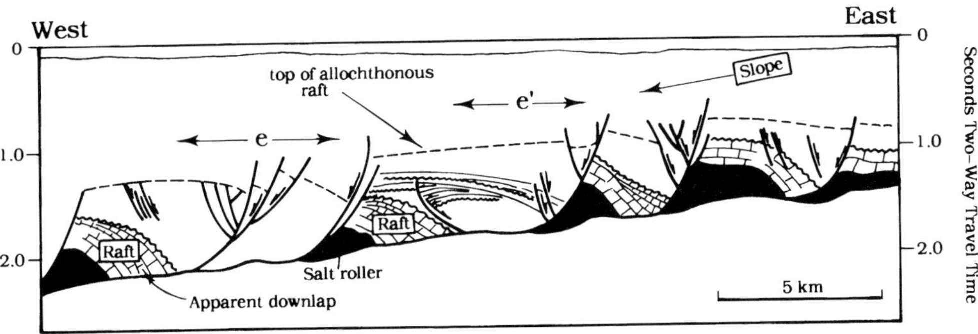

Fig. 2. Interpreted seismic profile through gravity-driven 'rafts'. They comprise Middle and Upper Cretaceous limestone and Paleogene clastics in pronounced rollover anticlines cored by Lower Cretaceous salt. e + e' = extension based on fault heave. From Duval *et al.* (1992).

Atlantic margin basin system is from the Apto-Albian syn-rift of the Kwanza basin, offshore Angola (Duval *et al.* 1992; Lundin 1992). Here, allochthonous units composed of Apto-Albian shelfal carbonates are emplaced in a relatively deep basin setting by gliding on an intra-salt detachment. Duval *et al.* (1992) have coined the term 'raft tectonics' to describe extreme thin-skin extension in which 10–20 km long, fault-bounded blocks separate into rafts that are no longer in mutual contact (Fig. 2). They estimate that 200 km of space was needed to accommodate the rafting which took place by downslope gravity-gliding and spreading into the actively opening South Atlantic Ocean.

Comparable structural geometry to the Kwanza basin rafts are seen in the Rio Muni basin, offshore Equatorial Guinea. A map of the basin (Fig. 3)

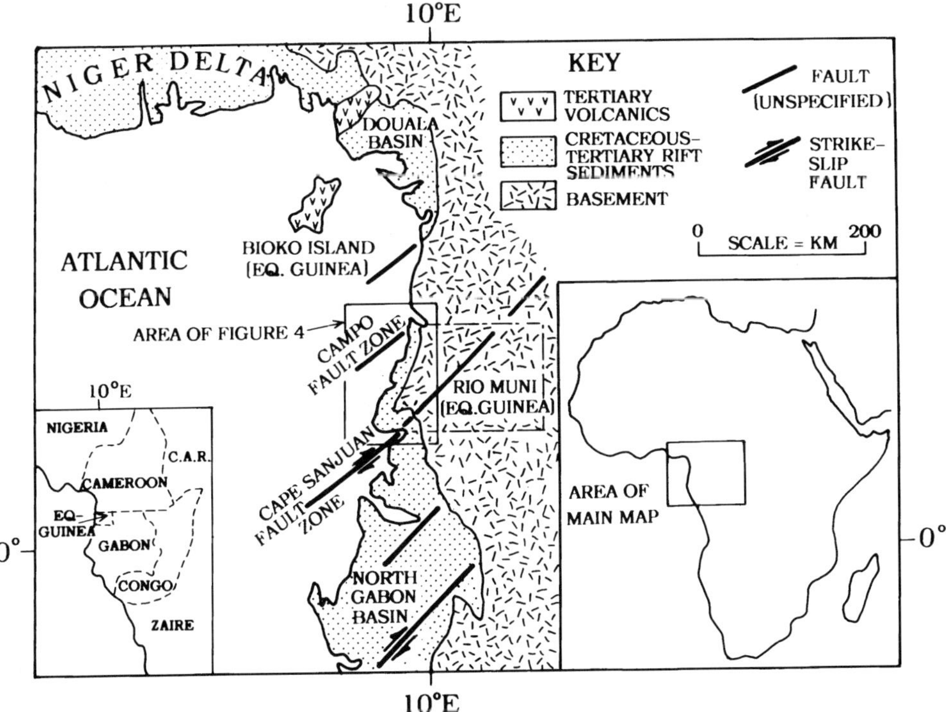

Fig. 3. Regional geology and setting of the Rio Muni basin, Equatorial Guinea, West Africa.

shows it bounded to the north and south by transversely oriented structural highs, interpreted as the continental extensions of oceanic transform faults. These culminations separate the Rio Muni basin from adjacent basins in Gabon, to the south, and Cameroon, to the north. Structurally, the Rio Muni basin is dominated by horsts and graben defined by relatively steep and planar normal faults that accommodated roughly east–west extension during Apto-Albian continental rifting.

The Mesozoic-Cenozoic evolution of the basin is represented by four megasequences (*sensu* Hubbard *et al.* 1985) shown in Table 1. Collectively, they record extension and subsidence associated with the northward 'unzipping' of the South Atlantic Ocean. The megasequences were identified from well data, using biostratigraphy and petrographic interpretation of wireline logs to map them on seismic profiles. A more complete discussion of their identification and mapping is given by Turner (1995).

Stratigraphy

The four-part division of Rio Muni stratigraphy outlined in Table 1 represents a somewhat unorthodox division compared to the conventional pre-, syn- and post-rift schemes applicable to most other rift basins. The Late-Rift Sequence is separated from its encasing Main- and Post-Rift Sequences by a subjacent basal detachment and a suprajacent discordant unconformity, the Drift unconformity. It is composed of regular-bedded shallow water limestone containing abundant oolites. These shelf carbonates are interpreted as the products of deposition on a now-eroded shallow shelf lying to the east of the present Rio Muni basin margin.

The shelf carbonates overlie a series of deep to shallow lacustrine clastics and salt precipitates that belong to the Main-Rift Sequence. They accumulated in narrow seaways along the actively extending rift basin floor and they vary locally, according to variations in water depth. The lower part of the Main-Rift Sequence comprises a heterolithic series of intercalated organics-rich shale and salt precipitates, typical of relatively slow deposition on the floor of a clastics-starved basin in which subsidence outpaced sedimentation. The upper part of the sequence is dominated by *c.*150–200 cm thick sandstone bodies that record turbiditic influxes to the basin. Their favourable porosity–permeability characteristics, and their suprajacent position with respect to the potential hydrocarbon-sourcing shales, makes them the prime target of most oil exploration wells drilled in the Rio Muni basin to date (Ross & Hempstead 1993).

Structural geometry

The structural-stratigraphic interpretations of seismic data illustrated in Figs 5, 6 & 7 are tied to five exploration boreholes (shown on Fig. 4) using a full suite of wireline logs (Ross & Hempstead 1993).

The profile in Fig. 5 illustrates features typical of the structure and stratigraphy of the Rio Muni basin. It shows a low-angle, ramp-flat detachment which forms the lower boundary to the Late-Rift Sequence, emplacing it in a relatively deep

Table 1. *Summary of the features of the principal stratigraphic divisions of the Rio Muni basin*

Megasequence	Age	Sedimentary facies	Depositional setting	Average sonic velocity (m s^{-1})
Post-Rift	Turonian–Cenozoic	Fine-grained clastics	Post-drift unconformity; sustained thermal subsidence and passive margin development	2125
Early Drift	Albian–Cenomanian	Oolitic shelf carbonates	Pre-continental splitting; syn-gravity-gliding	3900
Late-Rift	Aptian–Albian	Shales and precipitates coarsening upward to turbiditic sandstone	Starved lake systems and narrow seaways along the rift basin floor	3050
Pre/Main-Rift	Barremian–Aptian	Little-known fluvial and paralic clastics	Rift basins	no data

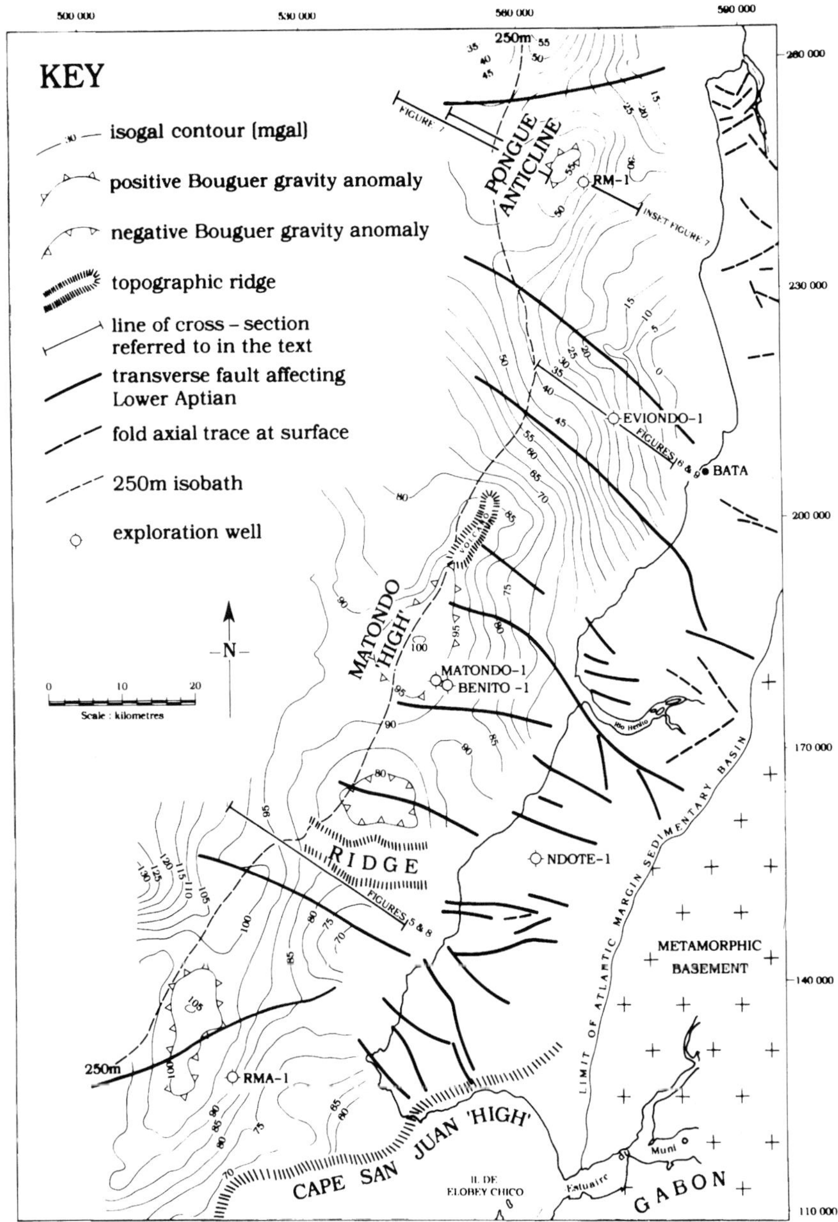

Fig. 4. Major fault structure of the Rio Muni basin, and the location of wells and cross-sections referred to in the text. Bouguer gravity isogals are from Elf Aquitaine Internal Report (1986).

basin setting. The gravity-driven origin of the detachment and associated hanging wall deformation is suggested by several observations. First, an extremely thin-skin style of deformation, responding sympathetically to the tectonic relief of the footwall (i.e. the positions of major ramps are controlled by fault zones in the immediately underlying Syn-Rift Sequence), and with a hanging wall aspect ratio (i.e. down-dip length : average thickness) of eight. Secondly, upslope extension is 'balanced' by downslope contraction. In spite of the lack of direct well evidence for stratigraphic duplication at the toes of the nappes (none of the Rio Muni basin wells have drilled through these

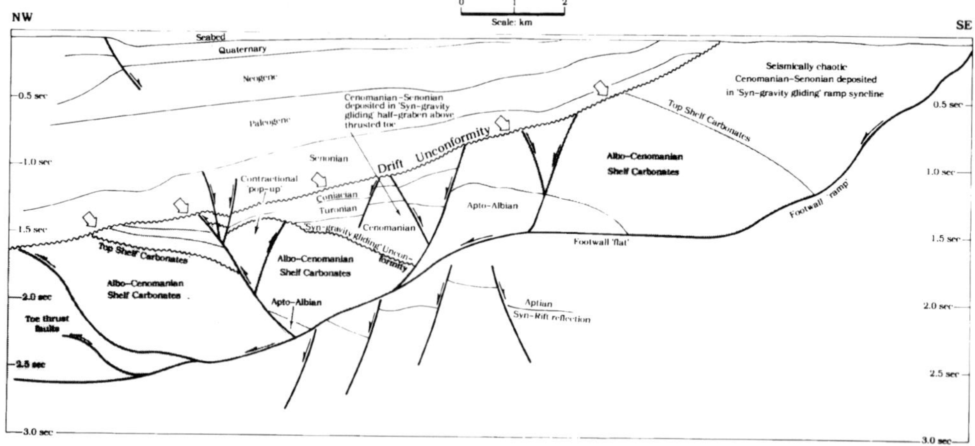

Fig. 5. Interpretation of a migrated seismic profile whose location is shown in Fig. 4. Note that the gradient of the basal detachment is exaggerated in seismic two-way travel time due to the landward tapering prism of sonically slow Upper Cretaceous and Tertiary sediment. Allochthonous shelf carbonates of the Early Drift Sequence are stippled. Open arrows signify onlap.

toes), the interpretations in Figs 5 & 6 provide the following indications of the accommodation of upslope extension by downslope contraction:

- Fold geometry typical of compressional hanging wall folds above landward dipping detachments.
- Progressive landward steepening of east-dipping detachments at the toes of the nappes. These are interpreted in terms of a 'piggyback' fault propagation sequence in which reverse displacement is transferred to subsequent fault planes that develop in the footwalls, thereby causing progressive steepening of the precedent, inactive faults.
- By analogy to smaller scale slumps and landslides, where detached units are defined by a basal fault that connects to the surface, the basal detachment to the Rio Muni basin examples appears, likewise, to connect to the palaeosurface by cutting up-section at the downslope, contractional end of the detachment.

The profile in Fig. 6 is structurally more complex, but otherwise it shares the same features of the

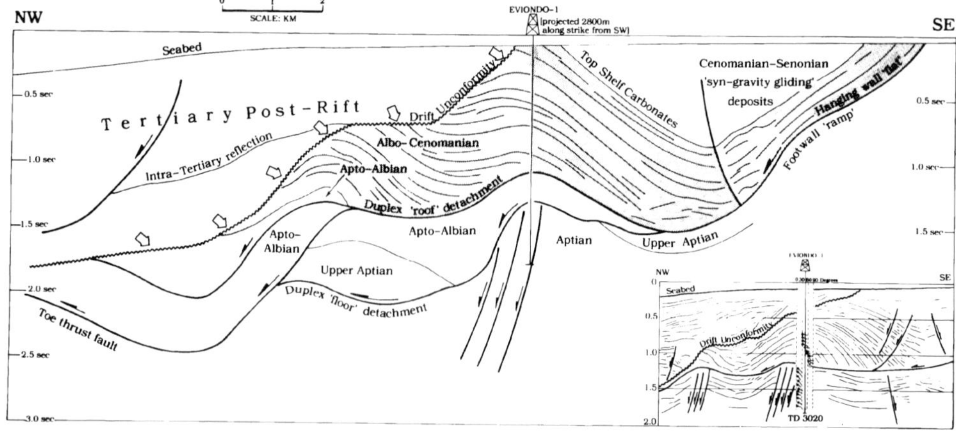

Fig. 6. Interpretation of a migrated seismic profile whose location is shown in Fig. 4. In seismic two-way travel time, the undulating geometry of the detachments is accentuated by locally preserved pockets of sonically slow Upper Cretaceous and Tertiary sediment. Allochthonous shelf carbonates of the Early Drift Sequence are stippled. Open arrows signify onlap.

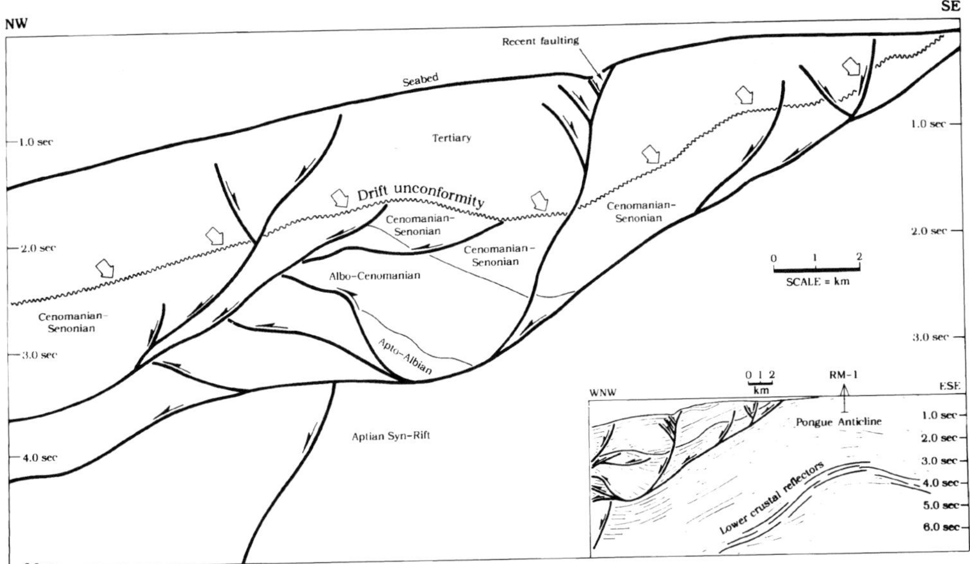

Fig. 7. Interpretation of a migrated seismic profile across the Pongue Anticline whose location is shown in Fig. 4. The inset is from a deep (16 s two-way travel time), unmigrated seismic profile showing the setting of the Pongue Anticline in the context of deep crustal structure. Open arrows signify onlap. See text for discussion.

somewhat simpler geometry illustrated in Fig. 5. A thin-skin detachment with a ramp-flat profile defines the base of the allochthonous shelf carbonates, again sandwiched between the Syn-Rift and Post-Rift Sequences. The shelf carbonates here attain a cumulative thickness of at least 4000 m. Hanging wall deformation is expressed as an upright anticline-syncline pair, interpreted to have formed in response to the progressive rotation of hanging wall ramps. On seismic profiles, the tightness of these folds is exaggerated due to the 'push-down' effect of the sonically slow Cenomanian sequence in the core of the syncline, and the 'pull-up' of the sonically fast shelf carbonates in the core of the anticline.

Greater geometric complexity is also introduced by the existence of a lensoid sliver of Syn-Rift Sequence rocks bounded both above and below by a low-angle detachment. This fault-bounded sliver comprises one of several horses forming part of an extensional duplex. It accommodated the transfer from an upper Syn-Rift Sequence detachment level, to a deeper detachment occuring at the upper limit of the condensed salts-and-shales interval of the lower Syn-Rift Sequence. The progressive development of the duplex is considered in greater detail by Turner (1995) who notes many parallels between the geometry and mechanisms of extensional (e.g. Gibbs 1984) and contractional (e.g. Boyer & Elliott 1982) duplexes.

Qualitative evidence for deep palaeobathymetry

Gravity-driven deformation in the Rio Muni basin occurred in response to major intra-basin, shelf-edge relief generating a significant gravitational potential. The existence of such relief is best accounted for by invoking deep late-rift bathymetry. Two main features of the Rio Muni basin may allude independently to the existence of deep bathymetry during the Late-Rift stage.

(1) The under-filled nature of the Late-Rift Sequence condensed clastics supports the notion of subsidence outpacing sedimentation, and thus, of increasing bathymetry. The assertion that the Late-Rift Sequence was deposited in a 'starved' basin is based on sedimentological interpretation of this sequence, and its corollaries in adjacent basins in West Africa and Brazil. Koutsoukos *et al.* (1991) and Natland (1978) elaborate on the widespread anoxic conditions that affected the northern South Atlantic during the Late Aptian to Albian. They interpret anoxic conditions as a product of a compartmentalized South Atlantic basin system in which warm waters and deep, narrow seaways led to restricted bottom water circulation and, implicitly, a limited clastic input.

(2) The geometry of the Drift unconformity in the Rio Muni basin can be interpreted as being partly a consequence of deep, late-rift bathymetry. The phase of uplift and erosion represented by drift

unconformities is a relatively long-term isostatic response of the lithosphere to geologically rapid crustal extension and thinning (Le Pichon *et al.* 1982; McKenzie 1978; see also Roberts & Yielding 1994). Assuming that syn-extensional sedimentation in the Rio Muni basin was exceeded by to the rate of subsidence, the deeply erosive Drift unconformity can be interpreted as reflecting major uplift and erosion following minimal basin subsidence, in which the component of sediment load-generated subsidence was negligible. Further, McGinnis *et al.* (1993) have suggested that the magnitude of Late Oligocene eustatic sea level fall, as predicted from sequence stratigraphy, has been overestimated due to failure to take into account the effect of flexural rebound following erosional unloading of the shelf. It is worth noting that gravity-driven translation of shelfal material to a deep setting, as described here, will compound the effect of erosional unloading, thus accentuating the magnitude of erosion represented by the Drift unconformity.

In comparison with other sub-basins of the West African basin system, in which the characteristic gravity-driven deformation described here is absent, the discordance of the Drift unconformity in the Rio Muni basin is spectacular (e.g. compare seismic sections from Gabon: Teisserenc & Villemin 1989, and SW Africa: Gerrard & Smith 1982, where the unconformity is less pronounced). Significantly, a similarly erosive Drift unconformity to that in the Rio Muni basin is observed on seismic profiles through the Kwanza basin of Angola, where comparable gravity-driven structures also occur (e.g. Brice *et al.* 1982; Duval *et al.* 1992; McGinnis *et al.* 1993).

Structural restoration and palaeobathymetry

Before describing the restoration method, it should be noted that the restoration technique used here requires: (i) negligible lateral spreading of the gravity-driven structures, and (ii) that they moved as discrete units in which deformation at their upslope head and downslope toe was penecontemporaneous (Fig. 1). First, from seismic profiles oriented parallel to the basin axis (Turner 1995), it is clear that lateral spreading was not a significant process accommodating the nappes' basinward displacement. Basin axis-parallel (i.e. coast-parallel) sections do not exhibit transversely oriented reverse and thrust faults, the chief mechanism by which lateral spreading would be accommodated. The nappes are infact constrained laterally by the transverse faults in the sub-nappe, Syn- and Pre-Rift Sequences (Fig. 4). Secondly, there is good evidence indicating the nappes did indeed move sychronously, and that in effect they behaved as small coulomb wedges, more or less on the point of failure throughout. Most notable among this evidence are the small basins located at both the extensional head-end, and at the contractional toe, of the nappe in Fig. 5. These basins contain a distinctly onlapping, seismically chaotic fill of Cenomanian-Turonian and Senonian clastics which were deposited in small ramp synclines (at the head) and half-grabens (at the toe) perched above the actively deforming nappes. The divergence of seismic reflectors within the fill of these basins, and their increased bed length with ascending stratigraphy, record syn-extensional deposition.

Figures 8 and 9 summarize the structural restoration procedure applied to the gravity-driven nappes described from the profiles shown in Figs 5 & 6. Conversion of the seismic data, from a vertical scale of seismic two-way travel time, to depth uses seismic velocities derived from borehole sonic log measurements (Table 1). In order to restore the Syn- and Late-Rift Sequences to their syn-deformational thickness, the post-deformational Post-Rift Sequence is removed and the underlying syn-deformational succession decompacted. Retro-deformation of the two profiles differs slightly according to structural complexity.

In view of the shallow depth (and hence brittle rheology) of the hanging wall rocks throughout nappe movement, and their regular-bedded nature, a flexural slip mechanism of hanging wall deformation is applied during retro-deformation of the nappes. Flexural slip restoration of hangingwall deformation allows for the greatest precision in determining the magnitude of fault displacement. Previous attempts to restore faults as a means of reconstructing palaeogeographic attributes have been achieved using rigid body rotations of faulted terrains in which no internal deformation of the footwall is allowed for (seen as gaps and overlaps within a restored template, e.g. Young 1992 fig. 11). The flexural slip procedure applied here more realistically simulates the mechanism of faulting and associated deformation at shallow depth in stratified rocks.

In Fig. 8, retro-deformation is achieved by simply reversing the nappe back up the basal detachment to a point at which the chosen sea level marker horizon is more or less flat, and thus in its pre-deformational position. The vertical difference in elevation of the hanging wall cutoff of the marker (literally the throw of the basal detachment) following nappe emplacement, and hence a minimum estimate of syn-deformational bathymetry as taken here, is 2700 m.

The greater structural complexity of the extensional duplex interpreted in Fig. 6 makes

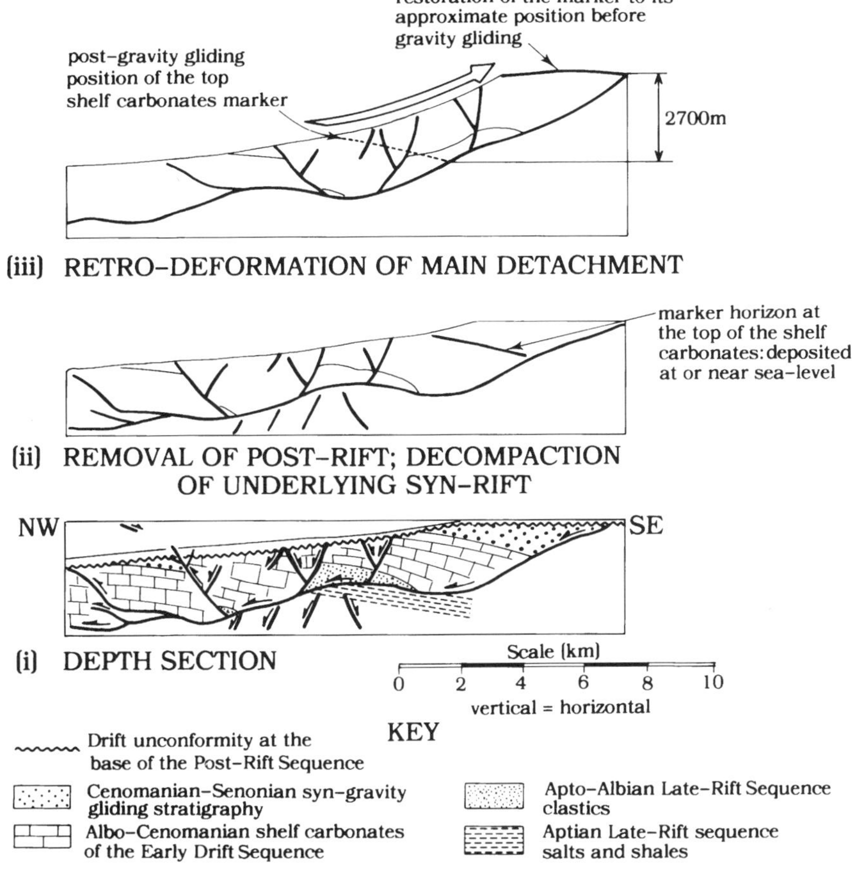

Fig. 8. Line drawing of the two main stages in the sequential restoration (retro-deformation) of the structure interpreted from the seismic profile in Fig. 5. Time-to-depth conversion uses sonic velocities given in Table 1. Dashed line in (iii) indicates the post-deformational position of the sea level marker horizon.

for a slightly more involved, three-part retro-deformation procedure (Fig. 9). Following depth conversion, and removal of the Post-Rift Sequence (allowing for decompaction), it is necessary to restore separately the roof and floor detachments of the duplex, in the reverse order to that in which they developed. Since duplexes develop by sequential propagation of subsequent faults into the autochthonous footwall, the upper, roof detachment is the first fault to be restored, followed by the lower, floor detachment. Figure 9 shows that restoration of an arbitrary upper Late-Rift Sequence marker to its pre-nappe configuration gives an elevational difference, and thereby a minimum estimate of palaeobathymetry, of 1600 m. Several observations emerge from this restoration of the duplex.

(1) The succession in the hanging wall of the nappe accumulated during active basin extension, as recorded by the fact that older markers, deeper in the succession, remain folded and require greater unravelling to fully restore them.

(2) The restored geometry of the duplex does not display a smooth fault profile. This suggests that it is structurally more complex than interpreted, perhaps comprising a greater number of imbricate faults than are shown in Fig. 6, and/or that the fault plane itself may have undergone deformation during the downslope translation of the nappe. A detailed investigation of the degree of syn-nappe deformation of the fault planes is beyond the scope of this paper. However, the presence of a major extensional duplex (recording brittle failure of the footwall during nappe emplacement), and several fault zones that propagate through the nappes' overburden (interpreted to be related to compaction of the nappes' hanging walls following their emplacement; e.g. Figs 5, 6 & 7) suggests that it is

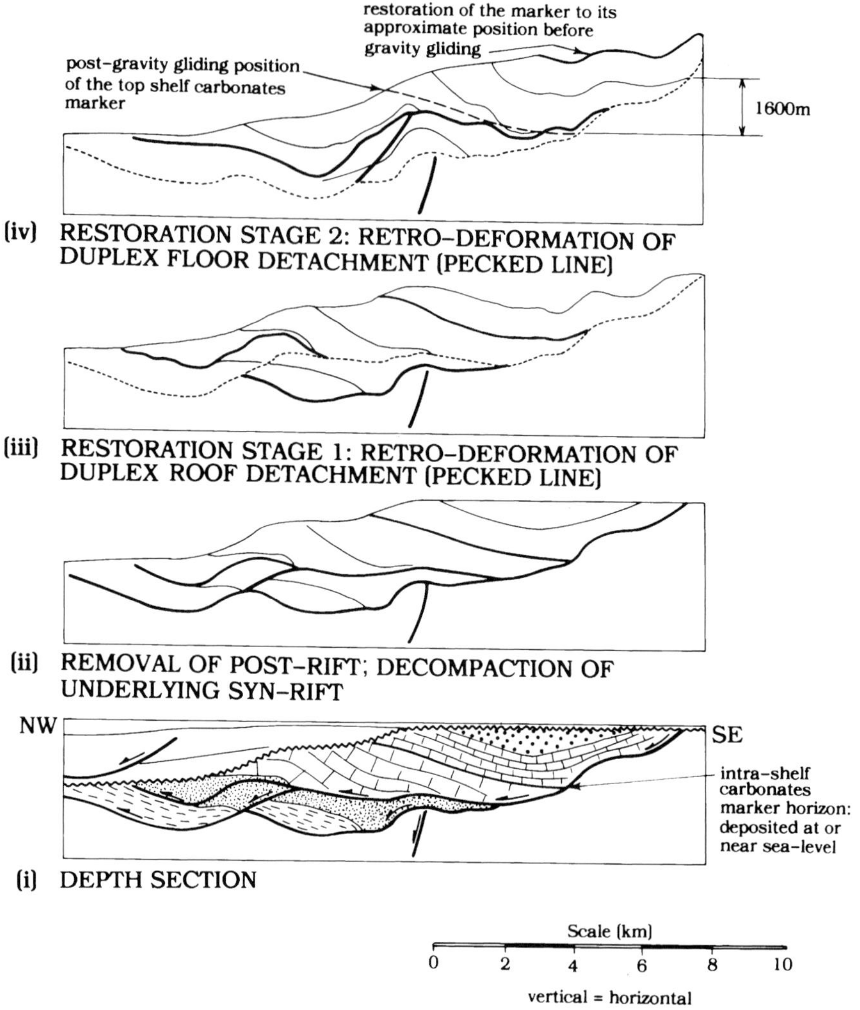

Fig. 9. Line drawing of the three main stages in the sequential restoration of the duplex structure interpreted from the seismic profile in Fig. 6. Time-to-depth conversion uses sonic velocities given in Table 1. Dashed line in (iv) indicates the post-deformational position of the sea level marker horizon. Fine-pecked lines in (iii) and (iv) indicate the active fault segment undergoing restoration.

reasonable to invoke a degree of modification of the fault planes during nappe emplacement.

Bouguer gravity

The profiles in Figs 5 and 6 are located to the north and south of the prominent Matondo High and, from the Bouguer isogals in Fig. 4, they are seen to traverse parts of the basin with contrasting gravity signatures. The Matondo High divides relatively shallow (*c.*2000 m) basin and thick crust to the north, with a Bouguer gravity of 30–50mgal, from deeper basin (*c.*3500 m) and thinner crust to the south, with a correspondingly higher Bouguer gravity of 90–110 mgal.

A positive Bouguer gravity anomaly correlates directly with thin crust above the Pongue Anticline (Fig. 4), in the northern-most part of the Rio Muni basin. Figure 7 is a line sketch taken from the upper part of a deep (16 seconds seismic two-way travel time) seismic profile across the crest of the Pongue Anticline. Interpretation of the Cretaceous and Tertiary succession is constrained by well RM-1, located at the left-hand end of the section. The interpretation depicts thin-skin gravitational detachment of Upper Cretaceous rocks on the

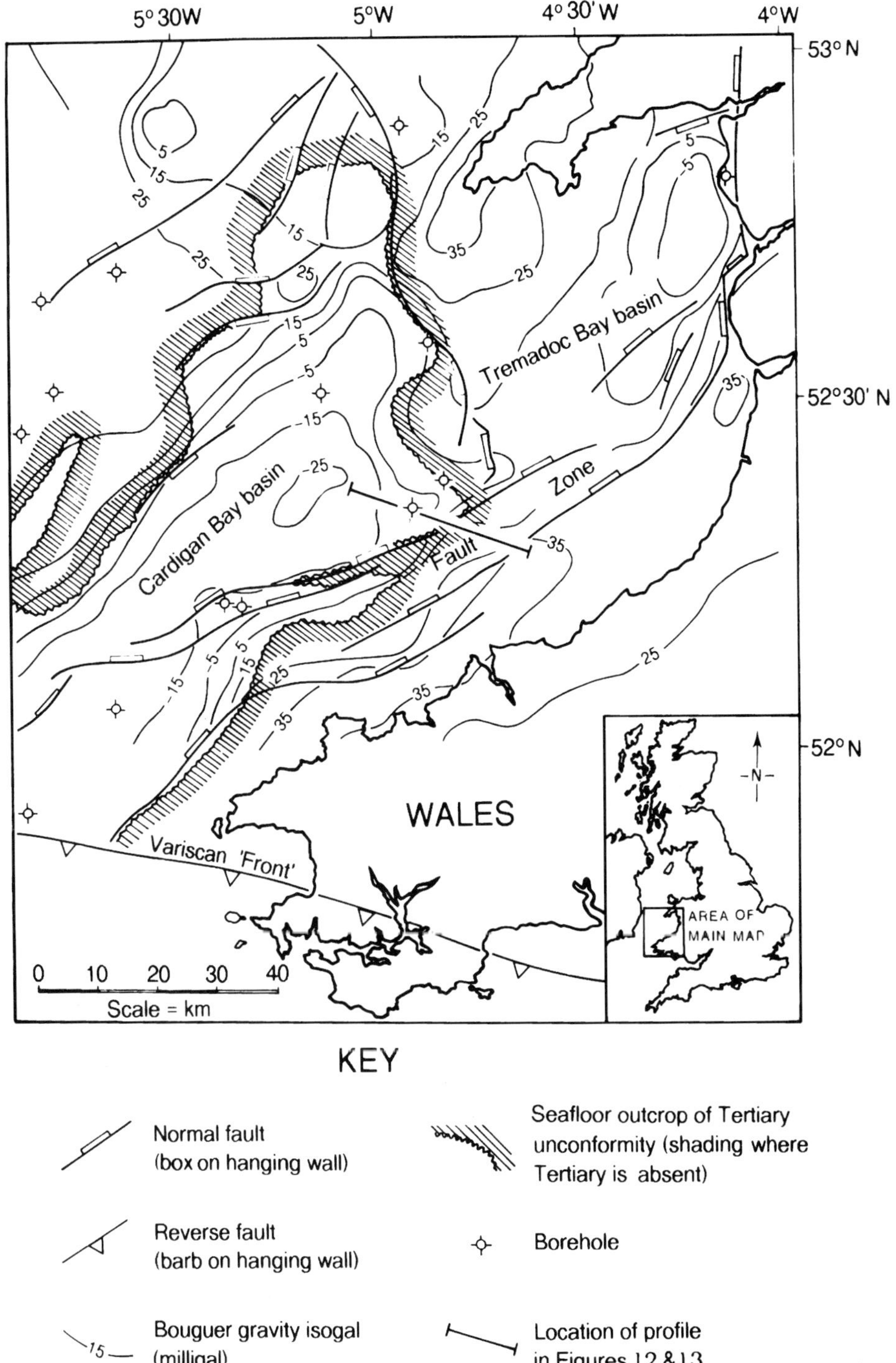

Fig. 10. Regional geology and setting of the Cardigan Bay basin and location of cross-sections referred to in the text. The map emphasizes the coincidence of the seafloor trace of the basal Tertiary unconformity, marking the principal Mesozoic (especially Liassic) depocentre, and the contour pattern of Bouguer gravity isogals. Isogals from BGS 1:250 000 Bouguer Gravity Anomaly Sheet 'Cardigan Bay' (Provisional Edition).

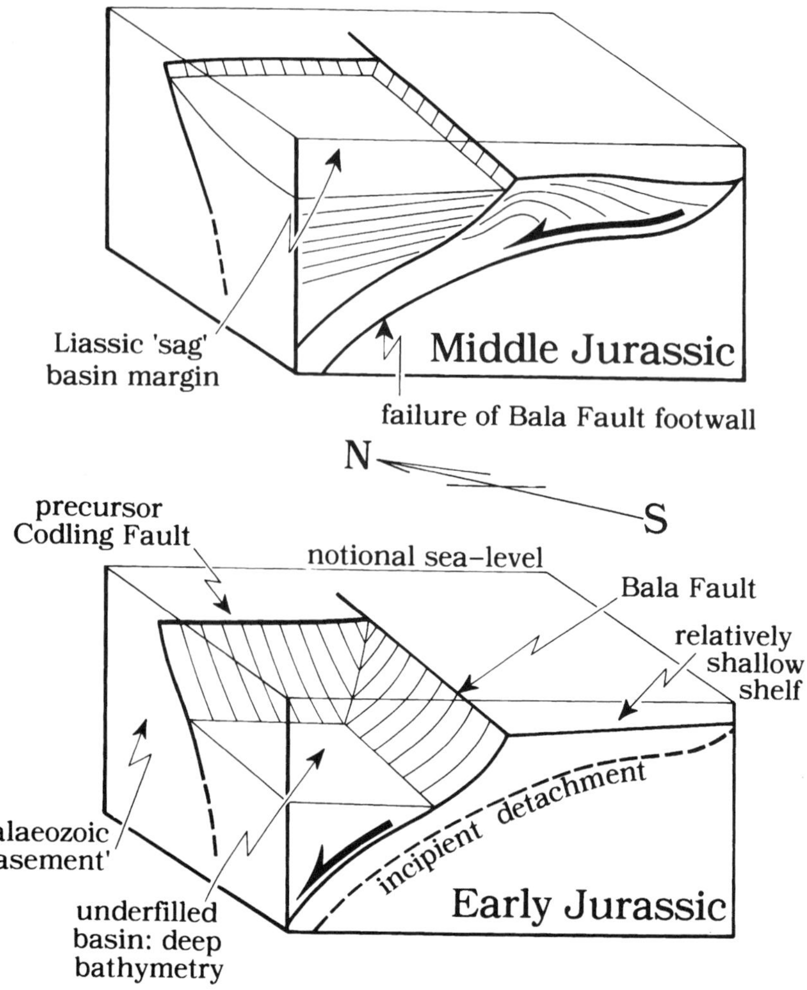

Fig. 11. Schematic block diagram placing the area of the Cardigan Bay basin shown in Fig. 12 in its palaeogeographic context. The diagram emphasizes the under-filled nature of the basin in the Mid Jurassic, interpreted here as leading to the collapse of the Bala Fault footwall. Rapid Mid Jurassic subsidence may have been at a maximum at this corner of the basin due to the combined effect of the basinward downthrow of the Bala and (precursor) Codling faults.

western, basinward flank of the anticline. The inset places the section in its deeper context, revealing the anticline to be located above lower crustal reflectors, expressed seismically as high amplitude/low frequency seismic events, occurring at a shallow depth of *c*.3.5 s TWT. They record thin crust beneath the Pongue Anticline and suggest the origin of the anticline is related more to thermal doming, than compressional folding. Consistent with the shallow lower crustal reflectors and thin crust, are the high geothermal gradients inferred directly from vitrinite reflectances recorded from Apto-Albian rocks in well RM-1. Compared to the average regional VR of 0.9%, VRs between 1.5% and 2.5% are recorded from the well (Turner 1995).

The two restored sections from the Rio Muni basin indicate varying palaeobathymetry that accords with the Bouguer gravity structure of the basin. The largest palaeobathymetry is inferred from a structure that occurs above a relative Bouguer gravity high, of 75–105 mgal. Conversely, lesser palaeobathymetry is indicated by a structure that occupies a relative gravity low of 20–40 mgal. In the north of the Rio Muni basin, a positive gravity anomaly correlates with demonstrably thin crust (Pongue Anticline). On this basis, high gravity signature is interpreted as an expression of thin crust and correspondingly rapid tectonic subsidence. Conversely, negative gravity anomalies will correspond with those parts of the basin where lesser amounts of crustal thinning led to slower rates of tectonic subsidence and sediment accumulation.

Interpretations of gravity maps generally associate negative anomalies with thick sedimentary successions, and vice versa. However, in the Rio Muni basin the seismic profiles indicate that subsidence outpaced sedimentation. In this case, therefore, areas identified as having experienced relatively rapid subsidence will not necessarily equate with those parts of the basin which have received the thickest sedimentary accumulation (cf. Bertram & Milton 1989). Consequently, the gravity structure of the basin provides a means of calibrating qualitatively the palaeobathymetric results obtained from structural restoration of the gravity-driven nappes. The largest palaeobathymetry occurs where greater crustal thinning generated rapid and sustained subsidence, leading to the generation of major relief between the basin margin and the basin centre.

Cardigan Bay

Setting

Comparable structures to the gravity-driven nappes of the Rio Muni basin occur in Jurassic rocks of the Cardigan Bay basin, offshore West Wales, UK (Fig. 10). The basin experienced a varied tectonic history of repeated cycles of subsidence and uplift since at least the Carboniferous. Its Jurassic evolution was dominated by prolonged subsidence leading to the accumulation of the most complete Liassic succcession in the British area (Hallam 1992).

Structure

The block diagram in Fig. 11 shows schematically the setting of the seismic profile interpreted in Fig. 12. The roughly WNW-trending profile is located at the junction of the Bala Fault and the so called Codling Fault which extends as far as the Kish Bank basin, to the northwest of the study area. Greatest Liassic subsidence is interpreted at this corner of the Cardigan Bay basin as a consequence of the combined effect of the two faults' basinward downthrow. Failure of the Bala Fault footwall is interpreted as a response to rapid subsidence outpacing sedimentation, thereby creating an underfilled bathymetric deep. Several features of the seismic line interpreted in Fig. 12 record qualitatively the existence of deep bathymetry during the deposition of the Liassic.

(1) A strongly onlapping Liassic succession comprising *c.*2500 m of fine-grained clastics interpreted as the products of deposition in an open marine environment of uncertain water depth (Barr *et al.* 1981).

(2) A low-angle, listric normal fault detaching within Permo-Triassic evaporites, with an associated 10 km-wavelength rollover anticline. The anticline has a locally N–NE striking axial trace which changes to the south of the profile where it becomes parallel with the trend of the Bala

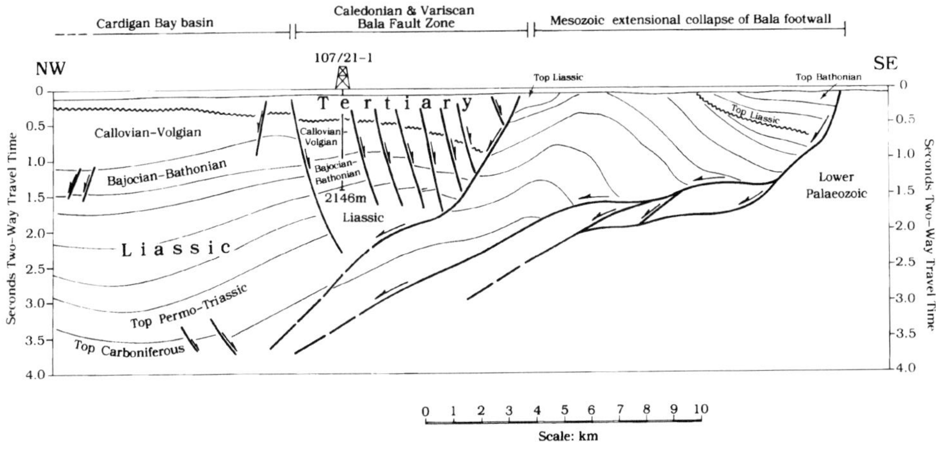

Fig. 12. Interpretation of a migrated seismic profile across the margin of the Cardigan Bay basin, whose location is shown in Fig. 10.

Fault. The hanging wall is occupied by a syn-kinematic sequence of Liassic and Dogger (Middle Jurassic) rocks recording their accumulation during fault displacement. This listric fault system accommodated Jurassic gravitational collapse along a 50 km length of the southeastern margin of the Cardigan Bay basin in response to the generation of significant relief between the deep water basin centre and the autochthonous, elevated Bala Fault footwall. Truncation of the hangingwall rollover anticline by the seafloor unconformity (Fig. 12) is interpreted as the medium- to long-term isostatic response to the unloading of the basin margin by the listric fault system (cf. the drift unconformity in West Africa). Unlike the West African examples, however, the Cardigan Bay basin experienced a major phase of uplift and erosion during the Early Tertiary.

Restoration

Like the West African examples, it is necessary first to establish a suitable sea level marker within the deformed sequence. This is provided in Cardigan Bay by a Bajocian-Bathonian succession of beds of shale with limestone and sandstone. The sandstones are clearly of shallow water deltaic to marginal marine lithofacies, containing *Planolites* burrows, rootlets and carbonaceous debris (Penn & Evans 1976). They form a seismically distinctive package of relatively high-amplitude reflectors that locally onlap the topmost Liassic unconformity.

In Fig. 13, restoration of the gravity-driven fault displacement of the Cardigan Bay basin margin succcession is achieved by reversing the Bajocian–Bathonian marker interval back up the listric fault plane to the point at which it is more or less flat-lying, and therefore, in its pre-deformational elevation. In this case, a minimum estimate of the relief beween basin margin and basin centre is gained from the 2100 m difference in the pre- and post-deformation elevation of the marker.

Bouguer gravity

In the Cardigan Bay basin, a very thick (*c*.4500 m) Mesozoic–Cenozoic sedimentary succession

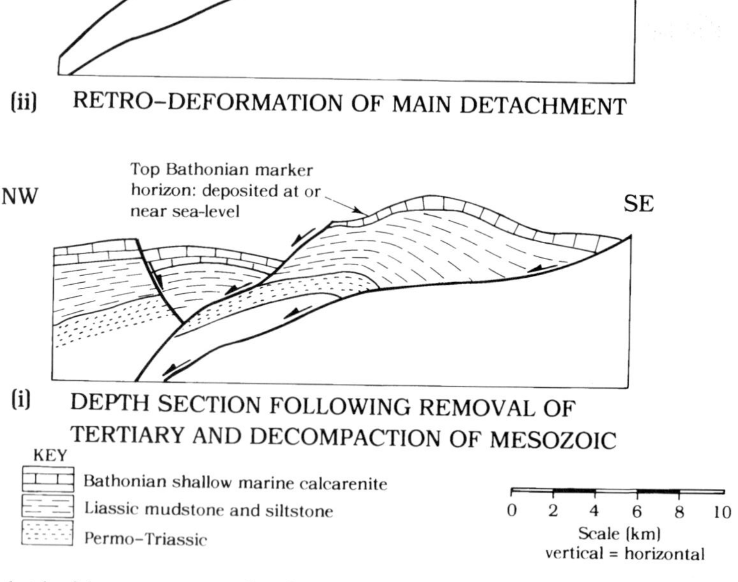

Fig. 13. Line sketch of the two-stage procedure for restoring (retro-deforming) the structure interpreted from the seismic profile in Fig. 12. Time-to-depth conversion uses the following borehole sonic-derived velocities: Permo-Triassic 4930 m s^{-1}; Liassic 3600 m s^{-1}; Middle Jurassic 2750 m s^{-1}; Upper Jurassic 3200 m s^{-1}; Tertiary 2220 m s^{-1}. Dashed line in (ii) indicates the post-deformational position of the sea level marker horizon.

correlates directly with a strongly negative Bouguer gravity anomaly of 5–25 mgal (Fig. 10). The mapped seismic structure of the basin fill, especially the seafloor trace of the basal Tertiary unconformity (recording inversion of the Liassic depocentre), accords with the contour pattern of the isogals. Gravitational collapse of the basin margin is interpreted as a response to deep bathymetry generated by rapid subsidence which was only partially filled by sedimentation.

Discussion

Implications and limitations of the restoration method

In this paper, it has been suggested that the presence of geometrically distinctive, gravity-driven structures in basins offshore West Africa and Wales may record late-rift water depths of between 1600 m and 2700 m. The following factors may contribute to the inaccuracy of palaeobathymetric estimates using the restoration procedure described above.

(1) Underestimates will result from the fact that some of the gravitational potential is accommodated by non-plane strain deformation not easily detectable on seismic sections. Ideally, all the gravitational potential energy generated by deep shelf-edge bathymetry will be expended in driving the downslope deformation of the nappes. However, in practice small-scale (i.e. below a scale that may be readily identified on seismic sections) lateral spreading and other mechanisms of distributed strain will absorb a fraction of the gravitational potential that is difficult to quantify. Such gravity spreading is promoted by low strain rate, ductile lithology and rapid sedimentation of the hanging wall succession in which the nappes' hanging wall successions remain uncompacted (un-dewatered) during initial sedimentation and burial. Further, as the nappe movement takes place in sediment which is less than fully compacted, lateral compaction in the downslope part of a nappe will accommodate a portion of the contraction that cannot be taken into account using the method described here.

(2) Underestimates of pre-deformational potential, and hence palaeobathymetry, may result if the downslope movement of the nappes is 'frozen' before the gravitational potential has been used. The degree to which the nappes' displacement continues until all original potential is expended will depend largely on: (i) the lateral persistence of easy-glide lithologies with low basal shear strength, and (ii) the capacity of the hanging wall succession to maintain high, near-lithostatic fluid pressures (cf. Hubbert & Rubey 1959).

(3) The method assumes that the nappes deform as a closed system, in which their total displacement is controlled by the mass balance between the upslope head and the downslope toe. This takes no account of the possibility that the toe of the nappe may be eroded as it is extruded above the palaeo-surface. Such a scenario may lead to overestimates of gravitational potential and palaeobathymetry because it means that the downslope mass that is balancing the movement of the upslope part of the nappe is being reduced, thereby allowing more downslope movement than would be possible if the toe remained intact.

Bearing in mind that the structures analysed here developed in basins floored demonstrably by continental crust (in West Africa, the structures are truncated by the suprajacent drift unconformity, widely interpreted as signifying the onset of continental splitting and oceanic crust production) such water depths are significant and indicate a special set of geodynamic circumstances. Assuming average sediment density and lithospheric physical parameters, gradual filling of a 2000 m deep basin would depress the basin floor by *c*.250%, to 5000 m (that is, a 5000 m deep basin overlying lithosphere with average physical parameters is generated by a tectonic or driving subsidence of *c*.2000 m, the remainder of the basin depth is due to sediment load-related subsidence).

Analogous gravity-driven deformation

Gravity tectonic processes are promoted by the contrast between the rapid accumulation of thick, coarse clastic successions at the shelf edge, and the condensed, muddy sequences that typify the lower slope area (cf. Galloway 1989). The most common setting of modern gravity-driven faulting is therefore in actively prograding, passive margin basins where abrupt shelf-edge relief maximizes gravitational potential. The gravity tectonics described here pre-dates the onset of post-rift thermal subsidence and it cannot, therefore, be ascribed to gravitational instability caused by the excessive build-up of shelf-edge clastics. No examples of large-scale gravity tectonics in a modern syn-rift setting are known to the author. However, an instructive example of analogous structures developing in a modern post-rift setting are described by Dingle (1977) from the Agulhas Bank of the continental margin offshore southeast Africa (Fig. 14).

Dingle describes one of the largest documented submarine slumps, with an average along-strike length of 750 km, a width of up to 220 km, and covering an area of 79 500 km^2. On the basis of the geologically instantaneous timing of movement of

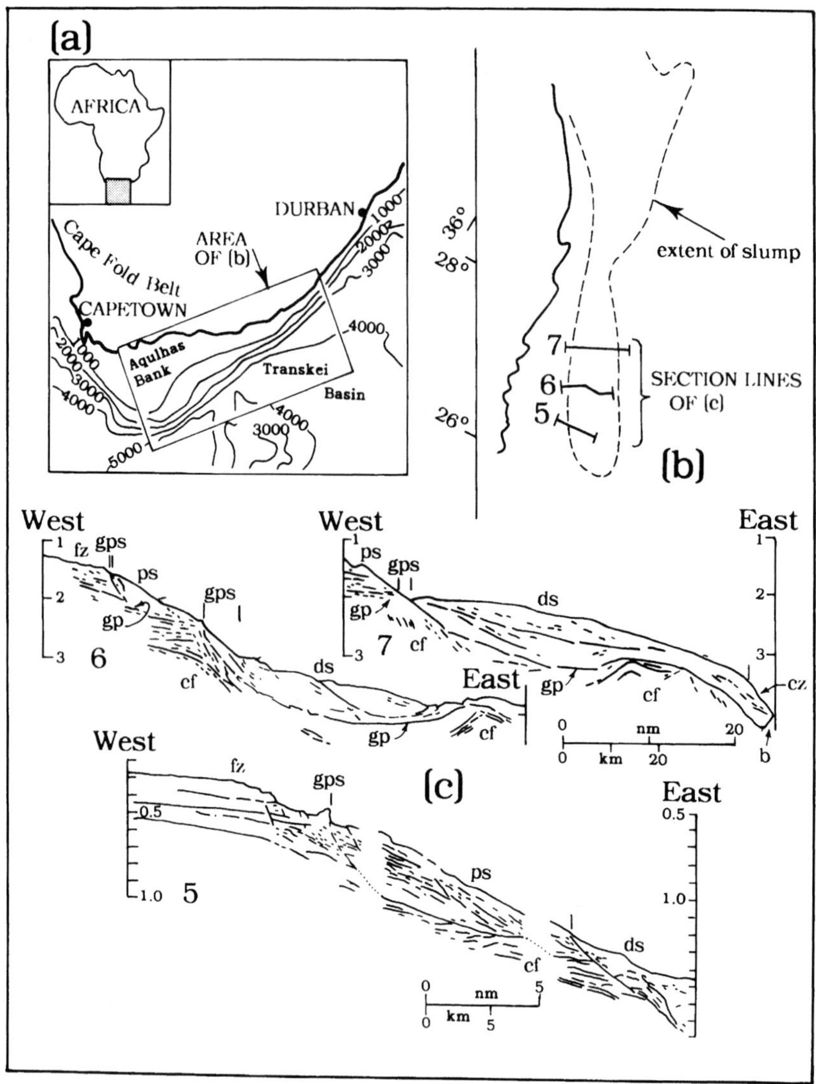

Fig. 14. Bathymetric setting, map and tracings of seismic profiles through the Agulhas slump, offshore Southeast Africa. Abbreviations used on the numbered profiles: fz, fissured zone; gps, glide plane scar; gp, glide plane; ps, proximal slump; cf, structures in Cretaceous strata; ds, distal slump; cz, compressed zone. Modified from Dingle (1977).

the entire length of the feature, and the lithified state of the slumped material (thereby suggesting it was unlikely to have been triggered by abnormal river discharge or storm induced movement), he interprets it as being triggered by a post-Pliocene earthquake.

In spite of the clearly post-rift setting of the Agulhas slump, it shares many similarities with the structures described here from West Africa and Cardigan Bay. From seismic, core and bathymetric data, it comprises a tensional, fissured head region which connects with a compressional toe via a concave, rotational basal glide plane. The internal structure of the slump is dominated by listric faults that splay from ramps along the basal glide plane. The structure in the sub-slump 'basement' exerts a strong control over the location of ramps and compressional zones within the slump sheet. In particular, the presence of the Marginal Fractured Ridge, a prominent topographic, structural and magnetic feature that defines the Cretaceous syn-rift basin margin, determines the point at which the slump cuts up-section to emerge at the depositional surface.

Table 2. *Summary of the relation between the magnitude of nappe displacement and syn-kinematic bathymetry in the Rio Muni basin and Cardigan Bay*

	Bouguer gravity (mgal)	Allochthonous section thickness (m)	Along-fault displacement (m)	Heave (m)	Throw (bathymetry) (m)
Rio Muni North (Duplex)	30–50	*c*.4000	*c*.4000	3100	1600
Rio Muni South	90–110	*c*.2000	*c*.6300	6000	2700
Cardigan Bay	−25–5	*c*.4000	*c*.4200	4800	2100

Dingle (1977) attributes along-strike variations in the total displacement along the basal glide plane largely to the structure of the Cretaceous basement and the rheology of the slumped material. However, total movement of the slump also accords systematically with along-strike changes in present bathymetry. From north to south, Dingle reports an increase in displacement of the Agulhas slump from *c*.12 km to 200 km; this is accompanied by an increase in average submarine gradient from between 0.57° and 0.86° in the southern part of the slump, to between 1.23° and 2.90° in the north (author's calculations from fig. 1 of Dingle 1977). Interestingly, these angles concur with the theoretical calculations of Price (1977, table 1)) which suggest that, along incipient glide planes with a low frictional resistance (i.e. high, near-lithostatic fluid pressure) (Hubbert & Rubey 1959), a *c*.100 km long slump 'block' may become unstable above a slope angle of around 0.57°.

Conclusion

(1) Large-scale, allochthonous nappes are characteristic of late-rift stage basins with deep bathymetry ('underfilled') e.g. the Apto-Albian of Rio Muni, West Africa; the Jurassic of Cardigan Bay, West Wales, UK.

(2) If lateral spreading is unimportant in their development, and it can be demonstrated that the nappes deformed synchronously at their toe and head, then the vertical displacement of a sea level marker provides an estimate of syn-kinematic bathymetry.

(3) In the Rio Muni basin, West Africa, different bathymetric estimates from separate basin compartments equate with differences in their Bouguer gravity signature (see Table 2).

Grateful thanks are extended to LASMO plc for allowing free access to data and computing facilities, and also Exploration Mapping Services, Western Geophysical, BP, Jebco, Elf, Agip and Murphy for contributing data to this study. Alex Maltman is thanked for drawing the authors attention to the Dingle paper. Reviews from James Buchanan and Jenny Urquhart, and comments from volume editor Dick Nieuwland, led directly to improvements in the content of this paper.

References

ALLEN, J. R. L. 1967. Depth indicators of clastic sequences. *Marine Geology*, **5**, 429–447.

ALLEN, P. A. & ALLEN, J. R. 1990. *Basin Analysis*. Blackwell Scientific Publications.

BARR, K. W., COLTER, V. S. & YOUNG, R. 1981. The geology of the Cardigan Bay–St George's Channel Basin. *In*: ILLING, L. V. & HOBSON, G. D. (eds) *Petroleum Geology of the Continental Shelf of North-West Europe*. Institute of Petroleum, London, 432–443.

BERTRAM, G. T. & MILTON, N. J. 1989. Reconstructing basin evolution from sedimentary thickness; the importance of palaeobathymetric control, with reference to the North Sea. Basin Research, **1**, 247–257.

BOYER, S. E. & ELLIOTT, D. 1982. Thrust Systems. *American Association of Petroleum Geologists Bulletin*, **66**, 1196–1230.

BRICE, S. E., COCHRAN, M. D., PARDO, G. & EDWARDS, A. D. 1982. Tectonics and seeimentation of the South Atlantic rift sequence: Cabinda, Angola. *In*: WATKINS, J. S. & DRAKE, C. L. (eds) *Studies in Continental Margin Geology*. American Association of Petroleum Geologists Memoir, **34**, 5–18.

CHADWICK, R. A. 1985. Permian, Mesozoic and Cenozoic structural evolution of England and Wales in relation to the principles of extension and inversion tectonics. *In*: WHITTAKER, A. (ed.) *Atlas of Onshore Sedimentary Basins in England and Wales*. Blackie, Glasgow, 9–25.

DAVIS, D., SUPPE, J. & DAHLEN, F. A. 1983. Mechanics of

Fold-and-Thrust Belts and Accretionary Wedges. *Journal of Geophysical Research*, **88**, 1153–1172.

DINGLE, R. V. 1977. The anatomy of a large submarine slump on a sheared continental margin (SE Africa). *Journal of the Geological Society, London*, **134**, 293–311.

DUVAL, B., CRAMEZ, C. & JACKSON, M. P. A. 1992. Raft tectonics in the Kwanza basin, Angola. *Marine and Petroleum Geology*, **9**, 389–405.

GALLOWAY, W. E. 1989. Genetic Stratigraphic Sequences in Basin Analysis II: Application to Northwest Gulf of Mexico Cenozoic Basin. *American Association of Petroleum Geologists Bulletin*, **73**, 143–154.

GERRARD, I. & SMITH, G. C. 1982. Post-Palaeozoic Succession and Structure of the Southwestern African Continental Margin. *In*: WATKINS, J. S. & DRAKE, C. L. (eds) *Studies in Continental Margin Geology*. American Association of Petroleum Geologists Memoirs, **34**, 49–77.

GIBBS, A. D. 1984. Structural evolution of extensional basin margins. *Journal of the Geological Society, London*, **142**, 609–620.

HALLAM, A. 1967. The depth significance of shales with bituminous laminae. *Marine Geology*, **5**, 481–495.

—— 1992. Jurassic. *In*: McL. D. DUFF, P. & SMITH, A. J. (eds) *Geology of England and Wales*. Geological Society, London, 325–349.

HUBBARD, R. J., PAPE, J. & ROBERTS, D. G. 1985. Depositional Sequence Mapping as a Technique to Establish Tectonic and Stratigraphic Framework and Evaluate Hydrocarbon Potential on a Passive Continental Margin. *In*: BERG, O. R. & WOOLVERTON, D. G. (eds) *Seismic Stratigraphy: An Integrated Approach*. American Assiciation of Petroleum Geologists Memoir, **39**, 79–93.

HUBBERT, M. K. & RUBEY,W. W. 1959. Role of fluid pressure in mechanics of overthrust faulting. *Bulletin of the Geological Society of America*, **70**, 115–166.

KOUTSOUKOS, E. A. M., MELLO, M. R., DE AZAMBUJHA FILHO, N. C., HART, M. B. & MAXWELL, J. R. 1991. The Upper Aptian–Albian succession of the Sergipe Basin, Brazil. *American Association of Petroleum Geologists Bulletin*, **75**, 479–498.

LE PICHON, X., ANGELIER, J. & SIBUET, J. C. 1982. Plate boundaries and extensional tectonics. *Tectonophysics*, **81**, 239–256.

LIVERA, S. E. & GDULA, J. E. 1990. Brent oil field. *In*: *Structural Traps II: Traps associated with tectonic faulting*. American Association of Petroleum Geologists Treatise, **A-017**, 21–63.

LUNDIN, E. R. 1992. Thin-skinned extensional tectonics on a salt detachment, northern Kwanza basin, Angola. *Marine and Petroleum Geology*, **9**, 405–412.

MCGINNIS, J. P., DRISCOLL, N. W., KARNER, G. D., BRUMBAUGH, W. D. & CAMERON, N. 1993. Flexural response of passive margins to deep-sea erosion and slope retreat: Implications for relative sea level change. *Geology*, **21**, 893–896.

MCKENZIE, D. P. 1978. Some remarks on the development of sedimentary basins. *Earth and Planetary Science Letters*, **40**, 25–32.

NATLAND, J. H. 1978. Composition, provenence and diagenesis of Cretaceous clastic sediments drilled on the Atlantic continental rise off southern Africa, DSDP Site 361. Implications for the early circulation of the South Atlantic. *Initial Reports of the Deep Sea Drilling Project*, **40**, 1025–1061.

PENN, I. E. & EVANS, C. D. R. 1976. *The Middle Jurassic (mainly Bathonian)* of *Cardigan Bay and its palaeogeographical significance*. Report of the Institute of Geological Sciences, **76/6**.

POTTS, G. J. & HENDERSON, G. 1993 The geometry and evolution of syn-sedimentary faults observed in cores from the Magnus Field. *In*: *Tectonic Studies Group Abstract Manual*, Annual Meeting 1993, 33.

PRICE, N. J. 1977 Aspects of gravity-tectonics and the development of listric faults. *Journal of the Geological Society London*, **133**, 311–329.

RAMBERG, H. 1981. *Gravity, deformation and the earth' crust* (2nd edn). Academic Press.

ROBERTS, A. M. & YIELDING, G. 1994, Continental extensional tectonics. *In*: HANCOCK, P. L. (ed.) *Continental Deformation*. Pergamon Press, 223–250.

ROSS, D. & HEMPSTEAD, N. 1993. Geology, hydrocarbon potential of the Rio Muni area, Equatorial Guinea. *Oil & Gas Journa*l, **91**, 96–100.

SPEKSNIJDER, A. 1987. The structural configuration of Cormorant Block IV in context of the northern Viking Graben structural framework. *Geologie en Mijnbouw*, **65**, 357–379.

TEISSERENC, P. & VILLEMIN, J. 1989. Sedimentary Basin of Gabon–Geology and Oil Systems. *In*: EDWARDS, J. D. & SANTOGROSSI, P. A. (eds) *Divergent/Passive Margin Basins*. American Association of Petroleum Geologists Memoirs, **48**, 117–200.

TURNER, J. P. 1995. Gravity-driven structures and rift basin evolution: The Rio Muni basin, offshore Equatorial West Africa. *American Association of Petroleum Geologists Bulletin*, in press.

YOUNG, R. 1992. Restoration of a regional profile across the Magnus Field in the northern North Sea. *In*: LARSEN, R. M., BREKKE, H., LARSEN, B. T. & TALLERAAS, E. (eds) *Structural and Tectonic Modelling and its Application to Petroleum Geology*. Norwegian Petroleum Society, **1**, 221–229.

Index

References to Figures and Tables are in *italics*